Marty Biermeier

Television Simplified

DELMAR PUBLISHERS

Library of Congress Catalog Card Number: 72–11273

PRINTED IN THE UNITED STATES OF AMERICA

PUBLISHED SIMULTANEOUSLY IN CANADA BY
DELMAR PUBLISHERS, A DIVISION OF
VAN NOSTRAND REINHOLD, LTD.

Television Simplified

Seventh Edition

Milton S. Kiver and Milton Kaufman

DELMAR PUBLISHERS, ALBANY, NEW YORK 12205
A DIVISION OF LITTON EDUCATIONAL PUBLISHING, INC.

Preface

The state of the television art has advanced considerably since the publication of the sixth edition of TELEVISION SIMPLIFIED. Much of this progress has been made possible by the use of solid-state components and printed circuit construction. Integrated circuits and modular construction are finding increasing use. The overall effect of these devices in television receivers is the construction of a receiver which is far more compact, more efficient, and more reliable than the vacuum-tube receivers.

The use of solid-state components has made possible the innovation of circuits which would not have been practical with the use of vacuum tubes. An example of this is the varactor tuner, which has no tubes and no selector switch but which may be push-button tuned to any VHF or UHF channel. As an example of the use of integrated circuits, we may find an entire audio section in one integrated circuit and a color-demodulator section in another.

There are now a great many color television receivers in operation and their number is growing rapidly. These receivers, of course, also make wide use of solid-state techniques, some being currently of 100% solid-state construction, while others are still manufactured exclusively with vacuum tubes or with a combination of vacuum tubes and solid-state devices (hybrid receivers).

In addition to the newer receivers, there still exist many millions of the older vacuum-tube receivers in service, and these must be expected to remain in service for many years. In order that this book may be a reference volume for all types of television receivers, the vacuum-tube types, as well as the newer hybrid and solid-state types are given extensive coverage, both in their monochrome and color-receiver designs.

Various changes have been made in the content of this edition, both to update the coverage and to make the book more useful to students and instructors. Some of the more important differences are:

1. There are now 24 chapters rather than the previous 19. This represents not only the inclusion of new material but also the reorganization of some of the original material and the deletion of that which is now obsolete. For instance, separate chapters are now devoted to the following important topics: "Principles of Scanning and Synchronization," "Television Camera Tubes," "Automatic Fine Tuning and Remote Controlled Tuners," "Deflection Oscillators and Horizontal AFC," as well as others of importance.

2. Wherever applicable, each chapter now has material covering both vacuum-tube and solid-state circuits for both monochrome and color

receivers. The coverage of color television is also considerably enhanced by individual chapters covering the theory, circuit operation, and troubleshooting procedures. New color slides (Chapter 24) were especially photographed for this book to improve the students' understanding of defects which might occur in color receivers.

3. The very latest techniques in circuitry and construction made possible by the newer solid-state devices and circuit designs have been included. Among these are the new automatic circuitry controls, such as automatic tint control, automatic color control, and automatic brightness and contrast control. The latest remote control techniques are also described.

4. The chapters on receiver analysis and troubleshooting (Chapters 22–24) have been completely modernized and discuss the very latest solid-state color receiver and the most recent troubleshooting and alignment techniques. The latest types of monochrome and color-receiver test equipment are also reviewed.

5. Many new circuit schematic and block diagrams have been keyed to the text to enhance the readers' understanding of the material. Among these are a number of simplified diagrams, presented wherever it was felt a more detailed analysis of the information was required.

The reader will find that this new 7th Edition represents a very comprehensive coverage of all of the types of television receivers now in service, from the standpoint of theory, operation, troubleshooting, and alignment. In addition, sufficient information regarding television transmission and television transmitters is presented to enable the student to complete his understanding of the overall television system.

The authors wish to gratefully acknowledge all the individuals and companies who were instrumental in providing helpful assistance and information. Credits are given for photos and drawings furnished by the various companies. Particular appreciation is extended to Mr. El Mueller of Motorola, Inc., Sylvania Electric Products, Inc., Varo Inc., General Electric Co., Heath Co., Admiral Corp., Sony Inc., Zenith Radio Corp., Fairchild Semiconductor, Sencore Inc., Curtis Mathes Sales Co., Philco-Ford Corp., Radio Corporation of America, and Oak Electronetics Corp. We also wish to cite Mr. David F. Stout, Mr. Burton D. Santee, Mr. Bob Barkley, and Dr. J. C. Prabhakar for their valuable cooperation.

MILTON S. KIVER
MILTON KAUFMAN

Contents

7 AUTOMATIC-FINE TUNING AND REMOTE CONTROL 136

8 VIDEO-IF AMPLIFIERS 156

9 VIDEO DETECTORS 183

10 AUTOMATIC GAIN CONTROL (AGC) CIRCUITS 192

15 TV POWER SUPPLIES 299

16 SYNCHRONIZING CIRCUIT FUNDAMENTALS 310

17 DEFLECTION OSCILLATORS AND HORIZONTAL AFC 330

The author and editorial staff at Delmar Publishers are interested in continually improving the quality of this instructional material. The reader is invited to submit constructive criticism and questions. Responses will be reviewed jointly by the author and source editor. Send comments to:

Editor-in-Chief
P.O. Box 5087
Albany, New York 12205

Television Simplified

1 Concepts of the TV System

1.1 INTRODUCTION

Television is the science of transmitting rapidly changing pictures from one place to another. Radio waves are usually used for the transmission of television pictures, but in some applications, such as *closed-circuit television* (CCTV), wires or cables carry the signal from one point to another. Television is not only one of the most significant factors in the home-entertainment industry, but it is also used extensively in science, industry, education, and military applications.

New types of components, new circuits, and new concepts are continually being introduced in the television industry. For this reason, it is very important for the technician to be knowledgeable about the technical aspects of television systems. Since 1946 when the complete television receiver consisted only of *monochrome* (that is, black and white) circuitry, we have seen the introduction and growth of such innovations as color television, battery-operated portable television receivers, video-tape recorders, closed-circuit television, and pay television. Early television receivers employed as many as thirty vacuum tubes, but the trend is now toward solid-state components such as NPN and PNP transistors, field-effect transistors, silicon-controlled rectifiers, and integrated circuits.

Figure 1-1(A) shows a modern transistorized television chassis. This chassis employs ten *plug-in modules* to simplify the problem of servicing. The chassis slides out in front of the cabinet, as illustrated in Fig. 1-1(B), and the modules are readily removed for servicing or replacement.

Because of the considerable complexity of the television receiver, a whole new array of test equipment and of servicing methods has evolved. However, these have not completely supplanted the instruments and service techniques employed in connection with broadcast radio receivers, since much of what is done to a radio receiver to isolate a defect can be applied as well to a television receiver. Rather, these basic service methods have been expanded to encompass the newer circuits which television receivers possess, and which have no counterpart in radio. This situation in the field has led to a natural upgrading in the knowledge and skill of the service technician and a corresponding increase in his financial earning power. Thus, television happily is more enjoyable to the user and more profitable to the technician.

Television receivers may be housed in console cabinets like the one shown in Fig. 1-2(A), in table-model cabinets like the one shown in

A

B

Fig. 1-1. A transistorized, color-TV set, employing modular construction. Part A shows the ten removable modules. Part B shows how the entire chassis can be pulled forward for ease of servicing. (*Courtesy of Motorola, Inc.*)

A

Fig. 1.2. Two styles of cabinets for housing color-TV sets. A console model is shown in part A and a table model in part B. (*Courtesy of Zenith Radio Corp.*)

Fig. 1-2(B), or in small-screen portable packages. Regardless of the type of enclosure, the circuitry used in all of these receivers is basically similar. The large models may possess a greater number of transistors or tubes and perhaps additional speakers for better sound quality, but they do not differ in their theory of operation from the smallest battery-operated portable television receiver.

There are two types of receivers in general use: monochrome (black and white) and color. There is a considerable difference between the basic circuitry used in these two types of receivers, although there is relatively little circuit difference between monochrome receivers or between color receivers.

Picture-tube screens range from a few inches in the smaller sets (found only in portable receivers) to 24 inches for larger sets. At one time, 27- and 30-inch picture-tube screens were available, but these are no longer used. They were too large and bulky, and they presented many manufacturing difficulties. In addition, the larger picture was not found to be appreciably more desirable than the picture obtained with 23- and 24-inch screens.

Use of Semiconductors in Television Receivers There are a number of reasons why semiconductor devices have replaced tubes in the design of many television receivers. For signal amplification, the transistor is more efficient than a tube because it does not require a heated filament. The fact that there are no filaments also means fewer cooling problems with the solid-state operation. Since filament burn-out is the most frequent cause of tube failure, solid-state circuitry means greater reliability. Solid-state sets are also smaller and lighter than comparable tube sets.

The smaller physical size of semiconductor circuits and the lower cost of printed circuit fabrication makes it possible to have more automatic circuitry that simplifies the operation of television receivers. Figure 1-1(A) shows how modern transistor circuits are fabricated with compact circuit boards.

One of the more recent developments in the use of semiconductors is the *integrated circuit*. This is a method of making a number of circuits on a single semiconductor unit; the unit is called a *chip*. Figure 1-3 shows an integrated circuit. The inset of this illustration shows the chip with connections to the various individual circuits. One chip may contain a complete audio section, an IF section, or other section in the receiver. The chip shown in Fig. 1-3 contains a complete color demodulator.

The advantages of integrated circuits include an increased reliability over transistor circuits. This is because fewer soldered connections are needed for a complete circuit. The excellent reliability of integrated circuits had been proven many times in computer applications before they were first used in television receivers.

Another advantage of integrated circuits over transistor circuits is their lower cost when comparing the cost of fabricating a large number of integrated circuits with the cost of fabricating a large number of identical transistor circuits. Integrated circuits are much more compact than equivalent transistor circuits.

Fig. 1-2(B).

Monochrome Receiver Controls Controlwise, monochrome television receivers are only slightly more complicated than radio-broadcast sets. For the sound section of the television receiver, a conventional *ON-OFF, VOLUME CONTROL*, supplemented occasionally by a *TONE CONTROL*, is used. Instead of turning the ON-OFF volume control knob fully counterclockwise to turn the set off, some manufacturers prefer to use a push-pull type of control. With this control, the receiver is energized by pulling the knob out and de-energized by pushing the knob in Volume variation is obtained by turning the knob clockwise or counterclockwise. The advantage of this arrangement is the fact that the same volume control setting can be maintained when the receiver is turned off and then turned on again.

For the *video* (or, image-producing) section of the television receiver, some additional controls are required. These are few in number and can be easily manipulated. A primary control is the *CHANNEL SELECTOR*. It is often made in two parts: one for selecting VHF stations (Channels 2 to 13) and the other for selecting UHF stations (Channels 14 to 83). In the VHF range there may be as many as seven channels used in one community, but not more. If there are more than seven stations, the rest with be in the UHF range. The VHF channel selector is usually a 12-position switch with an associated *FINE TUNING* control, which is

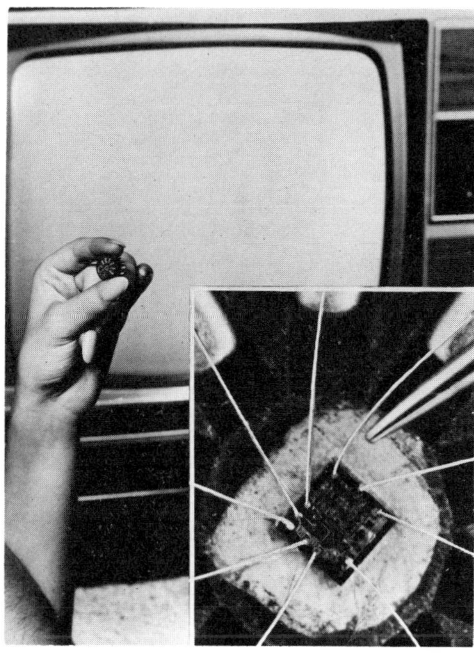

Fig. 1-3. An integrated circuit color demodulator. The monolithic chip, shown in the inset at the lower right, contains the equivalent of 19 transistors, 2 diodes, and 24 resistors, and is only 1/16 inch square. (*Courtesy of Zenith Radio Corp.*)

basically a vernier-tuning adjustment. The VHF tuning circuits used in television receivers are relatively fixed, and the desired station is obtained by means of the selector switch.

In addition to other circuits, the proper oscillator coil and capacitor are selected when the channel selector switch is moved into position. If a change should occur in the resonant properties of these latter circuits (due to heating or aging) and no fine-tuning adjustment were provided, the sound would possibly become distorted and the images would not be faithfully reproduced. To prevent this, a fine-tuning control is placed on the front panel. Within limits, this control permits the observer to *tune in* the signal, so that the proper frequencies are delivered to the receiver video and audio IF amplifiers.

Most UHF tuners are of the *continuous* type. Detent-type tuners, such as the VHF tuners with 12 selections, are considered easier to operate than the continuous type tuners, and legislators have applied pressure to TV manufacturers so that they will provide a detent-type tuning for UHF that is no more difficult to operate than VHF types.

Another control associated with the picture is the *CONTRAST CONTROL* which adjusts the relationship between the light and dark areas on the screen. Turning the control clockwise causes the darker areas of the picture to turn from gray to black. Rotating it counterclockwise causes the dark areas to lighten from black to gray. The contrast control is sometimes called the *PICTURE CONTROL* or *PIX CONTROL*.

Still another control is the *BRIGHTNESS CONTROL*, which regulates the overall shading of the picture and establishes the brightness of an image, that is, whether the overall image is light, medium, or dark. Controls which deal with the vertical and horizontal synchronization of the picture may also be found on the front panel of a television receiver. How these controls are manipulated will be discussed later. Generally, only those controls which *must* be manipulated by the customer are placed on the front panel. Others that require less frequent adjustment are positioned either on the side or back panel of the chassis or in a small covered recess on the front of the receiver. In any event, while the mechanism of a television receiver may be quite complex, use of a few necessary controls can be readily learned even by those entirely unfamiliar with the technical aspects of the system.

If the picture on the television receiver screen gets out of step (that is, out of *synchronization*) with the transmitted picture, then it will roll vertically or horizontally. The *VERTICAL HOLD* and *HORIZONTAL HOLD* controls are used for keeping the receiver synchronized with the transmitted picture.

Color Receiver Controls All of the controls discussed for monochrome receivers are usually included with color-television receiver controls. Two additional controls, *TINT*, or *HUE*, and *COLOR*, are also included. Figure 1-4 shows a typical control panel for a color television receiver.

Fig. 1-4. A typical control panel for a color-television receiver. (*Courtesy of Curtis Mathes.*)

The tint control selects the correct colors to be displayed. The most difficult colors to reproduce, and probably the most important, are the flesh tones. Because of the way color receivers are designed, if you obtain the proper skin colors with adjustment of the tint control, the other colors will be correct.

There are two things that can be accomplished by adjusting the color control. By rotating it fully counterclockwise, the color circuits are rendered ineffective, and the receiver reproduces the picture in black and white. Turning the control clockwise adjusts the amount of color for the picture. For example, this control determines whether the grass in the picture is dark green or light green, and whether the sky in the picture is dark blue or light blue.

The adjustment of the fine tuning control is more critical for color receivers than it is for monochrome receivers. A slight misadjustment of this control can result in a complete loss of color in the picture. Because of its importance, some color receivers are equipped with special circuits to simplify the adjustment of the fine-tuning control. Automatic-frequency control circuits are included in many receiver designs so that the receiver oscillator will not drift after it is properly adjusted. Some color sets also offer automatic control of flesh tones.

It is unfortunate that different manufacturers have used different names for the color and hue controls. Table 1-1 shows some of the names used by different manufacturers. Despite their different names, these controls perform the same functions from the viewer's standpoint as the control is adjusted as well as from the standpoint of what the control does electrically within the receiver.

As shown in Table 1-1, the *HUE CONTROL* may also be known by the name *TINT CONTROL*. However, the name *TINT CONTROL* is

TABLE 1-1 Color television controls.

Name of Control	Other Names Used	Its Function
Color	Chroma Color gain Color intensity	Controls the saturation of the color.
Hue	Color phase Tint Color tone	Primarily used to set the correct flesh tones. If these are correct, the other colors should also be correct.
Colorfast (or Tint)	Chromatone	Makes the monochrome pictures brown and white instead of black and white.

also used in some color receivers to operate a special circuit. When a black-and-white picture is being received, the viewer may change the picture to brown and white by operating this so-called tint control.

Thus, it is not only possible that different names may be used for the same control, but also, the same name may be used for two completely different controls. A good technician will take a moment to become acquainted with the controls of the individual receiver if he is not familiar with the particular model that he is working on.

1.2 PICTURE DETAIL

Consider an ordinary photograph, such as is shown in Fig. 1-5. This picture was obtained from a negative that contained a large number of chemical grains which were originally sensitive to light. When a photograph is printed in a newspaper or book, the usual procedure is to divide the picture into very small parts called *picture elements*. Dividing the picture into tiny elements permits one shade of ink (usually black) to be used for producing all of the different shades of black, gray, and white shown in the picture. When a picture is divided into tiny picture elements it is said to be *half toned*. With the aid of a strong magnifying glass you can easily see the picture elements in Fig. 1-5.

In order to give a smooth, continuous picture, the picture elements of a photograph, or a half-toned illustration, should be so small that they are not readily visible with the naked eye. A fine-grain illustration, with many picture elements per unit of area, can be viewed more closely than a coarse-grain picture without these elements being discernible.

With television images, the same kind of situation prevails. The television picture is displayed on the screen of a picture tube like the one shown in Fig. 1-6. This is a cathode-ray tube somewhat similar to the ones used in oscilloscopes. The electron beam from the gun produces light when it strikes the screen. The amount of light depends upon the

Fig. 1-5. A television studio scene developed from a negative. This picture has been half toned.

beam current. The beam can be moved to any point on the screen by using the magnetic field of a deflection current.

In the receiver, each picture element is just as large as the area of the circular electron beam striking the fluorescent screen of the picture tube. The light that is seen when observing a picture-tube screen is derived from the energy given off by the electron beam striking particles of the fluorescent coating on the inner face of the tube. If the points of light are closely spaced, the eye of the observer will integrate them, and their individual character as separate points will disappear. Hence, an electron beam of small diameter is one of the first requirements for a television picture that is to reproduce any amount of fine detail. This requirement is important for the receiver picture-tube, and for the camera tube.

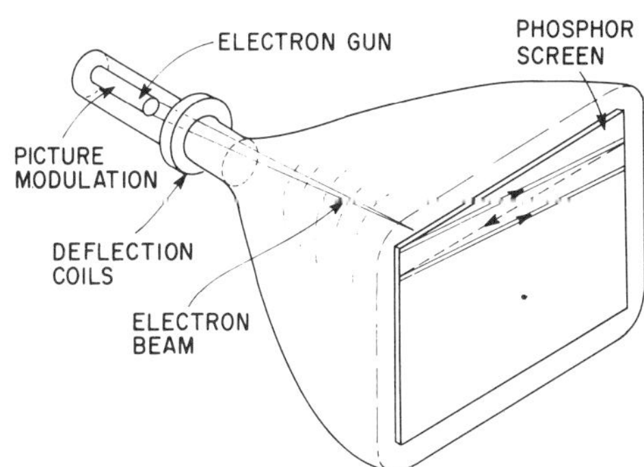

Fig. 1-6. The television picture displayed on a picture tube.

The television picture is produced by moving the electron beam rapidly back and forth across the screen to produce about 500 lines. This procedure is called *scanning.* As the lines are scanned, the beam is made brighter and dimmer in order to "paint" the picture. Ideally, the individual lines making up the picture should not be visible, just as the picture elements in the photograph of Fig. 1-5 are not easily visible. When the picture elements are seen, they distract the viewer and they make the picture appear to be blurred.

If you crowd 500 lines together to make a picture that is eight inches high, you cannot see the lines easily. You must make a close inspection of the screen before they are distinguishable. If, on the other hand, you use the same number of lines to make a picture twenty inches high, the lines making up the picture are easily discernible unless you move back some distance from the screen. If you could increase the number of lines in the larger picture, you would not be able to see the individual lines. However, the allowable number of lines is established by the FCC. Therefore, there is a limit to the size of the picture you can make. More important, the correct size of the picture on your receiver is determined by the distance from which you are going to view the picture. When you view a picture at a distance, it is difficult to see the lines.

1.3 INTRODUCTION TO SCANNING

In the last section, it was mentioned that the picture on the picture tube screen is "painted" by moving an electron beam back and forth as it scans individual lines, each one separated from the other by a very narrow margin.

Figure 1-7 shows the basic principle involved in scanning a picture. Starting at the top left side, the electron beam moves from left to right across the screen along line 1-1. (In this illustration, it is presumed that you are facing the screen of the picture tube.) After the beam reaches the right side of the screen at point 1, it is rapidly moved to the left side of the screen again along dotted line 1-2. During its motion from right to left, no picture information is transmitted. The line marked 1-1 is called a *trace*, and the line marked 1-2 is called the *return trace*, or *retrace*.

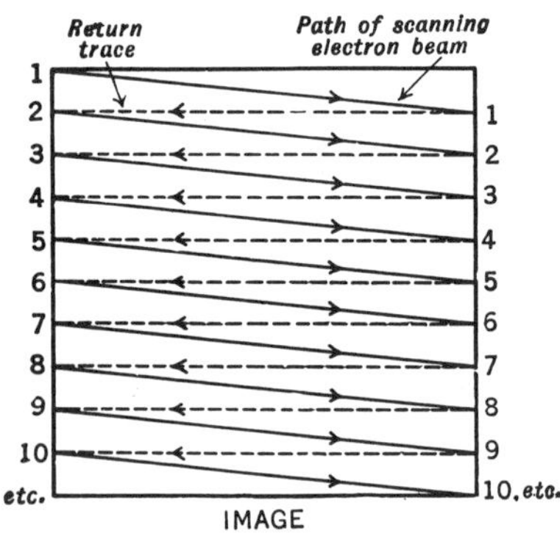

Fig. 1-7. The simplified procedure for scanning a television picture. Solid line 1-1 is called a *trace*, and dotted line 1-2 is called a *retrace*.

After retrace, the beam has arrived at the left side of the screen at point 2. It then moves from left to right across the screen for the next line. This procedure is repeated throughout the picture.

It is important to know that this is a simplified illustration of the procedure for scanning. In actual practice, the scanning procedure is modified somewhat. The important thing to note here is that the scanning procedure is the same as used when reading a line on a page of paper. The reader starts at the top left side of the page and moves his eyes across a line of type from left to right. Then, after reaching the right side of the page the eye quickly moves to the left side to begin the next line.

The electron beam moves back and forth across the screen of the picture tube regardless of whether or not there is a picture being displayed. In the absence of a picture, the beam will trace out a white rectangle on the screen of the picture tube. This lighted rectangle is called a *raster*. When the standards for television were first established, it was decided that a rectangular picture (or rectangular raster) would be most desirable, and the relationship between the height and width of the rectangle was established. The width of the rectangle divided by its height is called the *aspect ratio*, and in television the aspect ratio of the picture (and the raster) is four to three. In other words, if the picture is four inches wide, it must be three inches high in order to have the correct aspect ratio. Likewise, if the picture is twelve inches wide it must be nine inches high.

By adjusting the voltages to the picture-tube circuitry, it is possible to produce a rectangle with different aspect ratios. However, it is important to remember that the picture is *transmitted* in an aspect ratio of four to three. Therefore, in order to reproduce the transmitted picture with a minimum amount of distortion, it is necessary that the receiver picture also have an identical aspect ratio of four to three.

1.4 SCANNING RATE

After the trace reaches the bottom of the raster, it is quickly moved back to the top of the screen where it starts the production of scanning lines again. What actually appears on the television screen is a number of complete rasters (or fields) each second.

When the requirements of television were first established, it was necessary to decide on a standard repetition rate for showing complete pictures, that is, the number of complete frames per second. The actual choice of frame frequency involves a tradeoff. Ideally, a high frame frequency should be used so that the eye cannot perceive the individual pictures making up the continuous motion. However, a high frame frequency produces electronic problems in the scanning. For one thing, the amount of brightness on the screen is definitely affected by the rate at which the beam is moved across the screen. Specifically the faster the beam is moved, the more difficult it is to get sufficient brightness. Also a higher frame rate requires a wider transmission bandwidth.

Thus, to obtain the desired amount of brightness, to utilize a reasonable transmission bandwidth, and at the same time give the effect of continuous motion, a compromise was needed. It was finally decided that thirty complete pictures would be displayed on the screen each second. These complete pictures are referred to as *frames*. A frame frequency of only thirty pictures per second would result in a flicker discernible to the eye, so each picture is divided into two parts called *fields*. Two fields must be produced in order to make one complete picture or frame. The field frequency, then, is obviously sixty fields per second and the frame frequency is thirty frames per second. Each field contains one half of the total picture elements.

A situation similar to this occurs in motion pictures. Each individual picture is moved into the projector and flashed onto the screen. Then a shutter comes down in front of the picture momentarily and the same picture is shown again. The purpose of using the shutter dividing the picture into two displays is to reduce the flicker rate. In motion pictures, each frame is shown twenty-four times per second and each frame is divided into two parts, or fields, so that the actual repetition rate is 48 fields per second. By dividing the frames into fields, the eye is tricked into believing that the motion is continuous and no flicker is observed by the eye.

The picture on the television screen is also divided into two parts, but in a different manner, which is known as "interlaced scanning," to be described in the next chapter. As is the case with motion pictures, two fields are used to produce a complete frame, or picture.

The choice of sixty fields per second and thirty frames per second was deliberate, in order that the television picture would be synchronized to the standard United States power-line frequency of 60 Hertz. At the transmitter, all scanning frequencies are derived from the 60-Hertz line frequency, and the receivers are automatically synchronized to the transmitted synchronizing pulses. As a result of this power-line frequency synchronization, the effects of hum on the picture, which might be caused by imperfect power supply filtering, will be stationary on the TV receiver screen. If this synchronization did not occur, the hum effects would cause vertically moving patterns to pass through the picture. Adjacent cities, having TV transmitters frequently use synchronized power-line frequencies to prevent this unwanted type of interference.

It is interesting to note that when color television programs are broadcast, the field frequency is not 60 Hertz, but is reduced slightly to 59.94 Hertz. Thus, while hum interference patterns are not perfectly synchronized for color broadcasts, they move vertically at the rate of only 0.06 Hertz, which is a slow rate and not readily visible.

1.5 SUMMARY OF DESIRABLE IMAGE CHARACTERISTICS

Since the image is the final product of the television system, and everything centers about the production of this image, a summary of its most important characteristics will now be given. In order for a picture to be satisfactory from the observer's point of view, the following minimum requirements should be obtainable:

1. The composition of the image should be such that none of the elements that go into its make-up are visible from ordinary viewing distances. The image should have the fine, smooth appearance of a good photograph.

2. If a color picture is being displayed, the colors should be realistic. There are many different shades of blue, but most of them would not be suitable for the color of sky. A good picture will be a reproduction of the colors as they exist in real life.

3. The eye must not be able to perceive a flicker in the reproduced picture. To accomplish this, the television receiver displays sixty fields per second. An advantage of using a high number of fields per second is that it makes motion on the screen to appear to be smooth and continuous.

4. If the picture is to be viewed by more than one person, then it should be large enough so that it can be viewed comfortably by everyone. On the other hand, the picture must not be large enough to permit the viewer to see the lines that make it up.

5. To meet the changing requirements for viewing the screen either by day or by night, an adequate amount of light must be available from the picture tube. Less image brightness is necessary when the room illumination is low than when it is high, so a brightness *control* is needed.

6. An effective contrast range is desirable. Contrast refers to the ratio between points of maximum to minimum brightness on the same screen. In broad daylight, for example, the contrast ratio between areas in bright sunlight to shaded areas may run as high as 10,000 : 1. On fluorescent screens, however, the amount of light that can be emitted is definitely limited, and only contrast ratios up to 100 : 1 are obtainable ordinarily. These, however, prove quite satisfactory. In order to be able to take advantage of the maximum range of contrast values, a contrast *control* is needed.

The foregoing requirements have been listed with only a short explanation advanced for each. Each of these characteristics will be discussed again, in more detail, when the applicable circuitry is studied.

1.6 BLOCK DIAGRAMS OF THE TELEVISION SYSTEM

Before beginning a detailed discussion of television circuitry, it will be useful to get an overall view of the television system in block diagram form.

Figure 1-8 shows a simplified block diagram of a television transmitter. There are actually two transmitters involved in this system. One is used for transmitting the picture (or *video*) signal. The other is used for transmitting the sound (or *aural*) signal.

The TV camera converts the scene being televised into electrical impulses which can be amplified and transmitted. The camera tube scans the scene in a manner that is the same as the way the picture is scanned in the picture tube of the receiver. Therefore, scanning and synchronizing voltages must be delivered to the camera. A *synchronizing signal* must also be transmitted so that the scene in the receiver can keep in step with the scene at the transmitter. The required scanning and synchronizing signals are also delivered to the video amplifiers and mixers in the transmitter. The output signals from the video amplifier and mixers are delivered to an amplitude-modulation system. This is simply an AM transmitter, but it does differ from the transmitters used in AM broadcast radio. Instead of both sidebands being transmitted, only one full sideband and a part of the other is contained in the output signal. The purpose of eliminating most of one sideband is to conserve the electromagnetic spectrum, so that more stations can transmit in a given range of frequencies. Transmitting with one full sideband and one partial sideband is called *vestigial sideband transmission*. It is not to be confused with single sideband transmission in which one of the sidebands is completely removed.

If a color television scene is being transmitted, then the output from

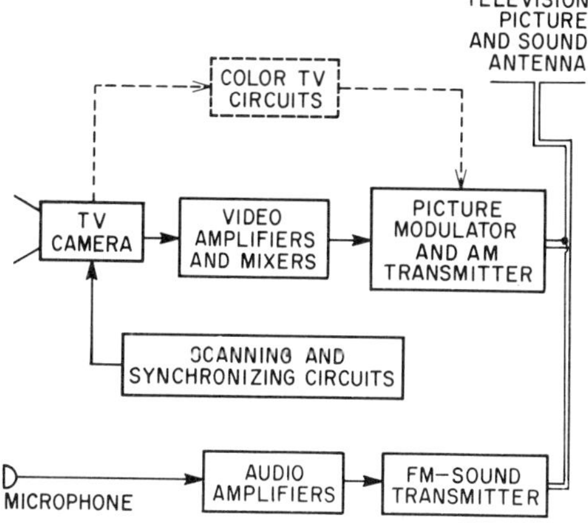

Fig. 1-8. Simplified block diagram of a television transmitter.

the color TV camera goes through special color processing circuits. This is shown in dotted lines in Fig. 1-8. It is important to understand that when the color signal is being transmitted, an equivalent monochrome signal is also being transmitted. This is a requirement which was made when color television was first introduced to the American public. It was necessary that the already-existing (monochrome) receivers would be able to receive and display the signal from a color transmission with no change in quality. In other words, it was necessary that the color and monochrome systems be *compatible*. Basically, color transmission involves the use of special color TV circuits to superimpose the color signal onto the already-existing monochrome signal.

The audio system in a TV transmitter is completely separate from the video section. It consists of a microphone, audio amplifiers, and a standard FM transmitter. Both signals are transmitted from the same antenna by the use of a special coupling device called a *diplexer*.

Figure 1-9 shows a simplified block diagram of a television receiver. The blocks drawn with the solid lines represent the monochrome circuitry necessary for reproducing the picture in black and white. If a color signal is present, a special color demodulator (shown in dotted lines) may be used. Of course, the picture tube used must also be capable of reproducing the colors.

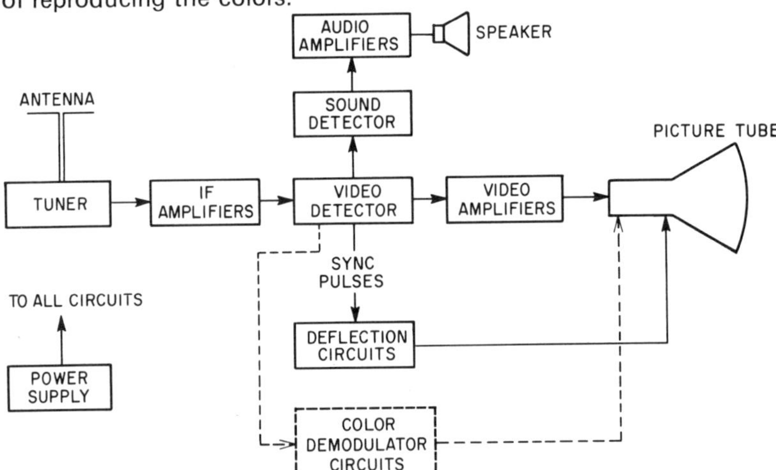

Fig. 1-9. Simplified block diagram of a television receiver.

There are four sections common to all receivers, and these sections are present in the block diagram of Fig. 1-9. The four basic sections are: an antenna system, a method of *selection*, a method of *detection*, and a device for reproducing the intelligence.

For television receivers, the antenna is considerably more elaborate than those used for radio receivers. This is because the signal is affected much more by reflections from large objects, and also because the signal has a more limited distance over which it can travel.

The selection portion of the receiver is accomplished in the tuner. It makes possible the selection of one channel and rejection of all others.

All television receivers in use today are superheterodyne types, so the tuner section includes a local oscillator for converting the RF signal to an intermediate frequency.

The output of the tuner is delivered to the IF amplifiers and then to a detector. This is the third section required of all receivers. Detection in the television receiver actually involves two steps. The video signal is amplitude modulated, and therefore, an AM detector is needed for that part of the signal; and, the sound portion is frequency modulated, so an FM detector is needed for that portion of the signal. The picture and sound signals are amplified by the video and audio amplifiers respectively. The picture tube reproduces the video signals, and the receiver speaker reproduces the sound signals.

You will remember that special signals were combined with the television signal in order to make it possible to synchronize the received picture with the one that is transmitted. These special signals are called *synchronizing pulses*, or *sync pulses* Sync pulses control the frequency of the deflection circuit currents. These currents cause the electron beam to scan the picture-tube screen.

If a color signal is present, special color circuit demodulators are present as shown in dotted lines in Fig. 1-9. There is an automatic circuit in a color television receiver that prevents the color demodulators from operating when there is no color signal present. This is done automatically without the need for switching by the person viewing the signal on the receiver. Thus, the transmitter may switch back and forth between color pictures and monochrome pictures, and the receiver control circuits will automatically determine which circuitry will operate to produce the monochrome or color picture.

1.7 TELEVISION SYSTEMS THAT USE CABLES

Instead of transmitting the television signal from one point to another by using electromagnetic waves, it may be transmitted along cables. One system that uses cables is called *closed-circuit television*. With this system the signal is transmitted from the camera to one or more receivers along a coaxial cable, and the signal is not receivable by the general public. In a typical application, the camera is mounted at some point where it would be inconvenient or impossible for a human to be, such as near an atomic energy experiment.

Since the signal for closed-circuit television is not radiated into space, it does not come under the rules and regulations of the FCC. Furthermore, much less power is required for operating a closed-circuit television system. The large RF power amplifiers are not required, so the system has a lower initial cost and lower operating cost. Another advantage sometimes listed is the privacy of the system. No set that is not connected to the cable can receive the signal.

The development of transistor video circuits has led to widespread educational, industrial, and private use. It would be impossible to list all of the uses, but a few will be discussed as representative.

Educational Uses A small closed-circuit camera may be located close to an operation or laboratory experiment, so that a large number of students may get a close-up view of what is going on. A lecturer, or the school principal or president, can address a large number of people in many different locations simultaneously. Students can practice television acting, programming, and directing with a low-budget television system.

Industrial Uses Closed-circuit television cameras can be mounted to monitor many meters and gauges from a single remote position. Close-up views of hazardous operations, such as very high temperature or very low temperature experiments and machine handling of radioactive materials, can be continuously monitored from a remote position. Detection of burglary, fires starting in remote positions, assembly line failure, etc., is an easy matter with closed-circuit cameras strategically located.

Private Uses As in industry, protection of personal property against burglary and trespassing can justify the relatively low cost of a closed-circuit television system. A camera in the baby's bedroom will allow a busy housewife to keep an eye on her child while doing her housework, and one outside the door will allow her to identify callers before opening the door.

The almost unlimited applications of closed-circuit television means more work for the television technician. The circuitry for both the camera and the receiver is less complicated than the commercial systems used in regular television.

Cablevision is another type of system in which television signals are transmitted over coaxial cables. In this system, signals from stations that are normally out of range of the viewer's receiver are delivered to the house. Cablevision permits the viewer to select from a larger number of programs. Subscribers pay for this service monthly much in the same way as the telephone bill is paid.

In large apartment buildings it is often difficult to obtain a television picture without using an outside antenna system. With 150 or 200 apartments in the building, the roof would be too cluttered if each apartment has its own antenna system. To get around this problem, a *master antenna television* system—abbreviated MATV—is used. With MATV, a single antenna system is mounted on the roof and the signal is distributed to each apartment through coaxial cables. The signal may be amplified before it is distributed.

Some communities are located in such a way that it is difficult for them to receive a television signal. An example is where the community is in a valley, and is shielded from the signal by nearby mountains. A master antenna system can be used to distribute the signal by coaxial cable to each house in the community. Such a system is called *community antenna television*, or simply, CATV.

1.8 EXTENDING THE COVERAGE OF TELEVISION

Television signals are transmitted in the range of frequencies between 54 and 890 MHz. This range includes frequencies in the very high frequencies (VHF) and in the ultra-high frequencies (UHF).

Unlike the lower frequencies used for AM broadcast radio, these frequencies are not reflected from the ionosphere. Instead, their transmission is limited to *line-of-sight distances,* which means a distance of about 45 miles from the transmitter. The value of 45 miles is an average because the terrain between the transmitter and the receiver antennas, and the heights of these antennas, must be taken into consideration.

By the use of cablevision and community antenna systems, transmission well beyond the line of sight can be achieved. Both of these systems use transmission lines. There are also methods of extending the coverage of television by ways other than by the use of cables.

Transmission across the ocean can be achieved by the use of *active satellites.* An active satellite receives a signal, amplifies it, and then retransmits it. It is different from a passive satellite that only echoes the signal by reflecting it back to Earth.

Figure 1-10 shows how the active satellite works. The televison signal that is transmitted from point A is received by the satellite and retransmitted to the station at point B. Note the line-of-sight distance over which transmission is normally achieved.

For transmitting signals coast-to-coast, *microwave relay stations* (Fig. 1-11) are used. These are simply repeater stations that receive the signal, amplify it, and then retransmit it. The relay stations are located at intervals of 50 to 100 miles.

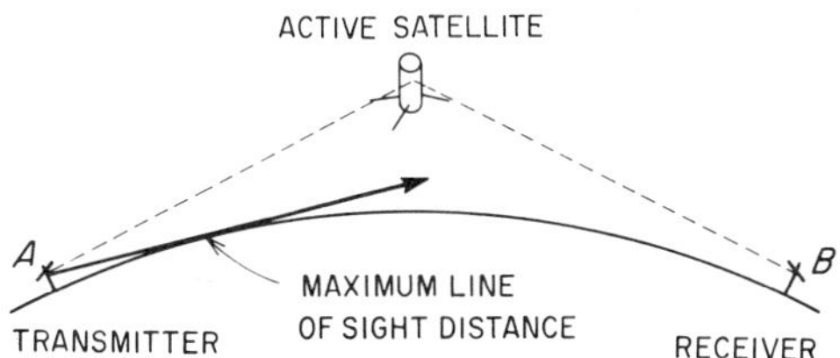

Fig. 1-10. Active satellites are used for extending television coverage beyond the line of sight.

Fig. 1-11. Microwave relays are used for transmitting signals over long distances.

1.9 SUMMARY OF TELEVISION STANDARDS

The transmitted television signal must comply with strict standards established by the Federal Communications Commission (FCC). As an introduction to a more detailed study of television circuitry, it will be useful to know what some of these standards are.

Each VHF and UHF television station is assigned a channel that is 6 MHz wide. The complete signal—called the *composite signal*—must be fitted into this 6 MHz bandwidth. The composite signal includes: the video carrier, one vestigial sideband for the video signal, one complete sideband for the video signal, the color signals, the synchronizing pulses, and the FM sound signal.

The video part of the composite signal is amplitude modulated. In order to obtain the same number of frequencies for the video signal, using frequency modulation, the required bandwidth would be prohibitive. The synchronizing pulses are also transmitted by amplitude modulation, and these pulses are superimposed onto the video signal. The color information is transmitted by a clever method of slipping the color signals in between spaces in the monochrome video signals. This process is called *interleaving.*

TABLE 1-2 Frequencies of television channels.

Channel No.	Channel Freq. (MHz)	
2	54–60	lower VHF band
3	60–66	
4	66–72	
5	76–82	
6	82–88	
7	174–180	upper VHF band
8	180–186	
9	186–192	
10	192–198	
11	198–204	
12	204–210	
13	210–216	

UHF Band

Channel No.	Channel Freq. (MHz)	Channel No.	Channel Freq. (MHz)	Channel No.	Channel Freq. (MHz)
14	470–476	38	614–620	62	758–764
15	476–482	39	620–626	63	764–770
16	482–488	40	626–632	64	770–776
17	488–494	41	632–638	65	776–782
18	494–500	42	638–644	66	782–788
19	500–506	43	644–650	67	788–794
20	506–512	44	650–656	68	794–800
21	512–518	45	656–662	69	800–806
22	518–524	46	662–668	70	806–812
23	524–530	47	668–674	71	812–818
24	530–536	48	674–680	72	818–824
25	536–542	49	680–686	73	824–830
26	542–548	50	686–692	74	830–836
27	548–554	51	692–698	75	836–842
28	554–560	52	698–704	76	842–848
29	560–566	53	704–710	77	848–854
30	566–572	54	710–716	78	854–860
31	572–578	55	716–722	79	860–866
32	578–584	56	722–728	80	866–872
33	584–590	57	728–734	81	872–878
34	590–596	58	734–740	82	878–884
35	596–602	59	740–746	83	884–890
36	602–608	60	746–752		
37	608–614	61	752–758		

There are 30 complete pictures, or frames, sent each second. Each frame is divided into two parts, or fields; and therefore, there are 60 fields transmitted every second.

There are 525 scanning lines generated for each frame, but only about 480 of these are actually used to make up the picture. Those lines that are actually used for producing the picture are called active lines. The lines that are not used for actually making the picture are generated during the time that it takes to get the beam from the bottom of the picture (at the end of the field) back to the top of the picture (at the start of a new field).

The sound signal is frequency modulated, and as with FM broadcast systems, it may also be produced by the *indirect FM method.* This actually produces phase modulation (PM), but any FM detector that can demodulate an FM signal can also demodulate a PM signal. The maximum deviation for TV, is ± 25 kHz.

The aspect ratio of the transmitted picture is 4 : 3, but the picture is sometimes purposely "stretched" at the receiver in order to give the appearance of a larger picture area.

Table 1-2 shows the frequencies assigned to the VHF and UHF channels.

REVIEW QUESTIONS

1. In what ways are the television sets of today different from the first receivers made?
2. In what way is the process of scanning like reading a page in a book?
3. Why is the video part of television transmission sent by amplitude modulation rather than by frequency modulation?
4. Which circuits in a TV receiver are controlled by the sync pulses?
5. What does the tuner section of a receiver do?
6. What are some advantages of using transistors for television circuits instead of using vacuum tubes?
7. What is an MATV system?
8. What are active satellites used for in television?
9. What is the bandwidth of a television channel?
10. What controls are used on color-television receivers, in addition to those used on monochrome receivers?
11. What is a raster?
12. What are some uses for closed-circuit television?

2 Principles of Scanning, Synchronizing, and Video Signals

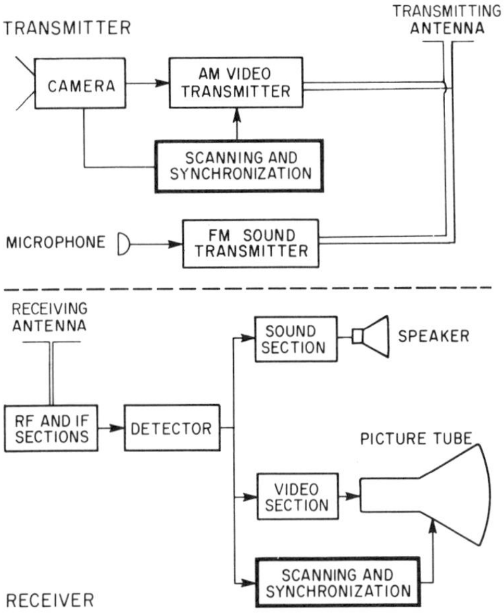

TRANSMITTER
TRANSMITTING ANTENNA
CAMERA
AM VIDEO TRANSMITTER
SCANNING AND SYNCHRONIZATION
MICROPHONE
FM SOUND TRANSMITTER
RECEIVING ANTENNA
SOUND SECTION
SPEAKER
RF AND IF SECTIONS
DETECTOR
PICTURE TUBE
VIDEO SECTION
SCANNING AND SYNCHRONIZATION
RECEIVER

Fig. 2-1. Simplified block diagram of a television transmitter and receiver.

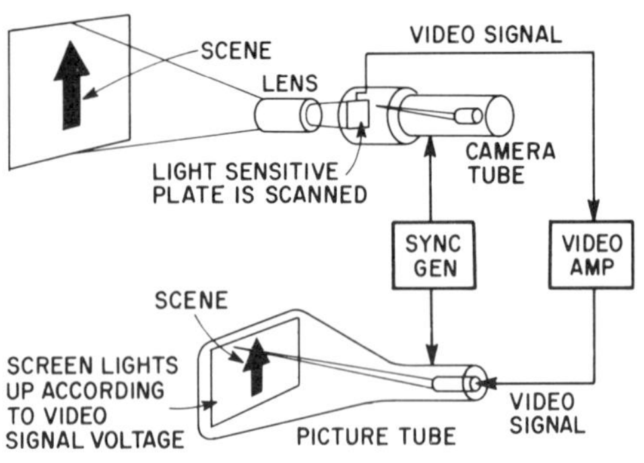

Fig. 2-2. The scene is focused onto a light-sensitive plate in the camera tube. The same scanning process is used in the receiver to reconstruct the scene.

2.1 INTRODUCTION

Figure 2-1 shows the basic components of a system for transmitting and receiving television programs. The sections related to scanning and synchronizing are shown in heavy outlines.

Scanning and synchronizing are needed for keeping the reproduced picture in step with the picture originated at the transmitter. The synchronizing *signals*, which will be referred to simply as the *sync signals*, are generated at the transmitter where they amplitude-modulate the transmitted signal. The receiver uses the sync signals to synchronize the scanning of the electron beam of the picture tube with that of the transmitter camera tube. As will be shown later, the sync signals are actually used for controlling the frequency of oscillators used in the scanning and synchronizing section of the receiver.

Another function of the scanning and synchronizing sections is that of generating *blanking signals*. For reasons to be described later in this chapter, it is desirable to turn off the electron beam in the camera, or in the receiver picture tube, during retrace periods. This allows the beam to be repositioned for the start of a new line or new picture. The blanking signals originate in the scanning and synchronizing section of the transmitter.

2.2 PRINCIPLES OF SCANNING

Figure 2-2 illustrates the basic elements of scanning in transmitter and receiver systems. The scene to be televised is focused by a lens system onto a light-sensitive plate located in the camera tube. This plate is made in such a way that the electric charge at any point on the plate depends upon the amount of light falling on that point. The camera tube video signal output current depends directly upon the amount of light falling on the plate at each point. The video signal, then, is comprised of variations of current which correspond to variations in brightness on the light-sensitive plate. This video signal is amplified and delivered to the gun of the receiver picture tube. In Fig. 2-2, the intervening transmitter and radio wave are not shown.

In order for the video signal from the camera tube to reproduce the picture at the receiver picture tube, the beam is swept back and forth across the screen in step with the beam that is sweeping the light-sensitive plate in the camera tube. The synchronizing generator provides the voltage for controlling the motion of the electron beam within the camera tube and also for controlling the beam in the picture tube. As the picture-tube electron beam scans any one point, its strength, that is, the number of electrons contained in its beam, is controlled by the amplitude of the video signal. At the same time, the location of the beam is controlled by the voltage from the sync generator.

To summarize: Scanning is employed in *both* the transmitter and the receiver. Synchronizing signals must be used to assure that the position of the electron beam in the receiver picture tube corresponds with its same position on the light-sensitive plate of the camera tube for any given instant of time.

There are many possible ways to scan a picture. For example, the beam could be started in the center and moved outwardly in a spiral. Also, the electron beam could be scanned vertically. However, television standards set up a horizontal scanning system, which means that the beam moves back and forth across the picture tube from left to right as you are viewing the picture tube.

Simplified Electron-Beam Scanning The following explanation of electron beam scanning is not the actual process used, but is a simplified method of scanning, presented as an introduction. While we are now discussing the action of the beam in a television camera tube, bear in mind that the action of the electron beam in a television picture-tube is identical.

An electron beam is formed and accelerated toward the upper left-hand corner of the camera tube image plate (point A in Fig. 2-3). From this point and under the influence of the currents in the deflection coils (positioned on the neck of the camera tube), the electron beam moves to the right, passing over the charged image plate which has been exposed to the focused rays of light from the televised scene. The beam continues along the first line until the end (point B) is reached. Here a generator connected to the camera tube will cut off or blank out the beam while the deflection coils bring it rapidly back again to point C at the left-hand side of the mosaic. This point is slightly below the first line. The blanking voltage is now removed, and again the cathode-ray beam moves toward the right. The sequence recurs until the end of the lowermost line is reached (at point D). The beam is blanked out and returned to the starting point A. The entire process is now ready to be repeated. In practice the beam passes over the entire image plate every 1/30 of a second. Hence 30 complete pictures are sent every second.

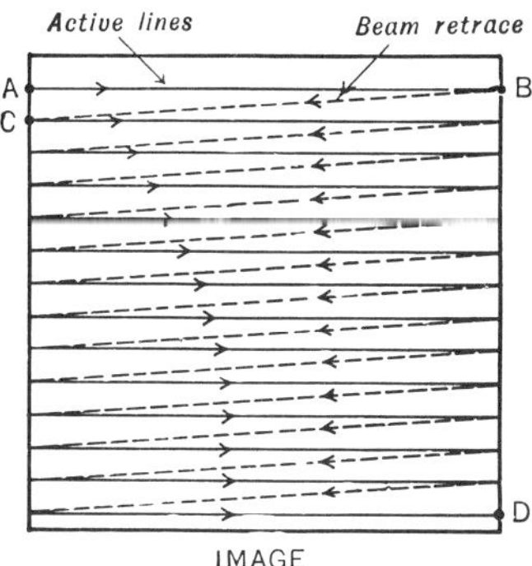

Fig. 2-3. The motion of the electron beam in one form of horizontal scanning.

In actual equipment, the motion of the scanning electron beam, as described previously, must be modified somewhat for two reasons. First, it is extremely difficult to generate a voltage that will cause the beam to drop suddenly from the end of one line to the level of the next

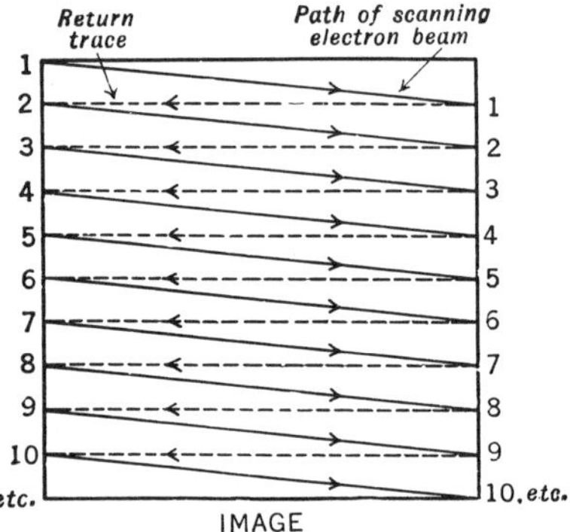

Fig. 2-4. In actual equipment it is easier to have the electron beam travel in the manner indicated above than that in Fig. 2-3.

one directly beneath it. It is simpler to have the beam move to the level of the second line gradually, as is illustrated in Fig. 2-4.

To obtain this type of motion for the electron beam, both the horizontal- and vertical-deflection coils in the camera tube are utilized. Without going into an extensive discussion at this time about the operation of the electron gun located in the neck of the camera tube, let us simply state that the horizontal-deflection coils can move the electron beams horizontally across the screen from left to right and back again, and the vertical-deflection coils can cause the beam to move vertically. When used together, and with different amounts of currents passing through each set of coils, it is possible to move the electron beam across the screen to reach any desired point.

In the foregoing type of motion (with the beam moving across the screen slantwise), we have the equivalent of a fast-acting current in the horizontal coils quickly forcing the beam across, while a slow-acting current in the vertical coils is forcing the beam down. The result of this is shown in Fig. 2-4. When the beam reaches the end of a line, it is quiclky brought almost straight across (with the blanking signals on) where it is in correct position to start scanning line 2 when the blanking voltage is removed. The remainder of the lines follow in similar fashion. At the bottom of the picture, after the last line has been scanned, a longer blanking signal is applied while the beam is returned to the top of the picture. The purpose of the blanking voltages is simply to prevent the beam from impinging on the screen when the beam is merely moving into position for the next scanning run.

A possible current that could be used for the horizontal- and vertical-deflection coils is the sawtooth wave illustrated in Fig. 2-5. This current rises gradually to a fixed level and then suddenly drops (almost vertically) to zero to begin the process over again. More will be mentioned about sawtooth-wave generators when the television receiver is discussed.

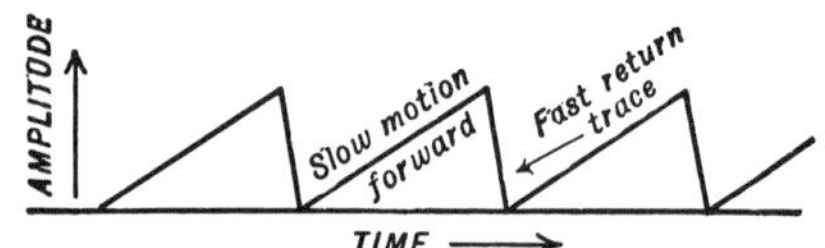

Fig. 2-5. A sawtooth current, as illustrated here, when passed through a set of horizontal deflecting coils, will cause the electron beam to move from left to right and to left again.

For the second reason why the horizontal scanning process must be modified, we must examine more closely the human eye and its action when observing motion on a screen.

2.3 FLICKER

If a set of related still films follow each other fairly rapidly on a screen, the human eye is able to integrate, or combine, them, and the motion appears continuous. The eye can do this because of a phenomenon called *persistence of vision*. Due to this property of the eye, visual images do not disappear as soon as their stimulus is removed. Rather, the light appears to diminish gradually, taking on the average, about 1/50 second before it disappears entirely. This situation is very fortunate, for otherwise, motion picture and television entertainment might be impossible.

It has been found that when theater films are presented at a rate of

15 stills per second, the action appears continuous. However, at this speed, flicker is still detectable and detracts from the complete enjoyment of the film. The flicker is due to the impression in the viewer's mind decreasing to too low a value before the next film is presented on the screen. Increasing the rate at which the stills are presented will gradually cause the flicker to disappear. At 50 frames per second—that is, 50 complete pictures per second—there is no trace of flicker, even under adverse conditions. The rate is not absolute, however, but depends greatly upon the brightness of the picture. With average illumination, lower frame rates prove satisfactory.

In the motion-picture theater, 24 individual still films (or frames) are flashed onto the screens each second. Since at this rate, flicker is somewhat noticeable, a shutter in the projection camera breaks up the presentation of each frame into two equal periods. The fundamental rate is therefore increased to an effective rate of 48 frames per second. This is accomplished by having the shutter move across the film while it is being projected onto the screen. Thus we are actually seeing each picture twice. By this ingenious method, all traces of flicker are eliminated.

In television, a fundamental rate of 30 images (or frames) per second was chosen because this frequency and the effective rate are related to the frequency of the ac power lines. Practically, this choice of frame-sequence rate necessitates less filtering in order to eliminate an ac ripple, which is called a *hum* in audio systems. With 24 frames per second for example, any ripple that was not eliminated by filtering would produce a weaving motion in the reproduced image. Less difficulty is encountered from an ac ripple when 30 frames per second are employed.

To eliminate all traces of flicker, an *effective* rate of 60 frames per second is utilized. This is accomplished by increasing the downward rate of travel of the scanning electron beam so that *every other line* is sent instead of every successive line. Then, when the bottom of the image is reached, the beam is sent back to the top of the image, and those lines that were missed in the previous scanning are now sent. Both of these operations, the odd- and even-line scanning, take 1/30 second; therefore 30 frames is still the fundamental rate. However, since all the even lines are transmitted in 1/60 second and the same is true of the odd lines, they add up, of course, to 1/30 second. To the eye, which cannot separate the two, the *effective* rate is now 60 frames per second, and no flicker is noticeable.

To differentiate between the actual fundamental rate and the effective rate, we say that the *frame frequency* is 30 Hertz, whereas the effective rate (called the *field frequency*) is 60 Hertz. This method of sending television images (Fig. 2-6) is known as *interlaced scanning.*

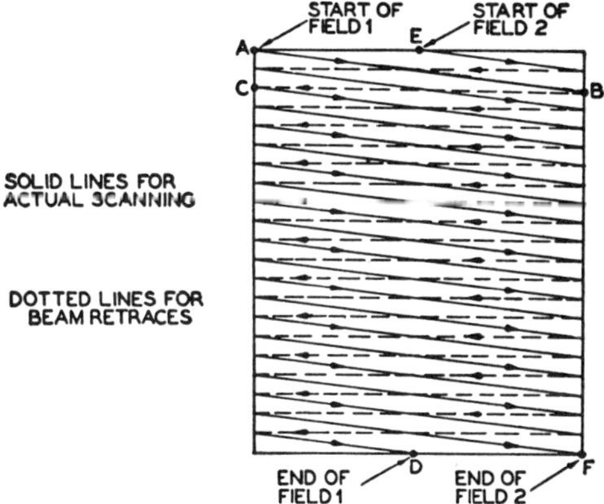

Fig. 2-6. The path of the electron beam in interlaced scanning.

Each complete scene is sent at a rate of 30 frames per second. To obtain the desired amount of detail in a scene, the picture is divided into a total of 525 horizontal lines. The technical reasons behind the choice of 525 lines are related to these requirements:

1. The frequency bandwidth available for the transmission of the television signals. As will be shown later, the required bandwidth increases with the number of lines.

2. The amount of detail required for a well-reproduced image.

3. The ease with which the synchronizing (and blanking) signals can be generated.

With each frame divided into two parts (because of interlaced scanning), each field will have one-half of 525 lines, or 262-1/2 lines, from its beginning to the start of the next field. As a matter of definition, a complete picture is called a *frame*. With interlaced scanning, each frame is broken up into an even-line field and odd-line field. Each field contains 262-1/2 lines, whereas a frame has 525, the full amount.

2.4 THE COMPLETE SCANNING PROCESS

From the foregoing discussion, it becomes possible to reconstruct the entire scanning process. Although only the movement of the electron beam at the television camera will be considered, an identical motion exists at the receiver screen.

At the start of the scanning motion at the camera-tube image plate, the electron beam is at the upper left-hand corner, point A of Fig. 2-6. Then, under the combined influence of the two sets of deflection coils, the beam moves at a small angle downward to the right. When point B is reached, the blanking signal acts while the beam is rapidly being brought back to point C, to start the third line, as required for interlaced scanning. The blanking signal then ends and the electron beam once again begins its left-to-right motion. In this manner every odd line is scanned.

When the end of the bottom odd line has been reached (point D), the blanking signals are applied while the beam is brought up to point E. Point E is above the first odd line of field 1 by a distance approximately equal to the thickness of one line. The beam is brought here as a result of the odd number of total lines used (525). Each field has 262-1/2 lines from its beginning to the start of the next field and, when the beam reaches point E, it has moved through the necessary 262-1/2 lines from its starting point A. From here the beam again starts its left-to-right motion, moving in between the previously scanned lines, as shown in Fig. 2-6. The beam continues until it reaches point F and from here is brought back to point A. From point A, the entire sequence repeats itself.

Thus, the electron beam moves back and forth across the width of the image plate 262-1/2 times in going from point A to point D to point E. The remaining 262-1/2 lines needed to form the total of 525 are obtained when the beam moves from point E to point F back to point A. The process may seem complicated, but actually it is carried out quite readily and accurately at the transmitter and receiver. A more detailed analysis, including the number of horizontal lines that are lost when the vertical blanking interval occurs, is given later in this chapter.

2.5 BLANKING AND SYNCHRONIZING SIGNALS

The cathode-ray beam at the receiver must follow the transmitter action at every point. For example, each time the camera-tube beam is blanked out, the same process must occur at the receiver and at the proper place on the screen. It is for this purpose that blanking-pulse signals are sent along with the video signals which contain the image details. These blanking pulses, when applied to the control grid of a cathode-ray tube, bias it to a large negative value, sufficient to prevent any electrons from passing through the grid and on to the fluorescent screen.

Blanking voltages, while preventing the electron beam from impinging on the fluorescent screen during retrace periods, do not cause the movement of the beam from the right- to the left-hand side of the screen, or from the bottom to the top. For this, another set of pulses, superimposed over the blanking signals, control the oscillators at the receiver and these, in turn, control the position of the beam. The pulses are called *synchronizing pulses.* A horizontal pulse at the end of each line causes the beam to be brought back to the left-hand side, in position for the next line. Vertical pulses, at the end of each field, are responsible for bringing the beam back to the top of the image.

2.6 THE VIDEO SIGNAL

In order to see how the picture detail, blanking signals, and synchronizing pulses are all combined to form the complete video signal, refer to Fig. 2-7(A). Here three complete lines have been scanned. At the end of each line the blanking signal is imposed on the beam and automatically prevents the electron beam from reaching the image plate at the camera or the fluorescent screen at the receiver. With the blanking signal "on," a synchronizing pulse is sent to cause (in this instance) the horizontal-deflection coils to move the position of the electron beam from the right side of the picture to the left. The sync pulse triggers a *sweep oscillator* circuit which provides the proper deflection coil currents needed for the retrace. This movement accomplished, the job of the synchronizing pulse is completed, and a fraction of a second later the blanking control releases its bias on the grid of the cathode-ray tube and the electron beam starts scanning again. This process continues until all the lines (odd or even) in one field have been scanned. Details involving the horizontal blanking and the sync pulse are shown in Fig. 2-7(B).

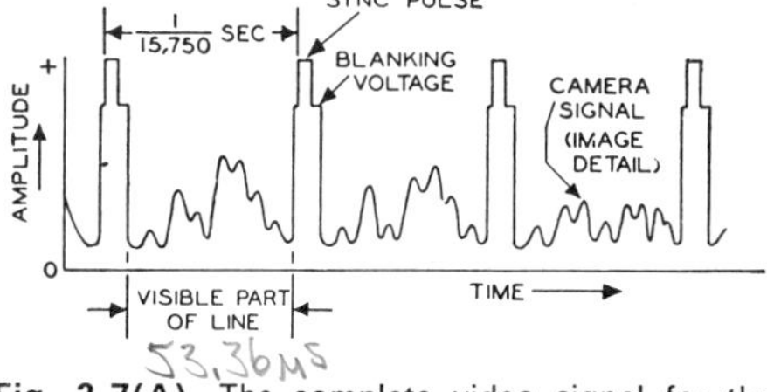

Fig. 2-7(A). The complete video signal for three scanned lines.

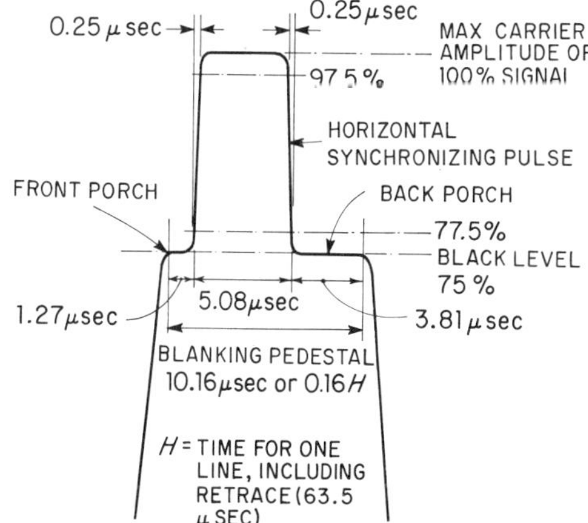

Fig. 2-7(B). The horizontal-blanking pedestal and horizontal sync pulse. The timing for these signals is very precisely controlled.

The vertical motion ceases at the bottom of the field, and it is necessary to bring the beam quickly to the top of the image so that the next field can be traced. Since the vertical-triggering pulse and the retrace require a longer period of time than the horizontal-triggering pulse and the retrace, a longer blanking signal is inserted. As soon as the blanking signal takes hold, the vertical-synchronizing pulse is sent. The form that this takes is shown in Fig. 2-8. Because the horizontal-synchronizing pulses must not be interrupted, even while the vertical-deflection coils

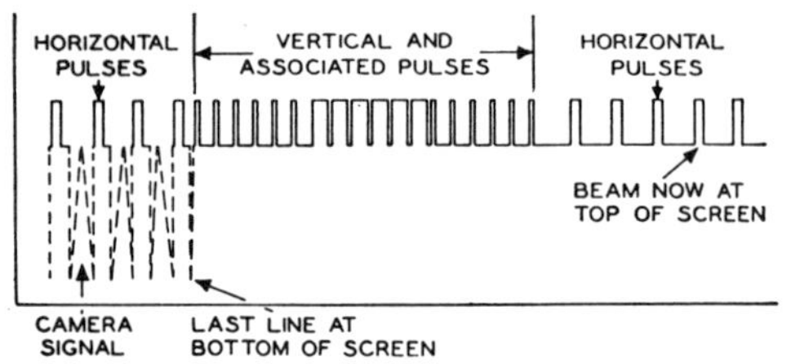

Fig. 2-8. The form of the vertical-synchronizing pulses.

are bringing the electron beam to the top of the field, the long vertical pulse is broken into appropriate intervals. In this manner it is possible to send both horizontal and vertical pulses at the same time, each type being accurately separated at the receiver and transferred to the proper deflection system.

By means of the vertical deflection coils, the electron beam is brought back to either point A or point E (Fig. 2-6), and then the scanning of the next field begins. The term used for the series of synchronizing pulses that combine to make up the total vertical synchronizing signal is, *serrated vertical pulses.*

2.7. NEGATIVE AND POSITIVE VIDEO POLARITY

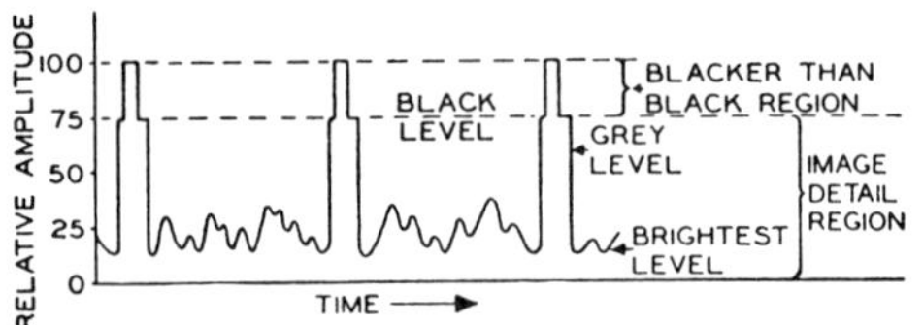

Fig. 2-9. The various proportions of a video signal (negative picture phase).

A closer inspection of a video signal (Fig. 2-9) reveals that of the total (100 per cent) amplitude available, from 75 to 80 percent is set aside for the camera-signal variations. At the level where the camera signal ceases, the blanking voltage is inserted. The remaining 20 to 25 percent of the amplitude is reserved for the horizontal- or the vertical-synchronising pulses. It will be noticed that, no matter where the camera signal happens to end, the blanking level and the synchronizing pulses always reach the same amplitude. This is done purposely at the transmitter, and several operations in the television receiver depend upon this behavior. It must be remembered, however, that this similar amplitude does not necessarily have to be used, but it is specifically employed because of the resulting simplicity at the receiver.

Figure 2-9 illustrates the form of the video signal as it is used in the United States. From the relative polarity marked on the ordinate (or vertical) scale, it is seen that the brightest portions of the camera signal cause the least amount of current to flow, or that their voltage has the least amplitude. The synchronizing pulses give the largest voltage and the greatest current of all.

Transmitting the video signal in this form is known as *negative picture phase*. It is claimed that less interference is visible on the viewing screen with negative picture transmission and that better all-round reception is obtained under adverse conditions.

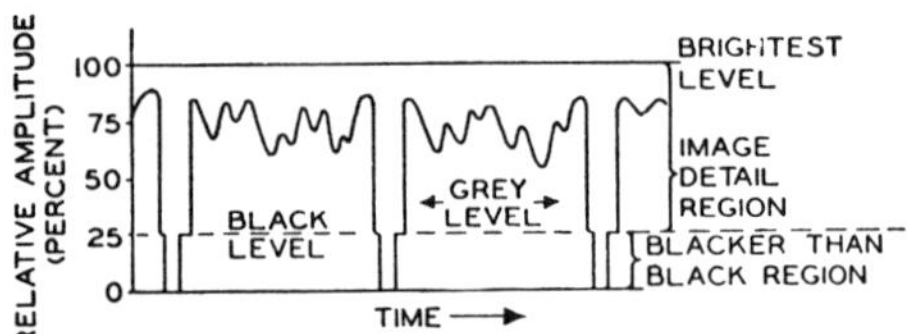

Fig. 2-10. The form of the composite video signal as applied to the grid of a TV receiver picture tube (positive picture phase).

In the receiver, before the video signal is applied to the control grid of the cathode-ray tube, the signal must possess the proper, or positive, picture phase (Fig. 2-10). The grid of the cathode-ray tube is then biased by enough negative voltage that, when the blanking voltage section of the signal is present, the electron beam is automatically prevented from reaching the fluorescent screen. With the positive picture phase, the camera-signal voltages are all more positive than the blanking pulse and, on these portions of the video signal, the electron beam is permitted to impinge on the screen with varying amounts of electrons. A bright spot in the received image causes the grid to become more positive than when the voltage of a darker spot is applied. More electrons in the beam mean that more light is emitted at the screen, and the various shades and light gradations of the image are formed by different voltages.

The purpose of the blanking voltage in the video signal is to prevent the electron beam from reaching the fluorescent screen. This fact is well known by now. The point in the video wave where the blanking signal is located occurs in the region where the currents corresponding to the very dark portions of the image are found. By the time the blanking voltage acts at the control grid of the viewing tube, the beam is entirely cut off and nothing appears on the screen. The blanking level is then properly called the *black region*, because nothing darker appears on the fluorescent screen. By nothing darker, we mean no light at all appears.

Now, consider the video signal of Fig. 2-10. Beyond the blanking level we find the synchronizing pulses. When applied to the viewing-tube control grid along with the rest of the wave, the pulses drive the grid to a negative voltage even greater than the cutoff voltage. The pulse region, for this reason, is labeled *blacker than black*, because the position of the blanking signal has been labeled black. The unwanted synchronizing pulses that ride through the video amplifiers with the necessary video signal need not be removed because they do not interfere in any way with the action of the control grid at the cathode-ray tube. As will be shown presently, the complete video wave is applied, after the detector, to the synchronizing- and video-amplifier circuits simultaneously. The synchronizing clipper tube permits only the pulses to pass through, whereas the video amplifiers allow the entire signal to pass.

2.8 WHY TELEVISION REQUIRES WIDE FREQUENCY BANDS

In dealing with television receivers, it will be found that extensive use is made of wide-band amplifiers designed to receive signals extending over a band 4 to 6 megaHertz (MHz) wide. The different forms which these amplifiers may assume and their characteristics are discussed in later chapters; however, the reason for the extremely wide bandwidth may be appreciated now.

In Chapter 1, it was brought out that the more elements in a picture, the finer the detail that is portrayed. The picture can also stand closer inspection before its smooth, continuous appearance is lost. Each 1/30 second, 525 lines are scanned, or a total of 15,750 lines in each 1 second. If each horizontal line contains 700 separate elements, then $15{,}750 \times 700$, or 11,025,000, elements or electrical impulses are transmitted each second. In order to attain full advantage of the use of this number of elements, it is first necessary to determine what relationship exists between two quantities: the number of elements and the bandwidth.

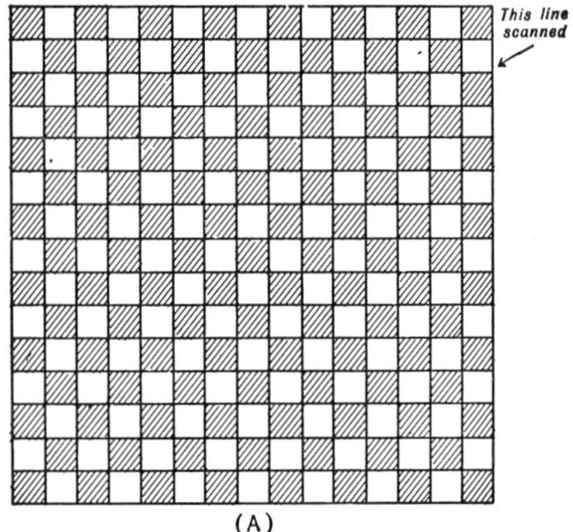

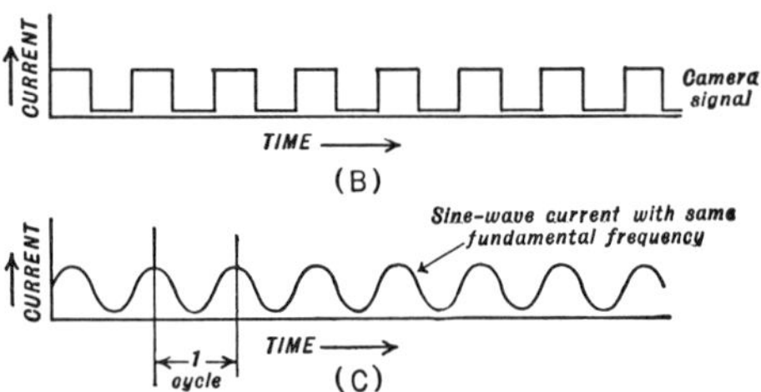

Fig. 2-11. The basic relationship between the number of elements in an image and the width of the frequency band required.

Consider, for example, that the image plate in the camera tube is broken up into a series of black-and-white dots, each dot representing one element. A portion of the resulting pattern is shown in Fig. 2-11(A). As the scanning beam passes over each element in turn, a pulse of current flows every time a white dot is reached, for this element has a large deficiency of electrons. At the next element, the current drops to zero, for theoretically a black dot represents an element that has received no light at all. It therefore requires no additional replacement of electrons. As one horizontal line is scanned, the electric pulses of current would have the shape shown in Fig. 2-11(B).

If one maximum point in the wave is combined with its succeeding minimum point, one complete cycle is obtained. The same situation prevails in any sine wave (see Fig. 2-11(C)). Since each white dot represents a maximum point and each black dot a minimum point, taking the total number of white and black dots on a line and dividing their sum by 2 gives the number of cycles the current goes through when one horizontal line is scanned. With 700 elements (dots, in this case) on a line, a fundamental frequency of 350 cycles per line is generated.

Under present standards, 525 lines are scanned in 1/30 second, or a total of 15,750 in 1 second. Employing 700 elements per line, 11,025,000 picture elements are sent each second, which, for our analysis results in a frequency of $\frac{11{,}025{,}000}{2}$ Hz, or 5.51 MHz. In actual practice, a bandwidth of 4 MHz is allowed. Thus, for the video-section requirements alone, this extremely large bandwidth must be passed by all the tuned circuits of the television receiver.

The above situation would seldom, if ever, be found in practice. However, the figures obtained by this reasoning yield results that have been found satisfactory, and the method derived from this viewpoint is justified.

Although 4 MHz are required to accommodate the video information alone, the transmitted bandwidth set in practice is 6 MHz. Of the extra 2 MHz, the FM audio carrier uses 50 kHz. Apparently some of the total bandwidth is not utilized. The reason for this seeming extra space is found in the process whereby the television video carrier is generated.

On ordinary AM radio broadcast frequencies, it is common knowledge that most stations occupy a 10-kHz bandwidth, or ± 5 kHz about the carrier position. Thus, if a station is assigned the frequency of 700 kHz, it transmits a signal that occupies just as much frequency space on one side of 700 kHz as on the other. With a modulating frequency of 5 kHz, the deviation is 5 kHz (or 5,000 Hertz) on either side of the carrier position of 700 kHz. Technically, we say that these side frequencies are *sidebands*. For the present illustration, each sideband may have a maximum deviation of 5 kHz about the mean, or carrier, position. The information of the signal is contained in the sidebands, since they are not generated until speech or music or other sounds are projected into the microphone. At the receiver, the variations in the sidebands are transformed into audible sounds and heard by the radio listener.

It can be shown that those sidebands that are generated with frequencies higher than the carrier frequency contain the same information as the sidebands with frequencies lower than the carrier. In other words, if one set of sidebands (either above or below the carrier) were eliminated, we could still obtain all the necessary information at our receiver. One reason that a sideband is not eliminated is for economic reasons. A transmitter naturally generates both sidebands, and it is often cheaper to transmit both than to try to eliminate one by expensive and complicated filters. However, single sideband transmission does exist for certain communication facilities.

Now, let us turn our attention to the video signal. This signal is generated by fundamentally the same type of apparatus that is employed at the AM sound broadcast frequencies. Since 4 MHz are needed for the picture detail, a signal would be generated that extended 8 MHz, or ± 4 MHz about the carrier. This spectrum does not include the sound. An 8 MHz band is undesirable because of the bandwidth occupied and the difficulties inherent in transmitting a signal of this bandwidth. Hence, the need arises for removing most of one sideband, since as noted previously, only one is required.

The undesired sideband is mostly removed by filters that follow the last amplifier of the television transmitter. But filters that will cut off one sideband sharply and completely and leave only the one desired are not easily constructed. Furthermore, in the process of elimination, nothing must occur that changes the amplitude or the phase of any of the components in the desired sideband. As a compromise arrangement, most, but not all, of one sideband is removed, and in this way the remaining sideband is not affected by the filtering. Thus part of the 6 MHz bandwidth is occupied by what may be called the *remnants* of the undesired sideband. This method is known as a *quasi-single-sideband or vestigial-sideband* operation.

In Fig. 2-12(A) the television-video signal appears with both sidebands present, and in Fig. 2-12(B) the signal is shown as it appears after passage through filters that partially remove one sideband. The frequency of the carrier is found 1.25 MHz above the low-frequency edge of the television signal. Then for 4 MHz above this, we have the television video signal with the desired picture information. A 0.5 MHz bandwidth separates the high-frequency edge of the video signal and the FM carrier. The space is left for the purpose of preventing undesirable interaction between the two, for example, cross-modulation, which would lead to the distortion of the video signal. In this manner, the allotted 6 MHz are distributed.

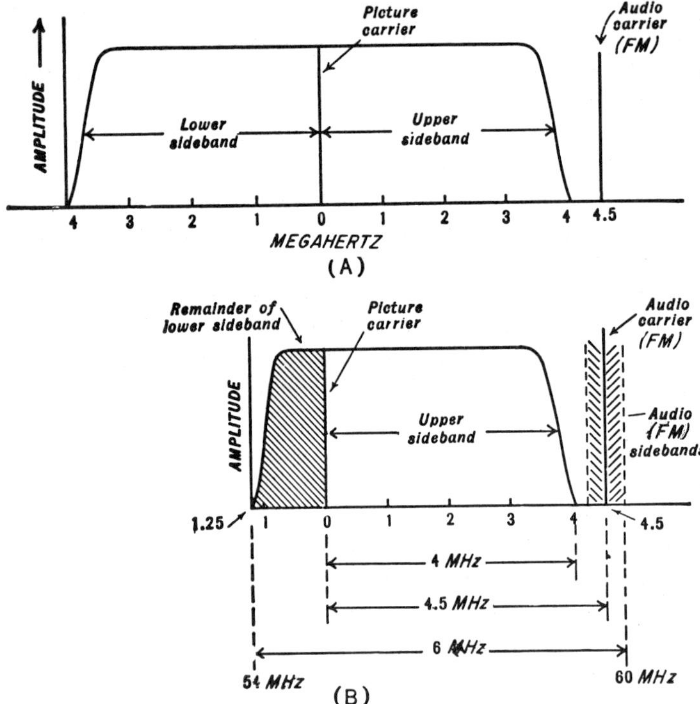

Fig. 2-12. Sideband transmission of television signals. (B) is standard in the United States and represents the signal as sent.

2.9 EFFECT OF LOSS OF LOW AND HIGH VIDEO FREQUENCIES

While uniform response over a 4-MHz band may be required in the picture IF and video amplifiers, this is not easily attained in practice. Special circuit designs must be resorted to which are more fully explained at the appropriate places in later chapters of this book. For the moment, it is only necessary to point out the effects of poor response at the high or low frequency ends of the band.

In the preceding analysis we have seen that a greater number of elements requires a greater bandwidth if advantage is to be taken of the increase. Since detail is determined mainly by the number of very small elements, any decrease in the response at the higher frequencies will result in less fine detail available at the receiving cathode-ray screen. The picture will lose some of its sharpness and may even appear somewhat blurred if the high-frequency response is degraded enough. In commercial monochrome television receivers, a video passband of from 3.3 to 4.0 MHz is generally considered good, while anything appreciably below 3.0 MHz is not too desirable. As we shall see, the frequency response of circuits in a color television receiver must not fall below about 4 MHz, or the color portion of the signal will be affected.

At the low-frequency end of the band, poor frequency response results in obliterating the slow changes that occur in background shading. This results in incorrect shading of the large areas in the reproduced picture.

REVIEW QUESTIONS

1. What type of signal is used for synchronizing a receiver-deflection system with the transmitter?
2. Explain the use of fields for reducing flicker.
3. What was the original reason for using a rate of 60 fields per second?
4. Explain what is meant by *interlaced scanning.*
5. What are the wave-forms needed (by the deflection system of the picture tube) for producing a raster?
6. Where are *equalizing pulses* found, and what are they used for?
7. Why is it necessary to focus the electron beam in a picture tube?
8. What is meant by the expression *active scanning lines*?
9. Compare vestigial-sideband transmission with double-sideband and single-sideband transmission (AM).

3 Television Camera Tubes

3.1 INTRODUCTION

In a previous chapter, we learned that in the telecasting process an optical image is converted into an electrical image for transmitting the picture to the receiver. We also learned that this important conversion is done in the TV camera by an image pickup tube. Now, the operation of the TV camera tube will be described in greater detail, for it is what this tube "sees" and converts into equivalent electrical impulses that determines the form of the image finally reproduced at the receiver.

For high quality images at the receiver, the camera tube must resolve the scene being televised into as many basic picture elements as possible, because the greater the number of these elements the higher the quality of detail in the reproduced picture. Also, the scanning beam in the pickup tube must produce electrical signals that faithfully represent each of these picture elements. To do this, the optical to electrical conversion must be achieved at a high signal-to-noise ratio to provide proper pickup sensitivity when low light-level scenes are being telecast. In other words, when there is no incident light on the face of the pickup tube, there must be little, if any, output signal. From these requirements, it can be seen that camera-tube characteristics and its electrical operation are important considerations.

Figure 3-1 is a simplified diagram of a typical monochrome-TV camera. An optical system focuses light reflected by the scene onto the faceplate of the camera tube. There, a photoelectric process transforms the light image into a virtual electronic replica in which each picture element is represented by an electrical potential. A scanning beam in the pickup tube next converts the picture, element by element, into electrical impulses and develops at its output an electrical sequence that represents the original scene. The output of the camera tube is then amplified to provide the video signal for the transmitter and to provide a sample of the video signal for observation in a viewfinder that is mounted on the camera housing.

Electronic circuits that provide the necessary control, synchronization, and power supply voltages operate the camera tube in the TV camera. A deflection system is included in the TV camera to control the movement of the camera-tube scanning beam. Many TV cameras receive synchronizing pulses from a studio-control unit, which also provides the sync pulses that synchronize the receiver with the camera; however,

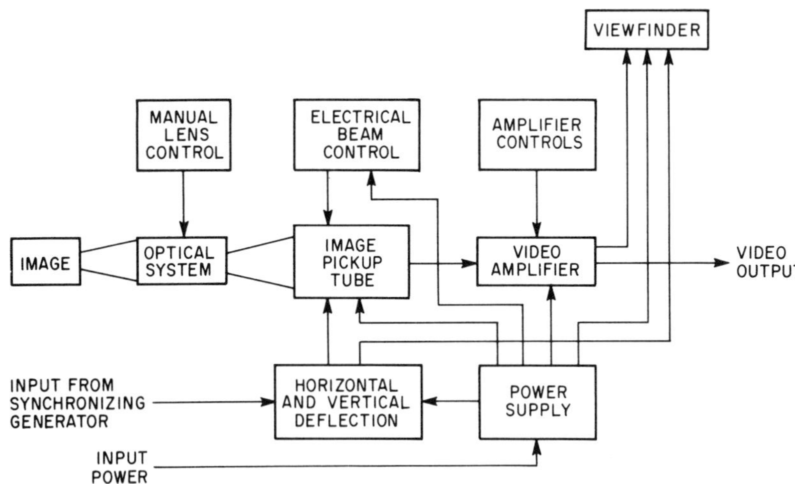

Fig. 3-1. Simplified diagram of a typical monochrome-television camera.

some of the newer TV cameras generate their own control signals and, in turn, provide output pulses to synchronize the control unit. Manual controls are also provided at the rear of the camera for setting the optical lens and for zooming. Because of the complex electronic circuits and controls, the TV camera was originally rather large and awkward to handle, however, recent innovations have revolutionized the camera's construction

In the last three decades, television has become the entertainment medium of the major population areas of the world. But television is not, by any means, limited to entertainment. In fact, television has found widespread use in education, medicine, industry, aerospace, and oceanography. Many of these applications require special camera tubes, some having widely varying sensitivities to light at different wavelengths throughout the spectrum. Much has been spent on image detection devices, and, today, whatever the need, there is a TV camera tube that can do the job. Some of these special purpose camera tubes will be discussed later on in this chapter.

Solid-State Camera Circuits Formerly, vacuum tubes were used in the electronic circuits of the TV camera. These tubes consumed a great deal of power in heating the filaments and produced a large amount of heat, which, in conjunction with the heat generated by studio and scene lighting, elevated the camera to a rather high temperature. As a result of this high-heat environment, the electronic circuits in the camera had to be aligned many times each day during regular program production. In general, the TV camera circuits were not stable and required at least one operator for each camera. Also, the vacuum tubes required spacious compartments, resulting in a camera which was both large and awkward.

Today, transistors, integrated circuits, and modular construction have made dramatic improvements in television broadcasting. Transistors have replaced the vacuum tube, eliminating the excessive power consumption and most of the heat. Integrated circuits have combined into one tiny block many of the circuits that previously included numerous capacitors, resistors, and transistors, all mounted on a large-area circuit board. The elimination of the vacuum tube and its heat made modular construction feasible and allowed a remarkable reduction in the overall size of the TV camera. Also, modular construction has virtually done away with troubleshooting to the component level. Now the trouble is merely traced to a particular module, and then the module is simply replaced, usually with no more difficulty than formerly encountered in replacing a vacuum tube. All of these advantages, together with temperature compensated deflection circuits, the use of feedback, and better regulated power supplies have resulted in a hundred-to-one improvement in stability. In fact, a common practice now is for one man to operate several TV cameras.

Color TV Cameras Thus far we have considered only cameras used for black and white, or monochrome, reproduction. To add color to the television system, a separate camera tube is used for each of the primary colors, red, green, and blue. A color filter system is also used to separate the incoming light from the image into the three colors and to focus each color onto the faceplate of an appropriate camera tube. The three camera tubes are identical except that the photosensitive material in each tube is more responsive to the particular color for which the tube is used. Remember though that each color is first determined by the filter system. The output from each tube consists of electrical impulses that represent the elements of the particular primary color reflected by the scene. These outputs are then amplified to provide three channels of video signals, one channel for each color. Color cameras will be discussed in more detail, after we have taken a closer look at the camera tubes which are in general use in the television industry.

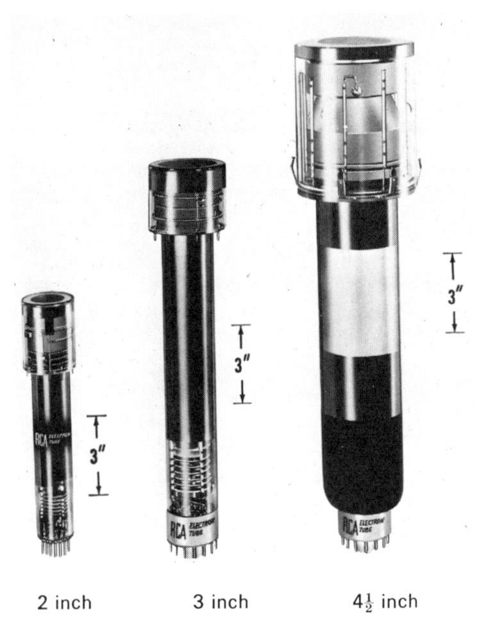

Fig. 3-2. RCA image orthicons. (*Courtesy of RCA.*)

3.2 TYPES OF CAMERA TUBES IN USE

Basically, TV camera tubes can be divided into two distinct classifications according to their method for producing an electrical image within the tube. The first method is by photoemission, whereby electrons are emitted from a photosensitive surface when light from the scene is focused onto the surface. The second method is by photoconduction, where the conductivity or resistance of the photosensitive surface varies in proportion to the intensity of the light focused onto the surface. By means of these two classifications, the camera tubes can be divided into the two types in general use: the image orthicon (see Fig. 3-2), which uses the process of photoemission; and the vidicon (see Fig. 3-3), which uses the process of photoconduction. Aside from these basic differences there are also several operational characteristics that are common to both types of camera tubes. These characteristics are important in considering a tube for a particular application.

At present, the vidicons in general use far outnumber the image orthicons. This has come about mainly because of the vidicon's small size and simplicity, which has made the tube especially desirable for industrial and aerospace applications. In fact, most of the color cameras today incorporate vidicons for the three color channels and, in some cases, a fourth vidicon for a luminance channel to provide better color registration and sharper monochrome pictures.

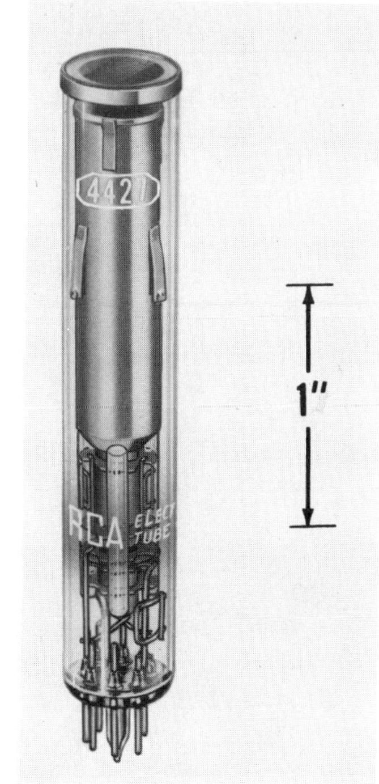

$\frac{1}{2}$ inch vidicon

Fig. 3-3. RCA vidicon. (*Courtesy of RCA.*)

Camera Tube Characteristics Among the characteristics common to all camera tubes is their light transfer capability. This is the ratio of faceplate illumination in footcandles to signal current in nanoamperes. This characteristic is referred to as the gamma of the tube, or simply the ratio of brightness variation in the reproduced image to the brightness variation in the original scene. Normally, the camera tubes are operated with unity gamma.

For a proper tonal rendition, the spectral response of a camera tube is an important parameter. The tube should have, as nearly as possible, the same spectral response as the human eye. This is necessary to render colors in their proper tones in reproducing black and white pictures, thereby producing the proper gray scale. Tubes designed to operate in a color camera have a greater response to each of the primary colors. Today, spectral response distribution has made possible the manufacture of camera tubes that are sensitive to the infrared, the ultraviolet, and even the x-rays. But variations in spectral response have had little effect on the other operating characteristics of the tube.

If the photosensitive material in a camera tube were able to emit an electron for each photon of light focused upon the material, the quantum efficiency of the material would be 100 percent: quantum efficiency being

$$Q_{\text{eff}} = \frac{\text{electrons}}{\text{photons}}$$

Quantum efficiencies this high are hardly possible, but quantum efficiency, nevertheless, is a practical way to compare photosensitive surfaces in the camera tube. In this comparison, photocurrent per lumen* is measured, using a standard light source. The source adopted for the measurement is a tungsten-filament light operating at a color

* A lumen is the amount of light that produces an illumination of one footcandle over an area of one square foot.

temperature of 2870° K. Since the lumen is actually a measure of brightness stimulation to the human eye, quantum efficiency is a convenient way to express the sensitivity of the image-pickup tube.

3.3 SCANNING AND PICTURE ELEMENTS

Scanning in the TV camera tube is identical to scanning in the receiver. The method of beam deflection, though, is slightly different in some camera tubes in that electrostatic deflection is used instead of, or in conjunction with, magnetic deflection. In electrostatic deflection, plates mounted inside the tube develop transverse electric fields that move the beam. In magnetic deflection, coils mounted around the camera tube produce transverse magnetic fields that move the beam fom the right to the left and from the top to the bottom of the scanned area. The purpose of scanning in the receiver cathode-ray tube is to "paint" the picture on the face of the tube. In the camera tube, scanning produces the picture elements that control the intensity of the painting beam in the receiver.

Photon-to-electron conversion is the means of changing an optical image to an electrical charge replica, in the camera tube. Because of the scanning beam, the charge replica becomes a sequence of electrical signals at the output of the tube. However, we have not yet determined how the scanning beam interrogates the charge replica of the original image and develops the sequential video signal. To better understand this electrical action, we will first define the picture element.

At this point, it is necessary to consider only the charge replica of the image. The exact way in which the replica is formed will be explained when we discuss the operation of the various camera tubes. In Fig. 3-4(A), the original image is focused onto a plate to form the charge replica. The plate contains a large number of photosensitive elements, or cells, each insulated from its neighbors. When the brightly illuminated portions of the original image reach the corresponding photosensitive elements, these elements assume an electrical charge (positive) which is different from that of the non-illuminated elements. The white areas of the plate are charged, while the black areas are not. The amount of the charge on any given element depends upon the intensity of the light impinging on that element. As a result, the electrical replica of the image, consists of a mosaic of electrical charges, which are distributed in individual elements, or cells, similar to the grains of a photographic plate or film. The individual elements are too small to be seen by the naked eye, and the finely focussed electron beam has a diameter equal to the combined size of several elements. When the electron beam scans the charge replica of the image, it will discharge (or sense) the average charge of several elements. Therefore, the resolution of the camera tube is limited by the diameter of the scanning beam. This is illustrated in Fig. 3-4(B).

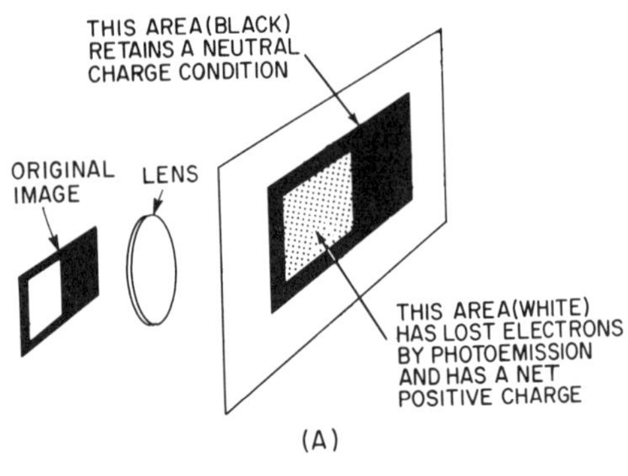

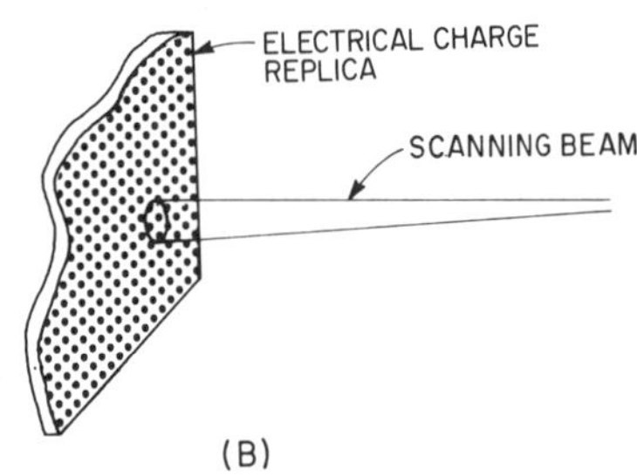

Fig. 3-4. Picture element charge buildup.

A fine-grain photograph, with many grains per unit of area, is capable of greater enlargement than a coarse-grain picture before these elements become discernible. With television images, the same kind of situation prevails. In the receiver, each picture element is just as large as the area of the circular beam impinging on the fluorescent screen of the cathode-ray tube. The light that is seen when observing a cathode-ray tube screen is derived from the energy given off by the impinging beam to the particles of the fluorescent coating on the inner face of the tube. If the points of light are closely spaced, the observer will integrate them in his eye and their character as separate points will disappear. Hence, one of the first considerations for a television picture that is to reproduce fine detail is an electron beam of small diameter. This requirement is as important at the receiver screen as it is at the camera tube.

The complete scanning process is covered in Chapter 2. Synchronization, blanking, and beam deflection are exactly the same at the camera tube as at the picture tube in the receiver. For that reason, synchronization of the two units is most critical. When the picture element gives up its intelligence to the scanning beam in the camera-pickup tube, the receiver-scanning beam must reproduce that information of the face of its cathode-ray tube.

3.4 OPERATION OF THE IMAGE ORTHICON

The image orthicon is a sensitive, stable TV camera tube intended for either outdoor or studio operation. It can produce high-quality commercially-acceptable pictures with incident light levels at the scene ranging from bright sunlight to deep shadows. Moreover, its sensitivity is equivalent to photographic film having an ASA exposure index of 8000. For comparison, the film ordinarily used by most newspaper photographers has an index of 100. Because of this sensitivity and its dependable service in almost every TV application in the field or studio, the tube has become known as the "workhorse of the industry."

The image orthicon is divided into three sections: an image section, where the electron replica of the image is formed and focused onto a target; a scanning section, where a scanning beam converts the image replica into electrical impulses; and a multiplier section where, through a process of secondary emission, more current is generated than is contained in the returning beam. Figure 3-5 illustrates all three sections of the tube, with its focus and deflection coils.

In operation, light rays from the scene being televised are focused by an optical lens system onto a semitransparent photocathode on the inner surface of the faceplate of the tube. Electrons are emitted from each point on this photocathode in proportion to the incident light intensity. Note that the light rays must penetrate the semitransparent photocathode to reach the photosensitive inner surface.

The emitted electron replica (in which, at each point, the density of the electrons corresponds to the light at that point) is directed toward

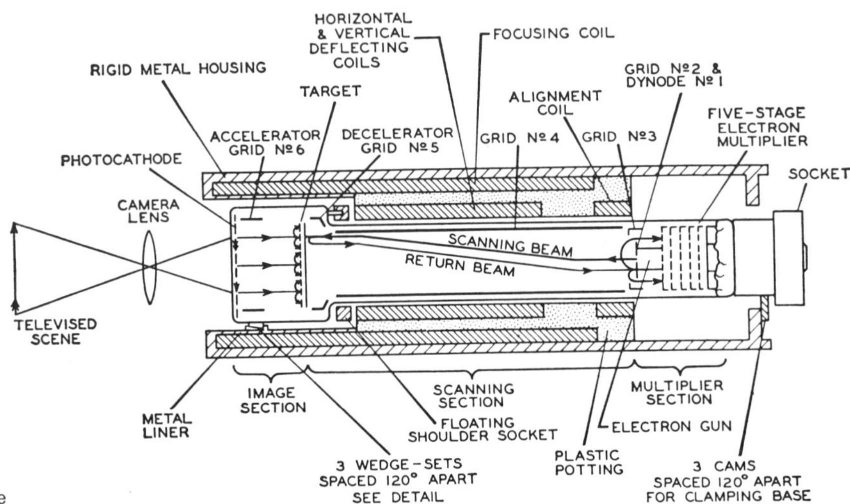

Fig. 3-5. The internal construction of the image orthicon.

the target by an accelerator grid. Focusing of the replica onto the target is accomplished by the external focus coil that surrounds both the image and the scanning sections. When the accelerated electrons of the replica strike the target, secondary electrons are emitted by the target material. These electrons are then collected by a fine-mesh screen that is spaced close to the photocathode side of the target. This screen is held at a definite potential with respect to the cutoff voltage of the target, thereby limiting target potential for all values of the incident light and stabilizing the operation of the tube.

Emission of the secondary electrons develops on the photocathode side of the target a pattern of positive charges directly proportional to the distribution of energy in the arriving electron replica. This pattern of charges, because of the target material, develops a similar potential pattern on the scanning-section side of the target. The target is not photosensitive but is made of a thin material that is capable of emitting many secondary electrons.

In the scanning section, an electron gun produces a low-velocity electron beam which scans the side of the target opposite the photocathode side. This gun includes a thermionic cathode, a control grid (grid No. 1 in Fig. 3-5, and an accelerating grid (grid No. 2 in the same figure). The beam is focused on the target by the external focus coil and an electrostatic field produced by grid No. 4. Grid No. 5 is used to shape the decelerating field between grid No. 4 and the target, so that electrons in the scanning beam will land uniformly over the entire area of the target.

During scanning, the electrons in the beam stop their forward motion at the surface of the target. Here, they are turned back toward the multiplier section, unless the particular point of target landing is a positive charge. In this latter case, enough electrons from the beam are deposited to neutralize the positive charge pattern on the target. This leaves a zero charge pattern on the scanning side of the target and a positive pattern on the photocathode side, but the two neutralize each other by conductivity through the target material in less than the time of one frame, or 1/30 second. The electrons turned back at the target form the return beam, which is directed into the multiplier section. This return beam has been amplitude modulated by the absorption of electrons at the target in accord with the charge pattern whose more positive areas correspond to the highlights of the electron replica.

It is evident that the most positive points on the target turn back to the multiplier section the least number of electrons from the original scanning beam. Hence, the voltage developed across the output load resistor is inversely proportional to the positive charge intensity on the target.

To function effectively, the target must be able to conduct between its two surfaces but not along either surface. The logic of this is evident. Whatever charge appears on one side of the target due to the focused image must likewise appear on the other side. It is this second side which is scanned, and it is from here that the video signal is obtained. Hence, a conducting path must exist between the front and back sides. However, nothing must disturb the relative potential that exists throughout the charge pattern, as deposited on the photocathode side of the target. Hence, no conduction is permissible between the various elements of any one side of the target. If this does occur, the charge differences between the various points on the electron replica disappear.

A major reason for the high sensitivity of the image orthicon stems from the current amplification that takes place in the multiplier section of the tube. In this multiplier, the electrons in the return beam are captured and then increased through a process of secondary emission to provide an output current that is several hundred times stronger than it would be without a multiplier.

The multiplier structure in the image orthicon consists of a series of circular screens set one below the other around the electron gun, as shown in Fig. 3-6. The return beam is directed to dynode No. 1 (this is also grid No. 2) of a 5-stage electron multiplier. For each electron that strikes this dynode, two or more electrons are dislodged from its surface. These secondary electrons are then directed (by the electric field of grid No. 3 and the higher voltage of dynode No. 2) to dynode No. 2. This latter element is a 32-blade pinwheel mounted directly below dynode No. 1. The arriving electrons strike the blade of dynode No. 2, causing secondary electrons to be emitted, which are drawn through the slots to the next stage. A schematic diagram of the image orthicon showing the arrangement of the dynodes is presented in Fig. 3-7.

This multiplying process continues at each successive dynode, with an ever increasing stream of electrons, until those emitted from the

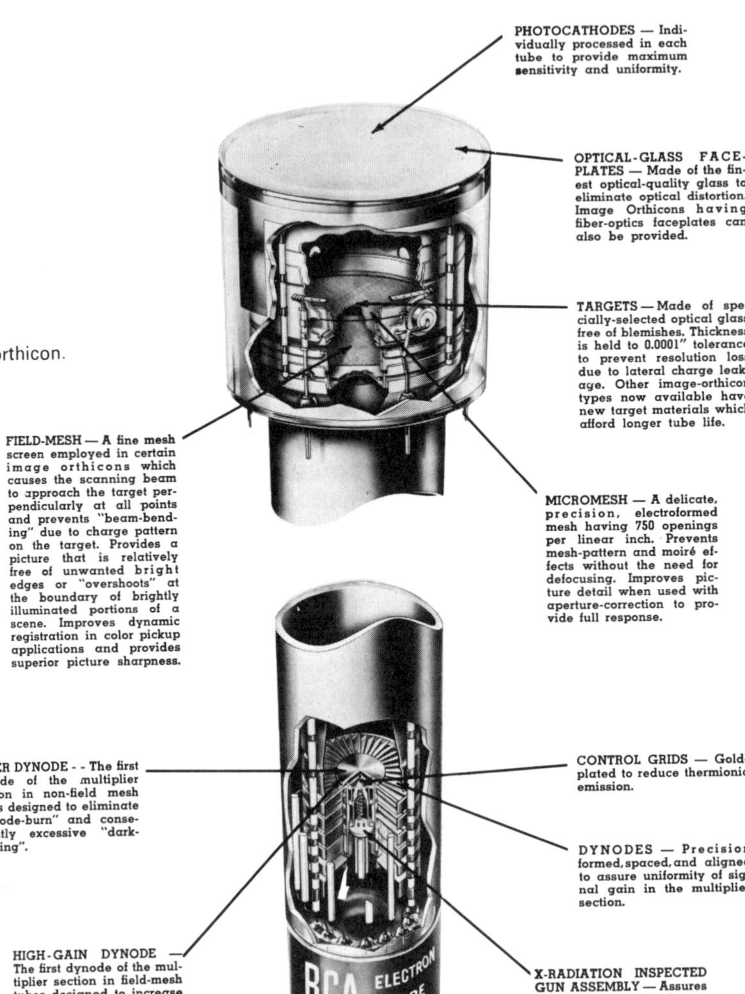

Fig. 3-6. Cut-away view of the image orthicon. (*Courtesy of RCA.*)

Fig. 3-7. The path of the electrons as they travel through the screens of the multiplier section (at left side of illustration).

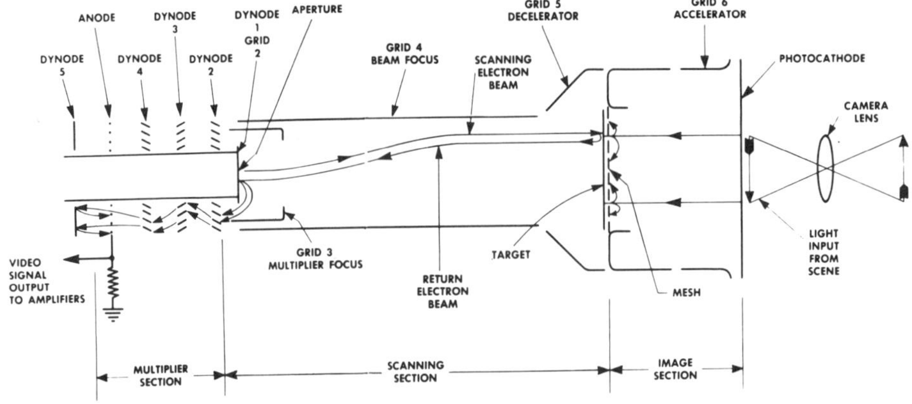

final dynode are collected by an anode and made available to the output circuit

An amplification of 500 or more electrons is achieved in this fashion. The multiplication so obtained maintains a high signal-to-noise ratio. The gain of the multiplier is sufficient to raise the output signal above the noise level of the video amplifiers to which the signal is fed, so that the amplifiers contribute no noise to the final video signal.

An alignment coil, mounted around the tube at the multiplier end of the focusing coil, provides a transverse magnetic field for aligning the scanning beam from the electron gun. Also, the horizontal and vertical deflection coils are mounted around the scanning section of the tube. This entire assembly, including all of the coils, is housed in a metal shield for installation in the camera.

Recently, many developments in photocathode materials have greatly improved the operating performance of the image orthicon and have greatly improved the operating performance of the image orthicon and have eliminated most of its disadvantages. Bialkali and multialkali materials, as used in the RCA image orthicon photocathodes, now provide a wide selection of spectral responses. These range from those which closely approximate the response of the human eye to extremely high sensitivities extending into the near infrared region of the spectrum.

To complement these recently developed photocathodes, several new target materials are now being used. Among these are electronically-conducting glass and thin-film semiconductive materials. Also, standard glass targets, made of the finest optical-quality glass to eliminate optical distortion, are used in some image orthicons.

Another feature of many image orthicons is a field mesh mounted in the scanning section near the target to prevent beam bending due to the charge pattern on the target. This device is a fine-mesh screen that causes the scanning beam to approach the target perpendicularly at all points on the scanning surface. Because of this perpendicular approach, the reproduced pictures are relatively free of unwanted bright edges or "overshoots" at the boundary of the brightly illuminated portions of a scene. Also, the field mesh improves the dynamic registration in color televising applications and provides greater picture sharpness.

3.5 OPERATION OF THE VIDICON

The vidicon is a simple, compact TV camera tube that is widely used in education, medicine, industry, aerospace, and oceanography. It is, perhaps, the most popular camera tube in the television industry. During the past several years, much effort has been spent in developing new photoconductive materials for use in its internal construction. Today, with these new materials, some vidicons can operate in exposure to direct sunlight or in near-total darkness. Also, these tubes are available with diameters ranging from 1/2 to 4-1/2 inches, and some of the larger ones even incorporate multiplier sections similar to those used in the image orthicon. The main difference between the vidicon and the image orthicon is physical size and the photosensitive material used to convert incident light rays into electrons.

The image orthicon depends on the principle of photoemission, wherein electrons are emitted by a substance when it is exposed to light. The vidicon, on the other hand, employs photoconductivity; that is, a substance is used for the target whose resistance shows a marked decrease when exposed to light.

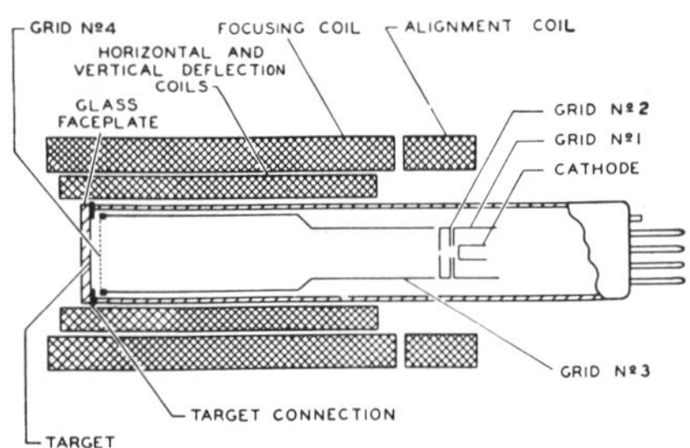

Fig. 3-8. Internal construction of a vidicon.

The operating principle of the vidicon camera tube is shown in Fig. 3-8. The target consists of a transparent conducting film (the signal electrode) on the inner surface of the faceplate and a thin photoconductive layer deposited on the film. Each cross-sectional element of the photoconductive layer is an insulator in the dark but becomes slightly conductive when it is illuminated. Such an element acts like a leaky capacitor, having one plate at the positive potential of the signal electrode and the other one floating. When light from the scene being televised is focused onto the surface of the photoconductive layer next to the faceplate, each illuminated element conducts slightly, the current depending on the amount of light reaching the element. This causes the potential of its opposite surface (i.e., the gun side) to rise toward the signal electrode potential. Hence, there appears on the gun side of the entire layer surface a positive-potential replica of the scene composed of various element potentials corresponding to the pattern of light which is focused onto the photoconductive layer.

When the gun side of the photoconductive layer, with its positive-potential replica, is scanned by the electron beam, electrons are deposited from the beam until the surface potential is reduced to that of the cathode in the gun. This action produces a change in the difference of potential between the two surfaces of the element being scanned. When the two surfaces of the element, which in effect form a charged capacitor, are connected through the external target (signal electrode) circuit and a scanning beam, a current is produced which constitutes the video signal. The amount of current flow is proportional to the surface potential of the element being scanned and to the rate of the scan. The video-signal current is then used to develop a signal-output voltage across the load resistor. The signal polarity is such that for highlights in the image, the input to the first video amplifier swings in the negative direction. In the interval between scans, wherever the photoconductive layer is exposed to light, the migration of the charge through the layer causes its surface potential to rise toward that of the signal plate. On the next scan, sufficient electrons are deposited by the beam to return the surface to the cathode potential.

The electron gun contains a cathode, a control grid (grid No. 1), and an accelerating grid (grid No. 2). The beam is focused on the surface of the photoconductive layer by the combined action of the uniform magnetic field of an external coil and the electrostatic field of grid No. 3. Grid No. 4 serves to provide a uniform decelerating field between itself and the photo-conductive layer, so that the electron beam will tend to approach the layer in a direction perpendicular to it, a condition that is necessary for driving the surface to the cathode potential. The beam-electrons approach the layer at a low velocity because of the low operating voltage of the signal electrode. Deflection of the beam across the photoconductive substance is obtained by external coils placed within the focusing field.

The Plumbicon The plumbicon, developed by Philips of Holland, is a small, lightweight television camera tube that has overcome many of the less favorable features of the standard vidicon. The tube has fast response and produces high quality pictures at low light levels. Its small size, together with its low-power operating characteristics, make it an ideal tube for transistorized television cameras designed to serve a particular purpose. Present-day color television cameras are making widespread use of the plumbicon because of its simplicity and spectral response to the primary colors. Typical plumbicon camera tubes are shown in Fig. 3-9.

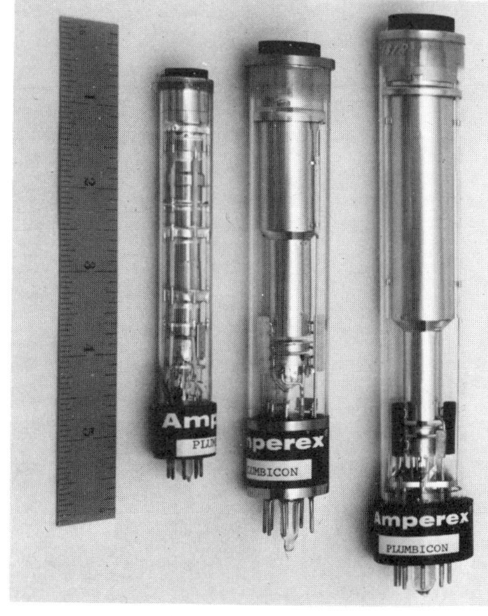

$\frac{5}{8}$ in (16 mm) 1 in (25.4 mm) $1\frac{1}{4}$ in (30 mm)

Fig. 3-9. Typical plumbicon tubes. (*Courtesy of Amperex Electronic Corp.*)

Functionally, the plumbicon is very similar to the standard vidicon. Focus and deflection are both accomplished magnetically. The main difference between the plumbicon and the standard vidicon is in the target.

Figure 3-10 is a simplified diagram of the plumbicon target. As shown in part (A) of this figure, the inner surface of the glass faceplate is coated with a thin transparent conductive layer of tin oxide (SnO_2). This layer forms the signal plate of the target. On the scanning side of the signal plate a photoconductive layer of lead monoxide (PbO) is deposited. These layers are specially prepared to function as three sublayers, each with a different conduction mode.

On the inner side of the faceplate, the tin oxide layer is a strong N-type semiconductor, commonly found in transistors (see Fig. 3-10(B)). Next to this N-type region is a layer consisting of almost pure lead monoxide, which is an intrinsic semiconductor. The scanning side of the lead monoxide has been doped to form a P-type semiconductor. Together, these three layers form a P-I-N junction diode.

The photoconductive target of the plumbicon functions similarly to the photoconductive target in the standard vidicon. Light from the scene being televised is focused through the transparent layer of tin oxide onto the photoconductive lead monoxide. Each picture element charge takes the form of a small capacitor with its positive plate toward the scanning beam, as shown in Fig. 3-10(C). The target signal plate becomes the negative side of the capacitor. When the low-velocity scanning beam lands on the charge element, it releases enough electrons to neutralize the charge built up on the element capacitor. The scanning-

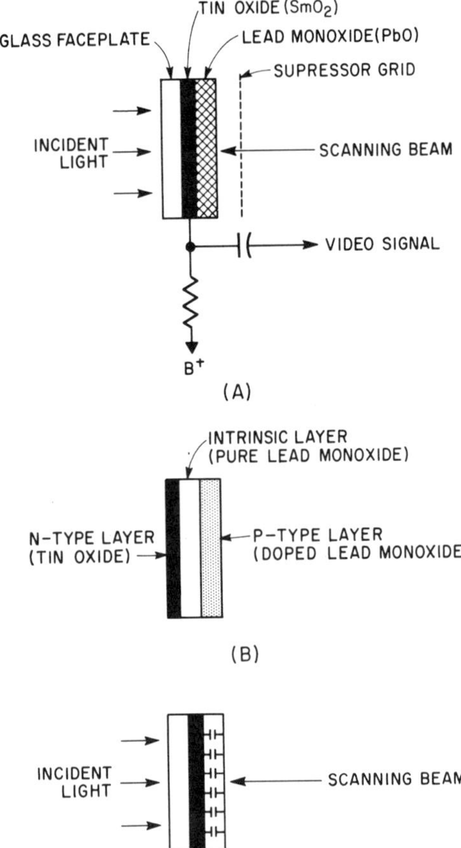

Fig. 3-10. The plumbicon target.

beam current through the external signal-plate load resistor develops the video-signal output.

The spectral response of the plumbicon can be varied during its manufacture to suit almost any particular application. Since the tube has gained wide popularity in color television cameras, it is now available with a spectral response suitable for any of the primary colors: red, green, or blue. The particular color response of the tube is designated by the letter R, G, or B following its type number, *e.g.*, a plumbicon for use in a red channel is designated type 55875R.

3.6 COLOR TELEVISION CAMERAS

In order to televise a color picture using the method of separating the three primary colors, three individual color cameras are required. In some color cameras, a forth tube is used to provide an improved monochrome picture and better detail for the color picture. This fourth tube will be described later on in this chapter.

Except for the number of camera tubes used, the distinguishing feature of the color camera is its optical system. It is here that all light entering the camera lens is divided, or "split," into separate color beams: one beam for each camera tube. This beam-splitting is accomplished by mirrors, or prisms, and lenses arranged in a unit called a beam-splitter. In the 4-tube camera, 4-way beamsplitters do the dividing, while 3-way units do the job in the 3-tube cameras.

Four-Way Beamsplitter A simplified diagram of a 4-way beamsplitter used in a typical 4-tube camera is shown in Fig. 3-11. Incident light from the scene being televised is divided into two parts. One part of the split beam is focused onto the faceplate of the fourth camera tube, in the luminance channel, to provide the "Y" voltage (monochrome signal) of the video signal. The other part of the split beam is reflected by a front-surface mirror through a relay lens into a system of dichroic mirrors. These mirrors are designed to reflect light of only one particular color and to allow light of all the other colors to pass on through. In the diagram of Fig. 3-11, the first dichroic mirror(1) reflects only the red light from the scene onto the faceplate of the red-channel tube to provide the R voltage of the video signal. Green and blue light pass on through the mirror and strike the front surface of the second mirror(2). This second mirror reflects the blue light into the blue channel and passes the green light on through into the green channel. These two latter channels provide the B and G voltages of the video signal.

Three-Way Beamsplitter The optical system in the 3-tube camera is somewhat simpler than the system found in the 4-tube camera. Figure 3-12 shows a simplified diagram of the compact 3-way beamsplitter in the Norelco* PC-70 3-tube color camera. In this color-splitting

* N. V. Philips of Holland.

system, dichroic layers are integral with, and enclosed within, a single-sealed assembly of small prisms. Color-trimming filters on the exit faces of the prisms provide suitable color characteristics, in conjunction with the spectral separation of the dichroic layers of the prisms and the spectral response of the plumbicon tubes used in the camera.

The plumbicon tube, positioned at the red, green, and blue exits of the beam-splitting optical system, has an antihalation faceplate which prevents light reflected back from the sensitive layer in the tube from generating an unwanted signal and impairing image quality. Each Plumbicon is mounted in an identical yoke assembly which is independently adjustable to center and align the tube in its optical axis. Each yoke also provides the precise adjustment for rotating the scanned area of the tube for angular registry.

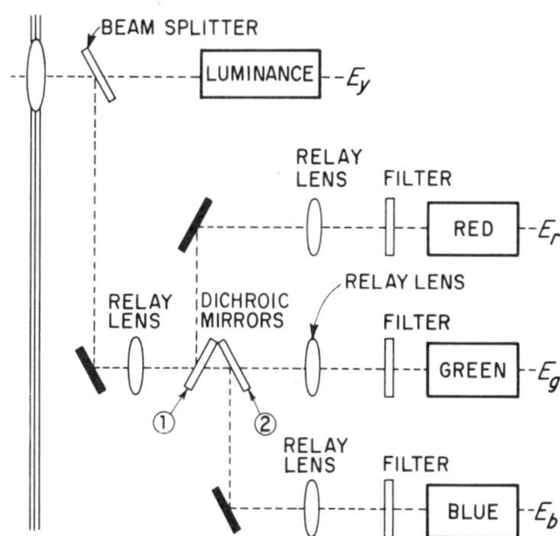

Fig. 3-11. Simplified diagram of a typical 4-way beamsplitter.

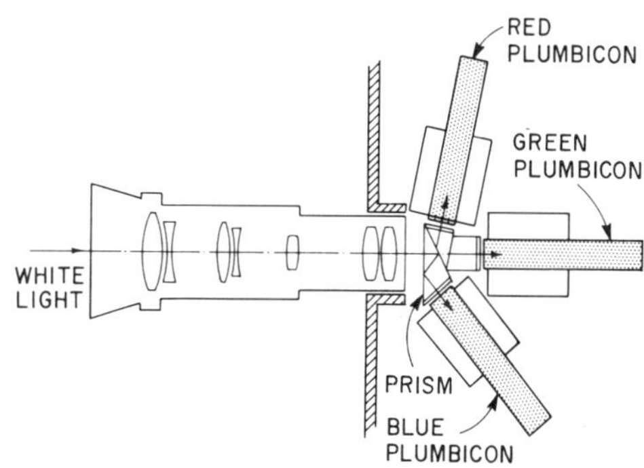

Fig. 3-12. Simplified diagram of the Norelco optical system.

3.7 SPECIAL PURPOSE TELEVISION CAMERAS

Over the past two decades, television has completely revolutionized home entertainment, and now it is making great inroads in education, medicine, industry, aerospace, and oceanography. In fact, there is hardly a field of endeavor today that does not use television in some particular application to perform tasks that heretofore were unsafe, inconvenient, or impossible. However, not all of these tasks can be performed by television equipment that is designed primarily for use in the broadcast studio, and, for this reason, many special purpose cameras have been developed.

Entertainment and Commercial Broadcasting In field applications that usually require the reproduction of high quality, on-the-spot television pictures, the cameras are set up wherever the action takes place. To facilitate the required mobility, these cameras should be small lightweight and easy to move about, yet they should retain all of the features of the larger studio cameras. In the field applications, they are usually connected to portable, field-type camera control units, which in turn are connected, by cables or microwave radio links, to the associated studio broadcast equipment. For studio use, the cameras are connected directly to the studio camera control units.

A typical portable color television camera is shown in Fig. 3-13. This camera is the Norelco PCP-90 "Minicam." As a color camera, it is designed for instantaneous news coverage and the on-the-spot pickup of current events. Unlike other portable color cameras which send separate red, blue, and green signals to their base stations for processing, the Minicam does all its signal processing in back-pack electronics carried by the camera operator. The encoded signals it produces are suitable for direct broadcasting without further processing. This technique reduces the possibility of noise pickup and cuts down on color errors caused by multipath effects.

Fig. 3-13. A portable color-TV camera. (*Courtesy of Philips Broadcast Equipment Corp.*)

As shown in Fig. 3-14, the Minicam system includes the color camera, back-pack electronics, a microwave radio link with a cable option, and a base station. This system uses a digital control technique, similar to that used by the Norelco PC-100 system, which permits radio control of all functions from the base station located as far as 10 miles away, depending, on the transmission path. If terrain features interfere with wireless communication, then the camera can be linked to its base station by a triaxial cable. However, cable losses due to stray capacitance limit the hookup to 1 mile between the camera and its base station. This system offers unrestricted mobility in areas where cables are prohibitive.

Education, Medicine, and Industry Television cameras that are designed specifically for education, medicine, and industry are typically small, simple to set up and operate, and are much less expensive than their studio counterparts. Mostly, they are operated by students or other personnel who are not specifically trained or experienced in television program production. Therefore, controls that must be set by the operator are reduced to a minimum, and, in many cases, to only an on-off power switch. Also, their method for mounting is determined by the particular purpose for which the camera is designed. Simple roll-around pedestals are often used for classroom or studio-type operation, while photographic-type tripods and simple shelf-type brackets serve in field or fixed applications.

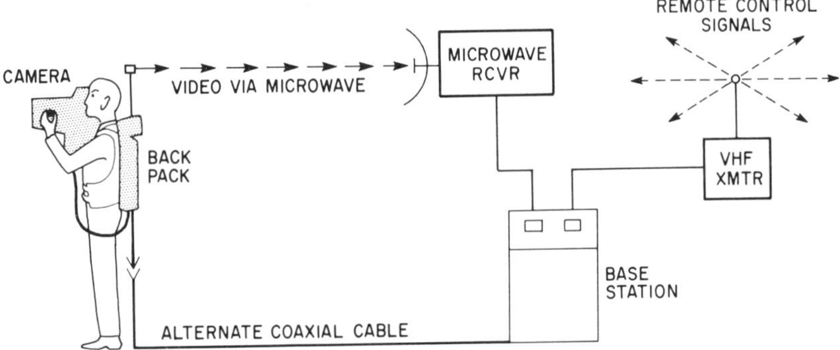

Fig. 3.14. Simplified diagram of Minicam system.

Figure 3-15 shows the RCA PK-730 color camera that is designed primarily for educational, medical, and industrial uses. This camera achieves full fidelity color reproduction with a single vidicon, in conjunction with a special color-detecting optical filter and innovative electronics. No revolving color wheels or other moving parts are used, and all circuit components are completely solid-state. Also, a zoom lens and a 5-inch viewfinder are included in the camera. It is 19-1/2 inches long, 8 inches wide, 13 inches high, and weighs less than 50 pounds. The operation of this camera is based upon the sequential, rather than on the simultaneous color system.

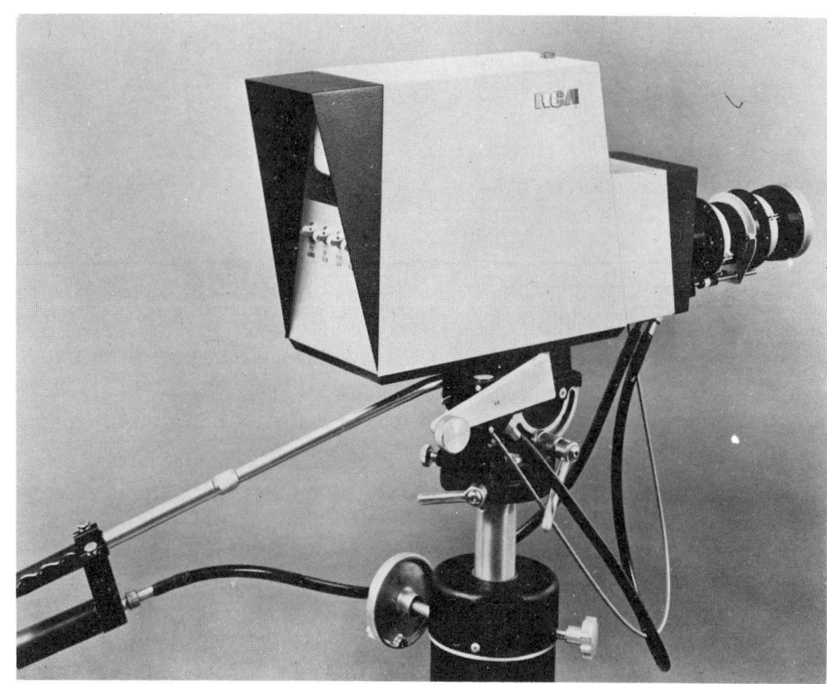

Fig. 3-15. The RCA PL-730 single vidicon color camera. (*Courtesy of RCA.*)

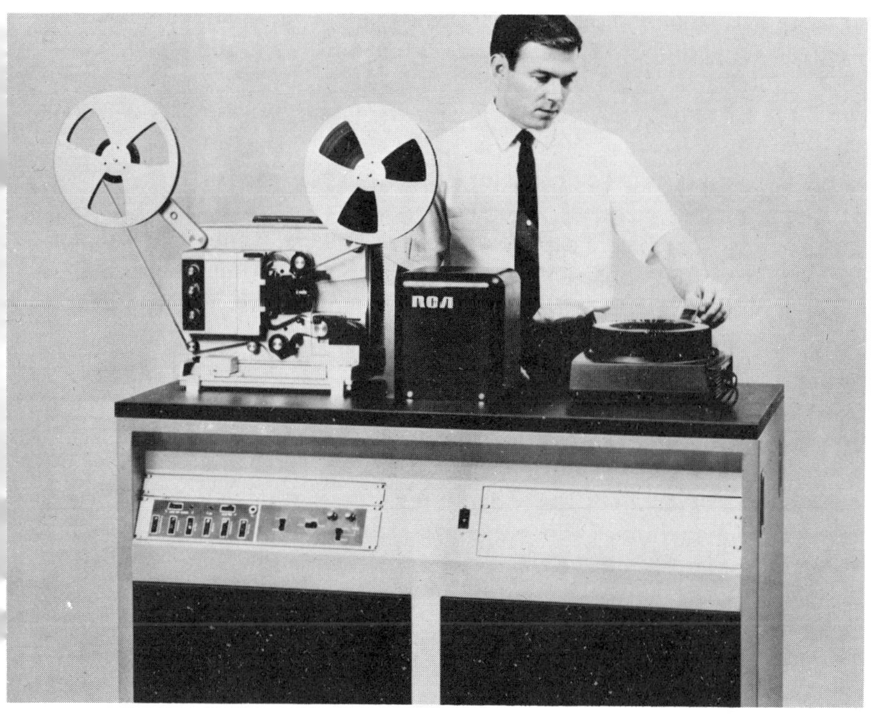

Fig. 3-16. A color-TV film system. (*Courtesy of RCA.*)

Figure 3-16 shows a specially designed color television film system. This completely self-contained system, the RCA PFS-710, includes a camera designed specifically for film reproduction, a 16 millimeter film projector, a slide projector, and a multiplexer. All of the units are contained in a mobile cart that is 4 feet long and 2 feet deep. The compatible NTSC signals are connected directly into the studio broadcast equipment through a cable.

3.8 CLOSED-CIRCUIT TELEVISION SYSTEMS

A closed circuit television system is a special application in which the camera signals are made available only to specially designated receivers. These systems, in recent years, have become an invaluable aid to education, medicine, industry, aerospace, and oceanography. Even entertainment television has found wide use in a closed circuit application where the viewer pays a fee to receive the programs in his home. Basically, these closed-circuit systems include one or more cameras with their required controls, one or more receivers, and some sort of transmission link between them.

The particular type of link used depends on the distance between the two locations, the number and dispersion of the receivers, and the mobility of either the camera or the receiver. In education, medicine, oceanography, and most industrial applications, the cable prevails, but in other industrial and all aerospace applications, the radio link is used exclusively. Figure 3-17 is a simplified block diagram that shows both types of links in typical applications.

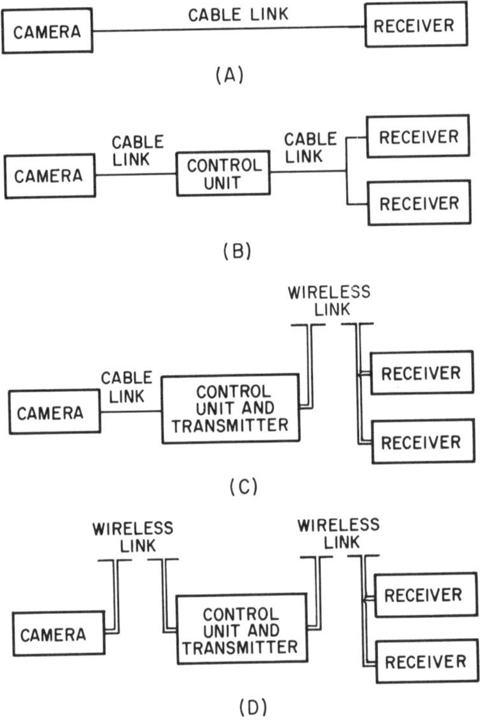

Fig. 3.17. Typical closed-circuit television systems.

In Fig. 3-17(A), the camera is shown connected through a cable directly to the receiver. This is perhaps the simplest form of a closed-circuit system and is used mainly for monitoring or surveillance in schools, shopping centers, estates, factories, warehouses, and crucial industrial installations. The camera used in this system is usually one of the simpler types that features only an on-off power control. In most cases, it is fixed mounted and requires no further attention. Its output signals, though, can be either video-coupled from an output amplifier in the camera, or an RF carrier onto which the video has been modulated.

The output signal determines the type of receiver that is connected to the other end of the cable. Those cameras that produce video outputs are connected directly to video circuits in the receiver: no tuner nor RF circuits are required. This video output, however, cannot be applied to the antenna terminals of a home-type television receiver, because the video is not superimposed on an RF carrier. Of course, the home-type receiver can be modified to provide terminals that connect into the video circuits, or an RF carrier can be provided at the camera.

Since there is a large number of economical home-type television receivers available for closed-circuit applications, many of the camera manufacturers have included an RF oscillator in their closed-circuit cameras to provide the necessary RF carrier. These oscillators, in many cases, are present at the factory to operate at some particular frequency to which the receiver can be tuned. A more versatile camera, though, is one in which the RF oscillator can be tuned to several different frequencies. In this way, a frequency that is not being used by a local commercial broadcast station can be selected. These variable or selectable frequency oscillators normally cover Channels 2 through 6. Separate camera control units and cable-driving amplifiers are seldom used in the single-receiver, closed-circuit television system.

A multi-receiver, closed-circuit system is shown in Fig. 3-17(B). These systems are used extensively in educational and training applications. They are particularly advantageous in medical schools where many students can observe a delicate operation in which attendance in the operating room is prohibitive. Also, a large group of students in many classrooms can view specially prepared slides mounted in a single microscope, motion pictures, or scientific experiments.

As shown in the figure, the multi-receiver, closed-circuit system includes a camera, a camera control unit, and the cable-connected receivers. The control unit in this system can vary from the simple desk-top type to the complex unit that was discussed in a previous section.

The wireless systems, as shown in Fig. 3-17(C), are used in most paid-television systems and in some industrial applications where cable links are either undesirable or impossible. These systems can range in complexity from a single receiver, with or without camera control, to the commercial broadcast-type facility that is used for paid-television. Their main feature is the wireless link, which is usually a microwave radio system. Even though the wireless link in Fig. 3-17(C) is shown between the control unit and the receiver, this link could also connect the control unit to the camera.

Figure 3-17(D) shows a closed-circuit television system in which the camera is remotely controlled over a microwave radio link. This is a versatile system that has been in use for a number of years in spacecraft, and, recently, has been introduced in a closed-circuit application. It is even being used in closed-circuit loops between the cameras and control units of several commercial broadcast systems to televise sports and news events. A typical system is the Norelco PCP-90 Minicam that was discussed earlier, along with other special purpose cameras. This system is noted for its computer-type digital circuits that are used in both the camera and the control unit to accomplish full control of the camera.

REVIEW QUESTIONS

1. How many different types of camera pickup tubes are in use? Name them and explain the differences between the types.
2. Can a monochrome-TV camera tube be used in a color camera?
3. Does the color Norelco TV camera have a separate lens system for each color? Draw a simple diagram of a lens system for a color camera.
4. Draw a diagram of a closed-circuit TV system in which a radio link is used. How many receivers can be used in this system?
5. Explain how the video signal is derived in the vidicon.
6. How is the video signal derived in the image orthicon?
7. What picture elements represent the minimum current in the output of an image orthicon?
8. Why must video amplifiers be used with a TV camera?
9. What is an electron multiplier? Where is it used?

TV Wave Propagation and TV Antenna Systems

4

4.1 INTRODUCTION

The antenna of a television receiver requires much more attention and care, especially with regard to placement, than the antenna of the ordinary sound receiver. In order to obtain a clear, well-formed image on the cathode-ray screen, the following requirements must be met:

(1) Sufficient signal strength must be developed at the antenna.
(2) The signal must be received from one source, not several.
(3) The antenna must be placed well away from man-made sources of interference.

In sound receivers, a certain amount of interference and distortion is permissible. For television, however, the standards are more strict, and added precautions must be taken to guard against many types of interference and distortion. Hence, there is a need for more elaborate antenna receiving systems.

The position of the antenna must be chosen carefully, not only to provide the maximum signal strength to the receiver, but also to avoid the appearance of so-called *ghosts* on the image screen. Ghosts are multiple images that are due to the simultaneous reception of the same signal from two or more directions. For an explanation of this form of interference, refer to Fig. 4-1 which shows a television receiver antenna.

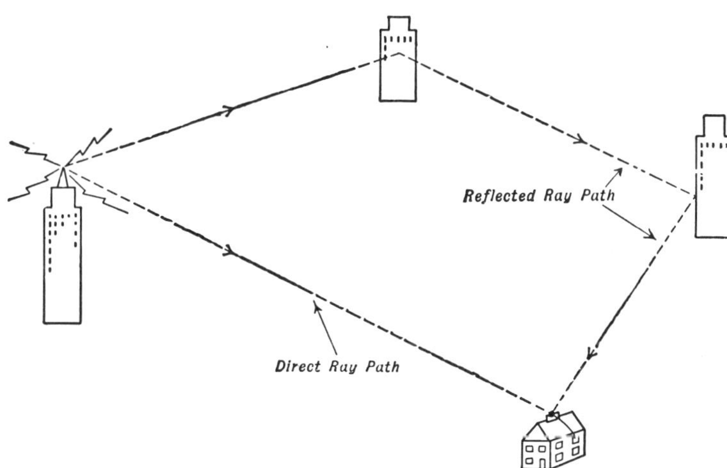

Fig. 4-1. The reflected ray and the direct ray arrive at the receiving antenna and form double images, or *ghosts*.

receiving one signal directly from the transmitting tower, while another ray strikes the same antenna after following a longer, indirect path. Reflection from buildings, water towers, bridges, or other large objects can cause an indirect ray to reach the antenna.

Because of the longer distance the reflected ray travels, it will arrive at the receiver antenna a small fraction of a second later than the direct ray. With sound receivers, the ear does not detect the difference in time between the direct and indirect signals. However, on a television screen, the scanning beam has traveled a short distance between the time the original wave arrives to the time the reflected ray arrives at the receiver. Hence, the image contained in the reflected ray appears on the screen displaced a short distance from similar detail contained in the direct ray. The result is shown in Fig. 4-2. When the effect is pronounced, a more distinct double image is obtained, and the picture appears blurred. To correct this condition, it is necessary to make changes in the antenna system so that only one ray is received. This is usually accomplished by using a highly *directional antenna*—that is, an antenna that will receive a signal from only one direction. The directional antenna is turned so as to eliminate the undesired signal. The antenna should not be turned to favor the reflected signal, unless it is impossible to obtain a clear image with the direct ray. The properties of reflecting surfaces change with time, and there is no certainty that a good reflected signal will always be received.

Fig. 4-2. A *ghost* image on a television viewing screen.

The placement of the antenna is one of the most important requirements of a television installation. To obtain optimum results, it is necessary to have a good understanding of the behavior of radio waves at the high frequencies.

4.2 RADIO-WAVE PROPAGATION

Transmitted radio waves at all frequencies may travel in either of two general directions. One wave closely follows the surface of the Earth, whereas the other travels upward at an angle which is dependent on the position of the transmitting antenna. The former is known as the *ground wave*, the latter as the *sky wave*. At the low frequencies, up to approximately 1,500 kHz, the ground-wave attenuation is low, and signals travel for long distances before they disappear. Above the broadcast band, ground-wave attenuation increases rapidly, and long-distance communication is carried on mostly by means of the sky wave.

The sky wave leaves the Earth at an angle that may have any value from 3 to 90 degrees, and travels in almost a straight line until the ionosphere is reached. This region, which begins about 70 miles above the surface of the Earth, contains large concentrations of charged gaseous ions, free electrons, and uncharged, or neutral, molecules. The ions and free electrons act on all passing electromagnetic waves and tend to bend these waves back to Earth. Whether the bending is complete (and the wave does return to the Earth) or only partial, depends on several factors:

(1) The frequency of the radio wave.
(2) The angle at which the wave enters the ionosphere.
(3) The density of the charged particles (ions and electrons) in the ionosphere at that particular moment.
(4) The thickness of the ionosphere at the time.

Extensive experiments indicate that, as the frequency of a wave increases, a smaller entering angle is necessary in order for complete bending to occur. To illustrate this, consider waves *A* and *B* in Fig. 4-3.

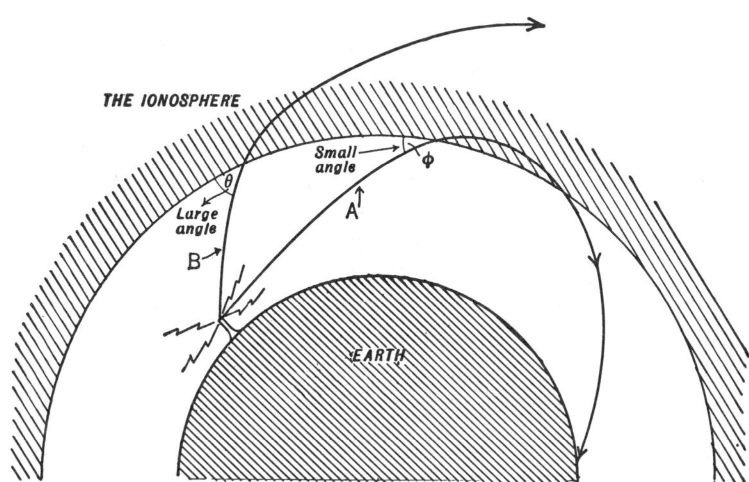

Fig. 4-3. At the higher frequencies a radio wave must enter the ionosphere at small angles if it is to be returned to Earth.

Wave *A* enters the ionosphere at a small angle (ϕ) and, hence, little bending is required to return it to Earth. Wave *B*, subject to the same amount of bending does not return to Earth, because its initial entering angle (θ) was too great. Naturally, the latter wave would not be useful for communicating between points on Earth.

By raising the frequency still higher, the maximum allowable incident angle at the ionosphere becomes smaller, until finally a frequency is reached where it becomes impossible to bend the wave back to Earth, no matter what angle is used. For ordinary ionospheric conditions, this frequency occurs at about 35 or 40 MHz. Above that frequency, the sky wave cannot be used for radio communication between distant points on Earth. Only the *direct ray* is of any use. Television bands starting above 40 MHz fall into this category. By direct ray (or rays), we mean the radio waves that travel in a straight line from transmitter to receiver. Ordinarily, at lower frequencies, the radio waves are sent to the ionosphere and from there to the receiver at a distant point. At television frequencies, the ionosphere is no longer useful, so the former sky waves must be concentrated into a path leading directly to the receiver. It is this restriction on the use of the direct ray that limits the distance in which high-frequency communication can take place.

There are present, at times, unusual conditions which cause the concentrations of charged particles in the ionosphere to increase sharply. At these times, it is possible to bend radio waves of frequencies up to 60 MHz. The exact time and place of these phenomena cannot be predicted and hence are of little value for commercial operation. They do explain to some extent, however, the distant reception of high-frequency signals that may occur.

4.3 LINE-OF-SIGHT DISTANCE

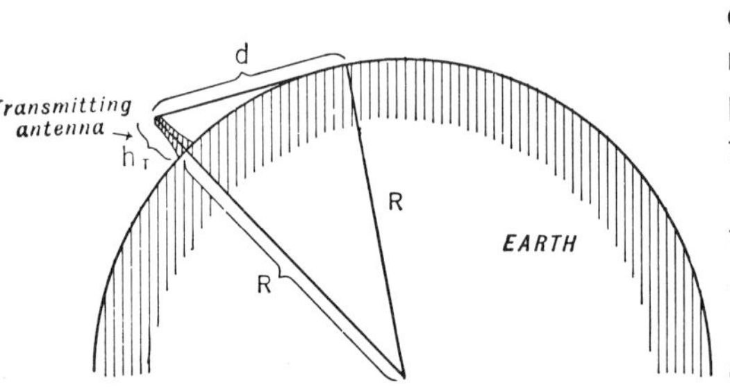

Fig. 4-4. Computation of the line-of-sight distance for high-frequency radio waves.

At the frequencies employed for television, reception is possible only when the receiver antenna directly intercepts the signals as they travel away from the transmitter. These electromagnetic waves travel in essentially straight lines, and the problem is resolved by finding the maximum distance from the transmitter where the receiver can be placed and still have its antenna intercept the rays. This distance, called the *line-of-sight distance*, may be computed as follows:

In Fig. 4-4, let the height of the transmitting antenna be called h_t, the radius of the Earth R, and the distance from the top of the transmitting antenna to the horizon d. These distances form a right triangle.

The Pythagorean Theorem in geometry states: *The sum of the squares of the sides of a right triangle equals the square of the hypotenuse.* Using this relationship for the distances shown in Fig. 4-4 gives the resulting equation:

$$R^2 + d^2 = (R + h_t)^2$$

By expanding:

$$R^2 + d^2 = R^2 + 2Rh_t + h_t^2$$

The value of h_t is very small compared to R, so h_t^2 can be eliminated from the equation without seriously affecting the accuracy. Also, R^2 can be subtracted from both sides of the equation. The equation then becomes

$$d^2 = 2Rh_t$$

The radius of the Earth is 4,000 miles, or 21,120,000 feet. Substituting this value into the equation, and taking the square root of both sides gives

$$\begin{aligned} d &= \sqrt{42{,}240{,}000\, h_t} \\ &= 6499\sqrt{h_t} \end{aligned}$$

Dividing by 5,280 to convert the distance to miles gives

$$d = 1.23\sqrt{h_t}, \text{ where} \qquad 4\text{-}1$$

d is the line-of-sight distance from the top of the transmitting antenna in miles, and
h_t is the height of the transmitting antenna in feet.

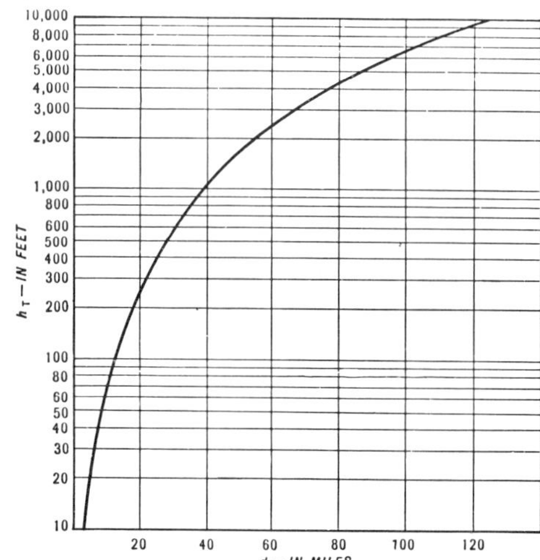

Fig. 4-5. The relationship between the height of the transmitting antenna in feet and the distance in miles from the antenna that the ray may be received.

The relationship between d and h_t for various values of h_t is shown in graph form in Fig. 4-5.

The ground coverage for any transmitting antenna will increase with its height. Likewise, the number of receivers capable of receiving the signals will increase. This fact accounts for the placement of television antennas atop tall buildings (for example, the World Trade Center, New York City) and on high plateaus.

Equation 4-1 can be used for computing the distance from the top of the transmitting antenna to the horizon. By placing the receiving antenna some distance in the air, it would be possible to cover a greater distance before the curvature of the Earth interferes with the direct ray. This condition is depicted in Fig. 4-6. By geometrical reasoning, the maximum line-of-sight distance between the two antennas can be arrived at from the distances shown in Fig. 4-6.

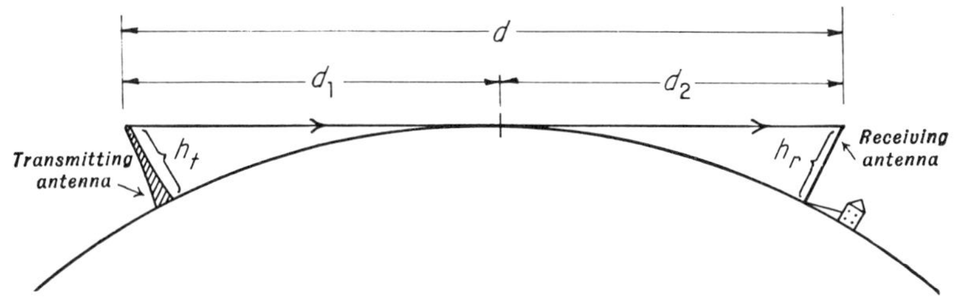

Fig. 4-6. The increase in the line-of-sight distance from the receiving antenna to the transmitter achieved by raising both structures as high as possible.

$d_1 = 1.23\sqrt{h_t}$ = the maximum distance from the transmitting antenna to the horizon.

$d_2 = 1.23\sqrt{h_r}$ = the maximum distance from the receiving antenna to the horizon.

$d = d_1 + d_2$ = the maximum distance from the transmitting antenna to the receiving antenna.

$$= 1.23\sqrt{h_t} + 1.23\sqrt{h_r}$$

$$d = 1.23(\sqrt{h_t} + \sqrt{h_r}), \text{ where}$$

h_t is the height of the transmitting antenna in feet,
h_r is the height of the receiving antenna in feet, and
d is the maximum line-of-sight distance between the two in miles.

The equations given are actually for the geometrical line of sight. In reality, electromagnetic waves are bent slightly as they move across the contact point at the horizon, and this increases the television line-of-sight distance by about 15% over the geometric line of sight. Thus, for a geometric line of sight of 30 miles, the television line-of-sight distance is $30 + (0.15 \times 30) = 34.5$ miles.

It has been observed that the received signal strength increases with the height of either *antenna*, or both. For television signals, this increase is most important. Placement of the antenna and utilization of its directive properties help in decreasing (and many times in eliminating) all but the desired direct wave.

4.4 UNWANTED SIGNAL PATHS

While the foregoing computed distances apply to the direct ray, there are other paths that waves may follow from the transmitting to the receiving antennas. These waves are undesirable, because they tend to distort and interfere with the direct-ray image on the screen. One type of indirect wave, reflection from surrounding objects, has already been discussed. Another type of indirect ray may arrive at the receiver by reflection from the surface of the Earth. This path is shown in Fig. 4-7. At the point where the reflected ray strikes the Earth, phase reversals up to 180 degrees have been found to occur. This phase shift places a wave at the receiving antenna which generally acts against the direct ray. The overall effect is a general lowering of the resultant-signal level and the appearance of annoying ghost images.

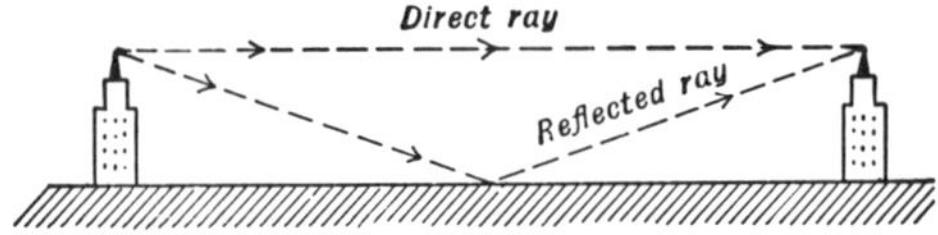

Fig. 4-7. The reflected radio wave, arriving at the receiving antenna after reflection from the Earth, may lower the strength of the direct ray considerably.

There are, however, two compensating conditions that reduce the problem of ground reflection. One is the weakening of the wave strength by absorption at the point where it grazes the Earth, and the other results from the added phase change caused by the fact that the length of the path of the reflected ray is longer than that of the direct ray. We see then, that there are two phase shifts affecting the reflected signal: (1) one at the point of reflection from Earth, and (2) one that is the result of the longer signal path. These two phase shifts are additive, so that the total phase shift is nearer to 360°. The worst possible phase shift would be 180°, because this would mean that the direct and the reflected waves are cancelling. Since 360° is equivalent to no phase shift, the fact that the individual phase shifts are additive reduces the problem of signal cancellation considerably. Of course, the direct wave is stronger than the reflected wave, so the two signals would not completely cancel, but the reduction in signal strength could be detrimental to the receiver picture quality.

4.5 WAVE POLARIZATION

The height of the antenna is one important factor that determines the quality of the reproduced picture. Another is the manner in which the antenna is held, that is, either vertically or horizontally. The position of the antenna is determined by the nature of the electromagnetic wave itself.

All electromagnetic waves have their energy divided between an electric field and a magnetic field. In free space these fields are at right angles to each other. Thus, if we were to visualize these fields and represent them by their lines of force, the wave front would appear as illustrated in Fig. 4-8. The squares represent the wave front, and the

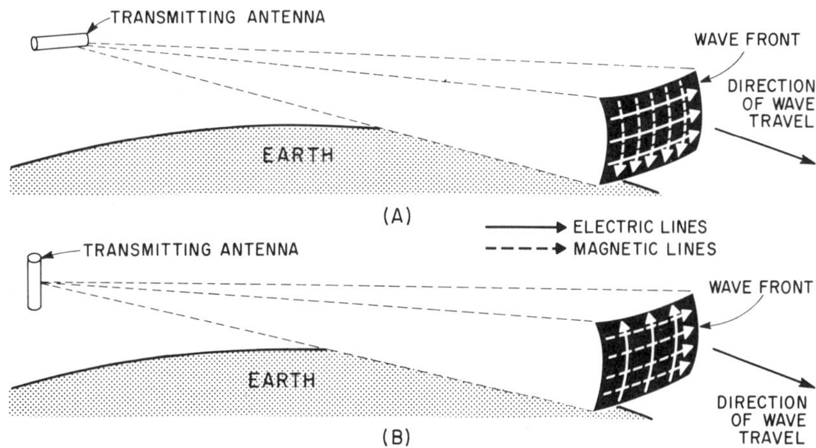

Fig. 4-8. The components of an electromagnetic wave showing their relationship with the direction of travel of the wave front.

arrows represent the direction in which the forces are acting. The direction of travel of these waves in free space is always at right angles to both fields. As shown in the figure, if the lines of the electric field are vertically directed upward and those of the magnetic field are horizontally directed to the right, then the wave travel is forward.

In radio, the polarity of a radio wave has been taken to be the same as the direction of the *electric* lines of force. Hence, a vertical antenna radiates a vertical electric field (the lines of force are perpendicular to the ground), and the wave is said to be *vertically polarized*. A horizontal antenna radiates a horizontally-polarized wave. In most cases, the signal that is induced in the receiving antenna is greatest if this antenna has the same polarization as the transmitting antenna.

There are different characteristics for horizontally- and vertically-polarized waves. For antennas located close to the Earth, vertically-polarized rays yield a better signal. When the receiving antenna is raised about one wavelength above ground, this difference generally disappears and either vertical or horizontal antennas may be employed. When the antenna is at least several wavelengths above ground, the horizontally-polarized waves give a more favorable signal-to-noise ratio. In television, the wavelengths are short, and the antennas are placed several wavelengths in the air. Horizontally-polarized waves have been accepted as standard for the TV industry. All television receiving antennas are mounted in the horizontal position.

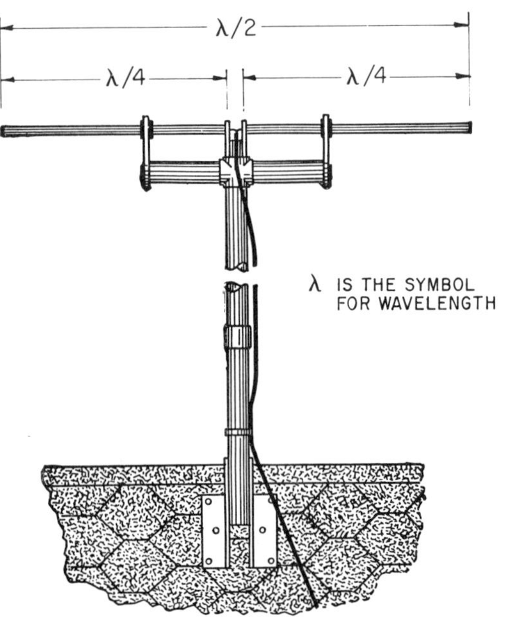

Fig. 4-9. A dipole-antenna assembly used extensively for television receivers.

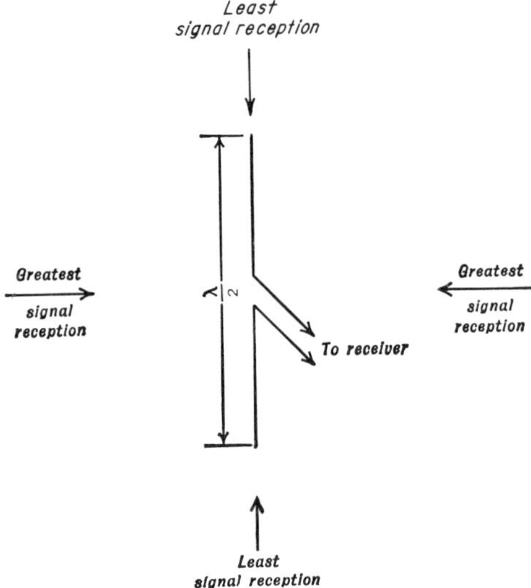

Fig. 4-10 (A). Dipole antennas of the type shown here receive signals best from the directions indicated.

4.6 TUNED ANTENNAS

Television antennas are made of wires or rods that are cut to a specific length. The particular length to which they are cut is the determining factor of their *resonant frequency*. An antenna has inductance and capacitance, and therefore it has a resonant frequency. Electromagnetic waves induce voltage in the antenna. The more closely the resonant frequency approaches the frequency of the electromagnetic wave, the greater the signal voltage generated in the antenna. For example, a 50MHz wave will induce more voltage into an antenna that is resonant at 50 MHz than in one that is resonant at 60 MHz. The greater the signal induced in the antenna, the greater the signal-to-noise ratio that is possible at the output of the receiver.

4.7 HALF-WAVELENGTH ANTENNAS

An ungrounded wire or rod which is cut to one-half the wavelength of the signal to be received is called a *half-wave antenna* or sometimes a *Hertz antenna*. This type of antenna is very popular since it represents one of the smallest antennas for its frequency, and consequently requires only a small amount of space. In troublesome areas it may be necessary to erect more elaborate arrays possessing greater gain and directivity than the simple half-wave antenna. They are, however, more costly.

A simple half-wave antenna may be erected and supported as indicated in Fig. 4-9. Two metallic rods are used for the antenna itself. They are mounted on a supporting structure and placed in a horizontal position. Each of the rods is one-quarter of a wavelength, the total of the two rods being equal to the necessary half-wavelength. This arrangement is called a *dipole antenna*. The transmission lead-in wires are connected to the rods, one wire of the line to each rod. The two-wire line then extends to the receiver. Care must be taken to fasten the line at several points to the supporting mast with stand-off insulators, so that it does not swing back and forth in the wind. Any such motion could weaken the electrical connections made at the rods.

An important property of dipole antennas is that they receive signals with the greatest intensity when the rods are at right angles to the approaching signal. This arrangement is illustrated in Fig. 4-10(A). On the other hand, signals approaching the antenna from either end are very poorly received. To show how waves at any angle are received, the graph of Fig. 4-10(B) is commonly drawn. It is an overall response curve for a horizontally held dipole antenna in the horizontal plane.

With the placement of the antenna as shown in the diagram, a strong signal will be received from direction *A*. As the signal angle made with this point is increased, the strength of the received signal decreases, until at point *B* (90 degrees) the received signal voltage is at a minimum

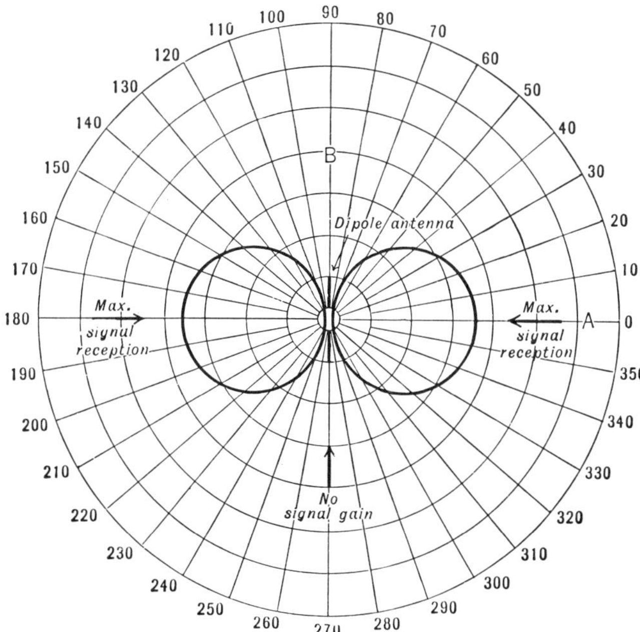

Fig. 4-10 (B). The directional response curve of a dipole antenna.

(or zero). By inspection of the graph the reader can determine the reception for waves coming in at other angles. Notice that good signal strength is obtained from two directions and, because of this, the dipole may be called *bidirectional*. Other systems can be devised that are *unidirectional* or *nondirectional*, or that have almost any desired properties. For each system, a response curve quickly indicates its properties in any direction.

It has been stated that an antenna must be tuned in order to have the strongest signal develop along its length. This is usually accomplished by cutting the wires (or rods) to a specific length. The length is determined by the frequency of the signal to the receiver; it is longer at the lower frequencies and shorter at the higher frequencies. It might be supposed, then, that a television receiver capable of receiving signals with frequencies ranging from 54 to 88 MHz would need several antennas, one for every channel. It is not usually necessary, however, to go to such extremes. In practice, one antenna is sufficient, if it is tuned to a middle frequency.

4.8 ANTENNA-LENGTH COMPUTATIONS

For the range of frequencies between 54 and 88 MHz, a middle value of 65 MHz might be chosen for the resonant frequency value to be used for the antenna design. While an antenna cut to this frequency will not give optimum results at the other frequencies, the reception will still be quite satisfactory in most cases.

To compute the length needed for the 65-MHz frequency half-wave antenna, the following formula is used:

$$L = \frac{468}{f}, \text{ where}$$

L is the length of the dipole in feet and
f is the frequency that the dipole is resonant to in MHz.

When f is 65 MHz, the correct dipole length would be equal to 468/65, or 7.2 feet. Practically, 7 feet might be used with each half of the half-wave dipole 3.5 feet long. For a full wavelength antenna, a length of approximately 14 feet would be needed.

An added advantage to be considered for half-wave antennas is the fact that their smaller physical size (compared with full-wave antennas) makes it easier to construct them so that they will be able to withstand high winds and the weight of ice or snow.

4.9 HALF-WAVE DIPOLE WITH REFLECTOR

The simple half-wave system provides satisfactory reception in many locations within reasonable distances of the transmitter. However, the signals reaching receivers situated in outlying areas are correspondingly weaker, so noise and interference have a greater distorting effect on the image. For these locations more elaborate arrays must be constructed —systems that have greater *gain* and *directivity* to provide better discrimination against interference. The *gain* of a receiving antenna is a measure of its effectiveness in receiving a signal divided by the effectiveness of a dipole in receiving the same signal at the same location. It is usually expressed in decibels (db). An antenna with a positive db will deliver a better signal to a receiver than a dipole would deliver. The *directivity* of an antenna is a reference to its ability to receive signals from one direction and reject signals from all other directions. Antennas with high directivities are often used to eliminate or reduce ghost problems. The gain of the antenna of Fig. 4-11 is about 5 db over the gain of a simple dipole.

The antenna of Fig. 4-11 is a dipole with an additional rod called a *reflector*. There is no electrical connection between the dipole—sometimes called the *driven element*—and the reflector which is some-

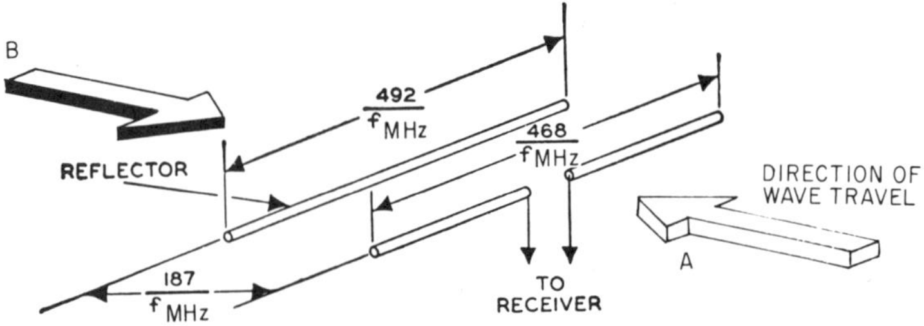

Fig. 4-11. Dipole receiving antenna and reflector.

times called the *parasitic element*. The spacing between the two elements is 0.15 to 0.25 wavelengths of the signal to which the dipole is resonant. This is an important factor in the antenna operation.

Let us assume that a signal arrives in the direction shown by arrow *A* in Fig. 4-11, and that the dipole is resonant to this frequency. The signal will generate a voltage in the driven element, and it will also generate a voltage in the reflector. Signal current flows in the reflector as a result, and the signal is reradiated. Now, the reradiated signal from the reflector also induces a signal voltage in the driven element. If the reflector is properly spaced from the dipole, the reradiated signal will be in phase with, and will reinforce, the original signal. The fact that the original signal is reinforced by the reflected signal accounts for the gain of this antenna over a simple dipole.

If a signal approaches from the direction marked *B* in Fig. 4-11, it will be reradiated by the reflector as before. However, the reradiated signal will now arrive at the driven element out of phase with the original signal, and the overall result will be a strong decrease in the received signal strength. You will note that in the case of signal *A*, the time required for the wave to travel from the dipole to the reflector and back to the dipole is longer than the time required for signal B and the reradiated signal from the reflector to arrive at the dipole. This accounts for the difference in phase between the original and reradiated wave for the two different cases.

Besides the greater gain that is observed with the addition of a reflector system, Fig. 4-12 shows that the angle at which a strong signal may

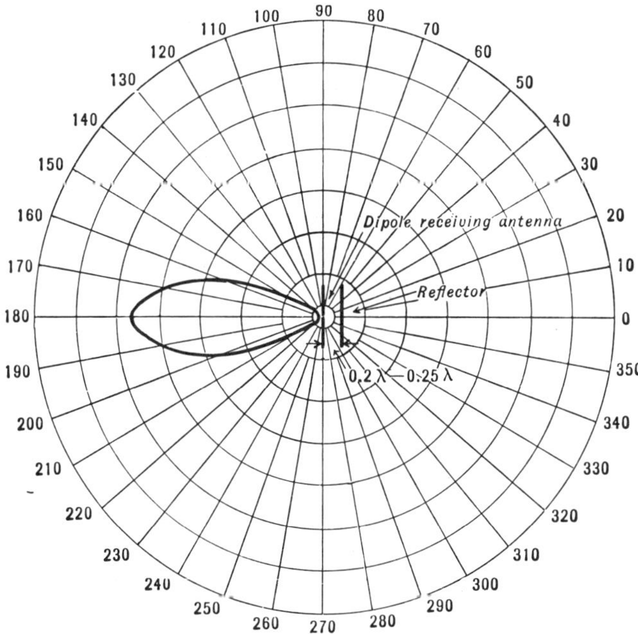

Fig. 4-12. The directional response curve for a half-wave antenna with a reflector.

be received is now narrower. This is also advantageous in reducing the number of reflected rays that can produce ghosts. Finally, partial or complete discrimination is possible against interference, man-made or otherwise.

The curve in Fig. 4-12 is an idealized representation. The response curve of an actual array would have small lobes extending in the direction of the reflector, which indicates that the reception of signals approaching the array from the rear can occur. This is understandable because the waves do not arrive at the driven element exactly 180° out of phase. The voltage induced in the driven element by the waves approaching from the reflector side is much less than for waves arriving from the front.

4.10 DIPOLES WITH REFLECTOR AND DIRECTOR

Additional gain and directivity can be obtained from an antenna if, besides the reflector, a *director element* is added. A director is a wire or rod that is approximately one-half wavelength. Like the reflector, it is a parasitic element. Normally the reflector is slightly longer than the driven element, and the director is slightly shorter. When the dipole is placed broadside to the direction of the incoming signal, the director is the first element of the combination to intercept the oncoming signal. Figure 4-13 shows an antenna with a director and reflector. The arrow points in the direction of greatest directivity. The director picks up part of the signal and then reradiates it with a phase relationship that strengthens the signal arriving at the driven element. The net result of this action is to make the response of the dipole more directive, and to reduce the ability of the array to pick up signals reaching the unit from the rear. Some antennas have more than one director. The *Yagi* type shown in Fig. 4-14 is an example. This type of antenna has a very high gain and a highly-directional response. Some Yagi antennas may have more directors than shown, which will increase the gain and directivity. A disadvantage of the Yagi antenna is that it is only responsive to a narrow range of frequencies. Therefore, it cannot be used as an all-purpose antenna in areas where there are a number of stations.

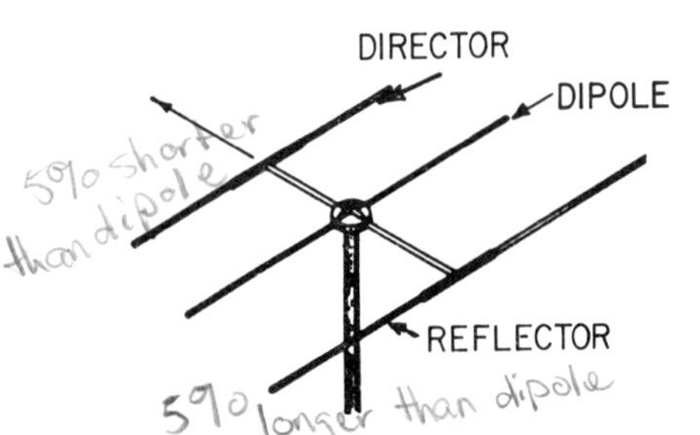

Fig. 4-13. A dipole with reflector and director.

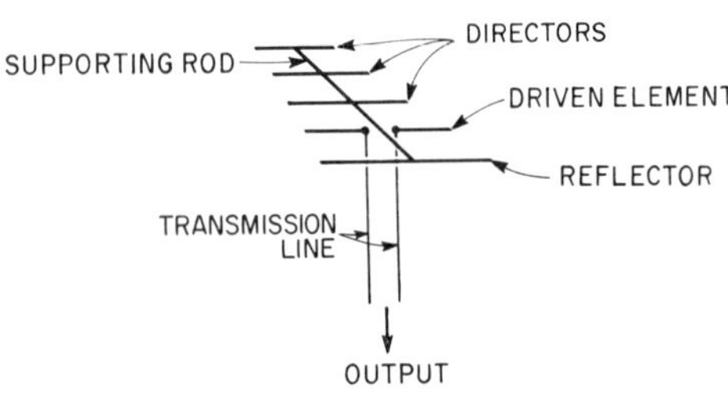

Fig. 4-14. A Yagi antenna with three directors and one reflector.

4.11 UHF TELEVISION ANTENNAS

In its original allocation plan issued in 1946, the Federal Communications Commission set aside 12 channels for commercial television broadcasting. At first 13 channels were allocated, but Channel 1 (44–50 MHz) was subsequently dropped. It did not take long to demonstrate that this was far too few channels for extensive nationwide coverage and, in 1952, an additional 70 channels in the UHF band (470–890 MHz) were added.

The problems facing the television technician in the UHF band do not differ in principle from those presently facing him in the VHF band.

They do, however, differ in degree. Thus, he must still erect an antenna system to capture as much signal as possible, except that the UHF signals reaching the antenna are often weaker than the VHF signals. Furthermore, the losses presented by the transmission lines are greater.

In order to send as much UHF signal to the receiver as possible, careful erection of the antenna is necessary. This means that the technician must not only have to find the best horizontal spot, but he must also have to determine the best height. In the choice of antennas, he has a considerable number of designs from which to pick, and, fortunately, high-gain arrays are more feasible for the UHF signal because antenna dimensions are smaller. A half-wave dipole at 550 MHz will have roughly 1/10 the overall dimensions of a half-wave dipole at 55 MHz. This means that more elements can be added to the UHF array without causing it to become unwieldy. Since the gain of an antenna generally rises with the number of elements, higher gain can be expected from the UHF arrays. We will now discuss a few representative designs.

Fan Dipole (Also known as a *bow tie* or *di-fan antenna.*) The half-wave dipole is the simplest type of VHF antenna, and also it is the simplest UHF antenna. Figure 4-15 shows an example of a dipole antenna with specially-shaped dipoles. By using triangular sheets of metal instead of rods, the unit becomes a broadband affair capable of receiving all those signals within the UHF band. The overall length of the fan for greatest gain is slightly higher than that of a rod dipole. The response pattern of a fan dipole is a figure -8, unless a screen reflector is placed behind the array as indicated in Fig. 4-16. In this case it becomes unidirectional. Even with a reflector these antennas are not highly directional, so they provide satisfactory reception only in strong signal areas where there are relatively few ghost problems. For greater gain, fan dipoles can be stacked two high and four high. This also provides better discrimination against ground-reflected signals and is useful in reducing the number of interfering signals capable of affecting the picture on the receiver.

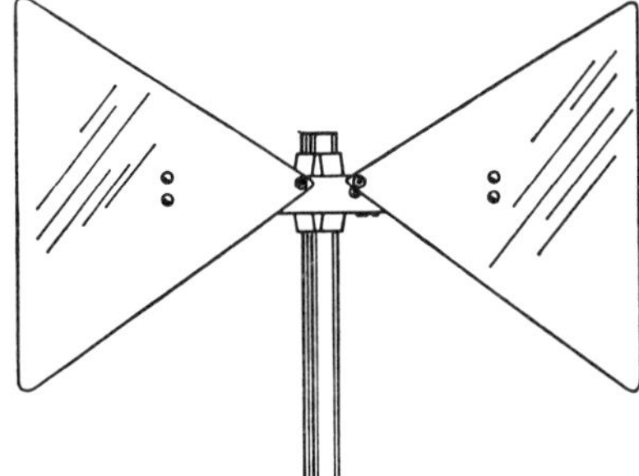

Fig. 4-15. A UHF fan dipole.

You will note that the di-fan reflector in Fig. 4-16 is a mesh screen instead of a rod, as is customary for VHF signals. Screens are considerably more efficient reflectors than rods, and the only reason screens are not used extensively for VHF is because they would prove too bulky and their wind resistance would be too great. Mesh screens are as effective as solid metallic sheets, provided the mesh openings are on the order of 0.2 of a wavelength or less at the highest operating frequency. Reflector dimensions are not critical, but the edges should extend a little beyond the dipole elements.

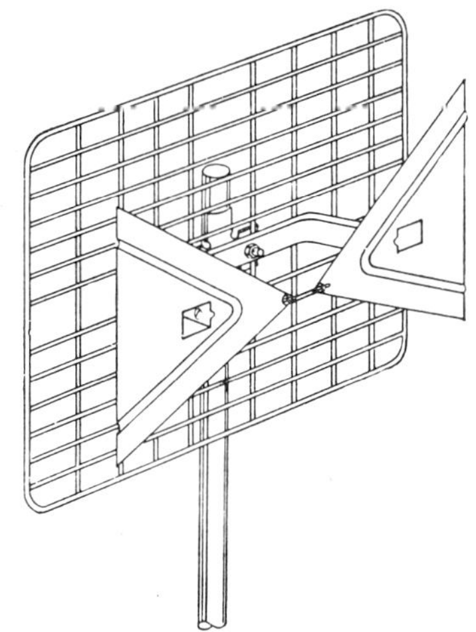

Fig. 4-16. A fan dipole with a screen reflector.

Parabolic Reflector Probably everyone is familiar with the fact that the headlights of a car have parabolic reflectors in order to provide a high concentration of light. In much the same fashion, parabolic reflectors

Fig. 4.17. A parabolic reflector. (*Courtesy of Taco.*)

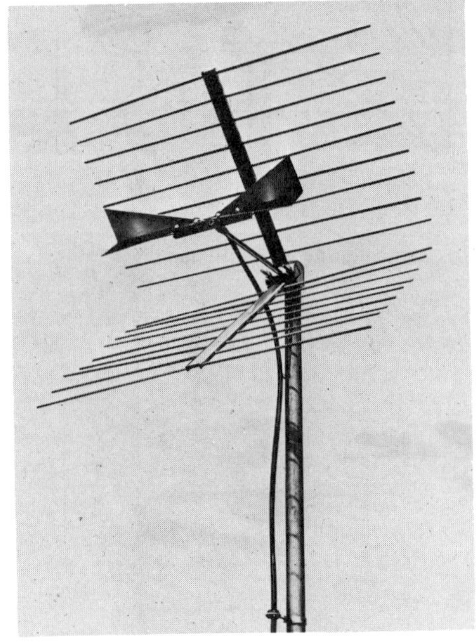

Fig. 4-18. A corner reflector for a UHF antenna.

are used to receive and transmit radio waves with a high gain and a sharp directivity.

Instead of using an entire parabolic reflector, it is possible to use only a section of one as shown in Fig. 4-17. Such a parabolic reflector can provide a gain of 8 db over that of a resonant half-wave dipole. The vertical directivity of this antenna structure is sharp, but the horizontal directivity is somewhat broad. Where high gain is desired and the ghost problem is not serious, this array will provide excellent results.

Corner Reflectors Instead of using curved surfaces as reflectors, it is possible to use two flat surfaces so positioned as to intersect at an angle, forming a corner. This type of reflector, depicted in Fig. 4-18, is known as a *corner reflector* antenna. The driven element, usually a dipole antenna, is placed at the center of the corner angle and at some distance from the vertex of the angle.

The response pattern of this antenna depends not only on the corner angle but also on the distance between the antenna and the vertex of the reflector corner. If the antenna is positioned too far from the vertex, the response pattern will have several lobes. If it is brought in too close, the vertical response will be broadened and the susceptibility of the array to ground-reflected signals will increase. The corner angle in the commercial array of Fig. 4-18 is 90 degrees, and a similar bend is placed in the fan dipole. Gain over the entire UHF-TV band is high, ranging from about 7 db at 500 MHz to 13 db at 900 MHz.

4.12 GENERAL-PURPOSE ANTENNAS

An antenna that has been widely used is the folded dipole illustrated in Fig. 4-19(A). This antenna consists principally of two dipole antennas connected in parallel with each other. The separation between the two sections is from three to five inches. The folded dipole has the same bidirectional pattern as shown for the simple dipole in Fig. 4-10(B). The two antennas (dipole and folded dipole) have approximately the same gain. However, the response of a folded dipole is more uniform over a band of frequencies than that of the simple dipole.

As with a simple dipole, the directivity of the folded dipole can be increased by the addition of a reflector. Figure 4-19(B) shows a folded dipole with a reflector. This antenna possesses the same directional pattern as the simple dipole with a reflector which was illustrated in Fig. 4-12. The length and spacing of the reflector can be determined by the same relationships as for a simple dipole and reflector. Figure 4-19(C) shows the correct relationships for the length of the elements and the spacing between elements for a folded dipole with a director and a reflector.

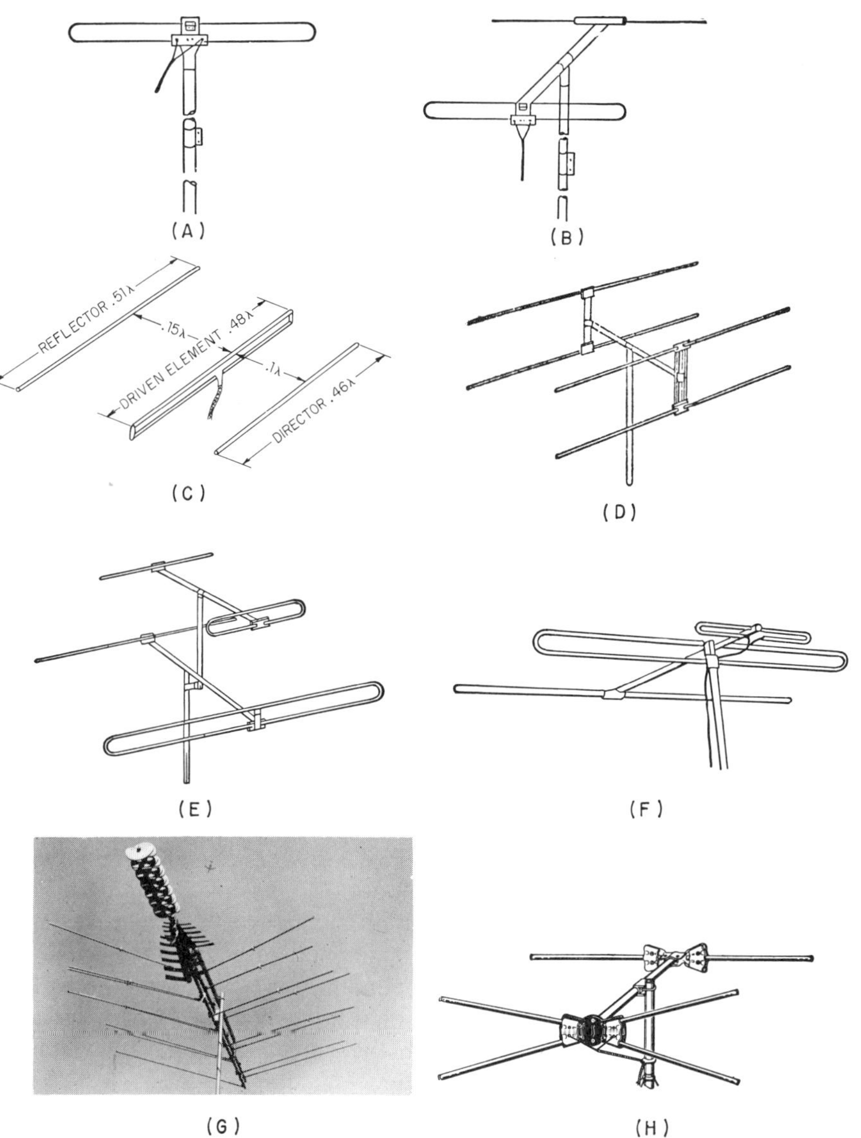

Fig. 4.19. Types of popular television receiving antennas.

In the remaining parts of Fig. 4-19 other popular types of television antennas are shown. These are elaborations of the basic dipole or folded dipole. In Fig. 4-19(D) we see a stacked-dipole array with reflectors. This antenna is sometimes referred to as a *Lazy H* because of its similarity to the letter H lying on its side. In it, two half-wave dipoles are placed at the front of the assembly, one mounted above the other. The center terminals of each dipole are connected by means of a parallel-wire line. Each conductor of the lead-in line to the television receiver is attached to a conductor of this connecting line at a point midway between the dipoles. A reflector is mounted behind each dipole.

In Fig. 4-19(E), there are two folded dipoles, with reflectors, mounted one above the other. The upper dipole is cut for a resonant frequency approximately in the center of the upper VHF-TV band (174–216 MHz,) and the longer folded dipole is resonated at the center frequency of the lower television band. A short length of a two-wire conductor connects the upper dipole to the lower one. From the lower antenna, a two-wire conductor feeds the signals to the receiver. With this assembly, each antenna can be oriented independently for best reception from stations within its band, thus providing the receiver with good coverage on both bands. Figure 4-19(F) is essentially the same arrangement as Fig. 4-19(E), except that the longer folded dipole acts as the reflector for the shorter folded dipole. The two folded dipoles are connected in the same manner as in the array in Fig. 4-19(E). However, the independent orientation of each folded dipole is not possible in the array of Fig. 4-19(F).

Figure 4-19(G) shows a type of antenna that is used extensively for color TV signal reception. It is called a *log periodic,* a name that will be explained later in section 4.16. The characteristic that makes this antenna so useful is its very wide bandwidth. In some versions the elements are bent forward, and a UHF section is included in front of the longer VHF section.

The array reproduced in Fig. 4-19(H) is one which has been used extensively because of its ability to receive low and high VHF band signals. The front elements are bent forward, while the rear elements, the reflectors, generally extend straight out. The response pattern of the antenna contains only one major lobe on all channels (see Fig. 4-12). This is an improvement over the conventional dipole where an element cut for the low frequencies will have a multi-lobed pattern on the high channels, and an element cut for the high frequencies will have a poor response on the low channels. With the conical antenna, one array may be adequate for all VHF channels.

Conical antennas may come either singly or stacked two or four high. The same is true of most other antennas.

4.13 COMBINATION ANTENNAS

In many parts of the country, both VHF and UHF stations are in operation. For those areas, combination antennas serve to simplify the installation problem. A number of combination VHF-UHF arrays are available, and the ones pictured in Fig. 4.20 indicate their range.

The array in Fig. 4-20(A) consists of a low-band conical antenna for VHF signals and a broadband fan dipole for UHF signals. A single lead-in

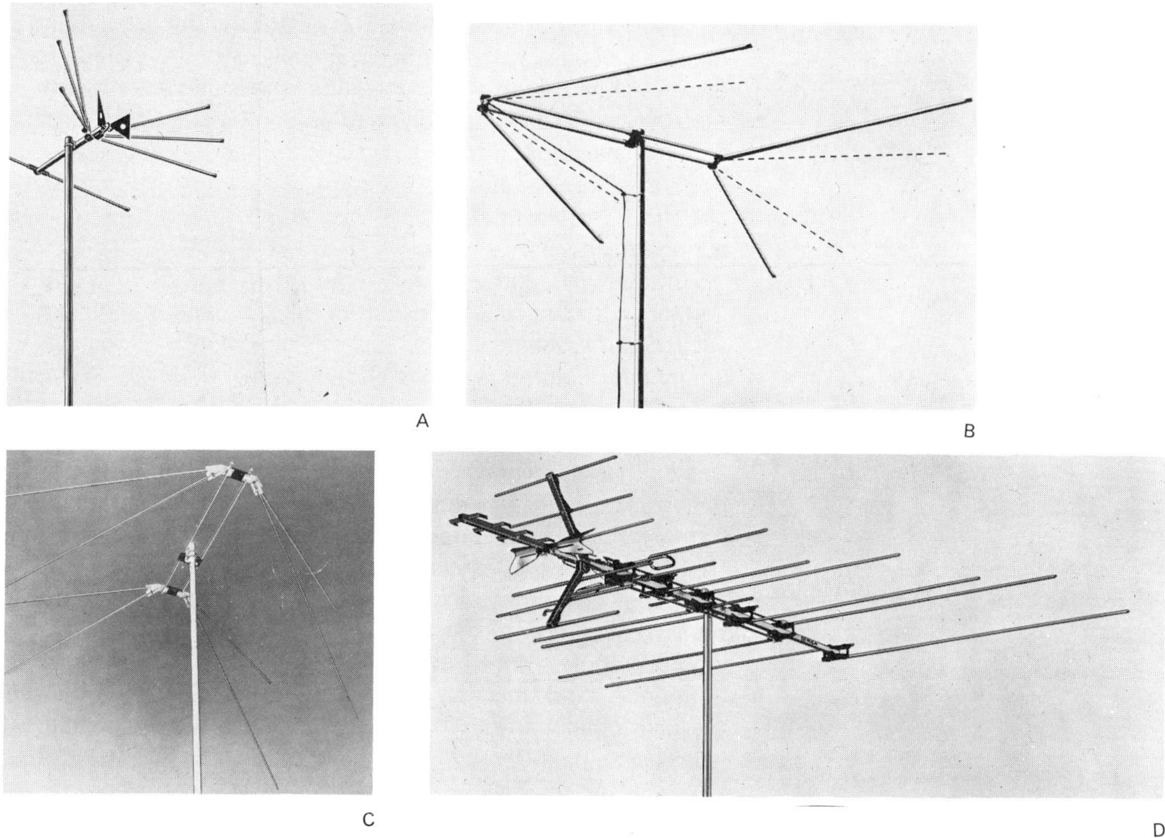

Fig. 4-20. Examples of combination antennas.

line delivers signals to the receiver through the use of a special coupling device which is mounted directly at the antenna itself.

V-Type arrays have appeared in a number of forms, of which the one in Fig. 4-20(B) is typical. Such an antenna contains four rods: two at the front and two at the rear. The two front rods may be considered the directors; the two rear rods are the driven elements supplying the signal to the receiver via a two-wire line. However, unlike the other arrays containing directors, the two front rods are electrically connected to the two rear rods. The purpose of the connecting rods is twofold. *First*, the rods serve as a transmission line to conduct to the two rear rods whatever signal is picked up by the two front rods. In addition, the two rear rods also pick up that portion of the signal which passed over the front rods and combine it with the energy received from the front rods via the transmission line. *Second,* the connecting rods support both front and rear rods and produce a mechanically sturdy array. It is characteristic of these antennas that the longer each side of the V becomes as compared to the wavelength of the operating frequency of the signal, the narrower the angle between the sides must be made for best gain and to obtain a single-lobe pattern. This means that for a low-band operation the sides are spread out farther, generally until the included angle is 90 degrees. For a high-band operation, the sides are brought in closer, generally to 60 degrees or even to 45 degrees. The best angle for a particular installation is determined at the time the antenna is erected. A good compromise angle is 60 degrees for both VHF and UHF signals, although if signals above 750 MHz are to be received, a smaller angle may be necessary.

Another variation of the V-type array is shown in Fig. 4-20(C). Stacking of the sections increases the gain about 2 db per additional section and sharpens the vertical directivity. A special VHF-UHF antenna for color reception is shown in Fig. 4-20(D). Note the corner reflector section for UHF.

Any of the VHF antenna designs can be used for UHF. Of course, the size of these antennas would have to be scaled down to the proper dimensions, but other than that no other changes are required. Design formulas remain unaltered. Because of the smaller dimensions of the UHF antennas, the stacking of such antennas is more common than with the VHF arrays. A four-bay array designed for reception at 60 MHz would be about 8 feet wide and 16 feet high. At 500 MHz, it would be only 1 foot wide and 2 feet high, certainly a considerable difference! And since the more elements used in an array, the greater the gain obtained from it, the trend toward more elaborate structures is understandable.

4.14 FRINGE-AREA ANTENNAS

In fringe areas where the signal level is very low, high-gain antenna arrays are needed. The gain of an antenna increases with the number of elements it possesses, and therefore fringe-area arrays have many more elements than the antennas used where the signal is strong. A gain of about 10 db is easily obtained, and with some, the gain may reach 15db or more. Most high-gain combinations are sharply directional. Such antennas must be carefully aimed, otherwise the captured signal will be much lower than it should be. Figure 4-21 shows four high-gain highly-directive antennas suitable for fringe-area reception.

Fringe Antenna Installation Hints After the antenna has been chosen, a number of points should be kept in mind before installing it:

1. It is usually true that the higher the antenna, the stronger the signal received. This is not *always* true, however, because there are heights at which cancellation due to the ground-reflected wave will occur. It is always a good idea to try changes in height in weak signal areas.

2. If possible, the antenna should be connected to its receiver before the supports are fixed in place permanently.

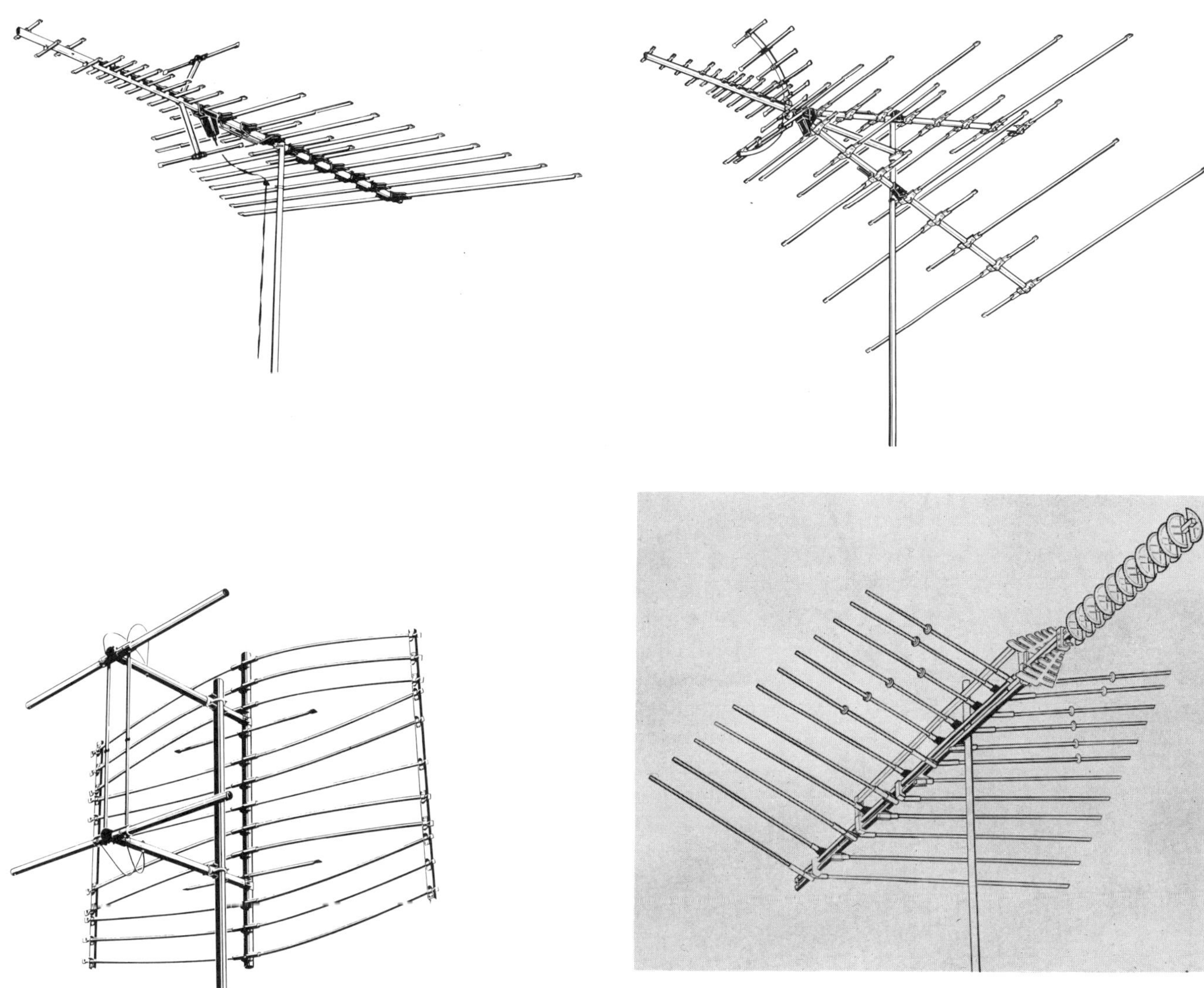

Fig. 4-21. Four representative high-gain antenna arrays designed specifically for weak-signal (fringe) areas.

3. When more than one station is to be received, the final placement of the antenna must, of necessity, be a compromise. In extreme cases, it may be desirable, or even necessary to erect several antennas.

4.15 INDOOR ANTENNAS

In strong signal areas it is sometimes feasible to use an indoor antenna for signal reception, provided the receiver is sufficiently sensitive. These antennas come in a variety of shapes, some of which are shown in

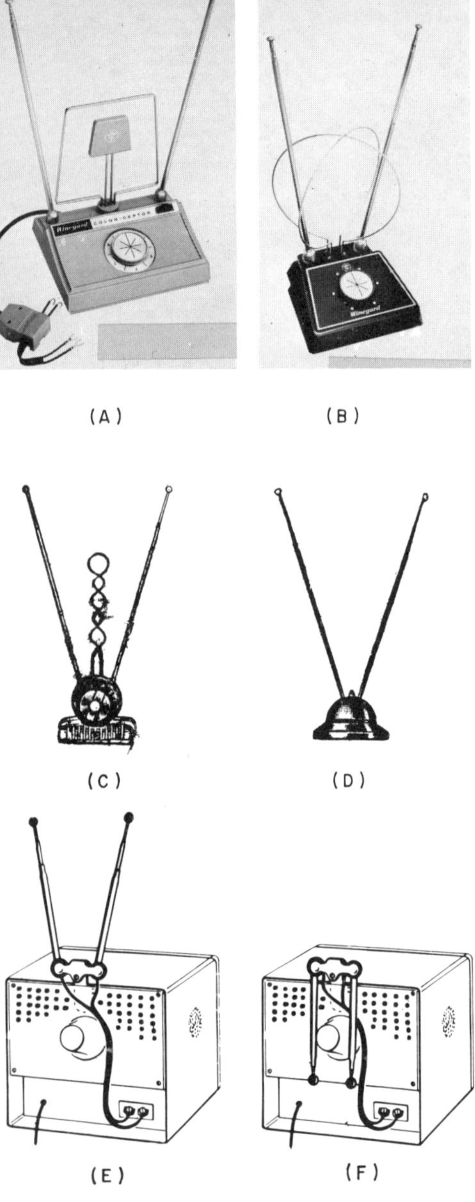

Fig. 4-22. Types of indoor TV antennas.

Fig. 4-22. The types shown in (A), (B), and (C) have selector switches which can modify the response pattern and raise the resonant frequency of the antenna so as to minimize interference and ghost signals. Generally, the switch is rotated (with the antenna connected to the receiver and with a signal being received) until the most acceptable picture is obtained on the screen. For all types shown, the dipole rods can be adjusted in length and position for best reception. When not in use, the rods can be telescoped.

In Fig. 4-22(E), the antenna is shown mounted directly on the back of the receiver. Its swivel base permits it to be manipulated into the correct position to capture the greatest amount of signal. When not in use, the arms are telescoped inward and then turned down behind the receiver cabinet as shown in Fig. 4-22(F).

4.16 COLOR-TELEVISION ANTENNAS

The requirements for color-television antennas are somewhat different than for monochrome television antennas. The emphasis on monochrome television has been in obtaining a *high gain* over a range of frequencies. In some installations a single antenna is used for the lower channels in the VHF range (2–6), and another antenna is used for the higher channels in the VHF range (7–13). When a law was passed in 1964 that required all television-receiver manufacturers to include UHF channels with every receiver sold, the use of these channels was assured. This means a third antenna must be included with some VHF antennas already discussed. The VHF antenna used for lower channels may be cut to resonate at Channel 3 or Channel 4; the VHF antenna for the upper VHF band may be cut to resonate at Channel 9; and, the UHF antenna is usually cut to resonate at the frequency of the local UHF station. Thus, with three antennas on the roof, the television reception for monochrome receivers can be satisfactory.

Figure 4-23 shows three typical response curves for the three antennas. You will note that the gain varies widely from one channel to another, and also from one end of a channel to the other. Suppose that in the lower VHF range the antenna is to be used for receiving Channels 3 and 5. An enlargement of the response curve of Fig. 4-23(A) is shown in Fig. 4-24. The color carrier portion of the curve is seen to be over 2 db below the video carrier portion in this illustration. While it is true that the sound carrier portion is also lower than the video carrier, this will not have any effect on the reproduced picture. The volume control and picture control of a receiver operate independently, so the viewer can compensate for the reduced sound intensity by turning the volume of the receiver to a higher point. However, the reduction in color carrier gain cannot be compensated for, and this reduction can result in a poor color picture quality. In fact, in color television it is desirable to have a total gain variation of less than + or − one db over the complete video channel. This accounts for the fact that when owners of monochrome receivers purchase new color receivers, they are immediately faced with the problem of installing a new antenna—an antenna capable of providing an even response and satisfactory gain for the channels to be received. For this purpose, a relatively new class of "color antennas" (more accurately called "broadband antennas") has been developed. The most popular type of broadband antenna is the *log periodic.* Figure 4-19(G) shows an example of this type of antenna. There is no single style of antenna that can be referred to as the log periodic. Instead, the term "log periodic" refers to a wide variety of antennas that have the same characteristics. In other words, two antennas may be completely different in appearance, but both may be log periodic antennas.

The term log periodic does *not,* as is sometimes mistakenly supposed, refer to the relative lengths of the elements or the relationship between their distances. Instead, it refers to their electrical characteristics. If a graph was made showing the antenna impedance against the *logarithm of the frequency,* it would be a curve similar to the one shown in Fig. 4-25(A). Note that the maximum value of impedance on this curve repeats itself *periodically* on the *logarithmic scale,* and hence the name log periodic. This repetition is not only true for the impedance of a log periodic antenna, but also for other electrical characteristics.

An important feature of log periodic antennas is the *geometric* relationship between the spacing of the elements and the length of the elements. Characteristically, these spacings and lengths vary in a geometric progression which results in the antenna getting larger and larger as the distance from a theoretical point—called the *apex*—increases. In a similar manner, the length of the elements also increases which can be demonstrated by the log periodic antenna shown in Fig. 4-25(B). This particular type of log periodic is sometimes referred to as a *dipole array log periodic antenna* because it consists of a number of dipoles which vary in distances and lengths. The antenna is designed in such a way that the distances between the dipoles and the length of the dipoles are in a constant ratio to each other.

When a television station broadcasts a signal, and the log periodic antenna is pointed in the direction necessary to receive this signal, it can be shown that only one or two of the dipole elements in the antenna will react to that frequency. All the other elements are inactive at that particular frequency; however, at some other frequency, they will become active. In other words, for any particular frequency being received only one or two of the log periodic elements are considered active.

4.17 TRANSMISSION LINES

The energy intercepted by the antenna's driven element must be delivered by a *transmission line* to the receiver with the least possible loss. The receiver may be located a considerable distance from the antenna, and since it is a conductor, it may act like an antenna, picking up signals of its own. These signals may be electrical noise impulses such as those generated by automobile ignition systems and machinery,

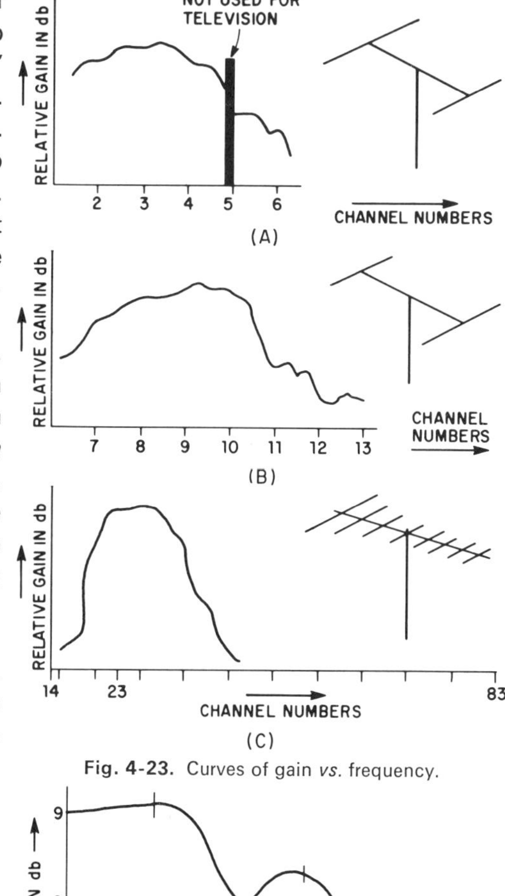

Fig. 4-23. Curves of gain *vs.* frequency.

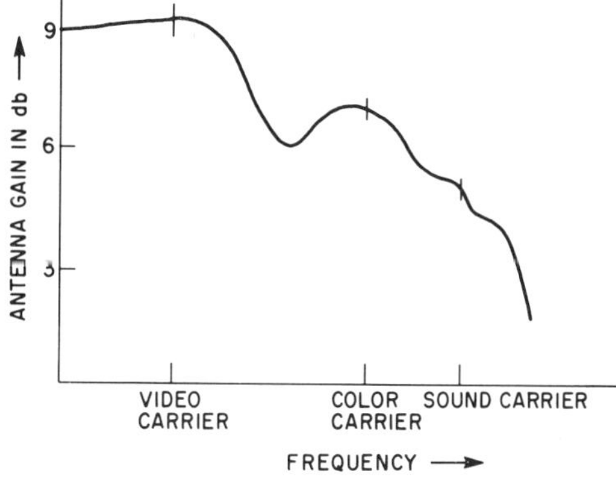

Fig. 4-24. An enlargement of Channel 5 of Fig. 4-23 (A).

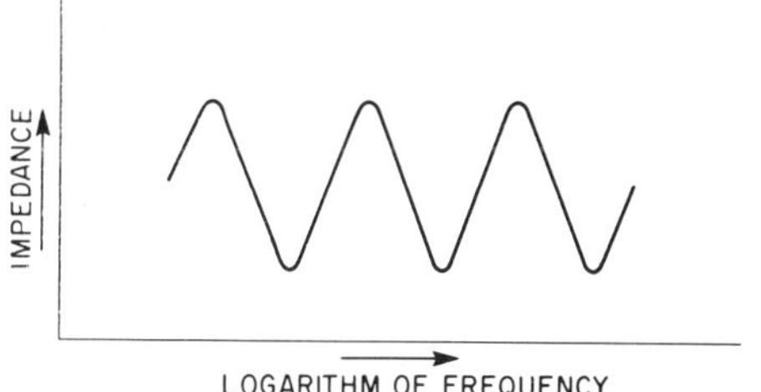

Fig. 4-25 (A). The periodic nature of antenna impedance when plotted on a logarithmic scale.

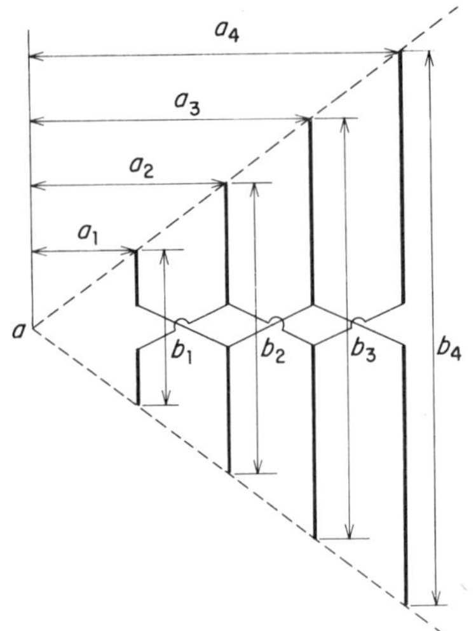

Fig. 4.25 (B). The relationship between the element length and the distance to the apex (a).

or they may be video signals from the transmitter. In any event, they are undesirable.

We will first discuss the types of transmission lines available, and then later we will discuss the methods of installing the transmission line so as to obtain minimum loss and minimum noise.

Types of Transmission Lines in VHF Although there are many different designs of transmission lines available, two general types are used most extensively in television installations: the *parallel-wire* types and the *concentric* or *coaxial cable* types. From the standpoint of convenience and economy, it would be desirable to use only one antenna that is capable of receiving all the VHF television stations. It should have, therefore, a fairly uniform response of 54 to 216 MHz. We have already shown that the log periodic antenna closely approaches this requirement. A resonant dipole presents an impedance at its center of about 73 ohms. To obtain the maximum transfer of power, the connecting transmission line should *match* this value, that is, the impedance of the transmission line should also be 73 ohms. When we attempt to use the same dipole for a band of frequencies, we find that the 73-ohm dipole impedance value is no longer valid. A dipole cut for 50 MHz presents a 73-ohm impedance at that frequency. At 100 MHz, the impedance has risen to 2,000 ohms. It is clear that the best transmission-line impedance would no longer be 73 ohms. A higher value is needed which will serve as a compromise between the 73 ohms and the 2,000 ohms. It is desirable to use as high an impedance value as possible, because the line loss is inversely proportional to the characteristic impedance. On the other hand, such factors as the size of the line and the wire gage must also be considered. It is a common practice to design the input circuit of the television receiver to be used with either a 73-ohm or a 300-ohm transmission line. It has been determined that a 300-ohm line used with a half-wave dipole produces a broad-frequency response without too great a loss due to mismatching, and satisfactory monochrome reception is achieved this way. A folded dipole has an impedance close to 300 ohms at its resonant frequency, and a much more uniform response is obtained with this type of antenna.

The flat, parallel-wire transmission line shown in Fig. 4-26(A) is one of the most popular transmission lines in use for VHF television today. The wire are encased in a plastic ribbon of polyethylene which is strong, flexible, and unaffected by sunlight, water, cold, acids, or alkalis. At 100 MHz, the line loss is on the order of 1.2 db per 100 feet of line. Characteristic impedance ranges from 75 ohms to 300 ohms are obtainable. The line is *balanced*, which means that both wires possess the same average potential with respect to ground. It is, however, unshielded, and therefore not recommended for use in extremely noisy locations.

A companion tubular twin-lead line pictured in Fig. 4-26(B), is also available, which, while somewhat more expensive than the flat twin-

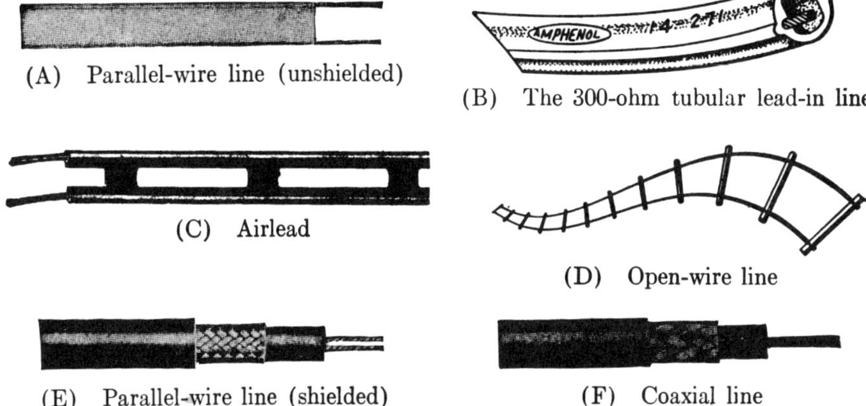

(A) Parallel-wire line (unshielded)

(B) The 300-ohm tubular lead-in line

(C) Airlead

(D) Open-wire line

(E) Parallel-wire line (shielded)

(F) Coaxial line

Fig. 4-26. Types of popular transmission lines used for FM and television istallations.

lead possesses the advantage of being less affected by adverse weather conditions than the flat line. Rain, sleet, snow, etc., may not physically affect the flat line, but electrically these serve to increase the signal loss in the line. Thus, at 100 MHz, flat and tubular lines under dry conditions possess equal attenuations of 1.2 db per 100 feet. When wet, however, the loss on a flat line rises to 7.3 db per 100 feet, whereas that on a tubular line is only 2.5 db per 100 feet. In strong signal areas this loss might not be important, but it would be excessive in a weak signal area.

A third type of parallel-wire line, known commercially as an *airlead,* is shown in Fig. 4-26(C). It has 80 percent of the polyethylene webbing removed, which is said to reduce the loss, or db attenuation per 100 feet, by at least 50 percent. The line impedance is still 300 ohms.

The fourth parallel-wire line, the one in Fig. 4-26(D), is almost completely open, being held together by small polystyrene spacers placed approximately six inches apart. The attenuation of this line is only in the order of 0.35 db per 100 feet at 100 MHz; it is relatively unaffected by changes in weather. The impedance of this line is 450 ohms.

A parallel-wire transmission line that is completely shielded is pictured in Fig. 4-26(E). The two wires are enclosed in a dielectric, possibly polyethylene, and the entire unit is shielded by a copper-braid covering. As a protection against the elements, an outer rubber covering is used. Grounding the copper braid converts it into an electrostatic shield which prevents any stray interference from reaching either of the inner conductors. Furthermore, the line is balanced with respect to ground. It can be built with impedance values ranging from 50 ohms up, but a 225-ohm line has found its greatest use in television installations. The attenuation of this line is 3.4 db per 100 feet at 100 MHz, and this value is considerably higher than the attenuation of any of the unshielded lines. Because of this, and because of its greater cost, the shielded line is used only where the surrounding noise is particularly severe.

An example of coaxial, or concentric cable, is illustrated in Fig. 4-33(F). It contains an insulated center wire enclosed by a concentric metallic covering which generally is a flexible copper braid. The inner wire is kept in position by a solid dielectric that is chosen for its low-loss properties. In some applications the signal carried by the coaxial line is confined to the inner conductor, with the outer copper-braid conductor grounded so as to serve as a shield against stray electromagnetic fields. This arrangement causes the line to be *unbalanced* with respect to ground, and the input circuit of the receiver must be connected accordingly. Coaxial cables are available in a range of impedances from 10 to 150 ohms.

At the receiver, the connections for balanced and unbalanced lines differ, as shown in Fig. 4-27. For a balanced line, the input coil is center-tapped and grounded at this center terminal. Stray fields, cutting across both wires of a balanced line, induce equal voltages in each line. The similar currents that flow because of the induced voltages are travelling in the same direction on the two conductors of the line and neutralize each other. Many receivers are made so that a simple change in the input circuit, such as removing or adding a jumper, changes the input impedance to the desired value. Special transformers, called *baluns*, are available for connecting balanced lines to unbalanced lines.

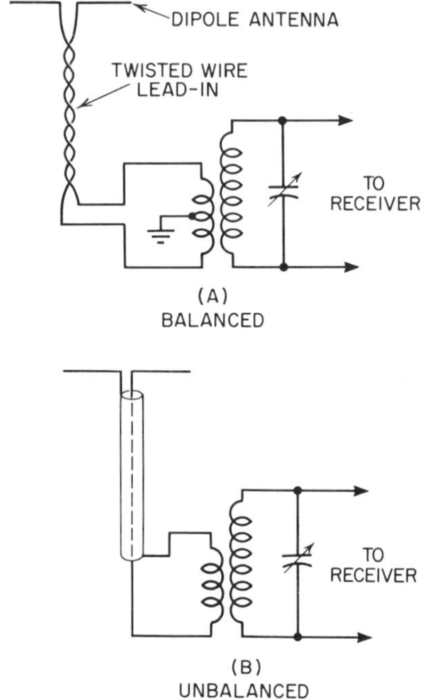

Fig. 4-27. Methods of connecting lead-in wires to the input coil of a receiver.

The open-wire line has, by far, the lowest attenuation and, thus, is frequently used for fringe-area installations. Its characteristic impedance is 450 ohms, and this high value may sometimes require a matching network between the line and the usual 300-ohm receiver-input impedance. That is does not always do so stems from the fact that receiver impedances may vary considerably from their stated value of 300 ohms. The 300-ohm line is the one which is most extensively employed in VHF because it is economical, it matches receiver input impedances directly, and its attenuation is low. However, in areas where the surrounding noise is high, use of the unshielded open-wire or 300-ohm lines becomes impractical and one of the coaxial cables must be employed.

Types of Transmission Lines in UHF With an increase in frequency, line db attenuation also rises. Thus, at 500 MHz the open-wire loss for 100 feet mounts to 0.78 db, the 300-ohm line loss becomes 3.2 db, RG-11U attenuation increases to 5.0 db, and RG-59U loss reaches the rather high figure of 9.4 db. Comparative figures at 100 MHz, and 1,000 MHz are given in Table 4-1, and it can well be understood why the amount of line needed should be figured closely in order that no more than absolutely necessary is used.

An interesting sidelight on line attenuation is the rapid rise in this value in unshielded lines when they become wet. The 300-ohm flat line appears to be most vulnerable, jumping from a value of 3.2 db (at 500 MHz) when the line is dry to 20.0 db when it becomes wet. What this

TABLE 4-1 Transmission line loss (db loss per 100 ft).

Type of Line	100 MHz		500 MHz		1000 MHz	
	Wet	*Dry*	*Wet*	*Dry*	*Wet*	*Dry*
450-ohm open-wire[a]	—	0.35	—	0.78	—	1.1
300-ohm flat	7.3	1.2	20.0	3.2	30.0	5.0
300-ohm tubular	2.5	1.1	6.8	3.0	10.0	4.6
RG-11U	—	1.8	—	5.0	—	7.6
RG-59U	—	3.8	—	9.4	—	14.2

[a] Estimated dry values—unknown for wet conditions.

rise will do even to a strong signal is quite obvious. This is an important thing to keep in mind when installing an antenna. The transmission line should not lie on the roof where it can become buried in accumulations of ice, snow, and water. Also, the line should not have long horizontal runs exposed to the elements that would permit a buildup of ice or moisture. Keep the line away from gutters, pipes, or other metal objects as much as possible. Avoid sharp bends in the line. If a bend must be made, have it occur gradually. Finally, secure the line tightly by means of stand-off insulators, so that it does not sway in the wind or otherwise alter its position. It is important to observe these precautions in both VHF and UHF installations.

The 300-ohm tubular line is considerably less affected by moisture, and hence would be more desirable for some installation purposes. No data are available on the attenuation increase in wet open-wire lines, although it is not considered to be appreciable. Shielded cables, such as the RG-11U and RG-59U are practically unaffected by inclement weather.

Impedance matching at the antenna and, more importantly, at the receiver will require careful attention in UHF installations. Mismatching at the receiver results in energy being reflected back along the line with the resultant standing waves. It has been found that the attenuation of a line may be increased by as much as 2 db over its normal rating when standing waves are present. In strong signal locations the additional loss may not be serious, but in moderate and weak signal areas it can mean the difference between usable and unusable television.

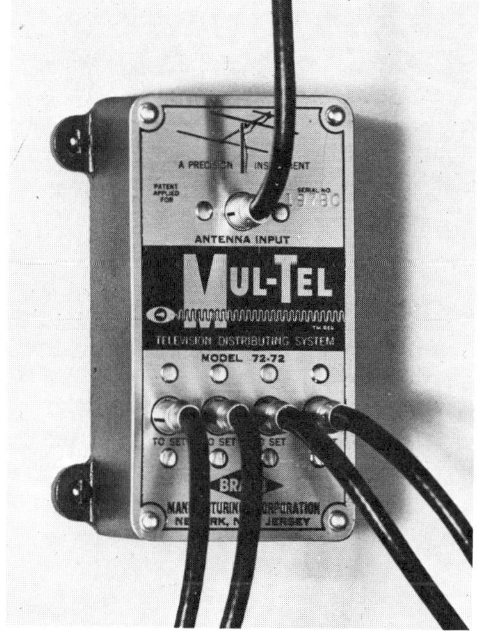

Fig. 4-28. A four-set coupler which enables one antenna to supply a television signal to four receivers. (*Courtesy of Brach Manufacturing Co.*)

4.18 MULTIPLE-SET COUPLERS

It is not unusual to find homes and apartments that have more than one television set. It is desirable to use the signal provided by one antenna array for all of these receivers. This can be accomplished with multiple-set couplers of the type indicated in Fig. 4-28. The lead-in from the antenna is connected to the input terminals of the coupler; the signal is then split into two, three, or four parts according to the design of the unit. Each receiver is then connected to an appropriate set of input terminals.

By this method, each set will receive less signal than it would receive if it were connected directly and solely to the antenna. However, in medium and strong signal areas, enough signal power is available to provide completely satisfactory signals for each receiver. In weak signal areas, separate antennas may be required, or a Master Antenna Distribution System—often called a MATV system—may be used. The initials MATV stand for *M*aster *A*ntenna for *TV*. MATV systems employ special high-frequency preamplifiers. This latter method is more expensive, but it can provide signals to an almost unlimited number of receivers.

4.19 MASTER ANTENNA DISTRIBUTION SYSTEMS

When a large number of sets are to be operated from a single antenna system, a master antenna television (MATV) system is used. Figure 4-29 illustrates a typical MATV system. It is comprised of three main sections:

(1) The antenna system(s).
(2) The signal processing system.
(3) The signal distribution system.

The Antenna System The problems encountered in mounting an antenna for an MATV system are similar to problems for other antenna systems. One significant difference is that a lower gain can be tolerated in favor of a greater bandwidth in an MATV installation. This is because

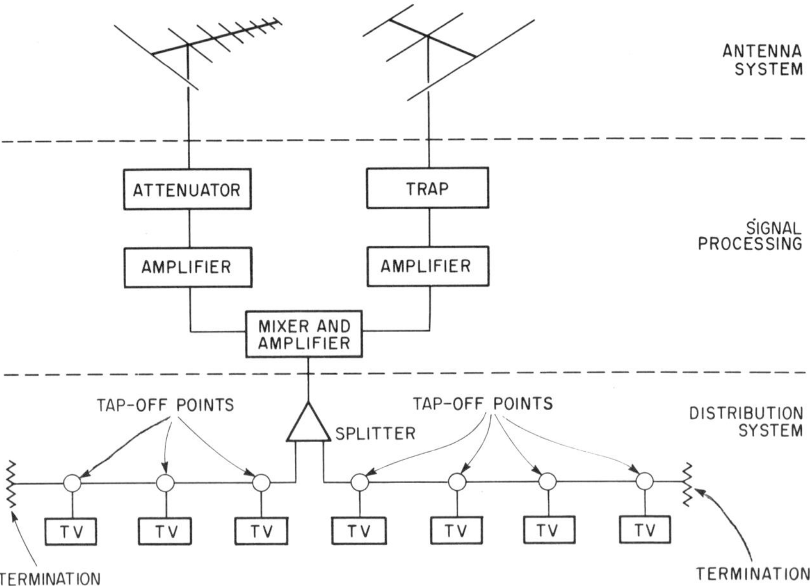

Fig. 4-29. An example of a MATV system.

an amplifier system is used in the signal processing section of the system. In some cases, when the signal amplitude is very low, a preamplifier may be mounted on the mast near the antenna terminals.

Coaxial cable is used almost exclusively in MATV systems because of its ability to reject noise signals, and also because it does not reradiate signals. If twin lead (balanced line) is used for the connection to the driven element, it must be converted to an unbalanced line. A special transformer, called a *balun*, is used for this purpose. The antenna lead-in may also contain traps or filters for eliminating interference at certain frequencies.

If a broadband antenna is used, it is conceivable that one channel may be very much stronger than the others. In such cases a tuned-circuit *attenuator* may be needed to reduce the strong signal, so that all frequencies presented to the receiver will be of the same strength.

Signal Processing The antenna and the signal processing sections of the MATV system are together referred to as the *head end*. The signal processing section combines the signals from each antenna system and amplifies them for distribution to the receivers. Figure 4-29 indicates the position of the mixer and the amplifier in the MATV system.

Distribution System In the *distribution system* shown for Fig. 4-29, the output of the amplifier and the mixer stage is delivered to a *splitter* which is like the two-set coupler previously discussed. Although there is a certain amount of loss in all such couplers, the signal from the amplifier is strong enough to overcome this loss. The output of the splitter is divided into two feeder lines. Each feeder contains a number of "taps." These taps may be located in different rooms in a house or different suites in an apartment building.

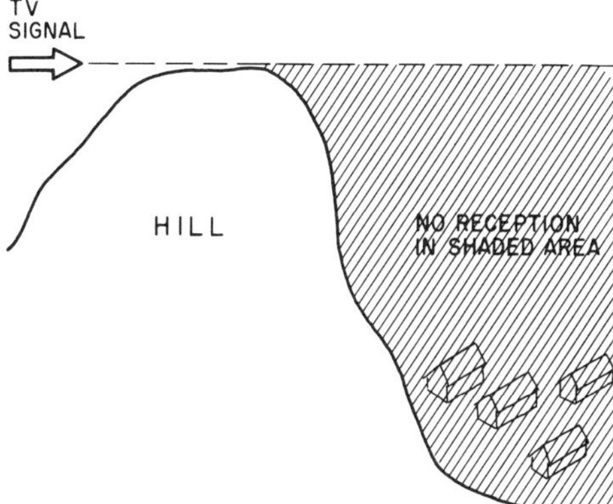

Fig. 4-30 (A). A situation where CATV is needed.

4.20 COMMUNITY ANTENNA TELEVISION SYSTEMS

Very often there are locations within the line-of-sight distance from the transmitter, where the television signal cannot be received because of the hilly terrain. Figure 4-30(A) shows such a situation. The houses in the valley cannot receive the line-of-sight television signal. This is the type of situation that leads to a *community antenna television system* (*CATV*) (see Fig. 4-30(B)). It consists of a receiving antenna which is mounted on the hill and a signal processing system which is similar to the ones used in the MATV system. As in the MATV systems, the antenna and the signal processing section together are called the *head end* of the system. The television signal is delivered to the town via a large coaxial cable. In the town, tap-off points are used for each house that desires reception. The houses that are connected to the system pay a monthly fee.

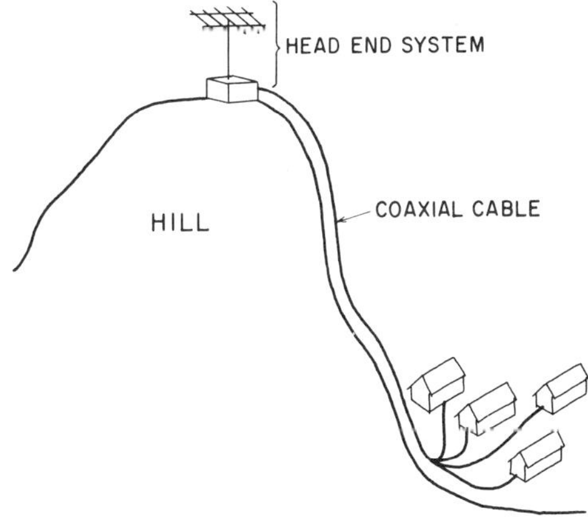

Fig. 4-30 (B). How a CATV system delivers the signal to the community.

Not only does the CATV system find applications where the television signal reception is normally very poor, but it is also used in locations where the viewer is usually limited to one or two channels. CATV systems with elaborate antennas and signal processing equipment can pick up very distant stations and deliver them to the customer, thereby increasing the number of channels from which he can choose.

In some cases, microwave links are used to transport a signal from one point to another more distant one. Suppose a local UHF station is at point a in Fig. 4-31, and it is desired to receive this signal at point b over 30 miles away. The signal *could* be transported via coaxial cable, but a less expensive system is illustrated in Fig. 4-31. The first step is to convert the signal to a microwave frequency. This microwave signal is then transmitted from point a to point b. Repeaters are used to receive

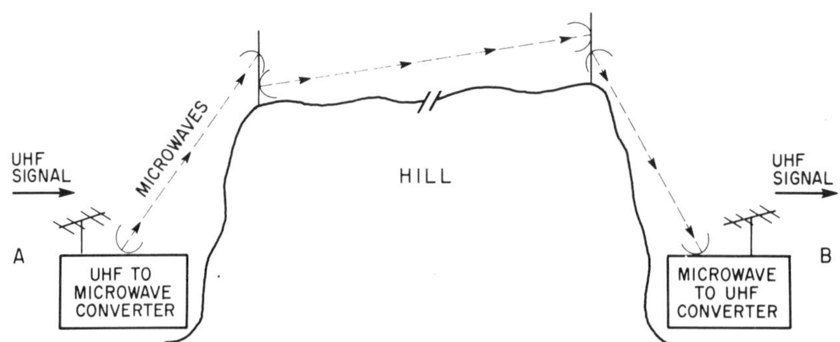

Fig. 4-31. Use of microwaves in a CATV system.

and re-transmit the signal at points along the way. At point b the signal is converted from microwave back to a UHF signal and delivered to the head end of the CATV system. In addition to saving miles of coaxial cable (and therefore reducing the cost), this system has the advantage of making use of the relatively uncrowded microwave band.

4.21 TROUBLESHOOTING ANTENNA SYSTEMS

Since the antenna is exposed to rain, snow, wind, and smog, it can be expected to deteriorate over a period of time. It is important for you to know the symptoms of a defective antenna. Obviously, if there is insufficient signal delivered to the receiver by a poor antenna system, no amount of work on the set will cure the trouble. A good technician will carefully observe the antenna system when he first arrives at a house where he is making a service call. Before he enters the house he will know whether or not: (1) The antenna has been damaged in high winds. (2) The antenna is oriented (pointed) in the right direction. (3) The transmission line is securely in place. Any of these factors can cause insufficient signal strength at the receiver.

Unshielded transmission line can become short circuited even though nothing is touching the wires. You can demonstrate this by wrapping a piece of metal foil around the line and moving it with your hand. In most cases you will notice that the picture is affected even though the foil is isolated from the wires. House painters sometimes get paint on an unshielded line. If the paint has a lead base, as many of them do, the paint can effectively short-circuit the line. (You can think of the paint as being a capacitive coupling between the lines which is the same as a short circuit.) Always inspect the transmission line in both new and old installations. Unshielded line should not be run close to rain gutters, aluminum siding, conduit, water pipes, furnace ducts, etc. All of these things can seriously affect the picture quality.

Inspecting Older Antenna Installations. If the antenna has been in service for some time, there are a number of features that should be inspected. Again, we will mention these as a checkoff list. An experienced technician will train himself to look for these things in a quick glance.

1. Note the orientation of the antenna. Strong winds can change the antenna's direction. Make a quick comparison of the direction in which the antennas in the neighborhood are pointing to see if the antenna is properly oriented.

2. Check the transmission line. Make sure it is fastened at the antenna, and there are no visible breaks. Make sure it has not been painted, and that it has not come loose from its fastenings. Watch for a fluttering transmission line. It may be a source of trouble.

3. If the customer has recently switched from a monochrome receiver to a color receiver, his old antenna may not be adequate for the job. The only solution here is to replace the old antenna.

4. White flashes in the picture—usually accompanied by static in the sound—are an indication that there is an intermittent break in the transmission line. The break may not be visible.

5. The ground connections for the mast and lightning arresters should be checked. Improper grounding will not affect the picture, but it represents a serious electric storm hazard.

6. Excessive snow in the picture, or a very weak picture, indicates that the antenna lead may be open or disconnected. Of course, the trouble *could* be in the receiver. A small portable receiver, which is known to be in operating condition, is useful for checking the signal from the antenna. If the portable can produce a satisfactory picture when connected to the antenna, but the customer's receiver cannot, then the trouble is in the customer's receiver rather than in the antenna. If a satisfactory picture cannot be obtained on the portable, then the antenna should be suspected. A field strength meter is useful if a large amount of antenna work is to be performed, especially if you are going to work on MATV and CATV systems.

REVIEW QUESTIONS

1. Explain the causes of *ghosts* on a TV receiver picture.
2. What factors determine whether or not a sky wave will be returned to Earth from the ionosphere?
3. What is the maximum line-of-sight distance between a transmitting antenna that is 900 feet above ground and a receiving antenna that is 144 feet above ground? (Assume that the terrain is flat and the bases of both antennas are the same altitude above sea level.)
4. The so-called line-of-sight transmission of television signals is greater than the calculated line-of-sight range. Why is this so? By how much is the transmission distance increased?
5. What is the length of a dipole antenna cut to resonate with the video-carrier frequency of Channel 49? (The video-carrier frequency of Channel 49 is 681.25 MHz.)
6. Name two kinds of parasitic elements used with Yagi antennas.
7. What property of a log periodic antenna makes it useful for color television reception?
8. A certain signal induces 100 microwatts of power in a standard dipole antenna, and 1000 microwatts of power in a Yagi antenna. What is the gain of the Yagi antenna in db?
9. At what point(s) on a half-wave antenna will maximum impedance be found?
10. What is an *unbalanced* transmission line?
11. What is the name of the transformer used to connect a balanced transmission line to an unbalanced transmission line?
12. What constitutes an MATV system?
13. Why is an attenuator sometimes used in an MATV system?
14. Why is a microwave transmission system sometimes used as part of a CATV system?

Wide-Band Tuning Circuits and RF Amplifiers

5

5.1 INTRODUCTION

The television signal occupies a 6-MHz bandwidth in the radio spectrum, a range far greater than anything we have to receive with the ordinary radio set. The problem must be met at the television receiver in the RF and mixer stages. The response of the tuned receiving circuit should be uniform throughout the 6-MHz band and yet be selective enough to discriminate against unwanted image frequencies or stations on adjacent bands. Before the circuits of the RF and mixer stages are considered, it will be helpful to discuss wide-band tuning circuits.

5.2 ORDINARY TUNING CIRCUITS

A single coil and capacitor, connected as shown in Fig. 5-1(A), form a parallel tuning circuit. At or near the resonant frequency, the variation of impedance which this combination presents is given by the graph of Fig. 5-1(B). At frequencies below the resonant frequency, the

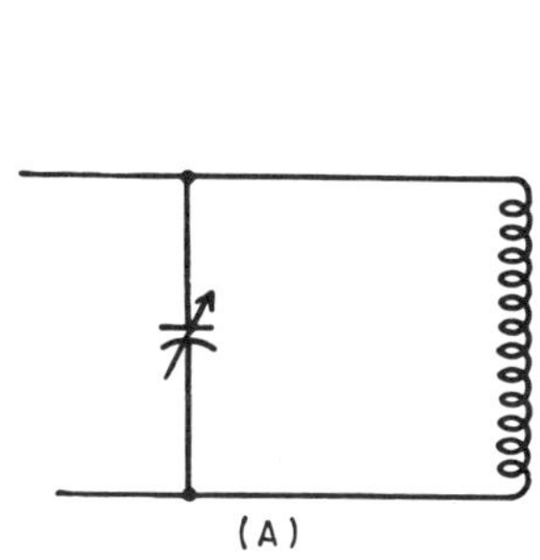

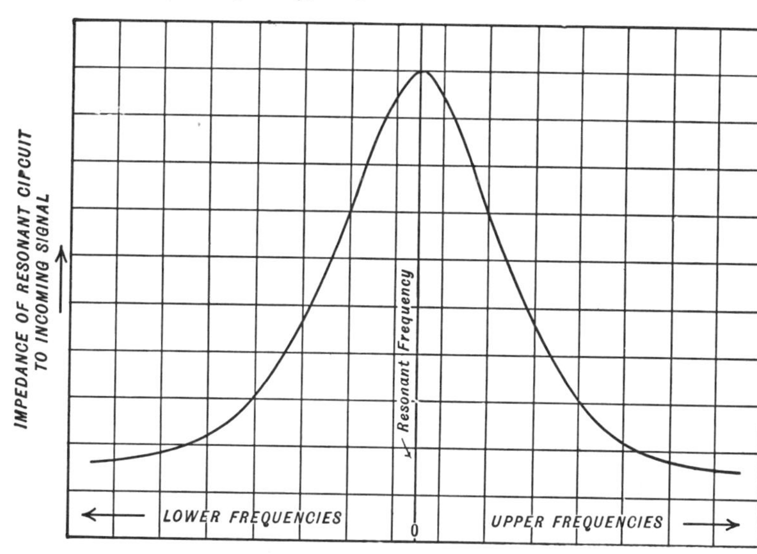

Fig. 5-1. A parallel tuning circuit and response curve.

parallel combination acts as an inductance with a lagging current, and impedance here drops off quite rapidly to a fairly low value. Above resonance, the effect is capacitive with a leading current. Again, the impedance decreases quite rapidly. At the resonant point, capacitive and inductance reactances cancel each other, the impedance becoming high and wholly resistive.

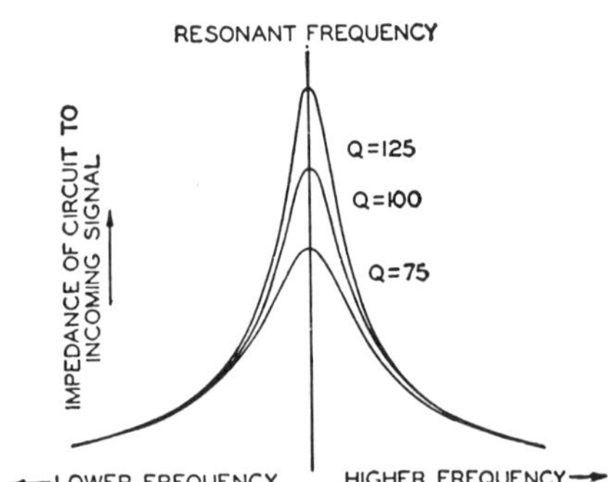

Fig. 5-2. The variation in the response curve with different values of *Q*.

While Fig. 5-1(B) shows the general shape of the resonant curve, more specific information is necessary. Hence, in Fig. 5-2 several resonant curves have been drawn, each for a circuit having a difference value of *Q*. The ratio of inductive reactance to coil resistance, *Q*, may be taken to indicate two things:

1. The sharpness of the resonant curve in the region about the resonant frequency. This, of course, is the selectivity of the tuning circuit.
2. The amount of voltage that will be developed by the incoming signal across the resonant circuit at resonance.

For any given circuit, the greater its *Q* value, the more selective the response of the circuit, and the greater the voltage developed. While these factors may be highly desirable, they are useful only if they do not interfere with the reception of radio signals. At the broadcast frequencies, each station occupies a bandwidth of 10 kHz. Within this region, uniform response is desirable. However, the sharply peaked curve of Fig. 5-1(B) does not produce equal response at all points within this region. The portion of the signal exactly at the resonant frequency, for example, would develop a greater voltage across the resonant circuit than those frequencies at the outer fringe, plus and minus 5 kHz away. A coil and capacitor combination having a lower *Q* would give a more uniform response and might be chosen over one with a higher value of *Q*. Less voltage would result from this change, but, with the advent of high-gain tubes, amplification is not too serious a problem. The emphasis now can be shifted to fidelity, which is also especially necessary for the reproduction of images in television receivers.

5.3 TRANSFORMER COUPLING

Whereas the simple circuit already described is sometimes used by itself for tuning, a more usual combination is the untuned primary coil inductively coupled to a tuned secondary coil (see Fig. 5-3). With this form of coupling, additional gain may result by having more turns in the secondary than in the primary coil. The stepped-up voltage applied to the grid of the next stage is larger than that obtained with only the single coil and capacitor by a value dependent upon the design of the coils.

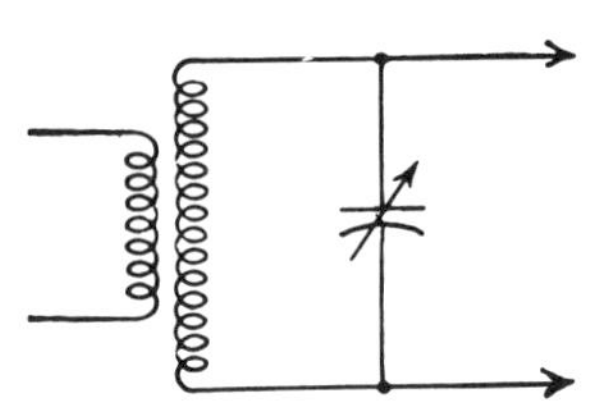

Fig. 5-3. A common form of transformer-coupled tuning circuit used in receivers.

The shape of the response curve of the primary circuit depends to a great extent upon the degree of coupling between the coils. When the coefficient of coupling *k* is low (*i.e.*, when the coils are relatively far apart), the interaction between the coils is small. The secondary response curve will retain the shape shown in Fig. 5-1(B).

As the coupling coefficient *k* is increased, the secondary circuit reflects a larger impedance into the primary. The primary current is affected more by variations in the tuning of the secondary capacitor. This, in turn, changes the number of flux lines which cut across the secondary coil, and the end result is a gradual broadening of both primary and secondary response curves. With very close coupling, the secondary response curve may continue to broaden and even develop a slight dip at the center. The dip, however, will never become too pronounced. It must be remembered that the discussion, so far, has dealt with coupled circuits where the primary is untuned. Hence, no matter how close a coupling is effected, the secondary will retain essentially the same curve shape given in Fig. 5-1(B).

On the other hand, with two *tuned* circuits coupled together, such as IF transformers, the effect of each circuit on the other becomes more pronounced. With close coupling, the familiar double-humped curve of Fig. 5-4 is obtained. The closer the coupling, the broader the curve and the greater the dip at the center.

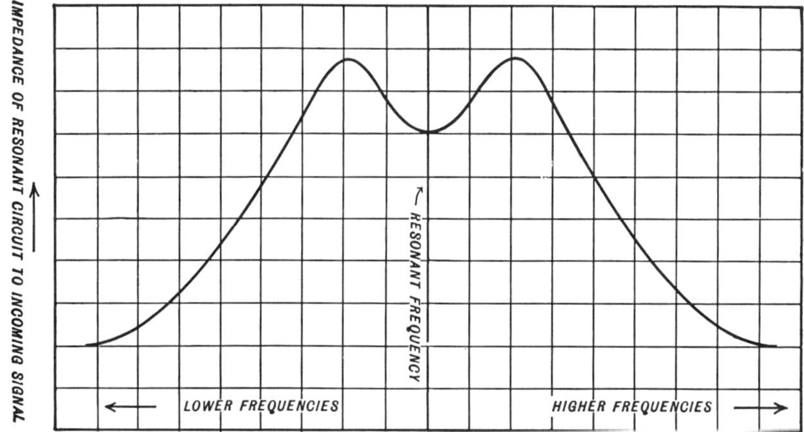

Fig. 5-4. Close coupling between two tuning circuits produces this type of response curve.

For television reception, none of these combinations provides the necessary uniform bandwidth. Loose coupling produces a curve that is too sharp and lacking in consistency over its range. Tight coupling tends to decrease the voltage of the frequencies near resonance because of the dip. Between these two extremes, we may obtain a semblance of uniform response about the center point of the curve, but never for a 6-MHz spread. However, if a low-valued resistor is shunted across the coil and capacitor, we can artificially flatten the curve to receive the necessary 6 MHz. The extent of the flat portion of the response curve will depend inversely on the value of the shunting resistor. The higher the resistor, the smaller the width of the uniform section of the curve. Hence, what we could not accomplish with a coil or a capacitor, we can do with a combination of these two with resistance.

One of the undesirable results of increasing the width of a response curve by the resistor method is the lowered Q that is obtained. As the value of Q decreases, the voltage developed across the tuned circuit becomes smaller for the same input. An inevitable reduction in output results. There are many ways of combining the tuned circuits and loading resistors to achieve the optimum gain and selectivity. Several of the more widely used circuits will be discussed in the section on RF amplifiers.

5.4 SPECIAL TUBES FOR TELEVISION RECEIVERS

A number of special tubes have been developed for the RF-amplifier stage in television receivers. These include pentodes, tetrodes, and triodes. All are of the miniature variety, and are used not only to achieve compactness in the tuner, but also because in a miniature tube, interelectrode capacitance is smaller and the connecting leads between the elements and the base pins are shorter. The latter introduces less inductance into the circuit, a factor which is particularly desirable since the inductance in the tuning network, itself, is small.

Pentodes From the standpoint of gain from a single tube, pentodes offer the best solution, and a number of them have been developed and extensively employed. When these pentodes are used, it is the usual practice to shunt resistors having values between 1,500 and 10,000 ohms across the tuning circuit to attain the necessary bandwidth. (The pentode, itself, is not very helpful in this respect, because it possesses a high internal resistance.) Because of the shunting resistor, however, stage gain is not very high: it is generally on the order of 20 to 25. The reason for this can be seen from the following.

A tuning circuit, when connected in the output of a tube, is essentially in series with the plate resistance of the tube. This is illustrated in Fig. 5-5, where the actual schematic and its electrical equivalent are shown. At resonance, the resistance of the tuning circuit itself may be high, but due to the low-shunting resistor, the total value of the combination becomes low. The plate resistance, on the other hand, is very high (in pentodes), and most of the output voltage is lost in the tube. Only a small portion of the total voltage appears across the tuning circuit to be transferred to the next stage.

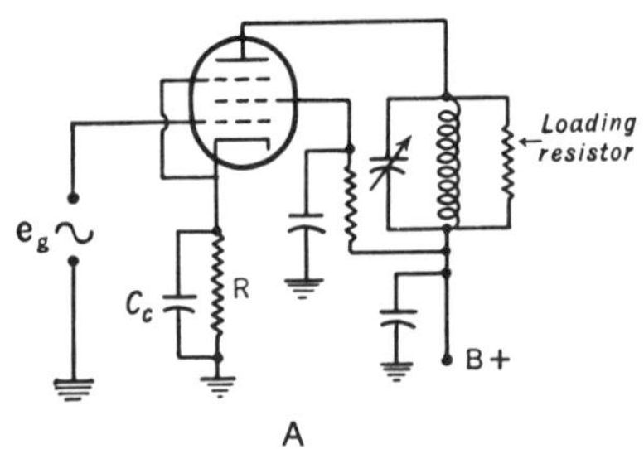

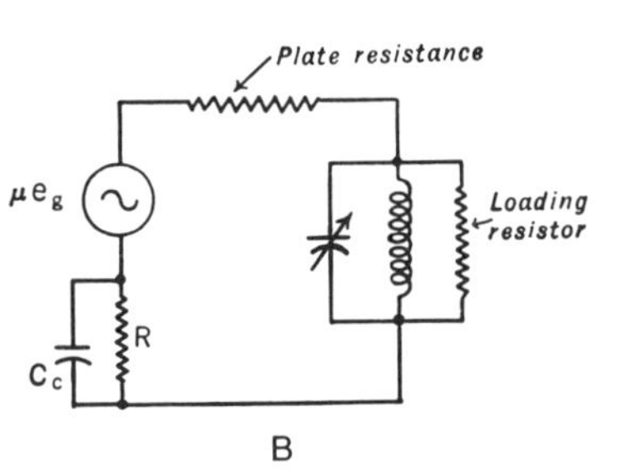

Fig. 5-5. An actual schematic and the electrical equivalent of an amplifier stage.

Mathematically, the gain of the pentode stage can be expressed closely by the relation:

$$Gain = g_m \times Z_L \text{, where} \quad (5\text{-}1)$$

g_m = mutual conductance of the tube, in mhos

Z_L = load impedance in output circuit, in ohms.

For a certain tube, g_m is 5,000 micromhos. With a plate load impedance of 2,000 ohms, we obtain

$$Gain = \frac{5{,}000}{1{,}000{,}000} \times 2{,}000$$
$$= 10$$

The 5,000 is divided by 1,000,000 to convert it from micromhos to mhos.

To obtain more amplification per stage, the mutual conductance of the tube must be increased. The mutual conductance of the tube, g_m, it will be recalled, represents the change in plate current caused by a change in grid voltage. To effect an increase in this ratio, tubes were designed in which the grid possesses greater control over the space charge near the cathode. This was done by moving the grid closer to the cathode. Although the change caused an increase in grid-to-cathode capacitance, it increased the mutual conductance even more. This design is exemplified in the 6JC6 and 6JD6.

As an example, the 6JC6 has a mutual conductance of 15,000, and the 6JD6 has a g_m of 16,000. Compare these values with older RF and IF pentode voltage amplifiers, for instance, the 6SK7, 6D6, 6S7, and the 6SJ7, which have mutual conductances of 2,000, 1,200, 1,750, and 1,600 micromhos, respectively. If the gain of the stage is computed and these values of mutual conductance are used, a voltage amplification much less than 10 is obtained.

Triodes The ability of a receiver to amplify a signal is governed not only by the amplification obtained from the tubes, but also by the noise generated in the tubes and in the associated receiver networks. Furthermore, the noise that is developed by the first stage, the RF amplifier, is actually the most important because at this point in the system the level of the incoming signal is more nearly on a par with the noise level than it is at any other point in the receiver. Once the signal becomes much larger than the noise, it can easily override the noise and hence mask its presence. Whatever noise voltage appears at the grid of the RF amplifier is amplified together with the signal. To obtain a picture as free of noise spots as possible, we need to have as much signal and as little noise as we are able at the front end of the set.

The best choice for a low-noise tube is a triode RF amplifier. This is because noise originating in a tube varies directly with the number of positive elements within that tube, and a triode has fewer such elements than a pentode. Hence, a triode is more desirable from a noise standpoint than a pentode. It is for this reason that high-frequency triodes have been employed in the RF amplifier.

A triode normally provides less gain than a well-constructed high-frequency pentode. As a result of this, special dual triodes have been developed for use in a circuit known as a *cascode amplifier*, in which the

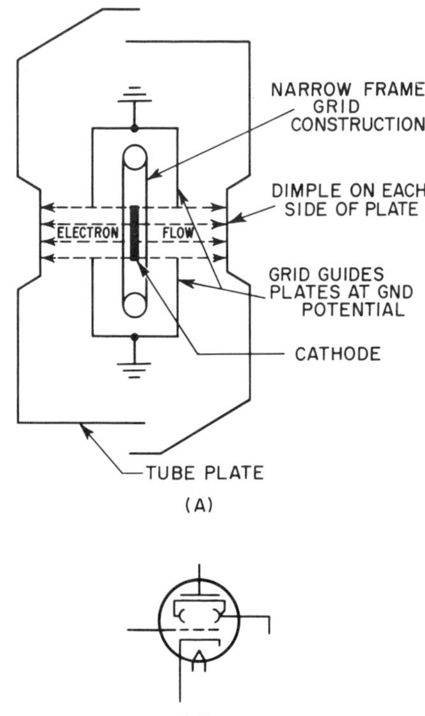

Fig. 5-6. An internal view and the schematic symbol of a 6ER5 RF triode.

two triodes are connected in a series. In this arrangement they are capable of providing about as much gain as a pentode. This enables the designer to achieve the desired amplification at a lower noise level. Cascode amplifiers will be discussed at length presently.

A significant change has taken place in high-frequency triode construction. In an effort to achieve higher efficiency and reduced plate-to-grid capacity, extra elements called *grid-guides*, or *shield plates*, have been inserted in the region between the grid and the plate. An internal view of the 6ER5, a tube of this group, is pictured in Fig. 5-6. The grid guides are U-shaped plates which surround the sides of the grid. By grounding the plates, a shield is inserted between the anode and the grid, reducing the capacity between these elements. In addition, the electron flow from cathode to plate is concentrated into a smaller area, striking the plate only at an indented section called a *dimple*. This concentration discourages sideway or random travel of electrons and reduces the noise which such undesired flow produces. Note that the grid-guide plates do not themselves intercept any electrons traveling from cathode to plate; hence they are not basically another element, which is why these tubes are still considered triodes. The dimples on the plate structures permit a closer spacing between this element and the grid. This reduces electron transit time, a desirable feature at VHF.

Higher gain is achieved by the use of a newly developed frame grid, which is formed by welding metal cross-braces between sturdy upright grid supports. This construction makes it possible to wind the grid with very fine wire under higher than normal tension. The result is a flatter (front-to-back) overall assembly which permits the grid to be positioned closer to the cathode. This closer positioning, with the closer spacing between the grid wire turns, enable the grid to exercise considerably greater control over the current flow, which, in turn, means a higher mutual conductance.

Very close spacing between the grid and cathode increases the tube mutual conductance, but this is not without disadvantages. Since the cathode and grid are both of metal, separated by a vacuum dielectric, they constitute a capacitor. You will remember that the capacitance of a capacitor *increases* as the spacing between the plates *decreases*. In order to reduce the input capacitance of the tube and still maintain very close spacing between the cathode and the grid, it is necessary to reduce the area of material. (The capacitance of a capacitor *decreases* as the area of the plate is *decreased*.) This is the theory upon which the triode nuvistor tube is based.

Figure 5-7 shows a cross section of a nuvistor tube. The overall length of this tube is only about 0.8 inch while the diameter is 0.435 inch at the widest point. All of the nuvistor electrodes are proportionately reduced in size from those of a conventional triode. This permits extremely close grid-to-cathode spacing with a reasonably small mutual conductance. A typical value of mutual conductance for a nuvistor tube is 15,000. Additional advantages are a very low filament current (typical value 135 milliamps), and a low plate voltage resulting in low tube noise.

Although the interelectrode capacitance is very small in a nuvistor, it must still be neutralized when used as an RF amplifier in tuner circuits. This is true of all triode amplifiers. The neutralizing circuits will be discussed later in this chapter.

Tetrodes High-frequency tetrode tubes, for example, the 6CY5, have also been developed specifically for use as RF amplifiers in television receivers. These tubes possess high mutual conductances (on the order of 8,000 to 10,000) and they do not develop much more internal noise than triodes. The high g_m is achieved by using fine (8-mil) wire for the control grid; the low noise level is partly due to the fact that the plate takes considerably more current than the screen grid (approximately in the ratio of 7:1). Since current division is one of the governing factors of internal noise in a tube, the relatively small percentage of the total current captured by the screen grid helps to keep the noise level low.

One of the principal reasons why tetrodes have been generally avoided in all types of circuits in the past has been their tendency to produce more current flow in the screen-grid circuit than in the plate circuit for certain low values of plate voltage. The oncoming cathode electrons strike the plate electrode with sufficient force to dislodge several electrons from the plate structure for each arriving cathode electron. Because of the close proximity of the screen-grid wires and their higher positive voltages, these electrons are drawn to the screen grid rather than to the plate, with the result that more current flows in the screen-grid circuit than in the plate circuit. The effect of this is to present a negative resistance in the plate circuit, leading to amplifier instability.

In the newer tetrodes, this tendency is largely reduced through element design. For example, in the 2CY5 and 6CY5 the interelectrode spacing has been shaped so that electrons arriving from the cathode develop a space charge in the region between the screen grid and the plate.* The negative electrostatic field of this concentration of electrons blocks the escape of the secondary electrons from the plate. It also prevents cathode electrons which have reached the vicinity of the plate from returning to the screen grid when the plate voltage swings below the screen-grid voltage in a normal operation. It is possible for the negative-resistance effect to appear if the plate voltage is driven to 30 volts or so. This is carefully avoided, however, so that no difficulty develops from this source.

Figure 5-8 shows a sectional view of a 2CY5. As indicated by the broken lines in the illustration, the stream of electrons is divided into

* In pentodes the suppressor grid serves this purpose.

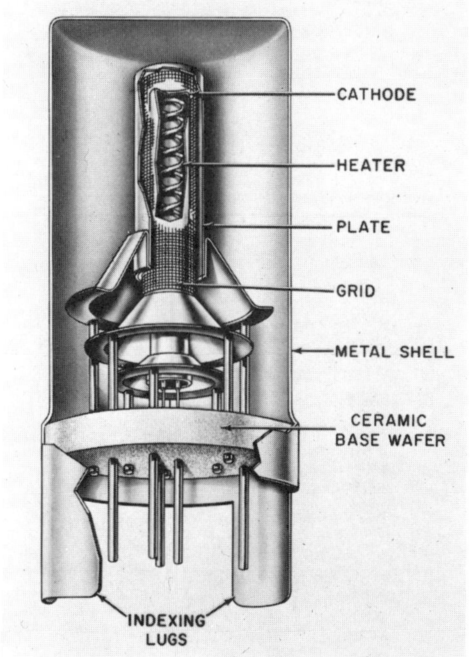

Fig. 5-7. A cutaway view of a Nuvistor tube. (*Courtesy of RCA.*)

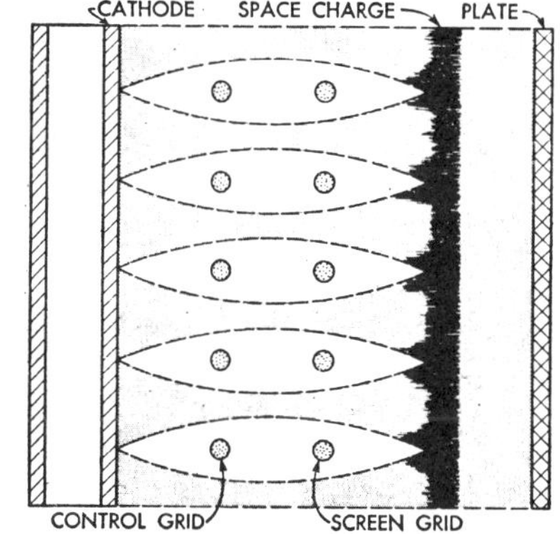

Fig. 5-8. Electron paths in the 2CY5 tetrode.

sheets or *beams* which tend to pass between the wires of the screen grid. Thus, relatively few electrons impinge on the screen grid. Also, by carefully spacing the various electrodes, a space-charge effect is created in the darkly shaded region.

5.5 TUBE-TYPE RF AMPLIFIERS

The typical television RF stage, indicated in Fig. 5-9, is very similar to the same stage in amplitude-modulated broadcast receivers. It has three functions. First, it provides signal amplification in a part of the set where the signal is at its lowest value. In outlying regions or noisy locations, this extra amplification may be the deciding factor in obtaining satisfactory reception. Second, it provides greater discrimination against signals lying in adjacent bands. This is especially applicable for image frequencies. A properly designed RF stage will help the signal to override any small disturbances that are produced in the tubes themselves. The internal tube disturbance is know as *noise*. In television receivers these disturbances are amplified along with the video signal, and if stronger than the received signal, they will appear as small white spots on the image screen. These spots are often referred to as *snow*, or *masking voltages*. Finally, the RF amplifier also reduces local oscillator radiation, which can be quite offensive to neighboring receivers.

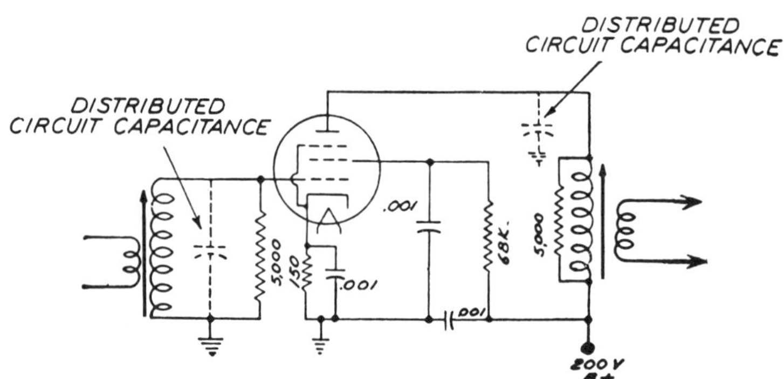

Fig. 5-9. A typical television RF amplifier.

The tube employed in the RF stage, besides having a low noise content and a high mutual conductance value, should also possess a remote cutoff characteristic. With remote cutoff properties, the stage does not distort as readily when large input signals are received. Furthermore, *automatic gain control* (AGC) voltages may be applied to the tube, materially aiding amplifier stability and tending to maintain a steady signal output. (AGC in a television receiver is similar to AVC in a radio receiver.)

Some of the forms that the RF stage may assume are shown in the accompanying diagrams. In Fig. 5-9, transformer coupling is used in the input and the output circuits of the RF amplifier. Each transformer is loaded down by a shunting resistor, so that its response will be fairly uniform over a 6 MHz bandwidth. The resistor value is chosen with the

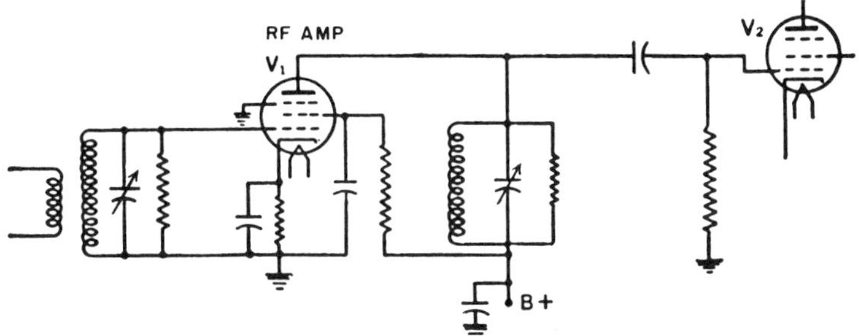

Fig. 5-10. Another RF amplifier stage. One tuned circuit is common to the plate of V_1 and the grid of V_2.

idea of maintaining the stage gain as high as possible. On the upper VHF channels (7-13) and throughout the UHF band, sufficient loading is usually provided by the tube itself so that external resistors are not needed. When triodes or tetrodes are used, loading resistors may be omitted, because the lower internal impedances of these tubes provide sufficient loading to achieve the desired bandwidth.

In Fig. 5-10 a single-tuned circuit instead of a transformer is used between the plate of the RF amplifier and the mixer tube. The tuning capacitance shown in each of these diagrams might be either a small variable trimmer capacitor or the stray-circuit wiring and tube capacitance always present in the circuit. In the latter instance, adjustment of the tuned circuit would not be accomplished by varying the capacitance (since the wiring and tube capacitances are not adjustable), but by using movable cores within the coil. Thereafter a selector switch or some other tuning arrangement is used. Although only one set of coils is indicated in some of these diagrams, there would be similar arrangements for each of the channels.

There are a number of variations of the coupling network between V_1 and V_2 of Fig. 5-10, two of which are shown in Fig. 5-11. In the first illustration, Fig. 5-11(A), the plate load for V_1 is a resistor. Coil L_1 is the resonant circuit between the two tubes, and it is placed in the grid circuit of V_2. In Fig. 5-11(B), the plate load for V_1 is an RF choke. It might also be another resonant circuit.

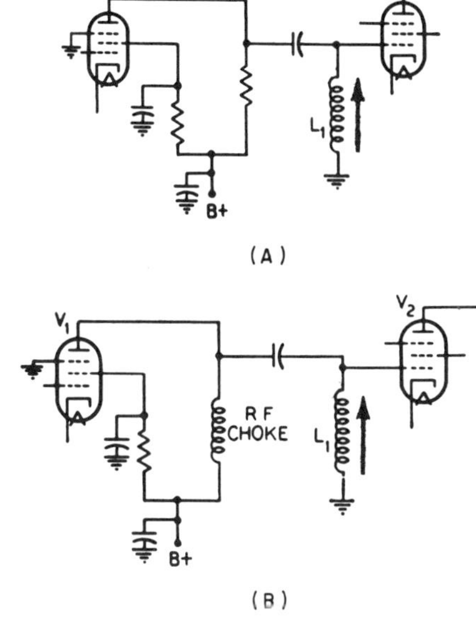

Fig. 5-11. Two additional interstage coupling networks between the RF amplifier (V_1) and the mixer (V_2). Either triodes or pentodes can be employed.

An approach sometimes practiced is to insert an overcoupled transformer in the input circuit and a single-peaked circuit in the plate circuit of the stage. One such circuit is shown in Fig. 5-12. The primary

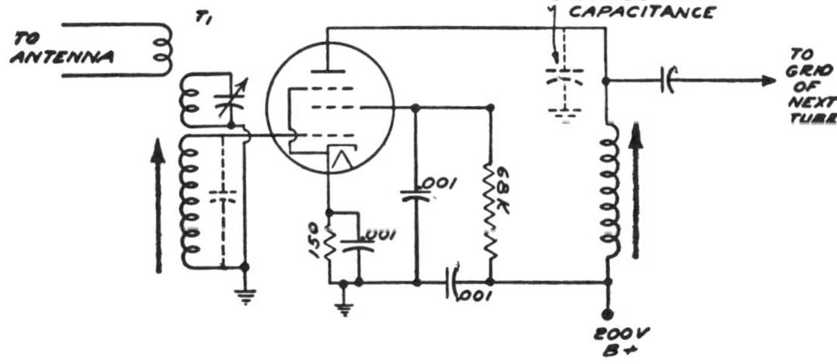

Fig. 5-12. An RF amplifier which combines the response characteristics of grid- and plate-tuned circuits to obtain a 6-MHz overall spread.

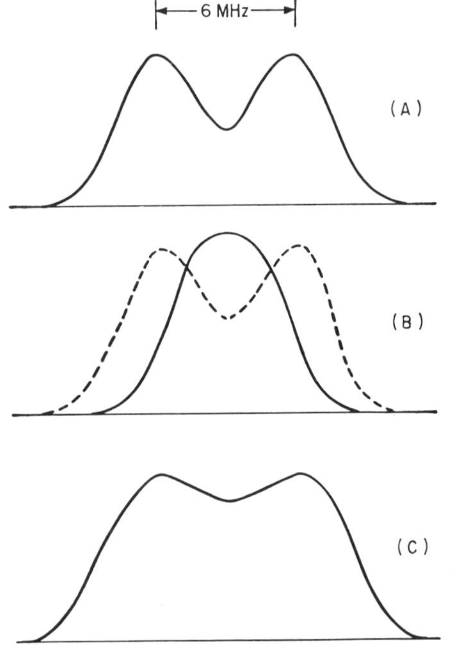

Fig. 5-13. The combination of two tuning circuits to produce a flat-topped overall response.

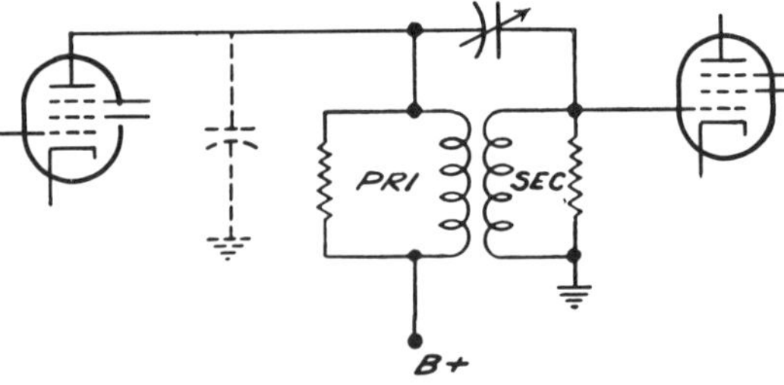

Fig. 5-14. One method of increasing the coupling between two tuned circuits to achieve a broad band-pass.

winding of T_1 is untuned and matches the transmission-line impedance. The grid winding is tuned by the grid-input capacitance of the tube, plus whatever stray capacitance is inevitably present in the circuit. The third winding contains a small trimmer to permit adjustment, although in some instances it is nothing more than a 1- or 2-turn winding which functions as a link coupling between the input and grid coils. The combination of these three coils results in a double-peaked response curve, as shown in Fig. 5-13(A). In the plate circuit of the stage, and serving as an impedance coupling between circuits, is a single-tuned coil. Its response is single-peaked, as illustrated in Fig. 5-13(B). By properly adjusting the peaks of these circuits, we can achieve an essentially overall flat response of 6 MHz for the stage, as is indicated in Fig. 5-13(C). (The word *essentially* is used because it seldom occurs that the RF response curve has an absolutely flat top. In practice, up to a 30 percent dip in the center of the curve is permissible.)

Another method of coupling between stages in order to achieve a broad band-pass is shown in Fig. 5-14. Here, a small capacitor connects the primary and secondary windings. The value of this capacitance is low (10 to 20 pF) and governs the extent of the bandwidth; increasing the capacitance increases the bandwidth.

In Fig. 5-15 mutual capacitive coupling between the RF amplifier and the mixer is achieved in still another way. Coils L_A and L_B are two coils which are coupled to each other *only* through the common capacitance C_4. In other words, their magnetic fields do not interact. Each coil is pretuned to the same frequency by means of a brass slug. Capacitor C_A represents the output capacitance of the RF amplifier plus other circuit capacitances; C_B is the input capacitance of the following tube plus the distributed wiring capacitance of the circuit.

In this type of tuned circuit, the bandwidth is determined by the degree of coupling and the Q's of L_A and L_B. The degree of coupling is controlled by the value of C_4. The smaller this capacitance, the greater the mutual impedance and the greater the bandwidth. The value of C_4 is chosen to provide a bandpass of approximately 6.0 MHz. To maintain a constant bandwidth, C_4 has a value of 250 pF on the lower channels and a value of 140 pF on the higher channels. This value com-

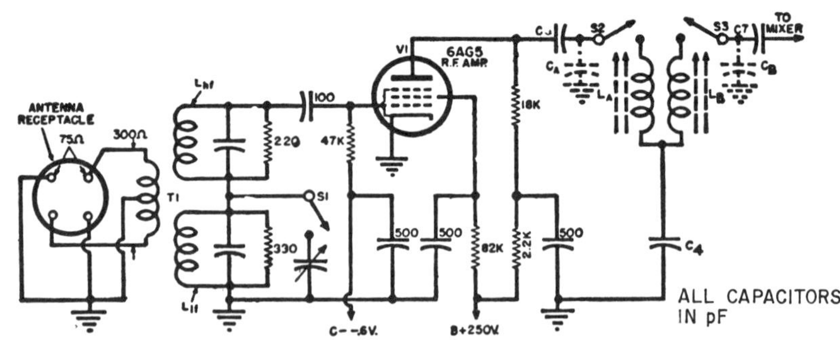

Fig. 5-15. Mutual capacitive coupling. C_4 is used between L_A and L_B.

pensates for the change in coil Q's with frequency. For each channel a new pair of coils is switched into the circuit.

Another feature of Fig. 5-15 is the provision for either 75- or 300-ohm input transmission lines. This is accomplished by using the full primary winding of the input transformer for the 300-ohm line and half of the winding for the 75-ohm coaxial line. Inductance of a coil is proportional to the square of the number of turns. Doubling the number of turns produces four times the inductance and, at the same frequency, four times the impedance; 300 ohms is four times 75 ohms.

Triode RF Amplifiers The low-noise qualities of triodes always make them attractive for the RF amplifier stage, and a number of different circuits have been designed which use these tubes. The simplest approach employs a single triode, as pictured in Fig. 5-16(A). The low plate resistance loads the tuning circuits sufficiently to achieve the desired bandwidth without the need for any external shunting resistors. A high g_m helps to achieve a fairly good gain even at these frequencies.

A triode tube, when used as shown in Fig. 5-16(A), requires a neutralizing network in order to prevent regeneration at high frequencies. This need stems from the relatively large capacitance that exists between the plate and grid elements inside the tube. At sufficiently high frequencies, the output signal can use this capacitance to travel from the plate to the grid and reduces the stage gain. By feeding an out-of-phase signal of the same amplitude back to the grid, the first signal can be effectively neutralized and the degenerative condition avoided.

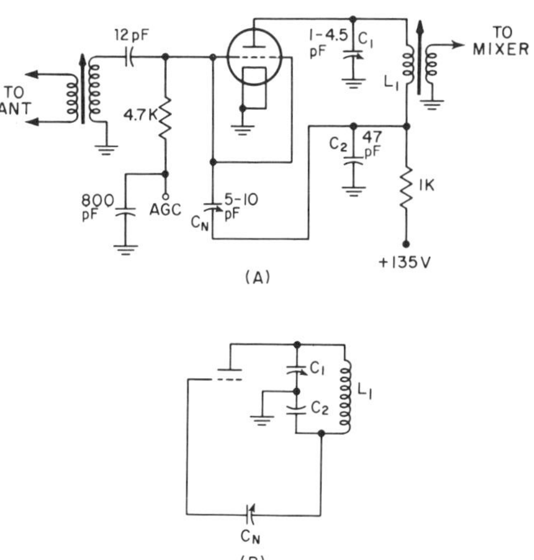

Fig. 5-16. A triode RF amplifier and how it is rearranged to better reveal out-of-phase voltage.

The required out-of-phase voltage for the grid is obtained at the bottom end of the plate coil (L_1). One way of looking at L_1 to see how the out-of-phase voltage is developed is to consider capacitors C_1 and C_2 as providing a capacitive network across L_1 which establishes an RF ground at some point on this coil (see Fig. 5-16(B)). With this condition, the RF voltage at the plate end of the coil is 180 degrees out of phase with the voltage at the other end. The capacitor C_N then feeds back to the grid as much out-of-phase voltage as is needed to neutralize the signal voltage reaching the grid by way of the interelectrode capacitance.

The tube actually possesses two grid leads and two cathode leads. This is done to minimize lead inductance in order to permit a single neutralization adjustment that will serve suitably throughout the entire VHF band. If such inductances were permitted to become high enough, operation on Channels 7 to 13 would require only one neutralizing adjustment and Channels 2 to 6 would require another. More will be said about tube lead inductance presently.

Push-Pull Triode Amplifiers Push-pull amplifiers have also been used in the RF section of television receivers. The circuit schematic is shown in Fig. 5-17. The transmission line from the antenna feeds directly into the grids of a push-pull triode amplifier. To match the

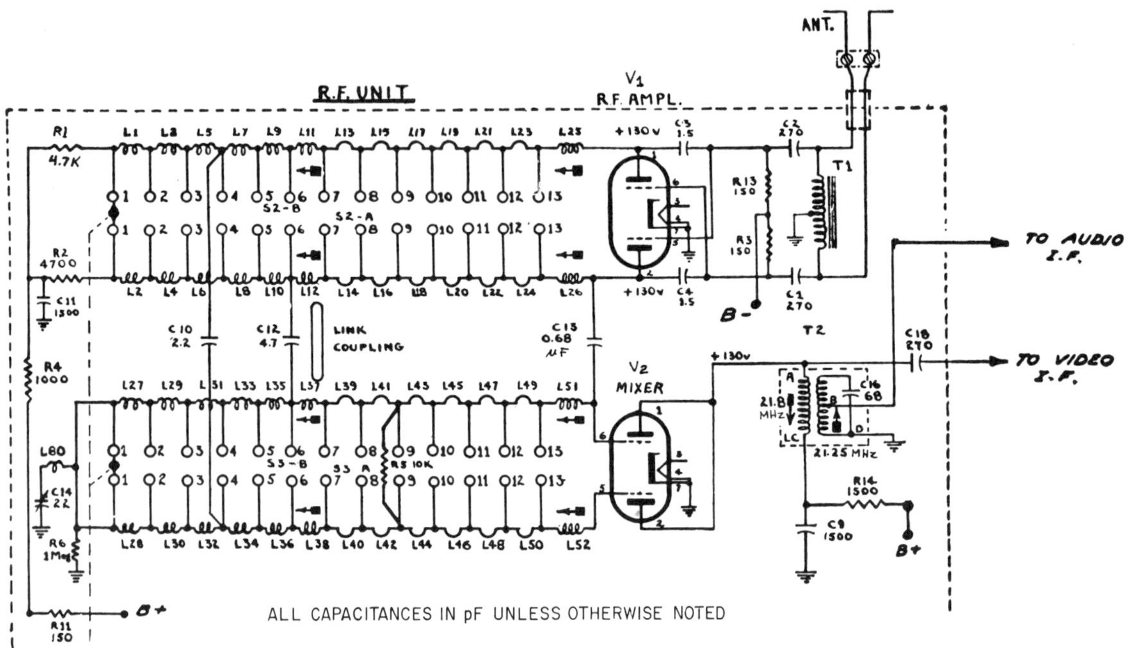

Fig. 5-17. A quarter-wave transmission line used for tuning the RF stages of a television receiver.

impedance of the line, two 150-ohm resistors are connected in series to provide the total of 300 ohms. Transformer T_1 is a center-tapped coil used to prevent low-frequency signals from reaching the grids of the RF amplifier. Capacitors C_1 and C_2 are antenna-isolating capacitors.

In the plate circuit of the RF amplifier, starting with L_{26} and progressing down to L_1, we have a series of inductances that may be considered as sections of a quarter-wave transmission line. The switch, as it moves progressively to the left, brings in more inductances, thus decreasing the channel frequency. In position 13, only L_{25} and L_{26} are in the circuit, and the receiver is set for the highest VHF-TV channel. At position 2, the set will receive the lowest channel. (Position 1 was for the now obsolete Channel 1.) At various points along the line, adjustments may be made by changing the position of the tuning slugs. The physical construction of each of the small inductances, L_{13} to L_{26}, is a small, fixed silver strap between the switch contacts. Each strap is cut long enough to introduce a 6 MHz change in frequency. In order to make the transition from the lowest high-frequency channel, 174–180 MHz, to the highest low-frequency channel, 82–88 MHz, adjustable coils L_{11} and L_{12} are used. Coils L_1 to L_{10} are more substantial in appearance than coils L_{13} to L_{26}, being wound in figure 8 fashion on fingers protruding from the switch assembly.

Since each section of the tube is a triode, neutralizing capacitors are necessary to counteract the grid-to-plate capacitance. This is the function of C_3 and C_4.

Coupling between the quarter-wave line of the RF amplifier and a similar section in the grid circuit of the mixer tube is twofold: by a direct capacitance connection and by link coupling. The response characteristic of these RF circuits extends the full 6 MHz. In addition, a 10,000-ohm loading resistor is placed across a portion of the mixer tuning circuits to provide the necessary bandwidth. (In Fig. 5-17 the resistor is effective only for channels below 9.)

Ground-Grid Amplifers Triode RF amplifiers are often employed in an arrangement known as the *ground-grid amplifier*. This type of amplifier is contrasted with the conventional amplifier in Fig. 5-18. Note that the grid of the tube is at a RF ground potential and that the signal is fed to the cathode. The tube still functions as an amplifier, because the flow of the plate current is controlled by the grid-to-cathode potential. Instead of varying the grid potential and maintaining the cathode as fixed, the grid is fixed and the cathode potential is varied. The net result is the same. In addition, the grid being grounded acts as a shield between the input and output circuits, thereby preventing the feedback of energy which is so essential to the development of oscillations.

The ground-grid amplifier also offers low input impedance, enabling the amplifier to match the antenna transmission-line impedance. The low impedance provides a broader band-pass response curve which is particularly desirable for 6 MHz television signals.

A commercial application of a grounded grid RF amplifier is shown in Fig. 5-19. The antenna is connected into the cathode circuit of the RF amplifier. Coil L_1 is a simple high-pass filter designed to reject all low-frequency signals, especially those at the intermediate frequency. The cathode chokes, L_2 to L_6, are placed in a series with the cathode resistor to prevent the input impedance from being lowered by the shunting effect of any stray capacitance to ground due to the cathode

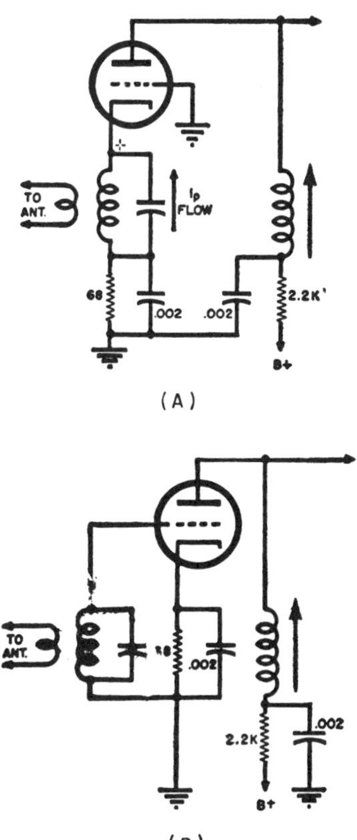

Fig. 5.18. A comparison between grounded-grid and conventional RF amplifiers.

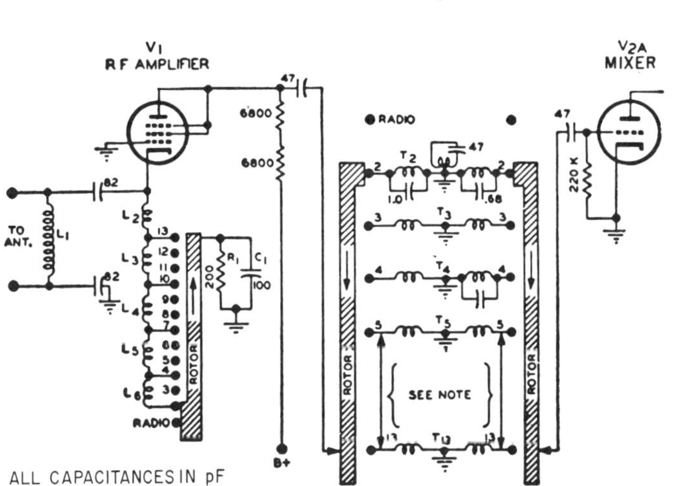

Fig. 5-19. The grounded-grid RF amplifier stage used in a television receiver. (Note: RF coils and switch points for Channels 6 through 12 are not shown. Coils l_6 through T_1 correspond to Channels 6 through 12 and are connected the same as T_5.)

of the tube. The choke value is changed with frequency. The resistor R_1 and the capacitor C_1 provide a cathode bias.

The RF amplifier is coupled to the mixer tube through a wide-band transformer. One such unit is provided for each channel. The windings are self-tuned by the distributed and tube capacitances to provide maximum gain through a high L/C ratio. The RF coils for each channel are placed physically near the oscillator coils of the same channel (which are not shown) in order that both voltages will combine at the mixer grid.

Cascode Amplifiers Still another RF amplifier arrangement that makes use of triodes is the cascode amplifier (see Fig. 5-20). Here two triodes are connected in a series, that is, the plate of the first section goes directly to the cathode of the second section. The same current flows through both tubes, and the amplitude of this current is controlled by the dc bias on the first triode.

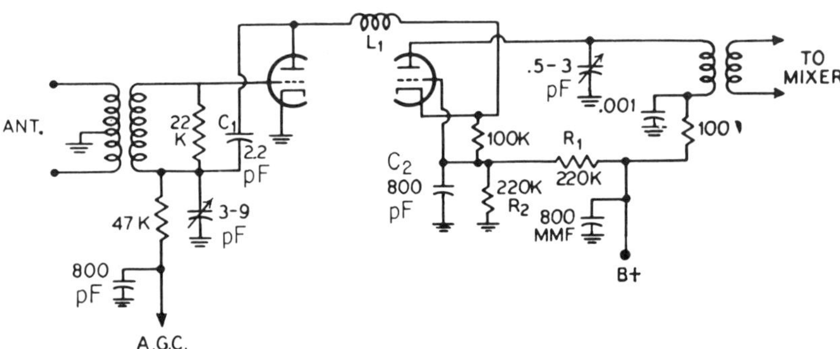

Fig. 5-20. A cascode RF amplifier.

The input-tuned circuit of this series amplifier connects to the control grid of the first triode; the output-tuned circuit is in the plate lead of the second triode. The first stage is operated as a conventional amplifier, that is, with the signal applied to the grid and the output signal obtained from the plate. The second stage is employed as a grounded-grid amplifier. The inductance, L_1, between both stages helps to neutralize the grid-to-plate capacitance of the first triode (with help from capacitor C_1), and it is designed to resonate with the grid-cathode capacity of the second section on the high VHF channels. While L_1 thus aids the stability of this combination, it is also largely responsible for the low-noise qualities of the cascode circuit.

The role that C_1 plays in helping neutralize the input triode to prevent it from oscillating can be seen perhaps more clearly by noting that it connects from the plate of the first triode to the bottom end of the coil in the grid circuit. Thus, it feeds its signal to the bottom end of this coil at the same time that the grid-to-plate capacity within the first triode feeds back its signal to the top of the coil. In this way we achieve the 180 degree phase reversal required for the two voltages to counterbalance and neutralize each other.

Direct coupling is used between the first triode plate and the second triode cathode. With cathode feed to the second triode, C_2 is used to place the grid at RF ground potential. Since the two triode sections are in a series across a common plate supply, the cathode of the second triode is 125 volts positive with respect to chassis ground. A divider across the plate supply consisting of R_1 and R_2 places the grid of the second triode at a sufficiently positive potential (with respect to its cathode) for a proper operating bias.

The cascode circuit is widely employed and a number of special twin-triode tubes have been developed for this particular purpose. All have basically similar electrical characteristics but different internal connections to facilitate the placement of the components and the layouts found in the various television tuners. The cascode arrangement gives an overall gain which is somewhat less than that obtainable from a well-designed, high-frequency pentode. However, the noise figure of a cascode combination is considerably better than that of a pentode.

Internal Tube Capacitances Just as important as the mutual conductance of a tube are its interelectrode capacitances. It has already been noted that the gain of a stage is equal to the product of the mutual conductance of the tube and the load impedance. The load impedance, in turn, is essentially equal to the value of the resistor shunting the tuning coil and capacitor. And, as we shall see in a moment, it is the value of the L-to-C ratio of the tuning circuit which determines how high a resistor can be used.

For the greatest gain over any band, a high L-to-C ratio should be maintained in each resonant circuit. The capacitance which shunts the coil includes the interelectrode capacitance of the tube. As we make this capacitance smaller, a higher L-to-C ratio is possible. In addition, the value of the resistance R needed to load (broadband) a tuned circuit is proportional to the resonant frequency reactance of the capacitance or coil. Thus, with a smaller capacitance we obtain a higher resonant frequency capacitive reactance and the loading resistor may be made higher in value for the desired value of Q. The end result is a greater gain since this means a higher resonant-circuit impedance.

For the RF input stage, the minimum capacitance is determined by

1. The grid-to-cathode capacitance, C_{gk}.
2. The grid-to-plate capacitance, C_{gp}.
3. The stray capacitance, C_s.

The total capacitance is equal to

$$C_{total} = C_s + C_{gk} + C_{gp}(1 + G), \text{ where} \quad (5\text{-}2)$$

G = the gain of the stage, usually about 15–30 in these amplifiers.

The total tuning capacitance across the tuned input circuit of the RF amplifier of Fig. 5-21 is comprised of the tuning capacitance (C) added

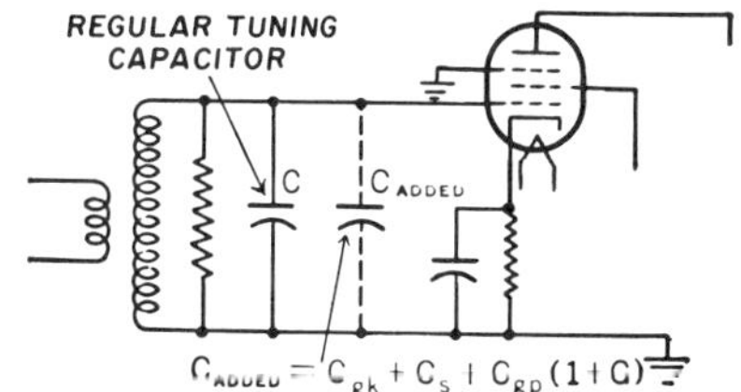

Fig. 5-21. At the higher frequencies, the stray-wiring and internal-tube capacitances represent an appreciable part of the total circuit capacitance and hence must be included in all computations.

to the tube and wiring capacitance. At the broadcast frequencies (500 to 1,500 kHz) in the ordinary home receiver, these tube and wiring capacitances are never serious when compared with the size of the tuning gang employed. However, when frequencies of 100 MHz or more are to be received, the tuning capacitor may be even smaller than these additional capacitances and they can no longer be disregarded.

Whereas the wiring and tuning capacitances remain fixed once the set has been completed, no such happy state of affairs exists for C_{gk} or $C_{gp}(1+G)$. The latter values will vary as the stage varies, which occurs every time the input voltage changes. The grid-to-plate capacitance will change its value as the electron current is altered. The effect of the variation, if great enough, is sufficient to detune the stage. Again, these small items, insignificant in themselves, may become very influential as the frequency increases and the size of the coil and capacitor decreases.

We have considered only the capacitance in the input circuit. A similar line of reasoning may be applied to the plate circuit, where the total minimum capacitance is composed of the following:

1. The output capacitance, C_0, as obtained in any tube manual.
2. The wiring capacitance.

The list is short because it has been assumed that the output circuit is inductively coupled to the next grid. This coupling tends to separate the input capacitance of the next tube from the plate circuit of the preceding tube. However, if a direct connection is made to the next tube, the additional input capacitances must be taken into account.

From the foregoing brief discussion, it is evident that in designing tube-type RF television amplifiers, tubes should be selected that have

1. High mutual conductance values.
2. Low input and output capacitances.

It has been suggested that the usefulness, or *figure of merit* of a tube may be determined by the ratio of (1) to (2), or

$$FIGURE\ OF\ MERIT = \frac{g_m}{C_{in} + C_0} \times 10^{-6}, \text{ where} \qquad 5\text{-}3$$

g_m = The transconductance in micromhos,
C_{in} = the input capacitance of the tube in picoFarads, and
C_0 = the output capacitance of the tube in picoFarads.

Large values for the figure of merit are desirable. It should be noted that both the numerator and denominator of the ratio are important at the high frequencies. At the low frequencies, the tube capacitances have less importance and only g_m needs to be considered.

Tubes with Two Grid and Two Cathode Terminals The tube in Fig. 5-16 possesses two cathode terminals and two grid terminals. It has been found that the input impedance of vacuum tubes, which is ordinarily so high as to be considered infinite, begins to decrease as we raise the frequency of the signal. For the television channels above 50 MHz, this tube loading on the attached tuned circuits causes a reduction in the gain and Q of the circuit. One of the causes for this reduction in tube input impedance is due to the inductance of the cathode leads within the tube itself. Why this is so can be seen from the following explanation.

The current of a tube must flow through the cathode lead wires and in so doing develops a voltage across the inductance of these wires. Note that this inductance is of importance only when the signal frequency is high. The average or dc component of the current does not enter into this consideration. The voltage produced across the lead inductance, although due to the plate current, is impressed between the grid and the cathode. As a result, the effective signal voltage acting at the grid of the tube is lowered because of the opposition of the cathode-lead voltage. The situation is analogous to inverse feedback, except that the lead-inductance voltage is present even though the cathode of the tube is grounded directly to the tube socket. This lead inductance occurs within the tube itself.

Note that the voltage which is developed across the cathode-lead inductance is due to the plate current. So far as the plate circuit is concerned, this voltage is of little significance. It is at the grid, where the signal is applied, that the voltage is important.

To eliminate the effect of the lead-inductance voltage on the input-grid circuit, it has been designed with two wires travelling directly from the cathode structure inside the tube to the tube base. In this manner one terminal is available for the grid-circuit return and one for the plate circuit and its current, and the two circuits are divorced from each other. In the circuit of Fig. 5-22, the RF amplifier tube possesses two cathode terminals. Even though both cathode terminals are grounded, pin 2 is connected to the grid coil and capacitor. Pin 7 is the cathode connection for the plate circuit. The screen-grid and plate bypass capacitors are connected to it. The dc plate current divides between both cathode terminals, but this is of no consequence since it does not contribute to the degenerative effect.

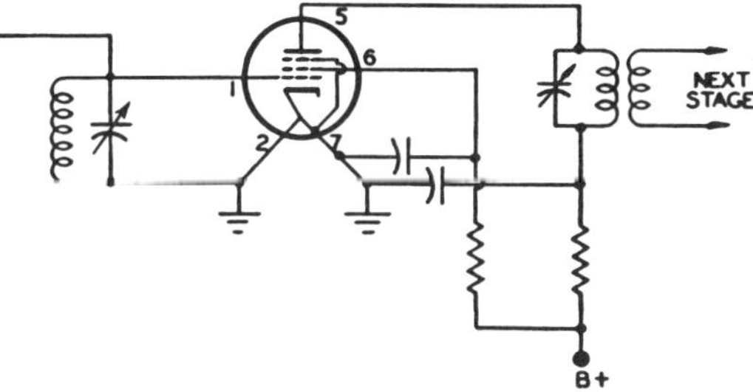

Fig. 5-22. The use of two cathode wires to eliminate the adverse effect of cathode inductance.

In RF triodes two control grids are also made available in order to reduce the lead inductance in this circuit as well. This reduction permits one neutralization adjustment to suffice over the entire VHF range. As indicated previously, if the control-grid inductance becomes too large, separate neutralization adjustments would be required for the low and high ends of this range. In tubes not requiring neutralization, such as tetrodes or pentodes, two control-grid leads are generally not employed.

5.6 TRANSISTOR RF AMPLIFIERS

Vacuum-tube triode-RF amplifiers may be operated as either grounded-cathode or grounded-grid amplifiers. The grounded-cathode amplifiers provide a somewhat higher gain, but have the disadvantage that neutralization is required to offset the plate-to-grid capacitive feedback. With grounded-grid amplifiers, the grid acts as a shield between the input signal at the cathode and the output signal at the plate.

Transistor-RF amplifiers may be operated as grounded-emitter amplifiers or grounded-base amplifiers. In this respect they are similar to the triode circuits. The grounded-emitter circuit (comparable to the grounded-cathode triode configuration) provides the higher gain and requires neutralization, while the grounded-base circuit (comparable to the grounded-grid triode configuration) can be operated without the need for neutralization.

The signal fed from collector to base via the collector-base junction capacitance may be either regenerative or degenerative depending upon the frequency of the signal and the nature of the collector load. If *regenerative*, the feedback signal *reinforces* the signal at the base; and, if *degenerative* the feedback signal partially cancels the signal at the base. In either case, the feedback signal may be highly undesirable in RF amplifiers. The purpose of neutralization is to deliberately feed a signal from the collector to the base which is 180° out of phase with the feedback signal traveling via the collector-to-base junction capacitance. The neutralizing signal cancels the effect of the undesired feedback signal. In the examples of neutralized RF amplifiers given here, both regenerative and degenerative feedback circuits are represented.

Figure 5-23 shows a typical grounded-emitter amplifier. Since the collector-signal voltage is 180° out of phase with the base-signal voltage, the collector-to-base capacitance will cause a degenerative feedback. In other words, the collector voltage, when fed back to the base via the collector-to-base capacitance, will partially cancel the input signal and cause a large reduction in the gain. To prevent this, an in-phase signal must be fed to the base to cancel (at least partially) the out-of-phase feedback.

The feedback voltage from the transformer secondary is 180° out of phase with the collector voltage at the point of takeoff. This feedback

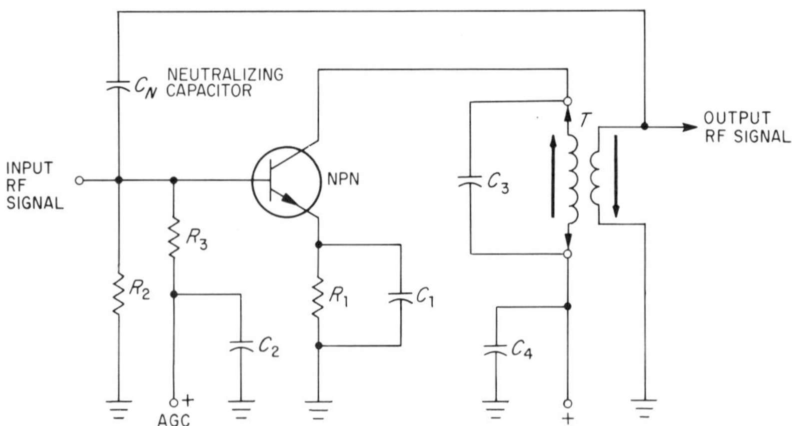

Fig. 5-23. A grounded-emitter RF amplifier with neutralization.

voltage is fed to the base via the neutralizing capacitor C_N. A *positive* AGC voltage is also fed to the base of the NPN transistor, but a *negative* voltage would be needed if a PNP transistor was employed. As with any AGC system, the AGC voltage is more effective if it is fed to the RF stage rather than to the IF stages only. However, the input and output impedances of a transistor vary considerably with changes in the operating voltages, and this is a decided disadvantage. Thus, in some transistor-TV receivers, there is no AGC voltage fed to the RF amplifier.

The primary winding of transformer *T* is changed whenever the channel-selector knob of the receiver is turned. This tunes the RF stage to the carrier frequency of the input signal. The primary is tuned by C_3, while C_2 and C_4 decouple the RF-amplifier circuit from the other stages.

Another method of neutralizing a transistor-RF amplifier is shown in Fig. 5-24. The basic circuit is shown in part A. It shows an NPN conventional amplifier biased with a voltage divider comprised of R_1 and R_2. The neutralizing capacitor C_N supplies the feedback-signal voltage to the base of the transistor. A decoupling filter (R_3 and C_4) is used to prevent the signal voltage variations from the other stages from affecting the RF amplifier.

The undesired feedback signal from the collector to the base via the base-collector capacitance is degenerative. To cancel the effects of this feedback signal, it is necessary that the feedback signal via C_N be regenerative. In other words, the neutralizing capacitor must be connected to a point where the signal is 180° out of phase with the collector signal. To show how the signal-phase reversal occurs between the collector of the transistor and point *a*, the circuit has been redrawn in Fig. 5-24(B). It will be noted that the junction of C_1 and C_2 is grounded, thus making the primary of the transformer winding also at ground potential. Therefore, the voltage at one side of the primary will be 180° out of phase with the signal at the other end.

A third method of neutralizing a transistor-RF amplifier is shown in Fig. 5-25. In this case the primary winding of the transformer is tapped and grounded. The voltage across the lower tap will, of course, be 180° out of phase with the voltage at the collector side, and therefore, the proper feedback voltage polarity is obtained. This circuit is more frequently used with IF amplifiers—a form of RF-amplifier circuit.

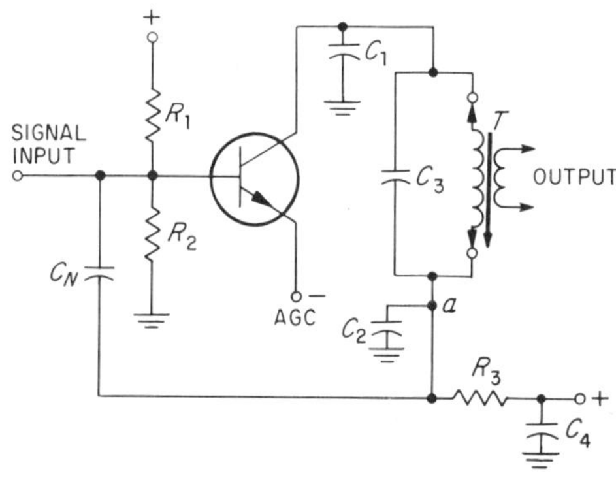

Fig. 5-24 (A). The basic neutralized RF amplifier circuit.

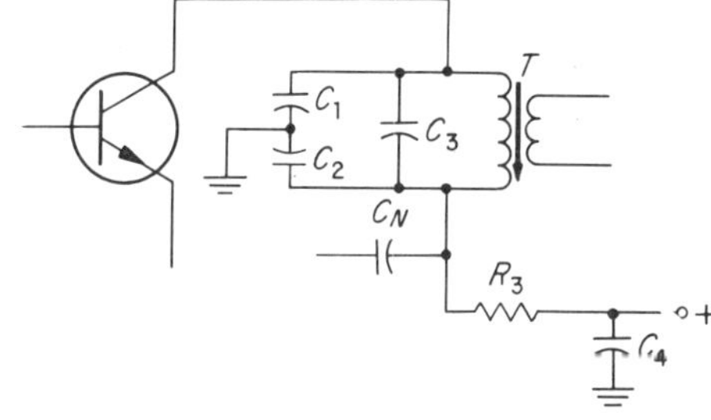

Fig. 5-24 (B). Circuit redrawn to show how phase reversal across a transformer primary is obtained.

5.7 FIELD-EFFECT TRANSISTORS

Because of the advantages of the field-effect transistor (FET), it was inevitable that it would be eventually used in television applications. In modern solid-state television receivers these field-effect transistors are used in a number of different types of circuits. At this time, we are especially interested in the use of the FET as an RF amplifier. Before discussing a typical RF-FET amplifier, we will first review the basic theory of the FET's.

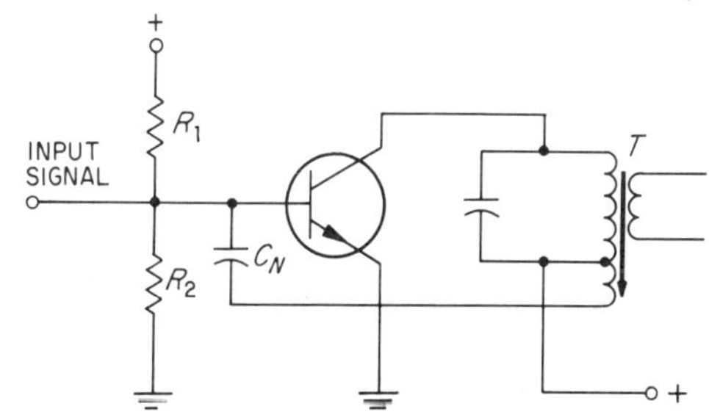

Fig. 5-25. A third method of neutralization.

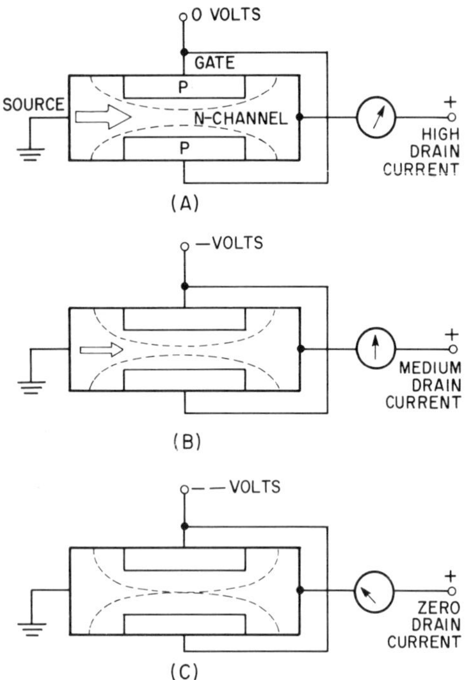

Fig. 5-26. The voltage and current relationships in an N-channel FET. (A) Current through FET with zero volts on the gate. The arrow indicates the electron flow. (B) Negative gate voltage reduces drain current. (C) Pinch-off voltage reduces drain current to zero.

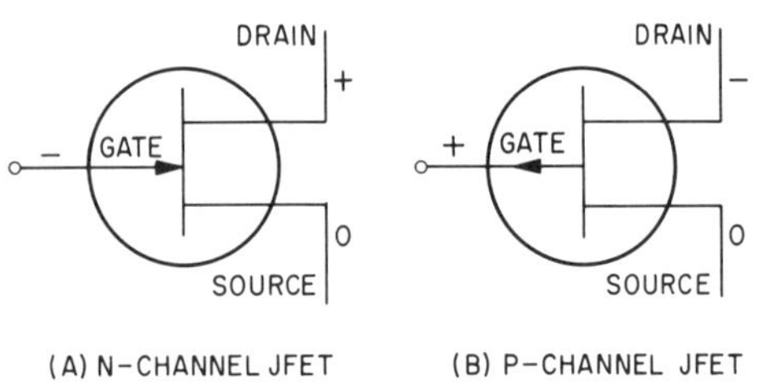

Fig. 5-27. Symbols for JFET's showing typical operating voltage polarities.

Two types of FET's are in popular use. One is called the *Junction FET* (*JFET*), and the other is the *Insulated-Gate FET* (*IGFET*). These latter are now called *MOSFET*'s (Metal Oxide Semiconductor Field-Effect Transistors). The newer term comes from the fact that a metal oxide is used for the gate insulation. Figure 5-26 shows the theory of operation for the JFET. The device depicted has an *N channel* through which current flows. There are two types of JFET's: the *N channel* and the *P channel*. They are named according to whether N-type material or P-type material is used for a conducting path through the device. In the N-channel JFET, current flow is by electrons. These electrons leave the *source* and are attracted to a positive voltage on the *drain*. Moving between the source and the drain, they pass through the gate area.

The dotted lines around the gate in Fig. 5-26(A) represent the boundaries of a depletion region that are present at the junction of the P- and N-type materials in this area. This depletion region will limit the number of electrons that can flow from the source to the drain. When no voltage is applied to the gate, the size of the depletion region is small. Under this condition, there is little opposition to the current flow through the device and the drain current will be high.

When a negative voltage is applied to the gate, the PN junction is reverse biased, and this increases the size of the depletion region as shown in Fig. 5-26(B). The result is a decrease in the drain current. The more negative and gate voltage for this type of FET, the greater the depletion region and the smaller the amount of drain current. Ultimately, a point will be reached where the depletion region is so large that it prevents any current flow through the channel. The negative voltage on the gate required to produce this situation is called the *pinch-off voltage* and it is shown in Fig. 5-26(C). A P-channel JFET operates in the same manner as just described except that the voltages on the gate and the drain are reversed. The more *positive* the voltage on the gate, the lower the amount of the current flow in a P-channel JFET. Current flow through a P-channel JFET is due to hole flow, and is toward a negative voltage on the source.

Figure 5-27 shows the two symbols that are used for N-channel and P-channel JFET's. Typical voltage polarities are shown on the symbols. The reason the N-channel JFET is often compared with the operation of the vacuum tube is indicated by the fact that the drain voltage is positive and the gate voltage is negative, while in the vacuum tube the plate voltage is positive, and the grid voltage is negative.

The junction FET of Fig. 5-26 operates by controlling the size of the depletion region with a gate voltage. The more negative the gate voltage the greater the depletion region, and vice versa. This type of FET is often referred to as a *depletion-mode junction FET*. Another mode of operation is possible. The junction between the gate and the channel can be made sufficiently large and with a sufficient amount of doping, so that the current through the channel is cut off whenever the voltage on the gate is zero volts. In order to make current flow through this FET, it is necessary to *forward bias* the junction in order to reduce the size of the depletion region around the gate. This type of a FET is referred to as an *enhancement-mode FET*. The important difference between the depletion-mode and the enhancement-mode FET's is that current will flow through the *depletion-mode FET* when the gate voltage is zero, whereas no current flows through an *enhancement-mode FET* when the gate voltage is zero.

The disadvantage to the type of FET's discussed so far (that is, of the junction-type FET's) is that a current will flow between the gate and the channel with a very small amount of forward bias. Even with reverse bias, the minority charge carriers flowing across the junction can result in an appreciable input current compared to that of a vacuum tube. To get around this problem, an insulated gate is sometimes placed at the junction between the gate and the channel. Figure 5-28 shows a cross section of a MOSFET. As with the junction FET's, MOSFET's may be either N-channel or P-channel types, and either type may be designed for operation in the depletion or the enhancement mode. The insulating layer—usually silicon dioxide—between the gate and the channel prevents the current flow in the gate circuit since neither majority or minority charge carriers can move across the dielectric layer. Therefore, the MOSFET has a very high-input impedance.

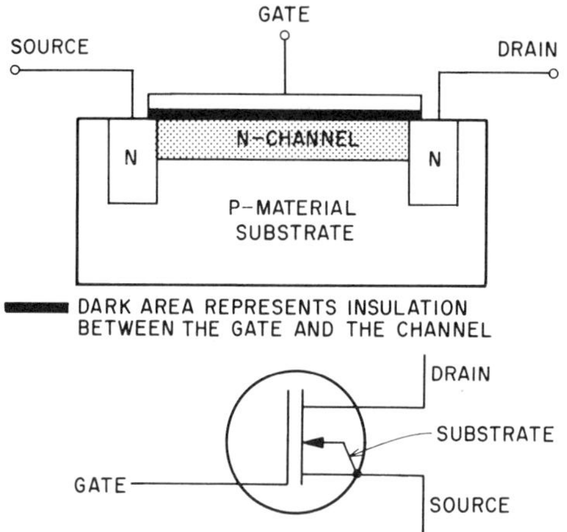

Fig. 5-28. A MOSFET and its symbol.

The insulated-gate layer is made very thin to allow the gate voltage to exercise the maximum amount of control over the current in the conducting channel. Because the layer is so thin, it can be easily punctured by any excessive voltage. For example, an electrostatic charge can destroy an Insulated-Gate FET. It is a common practice to ship MOSFET's with their leads tied together to prevent accidental static charges from destroying them before they can be placed into a circuit. Remember that a static charge from a person's hand or from the probe of a meter can easily destroy the very thin insulating layer, and special care must be taken when handling or measuring in circuits having MOSFET's.

Some FET's have more than one gate. One important example is the *Dual-Gate*, or *tetrode FET*. The symbol for this type of FET is shown in Fig. 5-29. Both of the gates exercise control over the current flow through the FET. Its characteristics are similar to those of a tetrode tube. The Dual-Gate FET is very useful as a mixer in which the oscillator and the RF-input signals are to be combined in the device. It is also useful as an RF amplifier because the AGC and RF voltages can be applied to the device at separate electrodes.

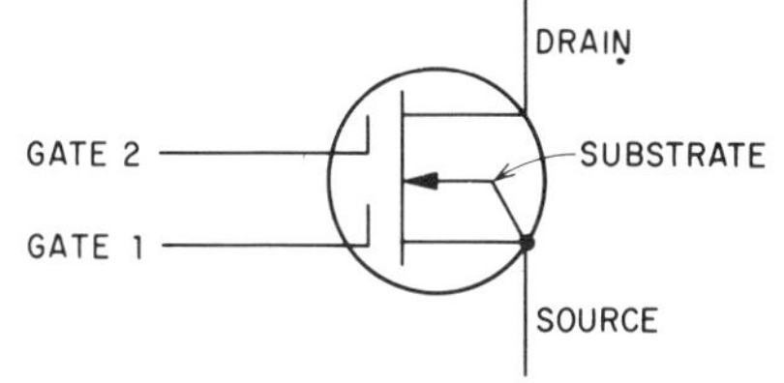

Fig. 5-29. The symbol for a dual-gate MOSFET.

Advantages of Field-Effect Transistors In an ordinary transistor, current flow through the device is dependent upon both the *hole flow* and the *electron flow*. Since both polarities of the charges are required for their operation, ordinary transistors are sometimes referred to as *bipolar devices*. In comparison a FET is a *unipolar device*, because the flow of current through it is either by hole flow *or* electron flow, but not by both. Since an electron-hole combination causes noise, a unipolar device generates less noise in its operation. This is an advantage in high-frequency operation.

Another advantage of the FET is its high-input impedance. In this regard it is similar in operation to the vacuum tube. Unlike transistor amplifiers which require current flow in the base circuit, and therefore require an input power, a FET circuit may be designed that has virtually no current flow in the gate circuit. By way of comparison, vacuum-tube amplifiers also have virtually no current flow in their grid circuits.

Another advantage of the FET is its *square law operation*. Current flowing through the FET is directly proportional to the *square* of the voltage on the gate, which means that is is a square law device rather than a linear device. With proper operating voltages, the output of a FET amplifier may be rich in second harmonics. Transistors and vacuum-tube amplifiers have a third and higher order harmonics in addition to the second harmonic, and this results in cross-modulation distortion of the output signal. Thus, the FET has an advantage over both the tubes and the transistors when it is used as a mixer

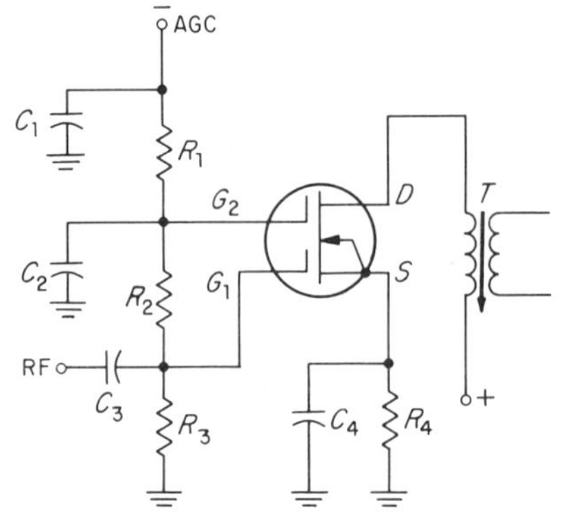

Fig. 5-30. A MOSFET RF dual-gate amplifier.

A Typical FET-RF Amplifier Figure 5-30 shows a simplified circuit with a Dual-Gate FET used as an RF amplifier. An AGC voltage is developed across a voltage divider comprised of R_1, R_2, and R_3. The capacitors C_1 and C_2 filter the AGC voltage so that only a pure dc operating voltage is applied to the gates. The largest part of the AGC voltage is applied to Gate No. 2 (G_2).

The RF signal is applied through C_3 and developed across R_3. This signal voltage appears on the first gate (G_1). Thus, the amount of current through the FET is controlled by two voltages: the AGC voltage and the signal voltage.

The source bias is supplied by R_4, and C_4 is used to prevent degeneration. The output signal is taken from the secondary of the transformer *T* in the drain circuit.

5.8 TELEVISION BOOSTERS

Owners of television receivers located in areas where the signal strength is weak often attempt to improve the quality of the pictures by adding an external booster to their sets. Boosters are basically nothing more than RF amplifiers, and when one is attached to a set, it means, in effect, adding one or more RF amplifiers to that already existing in the receiver.

The purpose in adding a booster is to strengthen the incoming signal to such an extent that it will produce a picture possessing the full contrast range, and at the same time, improve the signal-to-noise ratio so that the picture will be clear and free of annoying noise spots.

Of these two objectives, the improvement of the signal-to-noise ratio is the more difficult to attain, but it is the more important.

A booster capable of high gain but incapable of providing a good signal-to-noise ratio will give a picture filled with disturbing noise spots. A booster possessing a minimum of internal noise but capable of little gain will not amplify the signal sufficiently to permit it to override the noise of the set. Again, the picture will be covered with noise spots. The booster must have both attributes or it might as well have none.

Before we leave this subject of noise, it should be pointed out that nothing has been said about noise generated outside of the set or the booster. This noise, if present, comes down the transmission line with the signal and is indistinguishable from the signal so far as the booster is concerned. To overcome this noise, it must be attacked at its source, or if this is not feasible, try to keep as little of it as possible from reaching the signal via the antenna or the lead-in line. Standard methods of attack include increasing the antenna height, antenna replacement, and the use of a shielded lead-in line. It has also been found helpful to position the booster at the antenna (or at least as close to the antenna as possible), where the booster will strengthen the signal before it has been subjected to the noise and enables the signal with its amplified strength, to better overcome the adverse effects of the noise. In this way we can also attempt to improve the signal-to-noise ratio.

5.9 ANTENNA PREAMPLIFIERS

There are certain disadvantages to the use of a booster. One of the most important annoyances is that it introduces an additional unit to be connected to a TV set. Another complaint about boosters has been their rather unsightly appearance on top of an expensive television console.

Raising the antenna, using an antenna with a higher gain, and using two or more antennas in combination may, in some cases, eliminate the need for a booster, but not in all cases. Another solution is the use of an *antenna preamplifier*. This is simply a broadband transistor RF amplifier which is mounted on the antenna or antenna mast (near the antenna terminals). Figure 5-31 shows such a preamplifier. The higher reliability of transistors in comparison with tubes, and their lower operating voltages make the mast-mounted preamplifier possible.

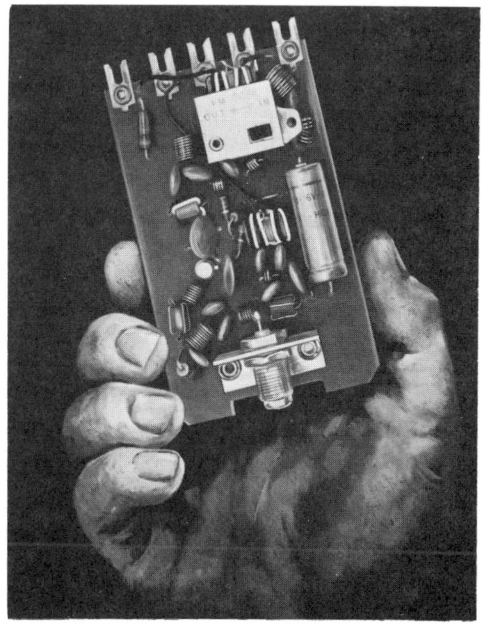

Fig. 5-31. An antenna preamplifier. (*Courtesy of Winegard Company.*)

Either a grounded-base or grounded-emitter circuit configuration may be used for the antenna preamplifier. The grounded-emitter circuit will give a somewhat higher gain, but replacement transistors must be carefully selected to avoid oscillation due to a change in the transistor parameter. The grounded-base configuration is often used because of greater stability, although it does have a lower gain. The grounded-base is also more stable under the widely varied temperatures that can be expected with a mast-mounted amplifier.

The amplifier is untuned, being capable of amplifying any frequency in

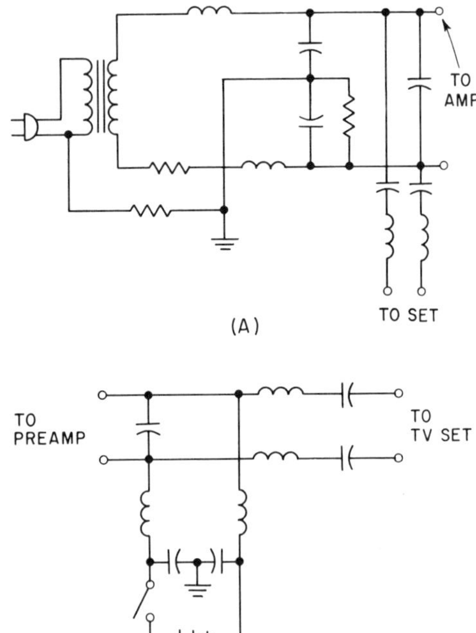

Fig. 5-32. Examples of power supplies used for operating preamplifiers. Power is obtained through the transmission line.

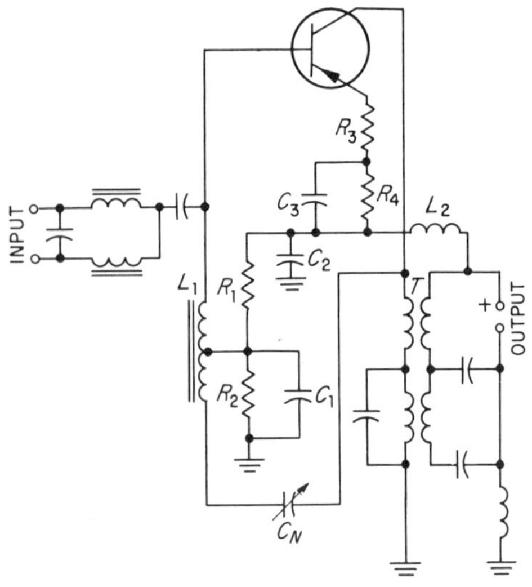

Fig. 5-33. Circuit of an antenna preamplifier.

the television and FM-broadcast range. Negative feedback is usually used for broadening the response and increasing the stability.

The small amount of power required for operating a transistor amplifier makes it possible to feed the required power through the transmission line that couples the antenna to the receiver. In this type of operation, an ac power supply can be located in the receiver cabinet, and the preamplifier may be energized whenever the receiver is being used. Figure 5-32(A) shows how the ac line is coupled to the antenna transmission line. A rectifier, located in the preamp at the antenna, converts the ac to dc for operating the transistor. In other versions the rectifier is located in the ac supply.

Batteries are also used for operating preamplifiers. If the battery supply is mounted on the mast, the replacement of the batteries would be inconvenient. In such cases, the supply may be mounted in the house or at some point that can be reached conveniently. Figure 5-32(B) shows an example of a battery supply for use with a preamplifier.

Figure 5-33 pictures a typical preamplifier circuit. The transistor is operated in the grounded-emitter configuration. The collector is at dc ground potential, while a positive voltage is supplied to the emitter and the base. The voltage divider R_1 and R_2 establishes the bias, and C_1, C_2, and L_2 filter the transistor-operating voltages. Neutralization is provided by C_N.

If it becomes necessary to service an antenna preamplifier, the power supply should be checked first. If the unit obtains its power from the ac line, the power supply should be disabled before touching it on the mast. It is not a good idea to try to service these units on the mast because a shortened power-supply transformer could result in a deadly voltage being present at the antenna.

REVIEW QUESTIONS

1. Should the tuned circuit of TV RF amplifier have a high Q or a low Q? Explain your answer.
2. In order to achieve a wide passband, should an RF or an IF transformer be tightly coupled or loosely coupled? Explain your answer.
3. Why is a loading resistor sometimes placed across a tuned circuit of an RF amplifier?
4. How is a low value of interelectrode capacitance achieved in a Nuvistor tube?
5. How is it possible to eliminate the tuning capacitor in some high-frequency tuned circuits? (See Fig. 5-9.)
6. For an RF amplifier, would you prefer a tube with a high input capacitance or one with a low input capacitance? Explain your choice.
7. For an RF amplifier, would it be better to have a tube with a high value of transconductance or a tube with a low value of transconductance? Why?
8. What is an advantage of the pentode over a triode for use as an RF amplifier?
9. What is a disadvantage of a pentode in comparison with a triode for use as an RF amplifier?

10. What is the purpose of neutralization as used in RF amplifiers?
11. Why is it usually unnecessary to neutralize a grounded-grid triode amplifier or a grounded-base amplifier?
12. What is the advantage of a cascode RF amplifier?
13. In a transistor RF amplifier, how is undesired coupling accomplished between the output stage and the input?
14. In a FET, what is the amount of drain current when the pinch-off voltage is applied to the gate?
15. What is a MOSFET and what is its advantage over an ordinary FET?
16. What type of tube is equivalent to a dual-gate FET?
17. What is an advantage of an antenna preamplifier over a booster?
18. What is the *figure of merit* of a tube?

6 TV Receiver Tuners

6.1 INTRODUCTION

All receiving systems have at least four basic sections: an antenna system for converting electromagnetic signals to electrical impulses; circuitry for selecting one station and rejecting all others; a detector for converting the RF (or IF) signal into signals that can operate a transducer; and a transducer to convert the electric signals into energy of some other form. Examples of transducers are loudspeakers that convert electric energy to sound energy and picture tubes that convert electric energy to light energy.

The tuner—sometimes called the *front end*—performs the job of selecting the desired station and rejecting all others. In addition, it performs the following:

1. It terminates the transmission line (between the antenna and the receiver) in the proper impedance, and presents a proper impedance match to the receiver IF system.
2. It amplifies the RF signal to present an acceptable signal-to-noise ratio.
3. It isolates the local oscillator signal from the antenna, thus preventing the oscillator signal from radiating and interfering with other receivers.
4. It converts the RF signal to an IF signal as required in a superheterodyne receiver. All television receivers employ the superheterodyne principle.

6.2 TYPES OF TV TUNERS FOR VHF AND UHF RECEPTION

Television tuners can be classified in more than one way. There are *VHF* and *UHF tuners, vacuum-tube* and *solid-state tuners,* and those with both tubes and semiconductors which are called *hybrid tuners.* In some of the earlier receivers, multichannel VHF tuning was achieved by the use of a ganged set of inductors wound spirally and with moving contacts. These were called *continuous tuners.* Tuning of VHF stations is now almost always accomplished by incremental steps for tuning from station to station. These types are sometimes called *step tuners,* or *incremental* tuners.

A vacuum-tube type of incremental VHF tuner is illustrated in Fig. 6-1(A). This tuner employs frame grid tubes and printed circuit wafer sections. This tuner is designed specifically for color sets. It also features preset oscillator frequencies. The tuner shown in Fig. 6-1(B) is a

A

B

C

Fig. 6-1. Three examples of VHF tuners. (A) A vacuum-tube type, using frame grid tubes and printed circuit wafer sections. (B) A transistor type, using printed circuit wafer sections. (C) Exploded view of transistor VHF tuner. ((*A*) *and* (*B*) *courtesy of Oak Manufacturing Company.*) (*C*) *Courtesy Zenith Radio Corp.*)

VHF tuner, which is completely transistorized. It also utilizes printed circuit wafer sections.

The tuners incorporating the incremental inductors are also termed *wafer type* and accomplish the change of channel by a progressive shorting of sections of the total inductance. Connections for the various sections are brought out at wafer contacts. In such layouts it is important to keep the length of wiring leads to the barest minimum, not only to avoid bringing in undetermined amounts of inductance, but also to avoid unwanted coupling effects to neighboring circuit elements. The

A

latter condition takes on an added importance when one recalls that modern tuners have a high *component density* compared to their predecessors of the late forties. Component density refers to the compactness of electronic circuits. Figure 6-1 (C) is an exploded view of the important details of another transistorized VHF tuner.

The manner in which a typical tuner wafer is constructed is shown in Fig. 6-2(A). Note that for Channels 2–6 individual coils are connected in a series and are progressively shorted out as the channel frequency increases. For Channels 7–13, a stamped inductance ring is utilized and sections of this are shorted out for higher channel frequencies. Oscillator wafers are generally tuned by individual screw adjustments which are accessible through the front panel of the set. There is a separate adjustment for each channel, and these generally function by varying the effective inductance utilized for each channel. The RF and mixer wafers are not tuned for each channel but have an overall tuning adjustment which is preset at the factory, but can be readjusted by the technician, if this should become necessary.

A newer type of tuner wafer construction utilizes a printed circuit wafer, which is shown in Fig. 6-2(B). In this method, all of the inductances, except those used for Channels 6 and 13, have been replaced

B

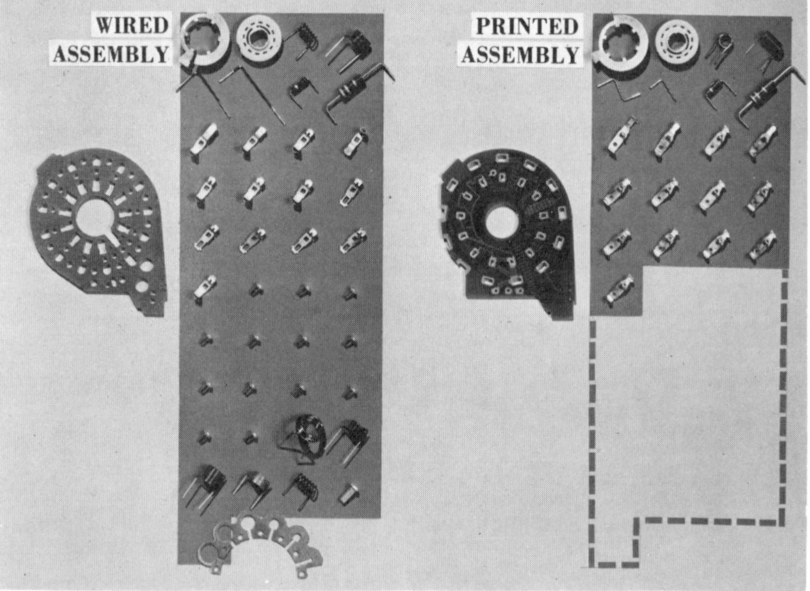

Fig. 6-2. Wafer construction for VHF TV tuners (A) Conventional wafer construction (B) Printed circuit wafer construction (C) A comparison between conventional and printed circuit wafer construction. (*Courtesy of Oak Manufacturing Co.*)

by printed circuit inductances. This method of construction has several advantages over that of Fig. 6-2(A). These are greater reliability, greater uniformity of alignment, lower cost, and greater versatility with regard to accommodating future designs. A comparison between the parts used in the wired assembly of Fig. 6-2(A) and the printed circuit version of part B is illustrated in the photo of Fig. 6-2(C). Note the simplicity of the printed circuit design. Tuning of the printed circuit wafer is accomplished by the use of fixed capacitors and by varying the inductances (coils) used for Channels 6 and 13. The complete tuner shown in Fig. 6-1(A) is constructed with printed circuit wafers.

Turret VHF tuners incorporate a bank of precisely wound coils with tight tolerances. The coils are switched into or out of the tuner circuit as the channel selector knob is turned from channel to channel. Both wafer switching and turret tuning have found wide acceptance and have shown remarkable performance. The *resetability* (that is, the ability of a tuner to repeatedly retune to exactly the same frequency after being switched) has been demonstrated to be better for turret tuners than for their wafer counterparts, and this is the reason why color tuners often incorporate their design. Fig. 6-3 illustrates a turret tuner.

All TV sets are required by FCC regulations to incorporate both VHF

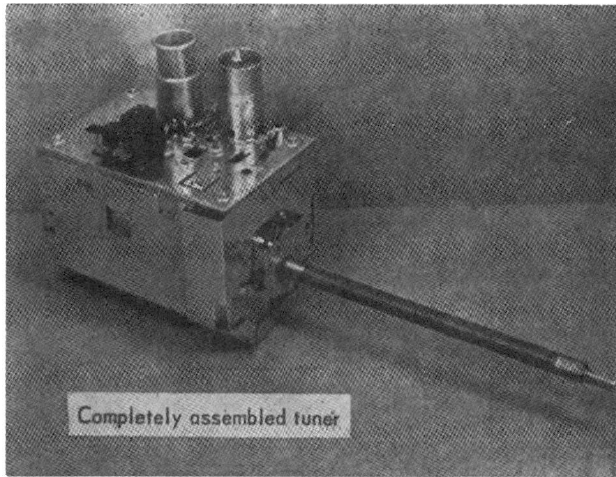

Fig. 6-3. A turret tuner. A strip holds all of the tuning coils for one channel.

and UHF tuners. The UHF tuner is separate from the VHF unit and is tuned by a separate control knob. While the VHF tuners are generally of the incremental type, most UHF tuners have been of the continuous-tuning type. However, it has been found that many viewers find it difficult to tune in UHF stations with a continuous tuner, and some designs permit selection of UHF stations either by switching or by push buttons. To accomplish this, special semiconductor devices are used in some tuners. These are called *varacters*, and they operate by providing a varying capacitance for tuning with a change in their operating voltage. The principles of operation of varactors and their operation in tuners are covered in detail later in this chapter. (Chapter 7 discusses the use of varactors for Automatic Fine-Tuning circuits.) The details of the theory and operation of UHF tuners are also covered later in this chapter.

6.3 ELECTRICAL CHARACTERISTICS OF TUNERS

In order to understand the electrical functions of the individual components of a television tuner, it is essential that we divide it into convenient and physically realizable sub-units. These subsections of a VHF tuner are shown in the functional block diagram of Fig. 6-4. It

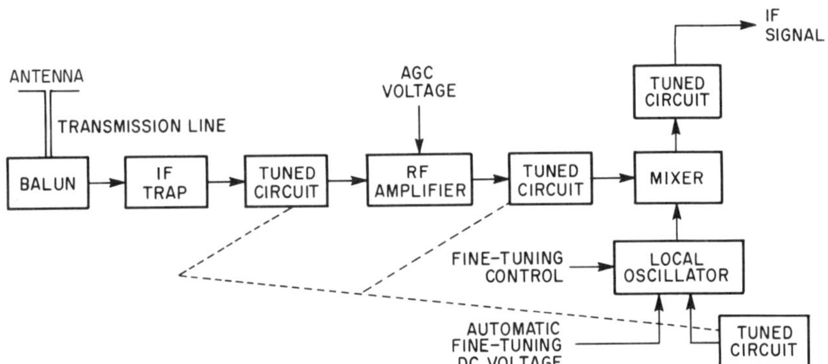

Fig. 6-4. Block diagram of a VHF tuner. The tuned circuits connected with a dashed line are varied simultaneously with the channel selector control.

shows a balun and an IF trap at the input. The former serves to match the 300-ohm transmission line (commonly used to bring the antenna signal to the receiver) to the 75-ohm input impedance level. Input impedances of most RF amplifiers are frequency-dependent, but a 75-ohm impedance to the receiver) to the 75-ohm impedance level. Input impedances of most RF amplifiers are frequency-dependent, but a 75-ohm impedance is a representative figure for the mid-VHF band. The IF trap prevents the injection of IF frequencies, generated by neighboring receivers, into the mainstream of signal flow. The balun and the IF trap are normally placed side by side on a separate mount and are externally located to the encased tuning mechanism of the tuner. This arrangement allows for the easy accessibility of the traps for alignment purposes. Often the mount also carries a pair of capacitors, one in each lead, to prevent any

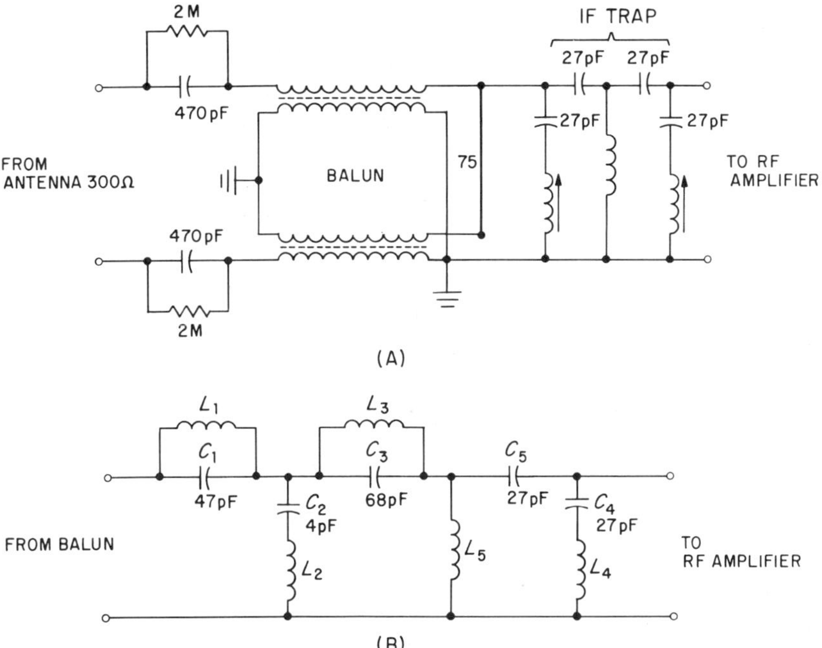

Fig. 6-5. Typical input circuitry for VHF tuners.

damage to the receiver by lightning, etc. Figure 6-5 shows the input circuitry of two VHF tuners.

The balun is a matching transformer with a ferromagnetic core upon which four tightly-coupled and evenly-spaced windings have been wound. The windings can be connected so as to provide an impedance transformation from a 300-ohm balanced source to either a 75- or a 300-ohm unbalanced load. Since the resistive component of the input impedance of most RF devices at VHF is close to 75 ohms, the balun is normally used to provide a 75-ohm output transformation. Figure 6-5(A) shows this configuration. The ferromagnetic core helps to keep the impedance transformation uniform throughout the range of frequencies of interest.

One of the sources of interference outside the signal passband occurs in the IF band when the tuner is in use for one of the lowest channels. The lowest channel (Number 2) with the video and sound carriers at 55.25 MHz and 59.75 MHz respectively is so close to the IF band of many receivers (41–46 MHz) that the natural attenuation of the tuner signal circuits may not be sufficient to give adequate rejection to sound IF frequencies. The IF trap, which follows the balun, consists of a number of sections of series or parallel tuned subsections in order to prevent interfering signals in the IF range from penetrating the RF amplifier and reaching the mixer. If the undesired signals reach the mixer or IF system, it is more difficult to reject them. Some of the inductors that make up the traps are made accessible for the purpose of achieving a proper IF response at the input of the IF amplifier during alignment. Two versions of the trap are shown in Fig. 6-5. One acts like a high-pass filter for frequencies beyond about 40 MHz and is pictured in part A. In part B additional tuned networks (L_1 and C_1, and L_2 and C_2) are built in to reject the FM broadcast signals which occupy the range 88–108 MHz.

Returning to the block diagram of Fig. 6-4, the RF amplifier boosts the incoming VHF signal before it is mixed with the local oscillator to produce an IF signal. The design of such an amplifier is one of the key factors in the overall performance of the tuner, since it is meant to handle the signals at their weakest level. Thus, in addition to magnifying the signal, the amplifier must not degrade the noise performance of the tuner as a whole. Also, the response should be as even as possible over the entire VHF range from Channels 2 through 13.

The input requirements for a UHF tuner are quite different from those of the VHF units. Figure 6-6 shows the UHF tuner in a block diagram form. The IF traps are almost never used, primarily because the lowest UHF channel (Number 14) frequencies are well above the highest IF channel frequency in use. Secondly, any passive network, such as a trap, will reduce the signal strength due to its insertion loss, thereby adding to the noise figure. The UHF tuner does not, in general, employ an RF amplifier because of its high cost and complicated design. The input circuit is thus led directly to the mixer through a balanced input transformer and a double-tuned, magnetically-coupled network.

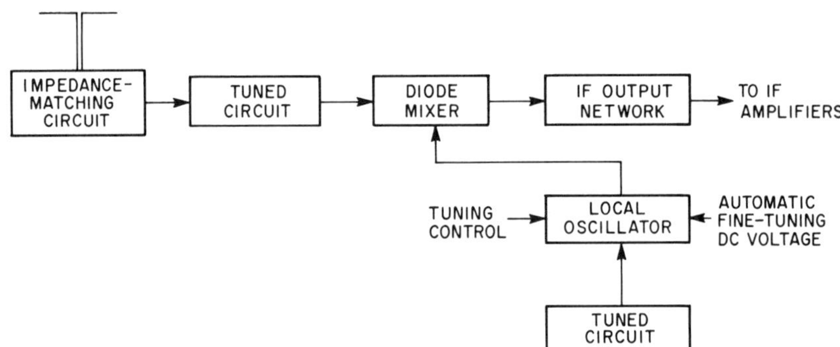

Fig. 6-6. Block diagram of a UHF tuner.

6.4 INTERFERENCE STABILITY AND NOISE PROBLEMS IN TUNERS

It has been pointed out that in addition to converting an RF signal into an IF signal, the television tuner must have an ability to provide immunity against undesired signals, and it must keep the overall noise level associated with sound and picture signals as low as possible.

Normally, the local oscillator frequency is above the incoming RF frequency by an amount equal to the IF frequency of the receiver. In other words, *the oscillator frequency minus the RF frequency equals the IF frequency*. As with all superheterodyne receivers, there is a problem with image frequencies. The image frequency is *above* the oscillator frequency by an amount equal to the IF frequency of the receiver. Thus, *the image frequency minus the oscillator frequency equals the IF frequency.* We see, then, that there are two different signals which can heterodyne (*i.e.*, mix) with the local oscillator frequency to produce frequencies in the IF range. The two signals are the *desired RF* from the station being tuned, and the *undesired image frequency* from a station having a carrier frequency above the local oscillator frequency by an amount equal to the IF frequency.

The combined selectivity of the RF input-coupling circuits and output circuits must be adequate to reduce the response of the tuner to image frequencies to a level that is below the visible and audible level. For IF frequencies of 41.25 MHz (for sound) and 45.75 MHz (for video), the corresponding image frequencies are removed from the desired signal inputs by 82.50 MHz and 91.50 MHz respectively. Since the television spectrum is quite wide, it is possible that the image frequencies represent another channel operating nearby. However, the frequency separation is large enough to ensure, without great difficulty, that the *images* are rejected to the extent of 60 db or better by the tuned circuits of the RF amplifier. In fact, one of the reasons for using higher IF frequencies is to achieve a better image rejection when using resonant circuits of moderate Q values.

Another source of interference is the IF band itself. The IF frequencies generated by a source external to the television receiver can find an easy passage through the tuner to the IF amplifier. This is especially true when the tuner is tuned to a lower channel. Special filter traps, tuned to the sound and video-IF frequencies, are placed at the input of the RF amplifier to reduce this interference. An overall IF attenuation figure of 60 db is considered quite adequate for acceptable tuner performance.

Electromagnetic radiation from the local oscillator of one television receiver can be the source of considerable interference at a neighboring receiver. At the early stages of television receiver development this fact was not fully understood, and after a time the Federal Communications Commission set limits on the field strengths of the allowable radiations both for VHF and UHF tuners. The RF amplifier used invariably in a VHF tuner acts to prevent these radiations from traveling to the antenna. The design of circuits used to prevent excessive radiations from UHF tuners is more challenging because they normally do not have RF amplifiers. Prevention of oscillator radiation in UHF tuners is achieved by a judicious placement of coupling coils and windows. Proper cable and case shielding can be a helpful deterrent to the radiation of interference by the local oscillator.

The need for a stable oscillator cannot be overemphasized. Any frequency changes that result from the inability of the step-tuning mechanism to resonate the circuits to exact channel frequencies can be taken care of by the fine-tuning adjustment. What is usually more of a problem is the oscillator drift which takes place during temperature stabilization of its environment. As an example of oscillator drift, take the case of a tuner set for Channel 13. The local oscillator could drift by as much as 1 megaHertz as a result of a temperature change within the tuner at 25°C. This can offset the sound and video performance of the succeeding circuitry, and result in a complete loss of the color signal. Fortunately, adequate design measures are available to compensate for such drift in both vacuum-tube and solid-state tuners. For instance, automatic fine tuning (AFT) circuits are used to lock the local oscillator onto the desired frequency regardless of changes in temperature. (These circuits are sometimes called automatic frequency control (AFC), but there is also an AFC circuit in the horizontal sweep circuit and AFT is preferred so that there will be no confusion of terms.)

Another factor which may cause oscillator drift is a change in the dc supply voltage resulting from poor regulation. A transistor oscillator is particularly sensitive to such effects. For example, when the dc voltage changes from seven to ten volts, a change of three volts, the oscillator frequency changes from 250 to 251 megaHertz. Many receivers now have regulated power supplies which deliver a constant dc voltage despite changes in load and changes in the line voltage input. This eliminates the problem of oscillator drift that is caused by voltage changes.

Receiver noise may originate from external sources such as poorly maintained electrical machinery in the receiver's vicinity, or from celestial bodies. Apart from these man made and natural causes, the noise may be produced in the circuit elements themselves (see Fig. 6-9). For instance, *thermal noise* is generated in the resistors due to temperature-dependent agitation of their atomic components. Also, *shot noise*, which is a serious problem in the vacuum tubes, is caused by the random fluctuations of the plate current. Thermal noise and shot noise are not frequency dependent. At frequencies where the corresponding wavelength is of the same order of magnitude as the dimensions of the electron transit, additional noise is produced. This suggests the need for miniature tubes with short transit times for RF amplifiers.

Since the tuner handles the RF signal at its lowest amplitude, great thought must be given to the design of the input stage of the RF amplifier (if one is used) so that the signal-to-noise ratio is optimum. A poor signal-to-noise ratio produces "snow" on the picture tube screen.

Having introduced the various elements that describe the performance of a television tuner, we can make a list of the "specifications" for one. The overall response for a receiver employing an intercarrier sound system is sketched in Fig. 6-7. The shape of the response curve depends a great deal on the load presented to the tuner, and the television receiver manufacturer almost always specifies the IF amplifier input conditions which the tuner is required to match.

The IF frequency ranges in Table 6-1 and elsewhere have been given as 41.25–45.75 MHz. Most of the recently manufactured receivers have IF frequencies in this range, but there are some older sets in operation that have lower IF ranges—usually in the 20–30 MHz range. It is necessary to consult the manufacturer's literature to determine the receiver's IF values, especially for the older receivers.

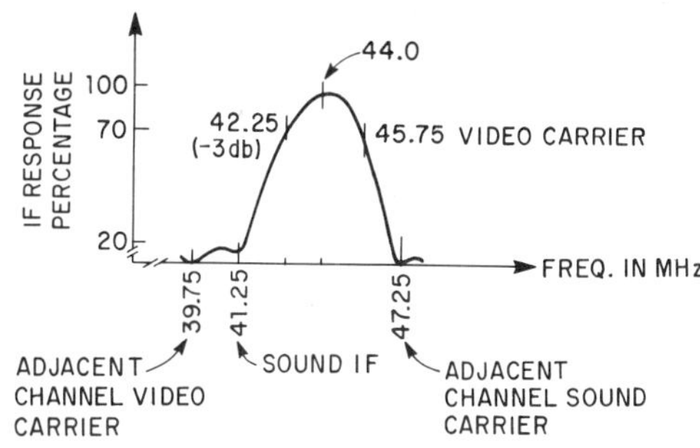

Fig. 6-7. Typical monochrome response curve for a tuner plus IF amplifier (intercarrier system).

TABLE 6-1 General Specifications for a Television Tuner

Video IF	45.75 MHz
Sound IF	41.25 MHz
Relative response	See Fig. 6-7. Depends on load conditions
Image rejection	VHF—Better than 60 db minimum UHF—Better than 70 db minimum
Adjacent channel rejection	Better than 60 db (min.)
Noise figure	VHF Tuners—5 to 8 db (typical) UHF Tuners—7 db (typical minimum) 11.5 db (typical maximum)
RF amplifier gain	Ch. 2– 6—28 db minimum Ch. 7–13—22 db minimum
Oscillator frequency resetability	Better than ±400 kHz for black and white units Better than ±100 kHz for color units
Oscillator frequency drift	For a temperature change of 22°C from ambient: 125 kHz maximum For a 10% change in the dc supply voltage: 50 kHz maximum
Fine-tuning range	±0.6 MHz (Average)

6.5 CHANNEL ALLOCATIONS

Table 6-2 shows the frequency allotments for the various television channels in the United States of America. The entire spectrum allocated for television transmission is divided into 12 Channels in the VHF (Very High Frequency) region and 70 channels in the UHF (Ultra High Frequency) region. The lower and upper limits of each channel are six MegaHertz apart, and this value is commonly referred to as the *bandwidth of the channel*. This should not be confused with the formal definition of the term *bandwidth*, which is the frequency region with a voltage (or current) response equal to or greater than 0.707 times the peak value.

Table 6-2 also lists the picture and sound carrier frequencies specified for all tabulated channels, the picture image frequencies (osc. freq. +pix IF freq.), and the location of spurious responses. While analyzing the response of a tuner, it is best to look for outputs which correspond

TABLE 6-2 U.S. Television Channel Frequencies (MHz)

Sound IF 41.25 Picture IF 45.75

	Channel No.	Freq. Limits	Center Freq.	Picture Carrier	Sound Carrier	Osc. Freq.	Picture Image Freq.	2 × Osc. + Pix IF	2 × Osc. − Pix IF	Channel No.	
VHF	2	54–60	57	55.25	59.75	101	146.75	247.75	156.25	2	VHF
	3	60–66	63	61.25	65.75	107	152.75	259.75	168.25	3	
	4	66–72	69	67.25	71.75	113	158.75	271.75	180.25	4	
	5	76–82	79	77.25	81.75	123	168.75	291.75	200.25	5	
	6	82–88	85	83.25	87.75	129	174.75	303.75	212.25	6	
	7	174–180	177	175.25	179.75	221	266.75	487.75	396.25	7	
	8	180–186	183	181.25	185.75	227	272.75	499.75	408.25	8	
	9	186–192	189	187.25	191.75	233	278.75	511.75	420.25	9	
	10	192–198	195	193.25	197.75	239	284.75	523.75	432.75	10	
	11	198–204	201	199.25	203.75	245	290.75	533.75	444.25	11	
	12	204–210	207	205.25	209.75	251	296.75	547.75	456.25	12	
	13	210–216	213	211.25	215.75	257	302.75	559.75	468.25	13	
UHF	14	470–476	473	471.25	475.75	517	562.75	1079.75	988.25	14	UHF
(Incomplete)	15	476–482	479	477.25	481.75	523	568.75	1091.75	1000.25	15	
	54	710–716	713	711.25	715.75	757	802.75	1559.75	1468.25	54	
	55	716–722	719	717.25	721.75	763	808.75	1571.75	1480.25	55	
	56	722–728	725	723.25	727.75	769	814.75	1583.75	1492.25	56	
	82	878–884	881	879.25	883.75	925	970.75	1895.75	1804.25	82	
	83	884–890	887	885.25	889.75	931	976.75	1907.75	1816.25	83	

Note: All frequencies are in megaHertz.

to input frequencies of $n \times$ osc. frequency $\pm$ pix IF, with $n = 1$, 2, or 3. Oscillator harmonics up to $n = 3$ can contribute substantially to the overall performance of the tuner, and such an analysis is the best means of determining whether or not the harmonics are offensive.

The television sound signal is transmitted by frequency modulation, and the video-signal amplitude modulates the channel picture carrier. The intermediate-Frequency for video (video IF) is 45.75 MHz, while the corresponding sound IF is 41.25 MHz. Thus, the IF pix and sound carrier spacing is 4.5 MHz. In the intercarrier sound system (monochrome receiver), the sound IF is frequently taken from the video detector. It is clear that the receiver IF section must pass the entire 4.5 MHz band in order to meet the requirements of this system. The reader is encouraged to complete the information for the missing UHF channels as an exercise. The UHF spectrum is continuous, unlike the VHF allocation which has two frequency gaps.

6.6 MIXER CIRCUIT OPERATION

The tuner has an RF amplifier, oscillator, and mixer circuit. In the last chapter the subject of RF amplifiers was covered. We will now cover the mixer circuit configurations.

As shown in the block diagram of Fig. 6-4, the mixer receives a signal from both the RF and the oscillator circuits and converts these to a difference frequency output, called the *intermediate frequency* (IF) signal. This conversion is achieved by applying both the RF and oscillator voltages to a nonlinear, vacuum-tube or transistor amplifier. If the amplifier is nonlinear, heterodyning will take place, and the output signal will consist of both the sum and difference frequencies. Assuming that the frequency difference between the RF and oscillator signals is appreciable, a selective network tuned to the lower-frequency component of the output will pass the desired IF only.

Vacuum-Tube Mixer Circuits In vacuum-tube tuners, energy from the oscillator may be either capacitively or inductively coupled to the mixer. Two frequently used methods are pictured in Fig. 6-8. Capacitive coupling is used in the circuit of part A, and inductive coupling is used in the circuit of part B. Interaction between the input signal and the oscillator outside the mixer tube is kept as low as possible in order to prevent any changes from occurring in the oscillator frequency, and to minimize oscillator radiations from appearing at the antenna.

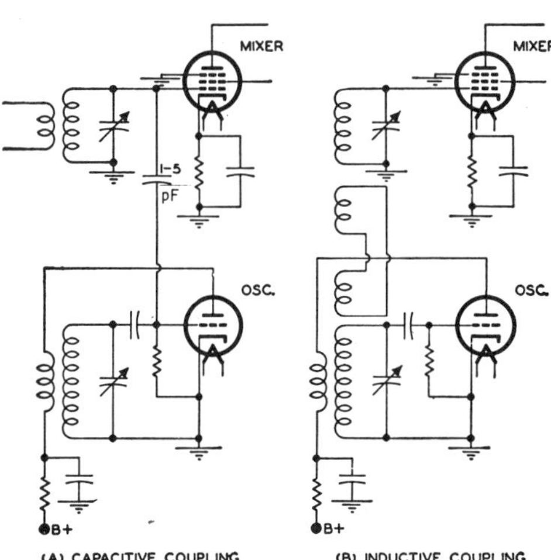

Fig. 6-8. Two types of circuits for coupling oscillator signals to vacuum-tube mixers.

In addition to the fact that there are FCC regulations covering radiated signals, such signals are troublesome because any considerable amount of radiated signal can produce a complete loss of contrast or even a negative picture in nearby television receivers. When the interfering frequency is close to the picture carrier of the station being received by the other sets, the *beat interference* produces vertical, horizontal, or slanted stripes across the screen. (Beat interference refers to the difference frequency signal obtained when the interfering signal and the picture carrier of the station being received mix with each other in the second detector stage of the receiver.)

In Fig. 6-8(B) an intermediate coil is employed to transfer the generated oscillator energy to the mixer. If it is possible to position the oscillator coil physically close to the mixer grid coil, the inductive transfer of energy may occur directly and we can dispense with the intermediate coil.

Within the vacuum-tube mixer of the television receiver, the received signal and the oscillator voltage both modulate the electron stream to form the desired audio- and video-IF voltages. In nearly all tube-type receivers, the signal from the RF amplifier is transformer- or impedance-coupled to the mixer. The oscillator voltage, as indicated, is transferred to the mixer tube either capacitively or inductively.

The mixing function for a UHF tuner is performed by a semiconductor, diode. This type of device requires less oscillator voltage. As an additional advantage, converter noise is reduced when diode mixers are used.

Transistor Mixer Circuits In a transistor mixer, the amplifier configuration, dc biasing and coupling to the RF oscillator, and the IF amplifier load should be considered. Adequate precautions must be

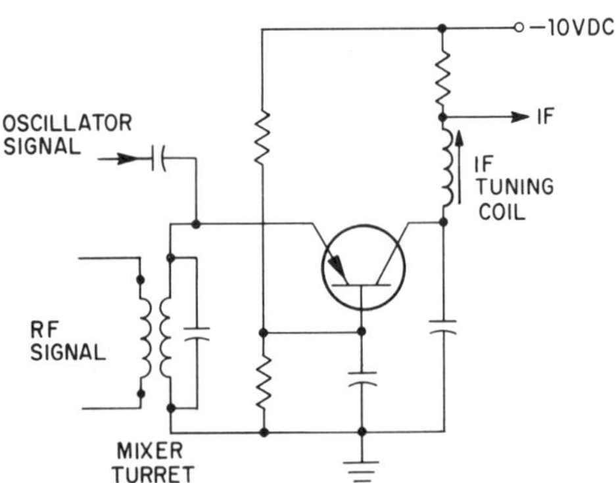

Fig. 6-9. A common-base mixer circuit with the RF and local oscillator signals injected at the emitter.

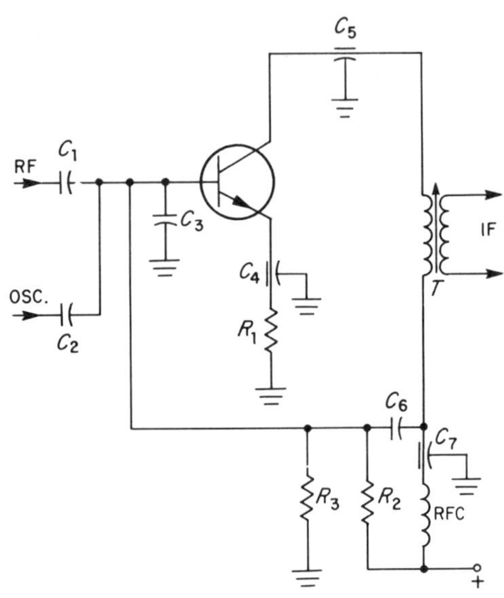

Fig. 6-10. A common-emitter mixer stage.

taken against the generation of IF frequencies from image frequencies and oscillator second harmonic frequencies. Also, since the mixer serves as an amplifier for IF frequencies, it must be stabilized in this range—either by neutralization or by the insertion of a series-resonant IF trap from input to ground.

There are many ways in which the RF and oscillator signals can be injected into a transistor mixer. The common-base configuration has found wide acceptance because of its stable operation, relative constancy of input impedance with input frequency, and immunity against cross-modulation. The choice of transistor type is governed by considerations of high gain at the IF frequency, base-emitter characteristics, and low base-emitter capacitance. The cutoff frequency need not exceed 50 MHz. The AGC voltage is usually applied to the RF amplifier. Thus, the transistor characteristics in this respect are not critical. A common-base mixer with oscillator injection at the emitter is shown in Fig. 6-9.

Common-emitter transistor amplifier circuits are also used for the mixer stage. The RF and oscillator signals may both be injected at the base, or both injected at the emitter. In some cases, one signal is injected at the base and the other at the emitter. In any case, heterodyning will take place provided, of course, that the stage is not linear.

Figure 6-10 shows a common-emitter mixer stage. The RF and oscillator signals are fed to the base via C_1 and C_2 respectively. Emitter stabilization is provided by R_1 and C_4. The base is biased by the voltage divider R_2 and R_3. Normally, this bias voltage is very low to assure that the transistor operates in the non-linear range. A filter comprised of C_7 and the RFC prevents the IF signal at the collector from entering the power supply.

A small IF signal voltage, developed across the RFC, is fed back to the emitter via C_6 for neutralization. Here you see an important difference between the common-base and common-emitter mixers, since neutralization is not required for the common-base configuration. This is especially true since the input and output circuits are tuned to different frequencies. In the common-emitter circuit, however, the stage is operating as a common-emitter IF amplifier, and the IF signal fed back from collector to base (via the collector-base junction capacitance) must be neutralized. The IF transformer (T) is used for two purposes: it transfers only the IF signal and rejects all other signals from the mixer; and, it matches the output impedance of the common-emitter mixer to the input of the first IF amplifier stage.

In Chapter 5 the cascode configuration for two triodes was discussed. It consists of a common-cathode circuit followed by a common-grid circuit. The plate of the common-cathode circuit is tied directly to the cathode of the common-grid circuit. Transistor amplifiers are also connected in a cascode configuration. Cascode transistor circuits have a common-emitter circuit followed by a common-base circuit, with direct coupling between the collector and emitter. Because of their high gain, good isolation between input and output, and low noise characteristics, transistor cascode amplifiers are used for both RF and mixer amplifiers in tuners.

Figure 6-11 shows a cascode mixer stage. Here the oscillator and RF signals are applied to the base of the common-emitter transistor (Q_2), and the IF signal is taken from the collector. When the tuner is switched to the UHF position, the IF signal from the UHF tuner is fed to Q_2 through the isolating diode D_1. Thus, the cascode amplifier provides an extra stage of IF amplification for the UHF stations. This partially offsets the fact that there is no RF amplifier in the UHF tuner.

The base of Q_1 is grounded by C_1, and the base bias for Q_1 and Q_2 is developed by the voltage divider R_1, R_2, and R_3. Emitter stabilization is obtained with R_4 and C_2. Note that neutralization is not required for this circuit.

The dual-gate MOSFET makes an ideal mixer amplifier. This type of transistor, which was discussed in the previous chapter, has two gates for input signals. One gate can be used for the RF input and the other for the oscillator input, or both inputs can be delivered to one gate, and the other used as a screen to isolate the input and output stages of the mixer.

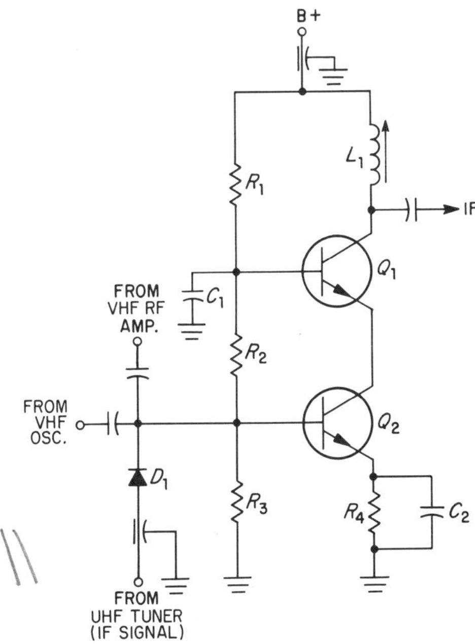

Fig. 6-11. A cascode mixer circuit. On UHF operation, VHF RF amplifier and VHF oscillator are disabled.

6.7 LOCAL OSCILLATOR CIRCUITS

The local oscillator provides a signal that can be heterodyned in the mixer with the RF signal. As in the case of radio superheterodyne receivers, the local oscillator frequency is normally *above* the incoming RF frequency by a frequency that is equal to the receiver IF frequency.

An important characteristic of the local oscillator used in tuners is its freedom from *drift*. The term drift, as applied to oscillator circuits, means any undesired change in oscillator frequency due to a temperature change, the aging of the circuit components, a change in the dc operating voltage, or a change in the oscillator load, etc. In monochrome receivers a small amount of oscillator drift can be tolerated. However, in color television receivers the color portion of the signal can be completely lost if the oscillator drifts too much. For this reason, automatic frequency controls (AFC) circuits may be used to maintain the oscillator frequency at a fixed value.

The local oscillator frequency is changed whenever the receiver is tuned from station to station. The *fine-tuning control* of the receiver allows the oscillator frequency to be varied over a narrow range. The setting of the fine-tuning control is very important if a satisfactory color picture is to be obtained. For this reason, manufacturers often include some kind of fine-tuning indicator (FTI) circuit to simplify the adjustment of the fine tuning control. The AFT circuit is also an aid to the proper adjustment of the fine-tuning control, since it will lock the oscillator onto the correct frequency when the control is anywhere within range of the AFT circuit limits.

Vacuum-Tube Local Oscillator Circuits Figure 6-12 shows what is perhaps the most frequently employed oscillator circuit in present-day television receivers. The circuit pictured in part A is known as the *ultraaudion*. Part B shows the ultraudion oscillator circuit as it is often drawn on schematic drawings. It is equivalent in its action to the well-known Colpitts circuit (see Fig. 6-12(C)). In the ultraudion, the voltage division across the tank circuit is accomplished through the grid-to-cathode (C_{gk}) and the plate-to-cathode (C_{pk}) capacitances within the tube. The feedback voltage which sustains the oscillations is developed across C_{gk}.

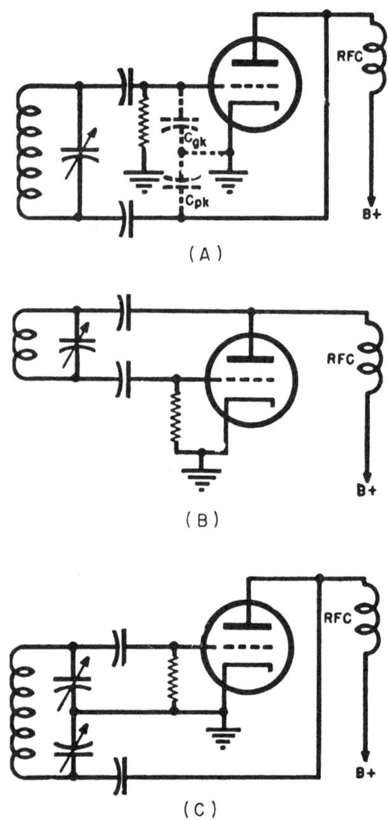

Fig. 6-12. The ultraudion oscillator.

The operation of the ultraudion oscillator can be understood from the circuit as it is used in television receivers. Figure 6-13 illustrates this circuit. As shown in part A, the inductance of the resonant circuit is changed as the receiver is tuned from station to station. In part B the equivalent circuit is pictured. The voltage-dividing capacitance network consists of the effective capacitance of C_{gk} in series with the parallel combination of C_{gk} and C_1, while C_t represents the combination of the grid-plate capacitance, the distributed capacitance, and C_2. Capacitor C_2 is a temperature-compensating capacitor and helps reduce oscillator drift. In spite of this, drift does occur, and C_1 is provided to permit the user of the set to adjust the oscillator frequency to the best sound output. Because C_1 is actually a vernier adjustment, it is labeled *fine-tuning control* and placed on the front panel. Any shift in oscillator frequency immediately alters the IF produced as a result of the mixing action. The effect is the same as detuning the receiver. By means of the fine-tuning control, the oscillator frequency can be readjusted to its proper value.

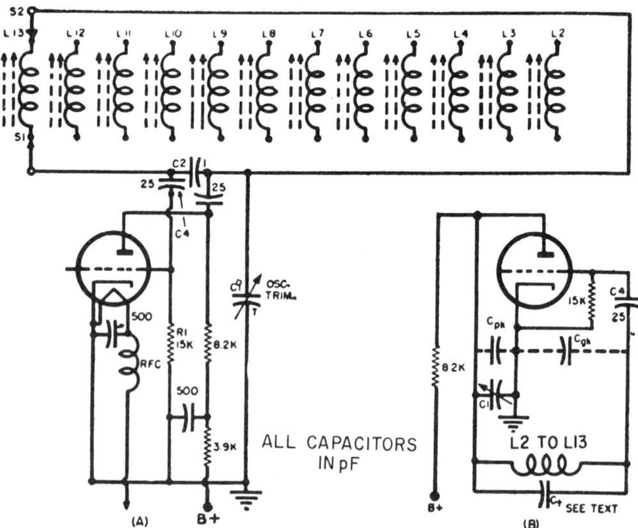

Fig. 6.13. A typical ultraudion oscillator.

The capacitors C_3 and C_4 keep the dc plate voltage off the exposed coils. Capacitor C_4 also makes it possible for the oscillator to develop grid-leak bias across R_1.

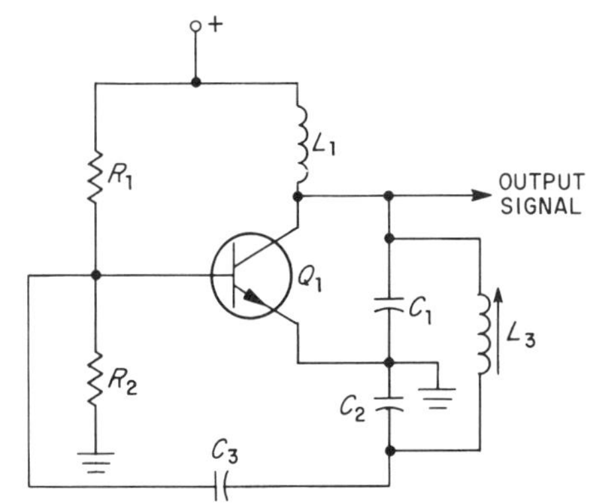

Fig. 6-14. The transistor-Colpitts oscillator is frequently used as a tuner local oscillator circuit.

Transistor Local Oscillator Circuit Figure 6-14 shows a Colpitts oscillator circuit employing a transistor amplifier. This circuit, including variations, is usually employed as a local oscillator in transistor television tuners. The transistor amplifier is Q_1. Its base is biased for conduction by a voltage divider comprised of R_1 and R_2. The junction of these resistors places a positive voltage on the base of the NPN transistor as required for conduction. Inductor L_1 is a filter which permits the dc power supply voltage to be applied to the collector of Q_1, but prevents the oscillator signal from entering the power supply circuitry.

The frequency-determining network in the circuit of Fig. 6-14 is comprised of C_1, C_2, and L_3. Often, one of these capacitors is made variable and serves as the fine-tuning control of the receiver. In another version, the inductance of L_3 is varied for fine-tuning control. Note that the capacitor C_1 is connected directly from the collector to the emitter. In most transistor oscillator circuits, this collector-to-emitter capacitor is present in one form or another. Capacitor C_3 is necessary to prevent a dc short circuit path from the collector to the base through coil L_3. It also serves to deliver the regenerative feedback signal required from the collector to the base circuit.

One important variation of the circuit of Fig. 6-14 employs the junction capacitance within the transistor for feedback, and therefore the external feedback capacitance network may not be immediately discernible. In another version, to be shown later, a common-base amplifier is used in a transistor tuner.

6.8 VACUUM-TUBE TUNERS

A schematic diagram of a pentode tuner is illustrated in Fig. 6-15. The input circuit is balanced with an impedance of 300 ohms. It is purposely designed to match the 300-ohm twin-lead line. However, by using a one-end terminal and ground, the input impedance becomes 75 ohms, and a shielded coaxial cable can be connected to the receiver without mismatch.

The secondary winding of the input circuit (L_2) is tuned by the input capacitance of the RF-amplifier tube in series with the parallel combination of C_1 and C_2. The trimmer C_2 is used for alignment. The 3900-ohm resistor R_8 across L_2 is inserted for the purpose of broadening the response of the input-tuned circuit to the necessary bandwidth.

The plate load of the RF amplifier is L_3 in conjunction with C_3 plus whatever tube and stray capacitance may be present here. The resistor R_6 is again a loading resistor designed to widen the band-pass characteristics of the tuned circuits. The gain of the stage is controlled by an AGC voltage fed to the control grid through a 47,000-ohm resistor.

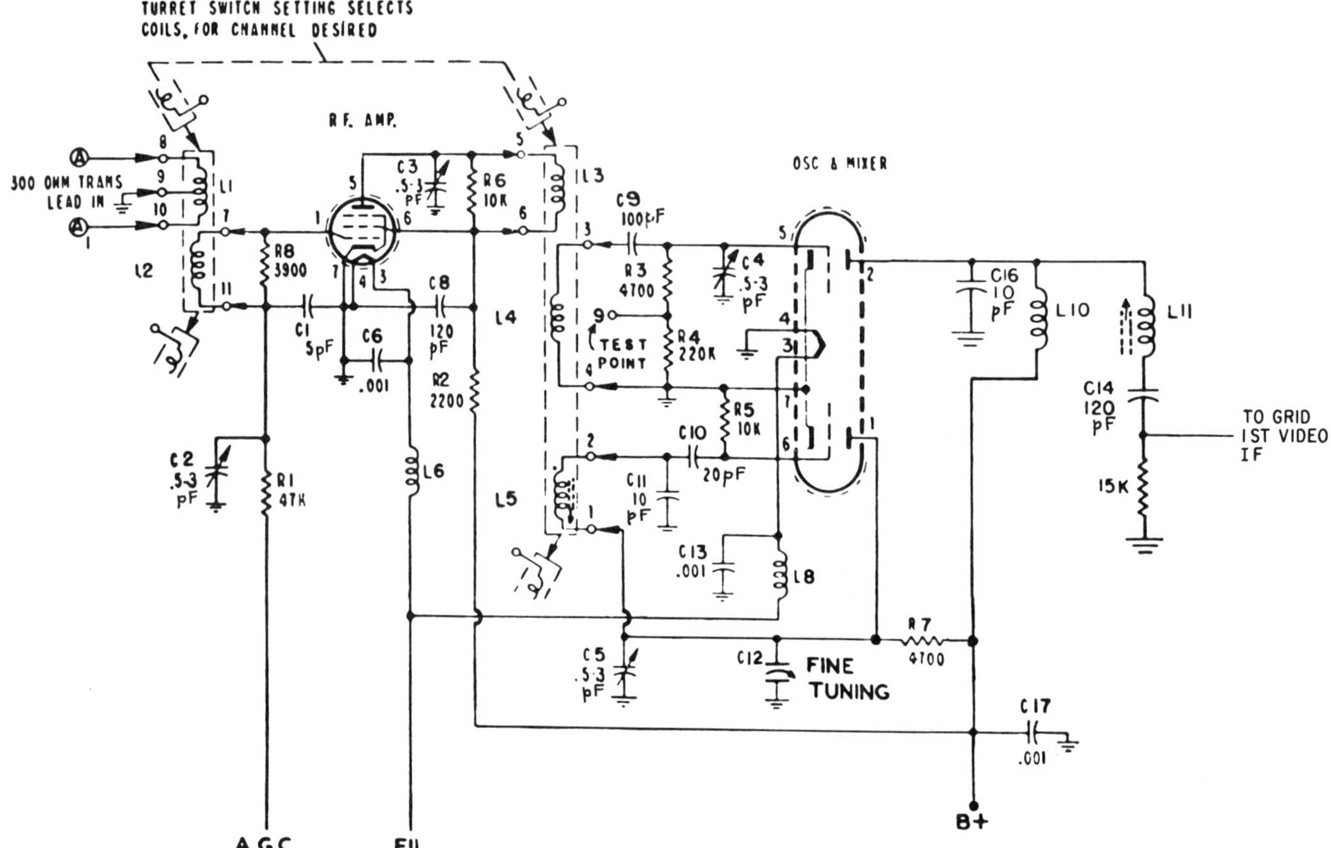

Fig. 6-15. Schematic diagram of a turret tuner using a pentode in the RF amplifier.

The signal is transferred from the plate of the RF amplifier to the mixer grid by inductive coupling between L_3 and L_4. The combination of R_3 and R_4 across L_4 is not for loading, but rather to provide a terminal (test point nine) where an oscilloscope can be attached to observe the response pattern of the RF-amplifier tuned circuits. Also, test point nine can be used as an injection point for video-IF-test signals. In the tuner, this test terminal projects above the top deck where it may readily be reached.

The resistors R_3 and R_4, together with C_9, develop a grid-leak bias for the mixer stage. The trimmer C_4 is used for alignment. The output of the mixer stage is coupled to the first video-IF amplifier by means of the low-pass network composed of C_{16}, L_{10}, L_{11}, and C_{14}. Capacitor C_{14} is a dc blocking capacitor.

The oscillator coil L_5 is inductively coupled to the mixer-grid coil L_4 to enable the oscillator signal to reach the mixer circuit. The capacitor C_{11} is in series with the parallel combination of C_5 and C_{12} to form the split-capacitor of a Colpitts oscillator. The trimmer C_5 is an RF-oscillator adjustment, while C_{12}, a variable dielectric-type capacitor, functions as the fine-tuning control. The grid-leak bias for the oscillator is developed by R_5 and C_{10}. The oscillator plate is shunt-fed by means of R_7.

Tetrode Tuner A tuner employing a tetrode in the RF stage would be practically identical to the circuit shown in Fig. 6-15. As a matter of fact, since the pentode RF amplifier in Fig. 6-15 contains no external suppressor connection, the two circuits would be exactly the same with the possible exception of a difference in component values.

Neutrode* Tuner An example of a neutrode RF amplifier as it would be employed in a turret tuner is shown in Fig. 6-16. It is followed by a pentode mixer and a triode oscillator. The pentode and triode sections are in one envelope.

The operation of a neutrode RF amplifier has already been discussed, but the input circuit of Fig. 6-16 contains a number of features which

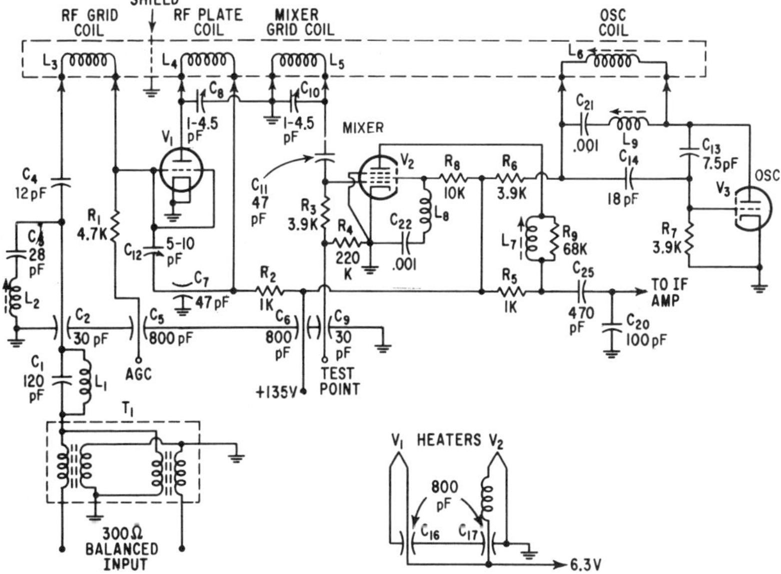

Fig. 6-16. An RF tuner using a neutrode RF amplifier.

have not been previously analyzed and which are quite common in present-day television receivers. For example, T_1 is a matching transformer designed to convert the 75-ohm input impedance of the RF-amplifier stage to 300 ohms in order to match the transmission line. It also provides a balanced input of 300 ohms, whereas the RF-amplifier input is unbalanced.

A parallel-resonant trap is formed by L_1 and C_1, while L_2 and C_3 form a series-resonant trap. Both traps are peaked for slightly different frequencies between 41 and 46 MHz. Their purpose is to prevent interfering signals in the IF range from penetrating the RF amplifier and

* Neutrode is a short term for *neutralized triode.*

reaching the mixer and, beyond it, the IF system where it is not possible to reject them. Inductor L_2 is adjustable and can be set for a specific frequency to be rejected. Inductor L_1 is factory adjusted and ordinarily is not touched in the field.

If we disregard the two foregoing traps, then the input-resonant circuit appears as shown in Fig. 6-17. It is essentially a low-pass, pi-type filter consisting of C_2, C_4, L_3, and C_{IN}, the latter representing the input capacitance of the tube. Since C_2 and C_4 are both much larger than C_{IN}, their impedance will be smaller than that of C_{IN}. Consequently, a greater amount of the signal voltage will appear across C_{IN}. This acts as a voltage step-up arrangement just as effectively as if we had employed a step-up transformer.

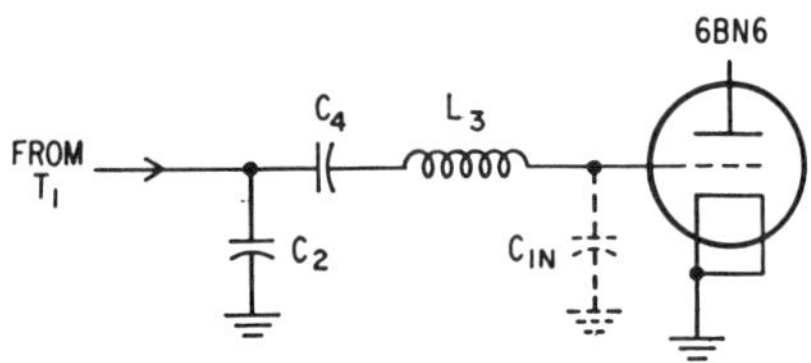

Fig. 6-17. Input-resonant circuit.

Inductor L_3 is changed for each channel, enabling the circuit to resonate to the required input frequency of that channel. Generally, the circuit is designed to peak at the mid-frequency of the specific channel.

The capacitor C_4, in addition to the function noted previously, also serves as a dc blocking capacitor. This is required because of the presence of a negative AGC voltage at the grid of the RF tube. This AGC voltage is brought in through R_1; C_5, in the AGC line, prevents any RF signal from reaching the AGC line beyond the tuner.

The output-tuning circuit of the RF amplifier consists of L_4, C_8, and whatever stray and output capacitance of V_1 might be present. Resistor R_2 is the plate-decoupling resistor, while C_7 functions as an RF-bypass capacitor. However, C_7 is small enough, so that it does not completely by-pass all of the RF voltage away from R_2. A small RF voltage does appear across R_2 and is fed back to the grid of V_1 by the neutralizing capacitor C_{12}.

The remaining two stages of the tuner, the mixer, and the oscillator, which have already been dealt with in our discussion of the preceding tuners, are conventional in design.

In Fig. 6-16 the capacitors C_2, C_5, C_6, C_7, C_9, C_{16}, and C_{17} are all somewhat different from the other fixed capacitors in the same diagram. The difference is that these are *feedthrough capacitors*. Such capacitors are constructed by taking a dielectric material and shaping it into the form of a small cylindrical tube. One metal plate is then deposited around the inner diameter of the cylinder, while the second plate is deposited over the outer surface. Wires are then attached to both ends of the inner plate. When the capacitor is inserted in a circular opening of one of the metal walls, or shields, of the tuner, the outer plate is automatically grounded. Filament and dc voltages can then be brought into the tuner by way of the inner plates of several feedthrough capacitors. If the capacitance values of these units are large enough, then the RF voltages traveling out along these same supply lines are bypassed to ground.

Feedthrough capacitors are employed extensively in TV tuners because of the convenient way in which they enable voltages to enter and leave the tuner housings. These capacitors are also useful between tuner sections separated by shield walls.

Nuvistor Tuners The advantages of a Nuvistor triode have been explained already in Chapter 5. Figure 6-18 shows the schematic diagram of a tuner that uses a Nuvistor for a RF amplifier.

RF amplifier V_1 is a conventional (that is, grounded-cathode) amplifier that employs a Nuvistor tube. The signal for the amplifier arrives at the tuner via the feedthrough capacitor C_{22}. Not shown in the illustration are an impedance-matching transformer and the filters for FM and IF signal frequencies. In addition to the signal input, the negative AGC voltage is also applied to the control grid of V_1. The AGC line is filtered by R_{1051} and C_{1050}, and a decoupling filter is comprised of C_{20} and R_3.

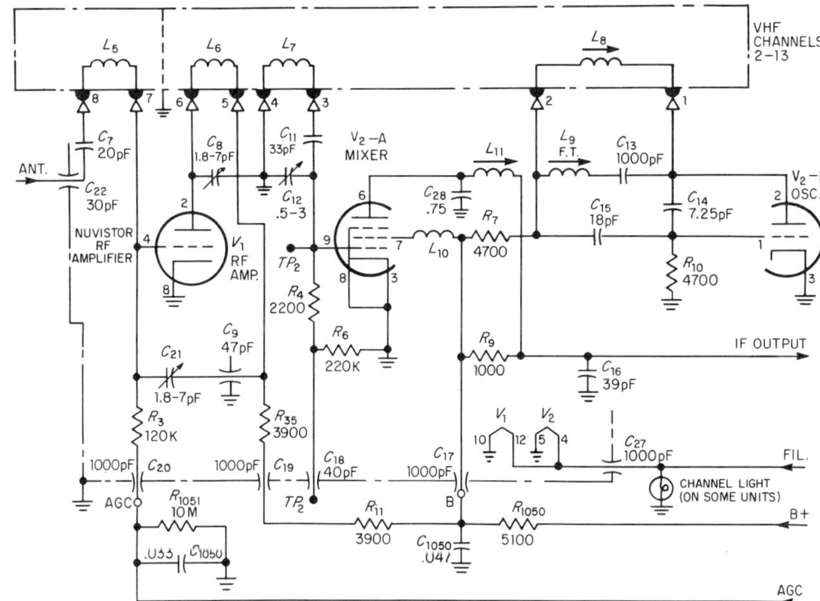

Fig. 6-18. A Nuvistor tuner. The Nuvistor is connected as a neutrode RF amplifier.

The output signal of V_1 is electromagnetically coupled from L_6 to L_7. The capacitors C_8 and C_{12} tune these coils. The RF amplifier circuit is neutralized by the feedback capacitor C_{21}.

There are two input signals to the mixer (V_{2-A}). The RF signal from L_7 is applied to the control grid at pin 9, and the oscillator signal from V_{2-B} is applied to the screen grid at pin 7. The difference signal—which is the IF frequency—is taken from the plate at pin 6.

6.9 TRANSISTOR TUNERS

Figure 6-19 pictures a typical transistor tuner. There are separate antenna input circuits for the VHF and UHF tuners. (The UHF tuner is not shown.) The RF amplifier is an NPN common-emitter circuit. The

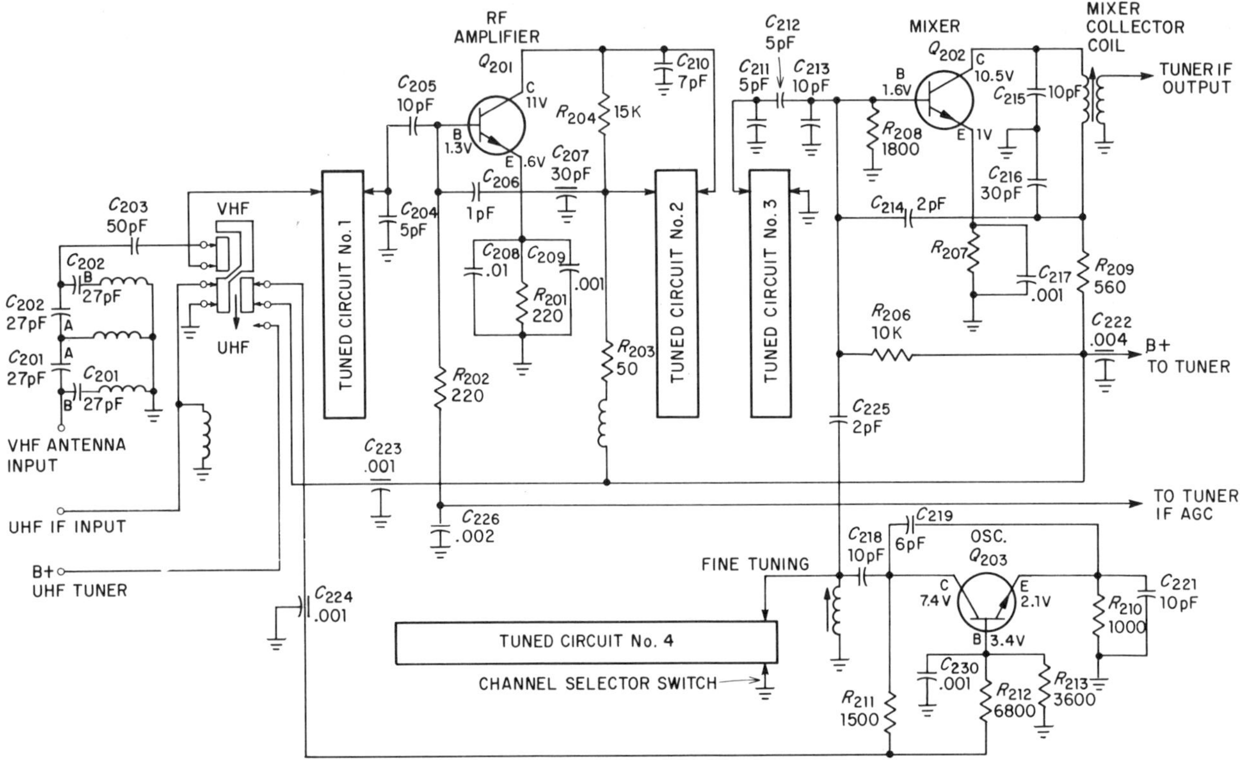

Fig. 6-19. A typical transistor tuner circuit.

input to the RF amplifier is tuned by tuned circuit number one. The output of the RF amplifier is tuned by tuned circuit number two, which is inductively coupled to tuned circuit number three. Thus, the RF signal is coupled to the mixer inductively, while the oscillator signal is injected into the mixer capacitively via C_{225}.

The mixer is a common-emitter circuit with a base bias established by a voltage divider consisting of R_{206} and R_{208} across the B+ source. Emitter stabilization is established by R_{207}, while C_{217} maintains the emitter at the RF ground potential.

The oscillator is a common-base circuit. Capacitor C_{230} grounds the base. When the tuner is set for a VHF station, the oscillator base is biased by R_{212} and R_{213} across the $B+$ source. Collector voltage for the oscillator is obtained through R_{211}. Regenerative feedback from the collector to the emitter is established by C_{219}. Tuned circuit number four sets the oscillator frequency.

The four inductive tuned circuits in Fig. 6-19 are adjusted simultaneously when the channel selector is moved from station to station.

Solid-State Tuning One of the more recent developments in television tuners is the solid-state tuning system. The solid-state tuning system is based upon the use of the so-called "varactor" diode. This diode is also known by the name voltage-variable capacitor and the trade names, Varicap and Selicap. The varactor is a special solid-state diode that acts as a capacitor whose capacitance varies inversely with the amount of reverse bias applied across the diode. All varactors operate only with reverse bias, and the tuned circuits to which they are connected have their resonant frequency changed merely by changing the amount of reverse bias across the varactor. The greater the reverse bias across the varactor, the less the varactor capacitance and the higher the tuned-circuit, resonant frequency.

In order to control the frequency of a tuned circuit, the varactor diode is placed across the tuning capacitor as shown in Fig. 6-20. In Fig. 6-20 (A) the amount of reverse voltage on the voltage-variable capacitor is dependent upon the settings of resistors R_1 and R_2, and of course, upon the position of the switch (*SW*). When the switch is in position one, the voltage at the arm of R_1 is picked off. This voltage establishes a certain amount of varactor capacitance. Capacitor C_1 is needed for isolating the varactor circuit so that the reverse bias voltage will not be grounded. In actual practice, the reactance of C_1 is negligible for the frequencies involved. Therefore, the varactor can be considered to be connected directly across tuning capacitor C_2.

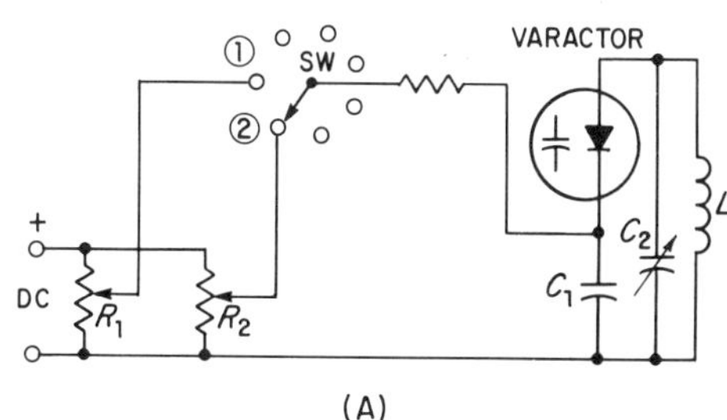

When the switch is turned to position 2, the voltage at the arm of resistor R_2 determines the capacitance of the varactor. The varactor, in parallel with the tuning capacitance, determines the frequency of the tuned circuit comprised of L and C_2. Only two positions are shown on the switch in Fig. 6-20(A), but it is presumed that all of the positions contain variable resistors which pick off different amounts of voltage from the applied dc. This type of tuning system has the advantage that only one inductance and capacitance is needed for all the different channel frequencies.

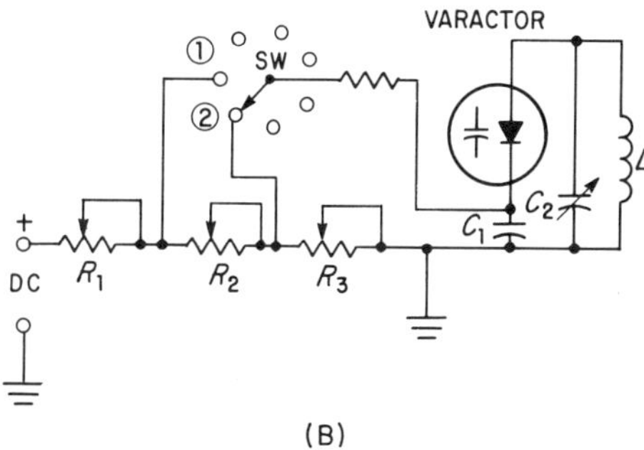

Fig. 6-20. Two ways of using varactors in detent tuners

In actual practice it is not possible to get a varactor diode to provide a complete capacitor range for tuning all VHF or UHF stations. Therefore it is common practice to switch in a different inductor for the upper VHF channel than is used for the lower VHF channels.

Another variation of the circuit in Fig. 6-20(A) uses push-button switches (rather than a rotating switch) to set the voltage on the voltage-variable capacitor. Both arrangements are useful, because they provide channel selection by switching. This is especially important in the UHF band due to the requirements of the FCC for providing the step tuning of stations in this range.

The circuit of Fig. 6-20(B) is slightly different, because the voltage picked off for the varactor is established by a number of resistors in series. Only three of the resistors are shown, but it is presumed that there is a variable resistor for each position on the switch. A single variable resistor in series with the resistor lineup can be used for fine tuning in this arrangement.

Figure 6-21(A) is a schematic diagram for a varactor tuner. The varactor diodes are D_1, D_2, and D_3. They tune the circuits for the output of the RF amplifier, the input of the mixer, and the local oscillator, respectively. Voltages for varying the varactor capacitance are obtained from the +2 *to* 24 *volt* line. A ganged switching system is used in the tuning

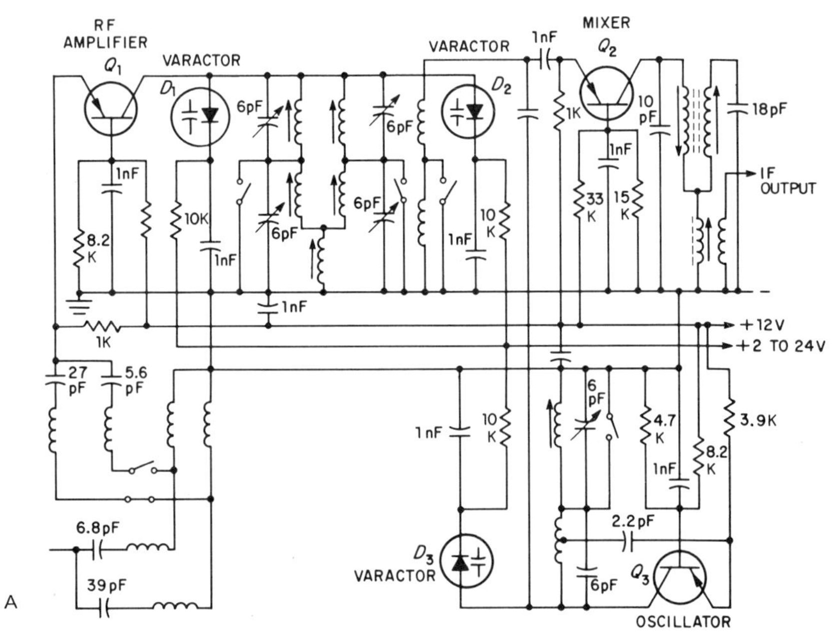

A

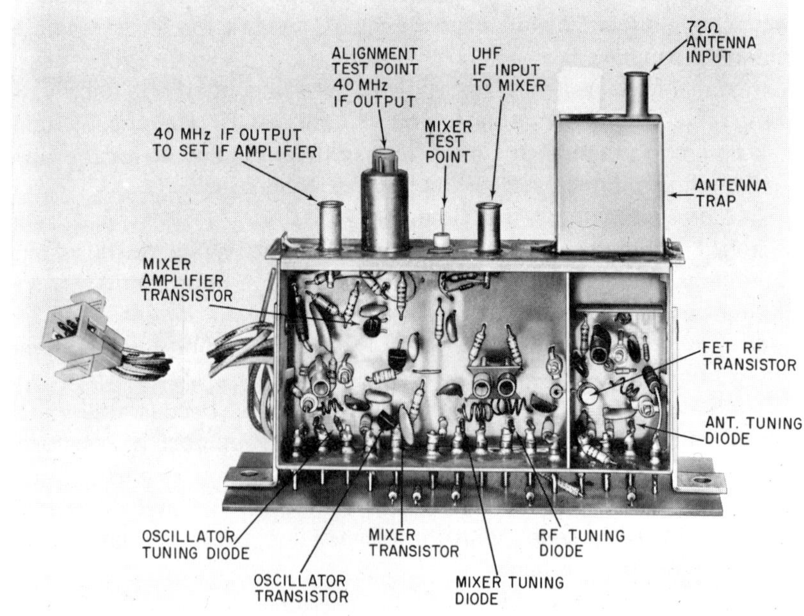

B

Fig. 6-21. (A) Schematic diagram of a varactor tuned VHF tuner (see text). (B) Bottom view of another varactor tuned VHF tuner. Note that four varactors are used in this tuner. (*Courtesy of Zenith Radio Corp.*)

circuits to provide a *low VHF* and a *high VHF* range. The varactor diodes are very carefully matched so as to provide an identical capacitance change when the 2 to 24 volt tuning voltage is varied. Except for the varactor diodes in parallel with the tuned circuits, this transistor tuner is similar to the others already discussed.

Another varactor-tuned VHF tuner is illustrated in the photo of Fig. 6-21(B). This differs from the one described before in that 4 tuning varactors are used. There is one each for the antenna, the RF, the mixer and the oscillator circuits. Note that no variable-tuning capacitors are used in a varactor tuner. The UHF IF input to the mixer comes from the UHF tuner which is described in Section 6.10. The mixer test point is used to feed in IF signals to align the IF amplifiers. The procedure employed to select channels for a combination UHF-VHF varactor tuner is also given in section 6.10 for the Zenith UHF tuner.

FET Tuners Television tuners are often identified by the type of RF amplifier used. Examples are *cascode tuners* and *Nuvistor tuners* which use cascode amplifiers and Nuvistor tube amplifiers respectively for RF amplifiers. As you might expect, then, a *FET tuner* is one which employs a FET as a RF amplifier.

Figure 6-22 shows the FET tuner used in an RCA solid-state chassis. (The RF amplifier for this tuner was discussed in greater detail in Chapter 5.) The RF signal is applied through the input filters and the input-tuned circuit to gate No. 1 of the FET amplifier Q_{201}. The receiver AGC voltage is applied directly to gate No. 2, and also to gate No. 1 from the junction of the voltage divider R_{201} and R_{202}. A filter in the AGC gate is comprised of C_{206} and the ferrite bead. (A ferrite bead behaves like a coil, but does not have the disadvantage of distributed capacitance.) The AGC line is also filtered by R_{203} and C_{227}.

The signal from the RF amplifier is delivered to the mixer stages Q_{202} and Q_{203} through a selective tuned circuit. The two transistor mixers are connected in a cascode configuration. This circuit was described earlier in this chapter.

The local oscillator Q_{204} is a Colpitts circuit. Its frequency is varied by adjusting the inductance of L_{210}. Transistor Q_{205} is connected across this inductance through the dc isolating capacitors C_{222} and C_{223}. The emitter of this transistor is open, and the base-collector junction is reverse biased. In this arrangement, Q_{205} is being used as a varactor. Voltage for this voltage-variable capacitor is obtained from the automatic fine-tuning control (AFT). The operation of AFT circuits is discussed in detail in Chapter 7. Basically, it involves obtaining a dc voltage that is proportional to the amount by which the oscillator drifts off frequency. This dc voltage—which is usually obtained from the receiver IF section—changes the capacitance of the voltage-variable capacitor in such a way that the oscillator-tuned circuit is brought back to the correct frequency value.

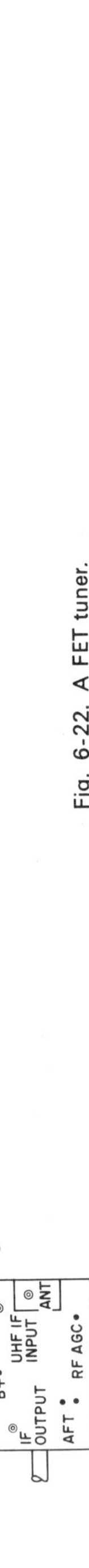

Fig. 6-22. A FET tuner.

A switch across Q_{205} disables the AFT system during the time when the manual fine-tuning control is operated. The receiver is designed so that it is necessary to push the manual fine-tuning control to operate it, and the action of pushing the knob operates the AFT disable switch.

6.10 UHF TUNERS

In the study of UHF tuners it is important to understand the distinction between a converter and a tuner. In principle, both devices convert the incoming signals in the UHF band (470–890 MHz) to a lower signal frequency. However, the converter steps down a frequency in the UHF channel to one of the VHF channel frequencies (usually Channel 6), and the VHF tuner, in turn, processes this channel as if it were the received signal. Since the relative position of video and sound IF is specified, the converter oscillator is tuned to a frequency lower than the incoming UHF signal to take into account the double-heterodyne action. For example, if the converter is to operate on Channel 54 which has the picture carrier at 711.25 MHz and the sound carrier at 715.75 MHz, the oscillator resonant frequency should be 628 MHz. This will transform the 710–716 MHz band to the 82–88 MHz band to correspond to Channel 6. Also, the picture and sound carriers of the UHF channel will be stepped down to 83.25 and 87.75 MHz respectively, as indeed should be the case (see Table 6-2). If the oscillator frequency were chosen on the higher side, the position of the two carriers would have been reversed. After conversion, the local oscillator of the VHF tuner steps down the Channel 6 frequencies to the intermediate frequencies in the routine way.

A UHF tuner converts the incoming UHF channel to an intermediate frequency and feeds this converted signal through the VHF tuner to the receiver IF stages. To summarize, the UHF converter changes the UHF signal to a VHF signal, while the UHF tuner changes the UHF signal to an IF signal. Except for this important difference, the converter and tuner actions are identical.

Since April, 1964, all television receivers manufactured in the United States must provide for reception of 12 VHF and 70 UHF channels. This has given impetus to the manufacture of UHF tuners. Gradually the converter units and channel strips are becoming obsolete.

Most UHF tuners fall into two general categories:

1. Those tuned by parallel-line tuned circuits.
2. Those tuned by coaxial tuned circuits (sometimes called cavities).

In addition to the above catagories, we also find UHF tuners which are varactor tuned. However, in these cases, the varactor is not the actual tuning circuit, but simply varies the total tuning capacitance of the tuned circuit, as previously described. The major advantage of varactor tuning in this case, is that it permits UHF stations to be selected easily by means of a switch or push buttons. It also makes it possible to tune to UHF stations by remote control.

Typical UHF Tuners Figure 6-23 is a schematic of a UHF parallel-line transistor tuner using variable ring inductors (L_2, L_3, and L_8) as tuning elements. The 300-ohm antenna impedance is converted by a balance-unbalance transformer T_1 to match the input impedance of the double-tuned circuit. An aperture between the antenna and the mixer cavities couples the two circuits magnetically. The tapping on coil T_2 provides an impedance match to the 200-ohm diode (D_1) impedance.

The diode IF output is coupled via a broadband filter L_5-C_6 which, together with the coaxial cable (connecting the tuner to the IF amplifier) and IF amplifier input circuit form part of a double-tuned circuit. The inductor L_5 also provides a dc return for the diode current.

The oscillator circuit uses a silicon planar epitaxial NPN transistor (Q_1) in a common-base configuration. Capacitors C_7, C_8, C_9, and inductors L_9, L_8, and L_7 form the tuned circuit with L_7 providing the coupling to the mixer diode. Capacitor C_{10} provides the oscillator feedback; resistors R_1, R_2, and R_3 establish the dc operation point for the transistor. Inductors L_{10} and L_{11} are isolating chokes.

Oscillator power from the oscillator cavity is injected into the diode mixer via the mutually-coupled coil L_6. The alignment of low- and high-numbered channels is achieved through the small variable capacitors C_1 and C_2 for the RF input, C_3 and C_4 for the mixer circuit, and C_7 and C_8 for the oscillator cavity.

The RF and mixer circuit of Fig. 6-23 can also be used in conjunction with a vacuum-tube Colpitts oscillator.

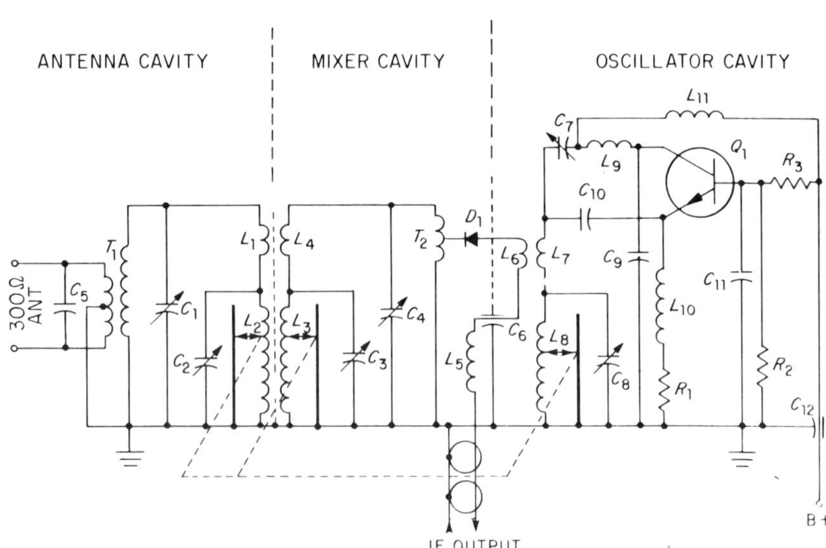

Fig. 6-23. A parallel-line UHF continuous-tuning tuner using ring inductors. (*Courtesy of Selectronics.*)

Figure 6-24 shows a version of the UHF tuner which uses a popular tuning arrangement. This circuit uses a coaxial tank with capacitor-end tuning. (The parallel-line tuner of Fig. 6-23 uses sliding contacts for tuning which can present service problems and are prone to higher noise figures.) In Fig. 6-24, antenna coil L_1 couples the RF power to a coaxial tank which, in turn, transfers the signal to the mixer diode (CR_1) through a suitably positioned window. The oscillator delivers its injection through the pickup coil (L_3), and the IF becomes available at L_6. This IF signal can be fed to the VHF tuner which then acts as a normal amplifier.

The circuit under discussion features a special automatic frequency control arrangement which helps to stabilize the oscillator frequency. At exact resonance, a varicap (varactor) diode, CR_2, is reverse biased by 2.8 volts across its terminals furnished from a 22 V regulated supply.

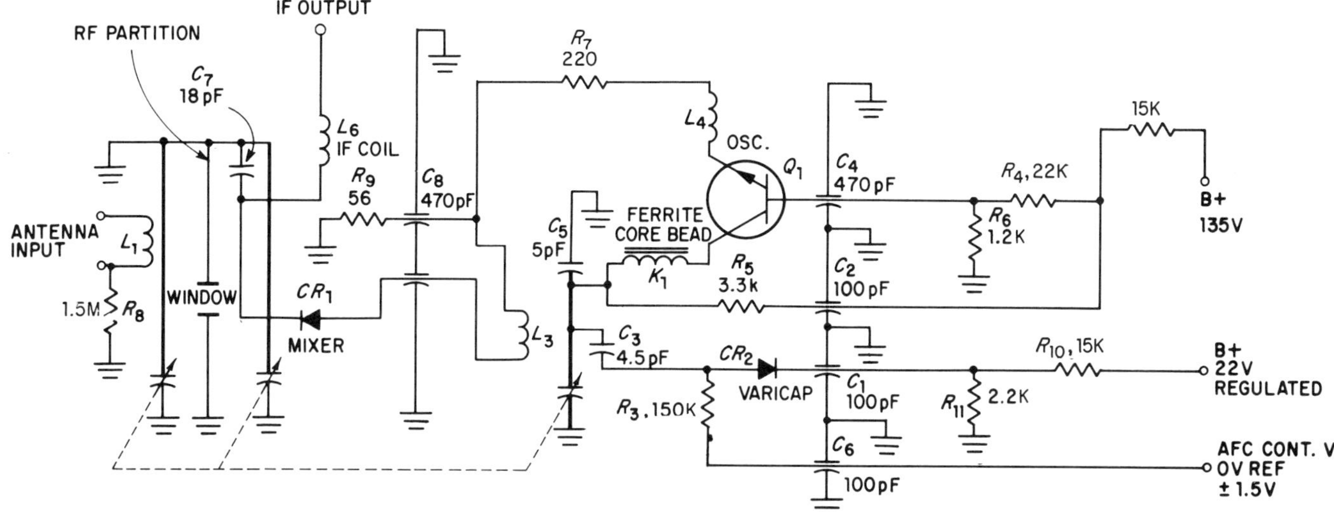

Fig. 6-24. A UHF tuner which uses coaxial tank circuits with capacitor-end tuning. (*Courtesy of General Electric Corp.*)

The resistors R_{10} and R_{11} are used as a voltage divider. Should the oscillator frequency drift, a voltage proportional to this drift will be generated, increasing or decreasing the reverse bias for the varicap. Since the varicap is a *voltage-variable capacitor,* the circuit elements can be so arranged that when the effective oscillator tank capacitance decreases causing a frequency rise, the varicap offers a higher capacitance across the coaxial tank. The effect is an opposite one with a decrease in capacitance. The arrangement, often used in high-priced tuners, checks the oscillator drift caused by the dc supply variations, as well as the temperature rise occasioned by the transfer of heat from the hotter parts of the receiver. A typical varicap is a silicon planar diode with a Q of 200 at 50 MHz, and a reverse breakdown voltage of 20 volts. The rest of the circuitry is conventional.

Another coaxial-tank UHF tuner with capacitor-end tuning is shown

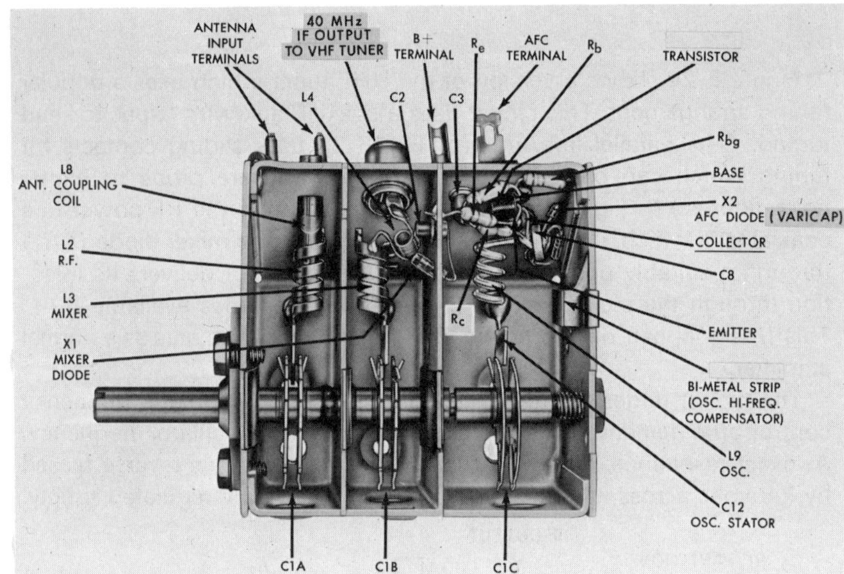

Fig. 6-25. A transistor, UHF coaxial-tank tuner. Note the individual coaxial cavities for the RF, mixer, and oscillator sections. (*Courtesy of Zenith Radio Corp.*)

in the photo of Fig. 6-25. There are three coaxial tanks (or cavities), one each for the RF, mixer, and oscillator sections. In each case, the inner walls of the metal partitions act as the outer conductors of the coaxial line, while the wound coils act as the inner conductors. The three end-tuning capacitors are ganged on a single shaft for ease of tuning. As is the case of the circuit shown in Fig. 6-24, this tuner also uses a varicap circuit for the oscillator to provide the automatic fine tuning. (See Chapter 7 for further details of AFT.) The antenna input signal is inductively coupled to the inner RF conductor L_2 by means of the antenna coupling coil L_8. Coupling between the RF and mixer, inner lines are accomplished by means of a window cut in the intervening partition. Note the bimetal strip in the oscillator section. This strip provides temperature-frequency compensation, thereby improving the oscillator stability. It also acts as a capacitor which varies with the temperature. This capacitor is formed between the strip and the oscillator inner conductor L_9.

The bottom view of a UHF varactor tuner is illustrated in the photo of Fig. 6-26. Note the absence of any variable-tuning capacitors. The tuning is accomplished by preset trimmers in conjunction with the variable capacity of the varactors. In the Zenith tuner shown in this figure, 4 varactor-tuning diodes are used. These are used in the antenna, RF, mixer, and oscillator sections. The varactor diodes are supplied in matched sets of 4 for proper tracking correlation. If it becomes necessary to replace one of these, a new matched set of all 4 must be installed.

In the Zenith Varactor-Diode Tuning System, of which the unit shown in Fig. 6-26 is a part, any combination of VHF and UHF channels can be set up, for example, Channels 2, 32, 5, 7, 44, 9, etc. All tuning is accomplished by means of a 3-position Band Selector Switch and a 14-position Channel Selector Drum. The Band Selector Switch is

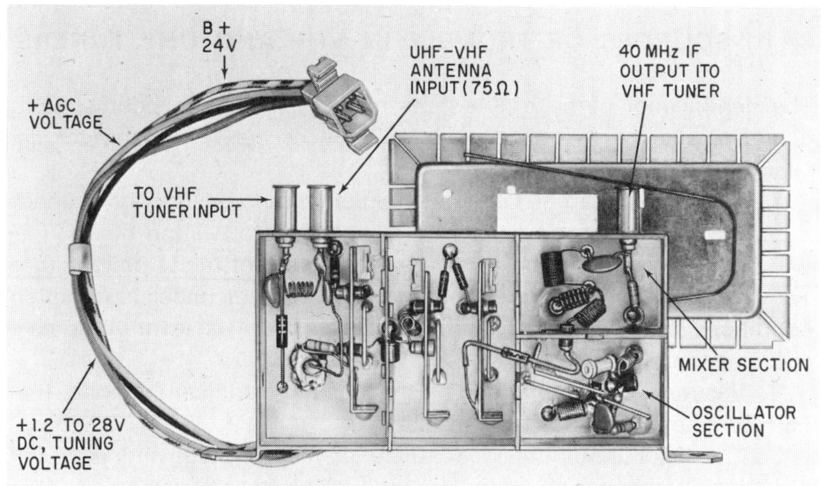

Fig. 6-26. The bottom view of a UHF varactor tuner. Note the absence of tuning capacitors. (*Courtesy of Zenith Radio Corp.*)

adjusted to either VHF-LOW (Channels 2–6), VHF-HIGH (Channels 7–13), or UHF. The exact channel desired is then obtained by turning the VHF-UHF FINE TUNING CONTROL, until the channel number appears in a window. Varying this latter control causes the resistance of a master potentiometer that changes in accordance each of the 14 detented positions of the Channel Selector Drum. Each potentiometer resistance change produces a corresponding voltage change which is applied to the varactor-tuning diodes. They, in turn, tune the resonant circuits by a variation of their capacity.

VHF-UHF Combination At this stage it would be appropriate to label various sections of the VHF-UHF signal flow in a receiver. Figure 6-27 gives an idea of the connections, both mechanical and electronic, that exist in a typical all-channel receiver. The VHF tuner in its UHF setting acts as an IF amplifier to boost the input to the video-IF amplifier, which thus receives nearly equal signal strengths for VHF and UHF channels. In this way it compensates for the absence of an RF amplifier in the UHF tuners. The circuitry beyond the tuner is identical for all channels.

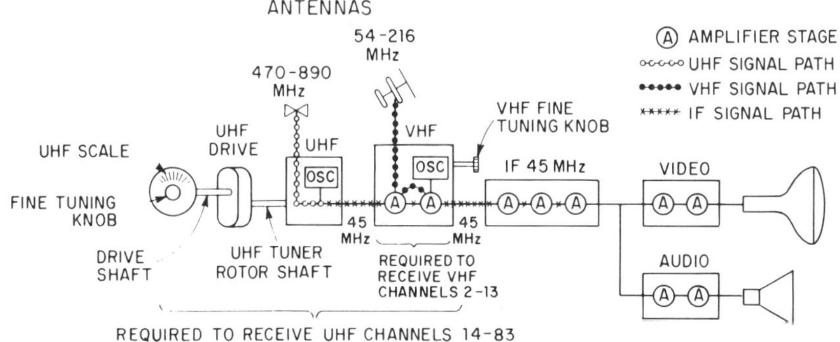

Fig. 6-27. A UHF-VHF tuning block diagram. (*Courtesy of Zenith Radio Corp.*)

6.11 SOURCES OF TROUBLE IN VHF AND UHF TUNERS

The identification of the trouble spots in a television tuner becomes easier if one recalls the following major differences between transistor and vacuum-tube characteristics.

1. Transistors used as voltage amplifiers are normally cool in operation, and unless these are working at abnormal operating points, they will not get warm enough to give an indication of faulty performance. Vacuum tubes, on the other hand, tend to get hot under high current conditions and may, in fact, burn themselves out. Transistor burnouts in equipment are relatively rare.

2. Vacuum-tube control grid bias values are significantly higher than base or gate bias values for transistors.

3. Transistor input impedances are markedly lower and a 20,000 ohms-volt voltmeter is adequate for making measurements in most transistor circuits. With tubes, a VTVM or FET meter is often required.

4. Large changes in ambient temperature affect transistor operation much more than they do vacuum-tube operation.

When the receiver does not function normally, the source of the fault is sometimes assignable from the kind of (sound or picture) symptoms it exhibits. The tuner is often to blame when one of the following conditions is present: there is a good raster but no sound or picture; good sound and picture quality is attained at markedly different settings of the fine-tuning control; a picture is available on only some of the channels operating in the vicinity; or the picture is snowy or intermittent. A step-by-step troubleshooting procedure is followed when one or more of these symptoms appear. The tuner contacts may require careful cleaning and lubrication if the picture is snowy or intermittent. Any broken coils on the turret sould be carefully replaced and, of course, realigned. However, the complete realignment of the unit should be undertaken only after the visual examination and voltage checks have revealed no abnormality.

In vacuum-tube tuners a quick change of tubes may sometimes highlight a faulty area. Transistors are in general more reliable in their operation and have a very long life. Thus, before ordering a transistor replacement, other avenues of search should be carefully weighed. A faulty active circuit element may be isolated by injecting a suitable sweep voltage at its input and observing the response on an oscilloscope. A faulty oscillator may be isolated if, on application of the proper supply voltage, it yields no output. Poor picture and sound are indications of a faulty RF amplifier or a faulty common IF amplifier circuitry.

A UHF tuner in its simplest form has only two active elements: the diode *detector* and the *local oscillator*. The transmission lines used for tuning are rigidly fixed to their supports, and the end-tuning variable air capacitors are almost always trouble free. It is no surprise, therefore, that these units give the least amount of trouble. When faults do occur, they are often attributable to mechanical gearing arrangements, or to an accumulation of dust on the capacitor plates. In replacing any components in the UHF circuitry, care should be exercised since a small amount of excessive lead length can misalign the tuner.

Should the picture become noisy on a specific UHF channel, the reduction of the oscillator injection voltage may be a cause. Coupling from the oscillator to the mixer may need attention. If the tuner should require complete alignment, it is advisable to turn to the manufacturer's instructions as detailed in his product manuals. This is particularly true in UHF tuners where a slight twist on the end-capacitors may offset the proper working of the entire unit. The importance of this fact cannot be over-emphasized.

Precautions to be undertaken while servicing any electronic product are also in order in the case of servicing tuners. Only in this case these are be followed even more rigorously. For instance, the transistor heat sinks should be adequate, soldering guns must not overheat the transistors, the solder joints must be clean, and so on.

REVIEW QUESTIONS

1. What is an incremental tuner as compared with a continuous tuner?
2. Why does the FCC set a limit on radiations from TV oscillators?
3. How does thermal noise generated in the tuner circuits show up on the reproduced picture at the CRT screen?
4. Explain the different ways that the RF signal and the oscillator signal can be injected into a mixer stage.
5. Which transistor amplifier configurations (*i.e.,* grounded emitter, grounded base, and grounded collector) are used for mixer stages?
6. When two signals are heterodyned in a mixer stage, there are four different output frequencies. How is the difference frequency chosen for use as the receiver IF frequency?
7. What is the most frequently encountered video IF frequency?
8. Compare the types of bias used on a vacuum-tube oscillator circuit and a transistor oscillator circuit.
9. In what circuit is the receiver fine tuning control located? What happens electrically when the fine tuning control is adjusted?
10. What is a "neutrode" circuit?
11. Where is a dual-gate MOSFET used in a tuner?
12. What type of oscilloscope probe is used for viewing the RF signal at the tuner?
13. What is a balun used for?
14. To what section of a tuner is the AGC voltage delivered?
15. What is a varactor diode and how does it operate?
16. What types of tuning circuits are used for UHF tuners?
17. Explain how channels are selected by a varactor tuner?

7 Automatic Fine Tuning and Remote Control

7.1 INTRODUCTION

This chapter discusses the requirements for automatic fine-tuning circuits and their operation. Circuits associated with both vacuum tubes and solid-state local (tuner) oscillators are covered. In addition, various types of remote control devices are explained and the use of tuning indicators is also described.

The fine-tuning control of a TV set adjusts the frequency of the local oscillator in the tuner. For a monochrome TV set, a moderate amount of misadjustment of this control can be tolerated without the viewer being particularly aware of this condition. However, in the case of a color receiver, the setting of the local oscillator frequency is more critical. In this latter case, even a small deviation of the oscillator frequency from its correct setting may result in serious deterioration, or complete loss of the color.

Even if the fine-tuning control was initially set correctly, the oscillator frequency may drift from its nominal frequency and cause color deterioration. Oscillator drift is caused mainly by temperature variations of the oscillator circuit parts.

In order to simplify the operation of a color set for the viewer, from the point of view of setting the fine-tuning control correctly and overcoming the problem of local oscillator drift, many color sets incorporate and *Automatic Fine-Tuning* (*AFT*) circuit. AFT circuits are designed so that any adjustment of the fine-tuning control within a predetermined range will cause the circuit to lock the oscillator onto the correct frequency. To simplify the job of getting the manual fine-tuning control adjusted to the required range, *tuning indicators* are used. In their simplest form, they are simply lights that are lit whenever the control is *not* correctly adjusted.

A number of remote-control devices for television receivers have been developed by their manufacturers. These devices offer an added convenience to the set user, who is thus able to change the volume, picture, and station without leaving his chair. The public has shown that they favor such controls and are willing to pay extra for their incorporation into the receiver. Manufacturers have responded by introducing a variety of remote-control methods and these are now in fairly wide use.

7.2 FUNCTIONS AND TYPES OF AUTOMATIC FINE-TUNING (AFT) SYSTEMS USED IN TELEVISION TUNERS

As pointed out in Chapter 6, the local oscillator for the tuner is tuned to frequencies which are *above* the center frequencies of the VHF and UHF channel by a frequency that is equal to the intermediate frequency. (The stability of sound and picture *carrier* frequencies is insured by an elaborate circuitry built as an integral part of the transmitter.) The automatic fine-tuning circuits therefore enable one to easily set the tuner oscillator frequency to the proper value—one which will insure that picture and sound IF markers from the tuner output appear at the proper positions of the IF amplifier response. It has also been pointed out that the requirements of the frequency stability of the local oscillator in a color receiver are much more stringent than those in a monochrome receiver. In fact, special design effort is expended in making provisions so that, in changing channels from one to the other, the oscillator sets itself precisely to the required frequency.

Oscillator drift is caused by such factors as ambient temperature change, component aging, power supply voltage fluctuations, and so on. A number of remedial measures may be taken in order to improve the stability of an oscillator circuit. The frequency-determining elements, such as inductors and capacitors, should be chosen so that their temperature coefficient is negligible or, so that a change in one is counterbalanced by a change in the other. Clearly, it is difficult to make such provisions in television tuners where a portion of the tuning components are often composed of stray capacitances and inductances. It must be noted, however, that capacitors have been designed with well-controlled positive and negative temperature coefficients.

The use of power supplies which are regulated to within a very small percentage of the nominal value has almost eliminated the voltage change as a serious cause for oscillator frequency drift in many receivers. The expense involved in incorporating such regulation is, of course, a manufacturer's concern.

The factors that contribute to the inherent oscillator drift can be compensated for by the addition of AFT circuitry, which will insure that the local oscillator frequency is always at the correct value. The basic principle of the operation of the AFT circuitry has been well known for many years and in fact appeared in monochrome TV sets decades ago. The AFT is basically an automatic frequency control circuit similar to those used in FM and other receivers. The basic principle is illustrated with the aid of the block diagram of Fig. 7-1.

An output from the third IF amplifier is applied to a frequency sensitive detector (discriminator) tuned to the picture IF carrier frequency of

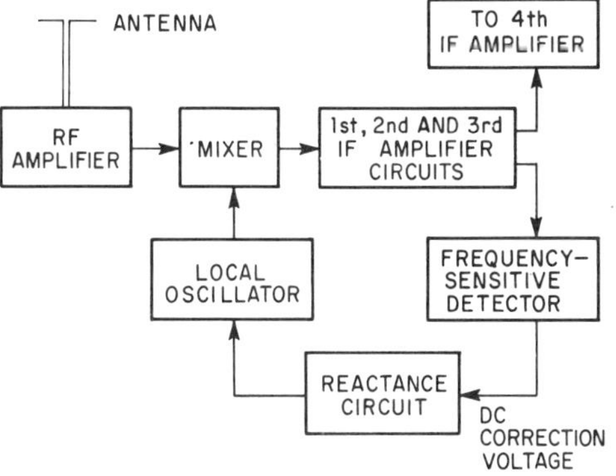

Fig. 7-1. Block diagram of an AFT system.

45.75 MHz. This IF carrier frequency is dependent upon the difference frequency between the incoming RF carrier frequency and the local oscillator frequency, which should be above the RF picture carrier frequency by exactly 45.75 MHz. If the local oscillator differs from this amount, either higher or lower, the frequency sensitive detector will provide a dc output having either a positive or a negative polarity, depending upon whether the oscillator is above or below its required frequency. In addition, the magnitude of the dc voltage will be proportional to the amount of error of the oscillator frequency. This dc voltage is applied to a "reactance circuit" (frequently a varactor, in more recent sets). The reactance circuit is actually a portion of the oscillator-tuned circuit and is varied by the dc voltage in such a manner as to correct the oscillator frequency. The AFT circuit makes it possible to mistune the fine-tuning control somewhat, without suffering any deterioration of picture quality. In addition, it also automatically compensates for any oscillator drift.

In the following section two types of frequency correction circuits are described:

1. A vacuum-tube arrangement using a triode which acts like a reactance in parallel with a resonant tank.
2. A solid-state arrangement using a reverse-biased P-N junction in order to make its capacitance dependent on the applied voltage. The former circuits are used in vacuum-tube television receivers, whereas most solid-state receivers use the P-N junction (varactor).

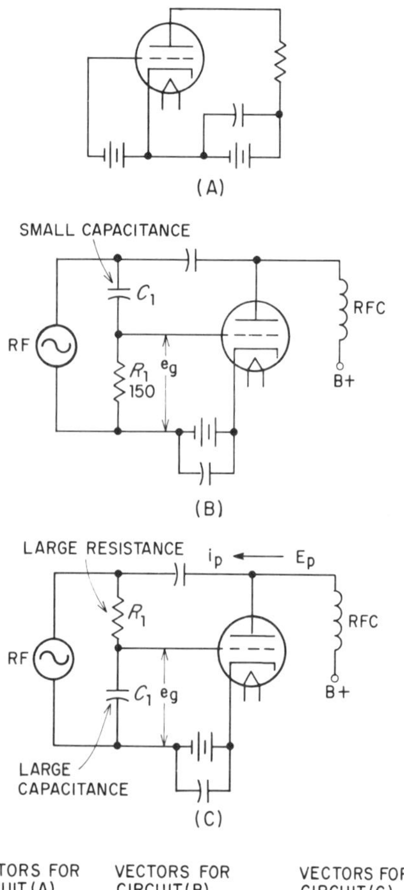

Fig. 7-2. A vacuum-tube may be connected so that it acts as a resistor, capacitor, or inductor. (A) In this circuit the tube behaves as a resistor. (B) In this circuit the tube behaves as a capacitor. (C) In this circuit the tube behaves as an inductance. (D) The vector relationship between the plate current (i_p) and the plate voltage (e_p) for the circuits in (A), (B), and (C).

7.3 TYPICAL AFT CIRCUITS

Reactance-Tube Circuits One method of correcting the frequency of the oscillator of an RF stage is by paralleling its resonant circuit with a reactance tube arrangement which has the characteristic of a variable reactance. Such a tube performs as a frequency controller when a dc voltage is applied to the grid to shift its bias. Figure 7-2 indicates the principle of operation of such a reactance tube.

In Fig. 7-2(A), a tube is connected with normal plate and grid voltages, so that the plate current is flowing through the tube. When we increase the plate voltage, the plate current will increase, and the two will be in step within fairly wide limits of the plate-current flow. The same is true of a plate-voltage decrease; the current will also decrease in step. Since the two follow each other directly, we obtain practically the same action as the voltage and current in a resistor, and consequently the tube is functioning as a resistor, although, at some voltages, it acts as a non-linear resistor.

Now, let us alter the circuit to the form shown in Fig. 7-2(B). A source of RF voltage is coupled to the plate of the triode, and the B+ voltage now reaches the tube through an RF choke. The choke prevents the RF component of the plate current from passing through the dc voltage source. In the same circuit, a small capacitor and resistor transfer part of the RF voltage to the grid of the tube. Let us investigate this section of the circuit more closely.

Since C_1 and R_1 are placed directly across the RF voltage source, an RF current will flow through both components, the amount of current being governed by the impedances of C_1 and R_1. In the diagram it is specified that C_1 is small, resulting in a large impedance. The resistor, on the other hand, is low in value. Thus, the circuit impedance will be largely capacitive, and the current flowing through C_1 and R_1 will lead the RF voltage by approximately 90 degrees. The RF current, flowing through R_1, will develop a voltage, e_g, which is in phase with the RF current and which leads the RF voltage from the generator by 90 degrees. It is also true that since e_g is the alternating grid voltage for the tube, the plate current will lead the applied RF voltage by the same 90 degrees. Whenever any electrical component exhibits the property of having the current through it lead the voltage applied across it by 90 degrees, it is said to act like a capacitance. Thus, by properly connecting the tube, we have made it appear as a capacitor to the circuit. The tube will have this effect on any circuit in which it is placed.

To indicate how a tube can be made to function as an inductance, we can employ the circuit of Fig. 7-2(C). Now we find that R_1 and C_1 have been interchanged, and the impedance of R_1 greatly exceeds the impedance of C_1, at the operating frequency of the RF-voltage generator. The current now through R_1 and C_1 will be in phase with the applied RF voltage. However, across any capacitor, the voltage always lags 90 degrees behind the current and, therefore, the RF grid voltage, in Fig. 7-2(C), will lag the current through R_1 and C_1 by 90 degrees. By the same token, it will lag the RF generator voltage by 90 degrees. Within the tube, the plate current, being in phase with the grid voltage, will also lag the RF voltage by 90 degrees. The tube will appear to the circuit as an inductance in which the current lags 90 degrees behind the applied RF voltage. If desired, a high resistance can be shunted across C_1 to provide a leakage path for any electrons accumulated at the grid. With a high value of resistance, the circuit operation remains as indicated.

In each of the foregoing instances, the amount of inductance or capacitance that the tube injects into the circuit will depend upon the amount of plate current flowing through the tube. If we increase the dc bias for the tube in the positive direction, the plate current will increase and its reactive effect on the circuit will likewise increase. With a greater negative dc bias, the opposite action and effect will occur. Figure 7-2(D) shows the vector relationships between plate current and plate voltage. In Fig. 7-2(D), if the capacitor C is not small enough, the current lead will be between 0 and 90°. The usual practice is to make the capacitive reactance larger than ten times the resistance of R_1. In the circuit of Fig. 7-2(C), the resistance of R_1 is made at least ten times greater than the capacitive reactance of C_1. It is interesting to note that a reactance circuit using a field effect transistor (FET) is identical to the ones shown in Fig. 7-2.

A typical tuner circuit (without the channel selection elements) is shown in Fig. 7-3. Here a single vacuum tube incorporates two triodes for oscillator and reactance tube functions. The mixer and RF amplifier are pentodes, and they perform the functions described in Chapter 6. (The turrets supply the requisite inductors for resonance.) If, for some reason, the local oscillator drifts, the color carrier frequency and the center frequency of the sound IF will drift. Unless an AFT system is used, the color proportions of the signal may be cut off or distorted severely. With automatic frequency control, a change in the local oscillator frequency causes a dc *correction voltage* to be generated by the *frequency-sensitive detector*. Frequency-sensitive detectors are also used for demodulating FM signals. Discriminators and ratio detectors are two examples of such circuits. They will be discussed in a later chapter.

If the oscillator frequency drifts to a value higher than normal (or is so tuned incorrectly), indicating a decrease in oscillator turret inductance (or in the resonating capacitance), the AFT control voltage from the frequency-sensitive detector will bias the reactance tube so as to cause it to look like a larger capacitance. This, in parallel with the frequency-determining network of the oscillator, will decrease the oscillator frequency thus compensating for the unwanted drift or mistuning. The opposite reaction will occur, if the oscillator frequency decreases.

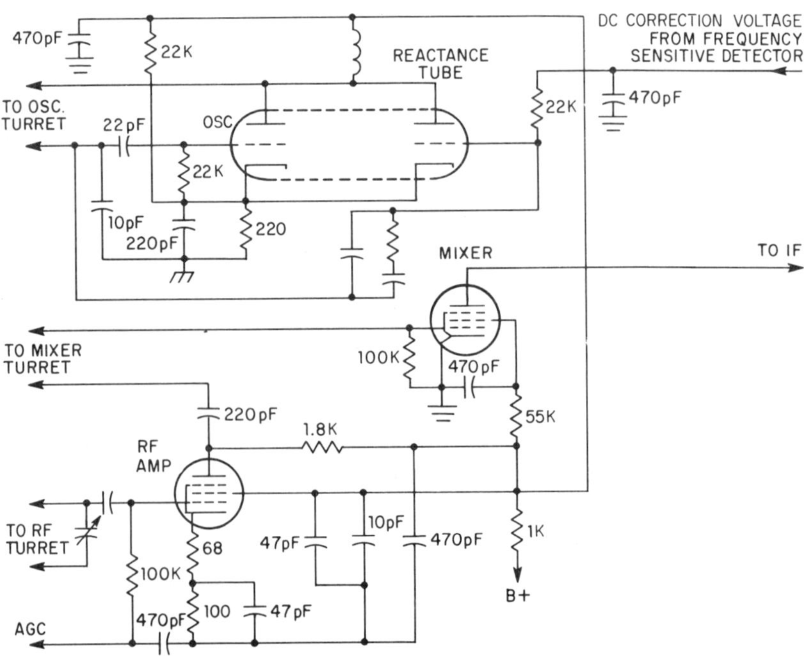

Fig. 7-3. A typical vacuum-tube VHF tuner arrangement using a reactance tube for AFT action.

Varactor-Tuned AFT System Corresponding to the function of a reactance tube, which acts as a variable capacitance, the *varactor* also acts as a variable capacitor.

Figure 7-4 shows a varactor diode as part of a typical oscillator circuit. The varactor diode is coupled to the main oscillator tuning circuit through the capacitor C and is, in effect, parallel to the $L_1C_1C_2$ combination which determines the oscillator frequency. Thus, should the oscillator frequency drop for any reason, the effective value of the shunting capacitance of the varactor circuit will decrease causing the local oscillator to reset itself to its original higher frequency. The required decrease in varactor capacitance is accomplished with a change in the dc bias voltage from the frequency-sensitive detector. In order to insure that the detector will cause the frequency correction to take place in the proper direction, the varactor is initially biased to provide a nominal capacitance for the oscillator circuit.

A more positive voltage from the detector dc output will *reduce* the bias across the varactor and its capacitance, while a less positive voltage will increase its capacitance.

As has been stated before, varactors are P-N junction diodes that are specifically made for use as voltage-variable capacitors. However, almost any junction diode that is reverse biased will exhibit a voltage-dependent capacitance. One manufacturer has used the base-collector junction of a silicon transistor in conection with this purpose. The base-collector junction of a transistor is normally reverse biased, so it can withstand the reverse voltage usually associated with varactor diodes. Figure 7-5 illustrates such a circuit. Note that the emitter is not connected.

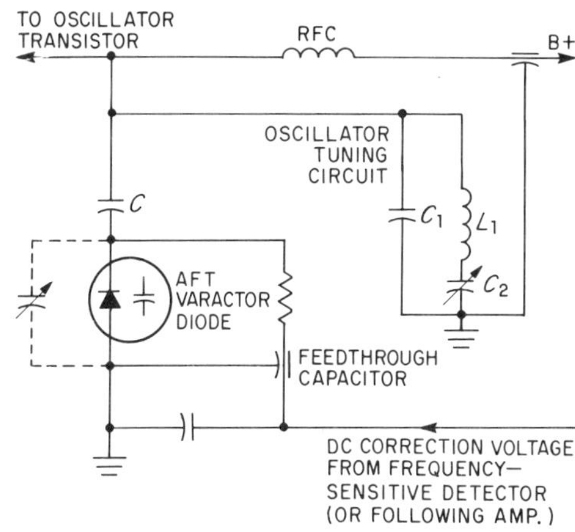

Fig. 7-4. The varactor connected across an oscillator tuning circuit.

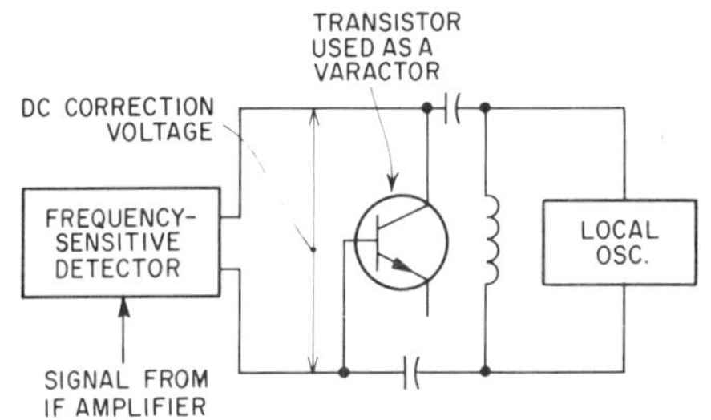

Fig. 7-5. In this circuit a transistor serves as a voltage-variable capacitor.

An AFT Frequency-Sensitive Detector The AFT detector circuit shown in Fig. 7-6 uses a transistor, Q_1, in a common-emitter configuration for the AFT frequency-sensitive detector and a second common-emitter transistor, Q_2 as an AFT amplifier. The input to the detector is taken from

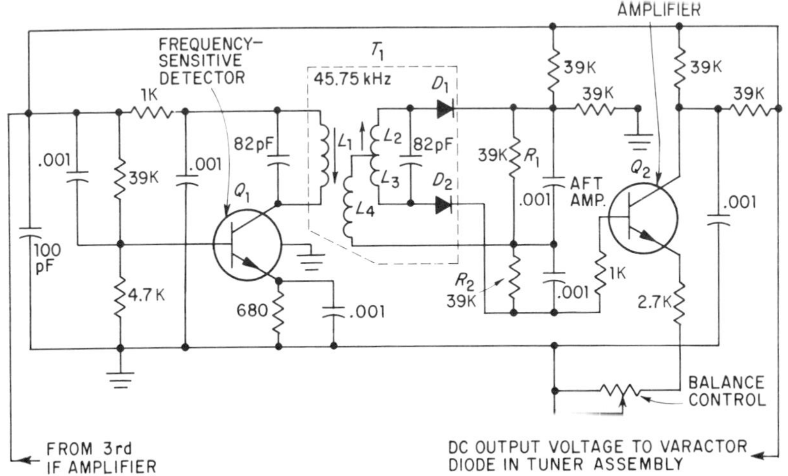

Fig. 7-6. A solid-state AFT, frequency sensitive detector, and amplifier circuit.

the output of the third IF amplifier, as indicated in Fig. 7-1. This input is applied to the base of Q_1. The basic circuit of the detector is that of a Foster-Seeley FM discriminator, which is explained in Chapter 20. The detector transformer T_1 is tuned to the picture IF carrier frequency of 45.75 MHz. The action of the detector is such that when the incoming carrier is exactly 45.75 MHz, the output as applied to the base of the amplifier Q_2 is zero volts. If the local (tuner) oscillator is not at its nominal frequency for the particular channel selected, the picture IF carrier frequency will be either above or below 45.75 MHz. In this event, the detector will have an output which is either a positive or a negative dc voltage. In addition, the farther off from the nominal frequency the local oscillator is, the greater will be the magnitude of the dc output voltage. In the circuit of Fig. 7-6, the dc detector output voltage is amplified by the transistor Q_2 and is then applied to the varactor in the tuner assembly, as pictured in Fig. 7-5.

The detector transistor Q_1 is normally operated at a very low quiescent emitter current (usually less than 50 microamperes) to improve the detection efficiency. This also provides amplification in the collector circuit.

All automatic frequency control circuits (including AFT) have a so-called "pull-in range" and a "hold-in range." The pull-in range is the frequency farthest from the nominal one that the AFT circuit can acquire and lock into the nominal frequency, when the AFT circuit is originally in the unlocked condition. An example of this is when we are switching channels. The hold-in range, on the other hand, is that frequency range, starting from the locked AFT condition, in which the AFT circuit can maintain a locked condition regardless of oscillator drift or manipulation of the fine-tuning control.

7.4 FUNCTIONS AND TYPES OF REMOTE-CONTROL DEVICES

The complexity of a remote-control device depends on the number of functions it performs. Its most important function is station-changing, since this is the major reason why most viewers leave their chairs and go to the receiver. Next, it is useful to be able to alter the sound level, preferably in a series of steps. Turning the receiver on and off is also a desirable control function, and finally, if not too costly, the control of contrast and/or brightness is useful. For color receivers, the tint and color circuits may be adjusted by the remote control. Obviously, the more functions to be altered from this remote position, the more complex the circuitry of the control device. This, in turn, directly determines the cost of the unit.

The most popular signaling method developed employs ultrasonic sound. In this method, a signal above the audible range is developed and directed toward the receiver, where it is picked up by a suitable microphone. This form of control is advantageous in that it can be confined by the walls of the room, with little possibility of interference to receivers in other rooms or beyond the receiver at which it is aimed. Furthermore, it is quite simple to generate acoustic signals by purely mechanical methods; there is no need for a battery or any other types of powering mechanism in the transmitter.

In the acoustic signaling system, the operating frequency has generally been kept in the 40 kHz range. This is sufficiently low so that air absorption does not seriously restrict the signal range. As the acoustic frequency increases, the air absorption also increases, and this can be a serious limitation on the system.

7.5 REMOTE-CONTROL SYSTEMS THAT EMPLOY ACOUSTIC SIGNALS

By far the most popular remote-control systems are those that employ ultrasonic signals. These signals can be generated in a number of different ways, but the overall principle is the same. The sound waves are picked up by a unit in the receiver and used for controlling some receiver function. In order to control a number of functions, a different frequency of sound wave is used for each.

Ultrasonic Transducers The operation of an acoustic type remote-control system depends upon the use of transducers. One transducer in the transmitter—that is, in the hand-held remote-control unit—converts a mechanical vibration into high-frequency sounds. Another transducer in the receiver converts the high-frequency sounds into electrical signals which can be used for operating a control in the receiver. In the systems to be described, the actuating transmitted signal is an ultrasonic sound that can be produced by vibrating cylindrical rods. There are other methods used for producing the ultrasonic sounds for remote control. For example, a vibrating column of air in a whistle, or a mechanical device, may be used. In any event, the systems are quite similar despite the difference in the method used for transmitting and receiving the sounds. The sound produced in the systems to be described is picked up by a microphone, converted into equivalent electrical signals, amplified, and then employed to actuate relays which perform the desired functions.

In the Zenith remote-control system of Fig. 7-7 and in several others, the transmitter consists of a cylindrical aluminum rod which vibrates in its longitudinal mode. If such a rod is struck on one end by a hammer moving along its axis, it emits a sustained note which has a definite frequency. For example, an aluminum rod 2 1/2 inches long has a fundamental resonant frequency of about 40 kHz. The internal damping of the aluminum is so slight that a large part of the vibrational energies, stored in the rod after the original blow of the hammer, is radiated.

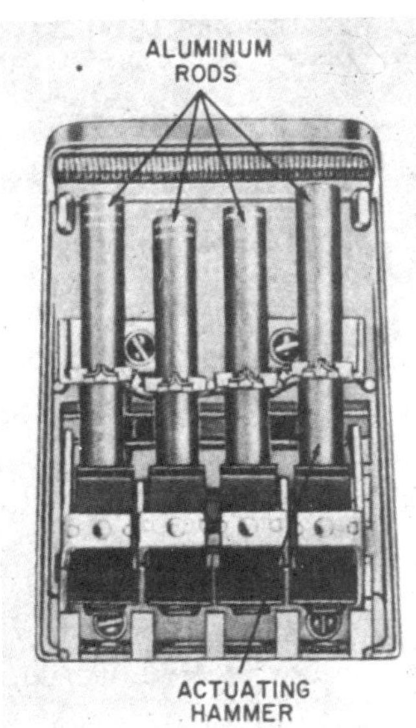

Fig. 7-7. The external appearance and internal construction of an ultrasonic transmitter for the remote control of a TV receiver. (*Courtesy of Zenith Radio Corp.*)

A single rod of a specified length will produce a certain resonant frequency. Therefore, to control three or four functions within a television receiver, three or four rods of slightly different lengths are employed (see Fig. 7-7). In the illustration we see the working parts of the transmitter. A hammer, a steel cylinder weighing about 2 1/2 grams, is located at one end of the cylindrical rod. When a button is pressed, the hammer is pushed away from the rod by the force of a spring. As the button is further advanced, the spring is suddenly released and the hammer strikes the rod. Generally, if more than one rod is employed, the difference in frequencies of the various rods may be on the order of 1,000 Hertz or so.

Once a rod has been struck and the energy transmitted, it is then desirable to dampen the remaining energy as quickly as possible. For this purpose, a mechanical damping method is employed. In the unit shown in Fig. 7-7, damping is achieved by a small piece of spring wire, covered by a plastic sleeve, which protrudes through the mounting and touches the rod. When the button is pressed, the damper is withdrawn; when it is released, the sleeve again makes contact with the rod.

At the receiver, the ultrasonic energy is picked up by a crystal-type microphone using a barium titanate crystal element. This material, when cut in the form of a small bar or plate and placed between two conducting electrodes, will generate a voltage when the bar or plate is mechanically strained by sound waves. Conversely, when a voltage is applied across the two electrodes, the barium titanate will be mechanically strained. This is the well-known piezoelectric effect.

Figure 7-8 pictures a typical microphone. Two thin rectangular wafers of barium titanate are combined, as shown in part A. Silver (the conducting electrode) is applied over a small section of each end of each wafer. These electrodes are indicated in part A by the dark segments at each end of the assembly. Between the two wafers, at the nodal points of vibration, two thin metal strips are cemented which serve as electrical contacts and as a mechanical suspension.

To broaden the response of this transducer, a small U-shaped piece of aluminum is added to the assembly, as shown in part B. This makes the microphone behave like two tuned circuits closely coupled together. It enables ultrasonic frequencies from 37.75 kHz to 41.25 kHz to be picked up and converted into their equivalent electrical signals.

Figure 7-8(C) shows the entire microphone in cross section. From left to right, there is a supporting piece which carries the barium titanate wafers by means of the two thin metal strips mentioned previously. Next is the aluminum bridge to which the wafers are cemented, together with a plastic piece and a rectangular window which fits closely around the bridge. Beyond the bridge, there is a space equal in length to a one-quarter wavelength at about 40 kHz and, beyond this, a rectangular horn two inches long. Both the one-quarter wavelength space and the horn serve to match the impedance of the barium titanate assembly to the air. Or, to state it another way, the horn and the one-quarter wavelength space help to couple the wafers to the air.

The combination of the mechanical transmitter, the microphone, and the amplifier in this system provides sufficient sensitivity to make the aiming of the transmitter quite unnecessary in most homes. Sound reflected from the floor, walls, ceiling, or furniture makes it possible to operate the receiver controls positively with the transmitter in almost any desired position—held in the hand or resting on a table or chair. The line of sight and the approximate aiming of the transmitter become important only at maximum range—a distance of 40 feet—which is rarely encountered.

Electronically-Generated Sound Signals Instead of using a mechanical device, such as a rod of aluminum or a column of air set into vibration, the ultrasonic generator may be comprised of an electronic oscillator and a speaker. Such a system is shown in Fig. 7-9(A). A typical remote-control panel is illustrated in Fig. 7-9(B).

The oscillator Q_1 is a Hartley type. The primary of transformer T is tapped to provide the regenerative feedback from collector to base by way of a feedback capacitor C_1. Base bias is established by the resistor R_1 connected to the negative side of the supply; and, the collector is connected directly to the negative side of the 9-volt battery when one of the function switches is closed. Note that the oscillator base and collector circuits are open, when all of the switches are open.

Each of the function switches connects a different capacitor combination across the secondary winding of the transformer T. Capacitors C_3, C_4 and C_{13} are permanently connected across this winding. The

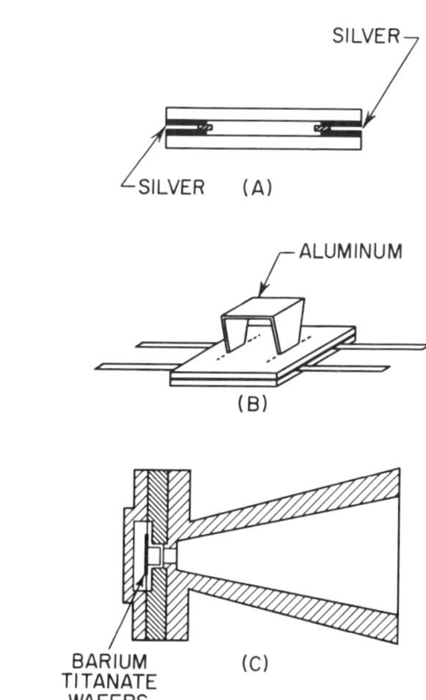

Fig. 7-8. The microphone utilized by Zenith to pick up the ultrasonic sound of the transmitter shown in Fig. 7-7.

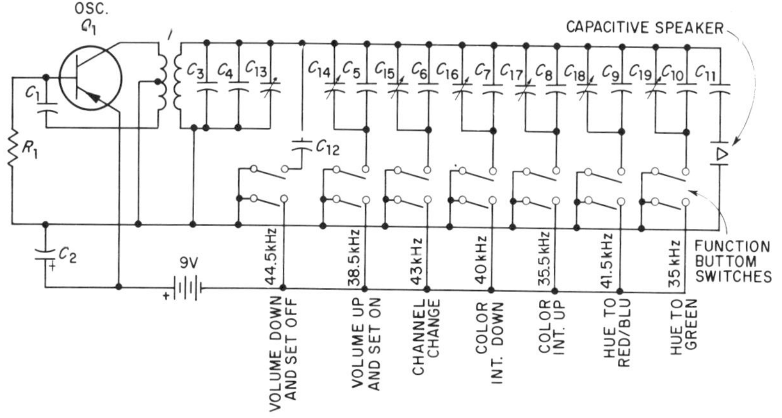

Fig. 7-9(A). With this type of remote transmitter, the ultrasonic signals are developed by an oscillator and converted to sound waves by a capacitive speaker. The control panel is shown in part B. (*Courtesy of Motorola Consumer Products, Inc.*)

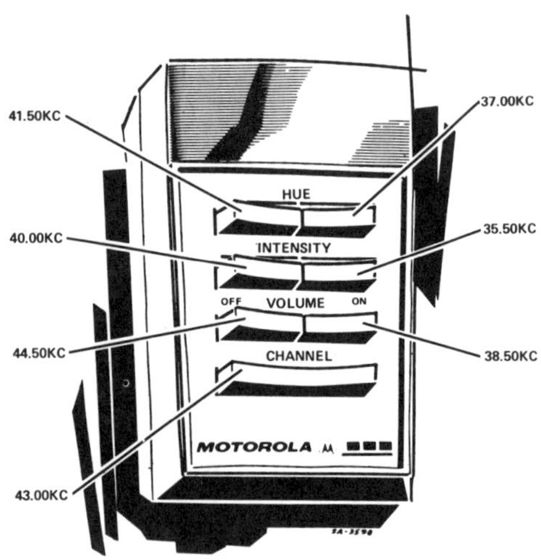

Fig. 7-9(B). A typical remote control panel for a color-TV receiver. Ultrasonic frequencies are transmitted to the receiver to provide the various indicated functions. (*Courtesy of Motorola Consumer Products, Inc.*)

oscillator frequency is controlled by the total amount of capacitance across the secondary winding. Closing a function switch places another capacitor combination across the C_3, C_4, C_{13} connection, thus increasing the capacitance and lowering the resonant frequency. The oscillator signal is delivered to a capacitive speaker that radiates the ultrasonic sound.

The capacitor C_2 is an electrolytic type placed across the 9-volt supply when the circuit is in operation. This prevents degeneration from occurring due to the internal resistance of the battery.

Let us assume that the "channel change" switch is pushed on the transmitter. (All of the switches are push-button types that are normally open.) The tracing of the circuit will show that this completes both the base and collector circuits for the transistor Q_1, and the circuit will oscillate. At the same time, capacitors C_{15} and C_{16} are connected in parallel with the secondary of transformer T, increasing the secondary capacitance and decreasing the oscillator frequency. This arrangement is quite common with remote-control circuits that employ electronically-generated ultrasonic signals. In the receiver, that circuit which is resonant to the 43 kHz signal will be actuated. This, in turn, will operate a motor that turns the channel selector switch from station to station.

Note that there are two function switches for controlling the volume. One is for turning the volume up, and the other is for turning the volume down. When the volume has been turned down all of the way, the set is turned off, but it can be turned on again from the remote position by pushing the "volume up" switch. Two "color intensity" switches are also provided for turning the color intensity up, or turning it down. Color intensity is sometimes referred to as the "color control" or "color amplitude." It sets the *amount* of color in the picture, but not the hue. In other words, varying the color control should change the greens from a light color to a dark color, but they should *not* change them from green to some other color. The hue control, which on some receivers is called the "tint control," sets the receiver for the proper phase relationship of the color signals. The hue control should be set for the proper flesh tones in the reproduced color picture. Operating the hue control, then, will change the flesh tones to a more greenish color, or to a reddish-blue color, depending upon which of the function switches is depressed.

Receivers for Ultrasonic Control Systems Three or four transistors are generally used for amplifying the ultrasonic signal from the microphone. This amplifier system is usually referred to as the *preamplifier*. Figure 7-10 shows an example of a transistor preamplifier. The signal from the microphone is coupled from J_1 to Q_1 through the coupling capacitor C_1. Resistance-capacitance coupling is used between the stages. The gain of Q_3 is set by R_{12}; the setting of this resistor determines the sensitivity of the system.

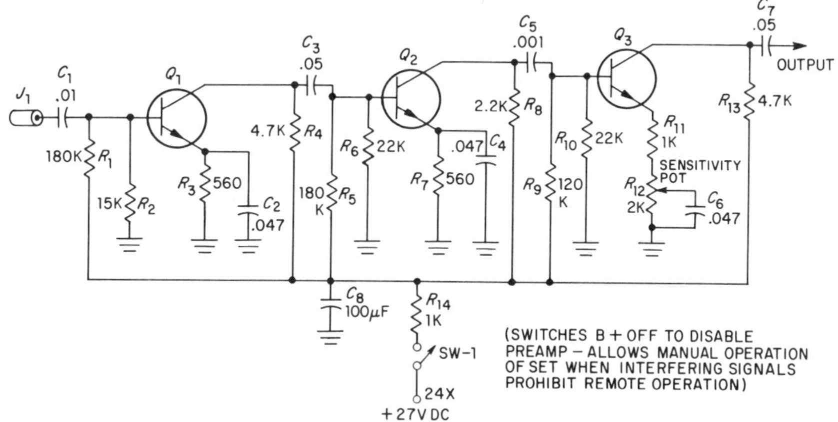

Fig. 7-10. This receiver amplifier is typical of those used with ultrasonic remote-control systems.

Figure 7-11 shows a block diagram of a typical receiver for an ultrasonic remote-control circuit. The ultrasonic transmitter, which is sometimes referred to as a *bonger*, is used to produce sounds at various ultrasonic frequencies according to the control being actuated. Different frequencies are used for the different functions that can be controlled by the hand-held transmitter.

The microphone receives the sound and converts it to an electrical signal that can be amplified. In most receivers there are a number of amplifier stages for this purpose. The output of the amplifier is delivered to a line which is used as the input to a number of *L-C* tuned circuits. Each tuned circuit is resonant to a particular frequency, and it will not respond to any of the other frequencies delivered by the ultrasonic transmitter. Suppose for instance, that the transmitter is sending a 37 kHz ultrasonic frequency. This signal is converted to an electrical signal and amplified and then fed to all of the tuned circuits shown in the block diagram. However, only the tuned circuit marked "37 kHz" will respond to this signal.

When an input of 37 kHz is present, the signal operates a control circuit which may be a relay, a stepping switch, a dc motor or an electronic control circuitry. Motors and stepping switches are often used to turn

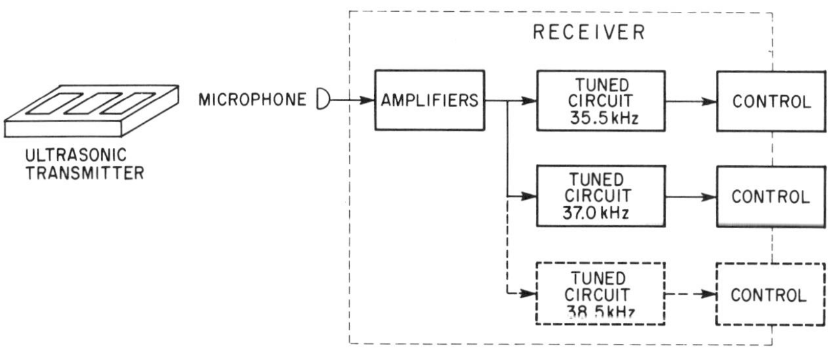

Fig. 7-11. Block diagram of a receiver used with an ultrasonic control system.

or vary the controls. Although there are only three tuned circuits shown in the block diagram of Fig. 7-11, it is possible to have as many as eight or ten, depending upon the number of functions being controlled.

Figure 7-12 illustrates a typical control circuit for an ultrasonic receiver. The input to this circuit comes from the amplifier, and is delivered simultaneously to a number of tuned circuits. (Only two of the tuned circuits are shown in Fig. 7-12.) Close inspection of the transistor circuits shows that the transistor bases are grounded through coils L_1 and L_2, and that there is no dc base bias present. When a transistor is not forward biased, it is in the cutoff condition. Therefore, all of the transistors in this line are normally at cutoff.

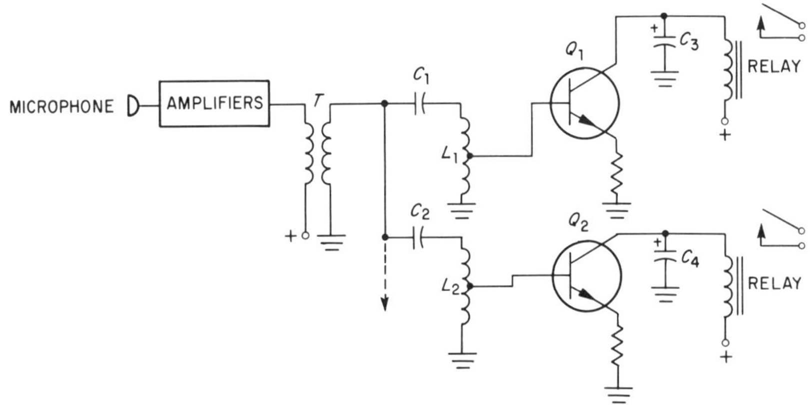

Fig. 7-12. Typical control circuitry.

Let us assume that an input signal of 37 kHz is present. The tuned circuit in the base of Q_2 is resonant to this frequency, and a signal will be delivered to the base. Whenever this signal goes in a positive direction, the base is forward biased and the transistor conducts. The output signal of the transistor is a series of pulses which are smoothed by the capacitor *C* and the coil of the relay. The relay is actuated, and the relay contacts are closed. These relay contacts are used to complete the circuit of a motor or a stepping switch.

7.6 AN ALL-ELECTRONIC REMOTE CONTROL SYSTEM

The Motorola remote-control system described here is an example of a truly solid-state system. The only electromechanical devices are an on-off relay in the power supply circuit, and a relay and a motor in the channel selector. The remote transmitter for this system is the one shown in Fig. 7-9. In the receiver, the electrical signal from the transducer is fed to a three-stage preamplifier similar to the one shown in Fig. 7-10.

The preamplifier feeds the *all-function driver* illustrated in Fig. 7-13. The all-function driver, in turn, drives the *function output* stage and the channel selection discriminator coil. The function output stage is a

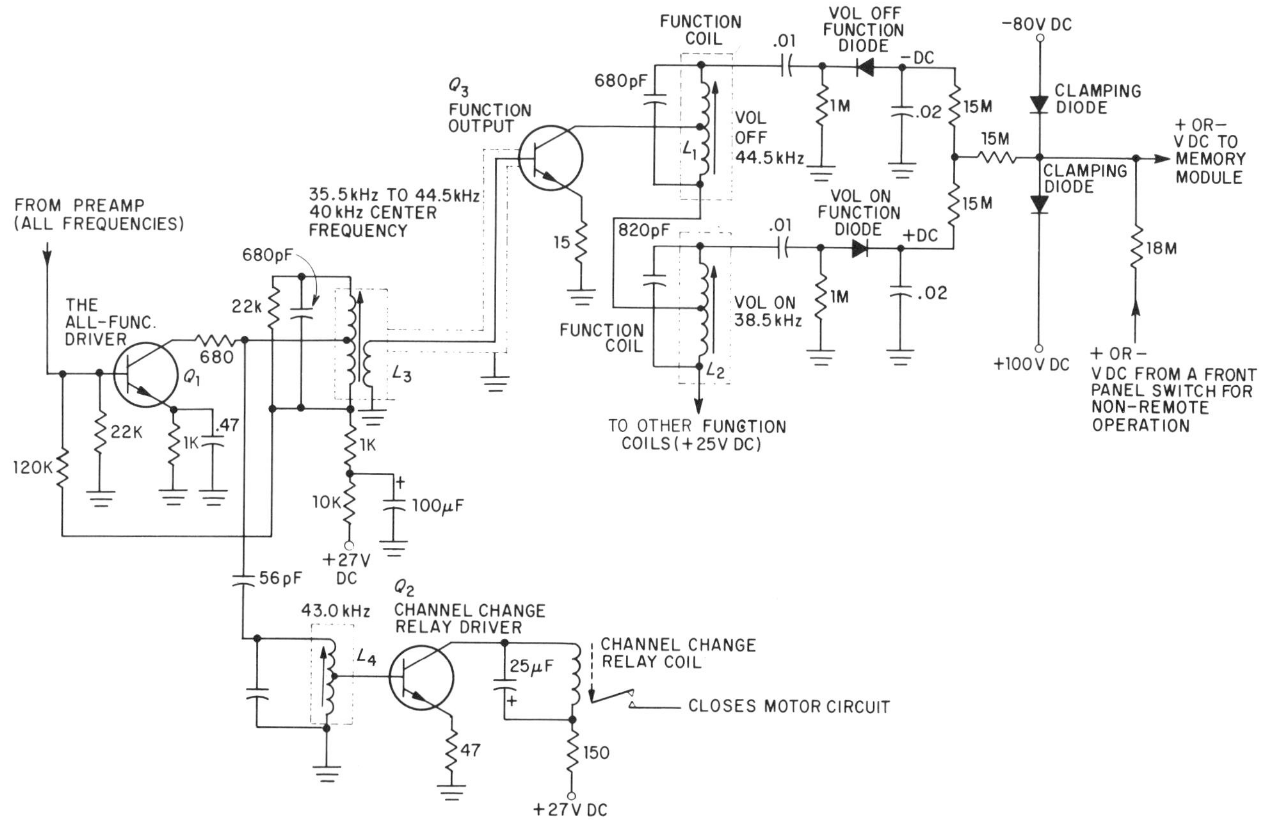

Fig. 7-13. The all-function driver and the function output stages of the Motorola remote-control system.

tuned input amplifier which accepts signal with frequencies ranging from 35.5 kHz to 44.5 kHz. Up to this stage the response is a wideband; but from this point on the receiver is very selective. The output stage feeds the function discriminators. The name *function discriminator* is somewhat misleading, because this is not the type of discriminator used for demodulating the FM signals. Instead, it is a frequency-selective system that accepts one frequency and rejects all others. Thus, it discriminates against the unwanted frequencies and is sensitive to only one frequency for each function. Therefore, its name "function discriminator."

The function discriminators consist of three pairs of coils and two function diodes. The six coils (two for each function) form a series load for the function output collector. Figure 7-13 shows the function discriminator circuitry for the *volume off-volume on* functions with the clamping diodes for the memory module. (The operation of the memory module is explained later in this section.) Each of the function coils is a high *Q* coil with a bandwidth between 500 and 1000 Hz. Under these conditions, the coil appears as a virtual short, unless the frequency of the incoming signal is resonant with the coil.

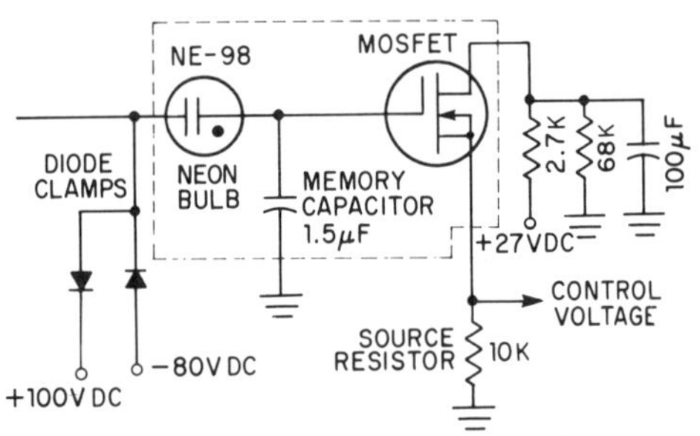

Fig. 7-14. A memory module circuit used in the Motorola remote-control system.

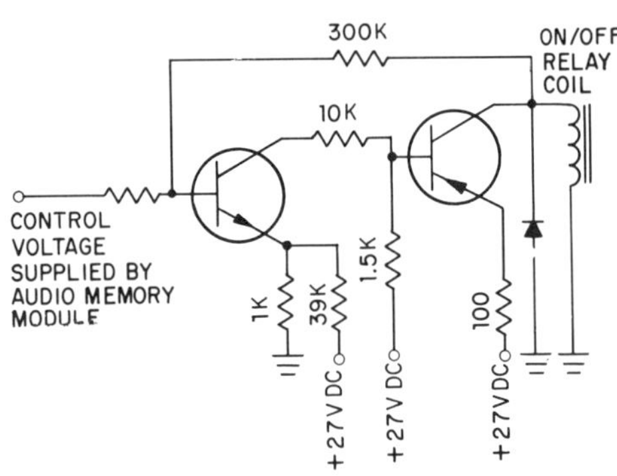

Fig. 7-15. Solid-state on-off switch (single-shot flip-flop).

At resonance, the coil develops a maximum ac voltage which is then rectified by one of the function diodes. In Fig. 7-13, if a 38.5 kHz amplified signal is fed to L_1, it is passed because this high Q coil is tuned resonant at 44.5 kHz. The system develops a maximum voltage across L_2 which is tuned to 38.5 kHz. The ac voltage from L_2 is half-wave rectified so that a positive voltage appears at the clamping diode of the memory module. The operation of the system beyond this stage depends on the working of three *memory modules* for intensity, volume, and hue functional controls.

The memory module pictured in Fig. 7-14 is a simple device composed of a neon bulb, a low leakage capacitor, and a MOSFET.

Recalling that a MOSFET has a large input impedance (of the order of 10^{14} ohms), and that a neon bulb will not conduct until the voltage across it reaches the ionization potential, both appear to be open. Once the memory capacitor is charged, it will remain so for a long while (more than 1000 hours). The charge on the capacitor holds a fixed potential on the gate of the FET, and thus fixes the drain and source currents. Voltages used for control purposes in later stages of the system are taken from a 10,000 ohm resistor in a source follower arrangement. Charging current for the capacitor goes through the neon bulb. Voltages applied to the bulb are clamped by the diodes external to the module. Should the function signal exceed + 100 V, the 100 Vdc clamping diode conducts, thereby limiting the applied voltage to + 100 Vdc. Since the ionization potential of the bulb is approximately 80 volts, the remaining 20 volts is used to charge the memory capacitor. The 80 volt clamp operates in the same manner with the exception of the direction of the current flow, which is opposite to the one in the previous instance. The negative clamp is lower than the positive clamp, so that the FET gate voltage will not exceed its cutoff (or pinchoff) voltage.

The operation of the circuits in the remote-control system has been uniform up to this point. That is, all functions have their signals generated, received, amplified, discriminated, and stored in exactly the same manner. However, the way in which the control voltage taken from the source resistor (10,000 ohms) is used varies from function to function.

The "on-off" function is accomplished, for instance, by the control voltage operating a one-shot flip-flop that energizes a relay. This circuit is shown in Fig. 7-15. The audio, on the other hand, is controlled by changing the bias of a two-stage audio amplifier as shown in Fig. 7-16. One of the two 3.58 MHz continuous wave color reference signals (that are 180° out of phase) has its amplitude varied by a controlled amplifier. Also, a single-stage controlled resistor decides the intensity. Similarly, a driver transistor which is used as a switch operates a relay to pull in the channel selection motor. Each function will now be discussed in detail. The control functions will be described in the order by which the discriminator coils are arranged.

The remote transmitter has two volume control buttons: *volume-off*

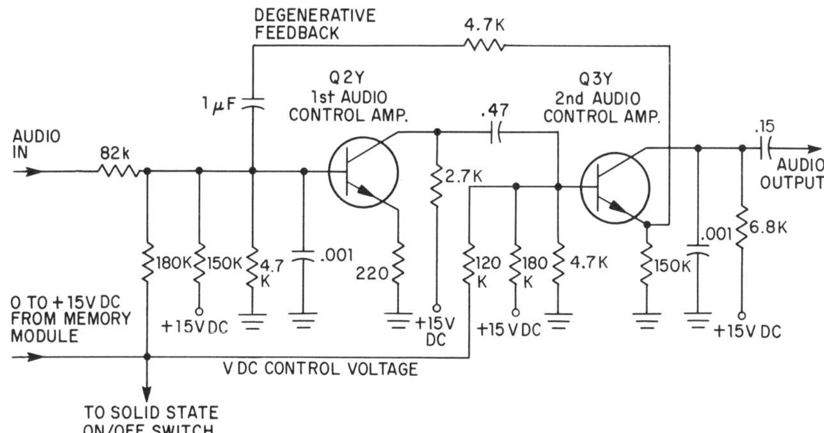

Fig. 7-16. Two-stage audio amplifier with a degenerative feedback. The control voltage from the memory module sets the gain by controlling the base bias voltage. (This amplifier is used in the Motorola remote-control system.)

and *volume-on* (see Fig. 7-9). Pressing the "on" button will charge the capacitor in the volume memory module (see Fig. 7-14). The potential at the gate of the FET rises above the pinchoff, and the FET conducts. A control voltage is thus developed across the source resistor. This control voltage splits between the audio and on-off controls and is thus used for two operations. One is to supply the turn-on voltage of a single-shot flip-flop, which Motorola calls a *Solid-State Switch*, pictured in Fig. 7-15. The load for the switch is a relay coil. When this is energized, the relay contacts that are in series with the main off-on switch are closed (see Fig. 7-17). The closing of these two sets of contacts applies power to the power transformer primary, the CRT filament transformer, and the power transformer. The skip and hold switches, along with the tuner motor and relay contacts shown in Fig. 7-17 will be explained later. Since the Motorola chassis has the *quick-on* capability, a quick-on defeat switch is in series with the CRT filament transformer.

The other path for the control voltage is to an audio preamplifier. An audio signal is received from the chassis, controlled in the remote-control unit, and returned to the chassis for further amplification. Figure 7-16

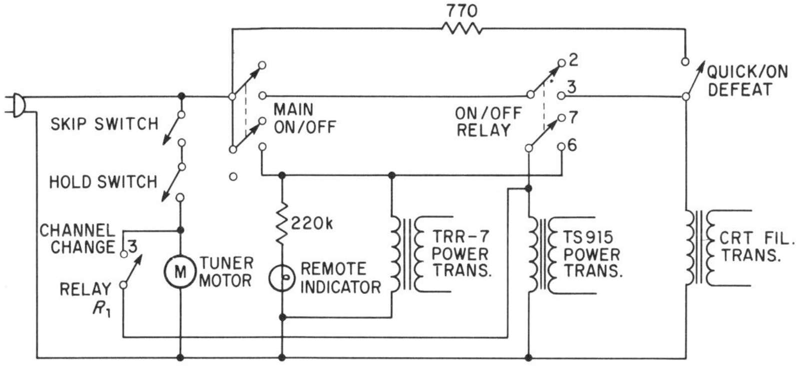

Fig. 7-17. This circuit shows how the on-off relay in the solid-state switch controls power to the receiver. (It is used in the Motorola remote-control system.)

illustrates the two-stage audio preamplifier located in the remote unit. The variable dc voltage received from the memory module shifts the bias of both of the transistors. Increasing the forward bias increases the volume. Pressing the *volume-off* control reverses the above process. Decreasing the forward bias of the two amplifiers decreases the volume level. When the control voltage falls below the firing point of the one-shot flip-flop, the switch snaps back to the equivalent position of *off*. The holding current for the relay (Fig. 7-17) is removed and the contacts open. The set is now off, but the filament voltage is still applied to the CRT through the quick-on switch.

Control of the hue reference phase is the result of adding two continuous-wave, 3.58 MHz signals that are initially out of phase by 180° and are generated in the appropriate circuitry in the receiver. Varying the amplitude of one of the signals controls the hue of the picture. In the remote system, pressing the hue-control button results in a voltage change in the hue memory module. This hue-control voltage is applied to a two-stage amplifier (Fig. 7-18). The signal to be

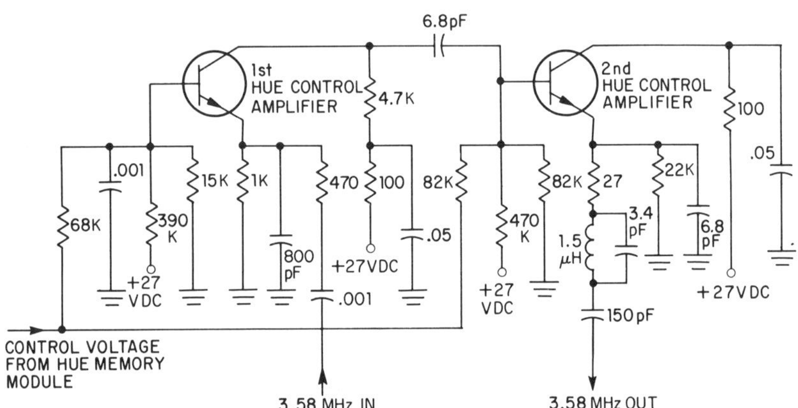

Fig. 7-18. The hue-control amplifier used in the Motorola remote-control system.

controlled is accepted from the main receiver chassis and fed to a common-base transistor amplifier. It passes through an emitter follower stage, resulting in a power gain without suffering a phase inversion or losses due to impedance mismatching. If the receiver did not have the remote control, the same objective could be accomplished by a potentiometer. The components used in the circuit illustrated in Fig. 7-18 that are not involved for biasing and load are used to eliminate amplified transients and harmonics which would interfere with the good picture reproduction. The more positive the voltage out of the function discriminator, the more positive the voltage on the FET gate. This causes an increase in FET conduction and an increase in the control voltage. Increasing the control voltage increases the bias on the hue-control amplifier resulting in an increase in the signal amplitude and the hue shift to red-blue. Decreasing the gate voltage has the opposite effect, *i.e.*, towards green.

The intensity control operates in a similar manner. As the control voltage from the module increases, the conduction of the intensity amplifier is increased. The use of the word "amplifier" is a misnomer in this application. The transistor is used as a variable resistor which simulates the action of a control potentiometer. A high intensity is achieved by an increased (positive-going) control voltage. Figure 7-19 shows the intensity *variable resistor*.

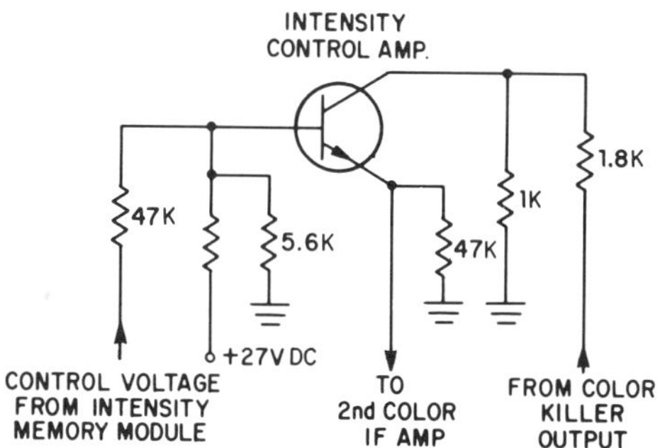

Fig. 7-19. The intensity control amplifier (which is really a transistor that behaves as a variable resistor. (It is used in the Motorola remote-control system.)

Channel selection is the only remote function that does not use a memory module. The all-function driver (Q_1) is loaded by two coils (Fig. 7-13). One is the low Q coil L_3 (35.5–44.5 kHz). The other is the narrowband, high Q coil L_4 which is resonant at 43.00 kHz. The voltage developed is used to turn the channel-change relay-driver transistor fully on. The collector current through transistor Q_2 pulls in the relay to apply power to the tuner motor. The location of the relay with respect to the power supply is shown in Fig. 7-17. When the motor starts to turn, it performs a solenoid action. The forward motion towards the front of the set is used to activate a *hold switch.* A *skip switch* is in series with the hold switch. The combination is in parallel with the relay contacts. *Skipping* is accomplished by detuning the fine-tuning coils. Rotating the fine-tuning knob counterclockwise at least ten turns will bring the tuning slug out so that the head of the slug will activate the skip switch (Fig. 7-17). The initial rotation of the motor is enough to carry the tuner drum past the channel which is to be changed. Each channel that has its slug out keeps the skip switch closed. The hold switch is closed, therefore, the motor keeps turning until a channel is in place which allows the motor to stop by opening the skip switch. Programming the tuner allows the operator to select one channel by pressing the button only once and then releasing it.

TABLE 7-1 Remote-Control Transmitter Functions and Frequencies for the Motorola System.

No.	Function	Oscillator/Transmitter frequency kHz
1	Volume-up, on	38.50
2	Volume-down, off	44.50
3	Intensity-up	35.50
4	Intensity-down	40.00
5	Hue-Red-Blue	41.50
6	Hue-Green	37.00
7	Channel change	43.00

7.7 TROUBLESHOOTING TECHNIQUES

As with any other section in a television receiver, good tools and instruments, and the manufacturer's literature are indispensable when you are troubleshooting. Chapter 24 covers troubleshooting techniques in detail, so the following will just be a discussion of general methods.

Troubleshooting AFT Circuits Automatic frequency controls, automatic gain controls, and automatic fine-tuning circuits are all examples of closed-loop systems. This means that part of the output signal is fed back, sometimes after it is modified in some way, to the input of the circuit. Locating troubles in this type of system is difficult, because the output signal depends upon the input signal and the input signal depends in part upon the output signal.

Suppose, for example, that a certain receiver with an AFT circuit drifts off frequency after it is tuned to a channel. This could be because the automatic fine-tuning control in not working, or it could be because the oscillator is drifting beyond the range of frequencies that the AFT circuit can control.

Manufacturers usually give typical operating voltages and test points for measuring these voltages. This is the first step in looking for a fault in an AFT circuit. For other than routine troubles, the only sure way to isolate a fault in a closed-loop system is to open the feedback loop. In some systems, it may be necessary to substitute a signal or a voltage to compensate for the open circuit. In an AFT system, the procedure is to open the part of the circuit that delivers a dc voltage to the oscillator control—that is, to the reactance tube or varactor diode. Very often receivers have an *AFC ON-OFF* troubleshooting switch to make it a simple matter to open the loop. A dc voltage from a battery or a bias pack is substituted for the dc. This procedure is illustrated in Fig. 7-20.

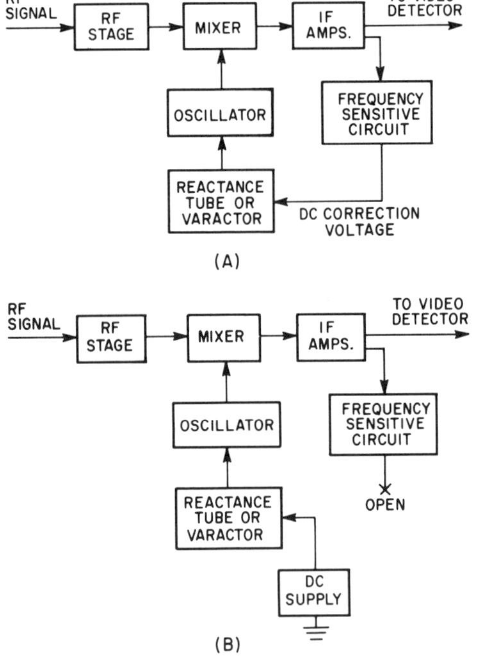

Fig. 7-20. The AFT circuit is a closed-loop system that presents a difficult troubleshooting problem. (A) The closed-loop system. (B) In order to troubleshoot a closed-loop system, some part of the system must be opened.

Figure 7-20(A) displays the AFT system in a block diagram form. There are several points in this closed-loop system that could be open circuited. The input to the frequency-sensitive circuit could be opened, but this is likely to affect the performance of the IF amplifier due to the change in its load. The reactance tube or the varactor circuit could be opened, but this would probably put the oscillator completely off frequency. Therefore, the most likely place to open the circuit is in the dc line. This is indicated in Fig. 7-20(B). The dc supply replaces the normal dc feedback voltage.

Once the feedback loop is opened, standard troubleshooting techniques can be used.

Troubleshooting Remote-Control Systems The ultrasonic sound from the remote-control transmitter may be produced by mechanical means. In that case, it is not likely to be a source of trouble. If the ultrasonic sound is produced by a transistor oscillator, the most likely problem will be a depleted battery. Usually the customer can replace this battery himself, but it should not be overlooked as a source of trouble. If the remote transmitter is the type that radiates a modulated RF signal, any point in the circuitry—including the battery—can be at fault. Voltage measurements are the first step.

One useful technique for determining if a remote-control transmitter is working properly is to tune an AM radio to a harmonic frequency of its radiation. (This technique will, of course, only work for transmitters that radiate electromagnetic waves.) If the transmitter is working, a buzzing sound should be heard in the radio loudspeaker.

If *one* of the remote-control functions, such as the " volume increase," does not work, but all other functions do work properly, then the transmitter and receiver circuitry is the least likely source of the trouble. Instead, one of the components in the section related to the inoperative function is probably at fault.

If none of the functions work, then the trouble may be in either the transmitter or the receiver. The first step is to check the power supplies and the operating voltages. The ultrasonic receivers can be checked by the same techniques used for troubleshooting audio amplifiers. Receivers designed for radio control are usually of the TRF type, and troubleshooting is accomplished with an RF and an audio generator.

REVIEW QUESTIONS

1. Explain why it is so important to have the fine-tuning control properly adjusted in a color receiver.
2. In what section of the tuner would you expect to find the reactance tube connected?
3. How is the dc correction voltage obtained in an AFT circuit?
4. Explain how the voltage across a varactor controls its capacitance value.
5. Describe two methods of producing ultrasonic waves for remote-control systems.
6. In what type of remote control system is a microphone used for picking up the remote control system?
7. What are the memory modules used for in an all-electronic remote-control system?

8 Video IF Amplifiers

8.1 FUNCTIONS OF VIDEO-IF AMPLIFIERS

The tuner and the common video-IF amplifiers are the only sections of the TV receiver that handle the entire band of frequencies required for the reproduction of the picture and sound. As discussed previously in Chapter 6, the tuner selects the desired TV channel and then converts the selected band of frequencies to the intermediate-frequency band of the video-IF amplifiers. These IF amplifiers then provide the signal amplitude necessary to drive the video detector, and the sound and video amplifiers. In this process, high-quality pictures and sound are dependent on the ability of the RF tuner and the IF amplifiers to select and amplify only the desired frequencies. All others must be attenuated to a very low level. Color video-IF amplifiers must also remove the sound IF frequencies before they reach the video detector. Figure 8-1 shows the relationship of the video-IF amplifiers to the tuner and the video detector, for both monochrome and color receivers.

In addition to its primary function of driving the video detector, the video IF amplifiers must compensate for the vestigial sideband method of transmission. This compensation is required to provide a uniform output at the video detector for frequencies near the video carrier as well as those in the higher frequency portion of the IF band. Also, the video-IF amplifiers must provide special tuned circuits, or traps, to prevent interference from the associated channel sound, the adjacent channel

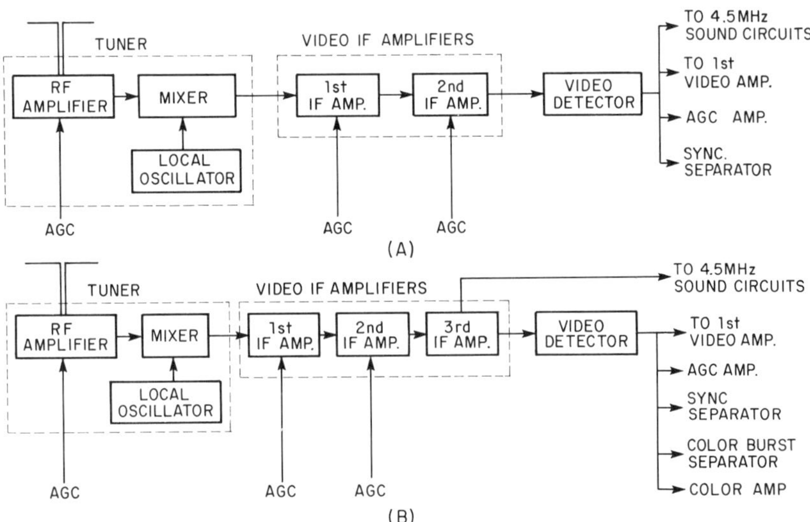

Fig. 8-1. The relationship of the tuner, video-IF amplifiers, and a video detector for (A) monochrome and (B) color sets.

sound, and, in some cases, the adjacent channel picture carrier. In addition, a color-set video IF amplifier must separate the sound from the picture frequencies before they arrive at the video detector, to reduce interference patterns on the picture tube caused by the heterodyning between the sound IF frequencies and the 3.58 MHz color subcarrier (see Fig. 8-1(B)). In monochrome TV receivers using intercarrier sound systems, the video IF amplifiers must attenuate the associated sound carrier so that its amplitude will be at a low level at the output to the video detector to prevent sound bar patterns on the picture tube.

From the foregoing discussion, it can be seen that the most important functions of the video-IF amplifiers are selectivity, gain, and elimination of unwanted frequencies.

In any discussion concerning selectivity and gain, it is important to speak in terms of bandwidth, or frequency response. In this light, anyone studying the functions of video-IF amplifiers must thoroughly understand frequency response curves and how they are related to the operation of the amplifiers. Figure 8-2 presents a typical response curve for an arbitrary amplifier. This curve represents the output voltage *versus* the frequency response of the amplifier. In solid-state circuits, the curve could represent the output current, or voltage, developed by the transistor amplifiers. Notice in the figure that as the frequency increases, the voltage remains at a low level until it is near 2 MHz. At that point, further increases in frequency cause the voltage to increase, until at around 5 MHz the voltage starts to decrease. Finally at about 8 MHz, the voltage, once again, is at its original low level. This curve then represents the response of our arbitrary amplifier. Its bandwidth is approximately the band of frequencies between the lowest and highest frequencies on the response curve that produces a specific output. This particular output is specified for video IF amplifiers and is discussed in greater detail in Section 8.5. The selectivity of the amplifier is determined by the steepness of the slopes, or skirts, of the response curve. The steeper the slope, the greater the selectivity because unwanted frequencies outside the curve will develop a much lower output voltage. The gain of the amplifier is represented by the highest point, or peak, on the response curve. This gain is usually expressed in decibels.

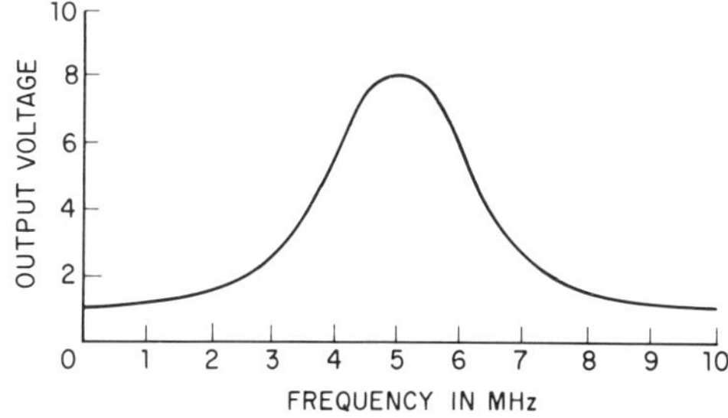

Fig. 8-2. The typical response curve of an arbitrary amplifier.

A brief summary of the major functions of the common video IF amplifier in both monochrome and color-TV receivers is as follows:

(1) It provides most of the radio frequency gain in the receiver.
(2) It provides most of the radio frequency selectivity in the receiver.
(3) It reduces accompanying sound interference.
(4) It reduces the adjacent channel interference.
(5) It separates the sound and picture IF frequencies.
(6) It supplies IF signals of sufficient amplitude to drive the AM video detector (and video amplifiers) and the FM sound detector (and audio amplifiers).

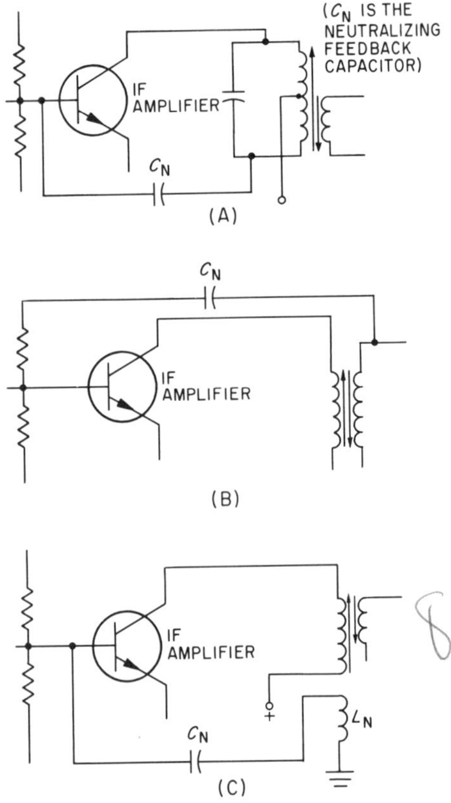

Fig. 8-3. Several neutralizing circuits for transistor-IF amplifiers. (A) An IF amplifier neutralized by using a tapped transformer winding. (B) The neutralization process without the need for a tap on the transformer winding. (C) Obtaining a neutralizing signal with a separate transformer winding.

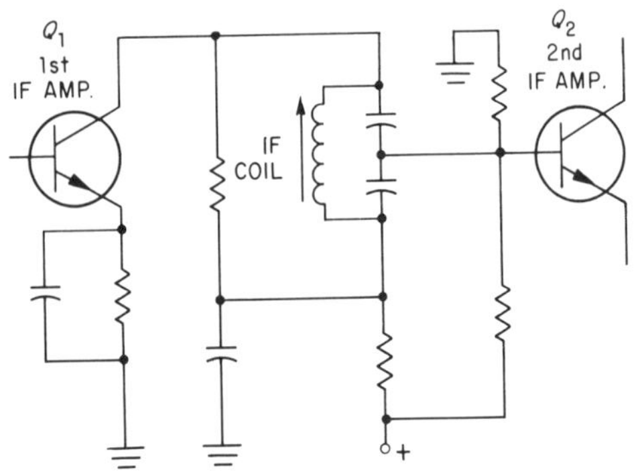

Fig. 8-4. The circuit used for impedance matching between IF amplifiers. Many variations of this circuit are possible.

As will be described in detail later in this chapter, the video-IF amplifiers used in color sets are somewhat more complicated in their circuitry and are more critical in their alignment. In general, they may be required to pass a greater bandwidth of frequencies, without attenuation, than the ones used for many monochrome sets, in order not to attenuate the color IF sideband frequencies.

8.2 COMPARISON OF VACUUM-TUBE AND SOLID-STATE VIDEO-IF AMPLIFIERS

Functionally, vacuum-tube and solid-state video IF amplifiers are identical. In construction, though, they are quite different. Vacuum-tube amplifiers are either hand-wired on a metal chassis or mounted on printed-circuit boards. Solid-state IF amplifiers are invariably constructed on printed-circuit boards. Also, some of the more recent solid-state receivers use integrated-circuit IF amplifiers mounted between tiny IF transformers on a printed-circuit board. In modular construction, though, both vacuum-tube and solid-state circuits have been used.

There are, however, several distinguishing features of the transistor amplifiers that should be noted. Whereas pentodes are used extensively in vacuum-tube IF amplifiers, and therefore neutralization is not needed, the transistor is more comparable to a triode. The feedback signal between the collector and base (through the junction capacitance) must often be neutralized. The neutralizing circuit may take several different forms. A few examples are shown in Fig. 8-3. Figure 8-3(A) presents the neutralization that takes a voltage from the tap on the IF transformer. Another method of neutralization is accomplished by taking a voltage of the proper phase from the secondary of the transformer. This method is presented in Fig. 8-3(B). It has the advantage of not requiring a tap on a transformer winding. In a third method of neutralization, presented in Fig. 8-3(C), a separate winding, L_N, on the transformer provides the feedback signal required for the neutralization. All solid-state video IF amplifiers are not neutralized in TV receivers. If the amplifier has a low gain and low impedance, and it is a broadband amplifier, then neutralization may not be required.

Another important feature of transistor IF amplifiers is the need for impedance matching between the collector circuit of one amplifier and the base circuit of the following stage. The input impedance of a transistor is quite low in comparison to that of a vacuum tube. This assumes, of course, that a common emitter circuit configuration is used, since this type of circuit is by far the most popular used for the IF amplifiers. Impedance matching can, of course, be accomplished by selecting the proper turns ratios for the transformers used in the IF stages. There is, however, some tendency to avoid the use of transformers in low-cost TV receivers, and, instead, impedance coupling is used extensively. When impedance coupling is used, a special impedance-matching network is needed, such as the one illustrated in Fig. 8-4.

In vacuum-tube circuits it is common practice to deliver a negative AGC voltage to one or more of the IF amplifier stages. In transistor IF amplifiers, either NPN or PNP transistors may be used, and, therefore, the AGC voltages may be either positive or negative. (AGC circuits are discussed in Chapter 10.)

Printed-circuit boards are employed in vacuum-tube as well as in all-solid-state video-IF amplifier circuits. Those boards used in conjunction with vacuum tubes are usually made of a thicker material than their solid-state counterparts. The extra material thickness is required to provide the strength and support for the heavier vacuum tubes and their associated sockets. Also, the vacuum-tube PC's require an extra printed circuit for the tube filaments. The solid-state PC's sometimes provide plug-in sockets for transistors, but on many boards, the transistors are soldered directly onto the circuits printed on the board. This latter construction technique, even though it is less convenient for the repairman and presents slightly more servicing difficulty, is by far the most efficient and reliable because the transistors are not connected into the circuit by spring-type contact. Another attribute of the PC is its adaptability to modular construction. For this application, contacts are mounted along the edges of the board for easy insertion in mating connectors on the master chassis or rack (see also Chapter 1).

Integrated circuits have been developed for both monochrome and color video-IF amplifiers. These components are about the same size as the transistors normally used for the IF amplifiers but contain all of the components, including the transistor, for one IF stage. The IC's do not, however, contain the IF transformer or the coupling-impedance device. Each IC unit has from 12 to 14 base pins and plugs into a socket in a fashion similar to vacuum-tubes and transistors. These IC's are finding use in both hybrid and all-solid-state receivers. IC's are particularly useful in modular construction and are finding widespread applications in high-quality receivers, such as Motorola's Quasar.

A major difference between the vacuum-tube and solid-state video-IF amplifiers is size. Even though vacuum tubes are used in modular construction, the all-solid-state circuits using transistors or IC's are more ideally suited for this technique. Most solid-state video-IF modules are no larger than an ordinary playing card, whereas the vacuum-tube module is two to three times that size. The solid-state circuits require no filament voltage and produce far less heat.

8.3 TYPES OF VIDEO-IF AMPLIFIERS

Even though all video-IF amplifiers perform basically the same function, there are several different types that must be considered. First, video-IF amplifiers are designated as monochrome or color types, depending on the TV receiver for which they are intended. Second, according to their circuit components, they are either vacuum-tube or solid-state circuits.

Third, within both the first and second classification, all video-IF amplifiers are classified according to their number of stages; one, two, three, or four. However, the two- and three-stage amplifier systems are the most common. Finally, all video-IF amplifiers are classified according to the method used for their interstage coupling. In the succeeding discussion, all of the classifications are presented separately.

Monochrome Video-IF Amplifiers Most monochrome-TV receivers employ either a two- or three-stage video-IF amplifier. An impedance-matching device is used to couple the signal from the tuner into the first IF stage. A sound trap is used to reduce the sound carrier amplitude to a low level, and all associated sound signals are passed on through the system to the detector, as shown in Fig. 8-1(A). Traps may also be provided to eliminate the adjacent channel sound. Some manufacturers also include traps to prevent adjacent channel picture interference. In these monochrome IF-amplifier systems, AGC is normally applied to two stages to provide a gain variation of at least 60 db, which is required to compensate for the wide difference in signal strength passed on from the tuner. This wide control latitude is difficult to achieve by controlling only one stage. Monochrome video-IF amplifiers generally do not have sufficient bandwidth for all color picture IF sideband frequencies.

Color Video-IF Amplifiers Color-TV receivers usually employ a three-stage video-IF amplifier system. Input to the color-IF amplifiers is similar to that of the monochrome amplifiers. An impedance-matching device is used to couple the first-IF stage to the tuner-coupling link, which is usually a length of coaxial cable. Sound traps are provided in the color-IF system to reduce the amplitude of the associated sound carrier and to eliminate adjacent channel sound frequencies. Since the associated sound carrier is not allowed to reach the video detector, a tuned circuit in the third stage removes the sound carrier (see Fig. 8-1(B)). Another sound trap is then placed between the third stage and the video detector to prevent associated sound interference from the video detection. As in the monochrome video-IF system, AGC is generally applied to the first and second stages. The bandwidth and selectivity requirements of color-IF amplifiers are more stringent than those of the monochrome-IF system, because of the color sidebands.

Vacuum-Tube Video-IF Amplifiers Vacuum-tube, video-IF amplifiers can be classified as either monochrome or color. In these systems, pentodes are normally used for the amplifiers, and transformers provide the interstage coupling. Two vacuum-tube stages are frequently employed for monochrome systems and three for many color applications. In the early days of color TV, some four-stage vacuum-tube, video-IF systems were used.

Solid-State Video-IF Amplifiers Transistor and integrated circuit, video-IF systems can also be classified as either monochrome or color.

Even though the functions of these IF systems are identical to those of the vacuum-tube systems, their circuits are different. The solid-state amplifiers have no filament circuits, and all components are considerably smaller. Transformers are normally used for interstage coupling but impedance coupling is often used. The transistors used are either of the NPN or PNP type, as individual components or in IC units.

8.4 INTERMEDIATE FREQUENCIES

To understand the importance of selectivity and gain in the video-IF amplifiers, all of the frequencies handled in the video-IF band must be considered, along with their amplitude and phase relationships. As shown in Fig. 8-5(A), there are two carriers involved in monochrome intermediate frequencies, an amplitude-modulated video carrier and a frequency-modulated sound carrier. As shown in Fig. 8-5(B), there are three carriers for color transmissions. These carriers are the AM-video and FM-sound carriers, which are identical to those used for the monochrome IF, and a phase/amplitude modulated, color subcarrier. Each carrier has sidebands associated with it. Notice that in Fig. 8-5 the carrier positions in the IF band are reversed from their normal RF transmitted order. The reasons for this reversal, and the various frequencies that make up the video-IF band for both monochrome and color, are discussed now.

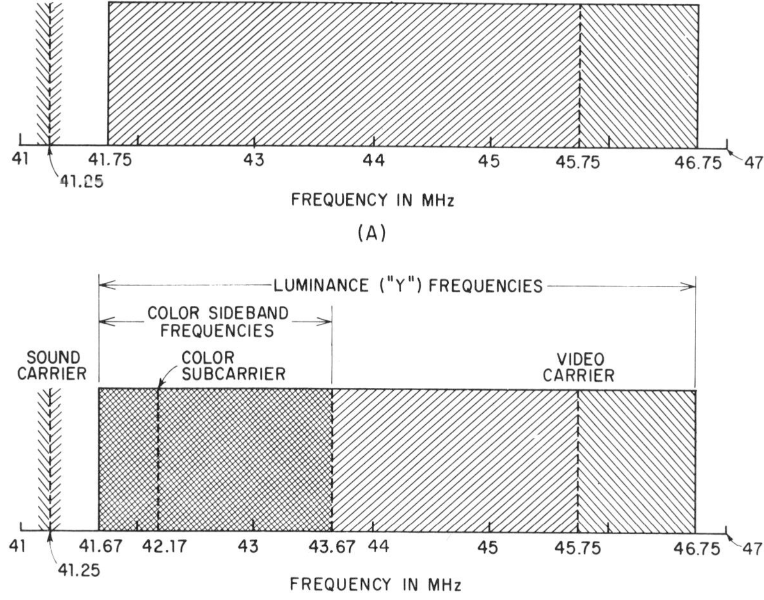

Fig. 8-5. The distribution of sound-and video-IF frequencies for (A) monochrome and (B) color IF passbands in an ideal receiver.

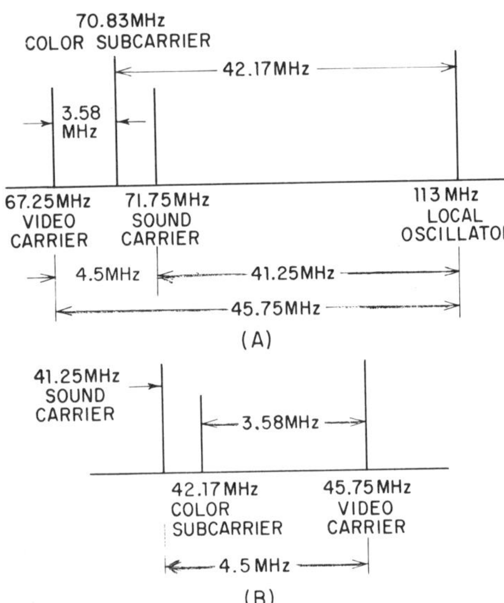

Fig. 8-6. The reversal of carrier frequencies as they pass through the mixer into the IF stages. (A) The distribution of frequencies for Channel 4 at the mixer stage input. (B) The distribution of frequencies in the receiver IF stages.

Frequency Reversal in the Video-IF Band. When the composite video signal is originally transmitted, the sound carrier is 4.5 MHz above the video carrier. In the receiver-IF stages, however, the sound carrier is 4.5 MHz below the video carrier as shown in Fig. 8-5. This signal reversal takes place in the mixer stage of the tuner. Figure 8-6 illustrates why the video and sound carriers exchange positions. (In color IF transmissions, the color subcarrier position is also reversed.) The frequencies shown in Fig. 8-6(A) represent the video, color, and sound carrier frequencies for Channel 4, and the local oscillator frequency required to produce the standard 45.75 MHz video-IF carrier, the sound carrier of 41.25 MHz, and the color subcarrier, of 42.17 MHz. In the figure, the transmitted sound carrier is shown at 71.75 MHz, which is 4.5 MHz above the 67.25 MHz video carrier, and the color subcarrier is shown at 70.83 MHz which is 3.58 MHz above the video-carrier frequency. The actual frequency difference between the color subcarrier and the video carrier is 3.579545 MHz; however, to simplify the discussion, the value is rounded off to 3.58 MHz.

Figure 8-6 indicates that there is a smaller difference between the sound carrier and the local oscillator frequencies than there is between the video carrier and local oscillator frequencies. This smaller difference also holds true for the color subcarrier, and is the reason why the sound carrier and color subcarrier are below the video carrier in the IF stages of the receiver, as explained by the following.

The relationship between the sound- and video-carrier frequencies in the receiver IF stages is shown in Fig. 8-6(B). Here, because of the mixing action, the sound carrier and the color subcarrier are below the video carrier. This frequency reversal would not occur in the local oscillator frequency was below the RF frequency of the broadcast station, but the normal procedure is to make the local oscillator frequency higher than the received signal frequency.

Monochrome Video-IF Frequencies In the monochrome-video IF stages, the sound carrier is established at 41.25 MHz and the video carrier at 45.75 MHz. All of those frequencies between 41.75 and 45.75 MHz, as illustrated in Fig. 8-5(A), make up the upper sideband of the amplitude modulated video carrier. Most of the lower sideband is eliminated in the vestigial sideband method of transmission, but the frequencies between approximately 45.75 and 46.75 MHz of the lower sideband are included along with the carrier. These lower sideband frequencies require special compensation in the IF amplifiers, as discussed in Section 8.5. This frequency range ideally permits a 4 MHz wide video-IF band that is sufficient for the reproduction of high-quality pictures. In practice, the video-IF bandwidth of monochrome receivers may vary between about 2.5 MHz and 3.6 MHz. The wide bandwidth is used for the larger screen and more expensive receivers.

Color Video IF Frequencies In the color-video IF band, the FM-sound and AM-video carriers are identical with the monochrome carrier, but chrominance, or color signals, are superimposed on the upper video sideband. After the mixer stage of the tuner, the color intermediate sideband frequencies appear as shown in Fig. 8-5(B). Monochrome-video frequencies, in color receivers, determine luminance, or brightness, of the picture and are referred to as the "Y" component.

Spurious Responses in the Video-IF Band. The picture carrier video-intermediate frequency of 45.75 MHz was recommended by the Electronic Industries Association (EIA), formerly the Radio and Television Manufacturers Association (RTMA). This particular frequency was derived after many years of study and research and is presently the standard for both monochrome- and color-TV receivers. The early TV receivers employed lower video-IF frequencies and, for that reason, were particularly prone to spurious responses. Even though these responses have been alleviated to a great extent by adopting the picture carrier 45.75 MHz standard IF frequency, it is important to understand how they may affect the TV receiver. The following spurious responses are among those found most troublesome:

(1) Image response;
(2) Response to two stations separated in frequency by the IF value;
(3) Direct IF response;
(4) Response to signals radiated from nearby TV receivers.

Image Response Image response is due to the mixing of an undesired signal with the local oscillator signal in the mixer stage to produce a voltage at the intermediate frequency. Since a frequency equal to the intermediate frequency is produced, this signal will be accepted and passed by the IF amplifiers. As an illustration, a TV receiver with a video-IF carrier value of 45.75 MHz has a video-IF band extending from 45.75 MHz to approximately 41.75 MHz. If the receiver is tuned to TV Channel 2, which has a frequency range of 54 to 60 MHz, the local oscillator will operate at 101 MHz. Remember that the video carrier is 1.25 MHz up from the low end of the TV channel frequency. This means that for Channel 2, the frequency of the video carrier is 55.25 MHz. The local oscillator frequency is then found by merely adding the picture-IF carrier value of the receiver to the frequency of the picture carrier in the TV channel, or

$$45.75 + (54 + 1.25) = 101.00 \text{ MHz}$$

From this discussion, it can also be seen that strong image signals in the frequency range of 142.75 to 146.75 MHz could beat with the 101

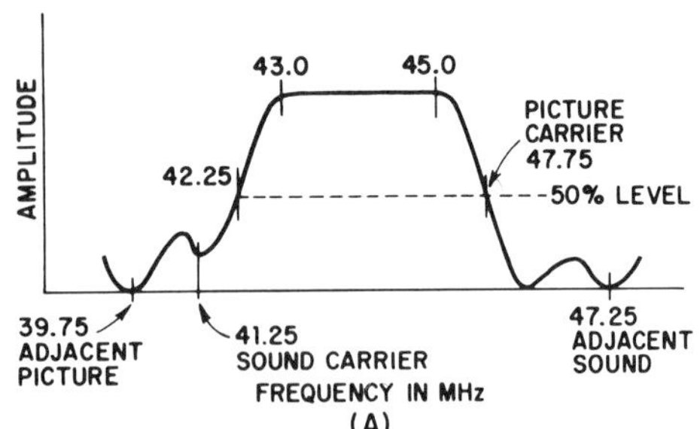

Fig. 8-7(A). The IF amplifier response curve of a typical monochrome-TV receiver.

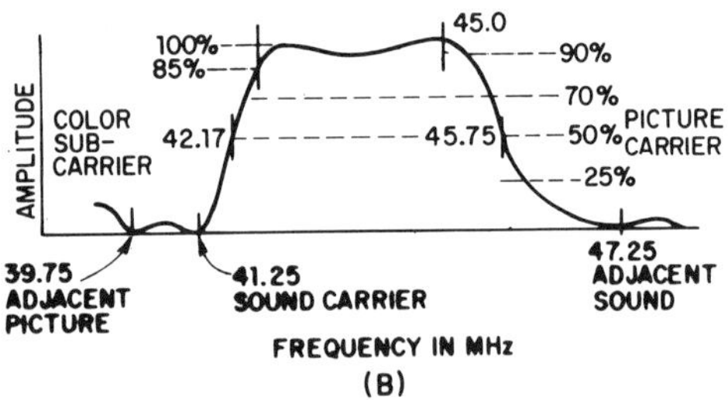

Fig. 8-7(B). The IF response curve of a typical color-TV receiver.

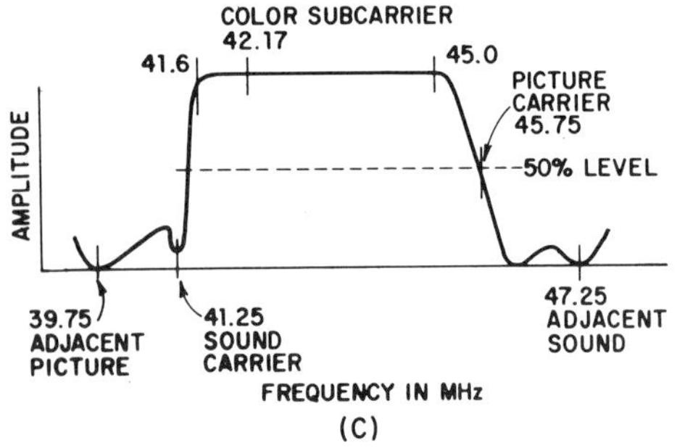

Fig. 8-7(C). Color IF amplifier response curve of the early color receivers.

MHz local oscillator frequency and produce voltages within the video-IF band of the receiver. These undesirable frequencies would, of course, have to get into the mixer stage of the tuner with enough strength to develop the spurious response. However, because of the high IF value of 45.75 MHz, the image frequency of 146.75 MHz is sufficiently removed in frequency from the desired signal at 55.25 MHz to be effectively eliminated from the mixer input by the RF resonant circuits in the tuner.

Stations Separated by the Intermediate Frequency A second source of interference is caused by two or more stations separated by the intermediate-frequency value of the receiver. These stations can be any sources of RF energy. In this situation, one incoming signal acts as the mixing oscillator for the other signal, or signals when several sources are involved. The result is a difference frequency that appears at the output of the mixer or converter stage at the IF value of the receiver.

Direct IF Response The third form of spurious response stems from the direct reception of signals which is equal to the intermediate frequency itself. To avoid the need of incorporating special filters, wavetraps, and shielding to prevent interference from this source, an IF value was chosen that is not the same frequency being used to any appreciable extent for commercial or amateur transmissions. This accounts for selection of the 45.75 MHz value.

8.5 AMPLIFIER BANDWIDTH

In the superheterodyne circuits of the TV receiver, a major portion of the overall gain and selectivity is provided by the IF amplifiers. For this reason, it is important for anyone working with TV receivers to fully understand the operation of the IF amplifiers, and especially their response characteristic curves. Figure 8-7 presents the typical response curves for both monochrome- and color-video IF amplifiers. The prime factors in determining the required shape of the curves are vestigial sideband compensation and the type of IF amplifier, monochrome or color. To achieve the desired bandwidth, the tuned circuits of the amplifier are usually stagger-tuned.

Vestigial-Sideband Transmission Compensation Figure 8-5 indicates that the video carriers in both monochrome- and color-IF bands are amplitude modulated, but differ from conventional AM carriers by having only one complete sideband. The other sideband, of which remnants are still present, has been effectively suppressed. This is known as vestigial sideband transmission and is the standard for both monochrome- and color-TV broadcasting. When any carrier is amplitude modulated, an upper and a lower sideband are generated; however, because identical information is contained in each sideband, only one sideband is required for demodulation in the receiver.

Complete suppression of the lower sideband is a desirable goal, but it is not economically achievable. It is impossible to completely eliminate one sideband by means of simple filters without, at the same time, distorting nearby portions of the remaining sideband. As a compromise between economy and the easily adjustable circuits on the one hand and the minimum distortion and bandpass on the other, it was decided to remove all but 1.25 MHz of the lower sideband of the video signal. The transmitted video signal, then, consists of this 1.25 MHz of the lower sideband, the video carrier, and all of the upper sideband. With the addition of the associated sound carrier and its sidebands, the full 6 MHz allotted to each television station is obtained.

Within the receiver, the IF amplifiers must use the upper sideband, together with the remnants of the lower sideband, to provide a response characteristic in which all sideband frequencies are amplified equally. In sound AM circuits this presents no particular problem because both sidebands are identical. In TV receivers, however, the sidebands are different because of the vestigial lower sideband. The lower video frequencies (those having frequencies close to the carrier) are contained in both the upper sideband and the remnants of the lower sideband. All video frequencies above 1.25 MHz, however, are present only in the upper sideband, having been suppressed in the lower sideband during transmission. If both the low and the high video frequencies are accorded equal amplification in the IF amplifiers, more low-frequency video voltage will be developed at the video-detector output than high-frequency voltage. It is to prevent this situation that the IF amplifier response characteristic shown in Fig. 8-7(A) is generally employed. At the video carrier the response of the IF amplifiers is down 50 percent, increasing linearly toward a maximum for the higher frequencies and decreasing for the lower frequencies. Roughly speaking, the lower video frequencies, for which there are two sidebands, receive half the amplification accorded all video frequencies above 1.25 MHz. In this way, the response for the low and the high video frequencies is equalized.

Monochrome Video-IF Amplifier Bandwidth Figure 8-8(A) illustrates the signal frequency distribution in a typical monochrome TV channel. The video carrier is set 1.25 MHz up from the lower band edge and the sound carrier is 4.5 MHz higher, or 0.25 MHz from the upper band edge. In the figure, the upper video sideband is shown to be 4 MHz wide. The width used in actual pratice, however, may vary from 2.5 to 4 MHz. In Fig. 8-7(A), the IF response curve shows the relationship between the various frequencies and the required amplitudes. On the curve, the frequencies indicated are marker frequencies that the technician injects into the IF band during the receiver alignment or testing. In this way, the response curve is observed on the screen of an oscilloscope, and the markers are used to spot particular frequencies along the curve. These markers are usually produced by crystal-

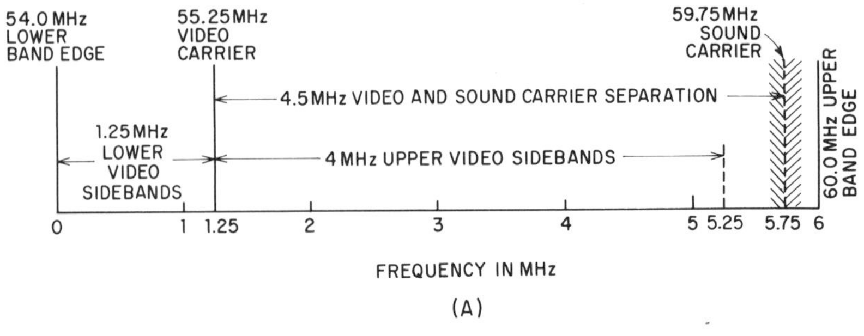

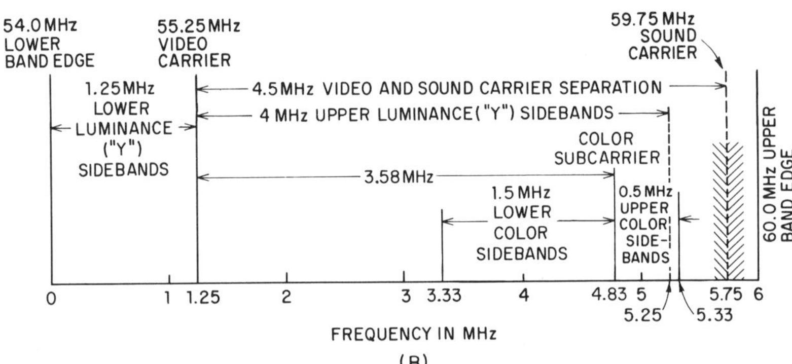

Fig. 8-8. The signal frequency distribution for monochrome and color transmissions for a typical TV channel (2).

controlled oscillators and are within a few Hertz of their intended frequencies.

Aside from the standard frequencies established for the sound, video, and color carriers, the marker frequencies can be any particular value specified by the manufacturer. In fact, there is a wide variation in the actual frequency of the markers used, as well as the shape of the response curve. For that reason, each manufacturer supplies, for his receivers, a special service bulletin that describes the shape of the curve, how to obtain it, and the marker frequencies to use.

Figure 8-7 shows marker frequencies at 39.75 and 47.25 MHz. These frequencies represent the adjacent channel video and sound carriers, respectively. Special trap networks in the IF amplifier circuits attenuate these frequencies to a low level to prevent interference in the reproduced picture. The associated sound carrier is also reduced by traps to a level that is from 5 to 10 percent of the total response curve amplitude. At this level, the sound frequencies can be removed easily from the overall IF band for application to the audio circuits and also to prevent these frequencies from reaching the video detector.

To the serviceman, the shape of this curve is important in his work on TV receivers. When aligning the IF stages, he must be careful to place the video carrier close to the 50 percent point to compensate for the vestigial sideband transmission. At the same time, he must tune the circuits to provide the maximum bandwidth. Detail in a TV image is dependent upon the strength of the high video frequencies present. When the response drops at the upper end of the curve, the fine detail becomes fuzzy and indistinct. Poor low-frequency response gives rise to the uneven shading of large areas on the screen, smearing, and a generally darker image.

Color Video-IF Amplifier Bandwidth The color IF amplifier response curve shown in Fig. 8-7(B) includes all of the frequencies present in the monochrome response curve of Fig. 8-7(A) plus the frequencies associated with the color subcarrier and its sidebands. Note the similarity between the two curves. The critical difference between them is that the color IF subcarrier of 42.17 MHz must be set at the 50 percent point of the curve opposite the IF picture carrier. In early color sets, the response curve was such that the color subcarrier and its sideband frequencies were positioned on the flat portion of the curve as shown in Fig. 8-7(C). To accommodate this latter bandwidth with adequate gain, the early sets required four IF stages. To reduce the number of IF stages required in color sets, the response curve of Fig. 8-7(B) was adopted. This narrower, overall bandwidth made it possible to provide the required gain in only three IF stages. The effect of placing the color subcarrier and its sidebands on the slope is compensated for in the Chroma amplifier (see Chapter 23) by a tuned network, which effectively places the color subcarrier and its sidebands on the flat portion of the Chroma amplifier response curve.

Trap networks are also provided in color-IF amplifier circuits to suppress the adjacent video and sound carriers. These are shown in Fig. 8-7(B), at 39.75 and 47.25 MHz, respectively.

Stagger Tuning Achieving the necessary bandwidth in a monochrome- or color-video-IF section with conventionally-tuned amplifiers is a difficult task and an expensive one. A simpler and more economical way to increase the bandwidth of the amplifiers is to tune each stage to a slightly different frequency, making the overall response of the amplifiers wider than that of any individual amplifier. This method, known as stagger tuning, is used extensively in both monochrome- and color-video IF stages. Because of its use, it is important to thoroughly understand stagger tuning and its effects on bandwidth. First, though, it is necessary to define bandwidth.

A typical resonance curve for a parallel-tuned circuit is shown in Fig. 8-9. The response is not uniform but varies from point to point. At the resonant frequency (labeled F_0 in the diagram), the response of the circuit is at its peak, or maximum impedance. From this point, in either direction, the response tapers off until it becomes negligible. With a characteristic of this type, the actual bandwidth is arbitrary. It could be said, for example, that all frequencies between the points *B* and *B′* on the curve are within the bandpass of the circuit. Note that this does not

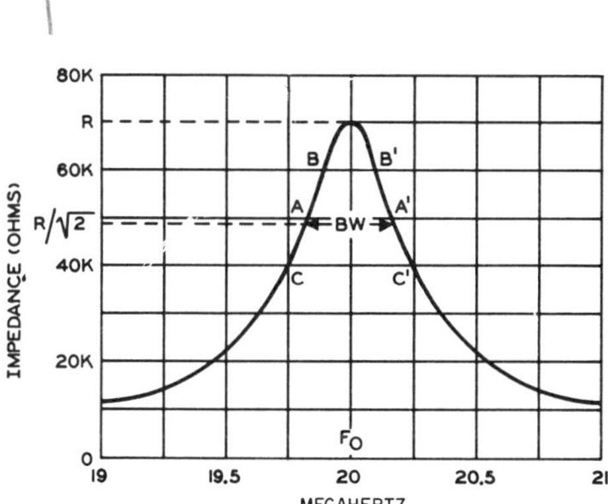

Fig. 8-9. The accepted definition for the bandwidth of a tuned system.

prevent other frequencies—those that receive less amplification—from passing through the circuit.

The arbitrary definition generally accepted for a bandwidth is illustrated in Fig. 8-9. The bandwidth of a circuit is equal to the numerical difference in Hertz between the two frequencies at which the impedance presented by the tuned circuit is equal to 0.707 of the impedance presented at F_0 (*i.e.*, the maximum impedance). Thus, in the response curve shown in Fig. 8-9, the impedance at points A and A' is 0.707 (or $1\sqrt{2}$) of the impedance offered by the circuit at F_0. In this particular illustration the bandwidth is approximately 0.4 MHz.

A further note of importance is that if the gain of the circuit is considered equal to 1 at F_0, it is down 3 db at points A and A'.

With this concept of bandwidth in mind, now consider two single-tuned amplifiers, both tuned to the same frequency. If these two amplifiers are in cascade (*i.e.*, connected so that the output signal of one is the input signal of the other), then the overall bandwidth is not equal to the bandwidth of either circuit, as might be expected, but to 64 percent of this value. The reason for the shrinkage in bandwidth will be apparent from the following discussion.

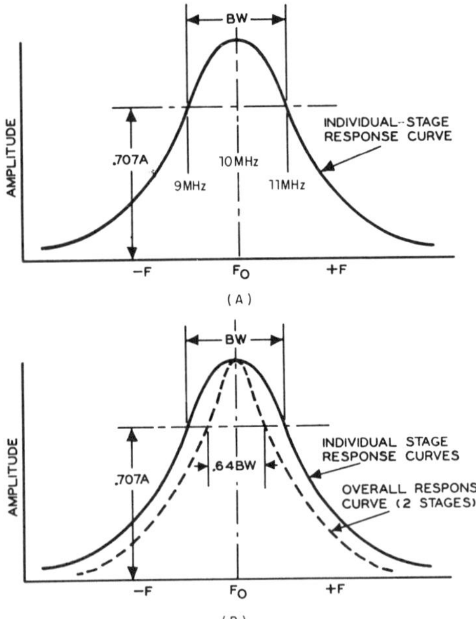

Fig. 8-10. Two tuned circuits, each peaked to the same frequency, produce an overall response in which the bandwidth is less than that of either curve taken separately.

The response curve of the first amplifier, illustrated in Fig. 8-10(A), has a maximum value of amplification of 1 at F_0, its peak, and 0.707 at the ends of the bandpass. Assume that the mid-frequency is 45 MHz, while the end frequencies of the bandpass are 44 and 46 MHz, respectively. If each of these three frequencies has an amplitude of 1 volt at the input to this tuned stage, then at the output they would possess the following values: at 44 MHz, $1 \times 0.707 = 0.707$ volt; at 45 MHz, $1 \times 1 = 1$ volt; and at 46 Mz, $1 \times H\, 0.707 = 0.707$ volt.

These same three frequencies are now passed through the second tuned circuit. Since this second circuit possesses the same characteristics as its predecessors, here is the result at its output: at 44 MHz, $0.707 \times 0.707 = 0.49$ volt; at 45 MHz, $1 \times 1 = 1$ volt; and at 46 MHz, $0.707 \times 0.707 = 0.49$ volt. After passage through the two amplifiers, 44 and 46 MHz are no longer within the 0.707 region about the resonant frequency of 45 MHz. The result, of course, is a narrower bandpass; more accurately, 36 percent narrower, as shown in Fig. 8-10(B).

Now consider two single-tuned amplifiers, each with the same bandwidth, but with their peaks separated (or staggered) by an amount equal to their bandwidth, as illustrated in Fig. 8-11. The result is a response in which the overall bandwidth (to the 0.707 point) is 1.4 times the bandwidth of a single stage. The overall gain, however, is now only one-half that of the two stages tuned to the same frequency. This is so because at the center frequency of the overall response curve, the individual stage responses are only 0.707 of their peak response. The product of the stage gains is approximately one-half ($0.707 \times 0.707 = 0.49$, or approximately 0.5). This is what is meant by stagger tuning.

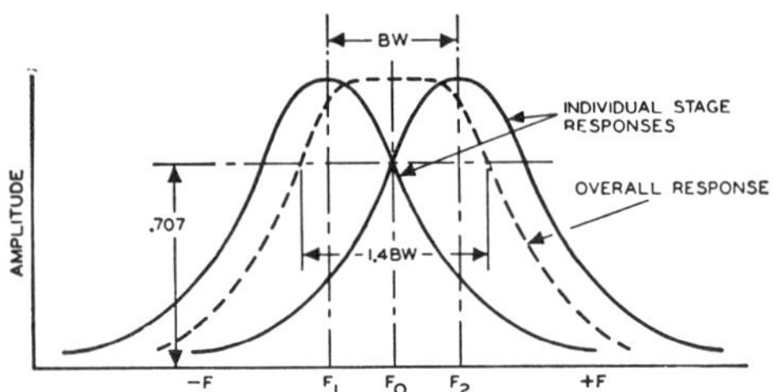

Fig. 8-11. By stagger tuning two tuned circuits we obtain a wider bandpass.

Now, to progress one step further. By stagger tuning two tuned circuits, 1.4 times the bandwidth of a single stage can be achieved, but with only one-half of the gain. Suppose, however, that stagger tuning is retained, but the bandwidth of each individual tuned circuits decreased. The overall bandwidth of the stagger-tuned system will still be 1.4 times the bandwidth of the individual stages. Since the bandwidth of the individual coil was decreased 1.4 times, however, this new figure will be less than 1.4 times the figure obtained when each individual bandwidth was greater. The advantage of this method is that a greater bandwidth is obtained than if the circuits had not been stagger tuned, and the overall gain remains high.

A simple illustration will demonstrate this point. Suppose that the bandwidth of each individual stage is decreased to 0.707 of its original value. To do this, the Q of the individual circuit is raised to 1.4 times its previous value, which will provide an increase in gain of 1.4 times. Now, when the stages are staggered by an amount equal to this reduced bandwidth, the overall gain is one-half the product of 1.4 times 1.4.

$$1/2 \times 1.4 \times 1.4 = 0.7 \times 1.4 \text{ or approximately } 1$$

Thus, the overall gain is now the same as that obtained with the previous amplifier with both circuits tuned to the same frequency.

An important relationship, and one to be remembered, is that the bandwidth of any paralleled-resonant circuit (or an ordinary resistance-coupled amplifier) is inversely proportional to the amplification of that system. Expressed a little differently, the bandwidth times the gain is a constant. This constant is known as the gain-bandwidth product. Thus, if the bandwidth of a system is increased 1.5 times, then its gain is decreased by that same amount. For any individual tuning coil,

$$\text{Bandwidth} = F_0/Q$$

where

F_0 = resonant frequency of the coil, and
Q = figure of merit of the coil.

This expression indicates that, for any given resonant frequency, increasing the bandwidth can only be accomplished by decreasing the Q of the coil by a proportional amount. If the resonant frequency of the coil is raised, however, maintaining Q constant, then the bandwidth will increase in like measure.

Remember that stagger tuning is a method of obtaining the bandwidth. It is not a form of coupling. Stagger tuning can be accomplished with either transformer coupling or impedance coupling.

8.6 SOUND-FREQUENCY SEPARATION

During the past decade, two basic methods were employed to separate the program sound from the video-IF band. One method often used in early TV receivers was a split-sound system in which the sound was taken from the output of the mixer stage. Filters (or traps) then prevented the sound carrier from entering the video-IF band. This method was disadvantageous, however, because it required critical adjustment of the fine-tuning control, often caused sound degradation due to oscillator drift, and was more expensive.

A second method, known as the intercarrier sound system, alleviated all of the disadvantages of the split-sound system. In this method, the sound frequencies were allowed to pass through all the video-amplifiers, but at a much reduced level. The program sound was separated in the video detector by beating the sound carrier against the video carrier, thus producing a 4.5 MHz sound IF. Filters, or traps, in the video amplifier section then separated the 4.5 MHz sound IF from the video signals. Sound IF filtering could be accomplished in the video amplifier section because the highest video frequency was 4.0 MHz, whereas the sound IF was 4.5 MHz.

With the advent of color TV, the intercarrier sound system had to be modified so that the sound-IF carrier could be removed before it reached the video detector. If this carrier were allowed to reach the detector, it would beat against both the video carrier and the color subcarrier. The result would be the development of two sound-IF frequencies: one at 4.5 MHz and the other at 920 kHz. It would then be impossible to filter the 920 kHz sound frequency from the video signals and this would cause an interference pattern on the color-tube screen. Today, monochrome receivers use the intercarrier sound system and separate the sound in the video detector. All color-TV receivers use the intercarrier sound system but remove the sound prior to the video detector.

Figure 8-12 is a simplified diagram of a typical sound-takeoff circuit in a color-TV receiver. Even though the circuit in the figure is shown as part of a solid-state IF system, it is also used in vacuum-tube systems. In the diagram, the 41.25 MHz sound carrier and the 45.75 MHz video carrier are coupled through C_1 to the sound detector, D_1, from the collector of the third IF amplifier, Q_3. D_1 detects the 4.5 MHz difference between the two carriers and feeds it into the sound-IF system.

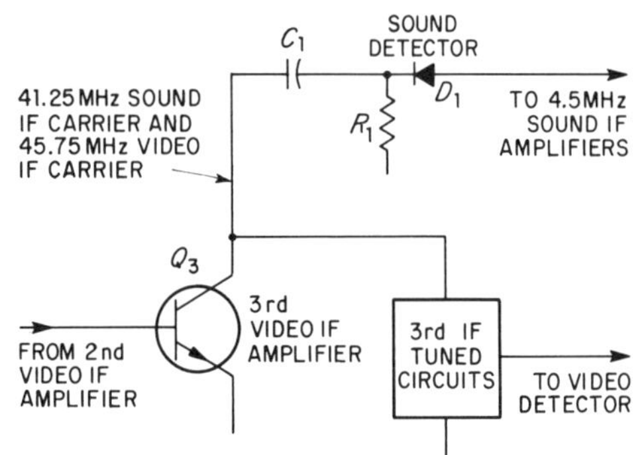

Fig. 8-12. The sound separation at the third video-IF amplifier in a color-TV receiver.

8.7 INTERSTAGE COUPLING

In present-day TV receivers, either transformer or impedance coupling is used to interconnect two or three IF amplifiers to make-up the video-IF system. For solid-state circuits, special lightweight, subminiature transformers and coils have been developed that can be mounted directly on printed-circuit boards. These small components also lend themselves well for use in modular construction and in small-screen portable TV receivers. Other coupling methods have been tried but none provide the same tuneability required in establishing the video-IF bandwidth, as transformer and impedance coupling.

Transformer Coupling The response of a transformer-coupled IF stage depends largely upon the coefficient of coupling between the primary and secondary windings of the IF transformer. The coefficient of coupling of a transformer is a measure of how well it is able to transfer energy from the primary to the secondary. Thus, a coefficient of coupling with a value of 1, or unity, means that all of the flux lines of the primary link with all of the turns of the secondary. The equation usually given for the coefficient of coupling is

$$\text{Coefficient of coupling} = \frac{M}{L_1 L_2}$$

where

M = mutual inductance between the primary and secondary,
L_1 = inductance of the primary, and
L_2 = inductance of the secondary.

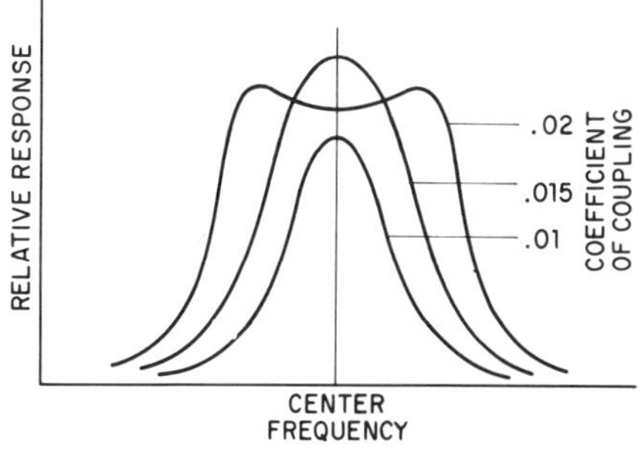

Fig. 8-13. Relationship between the coefficient of coupling and the IF response.

The value of the coefficient of coupling is always less than 1, but in power transformers it approaches values of 0.98 and 0.99. The windings of such a transformer are said to be closely coupled. As indicated by the equation, the greater the mutual inductance, the greater the coefficient of coupling, and hence the greater the numerical value of the coefficient of coupling. Figure 8-13 shows the relationship between the bandwidth and the coefficient of coupling. The coupling value of 0.02 is more desirable than that of 0.015, because it allows a greater range of frequencies to pass.

A bifilar-wound coil, which is also a type of transformer, can be used to obtain the necessary bandwidth when transformer overcoupling is desired. Shown in Fig. 8-14, it consists of two windings which are positioned so close together that the degree of coupling is near unity. The result is that practically all of the voltage developed across one winding, acting as the primary, is transferred to the other winding, acting as the secondary. A moveable iron core, inside the coil form, tunes both windings simultaneously. In most instances, the tuning rod extends above the top side of the chassis, where it can be reached for the alignment of the circuit.

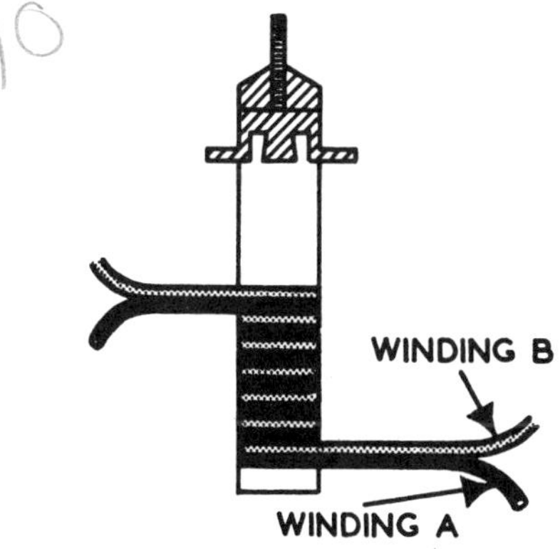

Fig. 8-14. A bifilar coil.

Impedance Coupling Figure 8-15 is a simplified diagram of an impedance-coupled vacuum-tube IF stage. This method is equally effective in solid-state circuits. In the diagram, the input signal is delivered to V_2 via C_1 and R_2. One side of R_2 is connected to the negative AGC source. In the absence of a signal, the AGC voltage drops to

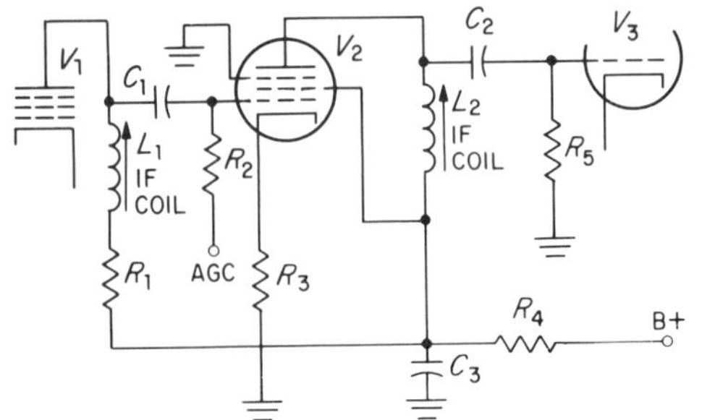

Fig. 8-15. An example of impedance-coupled IF amplifier stages.

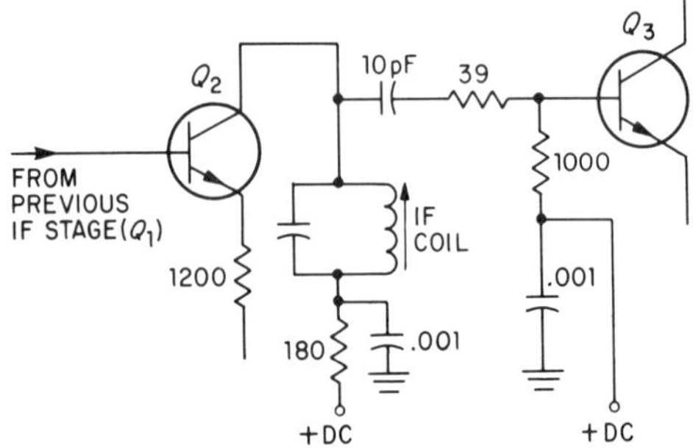

Fig. 8-16. Typical impedance-coupled IF stages used in a solid-state, color-TV receiver.

zero, so a small cathode resistor, R_3, prevents the tube from being operated without bias. The plate load for V_1 consists of a tuneable coil L_1. A decoupling filter comprised of C_3 and R_4 prevents any interstage coupling from occurring due to the common power supply. The output signal, taken across coil L_2, is delivered to the next stage via C_2 and R_5.

There are many circuit variations possible with impedance coupling. For example, a tuneable coil may be used in place of R_5 in the circuit pictured in Fig. 8-15. Also, the coupling capacitor may be replaced with an *LC* circuit to make the stage more selective. Figure 8-16 shows a typical impedance-coupling circuit used with solid-state IF systems.

8.8 IF-FREQUENCY TRAPS

The video-IF band in a TV receiver is particularly prone to interfering signals that, if allowed to pass into the video-amplifier section, could distort or destroy the image being reproduced on the screen. Fortunately, though, many of these interfering signals never get beyond the RF tuner and consequently are suppressed before they reach the first IF stage. Some undesirable signals, however, are so close to the channel frequency to which the receiver is tuned that they are able to pass through the RF tuned circuits and reach the video-IF system. The IF circuits present the greatest obstacle to undesired signals, but once these signals pass beyond this portion of the receiver, there is little that can be done to remove them. Every effort must be made to suppress all interfering signals before they reach the video detector.

To receive a 6 MHz band of frequencies, the RF and mixer tuning circuits are designed with a low *Q*, which means that the sides of the input response curve are not very steep, but instead tend to taper off gradually. Figure 8-17 illustrates the typical input response curve of a modern TV receiver. With a response of this nature, voltages at the frequency of the sound carrier of the next lower channel or the video carrier of the next higher channel could penetrate through the RF stages and reach the video-IF amplifiers. Unless trap circuits are inserted in the video-IF system, the effects of these interfering signals will be detrimental to the production of high quality pictures.

To eliminate interfering signals (by filtering or the use of traps), it is first necessary to determine their frequencies. Sometimes this can be done by simple subtraction, based on the assumption that the video-or sound-carrier frequency of adjacent channels is causing the interference. Suppose that the receiver is tuned to Channel 3 (60 to 66 MHz) and the video-carrier IF value is 45.75 MHz. The sound carrier of the next lower channel (Channel 2, 54 to 60 MHz) is at 59.75 MHz. A signal at this frequency when mixed with the local oscillator (107.00 for Channel 3) will develop a difference frequency of 47.25 MHz. This is the frequency of one interference signal (adjacent channel sound).

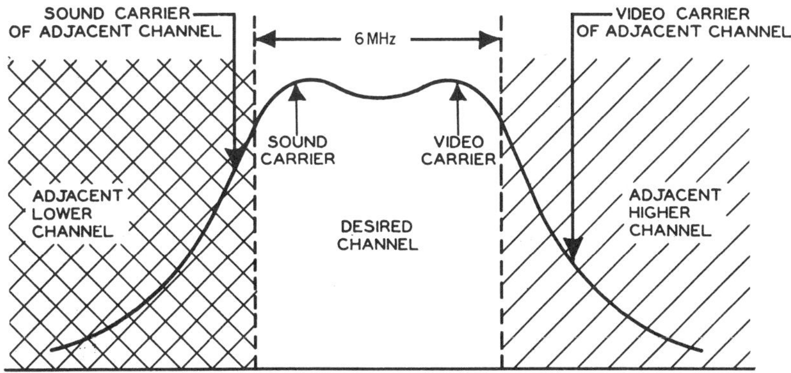

Fig. 8-17. The RF response curve of most television receivers. Note that signals from adjacent channels can be received.

Another possible interference signal is the picture carrier of the next higher channel (66 to 72 MHz). When this beats with the local oscillator, a difference frequency of 107.00 minus 67.25 MHz, or 39.75 MHz, is produced (adjacent channel picture).

In all cases where adjacent channels exist, the two interfering frequencies will be 47.25 and 39.75 MHz for a TV receiver with a 45.75 MHz video IF. There are, however, channels which are not subject to adjacent-channel interference. Thus, Channel 2 (54 to 60 MHz) does not have an adjacent lower channel; on the other hand, Channel 4 does not have an adjacent higher channel. Remember that adjacent as used here means channels that follow each other without any frequency separation. Channel 4 is followed by Channel 5, but the frequency of Channel 4 is 66 to 72 MHz and that of Channel 5 is 76 to 82 MHz. The 4 MHz separation is sufficient to prevent any of the frequencies in Channel 5 from adversely affecting Channel 4. Channel 2, however, is closely followed by Channel 3, and interference is possible. The same is true of many of the other channels; hence the use of traps is important. The trap frequencies are dependent upon the video- and audio-IF values employed in the receiver. The purpose of the traps, however, remains unchanged.

Even though adjacent channels are not assigned to TV broadcast stations within the same general area, there are cases where adjacent channels are assigned to stations in nearby areas. As an example, consider New York and Philadelphia which are only 90 miles apart. New York is assigned VHF channels 2, 4, 5, 7, 9, 11, and 13, whereas Philadelphia is assigned VHF channels 3, 6, and 10. Any TV receiver situated between these two cities would certainly be subject to considerable interference and would definitely require trap circuits. This same situation is true in many other parts of the country. Whether or not a receiver contains these adjacent-channel traps is largely a matter of design and economics. Some receivers have one or the other trap, some have both, and a few have neither.

Sound traps are always used in color-TV receivers because of the

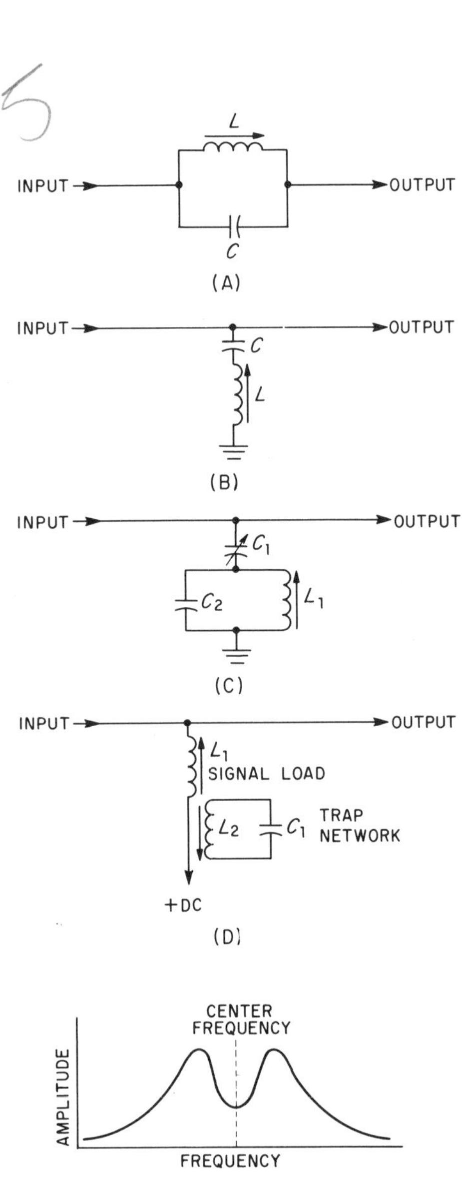

Fig. 8-18. Video IF-band traps.

narrow separation between the sound carrier and the color subcarrier. This interference signal has a frequency of 920 kHz and is virtually impossible to filter from the detected video signals. For that reason, sound traps in the video-IF section reduce the sound-carrier level to approximately 5 percent of the video-IF amplitude. After the sound is separated from the video-IF band, another sound-IF trap further reduces the sound signals to a very low level, thus preventing them from appearing in the video detector. In addition, a 4.5 MHz sound-IF trap may appear following the video detector.

Basically, there are five types of traps used in conjunction with video-IF amplifiers. These traps are known as series, parallel, absorption, degenerative, and bridged-T and, for the most part, are used in both vacuumtube and solid-state circuits for either monochrome- or color-TV reception. The bridged-T trap is the most widely used circuit and is indispensable in color-TV receivers.

Series Trap The series trap is a parallel resonant circuit configuration as pictured in Fig. 8-18(A). It is placed between two IF stages and tuned to the frequency to be rejected. This type of trap circuit is a sharply tuned network designed to reject one frequency or, at most, a narrow band of frequencies. When a signal voltage at the trap frequency appears at the input of the circuit, the impedance offered by LC is very high and almost all of the undesired voltage is dropped across the trap network. A negligible amount of the voltage appears across the input circuit of the following IF amplifier. At all other frequencies, the resonant circuit offers negligible impedance, and the desired signals pass easily.

Parallel trap These traps are tuned circuits which are placed across, or in shunt with, the circuit. Figure 8-18(B) shows a series-resonant circuit used in this manner. At the frequency for which it is set, the trap acts as a short circuit, bypassing the resonant frequency to ground and preventing further penetration into the circuit. At other frequencies the trap circuit presents a relatively high impedance, permitting these signals to proceed to the following stage. It is important that the parallel trap have a very high Q so that the circuit will bypass only a narrow range of frequencies.

A variation of the parallel trap is shown in Fig. 8-18(C). In this circuit, L_1 and C_2 form a parallel resonant circuit that is tuned to 42.25 MHz. The Q of the coil is 200, and a fairly large voltage is developed across the circuit at this frequency. Now, it is easily demonstrated that, for all frequencies lower than its resonant frequency, a parallel resonant circuit appears inductive. At the resonant frequency, of course, it presents a purely resistive impedance. For frequencies above resonance, the impedance presented by L_1 and C_2 is capacitive. This latter fact can be understood by noting that for the higher frequencies, the parallel capacitor offers less impedance than the coil. Consequently, most of the current flows through the capacitor, and the circuit current possesses a leading phase. Since the sound carrier, at 41.25 MHz, is below the 42.25 MHz resonant frequency of L_1 and C_2, the parallel combination appears inductive to the carrier. By resonating this inductance with C_1, a series-resonant path for the sound carrier is obtained, and the carrier is bypassed to ground.

By providing a parallel-resonant circuit (L_1 and C_2) for 42.25 MHz, a sharp rise in voltage is obtained just beyond 41.25 MHz on the IF response curve. Since the 42.25 MHz value is included in the range of desired video frequencies (they extend from 45.75 MHz down to 41.75 MHz), all of the desired video frequencies are passed by the trap with negligible attenuation while, at the same time, suppressing the undesired sound-IF carrier.

Absorption Trap The absorption trap, shown in Fig. 8-18(D), is a widely used type of rejection circuit. It consists of a coil and a fixed capacitor inductively coupled to the load inductor of an IF amplifier. When the IF amplifier receives a signal at the resonant frequency of the trap circuit, a high circulating current develops in the trap network as a result of the coupling between the trap and the load inductor. The voltage in the load coil, L_1, becomes quite low at the trap frequency. Consequently, very little of this interference voltage is permitted to reach the following stage. It is convenient to think of this kind of trap as being able to absorb all of the energy of the frequency to which it is tuned, and therefore no energy at that frequency is left available to pass on into the next stage. Absorption traps are also called "suckout traps."

The theory of absorption traps may also be explained on the basis of a fundamental transformer theory, with L_1 acting as the primary and the trap network, L_2 and C_1, acting as the secondary. Two resonant circuits closely coupled will produce a double-humped curve, such as the one shown in Fig. 8-18(E). Note the sharp decrease in primary current at the center frequency. In the case of the two tuned circuits of Fig. 8-18(D), L_1 is tuned to the desired band of frequencies while the trap, or secondary is tuned to the undesired frequency. Since the primary band coverage includes the undesired frequency to which the trap is tuned, there is a sharp drop in primary voltage at that frequency. It is this interaction between primary and secondary coils that produces the marked decrease in voltage at the trap frequency. The other frequencies in the IF band are unaffected by the trap.

Degenerative Traps Degenerative traps, as illustrated in Fig. 8-19, are designed to reduce the gain of an amplifier for frequencies to which the trap is tuned. These traps are used in the emitter circuit of a solid-state amplifier or in the cathode leg of a vacuum-tube circuit. In the latter application, the traps are often called "cathode traps." The two types of traps normally used to provide the degeneration are the absorption type and the series type. Figure 8-19(A) shows an absorption type in

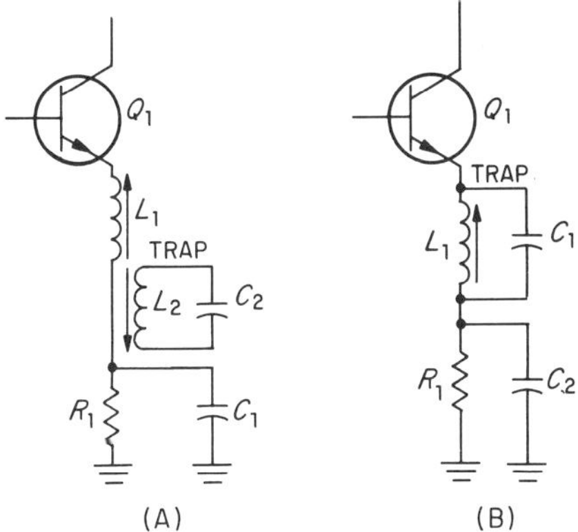

Fig. 8-19. Degenerative traps.

which the coil L_1 in series with C_1 forms a broadly tuned series-resonant circuit at the frequency to which the amplifier is tuned. This permits the amplifier to function normally for all signals within its frequency range. At the resonant frequency of the trap, however, a high impedance is reflected into the emitter or cathode circuit by the trap, and the gain of the stage is reduced by degeneration.

The series type of degenerative trap, illustrated in Fig. 8-19(B), places a parallel circuit directly into the emitter or cathode leg. At the resonant frequency of the trap, the impedance in this part of the amplifier circuit will be high, producing a large degenerative voltage and thus reducing the gain of the amplifier. At all other frequencies, the impedance of this parallel network is low. Only a small degenerative voltage appears, and therefore only a slight loss in gain occurs except at the undesired frequency.

Bridged-T Trap A trap that is more complex than any of the foregoing circuits, but also more effective, is the bridged-T trap shown in Fig. 8-20(A). In this circuit, L_1, C_1, and C_2 are resonated at the frequency of the signal to be rejected. Now, if the resistance of R is properly chosen, the attenuation imposed upon a signal to which L_1, C_1, and C_2 are resonated will be great. Ratios of 50 and 60 to 1 are easily attainable using standard components. This means that the strength of the desired signal at the output of the trap will be 50 to 60 times greater than the strength of the undesired signal.

Understanding the operation of this trap circuit can be simplified by transforming the bridged-T network shown in Fig. 8-20(B) into the

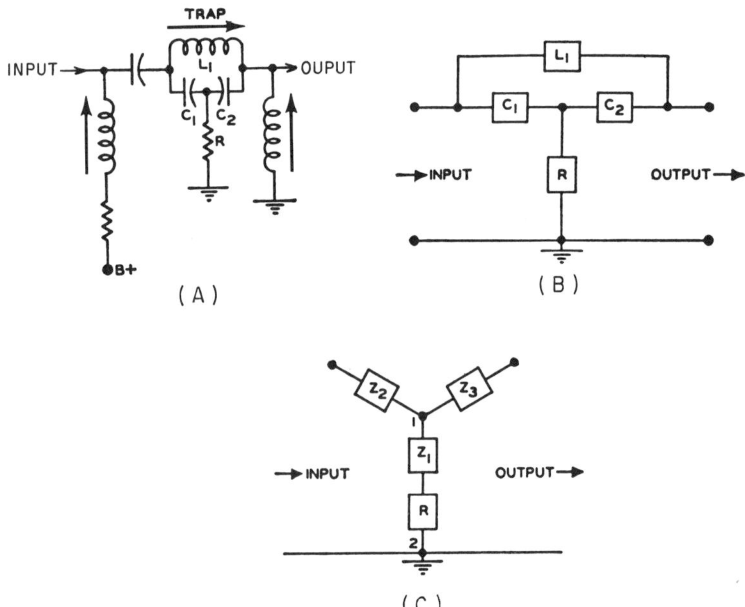

Fig. 8-20. The bridged-T trap.

equivalent network of Fig. 8-20(C). This is called a delta-wye transformation and can be readily accomplished with well-known electrical theorems. If the various components of the bridged-T network L_1, C_1, and C_2 are properly chosen, Z_1 will have a negative value. If R is made equal to Z_1, then the total impedance between points 1 and 2 will become zero, effectively short-circuiting signals of the frequency to which the network is tuned. To all other frequencies, the bridged-T network offers negligible attenuation.

8.9 TYPICAL VIDEO-IF AMPLIFIERS

As stated previously, the main functions of the video-IF amplifiers are to provide selectivity and gain. It was also pointed out that monochrome and color video-IF amplifiers are the two basic types and that each of these types can employ vacuum-tubes or solid-state components. With these facts in mind, then, it is important to examine several typical amplifier circuits. This examination should reveal that there is actually very little difference in the various amplifier circuits. The most important fact to remember is that color-IF circuits are more critical and, therefore, contain more traps or filters. Also, in color circuits the sound signals are removed prior to the video detector.

Figure 8-21 presents a typical transformer vacuum-tube, coupled, video-IF system for a color-TV receiver. In the diagram, each stage is shown as a separate tube; however, some manufacturers use two compactrons, with the first and second IF amplifiers in one and the third IF amplifier and a sound and sync amplifier in the other. Many of these circuits also use bifilar wound coils instead of transformers.

The video-IF amplifier system shown in Fig. 8-21 is coupled to the mixer stage of the tuner through a 41.25 MHz series-resonant sound trap. This trap reduces the associated sound carrier to the required 5 percent level, thus facilitating the intercarrier method of IF amplification. There is also a 47.25 MHz trap connected to the input of V_1 to reject the adjacent channel sound carrier. Some chassis also provide a 39.75 MHz trap at this point to reject the adjacent channel video carrier.

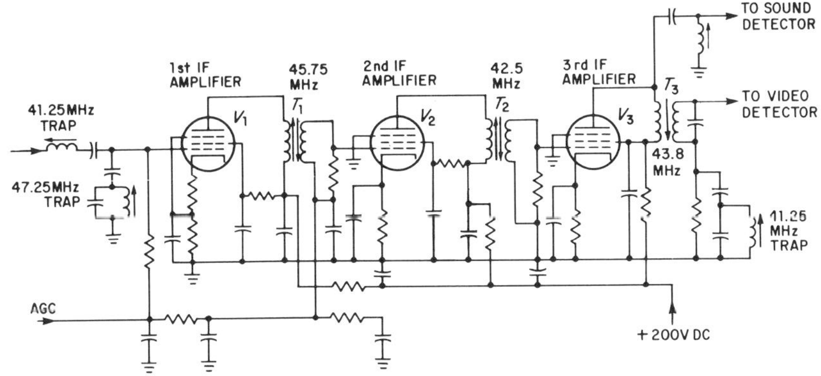

Fig. 8-21. A vacuum-tube, transformer-coupled color-video IF system.

From the plate of V_1, the amplified video signals are coupled to the grid of V_2 via transformer T_1. This transformer is tuned to 45.75 MHz and aids in placing that frequency at the 50 percent point on the IF response curve. T_2 couples the output of V_2 to the input of V_3. By tuning T_2 to 42.5 MHz, the 50 percent point on the opposite side of the response curve is established and places the color-IF subcarrier of 42.17 MHz at that point.

The plate circuit of V_3, the third video-IF amplifier, feeds a portion of the entire IF band voltage into the sound detector circuit. Here, the sound-IF carrier is beat against the video carrier to produce a 4.5 MHz sound carrier. Filters in the sound detector circuit then remove any color subcarrier frequencies, as well as the 920 kHz frequency that results from the sound carrier beating against the color subcarrier. T_3, also in the V_3 plate circuit, delivers the amplified IF voltage to the video detector. T_3 is then tuned to 43.8 MHz to flatten out the IF response curve near the center of the band-pass. A bridged-T trap connected to the detector side of T_3 is tuned to 41.25 MHz to reject all traces of the associated sound carrier.

AGC is applied only to the first two stages in this circuit. There are, however, some circuits in which AGC is applied to all three stages, but this practice is not popular. About 60 db of AGC control is usually desired, and this value can be achieved easily with two stages.

The interstage tuning circuits commonly employed in stagger-tuned IF systems, such as illustrated in Fig. 8-21, are either transformers, single coils, or bifilar-wound coils. Each transformer or coil contains a powdered-iron core that is adjustable within the coil, so that the circuit can be tuned to the desired resonant frequency. Because relatively high frequencies are employed in the IF band, actual capacitors are not always connected across each coil, although capacitance is present. It arises from the capacitance which is inherent between the turns of the coil, plus the capacitance that the vacuum tubes (or transistors) and the connecting parts and wires contribute.

It is a common practice when using single coils in a staggered arrangement, to place them in the plate circuit while the following grid circuit uses a coupling resistor. In short, each plate coil and the following grid resistor are in parallel, and by using low-valued resistors, the response of the tuning coils is broadened. This arrangement is needed to achieve an overall bandpass of 2.5 to 4 MHz, when the individual responses of all the coils are combined.

Solid-State, Color-IF Amplifier Figure 8-22 is a schematic diagram of the video-IF system used in an Admiral hybrid color-TV chassis. This system has three NPN transistor stages that are impedance coupled and pass a band of frequencies centered at 43.8 MHz. Input to the IF system is through a plug-in coax link from the mixer stage in the tuner. The mixer output coil is tuned to position 42.17 MHz at the 50 percent point on the IF response curve, and the impedance of the

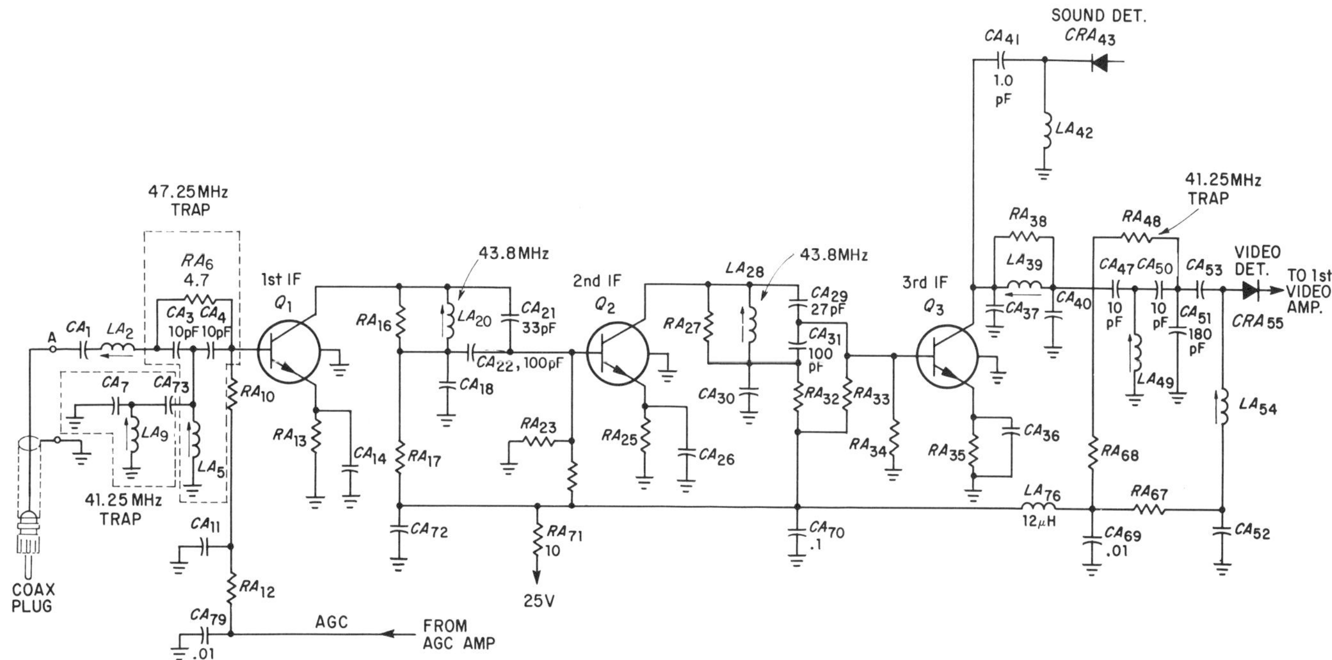

Fig. 8-22. An Admiral color chassis, video IF system. (*Courtesy of Admiral Corporation.*)

coil is tapped to match the impedance of the coax link. Q_1, the first IF amplifier, is coupled to the input coax link through LA_2 and CA_1. LA_2 is tuned to position 45.75 MHz at the 50 percent point on the opposite side of the response curve from the 42.17 MHz point. Both 50 percent points on the IF response curve are shown in Fig. 8-7(B). The mutual coupling of LA_2 and the mixer output coil provide a wide bandpass through the coax link.

At the input to Q_1 are two sound traps, one tuned to reject the adjacent channel sound frequency at 47.25 MHz and the other tuned to suppress the 41.25 MHz associated sound carrier. The 47.25 MHz trap consists of RA_6, CA_3, CA_4, and LA_5. This trap is a bridged-T configuration and delivers to the base of Q_1 two 47.25 MHz voltages that are equal but 180 degrees out of phase. One of these voltages is developed across RA_6 while the other is developed across CA_3, CA_4, and LA_5. At the base of Q_1 the voltages cancel each other, thus eliminating the adjacent channel sound carrier from the IF band. The 41.25 MHz trap is a series-resonant network consisting of CA_7, CA_{73}, and LA_9. This trap is loosely coupled to the input circuit and improves skirt selectivity and provides better fine tuning.

In this system, AGC is applied only to the first IF amplifier stage. As the AGC voltage increases, forward bias is developed at the base of Q_1. This bias increases the current flow through the transistor and, consequently, causes a greater voltage drop across RA_{16}, thus reducing the stage gain.

In the collector circuit of Q_1, CA_{21} and CA_{22} divide the voltage developed across LA_{20} and couple it to the base of Q_2. The junction of the two capacitors matches the impedance of the coil to the base of Q_2. LA_{20} is tuned to 43.8 MHz, which is the center of the IF system passband. Coupling between Q_2 and Q_3 is identical to that between Q_1 and Q_2, with LA_{28} also tuned to 43.8 MHz. The associated sound carrier is tapped off at the collector of Q_3, the third IF amplifier.

The output circuit of the IF system is, in effect, very similar to the input circuit. Here, LA_{39} is tuned to position 42.17 MHz at the 50 percent point on the IF response curve, as was indicated in Fig. 8-7(B), and LA_{54} positions 45.75 MHz at the same point but on the opposite slope. A bridged-T trap, consisting of RA_{48}, CA_{47}, CA_{50}, and LA_{49}, rejects the 41.25 MHz associated sound carrier, thus preventing it from reaching the video detector. The three sound traps in this system, two at 41.25 MHz and one at 47.25 MHz, in addition to suppressing interference, account for most of the selectivity in the IF system.

8.10 VIDEO-IF AMPLIFIER TROUBLES

The video-IF section of a TV receiver must pass the video information, the color signals, the synchronizing and blanking signals, and the sound signals. If the receiver-IF system is not operating properly, it cannot develop the voltages necessary for proper operation of the video, color, sweep, or sound sections.

Troubles in the video-IF section can be classified in three categories: complete or partial loss of signal, improper alignment, and AGC troubles. The symptoms are the same for both vacuum-tube and solid-state receivers, the difference being in voltage and resistance measurements. The alignment and troubleshooting of TV receivers is discussed in Chapters 23 and 24.

Complete Loss of Signal There can be a complete loss of signal in the IF system due to a lack of $B+$, or a faulty vacuum-tube, transistor, or other component. The picture-tube screen will have a raster, but there will be no sound or picture. The symptoms associated with an inoperative video-IF stage are similar to those for a defective tuner, and further tests are needed to isolate the defect. These symptoms can also be produced by defects in the AGC circuits. The reason for this is that any IF amplifier that is AGC-bias controlled, either fully or partially, can be driven to cutoff by an incorrect AGC voltage. This is especially true in transistor-IF amplifiers. Always keep in mind the fact that a defective IF amplifier may cause the same symptoms as a defective tuner or a defective AGC section.

Partial Loss of Signal The partial loss of a signal, due to a defect in the IF stages will result in a picture of insufficient contrast. Assuming that the tuner is operating normally, there shoud be no snow in the picture. Snow can be produced by a malfunction or incorrect setting of the AGC system, by a defective mixer in the tuner, by a defective video detector, or by a defective video amplifier. The isolation of the defective area must be accomplished and this is discussed in Chapter 24. With only a partial loss of signal caused by an IF defect, the sound may appear quite normal, or may appear with a 60 Hz sync buzz. In any case, sound will generally be present when there is only a partial loss of the video-IF signal.

Improper Alignment The misalignment of a TV receiver IF system may occur during normal usage or may occur when a component is replaced. In some cases, replacement of a tube or a transistor may make it necessary to realign the video-IF section. The aging of components, and vibration or shock when the receiver is moved, may also make alignment necessary.

Improper alignment of the video-IF amplifier section can cause a smearing of the picture. The adjustment of the fine-tuning control changes the amount of smear, because this adjustment varies the position of the video carrier on the IF response curve. The smear is due to a phase shift caused by an excessive low-frequency response. Any phase shift in the video-IF amplifier circuits can also upset the phase-modulated, color-hue signals. Troubles of this nature usually have little effect on the production of color, but the color appears in the wrong areas of the picture. If the smear is not tuneable with the fine-tuning control, it is more likely to be caused by a defect in the video-detector or video-amplifier circuits.

Improper alignment of the video-IF amplifiers may also be the cause of a loss of fine detail. This occurs when the higher video frequencies are attenuated. The color signal may be attenuated or completely lost because of an improperly aligned IF stage.

Sound interference can be caused by improper alignment in the video-IF system. This problem is often traced to trap adjustments which are used for obtaining the correct shape of the IF response curve.

An excessive high-frequency response in the IF section can cause ringing in the amplifier circuits. The symptom of ringing is when a ghost-like region appears following the vertical lines of the picture. These symptoms, however, are easily distinguished from the types of ghosts related to antenna problems, because they can be varied with the fine-tuning control of the receiver.

Some transistor IF amplifiers employ degenerative feedback to reduce the likelihood of oscillation, and to broaden the frequency response of the amplifier. If oscillation occurs, the picture will show a pattern of light and dark areas and will often display a "herringbone" design. This pattern will remain on the CRT screen even when the station selector is switched to an inactive channel. Also, the voltage across the detector load will be higher than normal, and the voltage will not be affected, as it should be, by tuning to another station or to an inactive channel.

The obvious cure for alignment problems in the receiver is to follow the correct alignment procedures (see Chapter 23). It is recommended that alignment not be attempted without the proper test equipment as recommended by the TV-receiver manufacturer.

AGC Troubles The symptoms described for a faulty video-IF system also may apply to trouble in the AGC circuits. Usually, AGC trouble appears on the CRT screen as a washed-out picture with poor contrast and weak color, or as an overloaded picture with too much cotrast. The first case is caused by too much AGC voltage, while the second case is caused by too little AGC voltage.

The difficulty in servicing AGC problems stems from the fact that it is part of a feedback loop. The amount of AGC voltage depends upon proper operation of the IF system, and proper operation of the IF system depends upon the amount of AGC voltage that is applied to the controlled stages. Since the AGC and IF circuits are interdependent, it is difficult to separate the problems and symptoms in the two circuits.

The best way to determine whether the IF of the AGC circuit is at fault is to disconnect the AGC circuit and substitute a dc bias voltage to take its place. When the bias of the tuner is supplied by a substitute source, the circuit will operate properly provided the trouble is in the AGC circuit. If the symptoms do not change when the AGC voltage is substituted for, then it follows that the trouble is in the tuner or the IF system. AGC troubles are discussed in Chapter 10 and Chapter 24.

REVIEW QUESTIONS

1. Draw a standard response curve for the IF system of a TV receiver. Indicate the position of the video carrier.
2. Name and explain two types of spurious responses.
3. Why are trap circuits used in video-IF amplifiers?
4. When is the audio signal of the adjacent TV channel not likely to cause interference? Why?
5. Draw three types of trap circuits commonly found in video-IF circuits.
6. Given that the order carrier IF is 45.75 MHz; list the oscillator frequencies for channels 2 and 4.
7. Explain the principle of stagger tuning.
8. Why are neutralizing circuits used more often in transistor IF amplifiers than in vacuum-tube IF amplifiers?
9. In addition to neutralizing circuits, name some other important differences between transistor and vacuum-tube IF amplifiers.
10. How does a bifilar coil differ from a conventional coupling?

Video Detectors 9

9.1 DETECTION OF THE VIDEO SIGNAL

In accordance with the general line-up that is peculiar to superheterodynes, the second detector follows the last IF amplifier. Detection in television receivers is carried out in much the same manner as in any ordinary sound broadcast receiver. The semiconductor single diode,

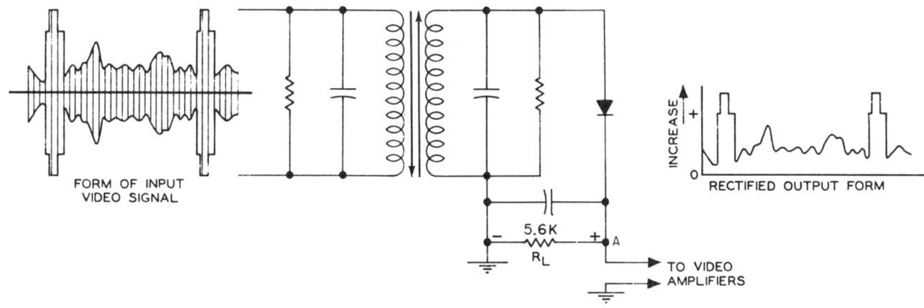

Fig. 9-1. A diode detector for a television receiver.

connected as shown in Fig. 9-1, is typical. The demodulated video signal with its blanking and synchronizing peaks is developed across R_L. The form of the IF signal when it enters the second detector is illustrated to the left of the figure. The rectified resultant is illustrated at the right.

As is true of the diode operation, anode current flows only when the anode is positive with respect to the cathode. The effect of this action is to eliminate the negative portion of the incoming signal. Since the positive and negative sections of the modulated video signal are exact duplicates of each other, either one may be used.

9.2 POSITIVE AND NEGATIVE-PICTURE PHASES

At this point it is necessary to consider the effect of the relative polarity of the voltage drop across the load resistor R_L. For American television systems, negative picture transmission is standard. This means that the brightest elements cause the least amount of current to flow, while maximum current is obtained when the blacker-than-black region of the synchronizing signal is reached. This method of transmission was adopted, because it was felt that better overall reception would be obtained.

The signal in the negative picture-phase form, as pictured in Fig. 9-2(A), however, cannot be applied directly to the grid of the picture tube. It first has to be reversed to the form shown in Fig. 9-2(B). That this is necessary is easily seen, for the blanking and synchronizing

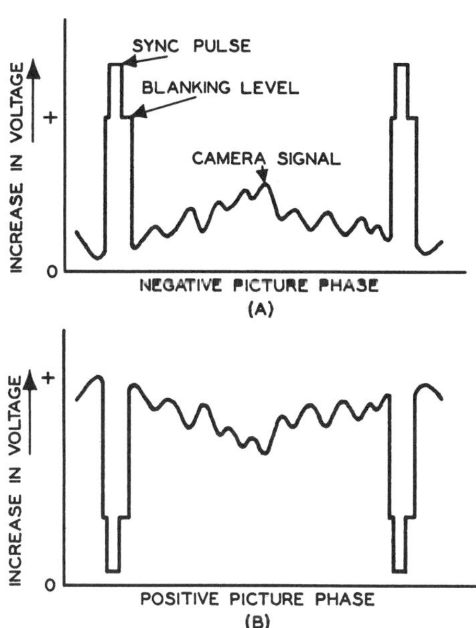

Fig. 9-2. Rectified video signals may be obtained from the output of the detector in either one of the two forms shown, depending upon how the detector is connected.

signals when applied to the control grid of a picture tube must bias it to cutoff. The objective can be attained only if the signal has the form given in Fig. 9-2(B). The television engineer calls this latter form of the television signal the positive picture phase. It is interesting to note that if the negative phase of the signal is applied to the control grid of the picture tube, all the picture values will be reversed and the observed scene will be similar to a photographic negative. (Another widely used method of distinguishing between the signal polarities is to employ the designation of a *sync pulse positive* (Fig. 9-2(A)) or a *sync pulse negative* [Fig. 9-2(B)]. The negative picture phase signal of Fig. 9-2(A) can be applied to the cathode of the picture tube, and the effect will be the same as applying the signal of Fig. 9-2(B) to the control grid of the picture tube.

In sound receivers, no attention is given to the relative phase of the audio signal, because our ears are insensitive to all but gross phase differences. Television, on the other hand, deals with visual images, and the reversal of phase produces noticeable effects. Possible ways of altering the phase of the video signal are discussed in the following paragraphs.

Turning to the half-wave detector circuit of Fig. 9-1, let us investigate the voltage developed across R_L. The incoming signal has the same form as at the antenna, with the synchronizing pulses giving rise to the greatest voltages. At the diode rectifier, these synchronizing signals cause the anode to become the most positive, resulting in a greater voltage drop across R_L and having the polarity as shown. On the other hand, those portions of the video signal representing the bright segments of the image will have the least positive voltage at the diode anode, with a smaller resultant voltage drop at R_L. Thus, with this circuit hookup, point *A* of resistor R_L will still give rise to a large positive voltage for the synchronizing signal, which means that the signal is still in the negative picture phase. The signal is unsuitable for direct application to the grid of the viewing tube.

The direction of the current flow through R_L may be altered quite easily to give the opposite polarity. Simply reverse the connections between the diode and the input transformer, as illustrated in Fig. 9-3. Rectification now eliminates the positive half of the modulated carrier

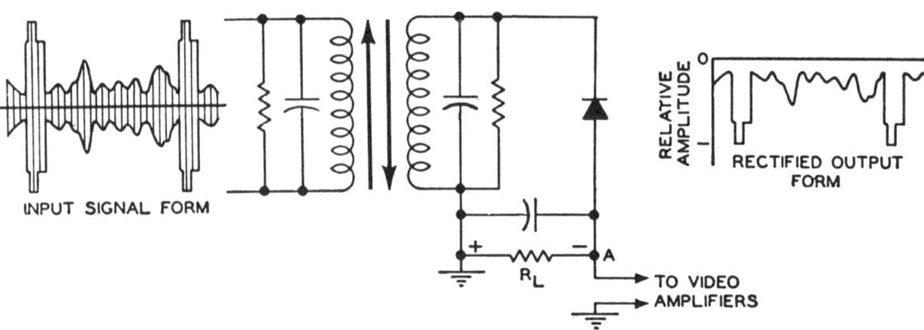

Fig. 9-3. A diode detector connected to give a positive picture-phase output signal. Note that, with an inverted diode detector, only the negative half of the input signal is rectified. In Fig. 9-1 the opposite is true.

and leaves the negative half. Since both contain the same information, nothing is lost. Point A becomes more strongly negative for the blanking and synchronizing portion of the video signal, while the bright elements cause A to become less negative. When the signal is applied in this form between the grid and the cathode of the image tube, the largest current will flow for the bright sections of the image and a bright spot will appear on the fluorescent screen. For the blanking and synchronizing parts of the signal, the voltage at the grid (from point A) will be highly negative and the electron beam will be cut off, as it should be.

The strength of the signal that is developed at the diode-load resistor is not strong enough to be used directly at the picture tube. Hence, further amplification is necessary. The following video amplifiers, which are generally of the *common cathode* and *common emitter* type, are capable of reversing by 180 degrees the polarity of any signal sent through them. Thus, if the video signal had a positive picture phase at the diode-load resistor, it would have a negative picture phase at the output of the first video amplifier. With another stage of amplification, the picture would be brought back to the positive phase again. As a general rule, then, an even number of video amplifiers is required if the picture phase across R_L in the detector is positive, and if the video signal is to be applied to the control grid of the picture tube. For a negative picture phase at R_L, an odd number of video amplifiers is needed, this time for a positive picture to appear at the grid of the image tube. These conditions are illustrated in a block form in Fig. 9-4.

The circuit of the video detector and the video amplifier in Fig. 9-5 appears at first glance to violate the foregoing rules. Examination of the video signal developed across R_1 reveals it to be positively phased. This would require 0, 2, 4, or some other even number of video amplifiers. Actually only one is present. The mystery is resolved when we note that

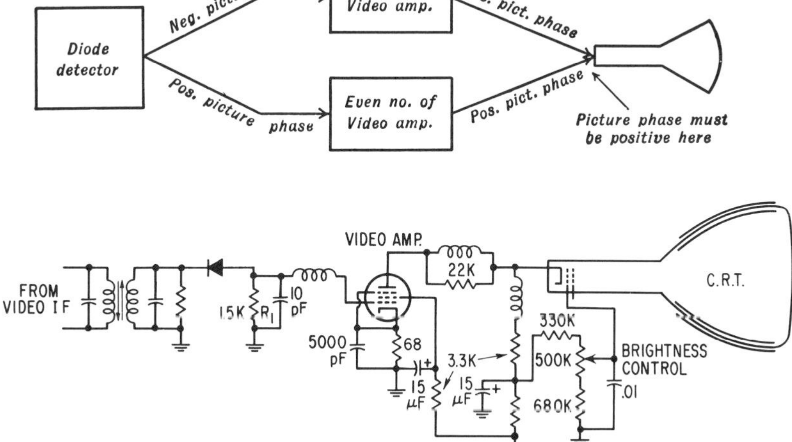

Fig. 9-4. An illustration of why the number of video amplifiers after the detector is dependent upon the polarity of the signal obtained from the detector.

Fig. 9-5. A video detector and a video amplifier feeding the signal to the cathode of the image tube.

the output of the video amplifier is fed not to the control grid of the cathode-ray tube but to its cathode. The foregoing rules were drawn up with the tacit understanding that all incoming signals are applied *to the control grid* of the picture tube! To produce similar results, signals applied to the cathode of a tube should differ by 180 degrees from the same signals applied to the control grid.

9.3 DETECTOR FILTERING AND PEAKING

The frequencies present in the detector circuit include the intermediate-frequency values and the actual video signals themselves, 0–4 MHz. The latter voltages are to be passed on to the video amplifiers and strengthened to the point where they are able to modulate the electron current in the picture tube to produce an image on the screen. At the detector output, the intermediate frequencies must be properly shunted around the load resistor to prevent their reaching the following video amplifiers. In the receivers currently being produced, the problem of filtering the IF voltages has been made comparatively simple because of the use of fairly high IF values. The rectified video signal has a maximum frequency of 4 MHz. In the early television receivers, the IF values ranged from 8.75 to 12.75 MHz, and considerable filtering was required because of the low order of separation between the desired frequencies (0–4 MHz) and those which were to be bypassed (8.75 to 12.75 MHz). However, by increasing the separation between the two, we have simplified the problem considerably. Current values for the video IF are between 40 and 46 MHz.

Adequate filtering can be obtained through the arrangement shown in Fig. 9-6. The rectified current passes through the low-pass filter composed of C_1, L_1, R_1, L_2, R_2, and C_2. Capacitor C_1 is a small fixed value of 5 pF, but actually there exists additional capacitance across this point produced by the diode junction and the wiring. At the other end of the filter, C_2 is shown in a dotted form because no such component is inserted. However, the sum of the stray-wiring capacitance plus the input capacitance of the following video amplifier produces the equivalent of an actual capacitor of 10 to 15 pF. The two coils, L_1 and L_2, while forming part of the low-pass filter, at the same time maintain a good frequency response to 4 MHz, thereby counteracting any tendency of the circuit to attenuate these higher video frequencies. More will be noted on this point in the chapter on video amplifiers. The 39,000-ohm resistor shunted across L_1 is used to prevent the response of the coil from rising abruptly at the higher video frequencies because of a natural resonant circuit formed by the coil and its inherent capacitance. The detector load resistor is R_2 (3,900 ohms).

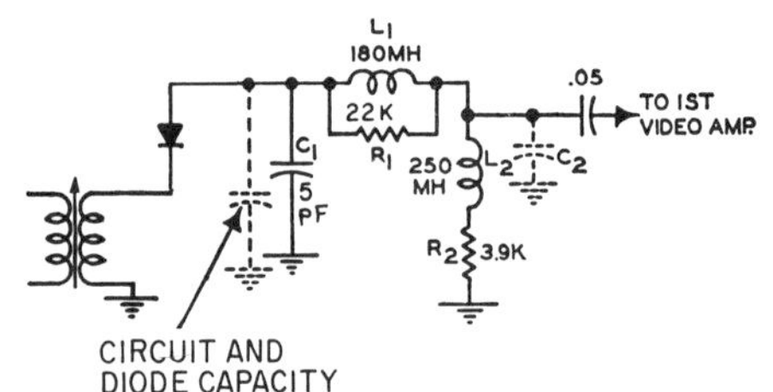

Fig. 9-6. A video-detector circuit with a low-pass filter and load resistor.

9.4 SHUNT-VIDEO DETECTORS

All of the video detectors discussed thus far are of the series type, in which the rectifier (tube or semiconductor diode), the input-tuned circuit, and the load resistor are all in series. It is also possible to achieve detection by placing the diode in a shunt with the tuned circuit and the load resistor. The basic circuit is illustrated in Fig. 9-7(A), and the complete circuit, with the low-pass filter, is shown in Fig. 9-7(B).

The circuit of Fig. 9-7(A) operates in the following manner. When the positive half of the IF video signal is active, the germanium diode conducts. Electrons travel up through the diode to the right-hand plate of C_1, charging this side of the capacitor negative. At the same time, an equivalent number of electrons leave the other plate of C_1 and travel down through L_1 and R_2 to $B+$ and thence to ground, completing the circuit. Because of the low resistance of the diode while it is conducting, C_1 will charge to the peak of the applied voltage.

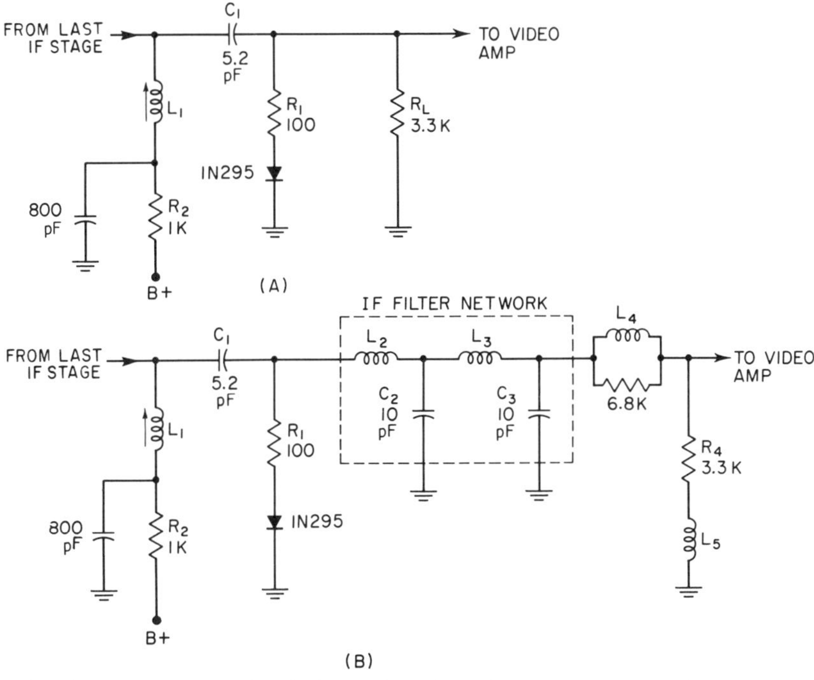

Fig. 9-7. Examples of shunt-video detection.

During the succeeding half-cycle, when the signal voltage is negative, the diode becomes nonconductive. Capacitor C_1 however, finds a complete discharge path through R_L, and electrons travel from the right-hand plate of C_1 through R_L to ground, then up through V_1 to the other plate of C_1. This flow places the full value of the voltage of C_1 across R_L. Thus, we obtain all of the IF variations across R_L, as well as the modulation signal (which represents the video information) and an average dc voltage, because rectification has taken place. Additional filtering is required to remove the IF portion of the voltage. What happens to the dc voltage depends on the succeeding circuit. If it is desired to retain the dc voltage, then a dc path is maintained to the next tube or transistor (a video amplifier). However, if this dc voltage is not desired, the signal across R_L can be capacitively coupled to the next stage.

A full shunt-detector circuit, complete with a low-pass filter, is shown in Fig. 8-7(B). Coils L_2 and L_3, and capacitors C_2 and C_3 form the filter, while L_4 and L_5 serve as the video-peaking elements. Resistor R_1 is a current-limiting resistor, designed to protect the diode. Note that in this arrangement the low-pass filter is placed before the load resistor, R_L. It could have been positioned after R_L, but the arrangement shown is the more efficient approach.

It will be instructive to see how a video signal would appear on the screen of an oscilloscope. A typical presentation is given in Fig. 9-8. To obtain a stationary pattern, the sweeping rate in the oscilloscope should either be 15,750 Hertz or, for two full lines (as shown in Fig. 9-8), 7,875 Hertz. How clearly the sync pulses appear will depend on the overall response of the vertical amplifiers in the oscilloscope. If this response is too narrow, the sync pulses will appear with rounded corners; however, even with the limited response of fairly low-cost oscilloscopes, the form of the video signal will be clearly discernible.

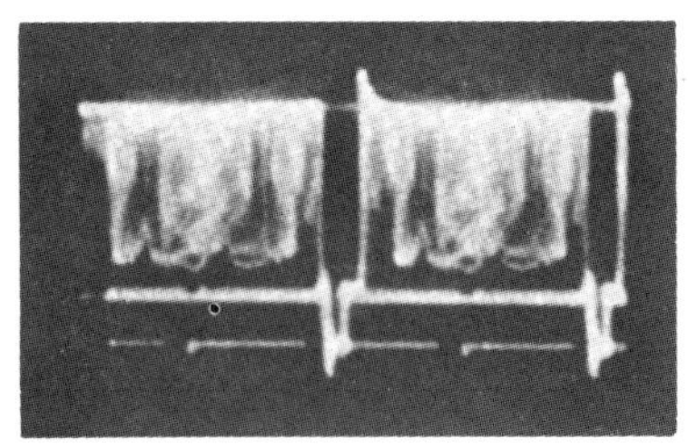

Fig. 9-8. Two lines of a video signal, as observed on the screen of an oscilloscope.

9.5 VIDEO DETECTORS IN COLOR-TELEVISION RECEIVERS

In monochrome receivers it is a common practice to feed the video and sound signals together through the video-IF amplifier stages, and through the video-detector stage. At the output of the detector, the two signals—that is, the video carrier and the sound carrier—are heterodyned and produce a 4.5 MHz sound IF frequency. This is a characteristic of intercarrier receivers.

In color television receivers, it is undesirable to allow both the sound signal and video signal to pass through the video detector. The reason for this is that a 920 kiloHertz difference signal would be generated as a result of heterodyning between the 3.58 MHz color carrier and the 4.5 MHz sound signal. The 920 kHz heterodyne signal would produce an interference pattern on the color screen.

To avoid the appearance of this pattern on the screen, the sound is obtained by tapping off the composite IF from the last video-IF amplifier stage and feeding it to a separate sound detector. The output of this sound detector is a 4.5 MHz FM sound signal which is then processed by the sound section of the TV set.

This method of obtaining the sound signal would not, in itself, prevent the aforementioned interference pattern. The pattern is suppressed by the use of two different traps, as indicated in Fig. 9-9. A trap, tuned to the sound IF of 41.25 MHz precedes the video detector. This greatly reduces the level of any sound IF signal prior to its introduction to the video detector and thus, in itself, would reduce the amplitude of the 4.5 MHz component contributing to the production of a

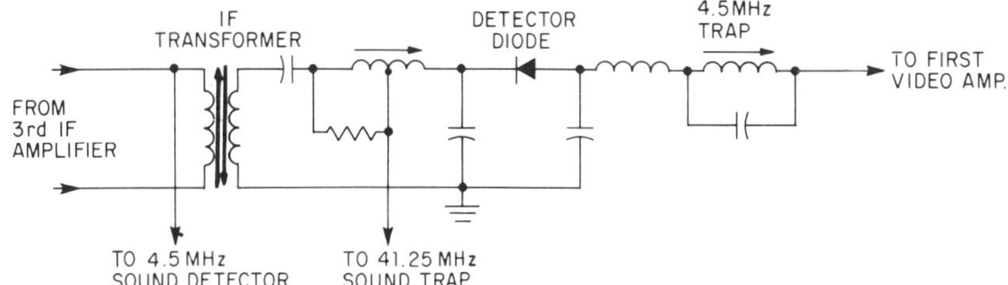

Fig. 9-9. A video-detector circuit in a color-TV receiver.

920 kHz signal. To further reduce the possibility of producing the interference pattern, a 4.5 MHz trap is placed after the video detector, to greatly reduce any remaining heterodyned 4.5 MHz signal produced at the output of the video detector.

Figure 9-10 indicates a typical IF response curve for a color-television receiver. This response curve is obtained by feeding a sweep generator into the tuner section and monitoring the output at the output of the video detector. (The complete procedure for obtaining this pattern is covered in a later chapter.) Since this response curve is obtained at the output of the video detector, the sound marker must be a point of minimum amplitude as shown in the illustration. The 41.25 MHz tuneable trap is adjusted to set the sound carrier at the minimum point. Note from the characteristic curve of Fig. 9-10 that the receiver circuits, including the tuner, the video-IF amplifiers, and the video detector, are capable of passing all frequencies in the video range including all frequencies in the color signal.

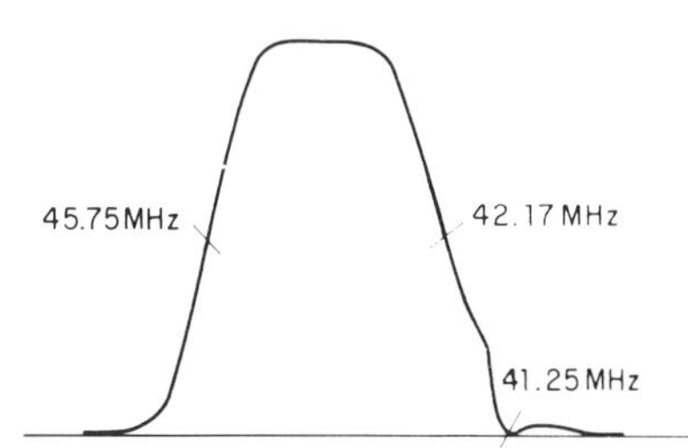

Fig. 9-10. Characteristic curve of a color receiver. The audio marker shows that the response at 41.25 MHz is extremely low at the output of the video detector.

Figure 9-11 illustrates the detector stage for an RCA solid-state color chassis. Capacitor C_1 at the output of the detector stage is used for filtering out undesired harmonics and for attenuating the video-carrier frequencies. A trap circuit, comprised of C_2, L_1, and R_1 is used for removing the 4.5 MHz beat frequency caused by the heterodyning of the audio- and video-carrier frequency. Inductors L_2 and L_3 remove

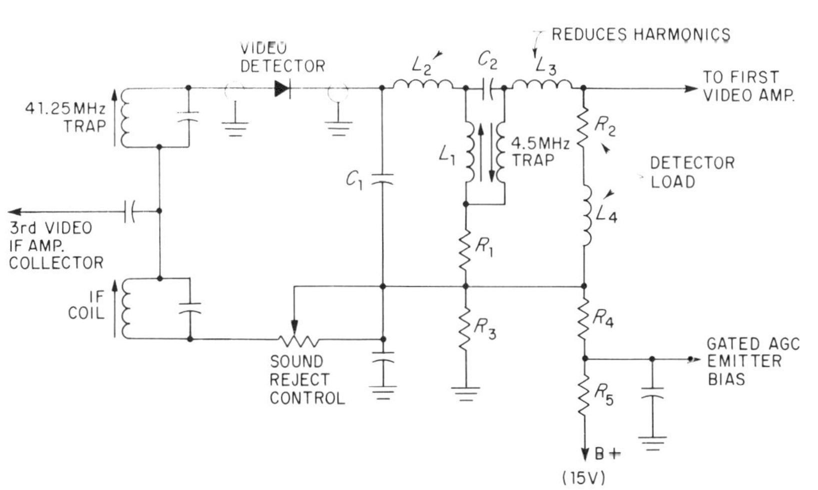

Fig. 9-11. The video-detector stage of an RCA color-TV receiver.

harmonic frequencies generated in the nonlinear detector stage. The resistor R_2 is the resistance load for the detector, while L_4 is a peaking coil.

You will note that in the circuit of Fig. 9-11 a dc bias voltage is applied to the detector stage from a voltage divider comprised of R_3, R_4, and R_5. This dc bias is actually needed for the first video amplifier which follows the detector stage. It brings up a very important point. Unlike detector stages in vacuum-tube receivers which are normally at dc ground potential, you will often find that the detector stage in a color transistor receiver has a dc bias voltage applied. Therefore, it is *not* true that you can completely ignore the dc voltage readings in the detector stage as far as the transistor receivers are concerned. If the dc voltage at this point is not correct, the first video amplifier will not be properly biased.

9.6 TROUBLES IN THE VIDEO-DETECTOR STAGE

In a monochrome receiver, both the sound and the video signals frequently pass through the video-detector stage, and therefore, a faulty detector can affect either or both of these signals. Since the sync pulses also pass through the detector stage, synchronization can be affected by a faulty component in the circuit. If the diode in the detector stage is not entirely defective, the circuit that is most affected by a weak signal will show symptoms first. Thus, if a receiver cannot tolerate the loss of the sync pulse amplitude, there may be a partial loss of synchronization and a weak picture. However, the sound may be acceptable.

The symptoms of no sound, no sync, and no picture might be an indication of a completely faulty detector stage provided that there is a raster present. Semiconductor diodes are used for detectors and when their front to back ratio decreases—that is, when the diode becomes faulty—a typical indication is a washed-out picture. A sufficient image cannot then be obtained with the contrast control.

Although the diode is a very likely source of trouble in the detector stage, it is also possible for one of the peaking coils to open. The best way to check for an open peaking coil in the detector stage, or for that matter in any stage, is to temporarily short it while watching the picture. If there is no effect on the picture when the peaking coil is shorted, it follows that the peaking coil is not open. If a short-circuit test indicates that the peaking coil is faulty, the next step is to replace it with an identical one.

If the video-detector stage becomes faulty in a color receiver, it is necessary to employ some common-sense troubleshooting. A look at the receiver schematic diagram will tell you which signals are present up to, and including, the detector stage. The sound is taken off at a point before the detector, so trouble with sound is not likely to be caused by the video-detector stage. The color signals are also taken off before the detector stage, so that trouble in the color section generally cannot be caused by a faulty video detector. Since sync signals pass through the detector stage and may receive additional amplification in one or two video-amplifier (or Y amplifier) stages, a faulty video detector may cause problems with synchronization as well as a loss in picture contrast or a total loss of the monochrome picture.

REVIEW QUESTIONS

1. Draw a response curve for a color receiver as observed at the video detector. Explain why the sound marker is at the point you have shown.
2. What are some differences between the video-detector circuit in a monochrome receiver and the video-detector circuit in a color receiver?
3. Why is there likely to be a dc-voltage level in the video-detector stage of a transistor receiver?
4. Draw a simplified shunt video-detector circuit showing the diode and the load resistor.
5. Draw a composite-video signal with a positive picture phase.
6. Assuming that the video signal is to be delivered to the cathode of the picture tube, and assuming also that there are three stages of conventional video amplification between the detector and picture tube, should the output signal from the video detector have a positive or negative-picture phase?
7. The video detector is called the second detector. Which stage is the first detector?
8. Which signals are present in the video detector of a color-television receiver?
9. What is the purpose of the filter circuit in the video-detector stage of a color-TV receiver.
10. What is the purpose of the trap circuits in the video-detector stage of a color-television receiver?

10 Automatic Gain Control (AGC) Circuits

10.1 COMPARISON OF RADIO AND TELEVISION AGC SYSTEMS

Automatic volume control (actually this should be called "automatic gain control") in a radio receiver serves to keep the output sound at a constant volume while wide variations occur in the strength of the input signal. Once the manual volume control has selected the output level that is desired, the AVC system tends to keep it there. In addition, when tuning to other stations, no adjustments are necessary to prevent a strong station from producing excessive volume. For television receivers, Automatic Gain Control (AGC) is advantageous in keeping the picture intensity fixed at one level even though the strength of the video signal at the input of the set may be varying. The eye is far more critical of changes than the ear, and anything that minimizes unwanted variations in image intensity is very desirable. AGC is advantageous when switching from one station to another because input-signal strengths may differ. Finally, more stable synchronizing is obtained if the signal fed to the synchronizing circuits is constant in amplitude.

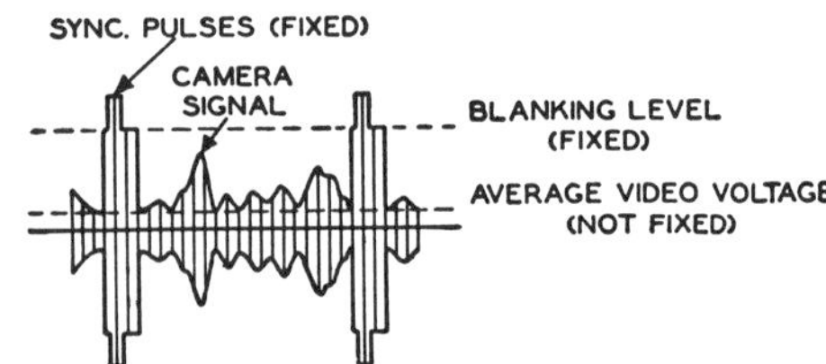

Fig. 10-1. An amplitude-modulated television signal. The fixed voltage levels are suitable for AGC control since these voltages vary directly with the signal strength.

A study of the television-modulated signal in Fig. 10-1 reveals that, so far as AGC is concerned, the rapidly varying camera signal is of little use to us. We desire some point which will be indicative of the strength of the carrier and which does not change with anything but the carrier.

With the present system of transmission, the carrier is always brought to the same level when the synchronizing pulses are inserted. Thus, so long as the signal being received is constant in strength, the level of the synchronizing pulses will always reach the same value. If something should affect the carrier level, these pulses would likewise change. With the change, the gain of the set would require adjustment to maintain the previous level at the detector. Hence, the strength of the synchronizing pulses will serve nicely as a reference level for the AGC system. It should be noted that the level of the blanking pulses (immediately below the top of the synchronizing pulse) is likewise fixed and may also be used.

To summarize, the function of the AGC circuit in a television receiver is to stabilize the output signal against input signal variations. Both the video and the audio signals are stabilized by using a feedback loop in the receiver. A dc voltage that is dependent upon the amplitude of the sync pulses is obtained and used for controlling the gain of the RF and IF stages. This is the essence of the AGC system in both monochrome and color receivers. However, there is a considerable difference between the method of controlling the gain of a vacuum-tube amplifier and the method of controlling the gain of a transistor amplifier. Therefore, this chapter will treat tube and transistor AGC circuits separately—beginning with the tube circuits.

At present, two general methods are employed to develop AGC voltage: *peak* and *keyed*. Each will be considered in turn.

10.2 PEAK-AGC SYSTEMS

In *peak AGC systems,* the designer uses the tips of the sync pulses to establish the AGC voltage. A separate diode rectifier is employed which receives the same video-IF signal as the second detector. Diode conduction occurs during a one-half cycle (positive or negative), and during this period a capacitor is charged to the full voltage of the arriving sync pulse. It is this voltage which is then employed for automatic gain control.

A typical peak-detector circuit is shown in Fig. 10-2(A). The AGC diode is one section of a twin diode. It receives the incoming signal from the video-IF system through capacitor C_1. The load for the AGC tube is R_1, a one-megohm resistor. To understand the operation of this circuit, consider the equivalent diagram shown in Fig. 10-2(B). (The cathode resistor and capacitor are omitted from the equivalent diagram, because they do not affect the AGC-voltage development.) The AGC tube will

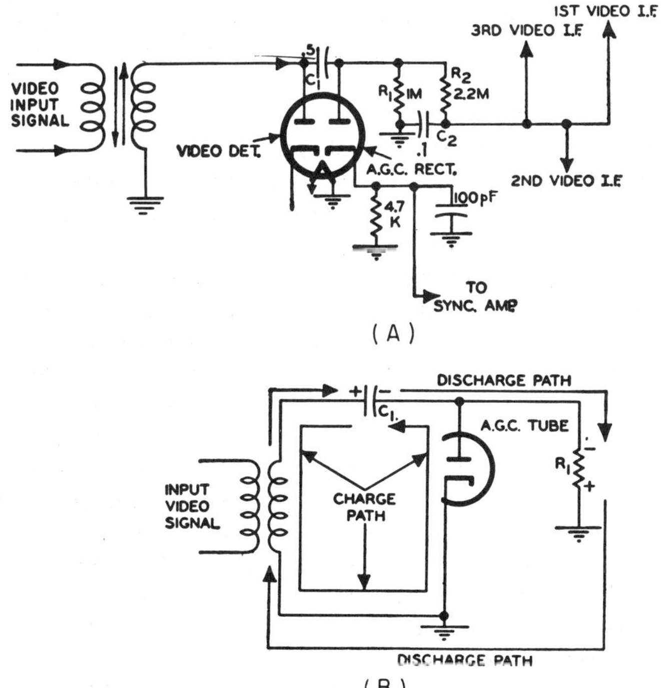

Fig. 10-2. One type of AGC circuit.

not conduct until its plate is driven positive with respect to its cathode. When this occurs, electrons flow from the cathode to the plate of the diode and into C_1, where the negative charge is stored. Very few electrons attempt to go through R_1 because of its high resistance value. On account of the low impedance offered by the tube when it is conducting, C_1 charges to the peak of the applied voltage, which is the value of the synchronizing pulses.

During the negative excursion of the incoming signal, the plate of the diode is driven negative with respect to its cathode and no conduction through the tube occurs. However, if we examine Fig. 10-2(B), we see that a complete circuit exists with C_1, R_1, and the input coil all in a series. Since a voltage exists across C_1 and a complete path is available, current will flow, making the upper end of R_1 negative with respect to ground. Because of the long time constant of R_1 and C_1, the charge accumulated across C_1 will discharge slowly through R_1—so slowly in fact, that only a small percentage of the voltage across C_1 will be lost during the interval when the tube is not conducting.

When the incoming signal becomes positive again, the tube does not immediately conduct because the applied signal voltage must first overcome the voltage that remains across C_1. Since C_1 has lost little of its voltage, tube conduction will occur only at the very peak of the positive cycle. These peaks, of course, are the synchronizing pulses. Thus, the voltage across C_1 is governed entirely by the sync pulses, which is what we desire. The negative voltage across R_1 is filtered by R_2 and C_2 to remove the 15,750-Hertz ripple of the horizontal sync pulse. The rectified dc voltage is then fed to the video-IF amplifiers as an AGC control voltage.

It is interesting to note that the same diode can also supply the sync pulses to the horizontal and vertical sync systems. Since current flows through this diode only during the sync pulses, voltage pulses will appear at these times across cathode resistor R_3. These pulses are tapped off and applied to the sync-separator circuits.

An AGC system which is more complex and involves the use of a special AGC amplifier is shown in Fig. 10-3. As before, the incoming video signal is received from the video-IF system and applied to the video detector and the AGC rectifier. In the cathode leg of the AGC diode section of the twin diode, we have the long-time-constant network of R_1 (4.7 meg) and C_1 (0.05 μf). Initially the capacitor charges up to the peak value of the incoming sync pulses and thereafter discharges slowly through R_1 during those intervals when the diode tube does not conduct. Since the discharge is very slow, because of the time constant of the circuit, much of the voltage established across C_1 will remain. Hence, current will flow through the tube only at the sync-pulse tips. The voltage thus established across C_1 and R_1 will be governed by the level of the sync pulses in the incoming signal.

As a result of the path of current flow through the AGC rectifier tube,

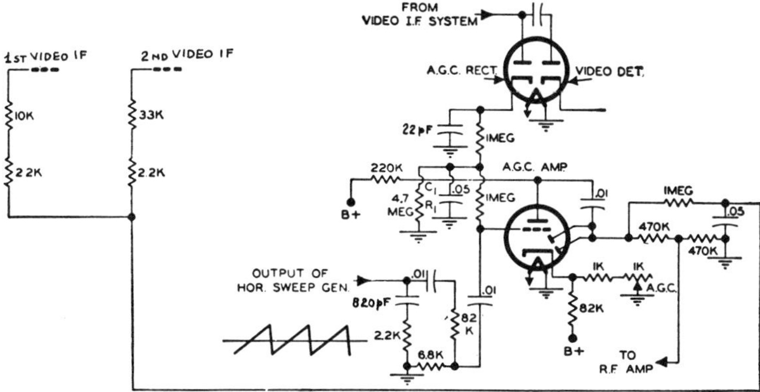

Fig. 10-3. An AGC network that uses an amplifier.

the voltage at the ungrounded end of R_1 is positive. This voltage is fed to the grid of the triode section of the AGC amplifier. The positive voltage, however, is offset by an even greater positive voltage that is obtained from the B+ power supply and applied to the cathode. Thus, the effective control-grid bias for the AGC amplifier is negative, although the value of this negative voltage varies with the voltage obtained from R_1 and C_1. The purpose of this arrangement is to vary the bias of the AGC amplifier in accordance with the strength of the incoming signal, and yet maintain the overall bias negative.

A second voltage, applied to the grid of the AGC amplifier, is obtained from the horizontal-deflection system. This voltage is amplified in the triode section of the AGC amplifier, and then rectified by the diode section of the same tube. It is this rectified voltage which is finally employed as the AGC voltage to control the gain of the RF, input IF, and first video-IF amplifiers.

When the signal level *increases*, the AGC rectifier develops more voltage across C_1 and R_1. This means more positive voltage for the AGC amplifier, and a greater output from this tube because of the increased gain. The horizontal deflection voltage fed into the tube remains constant at all times, and therefore, a more positive bias for the tube will produce a greater output and more rectified AGC voltage will be fed to the controlled stages. However, since the final AGC voltage is negative, the gain of the controlled stages decreases, counteracting the increased video signal and reducing the output to the normal level.

When the signal level *decreases*, less positive voltage appears across C_1 and R_1 and the AGC amplifier bias becomes correspondingly more negative. Less gain in this tube provides less voltage at its output, and the negative AGC voltage decreases. As a result, the gain of the controlled stages rises, and the signal level is again brought back to the normal level.

Whereas in the previous AGC system, the output of C_1 and R_1 was used as the controlling AGC voltage, here it is used to control the bias of the AGC amplifier. The final product of both systems is the same, but this arrangement is more sensitive to the carrier level changes because of the addition of the amplifier. A divider network at the output of the diode section of the AGC amplifier feeds less AGC voltage to the RF amplifier than to the IF stages. The potentiometer in the AGC system permits it to be adjusted for the desired sensitivity.

A number of receivers combine a *local-distant* switch with the AGC system to provide a degree of flexibility under varying signal conditions. The function of the switch is to have the full AGC voltage in use when the incoming signal is strong (that is, "local" conditions), and to reduce the control voltage when the received signal is weak (that is, "distant" conditions). There are a number of ways to achieve this flexibility, but the method shown in Fig. 10-4 is typical.

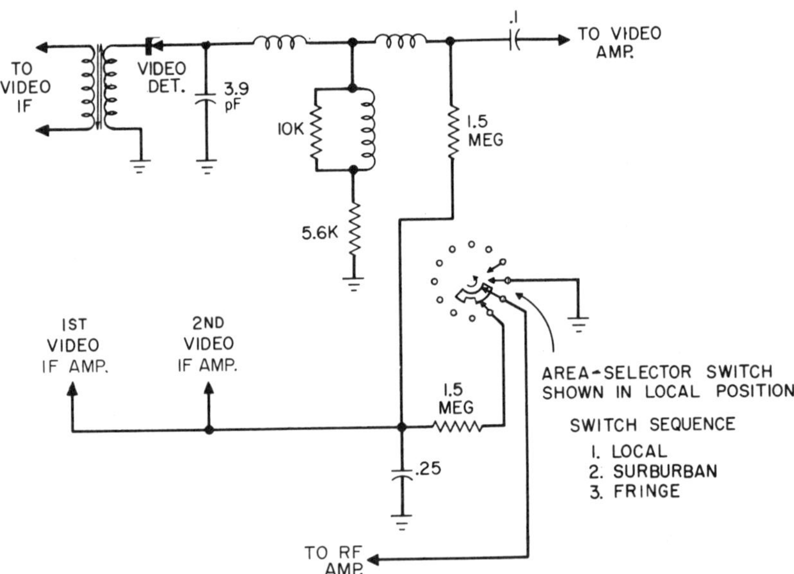

Fig. 10-4. An AGC system in which the control voltage can be varied to suit local reception conditions.

The AGC voltage and the detected video signal are both obtained from a semiconductor diode. The signal is fed via two 1.5 megohm resistors to the first and second IF amplifiers and to the RF amplifier. A three-position rotary selector switch is connected with this circuit so that it can alter the AGC-voltage distribution. When the switch is in the "local" position, all the controlled stages (IF and RF) receive the full AGC voltage. When the switch is in the "suburban" position, where the received signal is somewhat weaker, the AGC line to the RF amplifier is grounded. The two IF stages, however, still receive the full AGC voltage. Finally, in the "fringe" position, both IF amplifiers have their AGC voltage reduced by one half, and the AGC voltage to the RF amplifier remains at zero.

By the foregoing arrangement, the gain of the receiver can be regulated according to the strength of the incoming signal. This method of control can also be applied to the peak AGC systems.

10.3 KEYED-AGC SYSTEMS

Keyed-AGC systems represent still another approach to the problem of the adequate control of the receiver gain. They are very widely used.

By examining the AGC filter network in any of the circuits discussed thus far, it will be seen that high values of resistors and capacitors are used. This means that the capacitors charge and discharge slowly. High-value resistors and capacitors are required not only because they must smooth, or filter, a 15,750-Hertz ripple arising from the horizontal pulses which actuates the circuit, but because they must also smooth out the 60-Hertz ripple produced by the vertical pulses. If the 60-Hertz fluctuation is not removed, the bias on all controlled tubes will rise during the vertical sync interval and act to depress these pulses.

For slow changes in signal strength, these filter network arrangements are satisfactory. But what happens when a very fast change in signal level occurs? If such changes do not occur frequently, they will have very little noticeable effect on the AGC bias. However, if the changes occur rapidly and continuously (or frequently), they will affect the bias and, through this, the sound and picture signal strengths. As an example, when an airplane passes overhead, the picture and sound intensity will rise and fall (flutter). As a further disadvantage of the preceding AGC systems, when the noise level is high, the systems will react to large noise pulses by developing more bias than they will if they respond solely to the signal. The result will be less amplification for the signal when the noise is present. If the signal itself is quite weak, this decreased amplification may cause it to be lost altogether.

Both of these disturbances adversely affect the television image. In fringe areas, the weak signal and the accompanying noise are the most important considerations. In areas near airports, airplane flutter is important. Before we describe how a keyed AGC system overcomes both these annoyances, let us briefly determine the reason for airplane flutter.

Picture flutter occurs whenever an airplane passes overhead or nearby. The picture intensity rises and falls, becoming light and dark in turn at a fairly rapid rate. This effect might last from 15 to 30 seconds, depending upon how long it takes the airplane to pass.

The intensity pulsation is caused by the airplane acting as a reflector. Some of the television signals striking the metallic surface of the airplane bounce off and reach the television antenna. When these reflected television signals arrive in phase with the normal signal that the antenna receives, they will *add* to the desired signal and strengthen it. When the reflected signals arrive out of phase with the normal signal, they subtract from the desired signal, and the strength of the normal signal will be reduced. As the airplane moves, the reflected signal alternately adds to and then subtracts from the desired signal—depending upon the length of the reflected signal path. Since this rapid increase and decrease in the strength of the received signal cannot be counteracted by the usual slow-acting AGC filter, the picture on the screen will vary in intensity, producing the aforementioned flutter.

A keyed AGC system is able to overcome this flutter because of two facts: first, the AGC system is receptive to incoming signals only at certain specific times; second, the resistance and capacitances comprising the AGC filter are lower in value than the corresponding components in conventional AGC systems, and this means that their time constant is shorter. Therefore, a keyed AGC network can react instantly to a fairly rapid signal fluctuation (such as produced by airplanes) and can change the AGC bias fast enough to counteract this signal change.

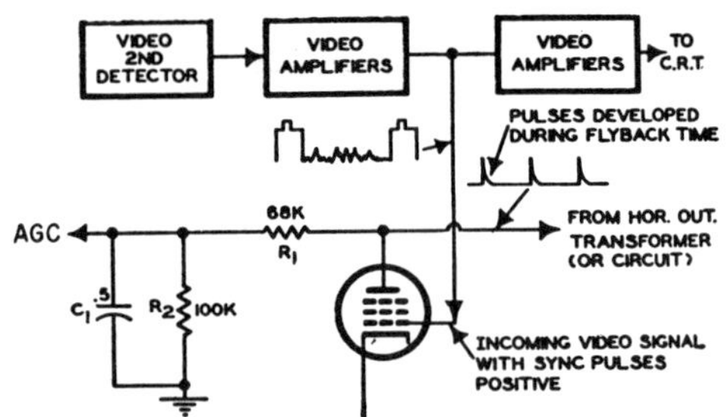

Fig. 10-5. A simplified diagram of the components of keyed AGC systems.

A simplified illustration of a keyed AGC system is shown in Fig. 10-5. A pentode is connected so that a portion of the detected video signal is applied to its control grid. The signal is in the negative picture phase, which means that the sync pulses are very positive. The plate of the pentode is connected to a winding on the horizontal-output transformer, and receives from this transformer a positive pulse of voltage at the end of each horizontal line.

Now, the pentode tube is so biased that it will not conduct unless the grid and plate are simultaneously active. If just one of these voltages is present, the tube does not ordinarily conduct.

The pulses applied to the grid are the horizontal-sync pulses. When these pulses arrive, the electron beam traveling across the face of the picture tube is about to start its retrace. At the moment it begins, a large pulse of voltage is developed in the horizontal-output transformer and a portion of this pulse is fed to the plate of the tube. With both positive pulses of voltage present (one at the plate and one at the grid), the AGC tube is keyed into conduction and the AGC bias voltage is established.

Note the foregoing sequence of events carefully, because they contain the key to the operation of this system. Positive pulses must be present at *both* the control grid and the plate of the tube in order for them to pass current and establish the proper AGC bias. The plate receives no positive voltage other than that furnished by the horizontal-output transformer.

Since the tube conducts only when the sync pulses are active at its grid, and it is inactive throughout the remainder of the video signal, it is evident that the AGC tube (and consequently the AGC network) is responsive to undesirable noise pulses for only a very short period of time. Actually, the sync pulses occupy only 5 percent of the composite video signal, and therefore, only 5 percent of the total noise can be effective.

Contrast this with AGC systems other than the keyed type. While they are supposed to be unresponsive to all but the sync-pulse tips, this is true only if the amplitude of the sync pulses is greater than the amplitude of any of the noise pulses present. Any noise signal possessing a greater amplitude than the sync pulses will cause current to flow in the AGC tube. Thus, noise pulses in such systems will develop a greater negative biasing voltage in the AGC network than that obtained from the sync pulses alone. Until this greater negative voltage diminishes and the normal sync pulses again resume control, the gain of the set will be down.

When the tube in the keyed AGC circuit of Fig. 10-5 conducts, current flows from the cathode to the plate, then through R_1 and R_2 to ground, and finally back to the cathode. Capacitor C_1 is charged to the voltage developed across R_2. The values chosen for the R_2-C_1 combination are designed to remove the 15,750-Hertz ripple of the plate-current pulses. No provision need be made to filter out a 60-Hertz ripple, since none exists in this AGC system. It is the use of small R and C values that largely accounts for the ability of a keyed AGC system to act fast in overcoming airplane flutter.

Before we progress further, there is probably one question which will occur to many readers: Why do the other AGC networks develop a 60-Hertz ripple, while the keyed AGC networks do not? To understand why, refer to Fig. 10-6 where horizontal-sync pulses and vertical-sync pulses are shown. Note that the duration of a horizontal-sync pulse is much shorter than the duration of the vertical-sync pulse. In the conventional AGC system, this inequality in pulse duration results in a larger charge being developed across the AGC capacitors by the vertical-sync pulses.

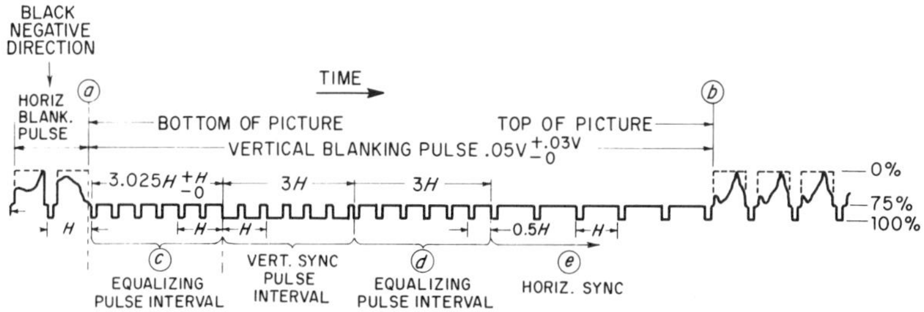

Fig. 10-6. The relative time intervals for the horizontal-sync pulses and the vertical-sync pulse interval.

Hence, every 1/60 sec, more current will flow through the AGC tube because of the greater width of the vertical sync pulses. As a result, the AGC biasing voltage will have a 60-Hertz ripple in addition to the 15,750-Hertz variation caused by the horizontal-sync pulses themselves.

In the keyed AGC system illustrated in Fig. 10-7, the pentode is *fired* by a *combination* of a positive plate pulse and a positive grid pulse. The plate pulse, however, is *constant* in duration, since it is obtained from the horizontal-output transformer where the pulses do not change. Hence, whether the grid pulse is a horizontal-sync pulse or a vertical-sync pulse, the tube conducts for the same length of time. It is because of this behavior that only a 15,750-Hertz ripple is present, and only this ripple frequency needs to be filtered in keyed AGC systems.

In applying the sync pulses to the grid of the AGC tube, care must be taken to see that they are all aligned to the same level. This alignment is necessary because the amount of current flowing through the AGC tube is determined in large measure by the amplitude of the sync pulses. As

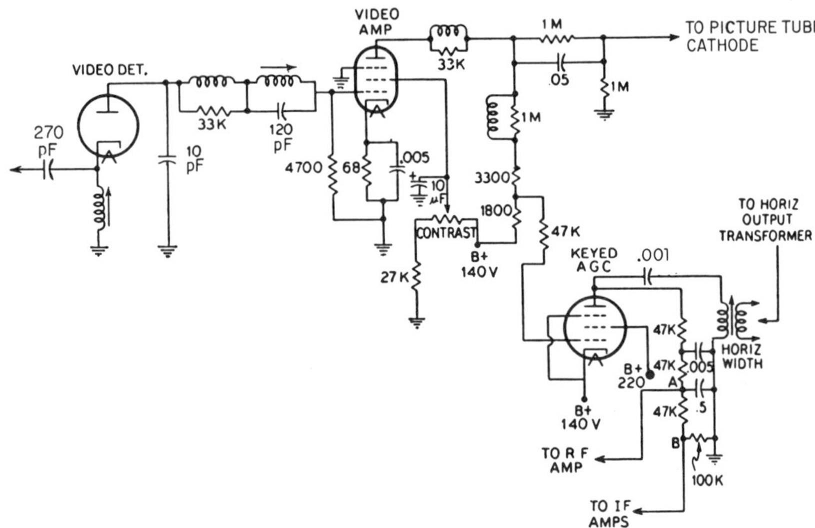

Fig. 10-7. A commercial application of the keyed AGC system.

the amplitudes vary, so will the tube current and the AGC bias. Hence the sync pulses will be aligned as required for the proper operation of the keyed AGC circuit. Inspection of the diagram reveals that the video signal which is applied to the AGC tube is obtained from the plate circuit of the video amplifier. At this point the sync pulses are positive. (This would have to be true, of course, since the signal from here is fed directly to the cathode of the cathode-ray tube.) A decoupling network consisting of a 3,300-ohm resistor, an 1,800-ohm resistor, and a 47,000-ohm resistor directs part of the video signal to the grid of the AGC tube. The decoupling network is designed to minimize the shunting effect of the AGC tube on the video-amplifier network. This is required in order to uphold the video response of the amplifier-coupling network.

The cathode of the AGC tube is connected directly to the 140V B + point. This is necessary because of the high positive potential present on the control grid of the tube. Actually, with the AGC tube in operation, the control grid is approximately 25 volts less positive than the cathode, and hence the tube does not conduct except when the horizontal-sync pulses are active.

The screen grid of the AGC tube has a positive potential of 220 volts applied to it. The flyback pulse for the plate is taken from a special winding placed over the horizontal width coil. This coil is connected across the secondary of the horizontal-output transformer. Whenever the AGC tube conducts, current flows through the three 47,000-ohm resistors and the 100,000-ohm resistor in its plate circuit. The polarity of the voltage developed at points *A* and *B* is thus negative with respect to ground, and this is the AGC-regulating bias sent to the grids of the controlled tubes.

Under normal conditions, the AGC voltage measured at the control grid of any of the video-IF amplifiers in this circuit will be approximately —4.5 volts. This voltage will vary slightly with signal-input and contrast control-setting. The latter control adjusts the screen-grid voltage of the video-amplifier tube between the limits of +64 and +140 volts. At +65 volts the video signal is cut off completely; at +140 volts it receives its maximum amplification.

The foregoing has indicated the advantage of a keyed AGC system and its manner of operation. There are certain disadvantages to this system, too. For example, when the horizontal sweep is not in step with the incoming signal, the pulses at the AGC tube will be also out of step and the AGC bias will vary rapidly. Hence it is most important that the horizontal-sync system be very stable.

Another factor which will influence the operation of this keyed AGC network is the timing of the horizontal-sweep pulse with respect to the incoming-sync pulses of the signal. The interval during which both must be active is very limited, being at most 10 microseconds, and this tends to make the operation of a keyed AGC system more critical than that of an ordinary AGC system. However, this critical timing factor cannot be reduced, since it is inherent in the operation of the keyed AGC circuit. Actually, to make the system less critical, the AGC tube would have to conduct for a longer period of time, and this, in turn, would cause the set to be more susceptible to noise pulses. Thus, a choice must be made and the operation of the circuit, as outlined here, provides very good results when it is functioning properly.

In place of a pentode-keying tube, a triode is occasionally employed. A pentode will provide more gain, but if the signal is first passed through the video amplifiers, the additional gain may not be required, and a triode will provide all the AGC voltage desired.

10.4 DELAYED AGC AND DIODE CLAMPING IN AGC CIRCUITS

Before we consider other keyed AGC networks, recognition should be given to a method in which the AGC voltage fed to the RF amplifier is varied differently from the AGC voltage applied to the video-IF stages. The reason for this difference stems from the need to operate the RF amplifier, under weak signal conditions, at maximum gain until the input signal has reached a value of about 500 mV. This is desirable in order to present as much signal to the mixer as possible, since the mixer is one of the greatest sources of noise in the receiver. With a large input signal, the output signal-to-noise ratio will be more favorable than it would be if the gain of the RF amplifier were less than maximum.

Actually, what we are seeking to do is to delay the application of an AGC voltage to the RF amplifier, at least until the input signal has attained

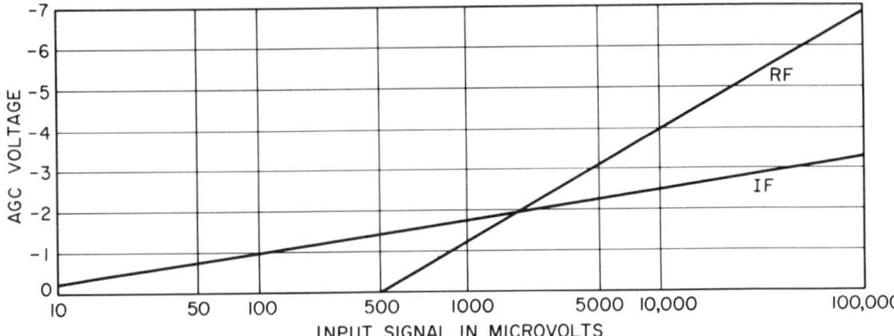

Fig. 10-8. The desired variation of the AGC voltage *versus* the input signal for the controlled RF and IF stages.

a level of 500 mV. Thereafter, the AGC voltage at this stage should rise fairly rapidly to avoid overloading the mixer. Figure 10-8 shows the desired variation of the AGC voltage *versus* the signal for the RF and IF stages. Note that the IF control is initiated immediately; the RF control is first delayed and then made to rise quite sharply.

One method of obtaining a RF delay is shown in Fig. 10-9. Tube V_1 is the keyed AGC tube which operates in the same fashion as the tube of Fig. 10-7. The AGC voltage which V_1 develops appears across C_1, and from this point it is distributed to the video-IF and RF stages through R_2, R_3, R_4, and R_5. Of these components, R_3 and C_2 are readily recognizable as a decoupling filter for the AGC voltage fed to the first video-IF stage. Resistor R_2 is an isolating resistor designed to keep the AGC variations at the RF tube distinct from the AGC variations occurring at video-IF tubes.

Diode V_2 and resistors R_4 and R_5 all connect to the line that goes to the RF amplifier at point A. If we first concentrate only on R_4 and R_5 and

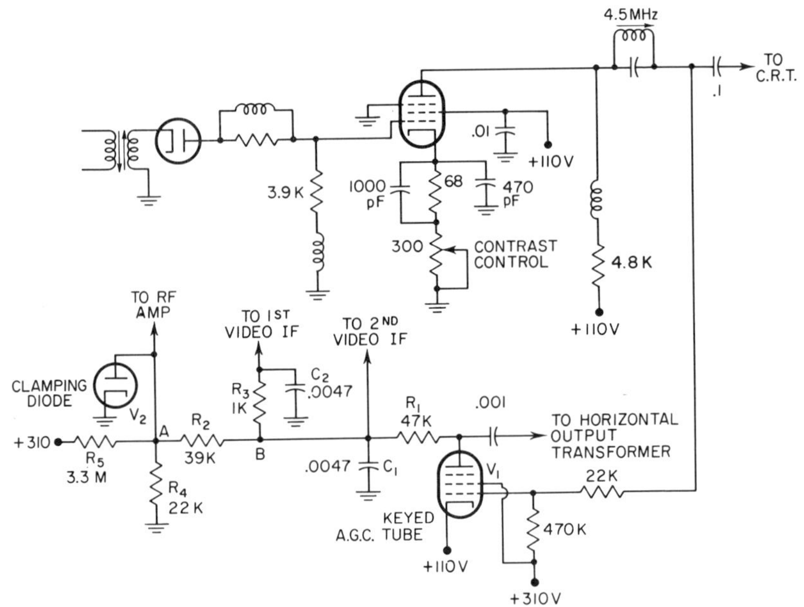

Fig. 10-9. An AGC system containing a clamping diode.

disregard V_2, then the $+310$ volts that are applied to one end of R_5 divides between R_5 and R_4. However, since R_5 is so much larger than R_4, all but 2 volts appears across R_5. Thus, point A becomes 2 volts positive with respect to the ground. If diode V_2 were not present, this $+2$ volts would be applied to the control grid of the RF amplifier. By connecting the plate of V_2 to point A, and the cathode of V_2 to ground, the tube conducts. The plate resistance of V_2 under these conditions is low enough so that resistor R_4 is practically shunted by a short circuit, and the voltage at point A drops to zero. Furthermore, point A remains fairly close to zero so long as the diode is conducting.

While all this is happening at point A, point B remains at whatever negative voltage V_1 develops across C_1. It is the purpose of R_2 to provide some isolation between points A and B. However, as the signal level rises and point B becomes increasingly negative, point A becomes less and less positive until the voltage across C_1 is strong enough to make point A negative. At this moment, V_2 stops conducting and the voltage at point A becomes more and more negative as the generated AGC voltage increases. By properly proportioning the resistive dividers in the RF and IF branches, the AGC voltage at point A can be made to rise faster than the AGC voltage reaching the video-IF stages. This results in the desirable situation illustrated graphically in Fig. 10-8.

An important thing to remember is that when the incoming signal is weak, the AGC negative bias is small, and the potential at point A, because of the presence of the clamping diode, is zero. The controlled video-IF amplifiers, under the same signal condition, have a bias close to -1 volt. But as the signal strength increases, so does the negative AGC bias, and part of it overcomes the slight positive potential at point A, driving the clamping diode into nonconduction, and raising the negative grid bias of the RF amplifier.

10.5 AGC SYSTEMS IN SOLID-STATE RECEIVERS

There are many similarities between the AGC systems used in tube receivers and those used in solid-state receivers. For example, the AGC circuitry in a solid-state receiver may, or may not be, keyed, although the keyed AGC system is strongly preferred. Also, the AGC voltage to the RF or IF amplifier may, or may not be, "delayed" so that the gain is not reduced for weak signals. As in the case of tube receivers, the AGC voltage in a transistor receiver may be delivered to the RF stage, or to IF stage, or to both. (Normally, only the first two IF amplifiers are controlled by the AGC voltage.) Also the AGC voltage in solid-state receivers may be amplified to obtain a greater change in gain for a small change in signal strength.

The AGC bias voltage is usually to the grid of a tube amplifier. In a solid-state receiver, the AGC voltage may be applied to either the base or to the emitter. It is preferable to apply the voltage to the base because with this scheme, the dc amplification of the transistor will permit a greater change in gain for a given change in the AGC bias.

While there are similarities between the AGC systems in the tube and the transistor receivers, there are also some very important differences. Tube circuits differ from transistor circuits in that different characteristics of the same type of transistor may strongly affect the operation of the AGC circuit. For this reason, transistors that are used for replacements in tuner and IF stages should carefully match the characteristics of the transistors originally employed. They not only must have similar characteristics, but they must also be similar types. As you will learn later, transistors of different types should not be interchanged.

Another important difference between the AGC systems in the solid-state receivers and those used in the vacuum-tube receivers is in the polarity of the AGC voltage, used. In transistor receivers, either PNP or NPN transistors may be used. Thus, in the case of the NPN transistor, the AGC voltage will be positive, and in the case of the PNP transistors, the AGC voltage must be negative. This is true regardless of whether forward or reverse AGC bias is employed. It is not uncommon for the AGC circuit in a solid-state receiver to supply both a positive and a negative dc bias voltage to the RF and IF stages. In a tube-type receiver the AGC voltage is employed to reduce the gain of the amplifier stage by increasing the negative bias between the grid and the cathode. This is true regardless of whether the voltage is delivered to the grid or to the cathode. The gain is always reduced by increasing the amount of negative bias voltage present, and therefore, reducing the plate current. However, in a transistor amplifier the AGC voltage may be used to reduce the gain by *reducing the forward bias* between the emitter and the base; or, it may be used to reduce the gain by *increasing the forward bias* between the emitter and the base.

When the AGC voltage is used to reduce the gain by reducing the bias, it is called a *reverse AGC bias*. When the AGC voltage is used to increase the forward bias, and hence reduce the gain by increasing the collector current, the system is called a *forward AGC bias*. The curve of Fig. 10-10 shows that it is possible to decrease the gain of a transistor amplifier by either increasing or decreasing the amount of the emitter-base forward bias of a transistor amplifier. The maximum gain of the amplifier is obtained by setting the bias at point *a*. This forward bias will be a negative voltage in the case of PNP transistors, and it will be a positive voltage in the case of NPN transistors. As shown on the graph, decreasing the amount of forward bias will reduce the gain of the amplifier, and the drop off is quite rapid. On the other hand, the gain will also decrease if the forward bias is increased, in which case the drop off is more gradual. From this curve it is obvious that the amplifier can be operated in such a way that either increasing or decreasing the forward bias will be used to decrease the gain.

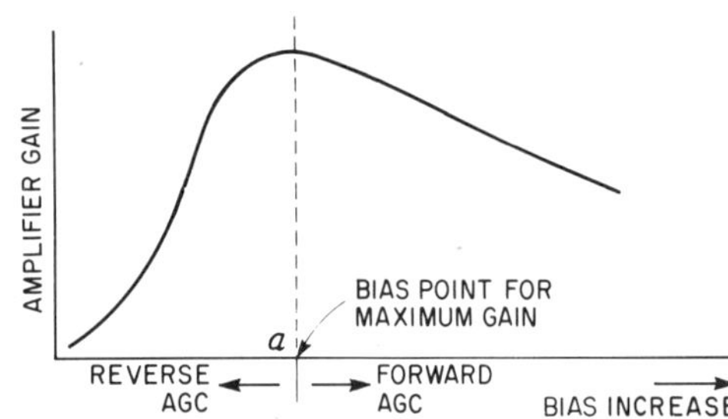

Fig. 10-10. This curve shows how the AGC voltage can reduce the gain by either increasing or decreasing the forward bias.

In most cases, the amplifier is not actually biased at the peak gain point, but it is biased on the slope in order to assure a stability of operation.

Advantages of Reverse AGC over Forward AGC In order to be better able to compare the advantages of forward and reverse AGC, we will first look at a typical amplifier system for each type.

Figure 10-11 shows two amplifiers with an AGC bias. In the circuit of Fig. 10-11(A) a reverse AGC is utilized. This means that the gain of the stage is reduced when the AGC voltage makes the base-emitter bias of the NPN transistor less positive. (This, or course, is only true for the NPN-type transistor shown. If a PNP-type transistor is used, the voltages will be the reverse polarity of those pictured.) Note that the collector circuit is connected directly to the B+ line through the output transformer primary.

In Fig. 10-11(B) the amplifier employs a forward AGC bias. In this case an increase in the positive dc voltage from the AGC line reduces the gain of the stage. Note that a more positive voltage increases the collector current, causing an increase in the voltage drop across the resistor R_1. The result of an increase in the collector current is to reduce the collector voltage, which decreases the gain of the transistor by operating closer to saturation.

The base-to-collector junction of a transistor is normally reverse biased. You will remember from your studies of the voltage-variable capacitor that the amount of reverse voltage across a P-N junction affects the junction capacitance in an inverse ratio. In other words, the greater the amount of reverse voltage, the smaller the amount of junction capacitance. In comparing the circuits of Fig. 10-11(A) and Fig. 10-11(B), it is apparent that a larger reverse voltage will appear across the base-collector junction, for the reverse AGC. Therefore, that junction capacitance is smaller than in the circuit of Fig. 10-11(B) because, the latter's collector voltage is lower, due to the drop across R_1. The lower value of the junction capacitance is desirable in RF amplifier stages, because it permits a higher gain. A change in the AGC voltage in the circuit of Fig. 10-11(A) cannot produce a large change in the base-collector junction capacitance. However, in the circuit of Fig. 10-11(B), a small increase in the base-emitter bias causes an increase in the collector current and reduces the collector junction voltage which, in turn, increases the base-collector junction capacitance. From the foregoing, it can be concluded that the base-collector junction capacitance will be larger, and will vary more in the circuit for a forward AGC. This is a disadvantage of this circuit as compared to the reverse AGC. Compared to the forward AGC, the input and output impedances vary less for changes in the reverse AGC voltage, and this is an advantage of this system.

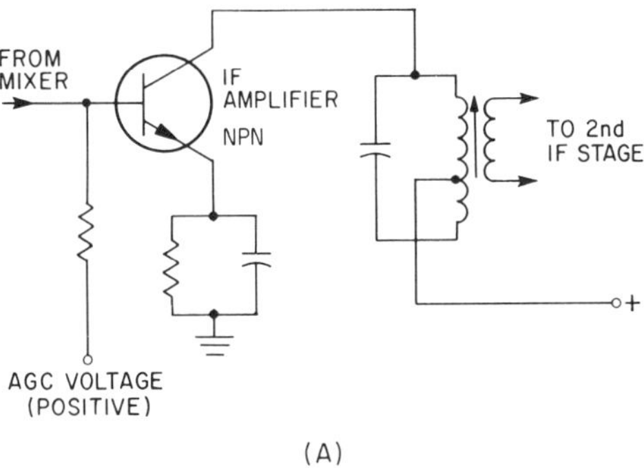

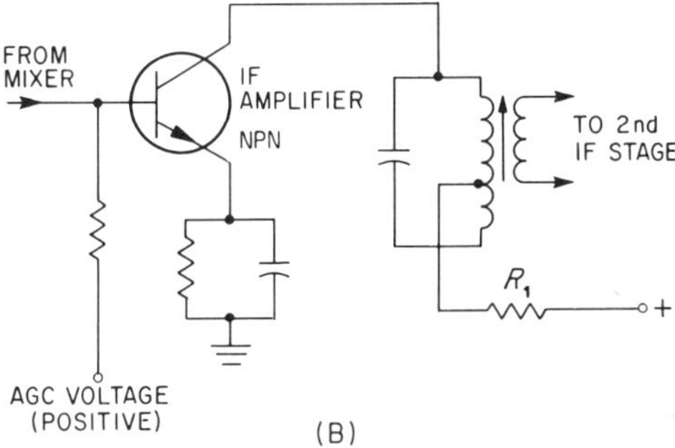

Fig. 10-11. Comparison of amplifiers controlled by AGC. In circuit (A) the AGC voltage becomes less positive. In circuit (B) the AGC voltage becomes more positive.

Advantages of Forward AGC over Reverse AGC One of the advantages of forward AGC can be understood from the curve of Fig. 10-10. Note that the region of the curve employed for the reverse AGC is non-linear, and therefore, any change in gain will not be in direct proportion to a change in the AGC voltage. Compare this with the linear slope on the portion of the curve employed for the forward AGC, and note that the change in gain is directly proportional to the change in the bias voltage.

It is also apparent from the curve of Fig. 10-10 that with a reverse AGC, a very strong signal could produce a sufficient decrease in AGC bias to drive the amplifier into the cutoff region. If this happens, severe distortion takes place. In order to avoid this, special circuitry which limits the amount of voltage applied to the stage must be used. When using the forward AGC, on the other hand, there is very little danger of an excess voltage driving the amplifier into the nonlinear region. As the forward bias is increased, the transistor tends toward saturation and the collector voltage approaches (but, of course, never reaches) zero volts with respect to the base.

We have briefly covered some of the advantages and disadvantages of forward and reverse AGC. It must be remembered that the choice of forward or reverse AGC is also influenced by the type of transistor employed for an amplifier. For example, transistors of the "mesa" type operate better with reverse AGC, but epitaxial planar transistors operate better with a forward AGC.

10.6 HOW AGC VOLTAGES ARE OBTAINED IN TRANSISTORIZED RECEIVERS

Figure 10-12 shows a block diagram of a typical keyed AGC system used in solid-state receivers. This is a popular type of AGC used in both the tube and the transistor receivers.

The input signal to the AGC keyer comes from a video amplifier. Remember that it is only the sync pulses which are of interest as far as the keyed AGC system is concerned.

An input pulse from the flyback transformer, or from some other point in the horizontal output stage, is also fed to the AGC keyer. The function of the keyer (which is also called the *AGC gate*), is to conduct only during that portion of the signal corresponding to the sync pulses. Since the amplitude of the sync pulses is the best indication of the amplitude of the input signal, the AGC keyer is turned on only during that period of

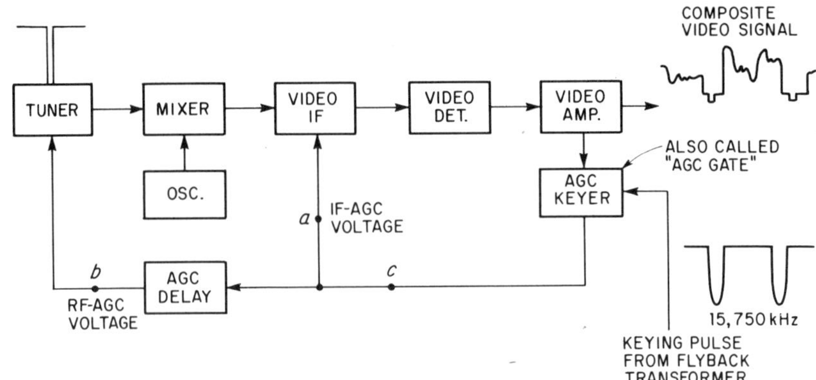

Fig. 10-12. Block diagram of keyed AGC system. The keyer conducts only during the period of the negative flyback pulses.

time. The pulses from the flyback transformer turn the keyer on. This system discriminates against the effect of noise impulses occurring between the sync pulse intervals.

The AGC voltage from the AGC keyer is filtered, and it may or may not be amplified. It is then fed to the video-IF stages and to the tuner-RF stage. Usually, only the first video-IF amplifier, or first *and* second video-IF amplifiers, are controlled by the AGC voltage. It is common practice to employ either direct coupling or stacked amplifiers in the video-IF stage, which means that an AGC voltage to one amplifier will affect the conduction to all other amplifiers which are directly coupled to it. This is important to remember if you are troubleshooting in a video-IF stage and cannot obtain the required amount of conduction through one of the transistors in that section.

The AGC voltage from the keyer is also fed to the tuner—usually through some type of AGC-delay circuit. The purpose of the delay is to permit the tuner to conduct at full amplification on weak signals. It is only on strong signals that the AGC voltage affects the tuner gain. If there is an amplifier in the AGC circuit, it is usually connected at point *c* so that it controls the dc bias to both the tuner and the video-IF amplifier. The usual practice is to make some provision for varying the gain of the AGC amplifier in order to regulate the amount of AGC voltage, and hence, regulate the gain of the RF and IF amplifiers. Normally, the AGC control is adjusted when the receiver is tuned to a very strong station. The object is to adjust the AGC bias voltage, so that the receiver will not be overdriven by a strong signal from a local station.

Figure 10-13 shows a transistorized AGC circuit. The AGC gate (keyer) is Q_1. The video input signal arrives at the base of this transistor,

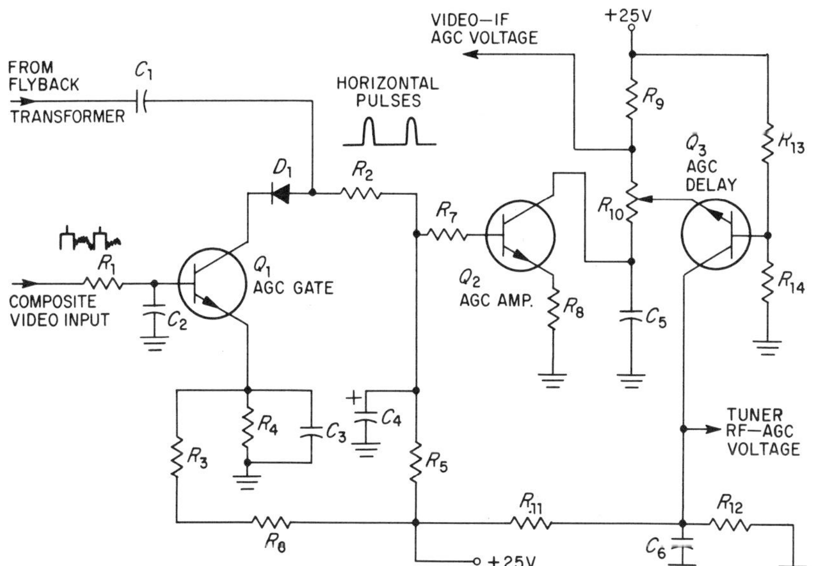

Fig. 10-13. A transistor AGC circuit.

while a positive pulse from the flyback transformer is coupled through capacitor C_1 and diode D_1 to the collector. When the sync pulse arrives at the base simultaneously with the pulse from the flyback transformer, transistor Q_1 conducts. Note that this transistor is normally held at cutoff with a positive voltage on its emitter provided by the voltage divider action of R_3 and R_4. A positive sync pulse at the base and a positive pulse at the collector are needed to override the positive cutoff voltage in the emitter stage. In some circuits the cutoff voltage at the emitter is adjustable, so that the threshold level of the keyer can be set. Noise spikes between the sync pulses cannot produce conduction in Q_1 due to the fact that its emitter is normally positive with respect to its collector, and the transistor can only conduct when the pulse from the flyback is present.

When the keyer conducts, the output signal is a series of pulses which are filtered by the electrolytic capacitor C_4 and the series load resistor R_5. The filtered signal, which is a dc voltage, sets the conduction level of the AGC amplifier (Q_2). Any remaining ac in the output of Q_2 is filtered by capacitor C_5. Transistor Q_2 conducts through R_{10} and R_9. The dc voltage at R_9 is used for the IF-amplifier AGC voltage.

Part of the dc output voltage of Q_2 is tapped at resistor R_{10} and fed to the AGC delay circuit. There is a positive voltage on the base and collector of this transistor, but there is also a positive voltage on the emitter. Transistor Q_3 cannot conduct until the conduction of Q_2 through R_{10} lowers the emitter voltage to the conduction value, that is, to the point where the emitter is negative with respect to the base. When this happens, Q_3 conducts through the load resistor (R_{11}). The voltage across R_{11} is filtered by the capacitor C_6, and the dc at that junction is the AGC voltage for the RF amplifier. Resistors R_{11} and R_{12} form a voltage divider to set the positive collector voltage of Q_3.

The circuit of Fig. 10-13 is very typical of the transistor AGC systems with one possible exception. Instead of the keying pulse from the flyback being delivered through a coupling capacitor, it is also possible for the flyback keying winding to be in series with the collector circuit. This variation is shown in Fig. 10-14.

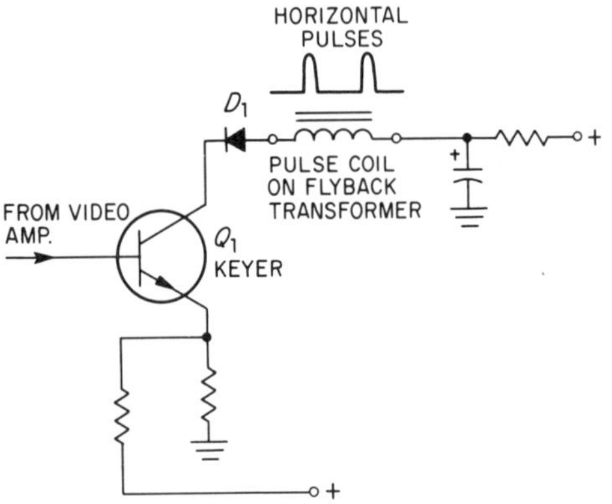

HORIZONTAL PULSES

D_1

PULSE COIL ON FLYBACK TRANSFORMER

FROM VIDEO AMP.

Q_1 KEYER

Fig. 10-14. Alternate method of connecting the flyback transformer keyer winding to the keyer transistor.

In some receivers a forward AGC is used for one IF amplifier and a reverse AGC is used for the next amplifier. The reason for using both forward and reverse bias in the IF-amplifier stages is that it increases the stability of the system and reduces the effect of a changing bias on the bandwidth.

When a dual-gate MOSFET is used for an RF or IF amplifier, the AGC voltage is used for biasing one of the gates. This type of circuit was discussed in Chapter 6. In that circuit, the AGC voltage was applied to one of the gates, and was stepped down in a voltage divider for application to the other gate. The circuitry for supplying AGC to a MOSFET tuner is similar to the one discussed in this section.

10.7 TROUBLES IN AGC CIRCUITS

Chapter 24 is devoted entirely to the methods of troubleshooting TV receivers, and we will not discuss troubleshooting methods in this section. What follows is merely a presentation of some of the most common troubles caused by AGC malfunctions and their effect on the picture.

Symptom: Negative Picture In a negative picture, the black-and-white values are reversed, similar to the appearance of a photographic negative. Basically, this trouble is caused by an insufficient AGC voltage supplied to either the RF or IF stages. This condition generally occurs only in tube receivers or in hybrid receivers that use tubes in the IF-, or video-amplifier stages. In order to produce a negative picture, grid current must be drawn in either the IF- or video-amplifier stages and one of these stages (or more) must be highly overdriven. If these conditions are present, the video signal is effectively inverted and the picture appears as a negative.

Possible causes of a negative picture are: improperly adjusted AGC control or switch; defective tubes (or transistors) in the AGC circuit; defective video-IF or video-amplifier tubes; or positive voltage on the AGC line.

Symptom: No Picture and No Sound (Raster Present) This symptom occurs when there is a complete lack of any signal reaching the picture tube. This is generally caused by an excessive AGC voltage applied to the RF-or IF-amplifier stages, or to both. The excessive AGC voltage causes one or more of these stages to be cut off. This type of trouble can appear in tube, hybrid, or solid-state receivers. The possible cause for this type of trouble is usually to be found in defective parts in the AGC circuit.

Note: This same symptom can also be caused by defects in the RF, mixer, IF, video-detector, or video-amplifier sections.

Symptom: Overloaded Picture With Intercarrier Buzz An overloaded picture is easily recognized by the dark overall appearance of the picture. If the overloading causes a sync compression in the IF stages, there may be horizontal picture pulling, or even vertical rolling. In addition an intercarrier buzz, caused by the modulation of the sound signal with the 60-Hertz sync pulses, may be present, particularly on strong signals.

This trouble is caused by an insufficient AGC voltage and may be the result of something as simple as an incorrect AGC control adjustment. If the control is set properly, tube, transistor, or parts defects will probably be present. Note that this same symptom may be caused by defective IF stages.

In general, it is frequently necessary to isolate the cause of AGC-type troubles between the actual AGC circuits and the RF, IF, and video-amplifier sections. This is necessary because many symptoms which appear to be caused by a defective AGC section, may actually be caused by defects in the other circuits mentioned previously. The manner of such isolation will be described in Chapter 24.

REVIEW QUESTIONS

1. In a peak AGC system, what prevents the AGC diode from conducting at periods other than during the sync pulse?
2. What is the purpose of an AGC amplifier?
3. Which circuits in a television receiver are controlled by the AGC voltage?
4. What is the purpose of the *area selector switch* used in some AGC systems? (Note this area selector switch also goes by other names, such as *local-remote.*)
5. What is another name for a *keyed AGC* system?
6. Where does the keying pulse come from in a keyed AGC system?
7. What are the advantages of the keyed AGC system over other AGC systems?
8. Explain what is meant by the expression *delayed AGC.*
9. Explain the differences between reverse AGC and forward AGC in transistor receivers.

Video Amplifiers 11

11.1 FUNCTIONS OF THE VIDEO AMPLIFIER

Up to this point, the television signal has been received and amplified by an RF stage, converted to another frequency by means of a mixer, further amplified by the IF stages, and rectified by the diode detector. The signal amplitude at the output of the video second detector is not great enough to drive the picture tube directly. Hence, further amplification is necessary, and this is provided by the video amplifiers. In monochrome receivers the amplifier stages following the video detector are called *video amplifiers* whereas in color receivers the comparable stages are called *the "y" channel, the luminance section*, or *the luminance channel*. Monochrome video amplifiers and color-receiver luminance stages will be discussed.

Video-Signal Requirements of Picture Tubes As a first step in determining the characteristics that a video amplifier must possess, let us look ahead to the monochrome picture tube and see what its requirements are. With these established, we can better determine how the video amplifiers should meet these needs.

In order for a video signal to produce a suitable image on the screen of the picture tube, it must possess certain attributes. *First,* it must maintain the video information in the same form that it had when it was originally developed at the studio. Any change in waveshape or any loss of frequency will result in an alteration in the image produced on the picture-tube screen. The video information contains frequencies from about 30 Hertz to 4 megaHertz to produce a high-quality picture. *Second,* the signal must possess the proper phase polarity, otherwise, it will produce a negative picture at the picture tube. *Finally,* the video signal must be strong enough to vary the intensity of the picture-tube scanning beam to produce a range from bright to dark values on the screen. Unless the amplitude of the video signal is large enough to fully drive the picture tube, the image on the screen will appear washed-out because it lacks sufficient contrast to provide a satisfactory image. These three requirements are basic to every picture tube, although the extent to which they are met will frequently vary from receiver to receiver. In order to appreciate more fully the relative importance of each characteristic, let us examine each in more detail.

Video-signal Bandwidth. The video signal, which is developed by the second detector and then passed through one or more video amplifiers before being applied to the picture tube, contains the blanking, synchronizing, and video information. The actual video bandwidth required to be passed by the video amplifiers depends upon the particular TV receiver involved. For inexpensive monochrome sets a bandwidth extending from about 30 Hertz to 2.5 MHz is adequate, particularly if a small size picture tube is used. The large screen and more expensive monochrome sets may extend the high-frequency bandpass to about 3.5 MHz. In the case of color receivers, the high-frequency bandpass depends upon whether the set is an older or newer type. For the older sets, a high-frequency bandpass of about 4.2 MHz was required (at least for the video amplifiers used to pass color information), while in the case of the newer sets, a high-frequency limit of about 3.6 MHz is adequate. The reasons for the difference in color-set responses will be explained in Chapter 22.

The video information deals directly with the detail which forms the picture. The low frequencies in the video signal produce the larger objects in the scene, while the high frequencies in the video signal produce the fine picture detail and may also contain color-video information. It is important for a video amplifier to have a uniform response over the entire range, otherwise either or both ends of the video spectrum will suffer. We shall learn in our subsequent study of video-amplifier circuits what precautions are taken to insure that the response does not fall off too soon at either the high or the low ends. We shall see also why certain manufacturers of monochrome receivers purposely restrict the video bandwidth, particularly at the high end because of economy or because of the small size of the picture-tube screen.

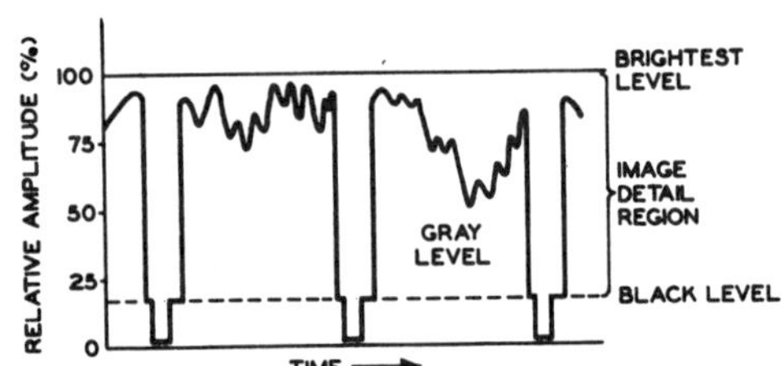

Fig. 11-1. Several lines of a typical video signal.

Video-signal Polarity. It was previously noted that the video signal must possess a certain polarity when applied to the picture tube, otherwise a reverse, or negative, image will be produced on the screen. Several lines of a typical video signal are shown in Fig. 11-1. The signal is drawn with the black level most negative. Whether it possesses this particular polarity at the output of the video second detector depends on the detector circuit. This we have already seen. When the video signal reaches the cathode-ray tube with this polarity, it must be applied to the control grid. The black level will serve to cut off the electron beam while the video variations, being relatively more positive, permit electrons to pass the control grid and reach the screen. The brightest portion of the video signal will be produced by the most positive voltages in this signal. These positive voltages represent the highlights in the image.

It is also permissible to apply the video signal to the cathode of the picture tube and, in this case, it is necessary that the video-signal polarity be reversed, that is, that the sync pulses will be more positive than the video-signal variations.

Direct-Coupled Video Amplifiers Between the video detector and the picture tube, direct-coupled (dc) amplifiers are sometimes employed. This is true, both for the vacuum-tube and for the transistor video amplifiers. The vacuum-tube dc amplifiers operate on the signal in the same manner as an ordinary resistance-capacity coupled amplifier; that is, a signal applied to the grid of a tube (or the base of a conventional transistor amplifier) will appear at the output inverted by 180 degrees. A typical vacuum-tube dc-coupled amplifier is shown in Fig. 11-2. Note the direct coupling between the plate of one tube and the control grid of the next without intervening coupling capacitors.

Direct-coupled amplifiers have certain advantages over amplifiers coupled by other methods. For example, the frequency response of the dc amplifier extends down to zero cycles per second. This is another way of saying that direct-coupled amplifiers can amplify changes in dc level. This is not possible with transformer-coupled amplifiers or with resistance-capacitance coupling. The average value of the video-signal voltage is important for properly reproducing a television picture. The direct-coupling of the video amplifiers preserves the average (dc) level and this is one of the important advantages of this type of coupling.

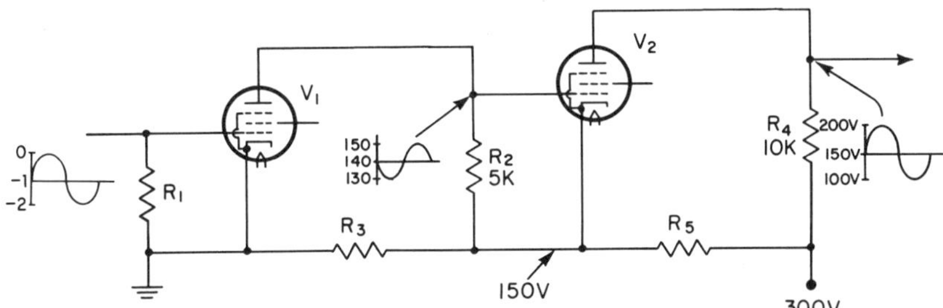

Fig. 11-2. A two-stage vacuum-tube dc amplifier.

Note, however, that when we employ a two-stage dc amplifier, the power-supply voltage must become increasingly large as we move through the system because of the direct connection between the plate of one tube and the grid of the next. For instance, since the plate of V_1 directly connects to the grid of V_2, we must return the cathode of V_2 to a point in the power supply where it will receive a higher voltage than its control grid in order to present this tube with an overall negative grid bias. Also, if we directly connect the plate of V_2 to the control grid of a following stage, then the cathode of this following tube will have to be placed at an even higher voltage in order to maintain the proper bias between the control grid and the cathode. This is one of the difficulties of using direct-coupled, vacuum-tube amplifiers. The same basic problem occurs in direct-coupled transistor amplifiers and FET amplifiers. However, this system is much more practical in such amplifiers, since the required supply voltages are much lower than in the case of vacuum tubes. Thus, in some solid-state TV sets, we will find direct-coupled video amplifiers.

Video-signal Amplitude The amplitude of the signal which is applied to the picture tube governs the contrast of the image which appears on the screen. To gain a better appreciation of this dependence, consider

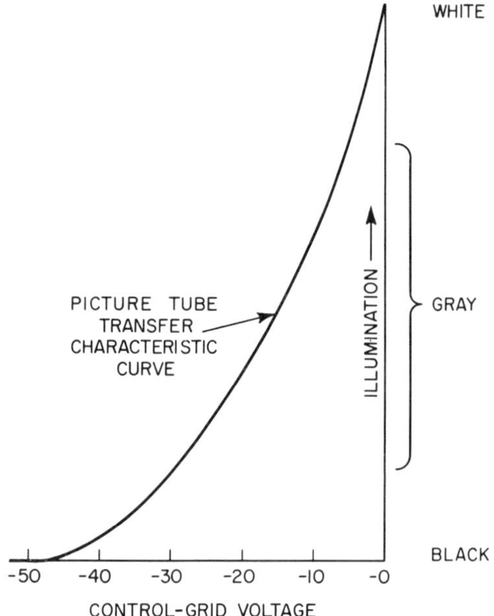

Fig. 11-3. A typical curve of a monochrome picture-tube transfer characteristic.

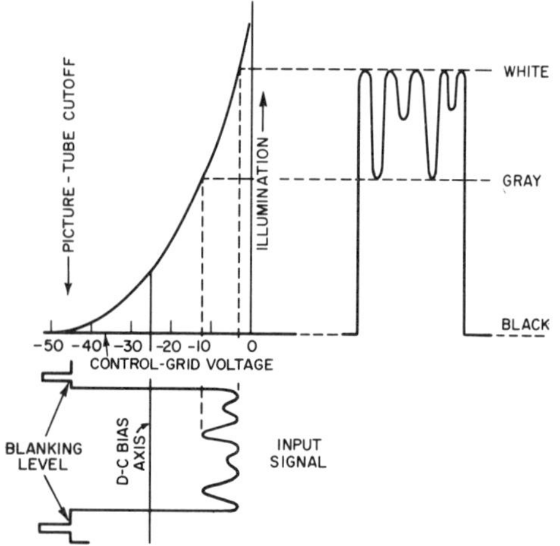

Fig. 11-4. The effect of a video signal at the control grid of a picture tube.

the typical monochrome transfer-characteristic curve shown in Fig. 11-3. This curve shows the relationship between the control-grid voltage and the intensity of any spot produced on the screen by the electron beam. For example, if the control-grid voltage is —50 volts (with respect to the cathode), the beam is completely cut off and nothing is seen on the screen. This is the condition when the screen is black. If we lower the grid bias voltage to —20 volts, the screen illumination produced can be determined in the following manner. Start at the —20-volt point on the horizontal axis and draw a straight line vertically until the curve is reached. Then draw a line to the right. A line so drawn will fall within the area marked *gray*. If we continue to reduce the control-grid bias voltage, perhaps to —10 volts, then, according to the previous procedure, we shall see that the screen will become brighter.

It is the purpose of the incoming video signal to vary the control-grid bias of the picture tube so that the desired variation of screen brightness is produced as the electron beam is moved back and forth across the screen. The first step is to establish the proper operating bias for the picture tube. Let us say that without any incoming signal this bias is adjusted to 25 volts (see Fig. 11-4). The video signal is now applied, and it will distribute itself about the operating point so that as much signal area appears on one side of the point as on the other. This distribution is also indicated in Fig. 11-4. Actually, the operating bias is adjusted so that the blanking voltage just reaches the cutoff level of the beam. If we examine the video-signal variations and the brightness which they produce on the screen, we see that the maximum white is produced on the screen for the video signal which extends the farthest to the right. For that portion of the video signal which does not extend quite as far, less screen illumination is produced and the signal falls, perhaps within the gray areas. Finally, whenever the blanking pulses appear, the voltage on the tube reaches the —50 volts cutoff point and the screen goes black.

If a smaller video signal is received, the situation shown in Fig. 11-5 prevails. First, the dc grid bias on the tube would be adjusted as before, until the blanking level of the incoming signal plus the dc bias voltage reaches the cutoff point. Once this is done, the signal variation to the right of the cutoff point will produce varying levels of illumination on the screen. In this case it is interesting to note that since this video signal has a smaller peak-to-peak variation than the preceding signal, we shall not obtain the range of screen illumination that we did with the preceding signal. The present signal extends from the cutoff to a point less than halfway between the maximum white and gray. The result is that there is less contrast in the picture between the dark and light areas. If the video-signal amplitude is further reduced, we shall eventually have a *washed-out* picture, which means that there will be little difference between the dark and light areas of the picture.

Screen Size and Video-signal Detail. It may not be immediately apparent, but the size of the screen on which an image is placed will also govern how much fine detail the image should possess. In a 525-line television system, there are only about 485 *active lines*. This means that only about 485 lines actually contain picture information, and the rest occur during the retrace. The 485 active lines of video information can be placed on a five-inch screen as well as on a 23-inch screen. The amount of detail necessary for the smaller tube is not as great as for the larger screen. The reason stems from the resolving power of the human eye.

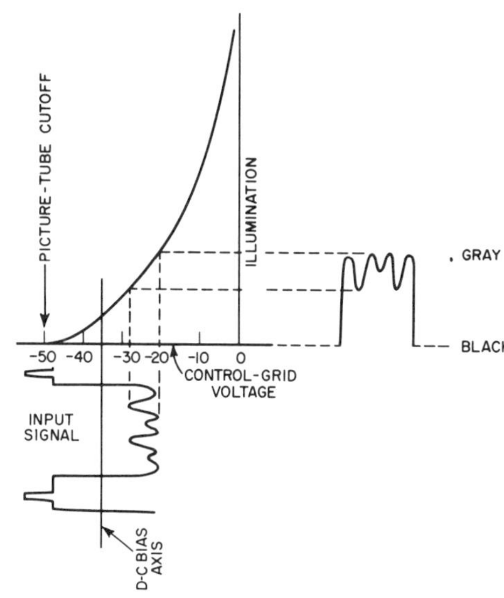

Fig. 11-5. The effect of a small video signal at the control grid of a picture tube.

The *resolving power* of the eye is the ability of the eye to distinguish between objects that are placed close together. As an example, consider the card shown in Fig. 11-6 with two narrow lines located side by side. So long as the card is held fairly close to the eye, it is possible for an observer to see each line separately. As the card is slowly moved farther and farther away, it becomes increasingly difficult to see each line distinctly. Eventually a point is reached where the eye is just capable of distinguishing between them. This point is the limit of the resolving power of the eye for these two lines.

Quite obviously, the farther apart the lines are, or the wider apart they are, the more easily they can be distinguished by the eye at any given viewing distance. For the average person, it is claimed that so long as the two objects subtend an angle of one minute or more at the observer's eye, they can be seen as distinct units. This angle is known as the *minimum resolving angle of the eye* and is illustrated in Fig. 11-6.

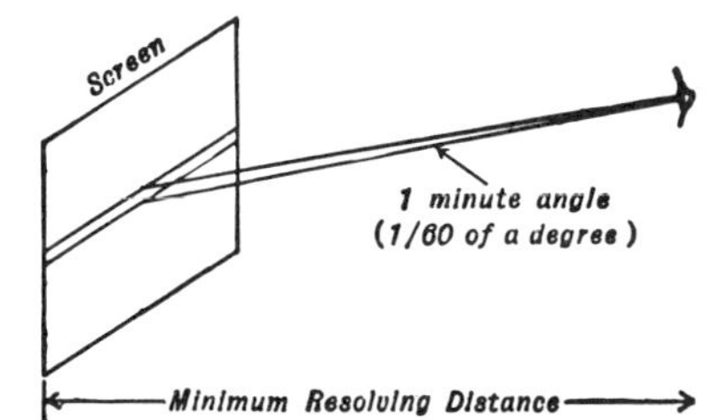

Fig. 11-6. The power of the human eye to resolve or separate two objects that are closely spaced depends upon the distance from the eye to the objects. If the objects subtend a 1-min angle at the eye, they may be seen separately.

The distance that the observer must be from the objects in order to have the one-minute angle subtended at this eye is known as the *critical resolving distance*. If the observer is farther away than this distance, the two objects merge into one. With television, it is necessary for the observer to remain outside the critical resolving distance. Coming closer only reveals the separate scanning lines, and this hampers the illusion of continuity.

From the foregoing line of reasoning, it would seem possible to calculate the exact viewing distance for an object of any size. Actually, with television images, an observer can approach the screen closer than the calculated figure and still be unable to distinguish one line from another. This is possible because the resolution of two lines depends not only on their separation, but also on the amount of light of the lines and their relative motion. The stronger the light, the more clearly they stand out. Under these conditions, the critical resolving distance increases.

On the other hand, the introduction of motion tends to make the line of demarcation less clear-cut and the objects blend into each other at much smaller distances than if they were stationary. The latter condition prevails for television images, and hence the observer may view the screen from closer distances than he could if the motion was absent. In addition, because of the impossibility of obtaining perfect synchronizing action, the positions of the lines of the picture tube tend to change slightly during each scanning run, and this further obscures any clear division between the lines.

Placing the same 500 active lines of picture information on a 19-inch screen as on a 7-inch screen means that the proper viewing distance for the larger screen is greater than that for the smaller screen. With the smaller screen, the ideal viewing distance is generally so short that the observer ordinarily never comes this close to the screen. Therefore, many of the finer details of the picture are not seen, even though they are present on the screen. Manufacturers of monochrome sets take advantage of this fact to design video amplifiers of small-screen receivers with their bandwidths less than 4 MHz. By the same token, as the picture-tube screens become larger, it becomes more important to have the bandwidths of the video amplifiers wider.

Although it is possible in the case of some monochrome sets to sacrifice some response at the high-frequency end of the 4 MHz signal, the amplifier should possess a flat characteristic at the low end. This provides a uniform response to 30 Hz. Since the amplifiers do not cut off sharply an any one frequency, but tend to decrease gradually, it is necessary for a flat response at 30 Hz in order to have the curve extend downward to 10 Hz, or even less.

To ascertain what a loss of response at the low frequencies would mean, let us examine the camera signal for it contains the information of the picture. A section of the signal which might be obtained from the scanning of one line is shown in Fig. 11-7. On either end of the line, we

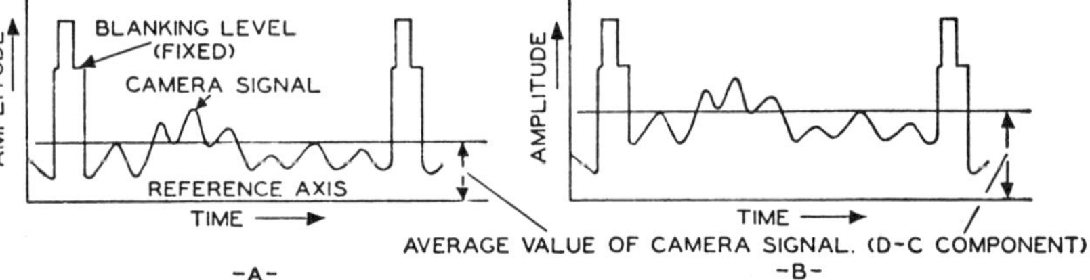

Fig. 11-7. The height of the camera-signal variations above the reference axis, represents the amount of background illumination that the line (or scene) will posses. This average value is known as the dc component of the video signal.

find the blanking and synchronizing pulses. These have a fixed level, always reaching the same voltage (or current) values whenever they are inserted into the signal. The elements of the image itself are represented by the varying voltages between the pulses, and naturally differ from one line to the next.

In addition to the ac variations that make up the video signal (such as synchronizing pulses, blanking pedestals, and video information) there is another important part of the signal called the *dc component.* Examine the two video signals placed side by side in Fig. 11-7. The blanking levels of both are of the same height, and the ac variations of each signal are identical. The only difference is in the average level of the ac variations of Fig. 11-7(A) as compared with the average level of the ac portion of the signal of Fig. 11-7(B). It is clearly shown that the average level of the signal in part B is higher. This average level represents the background illumination of the scene at that line and is the dc component of the video signal. The background illumination may vary from line to line, but this situation is unusual. Generally, it changes slowly over the entire scene, and adjacent lines will have almost equal dc components.

When the value of the dc component is high as in Fig. 11-7(B), the people and objects in the scene being televised will appear against a dark background. This is true because with negative transmission every value is reversed. The darker the scene (or element), the greater the amplitude. As the scene becomes brighter, there is correspondingly less amplitude, and the ac variations of the video signal move close to the zero axis. Hence, as the dc value is less in Fig. 11-7(A) than in Fig. 11-7 (B), the background illumination of Fig. 11-7(A) will be brighter. Neither the people nor the objects, however, will have changed. A lighted background will convey to the viewer of a television scene the impression of daylight, sunshine, and clear weather. A darker background, on the other hand, will give the viewer the impression of night.

At the transmitter, the dc component of the video signal can be inserted manually by an operator viewing the scene from a monitor, or automatically by using the average signal level derived from the viewing tube. If the latter cannot be accomplished, the light from the scene is allowed to fall onto a photoelectric tube at the camera position and the dc component is derived in this manner. Once obtained, it is inserted into the video signal, raising the ac component to the desired level.

From our discussion of the dc component, we can see that the average illumination of a scene may change with each frame, or 30 times a second. Of course, if the exact scene is televised without any variations, the average illumination remains constant. Actually, each frame scanned at the camera has a somewhat different average value. In order to obtain the correct shading of the image background at the receiver, it is necessary that all transmitting and receiving circuits be capable of passing 30 Hz without too great an attenuation. Any poor response would result in incorrect values for the background illumination and, as shown later, the left-to-right stretching or smearing of large objects.

11.2 VIDEO REQUIREMENTS FOR COLOR TV (GENERAL)

Although the subject of color television is covered more completely in Chapters 21, 22, and 23, a comparison of monochrome- and color-television video amplifiers is necessary at this point. In color-television receivers, the video-amplifier section accomplishes the same basic purpose as in monochrome receivers; that is, it raises the level of the video signal sufficiently to drive the picture tube to full contrast. In monochrome receivers, the video signal is applied to the single gun of the picture tube, to control the brightness of the various portions of the screen as the beam is scanned over the face of the tube.

In the case of a color set using a three-gun color tube and producing color pictures, there are *two* types of video signals simultaneously applied to the three guns. One of these is the so-called "Y" signal which contains all the information normally found in a monochrome-video signal. The other video signal consists of three color video signals to be described in a later chapter, which are individually applied to the three guns of the color tube. It is the combination of these *two types* of video signals which causes the picture tube to produce color pictures, as shown in Fig. 11-8.

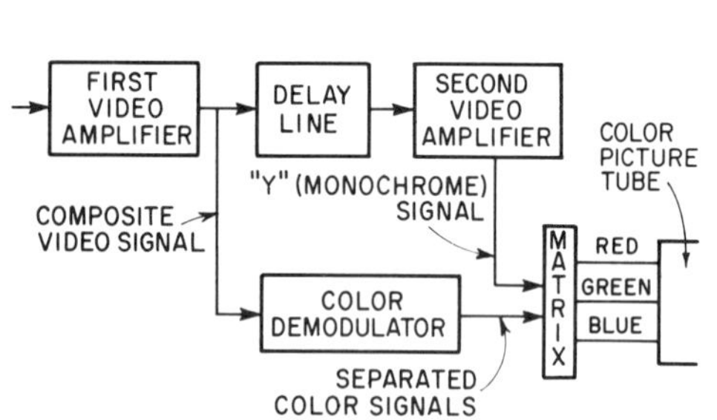

Fig. 11-8. A simplified block diagram of a video-amplifier section for a color-TV receiver.

It is important to note at this point, that while in a conventional monochrome set, a video bandwidth of about 3 MHz will produce a good monochrome picture, this bandwidth is totally inadequate for a color set and would, in fact, result in no color picture being produced. A color-set video amplifier must be capable of passing, without distortion, a bandwidth of at least 3.6 MHz. Any bandwidth appreciably less than this would result in a serious deterioration of the colors. When a color receiver is reproducing a monochrome picture, the color circuits are automatically disabled, and only the "Y" signal is applied to the three guns. This signal is so proportioned to each gun that the combination of the red, green, and blue phosphors, when illuminated, will produce a white light. A monochrome picture is now produced by the same general process used to create a monochrome picture in a strictly monochrome receiver. Figure 11-9 shows the video signal being applied simultaneously to the three guns of a color tube to produce a monochrome picture. This video signal is the "Y" signal, and *not* the color-video signals.

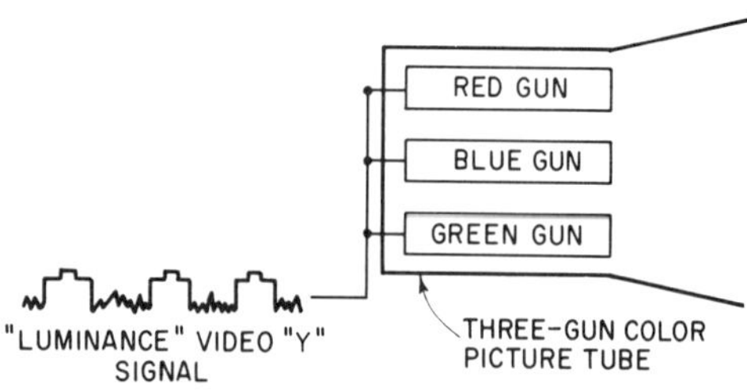

Fig. 11-9. When a three-gun color picture tube is used to produce a black-and-white picture, the video ("Y") signal is applied simultaneously to the three guns in the proper proportions.

In Fig. 11-8, the composite video signal that contains the color information is fed to the color-demodulator section, where individual color-video voltages are generated, to be fed to the correct gun of the color tube. In passing through the color-demodulator bandpass filters, a delay is introduced into the color-video signals. This causes them to be delayed with respect to the "Y" signal. To produce a correct color picture, it is necessary to introduce a compensating delay into the "Y" signal channel, so that the "Y" signal arrives at the three guns exactly in step with its related color-video signals. This delay is provided by the delay line, shown in Fig. 11-8.

Another name given to the "Y" signal is the "luminance signal," and the channel carrying this signal is sometimes called the "luminance channel." This channel, of course, is part of the video amplifier in a color set.

Having now studied the general requirements of the video section in a color set, we shall now look at some block diagrams of video amplifiers. This will give us a better understanding of the various signal paths and operational requirements of the video section.

11.3 TYPES OF VIDEO AMPLIFIERS FOR MONOCHROME AND COLOR SETS (GENERAL)

Monochrome Video Amplifiers. The video section of a television receiver comprises all of the stages between the video detector and the picture tube. We will first discuss a block diagram of a typical video amplifier in a monochrome receiver and then a block diagram of a typical luminance channel in a color-television receiver.

Figure 11-10 shows a block diagram of a video-amplifier section in a monochrome-television receiver. The video section is enclosed within the dotted lines. The complexity and design of this section varies from manufacturer to manufacturer, but the block diagram of Fig. 11-10 may be considered to be a typical video-amplifier section. In this discussion, the video-amplifier section is assumed to include all of the stages that handle the video signal, from the video detector to the picture tube. Although the dc restorer is included in the video section of the block diagram, it is of such sufficient importance that it is treated in a separate chapter in this book.

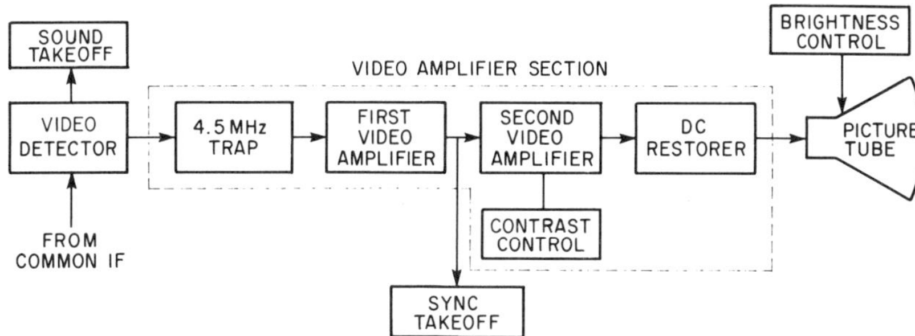

Fig. 11-10. Block diagram of the video section in a monochrome receiver.

The output of the video detector is delivered to the sound takeoff point and also to the first video amplifier, through a *4.5 MHz trap*. In some receivers the synchronizing pulses are also taken from the output of the video detector. Also, in some receivers the sound takeoff occurs after one or more stages of video amplification.

The first video amplifier accepts the composite video signal from the detector and amplifies it. In some cases a single stage of video amplification is sufficient to develop enough signal strength to drive the picture tube. If more than one stage of video amplification is used, careful attention must be paid to the coupling circuits between amplifiers, and also between the last video amplifier and the picture tube to insure that the full desired range of video frequencies are passed from stage to stage (see Chapter 12). In monochrome receivers, the actual full range of video frequencies that pass through the video amplifier may be somewhat less than the 0–4 MHz transmitted by the television station. This may be caused by the design limitations of the response of the IF amplifiers and/or the video amplifiers, as previously mentioned.

Contrast control is accomplished by controlling the gain of one of the video amplifiers. Changing the gain of the amplifier causes a change in the amplitude fo the signal delivered to the picture tube. The larger the difference in the amplitude between the maximum and minimum voltages, the greater the contrast of the picture. The ideal picture on the receiver CRT occurs when the light and dark areas in the reproduced picture are in the same ratio as the light and dark areas in the original scene. Television engineers have a measurement that compares the contrast in the original and reproduced pictures. They call this measurement the *gamma* (γ).

$$\gamma = \frac{\text{Ratio of bright to dark areas in the reproduced picture}}{\text{Ratio of bright to dark areas in the original picture}} \qquad 11\text{-}1$$

There are no units of measurement for gamma. If the contrast of the original picture is identical to the reproduced picture contrast, gamma is equal to one ($\gamma = 1$). Excessive contrast means a gamma greater than one ($\gamma > 1$), and a washed-out picture has a gamma value that is less than one ($\gamma < 1$). When a monochrome picture is being reproduced, the presence of colors in the original picture tends to reduce the contrast, so engineers recommend that the gamma be adjusted to a value greater than one. In such cases, a value of 1.4 ($\gamma = 1.4$) is considered to be ideal.

In the discussion on waveforms (see Fig. 11-7) it was shown that the dc level of the video signal changes from scene to scene. A picture with large white areas has a different dc level from a picture with large dark areas. If this dc level has been lost in the coupling circuits in the video section, then it must be restored. That is the purpose of the *dc restorer* illustrated in the block diagram of Fig. 11-10. If direct coupling is used from video detector to the picture tube, the the dc level will be preserved, and dc restoration is not necessary.

Color Video Amplifiers. Figure 11-11 shows a block diagram of a color-set luminance channel. At this time we are interested primarily in the circuits that produce the "Y" signal, since this is the equivalent of the monochrome video signal. However, a brief description of the related sections will also be given. As shown in Fig. 11-11, the sound takeoff precedes the video detector in color sets. At the output of the video detector the composite video signal is delivered to the first video amplifier through a 4.5 MHz trap. The color burst signal centers around a 3.58 MHz color subcarrier frequency and is present in the video-detector output. Although the color carrier is not acutally transmitted, its sidebands are present. These sidebands and the 3.5 MHz burst could heterodyne with the 4.5 MHz sound signal to produce frequencies that are within the video range. The heterodyne frequencies

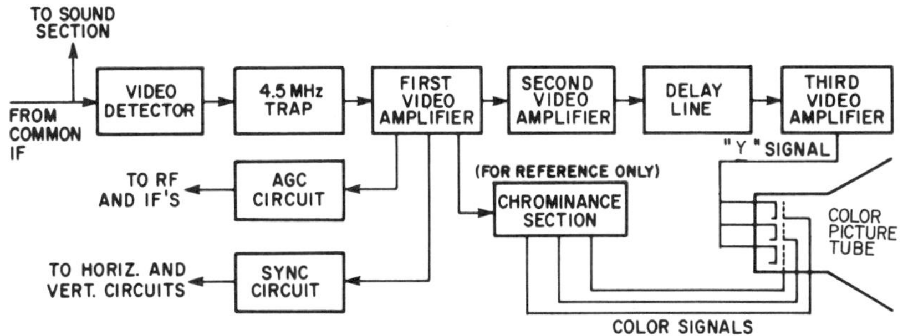

Fig. 11-11. Simplified block diagram of the luminance channel of a color-TV receiver.

(if present) would appear on the screen of the color tube as interference. The 4.5 MHz trap reduces the 4.5 MHz level and eliminates such interference.

The first video amplifier is similar to the first video amplifier in a monochrome receiver. As we mentioned previously, in older color receivers, the video amplifier(s) required to pass color information had to pass at least 4.2 MHz. However, in the newer color receivers this requirement is only about 3.6 MHz, since the compensation for the higher color-video sidebands is provided in the chrominance section of the receiver, as described in Chapter 22. Some color receivers have their color-video signals tapped off directly after the video detector, making it possible to restrict the video-amplifier bandwidth to less than 3.6 MHz. In the case of other newer color receivers, the color-video sidebands are frequently taken from the output of the first video amplifier, so that the bandwidth of the subsequent amplifier(s) can be somewhat reduced.

The output of the third video amplifier is the final "Y" (or monochrome) signal and is applied in many sets to the three cathodes of the color tube. As shown in Fig. 11-11, the three color video signals are applied to the appropriate color guns. In this case, the mixing (or matrixing) of the two signals actually takes place in the guns, producing the required color-video signals for red, green, and blue.

Some color-TV sets use a two-stage video amplifier, instead of the three-stage type described previously. In this case, the video outputs for the color section and the delay line are taken from the first video amplifier. The output of the second video amplifier will then go to the color-tube cathodes.

11.4 COMPARISON OF TUBE AND SOLID-STATE VIDEO AMPLIFIERS (GENERAL)

There are some advantages of transistors over vaccuum-tube amplifiers, and there are also a few disadvantages. Some of the obvious advantages include very small physical size, low power supply voltage requirement, and low cost. Greater reliability is another factor in the increased popularity of the transistor. On the other hand, it is more difficult to design transistors to accommodate large signal amplitudes. The required video-signal amplitude to a picture tube may be as high as 150 volts. While tube circuits can readily produce this much signal voltage, it is more difficult to obtain linear amplification from transistors for high-signal amplitudes.

Another consideration is the fact that it is desirable to have the video detector connected to an amplifier with a high-input impedance. While tube amplifiers have a characteristically high-input impedance, conventional transistor amplifiers do not. FET's have the high-input impedance advantage of vacuum tubes and the solid-state advantage of transistors. Of the four general methods of coupling tube amplifiers, direct coupling provides the best overall frequency response. However, this also is not without disadvantages. The power supply requirement for tubes that are direct coupled is quite rigid. Transistors are more easily direct coupled for two reasons. First, they operate at lower voltages, and the increased voltage requirement arising from cascading is not as difficult to obtain. Second, transistors can be put into special direct-coupled configurations that are not possible with tubes. The tube must always be connected into the circuit, so that the plate is positive with respect to the cathode. Transistors, on the other hand, can be connected so that the collector is positive (in the case of NPN transistors), or they may be connected so that the collector is negative (in the case of PNP transistors). The combination of one NPN and one PNP transistor in a direct-coupled configuration is sometimes referred to as a *complementary amplifier*. Figure 11-12 pictures a complementary amplifier. It should not be inferred that all transistorized video amplifiers are complementary amplifiers. As a matter of fact, the first stage following the video detector is more often an emitter follower because of the high input impedance of this configuration. The emitter follower delivers its output to a grounded base or grounded emitter amplifier. The important point is the fact that designers have a greater range of design choices when transistor amplifiers are used.

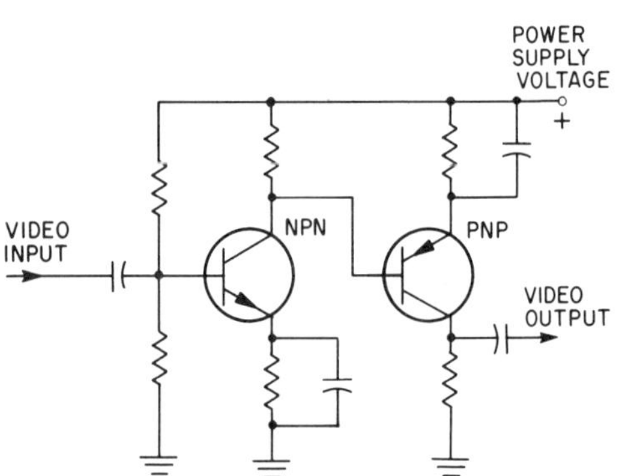

Fig. 11-12. Complementary amplifiers.

In some of the early transistorized television receivers the ability of tubes to handle high-signal voltages resulted in *hybrid* designs. A hybrid amplifier is one that uses both tubes and transistors. In the video-amplifier stages of these receivers, the output of the video detector is fed to a transistor amplifier which, in turn, delivers its output signal to a vacuum-tube, video-output amplifier.

11.5 PHASE DISTORTION

Phase distortion, which can be tolerated in an audio amplifier, is capable of destroying the image on the cathode-ray-tube screen and must be given careful attention when a video amplifier is designed. Since phase distortion is such an important factor, a brief discussion at this point may be helpful. Phase distortion is produced when the time or angle relationship of the electric waves to each other changes as they pass through any electrical system.

Results of Phase Distortion To correlate phase distortion and its effect on the television picture, let us study the dependence of phase distortion and time delay. It should be noted that at the low frequencies the phase angle between input and output voltages increases to a maximum of 90 degrees as the frequency decreases. Suppose that a video signal is sent through an *R-C* network containing (among others) two frequencies of 40 Hertz and 90 Hertz. We know that the 40 Hertz wave will receive a greater phase delay than the 90 Hertz wave. Let us assume that the 40 Hertz wave is shifted 45 degrees and the 90 Hertz wave, 10 degrees. Obviously the two waves will no longer have the same relationship at the output that they had at the input, and by simple mathematics it is possible to compute their difference.

A 40 Hertz wave takes 1/40 second to complete one full cycle, or 360 degrees. With 1/40 second for 360 degrees, it will take 1/320 second for the wave to change 45 degrees; 1/320 second is approximately 0.003 second. Thus there will be this time difference between a maximum occurring at the input to the next tube and that occurring at the output of the preceding tube. The appearance of the maximum at the next tube will lag behind the other by 0.003 second.

The 90 Hertz wave, we know, has a 10 degree phase angle introduced into it. One cycle, or 360 degrees, of a 90 Hertz wave occurs in 1/90 second. Ten degrees would require only 1/3240 second, or approximately 0.0003 second. Thus the input variations will differ by this time interval for the 90 Hertz wave.

At the cathode-ray screen, the electron beam moves across a 12-inch screen a distance of one inch from left to right in about 0.000007 second. The time interval is extremely short, and if waves containing the 40 and 90 Hertz receive the time displacements computed previously, the end result is a displacement of the picture elements that they represent.

At low frequencies a slight time delay causes certain parts of the object to be displaced from the correct position. The visible consequence of this displacement is smearing. Since the beam moves from left to right, the extended stretching of large objects will always be toward the right, or in the direction that the beam is moving. Only large objects are affected, because they are the only ones represented by the lower frequencies.

At the high-frequency end of the video signal, phase distortion results in the blurring of the fine detail of the picture. The larger the size of the cathode-ray-tube screen, the more evident this defect. This is another reason why the larger sets require more careful design and construction.

Phase distortion can be eliminated if the phase difference between the input and output voltages is zero, or if a proportional amount of delay is introduced for each frequency. Thus, a phase delay of 45 degrees at 60 Hertz is equivalent to a 90 degree delay at 120 Hertz, etc. The first introduces a delay of approximately 0.002 second, similar to 90 degrees at 120 Hertz. The net result is that all the picture elements are shifted the same amount, and correction is attained by positioning the picture.

Phase shifts introduced by the electrical constants of one stage are added to those of any other stage. The total phase delay of a system is equal to the sum of all the individual phase delays.

11.6 SQUARE-WAVE RESPONSE OF VIDEO AMPLIFIERS

There are a number of ways to check the response of a video amplifier. One method is to feed a wide range of frequencies into the amplifier, one at a time, and measure the output signal amplitude. A graph is then plotted showing the output signal amplitude *versus* the frequency. When this method is used, it is necessary to maintain the input signal voltage to the amplifier at a very precise value. Also, it is necessary that the instrument used for measuring the output signal amplitude be capable of responding equally to all frequencies used for testing. This method is very time consuming and is used only for very precise measurements.

Another method of checking a video amplifier response is to use a sweep generator. The generator sweeps back and forth throughout the range of video frequencies and an oscilloscope, sweeping in step with the sweep generator, displays the amplifier response. This technique is the same as that used for aligning the IF stages.

It has been shown that a square wave can be considered to be comprised of a fundamental frequency and a large number of odd harmonic frequencies. The greater the number of odd harmonics, the more nearly the wave approaches a perfect square or rectangle. If a square wave is passed through a video amplifier, it should appear at the output undistorted. If there is any frequency or phase distortion present, the square wave will be modified. The block diagram of Fig. 11-13 shows how a square-wave test may be performed on a video-amplifier stage.

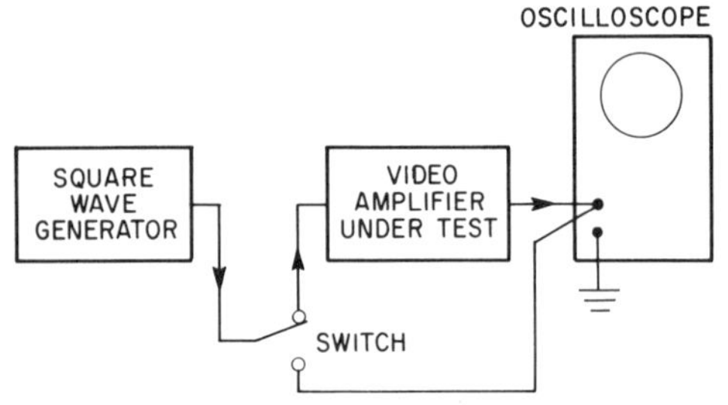

Fig. 11-13. Test setup for performing a square-wave test on a video amplifier. The switch makes it possible to re-check the amplitude of the square wave whenever the generator frequency is changed.

A little practice is needed in order to be able to correctly interpret the waveforms, but it is a quick method of determining video-amplifier response. It might be useful to remember that the horizontal portion of the square wave represents a period when the amplifier maintains a steady voltage. This steady voltage is like a dc voltage for the duration of the horizontal portion of each half cycle. Now, the lowest possible frequency is 0 Hertz, which is a dc voltage. Any change in the horizontal portion of the square wave represents a poor low-frequency response. The sides of the square wave represent a condition where the voltage changes almost instantly from one point to another. This rapid change corresponds to a high-frequency signal. The voltage of a high-frequency signal changes very rapidly from one point to another. This means that any distortion of the steep sides of a square wave represents problems with the high-frequency response.

A low-frequency square wave is fed to the amplifier to test the amplifier's low-frequency response, and a high-frequency square wave is used to test the high-frequency response. Figure 11-14 shows the various symptoms that may be expected with video amplifiers. One important precaution should be observed when making this test, or any of the previously-mentioned tests. The input signal to the video amplifier must not drive the amplifier into the nonlinear portion of its characteristic curve. To do so might cause the top and bottom of the wave to be cut off, and it would be impossible to tell if the amplifier is introducing distortion. Since transistor amplifiers can be damaged by overloads, it is especially important in solid-state receivers to hold the amplitude of the test signal within the prescribed values.

INPUT TO VIDEO AMPLIFIER	SHAPE OF OUTPUT WAVE SEEN ON CRT SCREEN	INTERPRETATION
60 HERTZ SQUARE WAVE		GOOD LOW-FREQUENCY RESPONSE AND NEGLIGIBLE PHASE SHIFT
		LEADING LOW-FREQUENCY PHASE SHIFT AND LOW-FREQUENCY ATTENUATION
		LAGGING LOW-FREQUENCY PHASE SHIFT
25 kHz SQUARE WAVE		GOOD HIGH-FREQUENCY AND TRANSIENT RESPONSE
		POOR HIGH-FREQUENCY RESPONSE
		EXCESSIVE HIGH-FREQUENCY RESPONSE(OSCILLATION) AND PHASE SHIFT DISTORTION
		EXCESSIVE OR INSUFFICIENT MID-FREQUENCY RESPONSE AND PHASE SHIFT DISTORTION

Fig. 11-14. Common symptoms of video amplifier troubles as indicated by the square-wave test.

11.7 CONTRAST CONTROLS IN VIDEO AMPLIFIERS

The purpose of the *contrast control* (also known as *picture control* or *pix control*) is to regulate the amount of video signal reaching the picture tube. It is manually operated, and is adjusted by the viewer until the relationship between the light and dark areas of the picture meets with his particular requirements. If the room is light, he may want to increase the contrast. If the room is dark, he may want to decrease the contrast. In either case, when he adjusts the contrast control, he is regulating the intensity (that is, the amplitude) of the video or luminance signal that reaches the picture tube.

Regulation of picture intensity may be accomplished in several ways. The bias of one or more video-IF amplifiers will vary the gain of these stages and this variation, in turn, will control the signal amplitude of the picture. The automatic gain control voltage adjusts the gain of the video-IF stages. For a manual contrast control, it is a more common practice to control the picture intensity in the video amplifier stage. This is normally accomplished in either of two ways: (1) control the gain of one of the video amplifier stages, or (2) control the amplitude of the signal delivered from one stage to the next.

Controlling Contrast by Controlling Gain Figure 11-15 pictures a circuit for varying the gain of a vacuum-tube video amplifier by controlling the amount of the cathode bias. Moving the arm of variable resistor R_1 all the way to the cathode side (point *a*) will provide maximum gain by reducing the cathode resistance to the value of R_2. Resistor R_2 is usually a relatively small resistance value used to prevent the tube from being operated without bias when the contrast control is adjusted to its maximum value. Capacitor *C* prevents degeneration due to the presence

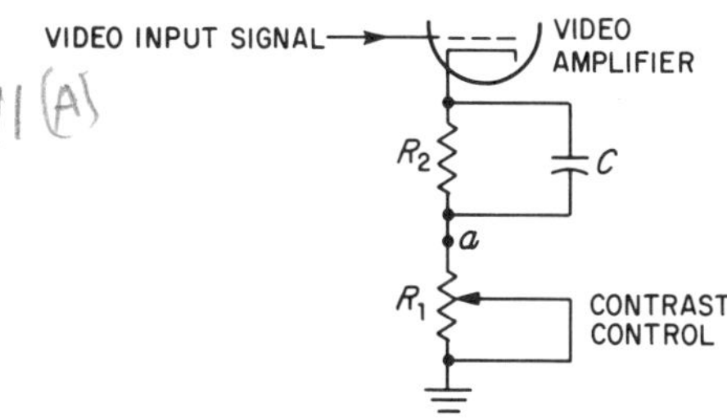

Fig. 11-15. Cathode control of video amplitude.

of R_2. When the amplifier is set for the maximum gain, the picture will have the maximum contrast.

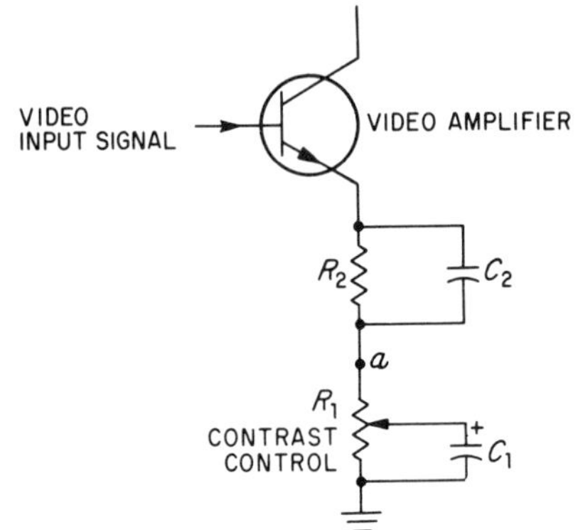

Fig. 11-16. Emitter control of a transistor-video amplifier.

Figure 11-16 shows the equivalent transistor-video amplifier with the contrast control in the emitter leg. In this case, adjustment of the contrast control does not change the dc voltage on the emitter. Instead, the arm of the control determines the gain of the stage by controlling the amount of degeneration introduced. When the arm is at point a, resistor R_1 is bypassed by C_1 and there is no degeneration. At this point the gain will be maximum. When the arm is moved toward the ground, the amount of degeneration is increased, and the gain is decreased. The capacitor C_1 is an electrolytic—a large value—that is capable of preventing degeneration even at the lowest video frequencies.

The bias circuits of Fig. 11-15 and 11-16 may be interchanged. In other words, the contrast control may regulate the amountof degeneration in a vacuum-tube cathode, or it may regulate the dc emitter voltage in a transistor. Except in cases where the screen-grid voltage of a tube is varied to introduce changes in gain, tube- and transistor-contrast control circuits are very similar.

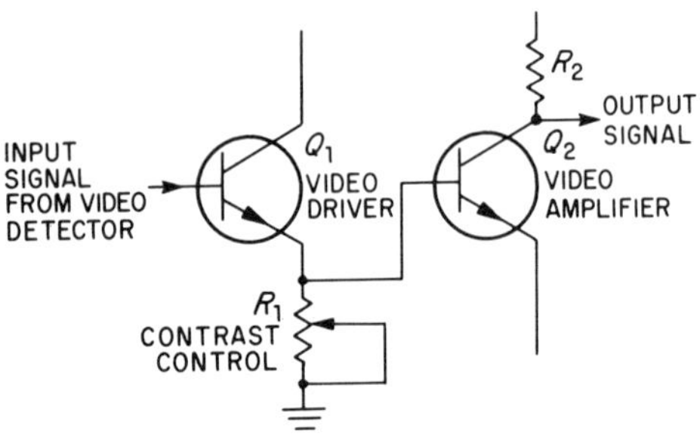

Fig. 11-17. Contrast control in a video-driver stage. Stage Q_1 is frequently called an emitter-driver.

A commonly-encountered transistor video system employs an emitter follower in the stage immediately after the detector. The emitter follower stage is often called the emitter driver. The emitter driver may be either direct coupled or capacitively coupled to the next stage. Figure 11-17 pictures a contrast control in the emitter-driver system. The adjustment of R_1 in this circuit will affect the dc base bias voltage on Q_2, and will therefore affect the gain of that stage. Changing the resistance of R_1 will also affect the amount of signal delivered to Q_2.

Controlling Contrast by Controlling Signal Amplitude Instead of changing the gain of the video-amplifier stage, it is possible to control the amount of signal delivered from one stage to another. Figure 11-18 illustrates such a circuit. Transistor Q_1 is a video driver that is R-C coupled to the conventional amplifier Q_2. The amount of signal delivered to Q_2 depends upon the setting of R_1. The base bias voltage on Q_2 is established by the voltage divider comprised of R_2 and R_3. The capacitor C_1 prevents the dc bias voltage from varying with changes in the setting of R_1. When the arm of R_1 is moved toward the ground, less signal amplitude is tapped off and the contrast decreases.

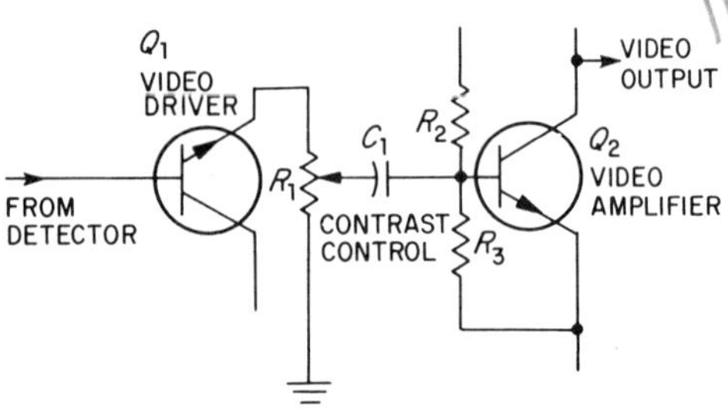

Fig. 11-18. Controlling the contrast by controlling the signal amplitude.

Figure 11-19 shows a vacuum-tube, video-amplifier circuit in which the contrast control regulates the amount of the video signal delivered to the picture tube. As the arm of the variable resistor R_L is moved toward the plate connection of V_1, a greater amount of video is delivered to the picture tube.

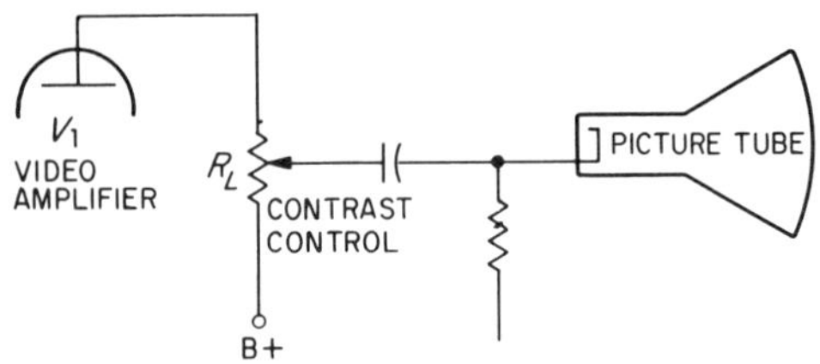

Fig. 11-19. Contrast control in the plate circuit of a video amplifier.

This chapter has presented a general discussion of the requirements for video amplifiers. A conventional R-C coupled amplifier cannot meet these requirements unless special compensating circuits are used. The next chapter will discuss the use of these compensating circuits, and the design of actual video amplifiers.

REVIEW QUESTIONS

1. Compare the requirements of a video amplifier for a black-and-white receiver with the requirements of a luminance channel in a color receiver.
2. What governs the minimum viewing distance for a television screen?
3. Which frequencies in the video composite signal produce the large areas of the reproduced picture, and which produce the fine detail?
4. How will changes in the dc, or average, level of the video signal affect the picture?
5. What relationship must exist between phase angle and frequency in order that no time delay distortion exists?
6. What is the relationship between the gain and bandwidth in an amplifier?
7. What is a delay line used for in a color-TV receiver?
8. What are the frequencies of traps that may be found in the luminance channel of a color receiver?
9. What does the term *y signal* mean?
10. In order to keep the bandwidth as large as possible, should the coupling capacitor between two *R-C* coupled video amplifiers be large or small?
11. Name two different ways that contrast can be controlled in a video amplifier.

12 Video Amplifier Design

12.1 AMPLIFIER GAIN AND EQUIVALENT CIRCUITS

In Chapter 11 the requirements necessary for the reproduction of television images were discussed. The methods whereby these requirements are met in practice represent an important consideration in television today.

The type of amplifier that can be used to give the necessary bandwidth is restricted, almost without exception, to the direct-coupled or the resistance-capacitance-coupled networks. *R-C* amplifiers have the advantage of small space and economy. Direct-coupled amplifiers have a perfect low frequency response due to the absence of reactive components in the coupling circuit.

From the discussion of conventional resistance-coupled amplifiers given in Chapter 11, the reader knows that a flat response is obtained in the middle range of frequencies (say from 200 Hertz up to approximately 2,000 Hertz) with ordinary circuits. This frequency response applies to any conventional *R-C* amplifier. The frequency and phase characteristics of the amplifier, throughout the middle range, are suitable for use in video amplifiers, and this section of the curve requires no further improvement. However, the responses at either end of the curve are far from satisfactory and correction measures must be taken. Fortunately, any changes made in the circuit to improve the high- or low-frequency responses of the curve will generally not react on each other (with one limitation which will be explained later), and each end can be analyzed separately and independently. However, before we see what can be done to extend the high- and low-frequency ends of the response curve, let us derive the gain equation for amplifiers.

Vacuum-Tube Amplifiers The gain of an amplifier, and by this we mean the voltage gain, is the ratio of the output-signal voltage to the input-signal voltage. The symbol A_e is used to represent voltage gain. We particularly stress the voltage gain, because it is also possible to obtain a power gain in an amplifier; however, in most video-amplifier applications, voltage gain is the figure of merit, and this is the value to be considered here.

The gain of a stage is dependent upon the amplification factor (μ) of the tube as well as the value of the load which is employed in the plate or output circuit. The μ value is governed largely by the tube construction, and also by the operating voltages applied to the various elements, particularly the screen grid and plate. The output load impedance of an amplifier is more than simply the resistance or impedance of any circuit to which the plate may be connected. In most instances, this load impedance includes the control-grid-input circuit of the following stage.

In order to appreciate better the roles which the two foregoing factors play in determining the gain of an amplifier, it is desirable to analyze the amplifier circuit by means of an equivalent diagram. The purpose of an equivalent diagram is to provide a circuit which contains only those components essential to the operation of the amplifier insofar as the *signal* is concerned. It is important that the latter distinction be carefully noted, because equivalent circuits do not utilize dc voltages, it being assumed that the proper voltages are applied and the tube is operating in the desired manner. Generally, this means that the stage operates as a linear amplifier.

Figure 12-1(A) contains the diagram of a conventional (grounded-cathode) triode amplifier in its simplest form, with the input signal applied between the grid and the cathode, while the output signal appears across Z_L. Grid bias is established by a battery inserted between the cathode and the grid. The plate voltage is obtained from a power supply which applies B+ to the bottom end of Z_L, while the negative side of the supply connects to the ground and the cathode.

In any amplifier, such as the one shown in Fig. 12-1(A), a small ac voltage inserted between the grid and cathode will develop a much larger voltage across the output-load impedance. In the equivalent diagram of Fig. 12-1(B), we can represent this amplifying action of the tube by a generator producing a voltage of μe_g. In this equation, μ is the amplification factor of the tube and e_g is the input ac signal voltage to the grid of the amplifier.

The μe_g generator is placed in series with the plate resistance of the tube (r_p) and the output-load impedance, Z_L. The plate resistance must be included, because we know such a resistance exists. When current flows through the tube, a certain voltage drop is developed across the resistance, *i.e.*, across the tube.

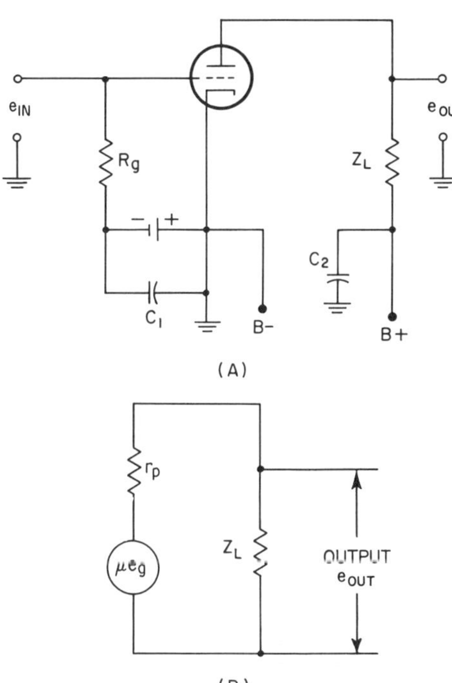

Fig. 12-1. (A) A triode amplifier and (B) its equivalent circuit.

The circuit of Fig. 12-1(B) thus becomes the equivalent circuit of the amplifier in Fig. 12-1(A). Note that capacitors C_1 and C_2 do not appear in the equivalent circuit, because their purpose is simply to provide a low-impedance path for signal current flowing around the bias battery in one case and the power supply in the other. Since it is not their purpose to impede the flow of the current, and since their reactance is, or should be, negligible, they need not be shown in the equivalent diagram. Also, we can disregard completely the grid circuit of the tube, because the signal voltage here is multiplied by the μ of the tube and is represented by the generator, μe_g. Thus, all we have left is the generator μe_g, the plate-load resistance, and the tube internal resistance. We know that Z_L and r_p must be in series with each other, because any current that flows through the tube, and hence through r_p must also flow through the load impedance, Z_L. Z_L is designated as an impedance rather than as a resistance, because resistors are not always employed as the load. To make the discussion general, the load is being considered as an impedance.

Conventional Amplifier Gain. The equation for the grain of the amplifier shown in Fig. 12-1 will now be derived. Since this is a simple series circuit, its governing equation is (from Ohm's Law)

$\mu e_g = i_p \times (r_p + Z_L)$, where
μ is the amplification factor,
e_g is the signal voltage on the grid,
i_p is the signal current in the plate circuit, 12-1
r_p is the dynamic plate resistance,
Z_L is the total plate circuit impedance.

Solving this equation for i_p, we obtain

$$i_p = \frac{\mu e_g}{r_p + Z_L} \qquad 12\text{-}2$$

The next step is to multiply the plate current, i_p, by the load impedance, Z_L, to obtain the voltage developed across Z_L. This gives us

$e_{\text{out}} = i_p Z_L$, where 12-3

e_{out} is the output voltage across the plate load impedance.

Furthermore,

$$e_{\text{out}} = \frac{\mu e_g}{r_p + Z_L} \cdot Z_L$$

The overall voltage gain of the amplifier (A_e) which is defined as the ratio of e_{out} to e_g, can now be obtained.

$$A_e = \frac{e_{\text{out}}}{e_g}, \text{ where} \qquad 12\text{-}4$$

A_e is the voltage gain, and

$$A_e = \frac{\mu e_g Z_L}{(r_p + Z_L) \cdot e_g}$$

$$= \frac{\mu Z_L}{r_p + Z_L} \qquad 12\text{-}5$$

From the equation 12-5, we see that the overall gain of the amplifier will always be less than μ, because of the fraction that it is multiplied by: $Z_L/(r_p + Z_L)$. Unless Z_L becomes much larger with respect to r_p, this fraction will always have a value which is less than one. Hence, the gain will always be less than the amplification factor μ of the tube that is employed. It can also be seen that in order to obtain the maximum gain, the plate-load impedance should be as high as possible. However, if this happens to be a resistor, then obviously, the larger this resistor, the lower the plate voltage for any given value of B+. With a lower plate voltage, less signal can be handled by the tube without distortion. Furthermore, it was shown in Chapter 11 that the upper end of the bandwidth is adversely affected when the value of the plate-load resistance is made too large. In a practical design, a compromise is reached between the power-supply voltage, the value of the plate-load resistance, and the desired bandwidth.

When a pentode tube is employed, we may use the same equivalent circuit as was used for the triode (Fig. 12-1). However, a better appreciation of the stage gain for this tube can be obtained in the following manner. In a pentode, the plate-load resistance is very high, certainly much higher than it is in a triode. Under these conditions, r_p is much larger than Z_L. By ignoring the relatively low value of Z_L compared to r_p in the denominator, we can rewrite equation 12-5

$$A_e = \frac{\mu Z_L}{r_p} \qquad 12\text{-}6$$

We see that μ is multiplied by the ratio of the load impedance to the plate resistance of the tube. Now, μ and r_p are values determined by the particular tube which is being used. Furthermore, these two quantities are related to each other by $g_m = \mu/r_p$. This relationship is true of every tube, whether it is a triode, tetrode, or pentode. If g_m is substituted for the ratio μ to r_p, the gain equation for an amplifier is given by

$$A_e = g_m \cdot Z_L, \text{ where} \qquad 12\text{-}7$$

g_m is the mutual conductance of the tube.

This equation is much more convenient to work with and can be used when the internal plate resistance of the tube is considerably greater than the load resistor. We also see why equation 12-7 would not be correct for triodes unless the same conditions can be assumed. In recognition of the preceding relationship, tube manuals are more likely to list g_m values for pentodes than μ values.

Thus, as a rough measure of the amplification of an amplifier, it is necessary simply to multiply the mutual conductance of the tube in question by the impedance or the resistance of the plate load. The reason this does not always give an accurate value stems from the presence of other impedances which affect the plate circuit and which frequently have a significant effect on the total value of impedance that the tube sees as a load. However, the foregoing procedure can be used as a rough indication. In the discussion that ensues, we shall consider the impedance of the plate as being purely resistive and use R_L in place of Z_L. This is permissible for video amplifiers in the mid-frequency range.

The g_m, or mutual conductance, of a tube is governed by the particular tube used and the plate current flowing through the tube. The latter is dependent upon the B+ applied to the plate. The second part of equation 12-7 is the load resistance into which the tube works. To see what fully constitutes this load, consider Fig. 12-2. Here we have the coupling network between the output of one amplifier and the input of the following stage. In addition to the plate-load resistor, R_L, we also see C_c, the coupling capacitor, and R_G, the grid resistor of the following tube. These three components are ordinarily wired into the circuit. Also present, but not physically wired into the circuit by the designer, is the output capacitance of the first tube, C_{out}, and the input capacitance of the following tube, C_{in}, plus two additional shunt capacities, C_s and C_M. C_s is the stray capacitance which exists across the circuit because of the wiring between the stages, the capacitance that R_L or R_G may have with respect to the chassis, and any capacitance that C_c itself may develop with respect to the chassis. This stray capacitance, while it is seldom greater than 5 or 6 picoFarads, must be taken into account when dealing with the high frequencies which pass through a video amplifier. C_M is a capacitance which is reflected from the plate of V_2 to its grid circuit. C_M is an effective capacitance at the grid of the tube, caused by the so-called *Miller effect.* C_M is equal to the grid-to-plate capacitance (C_{gp}), multiplied by $(1 + A_e)$, which is the voltage gain of the stage. Because of the Miller effect, the realized grid-input capacitance may be many times greater than the actual, physical capacitance.

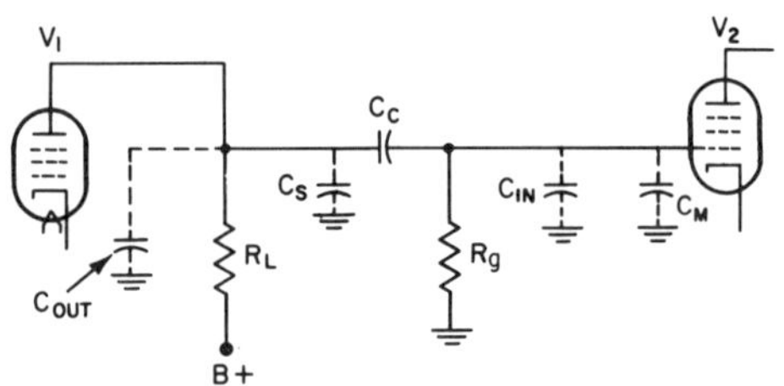

Fig. 12-2. The complete coupling network between the two amplifiers.

Thus, what initially appeared to be a fairly simple circuit, consisting of two resistors and a coupling capacitance, actually turns out to be a fairly complex network containing three additional capacitances which are ordinarily not visible. These capacitances, combined with the resistances, make a complex quantity, Z_L. As we shall see, however, not all of these components need be considered when dealing with any specific section of the overall video response. This fact will become evident as we consider the operation of these amplifiers, first at the high-frequency end, then over the mid-frequency section, and finally at the low-frequency end.

12.2 TRANSISTOR AMPLIFIERS

Common-Emitter Amplifiers Common emitter amplifiers are also known as *grounded-emitter amplifiers*. Unfortunately, transistor amplifiers cannot be analyzed quite as simply as vacuum-tube amplifiers. An approximate voltage gain can be easily determined and the method is given in this section. The circuit in its basic form, shown in Fig. 12-3 is used extensively as a voltage amplifier. It has a medium value of input impedance and a medium value of output impedance. The voltage gain and the power gain are both high in a common-emitter amplifier.

The voltage gain is the output-signal voltage (V_O) divided by the input-signal voltage (V_i). The output voltage is the signal current in the collector multiplied by the load resistance R_L.

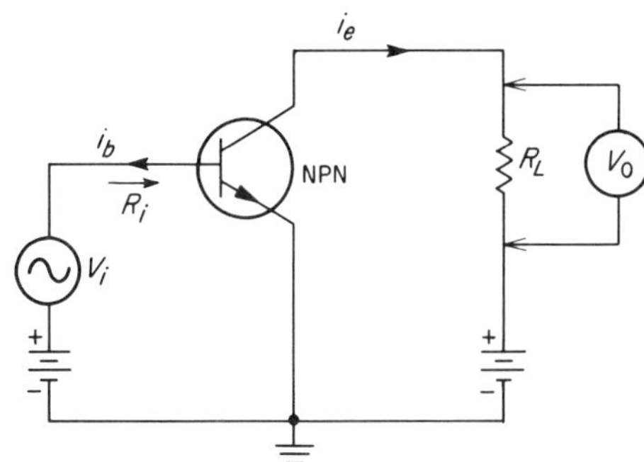

Fig. 12-3. Circuit for a grounded-emitter amplifier.

$V_O = i_e R_L$, where
V_O is the output signal voltage,
i_e is the collector signal current,
R_L is the load resistance. 12-8

The input voltage is equal to the base signal current (i_b) times the input resistance (R_i) of the transistor.

$V_i = i_b \cdot R_i$, where
V_i is the signal voltage applied to the input terminals,
i_b is the signal current flowing in the base, and
R_i is the input resistance of the transistor. 12-9

The approximate voltage gain equation can now be derived:

$$A_e = \frac{V_O}{V_i} = \frac{i_e R_L}{i_b R_i} = \frac{i_e}{i_b} \cdot \frac{R_L}{R_i} \qquad 12\text{-}10$$

If the ac base current is held to a relatively small value, then the relationship i_e/i_b is approximately equal to the transistor *beta* (β). In a grounded-emitter circuit, beta is called the *common-emitter forward current transfer ratio* represented by h_{fe}. The transistor *alpha* (α), often represented by h_{fb}, is the grounded-base forward current gain of a transistor, and it is related to β by the equation

$$\beta = \frac{\alpha}{1-\alpha}$$

Substituting $\alpha/(1-\alpha)$ for $i_e\, i_b$ in equation 12-10

$$A_e = \frac{i_e}{i_b} \cdot \frac{R_L}{R_i} - \frac{\alpha}{1-\alpha} \cdot \frac{R_L}{R_i} \qquad 12\text{-}11$$

The approximate input resistance (R_i) of a junction transistor is given by

$$R_i = \frac{r_e + r_b(1-\alpha)}{(1-\alpha)}, \text{ where}$$

r_e is the emitter resistance of the transistor, and
r_b is the base resistance of the transistor.

Substituting this fraction and performing a little algebra manipulation, the approximate voltage gain of a common emitter-transistor circuit is

$$A_e = \frac{\alpha}{1-\alpha} \cdot \frac{R_L}{\dfrac{r_e + r_b(1-\alpha)}{(1-\alpha)}}$$

$$A_e = \frac{\alpha}{1-\alpha} \cdot \frac{R_L(1-\alpha)}{r_e + r_b(1-\alpha)}$$

$$= \frac{\alpha R_L}{r_e + r_b(1-\alpha)} \qquad 12\text{-}12$$

This equation shows that increasing the value of R_L will increase the voltage gain. It also shows that the gain is greater if transistors with larger values of alpha are used.

As with vacuum-tube amplifiers, high-frequency circuits have a reduction in gain because of the input capacitance of the amplifying device. The alpha rating of a transistor varies with the frequency. It follows, then, that the beta rating also varies with the frequency. The relationship between the frequency and the alpha and the beta of a transistor is shown in Fig 12-4. The frequency where the alpha falls off to 0.707 (−3 db) of its 1 kHz value is called the *alpha cutoff frequency*. The point where beta falls off to 0.707 (−3 db) of its 1 kHz value is called the *beta cutoff frequency*. These curves show that the β cutoff, which is associated with a grounded-emitter configuration, is lower than the alpha cutoff which is associated with a grounded-base configuration. In other words, the grounded-base circuit has a higher cutoff point than the ground-emitter circuit. However, the gain bandwidth product is approximately the same for both configurations.

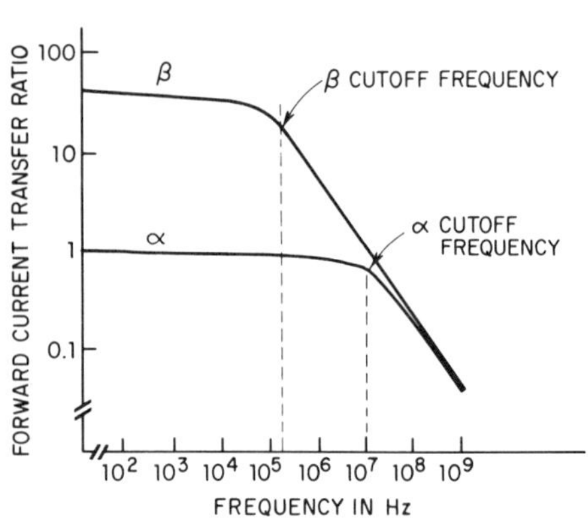

Fig. 12-4. A comparison of α and β cutoff points.

Emitter Followers. The emitter-follower (also known as the *common collector*) transistor configuration has characteristics that are similar to the cathode follower. For example, the voltage gain of an emitter follower is slightly less than one, but it may be considered to be equal to one in most cases. Also, it has a high-input impedance and a low-output impedance. We have already encountered the emitter follower as a driver for a video amplifier (see Fig. 11-23). Although an emitter follower may have a high-current gain, its power gain is the least of any of the three basic transistor circuits.

Grounded Base The grounded-base configuration is also called a *common-base* circuit. It has approximately the same gain as a conventional transistor amplifier, and it is similar to its counterpart—the grounded-grid amplifier. The input and output impedances of a grounded-base transistor circuit are the reverse of those given for the emitter follower circuits. That is, the input impedance is low and the output impedance is high. In applications where the ambient temperature may change over a wide range of values, the common-base circuit is more stable than the common-emitter circuit.

The grounded-base circuit is used extensively in high-frequency circuits. The high-frequency limit is set by the input capacitance of the transistor and the *transit time* of the charge carriers. The transit time is the time it takes an electron or hole to travel from emitter to collector through the semiconductor material of the transistor. Some transistors, such as Drift, and Mesa, types have been designed specifically to reduce input capacitance and transit time.

The characteristics of the common-emitter, the common-base, and the emitter-follower circuits are summarized in Table 12-1.

TABLE 12-1 Transistor Configurations—Comparison of Characteristics.

Characteristic	(Conventional) Common Emitter	Common Collector (Emitter Follower)	Common Base
Input Impedance	Moderate	High	Low
Output Impedance	Moderate	Low	Highest
Voltage gain	High	Less than 1	High—may be somewhat higher than for the common emitter
Current gain	High	High (Approximately the same as a common emitter)	Less than 1
Power gain	Highest of the three configurations	Lowest of the three configurations	Slightly lower than for the common emitter
Phase inversion between input and output signals	Yes	No	No

12.3 HIGH-FREQUENCY BEHAVIOR OF VIDEO AMPLIFIERS

The high-frequency response of a video amplifier depends upon two things: the nature of the amplifying device and the type of impedance between amplifiers. We have seen that the mutual conductance (g_m)

of the tube or transistor (α, β) will affect the gain of the amplifier. We have also seen that the input and the capacitance values are the determining factor of the highest frequency that can be amplified. We shall now direct our attention to the inter-stage impedances, and how they relate to bandwidth.

Vacuum-Tube Amplifiers When we considered the high-frequency operation of the circuit of Fig. 12-2, we did not have to include the coupling capacitor C_c. The reason for this is that C_c will generally have a value of approximately 0.1 μf, and the high-frequency end of a video-amplifier-response curve generally falls off at about 3 to 4 MHz. At these frequencies, C_c has a negligible reactance. All we need include are the two resistances, R_L and R_G, plus the shunt capacitances which are present in the circuit. The high-frequency version of the network between V_1 and V_2 is now as illustrated in Fig. 12-5. We could, if we wish, simplify this circuit even more by showing only R_L and omitting R_G. This simplification can be made because R_G is considerably higher in value than R_L, and the two resistances in parallel will provide a total resistance very close to the value of R_L. However, for the sake of those instances when R_G may not be negligible in its effects on the parallel circuit, we shall retain R_G.

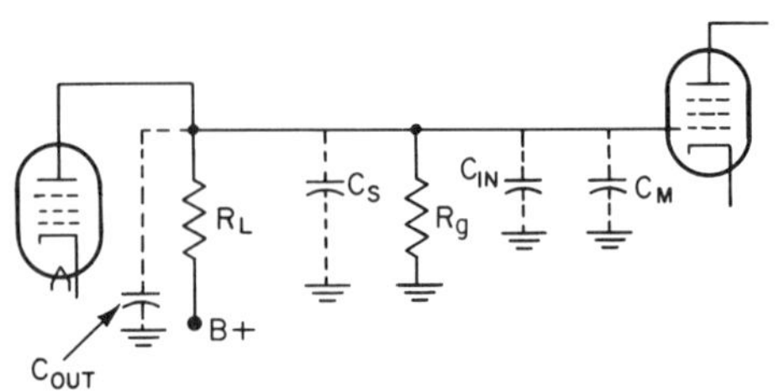

Fig. 12-5. The high-frequency equivalent of the interstage network.

To determine the high-frequency gain of an amplifier with the inter-stage network shown in Fig. 12-5 we must take not only R_L into consideration, but also the four shunt capacitances. These capacitances are in parallel and their total value is equal to the sum of the separate units, that is,

$C_T = C_{out} + C_s + C_{in} + C_M$, where
C_T is the total capacitance of the circuit,
C_{out} is the plate capacitance of the first tube, 12-13
C_s is the stray wiring capacitance of the circuit,
C_{in} is the input capacitance of the second tube,
C_M is the capacitance due to the Miller effect.

To evaluate C_T numerically, we must know the exact values of each of its four components. The values of C_{out} and C_{in} can be obtained from a tube manual. A typical value of C_{out} is 3 pF, and for C_{in} it is 7 pF. C_s will vary with the circuit, but generally it falls at about 6 pF. Still remaining is the determination of C_M.

Whenever two pieces of metal are separated by a dielectric, they form a capacitor. The metal grid and metal cathode of a tube are separated by a dielectric which in this case is a vacuum. The capacitance thus formed is the input capacitance (C_{in}) of the tube, but the actual measurement when the tube is amplifying will show the capacitance from the grid to the ground to be larger than the value of C_{in}. The additional capacitance is due to the *Miller effect*, which will now be explained. The triode circuit of Fig. 12-6. will be used for this explanation.

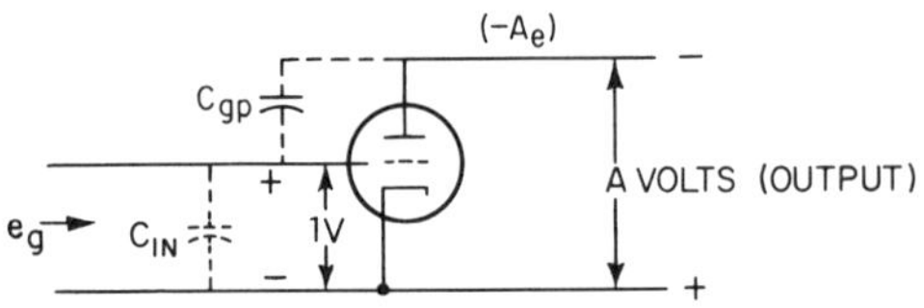

Fig. 12-6. Simplified circuit for explaining the Miller effect.

The plate signal voltage on a tube is A_e times larger than the grid-signal voltage. If we examine Fig. 12-6, we shall note that C_{gp} (the plate-to-grid capacitance) has the incoming signal voltage applied to it on the grid side. Let us assume that this is 1 volt. On the other side of C_{gp}, the amplified signal is present. If the stage gain is A_e, then $-A_e$ volts appears at the plate side of C_{gp}. The minus sign takes into account the phase reversal that occurs in the tube. Hence, the two voltages, so far as C_{gp} is concerned, are series aiding, or $(A_e + 1)$. Now, the grid-to-plate capacitance, C_{gp}, is in series with the normal-input capacitance of the tube, C_{in}, and the charging current that flows through C_{gp} will affect C_{in}. If we keep the voltage across C_{in} constant, which in this case is the input signal, e_g, but increase the effective charging current, then the overall result is equivalent to an increase in capacitance. This effect can be seen from the equation which governs the capacitor charge and the capacitance,

$Q = CV$, where
Q is the capacitor charge,
C is the capacitance, 12-14
V is the voltage across the capacitor.

If V is held constant, but Q increases, then C must become larger to maintain the equality of the equation.

Thus, because of the presence of C_{gp}, the input capacitance of a tube rises above the published value of C_{in}. If a pentode is used, C_{in} represents the capacitance between the control grid and the cathode, the heater, the grid number two, the grid number three, and any internal shield that may be employed. The input capacitance due to the Miller effect is equal to

$C_M = C_{gp}(1 + A_e)$, where 12-15
A_e is the voltage gain of the stage.

If we assume a value for C_{gp} of 0.05 and a stage gain of 20, then the Miller capacitance is equal to

$$C_M = 0.05(1 + 20) = 1.05 \text{ pF}$$

We can now compute the value of C_T.

$$C_T = C_{\text{out}} + C_s + C_{\text{in}} + C_M = 3 + 6 + 7 + 1.05 = 17.05 \text{ pF}$$

This capacitance is obviously not negligible when high frequencies are involved. It will certainly affect the total plate impedance at the high-frequency end of the video-response curve. The plate-load resistor, R_L,

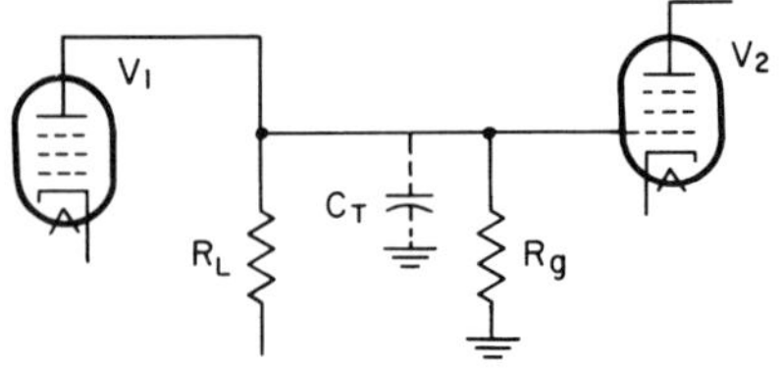

Fig. 12-7. The plate-load resistor R_L and C_T are in parallel with each other. The coupling capacitor C_C has been omitted, because it does not enter into the high-frequency gain calculations.

and the total capacitance C_T are in parallel with each other as shown in Fig. 12-7. It is useful to determine at what frequency the reactance of C_T (we shall call it X_{CT}) equals R_L. When this point is reached,

$$R_L = X_{CT}$$

Substituting $1/2\pi f C_T$ for X_{CT}:

$$R_L = \frac{1}{2\pi f C_T} \qquad 12\text{-}16$$

Solving this equation for f:

$$f = \frac{1}{2\pi C_T R_L}$$

At this frequency (f), the total impedance in the plate circuit is $1/\sqrt{2}$ of the value it has at lower frequencies.* At lower frequencies X_{CT} is so large it can be disregarded. The expression $1/\sqrt{2}$ is equal numerically to 0.707. In terms of decibel loss, 0.707 represents a decrease of 3 db. Hence, at the frequency f, the amplifier gain is 3 db down from its gain at lower frequencies.

It is apparent from equation 12-16 that in order to raise frequency f, either C_T or R_L must be decreased in value. The curves in Fig. 12-8 demonstrate how the bandwidth of a video amplifier is broadened by lowering the value of the plate-load resistor. The relationship between R_L and the frequency response was mentioned briefly in Chapter 11. Now we have given the mathematical explanation of why low-valued load resistors are used in video amplifiers. Unfortunately, Fig. 12-8 reveals that lower load resistors also provide less stage gain. Hence, while this method of increasing the response of a stage with regard to frequency is useful, it cannot be carried too far if any useful gain from the stage is to be achieved.

Transistor Amplifiers The coupling circuit between two *R-C* coupled transistor amplifiers is very similar to the vacuum-tube circuit

* This can be seen readily. Let us assume that R_L and X_{C_T} each have a value of 1 ohm. Then, since both are in parallel, and since one is a resistor and the other is a reactance,

$$Z = \frac{R \cdot X_{CT}}{\sqrt{(R)^2 + (X_{CT})^2}} = \frac{1 \cdot 1}{\sqrt{(1)^2 + (1)^2}} = \frac{1}{\sqrt{2}}$$

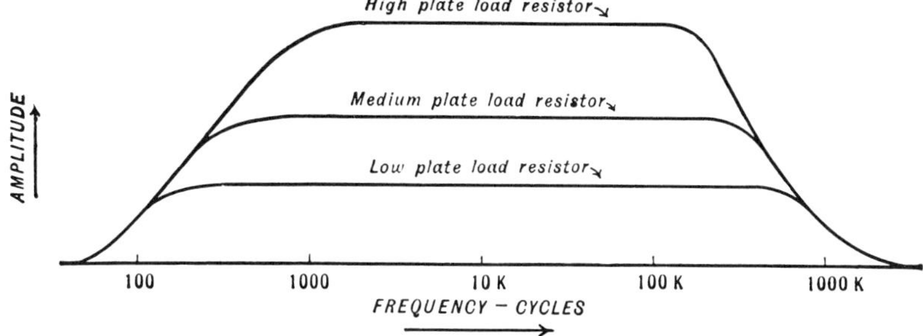

Fig. 12-8. By lowering the plate-load resistor value, it is possible to increase the extent of the flat portion of the response curve.

just discussed. However, the transistor input and output impedances are different due to the nature of its construction. It will be helpful to look at the characteristics that determine the transistor impedances, and compare them with the vacuum-tube impedances.

The grid of a vacuum tube presents an open circuit as far as dc is concerned, provided that the grid is maintained negative with respect to the cathode. Such is not the case with the transistors. The input resistance of a transistor is seen to be a combination of three resistors. Fig. 12-9 pictures the combination that is R_e, R_b, and R_c. These are the internal emitter, the base, and the collector resistances respectively. In this illustration the input capacitances are not shown. They will be considered presently. The resistors R_1 and R_2 form a voltage divider network to provide the correct emitter-to-base transistor bias. Now the collector-base junction of a transistor is normally reverse biased, so the collector resistance is high and can be disregarded in our discussion. The base and the emitter resistances are in series, and the combination is in parallel with R_2, the input resistor across which the signal is developed. The resistor R_1 is ac grounded by a filter capacitor C at the power supply, so it is effectively in parallel with R_2 and the input resistance $(R_b + R_e)$ of the transistor.

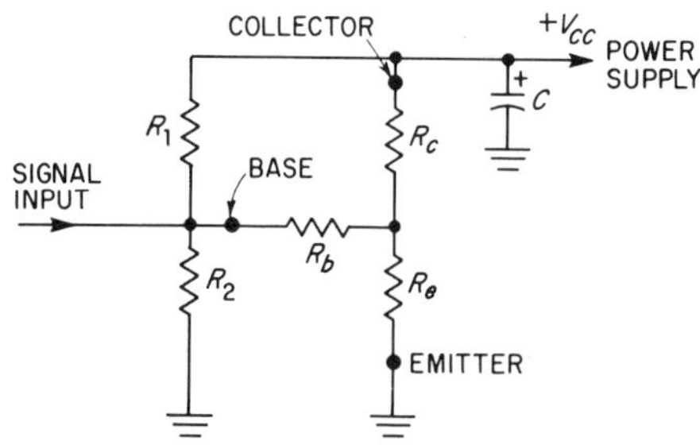

Fig. 12-9. Resistances seen when looking into the base of a common-emitter transistor amplifier.

The equivalent input resistance of the transistor amplifier as seen by the input signal is shown in Fig. 12-10. It is obvious from this illustration that the effect of the internal transistor resistances is to *lower* the total input resistance of the amplifier. This is an important point, because the input resistance of a transistor amplifier will be an influence on the high-frequency response.

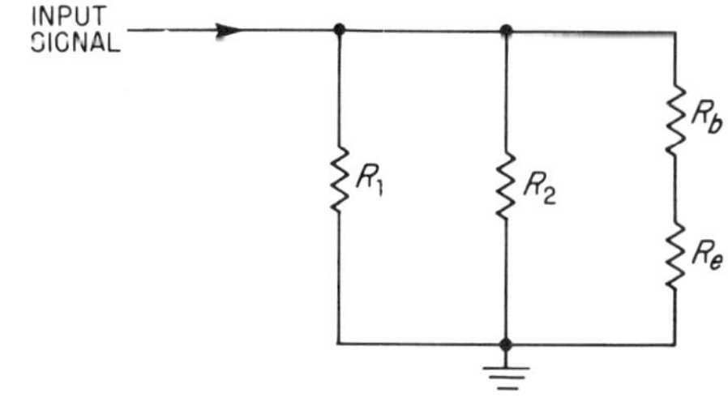

Fig. 12-10. A simplified circuit of resistance as seen by the input signal.

The circuit of Fig. 12-10 is a simplification given for the purpose of understanding the transistor amplifier. The actual input resistance is more complicated because the internal transistor resistances vary with the temperature and with the amount of junction current flowing.

The junction capacitance in a transistor introduces the same effect as the interelectrode capacitance in the tubes.

The collector-base junction of a transistor is normally reverse biased, while the emitter-base junction is forward biased. When a semiconductor junction is forward biased, it presents a greater capacitance than if it were reverse biased. The reason is that the forward-biased junction has a preponderance of charge carriers present at the junction, while the reverse-biased junction charge carriers are forced apart at the junction, introducing a wider, effective dielectric at the junction. (See also the discussion of the varactor in Chapter 6.) From the foregoing discussion, we can see that the emitter-base junction of a transistor will normally present a somewhat greater capacitance to the input signal than the collector-base junction.

It should be emphasized that the amplitude of the input signal to a transistor is very much dependent upon the emitter-base junction. There are other factors besides the junction bias that determine the amount of junction capacitance. The physical size of the junction is an important consideration, and the capacitance between the transistor leads must also be considered. These factors are controlled by the manufacturer in producing high-frequency transistors.

It is an interesting fact that the junction capacitance is made use of in certain applications. *The voltage-variable capacitors* operate on this principle. These capacitors are used in automatic tuning circuits. The amount of capacitance in such circuits is varied by controlling the amount of reverse bias present.

The input capacitance of a transistor amplifier is higher than would be expected from the presence of the junction capacitance alone. This increase in capacitance is due to the Miller effect, a problem that we have already encountered in vacuum-tube amplifiers. The Miller effect refers to the increase of input capacitance with amplification, and as in the case of the tube amplifiers, it becomes larger as the amplifier gain is increased.

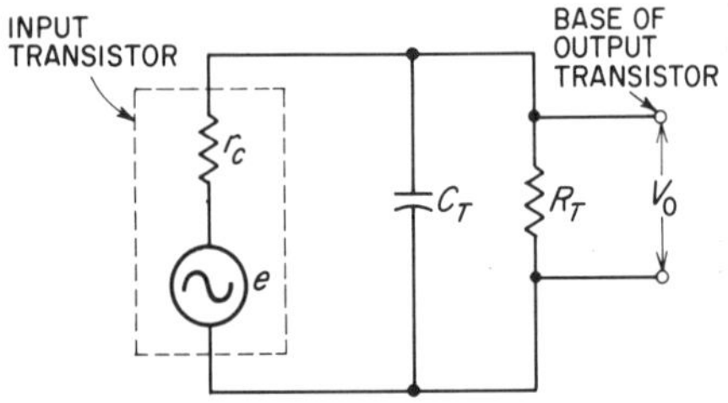

There are also wiring and stray capacitances that affect the high-frequency response of transistor amplifiers. These capacitances are identical to those discussed for the tube circuit. The circuit of Fig. 12-11 can be used to represent two R-C coupled transistor amplifiers. All of the capacitance values are combined to make C_T. The input transistor is represented as a voltage generator (e) and a series resistor (r). The voltage at the base of the following transistor is, V_0. The upper frequency response of this circuit is similar in shape to the response curve for a tube amplifier. In other words, as the frequency rises, the output voltage, V_0, decreases due to the shunting effect of C_T.

Fig. 12-11. A simplified circuit of a transistor amplifier at high frequencies.

12.4 SHUNT PEAKING

A method which is useful in extending the high-frequency response of an amplifier is the addition of a small inductance in parallel with the total capacitance (C_T) of the stage. This is usually accomplished by placing the inductance in series with the load resistor of the input stage. The inductance is designed to neutralize the effect of the shunting capacitances, at least to the extent that we can improve the amplifier response at the upper frequencies. This method is known as *shunt peaking.*

A circuit diagram using this type of compensating inductance is shown in Fig. 12-12. The impedance of the combination of L, R, and C_T is given as

$$Z = \sqrt{\frac{X_L{}^2 X_{CT}{}^2 + R_L{}^2 X_{CT}{}^2}{R_L{}^2 + (X_L - X_{CT})^2}} \qquad 12\text{-}17$$

It has been shown that when the reactance of C_T is made numerically equal to R_L, the gain of the stage is down 3 db. (See footnote in Section 12.3.) By letting $X_{CT} = R_L$ and making the value of $X_L = \frac{1}{2}R_L$ in equation 12-17, we get:

$$Z = \sqrt{\frac{\left(\frac{R_L}{2}\right)^2 R_L{}^2 + (R_L{}^2)^2}{R_L{}^2 + \left(R_L - \frac{R_L}{2}\right)^2}}$$

$$Z = \sqrt{\frac{R_L{}^2\left(\frac{R_L{}^2}{4} + R_L{}^2\right)}{R_L{}^2 + \frac{R_L{}^2}{4}}}$$

$$= R_L$$

Thus, under the conditions specified, the impedance of the plate circuit remains the same to frequency F as the impedance it possesses at mid-frequency. This is certainly a desirable situation, since it means that instead of falling off 3 db, the response of the circuit remains flat up to frequency F, the frequency at which X_{CT} equals R_L. In an uncompensated amplifier, at frequency F, the response is down 3 db from its mid-frequency value. By the addition of the inductance L, we have raised the gain 3 db at F.

The procedure for finding actual values for R_L and L is as follows. First, the highest frequency at which it was desired to have the response remain flat would be specified. In a video amplifier, this would usually be between 3 and 4 MHz. Once the type of tube or transistor to be used has been decided upon, the capacitance values that make up C_T can be obtained from manufacturer's specifications. The wiring capacitance, which is also needed for finding C_T, could be measured from a circuit layout, or it could be estimated. With both f and C_T known, the value of R_L could be obtained from the equation:

$$R_L = \frac{1}{2\pi f C_T}$$

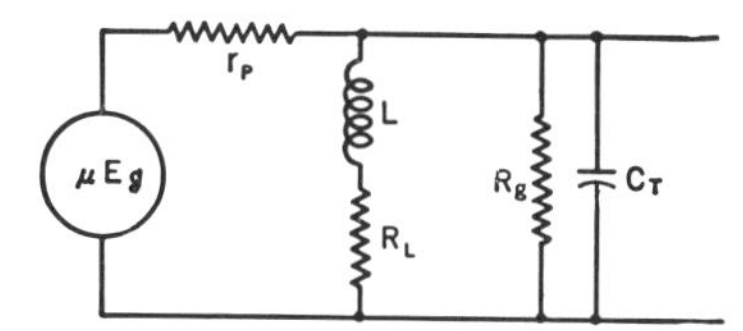

(A) EQUIVALENT HIGH-FREQUENCY CIRCUIT

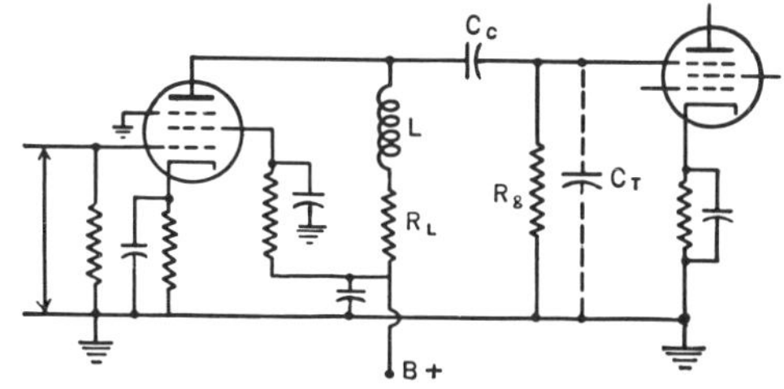

(B) ACTUAL CIRCUIT

Fig. 12-12. High-frequency compensation for an *R-C* amplifier.

Also, since X_L is to have a value equal to one-half R_L,

$$X_L = 0.5R_L = \frac{0.5}{2\pi f C_T}$$

By substituting $2\pi fL$ for X_L

$$2\pi f L = \frac{0.5}{2\pi f C_T} \qquad 12\text{-}18$$

or

$$L = \frac{0.5}{4\pi^2 f^2 C_T} \qquad 12\text{-}19$$

and

$$L = 0.5 C_T R_L^2$$

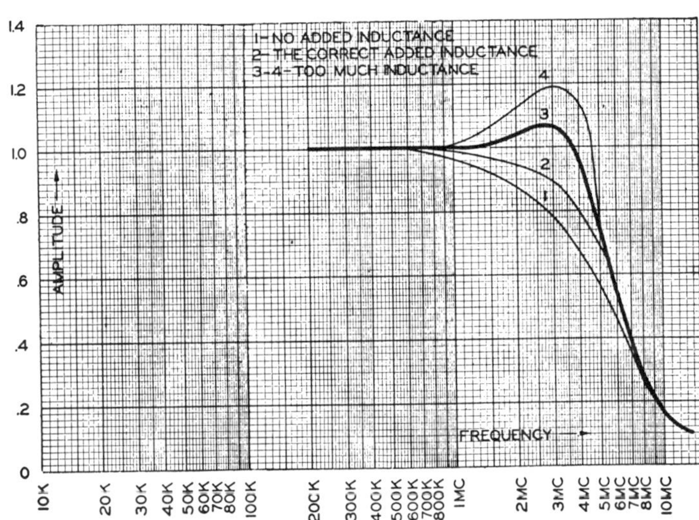

Fig. 12-13. The effect on an amplifier response of the insertion of various amounts of peaking inductance in the plate load.

Typical values of R_L range from 4,700 ohms to 6,800 ohms. Fig. 12-13 demonstrates the effect when L is too high an inductance value (curves 3 and 4) and when L is too low a value (curve 1). A small amount of overpeaking may sometimes be employed to sharpen the fine detail in the picture. Too much peaking, however, will lead to ringing, a condition where multiple lines follow the edge of an object. This condition is sometimes mistaken for ghosts. Figure 11-14 shows the symptom of ringing due to overcompensation as indicated by a square-wave test. The appropriate curve is marked "*Excessive High-frequency Response* (*Oscillation*)."

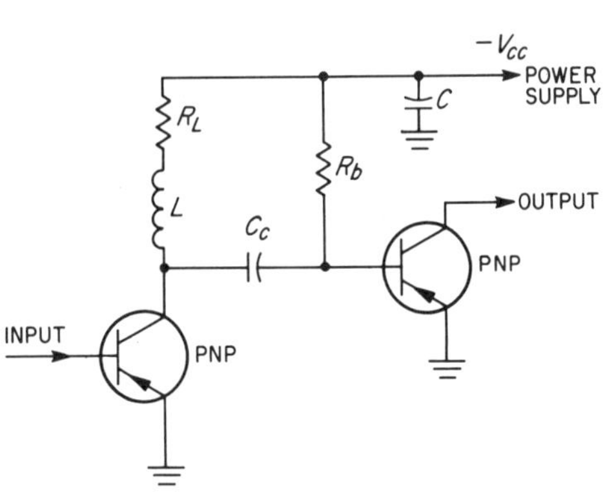

Fig. 12-14. Shunt-peaking compensation in a transistor-video amplifier.

Fig. 12-14 shows an R-C coupled transistor stage with shunt peaking compensation. The coupling, comprised of R_L, L, C_c and R_b, is the same as the one illustrated in Fig. 12-12, although it is drawn somewhat differently. It should be remembered that the B+ point to which R_L is tied in Fig. 12-12(B) is at ground potential as far as the signal voltage is concerned. That is why the equivalent circuit of Fig. 12-12(A) shows R_L and R_g tied to the same point. In the transistor circuit the base and the collector must both be negative (for PNP transistors), so they can both derive their voltage from the same power supply.

Figure 12-15 reveals another way of connecting a shunt peaking coil. In this case the coil is in the base leg of the second transistor stage, but it is still in parallel with C_T. An advantage of this circuit is that the amount of current through the coil is considerably less than when it is connected in a series with the load. However, there is no significant change in the shape of the response curve with the circuit of Fig. 12-15 as compared with the response for Fig. 12-14.

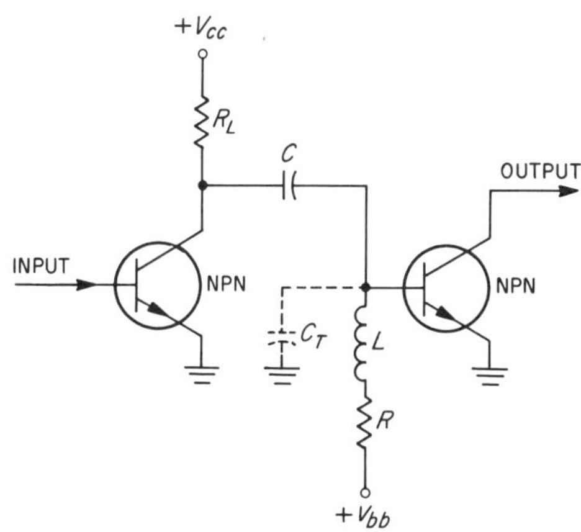

Fig. 12-15. A different circuit for shunt-peaking compensation.

12.5 SERIES PEAKING

A second method of improving the high-frequency response is to insert a small coil in a series with the coupling capacitor, as illustrated in Fig. 12-16. This method gives a higher gain and a better phase response than

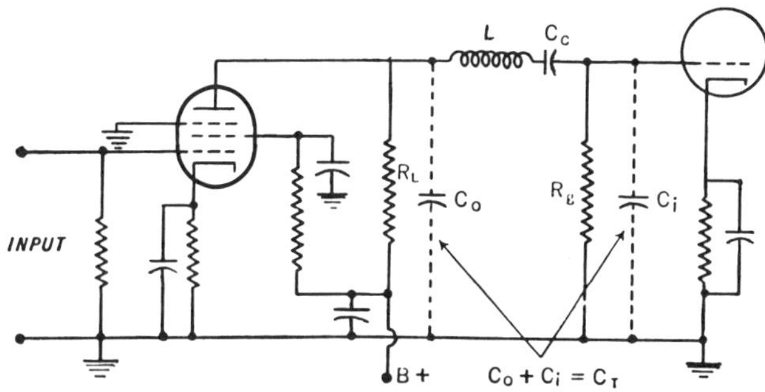

Fig. 12-16. High-frequency compensation by means of a series-peaking coil.

shunt peaking. The improved gain of this type of coupling is due to the fact that the components of C_T are no longer lumped together in one unit, but are separated. On the left-hand side of the series inductance is the output capacitance of the preceding tube, and on the other side is the input capacitance of the next tube. With this separation, the load resistor R_L may be higher in value, because only C_0 is directly across it and not the larger C_T. As C_0 is smaller than C_T, its capacitive reactance is greater, and it will have less of a shunting effect on R_L. A larger value of R_L is then possible, actually 50 percent larger. Thus,

$$R_L = \frac{1.5}{2\pi f C_T} \qquad \text{12-20}$$

In this case C_T is the *total* capacitance shunting R_L, that is, $C_0 + C_i$ in Fig. 12-16.

It has been found that the best results are obtained when the ratio of C_i to C_0 is approximately 2. This is accomplished (or closely approached)

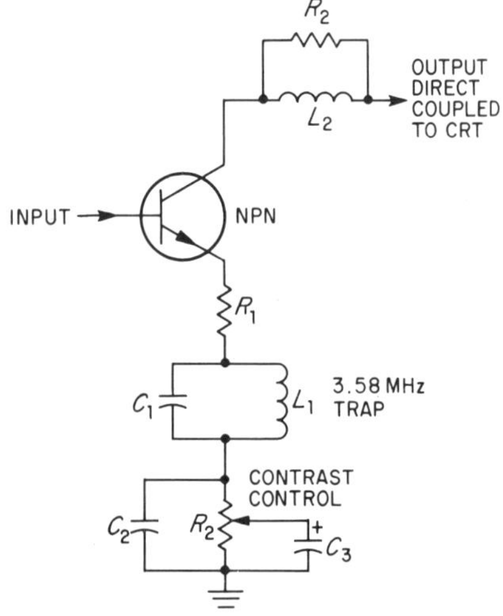

Fig. 12-17. Another peaking compensation circuit.

by the proper choice of the tubes used for the amplifiers. The value of the series coil, L, is given as

$$L = 0.67 C_T R_L^{\,2} \qquad \text{12-21}$$

Figure 12-17 shows a transistor video amplifier with a series peaking compensation provided by L_2. The resistor R_2, in parallel with L_2, is called a *swamping resistor.* The distributed capacitance of any coil in conjunction with the circuit capacitance will act with its inductance to produce resonance. This usually occurs at high frequencies. The overall effect is to cause the output circuit of the video amplifier to resonate, or overcompensate, over a narrow band of frequencies. This is undesirable in view of the fact that a flat response is needed. The swamping resistor reduces the Q of the tuned circuit formed by the peaking coil, and the distributed and circuit capacitances. The standard practice is to wind the peaking coil on the resistor, using the resistor as a coil form. The assembly is then covered with a compound to seal it against moisture.

12.6 SERIES-SHUNT PEAKING

It is possible to combine shunt and series peaking and obtain the advantages of both. Such combinations appear in both the tube and transistor circuits. The shunt coil is designed to neutralize the output capacitance of the preceding tube, while the series coil combines with the input capacitance (and stray-wiring capacitance) of the next tube. With this double combination, it is possible to extend the bandwidth to 4 MHz, which is more than can be derived through the use of the shunt peaking alone. Furthermore, the phase distortion of the combined compensation in the coupling network is lower than either of the two preceding types.

An amplifier using both shunt and series peaking combined is shown in Fig. 12-18. A swamping resistor is shunted across the series coil to

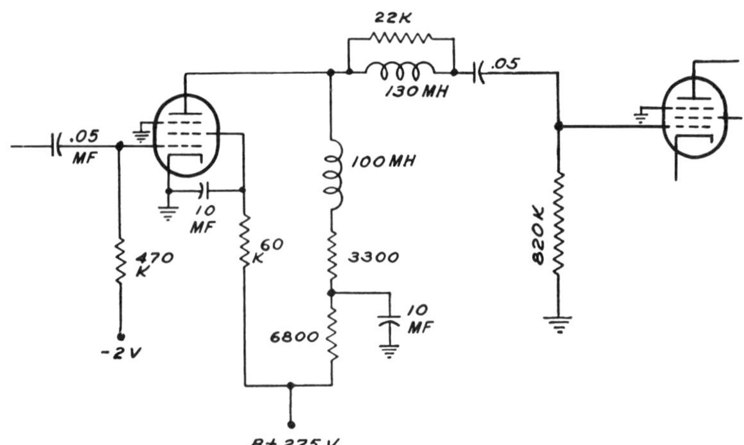

Fig. 12-18. A video amplifier employing series-shunt peaking.

minimize any sharp increase in the circuit response due to the combination of the series coil inductance and its natural or inherent capacitance. The coil is specifically designed to have a natural frequency considerably above the highest video frequency. In production, however, a certain number of coils will be produced with natural resonant frequencies within the range covered by the amplifier. The value of the swamping resistor is generally 4 or 5 times the impedance of the series coil at the highest video frequency.

For the combination circuit, the values of the shunt-peaking inductance (L_s), the series-peaking inductance (Lc), and R_L are obtained from the following relationships:

$$R_L = \frac{1.8}{2\pi f C_T}, \text{ where}$$

$$C_T = C_i + C_0 \qquad 12\text{-}22$$

$L_s = .12 C_T R_L{}^2$ (shunt coil)
$L_c = .52 C_T R_L{}^2$ (series coil), and
f is the highest frequency at which the response remains uniform.

These equations hold true only if the input capacitance of the second tube (C_i) and output capacitance of the first tube (C_0) are related by the ideal ratio:

$$\frac{C_i}{C_0} = 2$$

12.7 LOW-FREQUENCY COMPENSATION

With the high-frequency end of the response curve taken care of, let us determine what changes can be made to improve the low-frequency response. At this end of the band, it is possible to disregard the shunting capacitances since their reactance, given by

$$X_c = \frac{1}{2\pi f C}$$

is very high, and they do not affect the low-frequency signal voltages in any way. Now, however, it becomes necessary to include the coupling capacitor. Figure 12-19(A) shows the R-C coupled amplifier. The equivalent low-frequency circuit can be represented as illustrated in Fig. 12-19(B). The operation of the circuit, as explained in conjunction with Fig. 12-19, shows that the lower the frequency, the greater the effect of the coupling capacitor. The response gradually falls off because the reactance of C_c soon becomes dominant, and a large

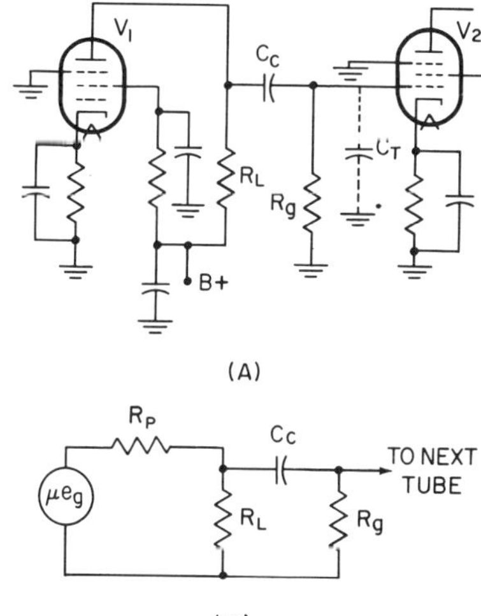

Fig. 12-19. (A) An RC-coupled amplifier and (B) its low-frequency equivalent circuit.

portion of the output voltage of V_1 is lost across C_c. The phase delay of the signal begins to change, eventually approaching 90 degrees. As a result, the background illumination of the reproduced image is affected.

In order to make the response more linear at the low frequencies, either C_c should be made larger (so that it will have less reactance) or R_g should be made larger. The limit of the size of either C_c or R_g is governed by several factors, some of which have already been discussed and now will be reviewed here:

1. Too large a value of C_c increases the stray capacitance to ground, and is certain to interfere with the high-frequency response.

2. Larger coupling capacitors generally have higher leakage currents. This permits the positive power-supply voltage on the preceding plate to affect the grid of the next tube and to bias it positively.

3. A large value of R_g can prove to be detrimental because of contact bias. This is especially true if there is even the smallest amount of gas present in the tube.

4. Finally, high values of R_g and C_c have a tendency to produce motorboating (oscillations), because of the slow building up and leaking off of the charge across the combination.

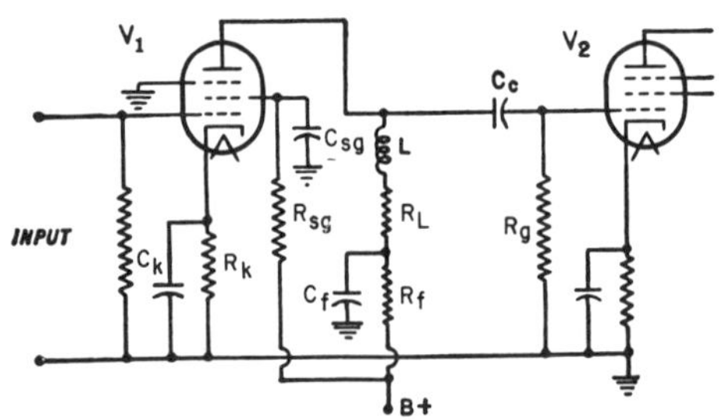

Fig. 12-20. Low-frequency compensation (R_f, C_f) of an R-C amplifier.

By inserting a resistor and a capacitor in the plate circuit of tube V_1, as indicated in Fig. 12-20, it is possible to improve the low-frequency response without making either R_g or C_c too large. R_f and C_f are the two added components. They form the *low-frequency compensation* circuit. The gain of an amplifier is directly related to the size of the plate-load resistor (or collector-load resistor). At low frequencies, the plate load in the circuit of Fig. 12-20 is comprised of R_L and R_f in series. The reactance of C_f is high at low frequencies, and it appears as an open circuit. At the higher frequencies, the reactance of C_f is low enough, so that it presents a short circuit to ground for the signal. This means that the plate load is comprised only of R_L at the higher signal frequencies and the gain is reduced. By amplifying the low frequencies with a higher gain, the loss of the signal at the coupling capacitor is overcome. An additional advantage of the low-frequency compensation is that it reduces the phase-shift distortion in both the tube and the transistor circuits.

The value of C_f in Fig. 12-20 is obtained from the expression

$$R_L C_f \doteq C_c R_g, \text{ where} \qquad 12\text{-}23$$

R_L, C_c, and R_g have previously been assigned values, and

C_f is the value of the low-frequency compensating capacitor.

Load resistor R_L will be determined by the highest frequency to be passed by the amplifier, and C_c and R_g will be as large as possible but within the limitations noted here. Finally, R_f should have a resistance which is at least 20 times larger than the reactance of C_f at the lowest frequency to be passed.

The network comprised of C_f and R_f provides the greatest amount of low-frequency compensation, but there are additional factors which influence the extent of the low-frequency response. One of these is the screen-grid dropping resistor and bypass capacitor. For best results, R_{sg} and C_{sg} should have a time constant which is at least 3 times as long as the period $(1/f)$ of the lowest video frequency to be passed by the amplifier. Another governing factor is the cathode resistor, R_k, and the cathode bypass capacitor, C_k. These should be chosen so that they satisfy the following expression:

$$R_k C_k = R_f C_f, \text{ where} \qquad 12\text{-}24$$

R_k, C_k, R_f, and C_s are as shown in Fig. 12-20.

By leaving the cathode resistance unbypassed, or by dividing it into two resistances with one unbypassed, the gain of the amplifier is decreased at all frequencies. However, the low-frequency response is improved by the addition of degeneration. This occurs because the bypass capacitor around the cathode resistor has a reactance related to the frequency. High frequencies are effectively bypassed, but low frequencies are only partially bypassed.

Admittedly, the screen and cathode circuits are not quite as important as the decoupling resistor and capacitor, the R_f and C_f circuit, but they should be considered in the amplifier design.

In the design procedure of video amplifiers, the values of the high-frequency compensating components are selected first. These include R_L, L_s, and L_c. Next, the low-frequency compensating components, C_f and R_f, are computed, then R_{sg} and C_{sg}, and, finally, R_k and C_k. The values of each of the latter three resistors are determined by the voltage needed for the proper operation of the tube as recommended by the manufacturer. This requirement imposes a limitation. However, since we are concerned with a time constant in each instance (as $C_f \times R_f$, $R_{sg} \times C_{sg}$, and $R_k \times C_k$) rather than the individual value of each part, we can usually satisfy all the required conditions.

When the high- and low-frequency compensating circuits are simultaneously applied to a video amplifier, the circuit appears as shown in Fig. 12-20. The frequency-and-phase response of this amplifier is plotted in Fig. 12-21.

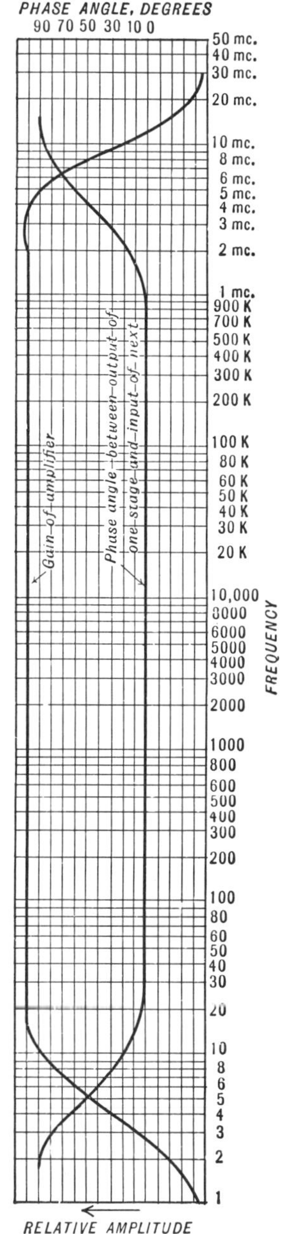

Fig. 12-21. The frequency response of a fully compensated resistance-coupled amplifier.

12.8 DIRECT-COUPLED VIDEO AMPLIFIERS

Direct-coupled amplifiers have been known and used for many years, even before the advent of television. The most important advantage of direct coupling in video amplifier stages is the excellent low frequency response which may extend down to zero Hertz.

A typical direct-coupled video amplifier system which provides a dc path from the video second detector to the picture tube is shown in Fig. 12-22. The video amplifier receives the signal direct from the video detector, amplifies it, and then feeds it to the cathode of the picture tube. Because of the direct coupling, the cathode of the picture tube has a fairly high positive potential. In order to offset this, the control grid of the picture tube is also given a positive voltage, but it is less positive than the cathode. (The exact value will be determined by the setting of the brightness control.) In this way we still maintain the control gird as negative with respect to the cathode. This is required for the proper operation of the tube.

Elimination of the coupling capacitor extends the amplifier response down to dc, or zero frequency. It was the presence of this capacitor, it will be recalled, that caused the response of an amplifier to drop off at the low frequencies and also causes phase-shift distortion. High-frequency compensation is still necessary. L_1 and L_2 of Fig. 12-22 serve this purpose.

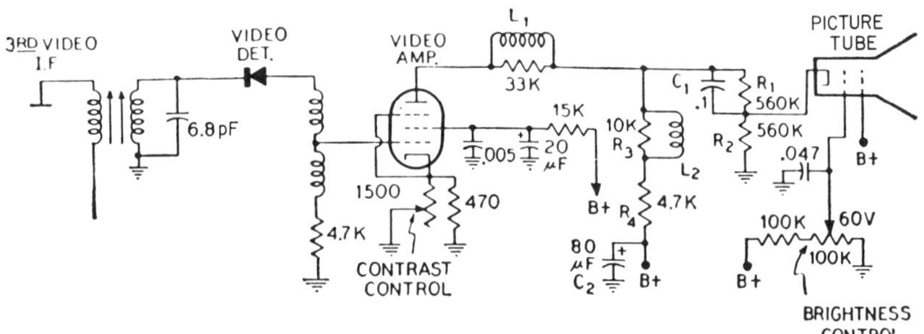

Fig. 12-22. A direct-coupled video-amplifier system.

The presence of C_1, R_1, and R_2 requires an explanation, because they are related to the fact that direct coupling is used between the video amplifier and the picture tube. If we tie the plate of the video amplifier directly to the cathode of the picture tube, the positive voltage at this cathode will be very positive (in this circuit). To offset this, the control grid of the picture tube will also have to have a high positive voltage. And finally, since the second grid (or first anode) of the picture tube should be more positive than the control grid, its voltage in this case may have to be as much as 500 volts positive with respect to the chassis To avoid using voltages this high (which are obtained from the low-voltage power supply), the following system is employed.

In the plate circuit of the video amplifier, the load presented to the medium- and high-frequency component of the video signal is a combination of L_2, R_3, and R_4. The load for the low-frequency portion of the signal, however, includes not only these three components but also the resistance formed by C_2 and the power supply. C_2, an electrolytic capacitor, may represent a short circuit to higher frequency ac, but to low frequencies its impedance may be very high. Therefore, this impedance in parallel with the power-supply impedance must also be considered in computing the load seen by the low-frequency component of the video signal.

Since the low-frequency load of the amplifier is greater than its medium and high-frequency load, the low-frequency component will receive more amplification. To equalize the response, a dc voltage divider consisting of R_1 and R_2 is formed. From the value of these resistances, it is seen that slightly less than half of the dc and low-frequency component reaches the picture tube. The network comprised of R_1, R_2, and C_1 is actually a frequency-compensating network that emphasizes higher frequencies. The 0.1 pF capacitor prevents the higher-frequency signals from being attenuated by shunting them around R_1. Thus, there will be a greater voltage across R_2 for the higher frequencies.

The voltage-dividing action of R_1 and R_2 also reduces the dc plate voltage reaching the cathode of the picture tube, and thereby reduces the amount of $B+$ required by the control grid and second grid of the CRT.

Series-peaking compensation is obtained in the direct-coupled video amplifier in Fig. 12-22 by virtue of L_1 and its parallel swamping resistor. Shunt peaking is accomplished by L_2 and R_3. Contrast control is obtained by varying the cathode-resistance value of the video-amplifier stage. Two filter capacitors are used in the screen circuit. The 20 μf electrolytic provides the filtering for low frequencies. But the self inductance of an electrolytic may be sufficient to introduce considerable reactance at high frequencies, so a small 0.005 μf capacitor is used as a high-frequency bypass.

A two-stage, direct-coupled, video-amplifier system is shown in Fig. 12-23. Note how the voltages start at -130 volts at the grid of V_1 and work up to $+225$ volts at the plate of V_2. This increase in the operating voltages from stage to stage is a characteristic of the direct-coupled amplifiers which is sometimes called *level shifting*. If we add another stage after V_2, the grid of this third stage will be positive because of its connection to the plate of V_2. This, in turn, means that the cathode and plate of this third tube will have to be even more positive. By adding enough stages, the B+ voltage required soon rises to extremely high

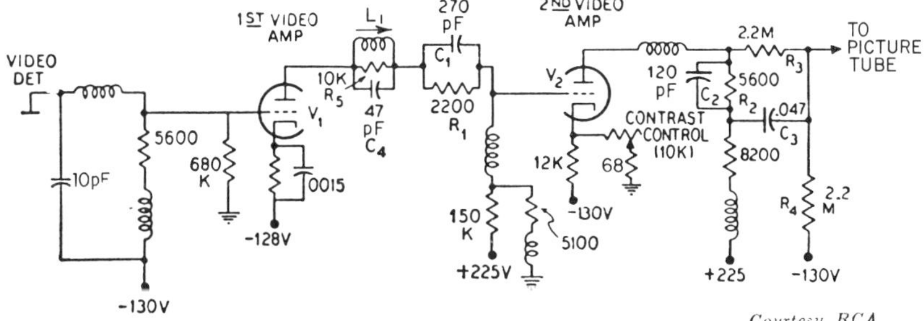

Fig. 12-23. A two-stage, direct-coupled video-amplifier system.

values. Fortunately, television receivers seldom require more than two stages of video amplification, and hence direct coupling can be successfully employed.

In the circuit of Fig. 12-23, the balance between the amplification accorded the *medium and high frequency* and the *low frequency and dc* components is carefully maintained. C_1 and R_1 form one equalizing network wherein the ac component is left alone while the low-frequency component is reduced. C_2, C_3, R_2, R_3, and R_4 form another such network. Finally, L_1, C_4, and R_5 constitute a 4.5 MHz trap to remove any 4.5 MHz signal that may be present in the circuit.

12.9 THE SELECTION OF TUBES AND TRANSISTORS FOR VIDEO AMPLIFIERS

Tubes The ability to provide a high gain and to handle signals up to 4 MHz are the primary considerations in the selection of tubes for use as video amplifiers. To achieve a high gain, pentodes and beam-power tetrodes are the favored types, although occasionally a triode or a dual triode is utilized. The higher mutual conductance values of pentodes compared to triodes makes the pentodes more useful in video-amplifier circuits. In addition, the pentode sections of these tubes are capable of handling relatively large signal swings without introducing excessive amplitude distortion.

The requirement of amplifying signals up to 4 MHz is met by a low interelectrode capacitance, particularly between the plate and grid. The figure of merit of a high-frequency tube is given by the expression

$$\frac{g_m}{C_{\text{in}} + C_{\text{out}}}$$

To the denominator C_M (or its equivalent $C_{gp}[1 + A]$) should be added, since this appears in parallel with C_{in}. When C_M is added, the expression becomes

$$\frac{g_m}{C_{\text{in}} + C_{\text{out}} + C_{gp}(1 + A)} \qquad \text{12-25}$$

The larger the value of this expression, the better the tube is for use as a video amplifier. Sometimes, the power need is the more important requirement. In such cases the interelectrode capacitance is sacrificed, because higher power requires a larger tube structure and this, in turn, leads to greater interelectrode capacitance values.

Transistors Transistors used for video amplifiers should have a high gain-bandwidth product. The beta of the transistor should be high at both dc and high frequencies, and of course, the input capacitance should be small. The high-signal voltages that are needed for driving the picture tube means that output transistors with a high-breakdown voltage rating must be chosen. It is difficult to make transistors with a high β, low-input capacitance, high-power gain, and high-breakdown rating all in the same device, so design engineers are faced with compromises between these characteristics.

12.10 TYPICAL TRANSISTOR VIDEO AMPLIFIER CIRCUITS

It would not be possible to include every variation of video amplifiers that has been used by the manufacturers. However, a few examples of the actual circuits will be instructive. We have already discussed some representative video amplifiers, and we have also discussed some circuits in relation to contrast controls.

Figure 12-24 shows an emitter-follower connected as a video driver. The input signal from the video detector is delivered to the base of the driver via capacitor C_1. This is an electrolytic capacitor with a relatively large value of 10 microFarads or more. The large value serves to extend the frequency response by reducing the reactance of the coupling network at low frequencies. The resistors R_1 and R_2 form a voltage divider network for obtaining the required dc base bias voltage. It will be noted that the collector of transistor Q_1 is connected directly to ground, and the output signal is taken from the emitter across resistor R_3. The emitter circuit, and the voltage divider network in the base, both obtain their operating voltage from the same source. This makes it necessary to supply a decoupling filter. Decoupling filters are used whenever more than one signal source is connected to the same point on a power source. The decoupling circuit has the same configuration as the low-frequency compensating network discussed in Section 12.7. In fact, such a circuit may serve the dual roll of frequency compensation and decoupling. If the purpose is decoupling only, then the resistance of R_4 will be quite low, usually less than 100 ohms. The important thing to remember is that you cannot interpret such a circuit by configuration alone, but must take the value of the component into consideration.

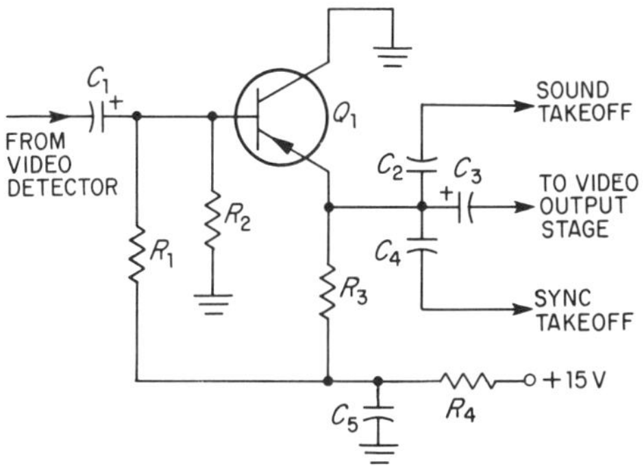

Fig. 12-24. An emitter follower used as a video driver.

The output signal of the emitter follower is delivered to the sound takeoff and the sync circuits through capacitors C_2 and C_4, respectively. Electrolytic capacitor C_3 delivers the video signal to the video-output stage. Note that no peaking compensation is used in the circuit which attests to the very wide freqeuncy response of the emitter-follower configurations. In actual practice it is possible to obtain the necessary 4 MHz bandwidth with such a circuit, even without compensating networks.

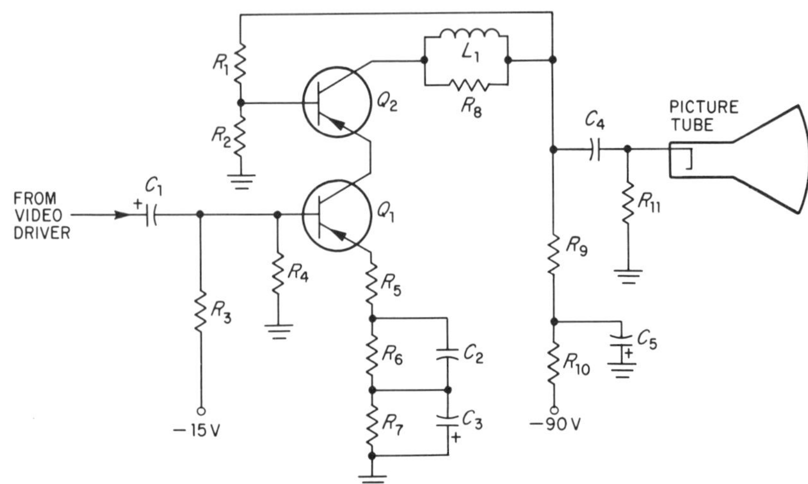

Fig. 12-25. Video-output stage using stacked amplifiers.

Figure 12-25 shows a transistor video-amplifier output stage that uses two transistors connected in series. This type of circuit is called a *stacked amplifier*. (It is also called a *cascode amplifier*.) It has the obvious advantage that the relatively large signal voltage needed to operate the picture tube is divided between the two transistors, thus reducing the drop across each transistor and decreasing the likelihood of a voltage breakdown. The input signal from the video driver is delivered through electrolytic capacitor C_1 to the base of Q_1. The voltage divider R_3 and R_4 establishes the base bias voltage for the stage. Three emitter resistors (R_5, R_6, and R_7) are connected in series in order to obtain the desired frequency compensation. Resistor R_5 is unbypassed and provides the current feedback for the amplifier at all frequencies. Current feedback, or degenerative feedback as it is sometimes called, serves to broaden the response of the amplifier. Resistor R_6 is bypassed for high frequencies, but not for low frequencies. This means that R_5 and R_6 are both in the emitter circuit as far as the low frequencies are concerned, but only R_5 is in the circuit for high frequencies. C_2 and C_3 effectively bypass the higher frequencies around R_6 and R_7. The overall result is that a different amount of degeneration occurs for low and high frequencies. The emitter circuit for Q_1 may be considered to be a form of high-frequency compensation. The series peaking coil L_1 and its swamping resistor R_8 are used to further improve the high-frequency response of the amplifier.

The necessary base bias for transistor Q_2 is obtained with the voltage divider R_1 and R_2. Now this voltage divider is different in principle than the one formed by R_3 and R_4. You will note that resistor R_1 is tied to the collector side of load resistor R_9. The signal voltage developed across R_9 is fed back to the base of Q_2 through R_1. This feedback is *degenerative* because the collector voltage is 180° out of phase with the voltage on the base. Earlier in this chapter we made a general statement that *the higher the gain at mid-frequencies, the narrower the bandwidth.*

By introducing the degenerative feedback, we will at the same time increase the frequency response of the amplifier. Compare this voltage divider with the one used to establish the base bias for Q_1. The high-voltage end of resistor R_3 is connected directly to the -15 volt source which is an unvarying voltage, and so no feedback is obtained in this case. Essentially, then, there are two feedback networks employed in the circuit of Fig. 12-25; the degenerative *current* feedback due to the unbypassed emitter resistor R_5 (and also R_6 at lower frequencies); and, the degenerative *voltage* feedback due to the connection of R_1 to the emitter side of the load resistor.

A decoupling filter comprised of R_{10} and C_5 is needed, because there are other amplifiers (not shown) connected to the -90 volt source. The input signal to the cathode of the picture tube is delivered by C_4, and the signal voltage is developed across R_{11}.

Integrated Circuits Integrated circuits are ideally suited for broad-band and high-frequency applications, but they were slow to appear in video-amplifier circuitry. Among the reasons is the relatively high output signal voltage requirement imposed by the picture tube. One of the first applications of integrated circuits in video amplifiers was in the video-driver stage where the signal amplitude is relatively low.

Many integrated circuits are comprised of monolithic circuits on a small single chip and include transistors, diodes, resistors, and capacitors. A description of this type of integrated circuit is given in Chapter 5 and again in Chapter 20, in greater detail. Some integrated circuits, however, contain only transistors (active elements) with the resistors and capacitors (passive elements), all being externally connected. The latter type of integrated circuit is shown schematically in Fig. 12-26, and the integrated circuit connected to its external passive elements to form a video-driver amplifier, is shown in Fig. 12-27(A). The external connection numbers for the integrated circuit are seen correspondingly

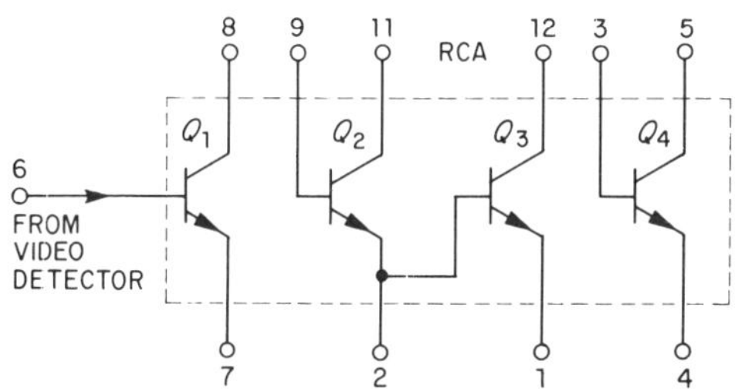

Fig. 12-26. An integrated circuit that can be used as a video-driver amplifier.

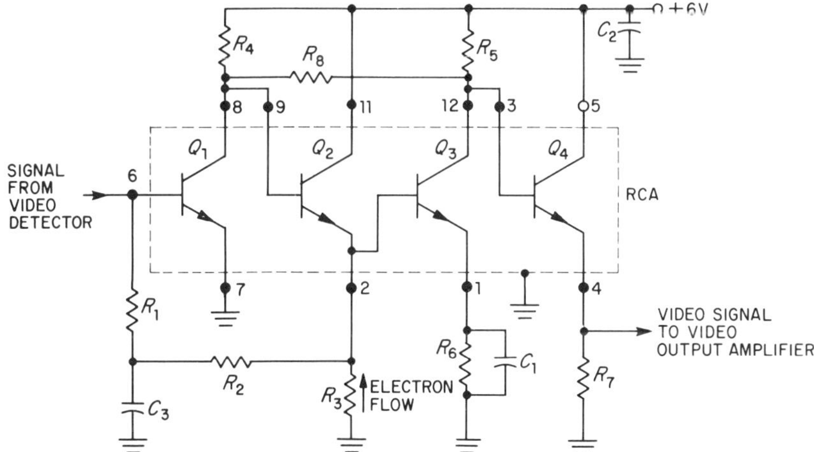

Fig. 12-27(A). The integrated circuit of Fig. 12-26 connected as a video driver. (*Courtesy of RCA.*)

on both of the aforementioned diagrams. The integrated circuit of Fig. 12-26 is that of the RCA, CA3018 unit. This consists of four silicon epitaxial transistors on a single chip, mounted in a 12 lead, TO-5 style (round) can. Two of the transistors are isolated, and the remaining two are emitter-base coupled. In addition to being adaptable for use as a video amplifier, this same IC may also be used to amplify RF, IF (through 100 MHz), and audio signals. Of course, the passive (discrete) components would be different for each type of use. Such passive circuits as resistors, capacitors, and inductors would be externally connected to the terminals of an integrated circuit. The circuit is designed to operate in a range of temperatures from −55°C to +125°C.

Figure 12-27(A) shows an RCA Integrated Circuit CA3018 connected as a video driver. The discrete components are located outside the dotted lines. The input signal from the detector appears as a voltage across R_1. This signal is applied to the base of Q_1. The dc operating voltage on the base of this transistor must be positive, since it is an NPN transistor. The required positive bias voltage is acquired from the emitter resistor of Q_2 as a result of an electron flow from the ground to the emitter. The positive signal voltage from the emitter is filtered by R_2 and C_3.

Transistor Q_1 is connected as a conventional amplifier with the output signal developed across R_4. This signal is direct coupled to the base of Q_2. Transistor Q_2 is connected internally in the integrated circuit as an emitter follower, the output being fed to the base of the conventional amplifier Q_3. The emitter resistor of Q_3 is bypassed for high frequencies, but not for low frequencies. This means that the high frequencies will receive a greater amount of amplification in this stage. The output of Q_3 is delivered to emitter follower Q_4, and the output of this stage goes to the video-output stage.

You will notice that all of the stages in the circuit of Fig. 12-27(A) are direct coupled. Thus, there is no need for low-frequency compensating networks. In addition, there is no phase distortion at low frequencies.

Three different kinds of integrated circuit packages can be seen in Fig. 12-27(B). The round can is similar to the one used in the video amplifier just described. The others illustrate two configurations of "flat-pack" circuits. The flat pack at the left is built to be wired into a printed-circuit board, while the one on the right is a plug-in unit.

Fig. 12-27(B). A TO-5 type (round can) and two "flat-pack" integrated circuit packages. (*Courtesy of Texas Instruments, Inc.*)

Grounded-base Video Amplifier You will recall that the grounded-base circuit has a low-input impedance and a high-output impedance. This characteristic, together with the fact that the grounded-base circuit has a higher-frequency cutoff, makes it useful in the circuit of Fig. 12-28. The low-output impedance of the emitter follower (Q_1) approximately matches the low-input impedance of amplifier Q_2, thus providing an optimum power transfer. The input signal for Q_2 is delivered to the emitter, and the output circuit is taken from the collector. The base is ac grounded by the capacitor C_4.

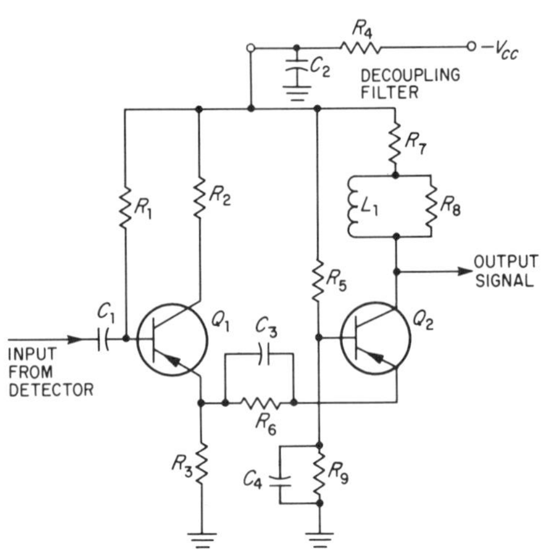

Fig. 12-28. A grounded-base video amplifier.

12.11 AUTOMATIC BRIGHTNESS CONTROL (ABC)

The amount of light in a room is a determining factor in the amount of brightness need to provide a satisfactory picture on a television-receiver screen. The brighter the room, the greater the amount of brightness required. Also, the amount of contrast needed is greater in a bright room. A circuit that automatically adjusts the brightness and the contrast settings to compensate for changes in the room brightness of the receiver is shown in Fig. 12-29. This circuit is more elaborate than some automatic brightness circuits that adjust only the brightness of the picture. The circuit operation depends upon the fact that a *L*ight-*D*ependent *R*esistor (LDR) changes resistance value as the amount of light falling on it changes. This type of resistor is sometimes known as a *photosensitive resistor.* The greater the amount of light, the lower the amount of resistance of this component.

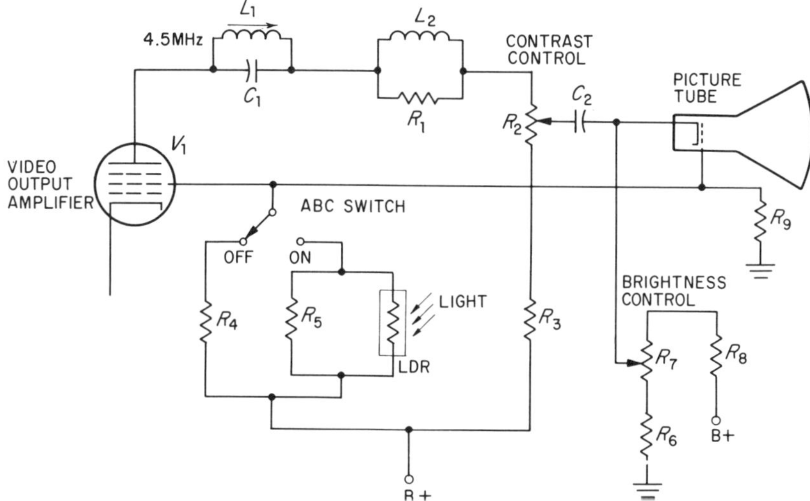

Fig. 12-29. A simplified version of an automatic brightness and contrast control.

The plate circuit of the video amplifier has a 4.5 MHz trap (L_1 and C_1), a peaking coil and swamping resistor (L_2 and R_1), and a contrast control (R_2). The resistor R_3 provides an additional load resistance for the video amplifier. The desired amount of video signal is taken from resistor R_2 and is ac coupled via C_2 to the picture-tube cathode. When the ABC switch is set in the off position, the grid voltage of the cathode-ray tube is set by the voltage-divider action of R_9 and R_4. The drop across R_4 is due in part to the electron current through R_9, and also to the screen current of V_1. The cathode voltage of the picture tube is set by the adjustment of the brightness control R_7. The resistors R_6 and R_8 establish the range of values through which the brightness control can be varied. With this arrangement, it is not possible to accidently set the brightness control so that the grid is positive with respect to the cathode. The brightness of the picture tube depends upon two dc voltages: one on the cathode and one on the grid.

When the ABC switch is in the "on" position, the screen current and the current through R_9 flow through the parallel combination of R_5 and LDR. The light-dependent resistor is located on the front of the receiver where it is exposed to the room lighting. Let us assume that the room becomes brighter. This causes the resistance of LDR to *decrease*. The resulting decrease in the screen-load resistance produces two effects. *First*, the gain of the amplifier is increased, producing a greater signal voltage across R_2 and a greater contrast in the picture. *Second*, the brightness of the picture is increased due to the increased positive voltage on the grid of the picture tube. In another variation of this circuit, the LDR is placed in parallel with the brightness control located in the emitter stage of a direct-coupled, transistor-video amplifier.

12.12 TROUBLES IN VIDEO AMPLIFIER CIRCUITS

This section is not intended to instruct the student in the methods of troubleshooting video-amplifier circuits. Troubleshooting methods are discussed in Chapter 24. In this section, we will become acquainted with the nature of the most common troubles, as they are seen on the picture tube. Only video troubles common to monochrome sets, or to the monochrome portion of color sets are presented here. Video troubles relating specifically to the color portion of color sets are discussed also, in Chapter 24.

Fig. 12-30. A normal test pattern as displayed on the screen of a typical monochrome-TV set.

Normal Test Pattern A normal test pattern, as seen on the picture-tube screen of a typical monochrome-TV set is pictured in Fig. 12-30. Note that the vertical lines in the top half of the pattern fade out a little above 3MHz, which is the overall high-frequency response (video) of the receiver. This response is typical of many monochrome sets. Note the good contrast throughout the pattern and the absence of smearing, following the letters and the large areas, which indicates correct low-frequency response.

Fig. 12-31. The appearance of the TV screen when the video amplifier is completely inoperative.

Symptom: No Video on Screen A completely inoperative video-amplifier section, with the remainder of the set operating normally, will result in a situation of a normal raster, but no picture. The appearance of this condition is shown in the photo of Fig. 12-31. Since there is no video at all, this condition is generally caused by a major malfunction of the video-amplifier section. This trouble also may be caused by a lack of operating voltages to one or more of the video-amplifier stages. It may also be caused by a defective tube or transistor. Other common causes are: a defective video-detector diode, an open-coupling capacitor, an open-peaking coil, or a break in the printed-circuit wiring.

Symptom: Poor High-Frequency Response Referring to Fig. 12-30, we note that the vertical lines in the upper half of the test pattern cease to be visible after about 3 MHz. If the high-frequency response of the video amplifier decreases substantially because of a defect in the video amplifier (or video IF) section, this would be apparent in the test pattern by the disappearance of the upper section vertical lines at a frequency appreciably lower than 3 MHz. For example the lines might disappear at 2 MHz. (The lower vertical lines would be similarly affected and would disappear closer to the lower end of the pattern.) Since the high-video frequencies are responsible for reproducing the smallest picture details, a loss of the high frequencies would manifest itself by a lack of fine picture detail in a TV picture. There would be an appearance of fuzziness, somewhat similar to the appearance of a photo which has been excessively enlarged. The viewer would note that the edges between the adjacent light and dark areas would not be sharply delineated. Also, small printed letters could not be easily read. A smeary picture may or may not also be present. This would depend upon the degree of the high-frequency phase shift that accompanies the loss of the high-frequency response. The greater the phase shift, the more the smeariness that would result. The picture may give the impression of being out of focus. However, in this case the scanning lines would be in focus, showing that the picture-tube focus is not at fault. Some possible causes for the poor high-frequency responses are: a defective peaking coil, a defective bypass capacitor, an incorrect type of replacement transistor, increased value of the load resistor, or a defective shunt capacitor in the video detector circuit.

Symptom: Poor Low Frequency Response The lower-video frequencies (below about 100 kHz) are responsible for reproducing the larger areas of the picture properly and also provide the correct overall contrast. If there is an insufficient low-frequency response (loss of the lower frequencies), the overall picture will give a weak appearance, with poor contrast. However, smearing will probably not take place and the picture, although weak in appearance will still appear sharp

An excessive low-frequency response (increased amplitudes at the lower frequencies), will produce excessive contrast and smearing caused by the resulting phase distortion. This condition is shown in Fig. 12-32 (compare with Fig. 12-30). Although not shown too clearly in the figure, the high-frequency response is not affected. However, the pattern shows excessive smearing to the right of all the large areas, and the overall picture is darker than normal. Note that the shaded, circular areas in the middle of the pattern tend to blend into each other. In addition, the horizontal wedges are darker than the vertical wedges, showing that there is an excessive low-frequency response.

Poor low-frequency response may be caused by a defective interstage coupling network, particularly a defective coupling capacitor. On the other hand, excessive low-frequency response is most likely caused by a defective low-frequency compensation network. In this case, the low-frequency compensation capacitors would be suspect.

Fig. 12-32. Excessive low-frequency response and phase distortion. Note the excessive contrast, the smearing of large areas, and the dark horizontal wedges.

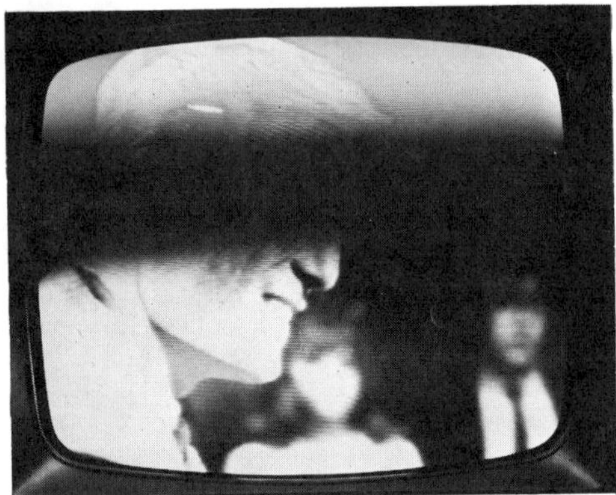

Fig. 12-33. A 60-Hz hum bar in the video signal. If this was caused by a 120-Hz power-supply hum, there would be two dark bars. Note that there is no bending of the picture, indicating that the hum voltage does not affect the sync circuits.

Symptom: Hum Bars on the Screen Figure 12-33 shows a typical case of a 60-Hz hum bar on the screen. This results from the introduction of a 60-Hz voltage into the video-amplifier section. In a tube set, this can be caused by heater-to-cathode leakage of a video-amplifier tube. Since transistor video amplifiers have no heaters, 60-Hz hum bars will not be produced in such amplifiers. However, it is possible to have 120-Hz hum bars in both transistor and tube sets, if the dc power supply voltages to the video-amplifier stages are poorly filtered. In this case, the appearance on the screen would be similar to that shown in Fig. 12-33, except that there would be two (rather than one) horizontal hum bars visible. Note that 60- or 120-Hz hum voltages may be produced by modulation in any of the RF or IF stages. This would be rectified by the video detector and appear on the screen as can be seen in Fig. 12-33.

REVIEW QUESTIONS

1. Draw a low-frequency compensated video amplifier.
2. Draw a series peaking coil in a typical tube-type, video-amplifier circuit.
3. Draw a shunt peaking coil in a typical transistor-type, video-amplifier circuit.
4. Why can we disregard all shunting capacitances when designing the low-frequency compensation network?
5. Without adding any additional components to an audio amplifier, how can we improve its low-frequency response? What limitations exist to this method?
6. What would be the visual effect of overpeaking? Underpeaking?
7. How is the gain of a tube-amplifier circuit related to the amplification factor of the tube?
8. What does the gain-bandwidth rating of a transistor indicate about its ability to amplify high frequencies?
9. How is the alpha cutoff frequency of a transistor determined?
10. Explain how the Miller effect influences the high-frequency response of an amplifier.
11. Why is a swamping resistor sometimes used with a peaking coil?
12. Why does level-shifting occur in direct-coupled amplifiers?
13. What is the advantage of using stacked amplifiers in transistor circuits?

DC Reinsertion* 13

13.1 THE DC COMPONENTS OF VIDEO SIGNALS

The composite video signal contains several distinct components, each of which serves a difinite purpose. There is, first of all, the ac component which represents the detail in the image. Second, there is the dc component which governs the overall background shading of the picture. Both components are separate, and each may be varied independently of the other. Finally, there are the blanking and synchronizing pulses. These are included in the signal for the purpose of synchronizing the scanning beam from one side of the screen to the other at precisely the correct instant of time.

The two preceding chapters dealt with video amplifiers which are primarily concerned with the ac component of the video signal. Here we are concerned with the dc component, its function within the video signal, what happens when it is removed from the signal, and how it is reinserted.

Several lines of a typical video signal are shown in Fig. 13-1. Between every two successive synchronizing and blanking pulses, we have the camera-signal variations, ranging from white (at the most positive value) to black at the level of the blanking pulse. The signals are shown in the positive picture-phase form. When applied to the control grid of a cathode-ray tube, each different value of video voltage produces a different spot intensity on the screen of the tube. From all these light gradations we obtain the image.

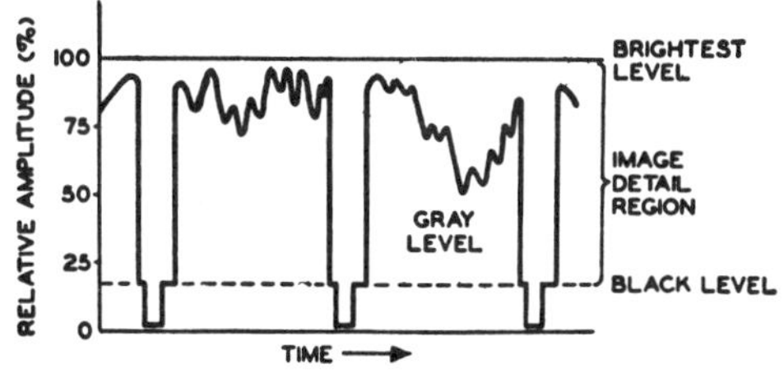

Fig. 13-1. Several lines of a typical video signal.

Suppose now, we take a video signal; and while maintaining the same camera-signal variations, we first move these variations closer to the blanking pulse level. This situation is shown in Fig. 13-2(A). Next, we shift the same variations as far away as possible from the blanking pulses as shown in Fig. 13-2(B). What will be the visual result in each instance? Since the blanking level represents the point at which

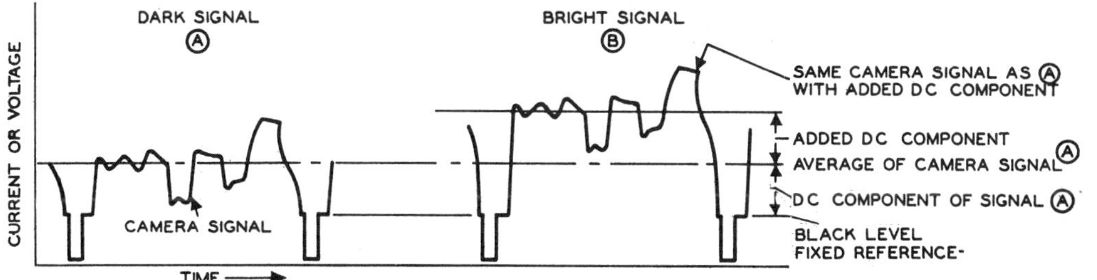

Fig. 13-2. Two video signals containing the same detail (ac component) but different background brightness (dc component).

* The name *dc reinsertion* circuit is common throughout the television field. However, the terms *clamping* circuit or *dc restorer* are also used. They refer to the same thing and may be used interchangeably.

the cathode-ray tube beam is supposed to be cut off, moving the video signal closer to this level means that the overall background of the image will become darker. On the other hand, when the video-signal variations are farther away from the blanking level, the background of the image becomes brighter. Note, however, that because the video-signal variations are identical in each instance, the same scene is obtained. The only thing we have altered by shifting the relative position of the video signal is the background brightness. In the first instance, it is dark; in the second, it becomes bright. We can simulate the same condition in a room by increasing or decreasing the intensity of the electric lights. This change does not affect the objects in the room; it merely affects the overall brightness of the scene.

To distinguish between the camera-signal variations and the average level of these variations (or the average distance of these variations from the blanking level), it has become standard practice to call the latter the dc component and the former the ac component of the video signal. The average level of the signal can be altered by the insertion of a dc voltage, thereby raising or lowering it and changing the background brightness of the image.

At the transmitter, the level of the blanking pulses is established as the dark level, at which point the electron beam in the receiver cathode-ray tube is cut off, and the screen for that point becomes dark. When the ac video-signal variations obtained from the camera tube are combined with this blanking voltage and the sync pulses, we have a complete video signal. At any point along the program line, the distance between the average level of the ac video signal and the blanking level may be varied (through the insertion of a dc voltage) to produce the desired shading or background brightness as dictated either by the program director or by the scene itself. Note that, since the dc voltage moves the video-signal variations closer to (or farther away from) the *blanking level,* we are using this level as a reference. Therefore, the level of the blanking pulses must always remain fixed, and the signal is transmitted with this relationship maintained.

The second-detector output in the receiver contains the full video signal, as shown in Fig. 13-1. The blanking pulse of each line is aligned to the same level. However, when the signal is passed through *R-C* coupled video-frequency amplifiers, the blanking pulses of the various lines are no longer aligned, because the coupling capacitors cause the video signal to possess equal positive and negative areas about the zero axis.

This situation is encountered in many circuits although in slightly different form. Suppose we take three 60 Hertz ac voltages and three dc voltages and combine them to form the signals indicated in Fig. 13-3 (A). Voltages of this type are frequently found in power supplies where the ac wave represents the ripple. For the sake of this discussion, we have provided enough dc voltage for each ac voltage, so that the positive

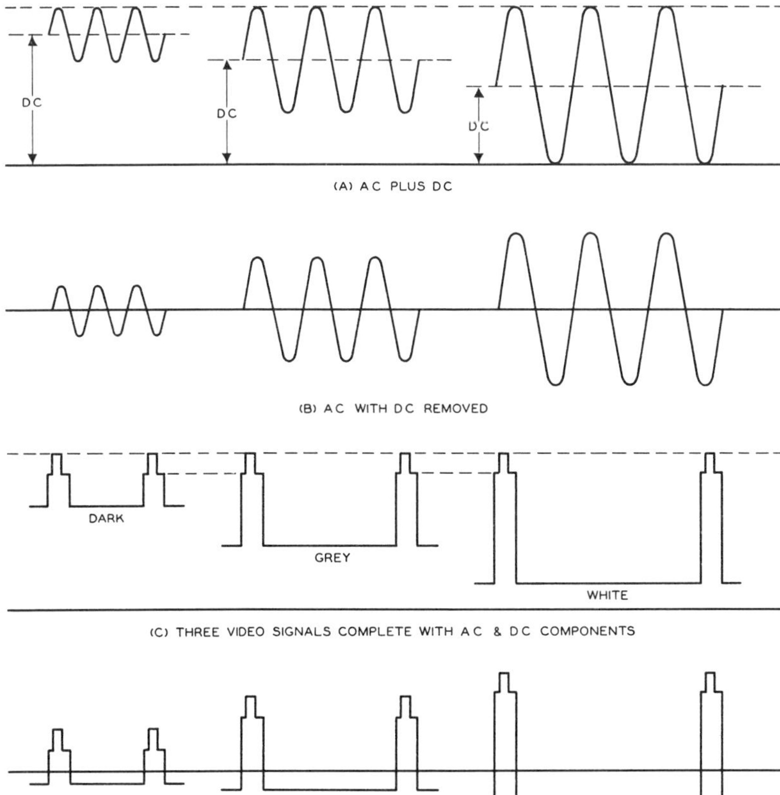

Fig. 13-3. An illustration of the effect of removing the dc component from the video signal.

peaks of all three waves reach the same level. Now, let us pass these voltages through a capacitor. The result is shown in Fig. 13-3(B). By removing the dc voltages, each wave has as much area above the axis as below it and, because of this, the positive peaks of the waves are no longer at the same level.

Let us look at the equivalent situation in a television system. In Fig. 13-3(C) there are shown three video signals taken at different moments from a television broadcast, which represent three lines. One line is almost white, one is gray, and one is dark or black. As they come out of the video-second detector, all the blanking voltages are aligned to the same level. After passing these three signals through a coupling capacitor, the signals possess the form indicated in Fig. 13-3(D). For each signal, the area above the axis is equal to the area below the axis. Because of this distribution, the blanking voltages of the signals are no longer at the same level, and we say that the dc component of the video signal is "missing." The question now is: What effect will this variation in the blanking level have on the image produced on the screen?

The top of each blanking pulse represents the darkest possible level of each line. Since all lines in an image should have the same reference (or black) level, the tops of all blanking pulses should have the same voltage value. This was true in the receiver just before we passed the detected video signal through a coupling capacitor in the video-frequency amplifier system. After passage through this capacitor, the blanking-pulse levels were no longer aligned to the same level. Now if we apply the three signals to a picture tube (reverse polarity of Fig. 13-3(D)) what happens?

First of all, when the signal corresponding to a white line reaches the cathode-ray tube grid, we manually adjust the brightness control (which controls the bias for the picture tube) to the point where the blanking-pulse level just drives the tube into cutoff. Thus, as long as the signal remains at the level it had when we adjusted the brightness control, the negative voltage of the blanking pulse—added to the negative bias set by the brightness control—will darken the screen at the blanking-pulse level.

Now, if the gray video signal comes to the cathode-ray tube, we see that its blanking-pulse level is less negative than the blanking pulse level of the previous video signal. In this case, then, the beam will not be cutoff by the blanking pulse, and the beam retrace will be visible. We could produce the proper cutoff conditions by increasing the negative bias on the image tube, but this procedure would be impractical for the continually-changing brightness levels. For scenes that are continually changing, the background shading varies too rapidly to be adjusted manually by the viewer. As a result, if the brightness control is set for a very bright picture, we will see the retrace lines when a darker picture arrives. Conversely, if the brightness control is set for a darker image, then, when a lighter image is viewed, part of the detail will be lost because of the greater cathode-ray tube-grid bias. The situation is aggravated even more when a dark video signal arrives. Now, we require an even greater negative bias and when the brightness control is set correctly for this signal, it is much too negative for any of the two previous signals. If either of these two other signals is viewed with the bias set for this last signal, the image will appear too dark. On the other hand, when it is correctly set for a white picture, a black picture will appear too light, with the retrace lines visible. The only solution to this state of affairs is to always return all blanking voltages to the same level just as we found them in the incoming signal. *This, then, is the function of the dc restorer in the receiver.*

Every cathode-ray tube has a definite characteristic curve. For a given amount of grid-bias voltage, a definte amount of light appears on the screen. All blanking pluses are purposely placed on the same level in order that the cathode-ray tube will react to them in the same manner throughout the entire reception of the signal. The same is true of white, gray, black, or any other shade that is transmitted to the scene. Any one shade must produce the same illumination on the cathode-ray tube screen each time its corresponding voltage is present on the control grid of the tube. However, this cannot occur unless all video signals have the same reference level. It is here that the usefulness of the dc component becomes apparent. Through the use of this inserted voltage, all blanking and synchronizing pulses are leveled off and the image detail attached to these pulses is likewise correctly oriented.

To operate the television receiver properly, then, a method must be devised whereby the ac video signals which appear at the cathode-ray tube are again brought to the same relative level that they had before the removal of the dc component in the coupling circuits of the video amplifiers. Of course, if direct coupling is used between detector ouptut and the picture tube, the dc level will not be lost. Consequently, dc insertion is not needed. However, for receivers that employ *R-C* coupling in the video stages, the problem resolves itself into one of reinserting a dc voltage that will take the place of the one removed. Consequently, there is a need here for special dc reinsertion networks.

It might be noted here that television receiver AGC systems (except the averaging type) use the sync-pulse level to establish the amount of AGC voltage. Here, too, it is important that the dc component is present, otherwise a change in the scene background will be taken by the AGC system to mean a change in the signal amplitude.

13.2 REINSERTING THE DC COMPONENTS

To understand how dc restoration is possible, it is necessary to know that removing the dc component from a video signal does *not* change its shape. It merely changes its reference level. This is evident when Figs. 13-2(A) and (B) are compared. The same variations in the ac components still occur, and the relationship of the ac signal to the blanking and synchronizing pulses remains the same, with or without the dc component. It is also seen that the brighter the line, the greater the separation between the picture-information variations and the pulses. As the scene becomes darker, these two components move closer together.

It is from these relationships that we are able to reinsert the dc component; for, if we can develop a variable bias that will affect each change in the blanking and synchronizing pulse voltage and act in such a manner that all pulses are brought to one common level, our purpose is achieved. It will mean, for example, that if a video signal in its ac form is applied to the input of a tube where the process of dc restoration occurs, a variable grid bias, developed here, will return the pulses to the same level again in the plate circuit of the tube. This bias will automatically adjust itself to accommodate each individual case. Then, with all the signals lined up again, they can be applied to the cathode-ray tube.

13.3 DC REINSERTION WITH A DIODE

The most common method of providing dc reinsertion is by the use of a simple diode circuit. This is illustrated in Fig. 13-4. The signal here is in its ac form until it reaches the input to the dc restorer, composed of capacitor C_1, resistor R, and diode D_1. The form of the signal, at this point, is the positive phase, since no further reversals take place before the grid of the cathode-ray tube is reached.

In the signal applied to the restorer, the blanking and synchronizing pulses are below the zero line. When applied to points 1 and 2, the signal will cause point 1 to become negative with respect to 2. This follows from the action of an ac wave. The other portion of the signal, which contains the image information, is above the line, and when applied across points 1 and 2, it will make 1 positive with respect to 2. The diode in the circuit conducts only when its anode is positive with respect to its cathode, or when point 2 is positive with respect to point 1.

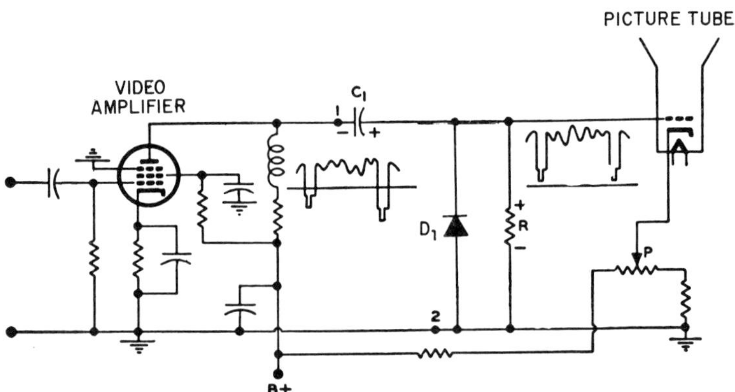

Fig. 13-4. DC reinsertion with a diode.

The action of the dc restorer is simple. When the polarity of the video signal at point 1 is negative, point 2 and the anode of the diode are positive with respect to point 1. A flow of current will occur through the diode, and capacitor C_1 will charge to a value dependent upon the strength of the signal acting at points 1 and 2. The polarity of the charge is indicated in Fig. 13-4. During the positive portions of the video signal at the input of the circuit, capacitor C_1 will discharge through R, since the diode anode is now negative, and the diode is nonconducting. The value of R is high, about 1 megohm, and C_1 discharges slowly.

The values of C_1 and R are so designed that the voltage on the capacitor remains fairly constant throughout an entire horizontal line, or during the time that the positive ac signal is acting on the picture tube grid. Note that this charge is between the grid and the gound (or cathode) and hence acts as a variable bias in series with the ac signal. When the negative portion of the signal (which is due mostly to the blanking and synchronizing pulses) acts at the input, the anode of the diode again becomes conductive. The charge on C_1 will now be automatically adjusted to the amplitude of the negative pulse. A bright line will place a larger positive voltage on the capacitor C_1 than a darker line (positive picture phase here). The positive voltage will cause the grid to become more positive, and the line will receive its correct value. The bias will raise each line until the blanking pulses are lined up again. Thus, in this instance, a bias develops which is proportional to the pulse amplitudes, which are, in turn, governed by the average brightness of the line, as previously explained. Potentiometer P is the brightness control, and its adjustment will cause the grid to cut off on the application of all blanking pulses.

13.4 THE BRIGHTNESS CONTROL

At the grid of the cathode-ray tube, a fixed bias between the grid and cathode is obtained from the power supply. This bias sets the operating point for the tube and, in conjunction with the video-blanking pulses, cuts off the electron beam at the proper moments. The setting of the CRT grid bias will depend upon the strength of the signal reaching the grid. A signal of small amplitude, say from some distant station, requires more fixed negative bias on the grid than a stronger signal.

The dependency of the cathode-ray tube grid bias on the strength of the arriving signal is illustrated in Fig. 13-5. For a weak signal, the bias must be advanced to the point where the combination of the relatively negative blanking voltage plus the tube bias drives the tube into cutoff. However, with a strong signal, the negative grid bias must be reduced; otherwise some of the picture detail is lost.

Since the bias of the cathode-ray tube may require an adjustment for different stations, or even for various conditions on the same station, a potentiometer is connected into the bias circuit, brought out to the front panel, and called the *brightness control*. By its use, the observer is able to adjust the bias on the grid of the picture tube in order that blanking pulses drive the grid only to cutoff and no retrace is visible on the screen.

The effects of the brightness control and the contrast control previously described overlap to some extent. If the setting of the contrast control is increased so that the video signal becomes stronger, the brightness control must be adjusted to meet the new condition, which means that no retraces are visible and that the picture does not look milky or *washed-out*. Too small a value of the negative grid bias allows the average illumination of the scene to increase and permits part of the return traces to become visible. In addition, the image assumes a thin, watery, washed-out appearance. Too low a setting of the brightness control, which will result in a high negative bias on the picture-tube grid, will cause some of the darker portions of the image to be eliminated, and the average illumination of the scene will decrease. To correct this latter condition,

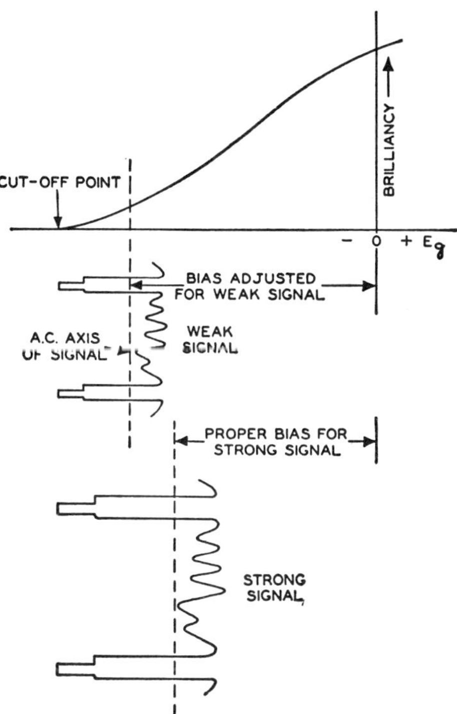

Fig. 13-5. Illustrating why the brightness control must be adjusted to suit the incoming signal.

either the brightness control can be adjusted or the contrast-control setting can be advanced until the correct position is obtained. If the brightness control is varied over a wide range, the focus of the electron beam in the CRT may be affected. However, in the normal range of brightness settings made by the viewer, changes in focus do not present a problem.

13.5 TELEVISION RECEIVERS THAT DO NOT EMPLOY DC RESTORATION

Cost is a strong determining factor in the disign of commercial television receivers, and if it is possible to reduce the cost of a set without compromising picture quality too much, this sacrifice is frequently made. A number of receivers have been manufactured that do not employ dc restoration, not do they possess a dc path between the video second detector and the picture tube. In other words, the dc component is removed from the signal and never reinserted.

As the preceding discussion has indicated, loss of the dc component will tend to make the overall picture darker. To counteract this, the viewer generally turns up his brightness control. This, in turn, frequently causes the vertical retrace lines to become visible. The continued presence of these lines during normal broadcasts will prove to be annoying. To rid the screen of these retrace lines, it has become standard practice to apply a negative pulse to the grid of the picture during the vertical retrace interval. (A positive pulse fed to the picture-tube cathode will achieve the same results. Generally the pulse is applied to the element not receiving the video signal.) The pulse biases the tube to cutoff, prevents electrons from passing through the tube, and effectively removes the vertical retrace lines for any normal position of the brightness and contrast controls.

It is true that removal of the dc component will reduce the contrast range of the image. However, this has been partly offset by the development of screen phosphors possessing wider contrast ranges, and it is doubtful whether any viewer can tell the difference when the dc component is missing.

When direct coupling is employed between all stages in the video amplifier section, including the coupling stage between the video detector and first video amplifier and between the last video amplifier and the picture tube, then dc restoration is not required. Figures 12-22 and 12-23 show examples of direct-coupled video stages that do not require dc restoration.

The disadvantage of direct coupling more than one tube-type video stage is that the power supply voltage must be higher than is required for a single amplifier stage. Also, any small amount of change in the dc level will be amplified by the first video amplifier and reamplified by the second video amplifier. The demands on the power supply, that is, the requirements of high voltage, good regulation, and exceptionally-good filtering, result in a preference for dc coupling, but only for the video-output stage.

It is an easier matter to direct-couple transistor amplifiers. The emitter-collector voltage is not very large to begin with, so three or four stages can be direct coupled without a voltage in excesse of 100 V. The low impedance of transistor circuits makes it easier to obtain a ripple-free dc voltage. Figure 13-6 shows a direct-coupled transistor video-amplifier circuit. The detector (D_1) is direct coupled to the first video amplifier (Q_1) through a peaking coil L_{212}. This is an emitter-follower stage which develops a signal across R_{248}. The base of the second

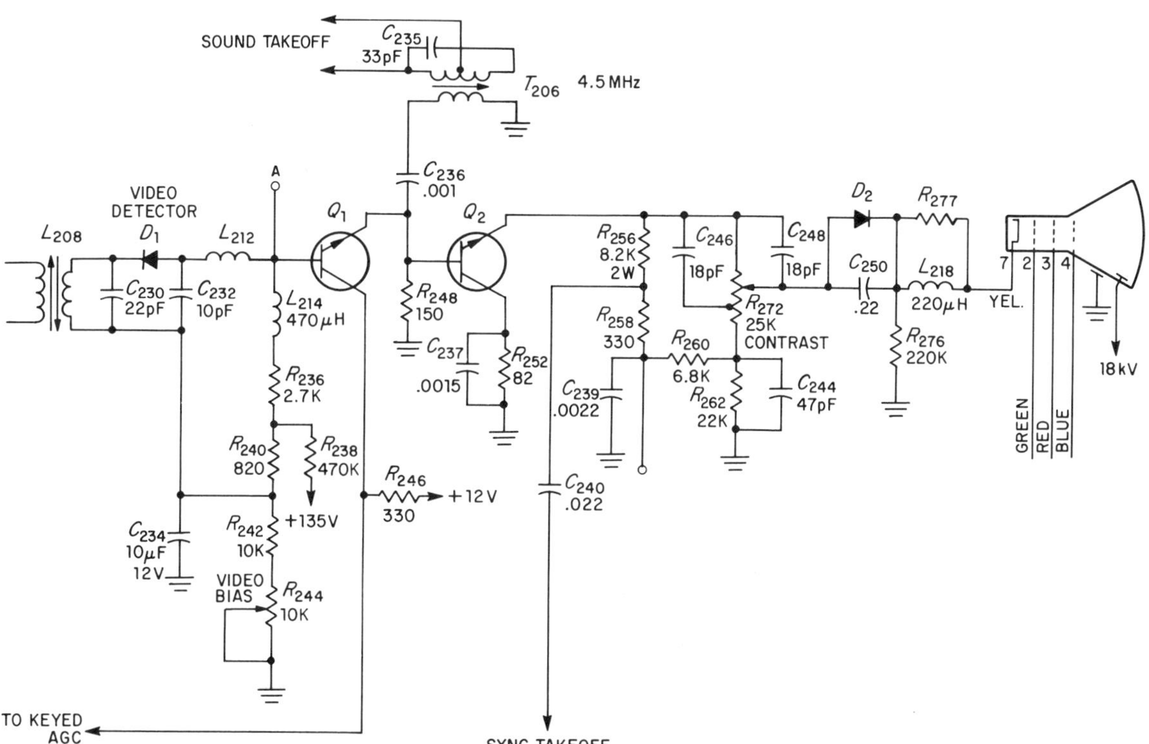

Fig. 13-6. Direct-coupled transistor, video-amplifier stages.

video amplifier (Q_2) receives its signal directly from the emitter of Q_1. The second amplifier is a conventional (common-emitter) amplifier stage direct coupled to the picture tube through the contrast control, diode D_2, and the peaking coil L_{218}. (Diode D_2 is used to compensate for nonlinearity in the transistor amplifier (Q_2) on high-amplitude signals.)

When direct coupling is used, the brightness control may be part of a video-amplifier circuit rather than part of the picture-tube bias circuit. Also, blanking signals may be fed to a video-amplifier stage rather than to the cathode or grid of the picture tube.

13.6 DC RESTORERS IN COLOR RECEIVERS

Although the subject of color receivers is covered in more detail in Chapter 22, the need for dc restorers in color-TV circuits is discussed here at this time.

As a general rule, whenever a video signal is delivered to the cathode-ray tube of television receivers, whether it is a color or a black-and-white receiver, then that signal must have a dc voltage reference if the maximum picture effectiveness is to be achieved. In this chapter it has been shown that the dc reference is lost whenever the signal passes through capacitive- or transformer-coupling networks. In a color receiver, the video signal (which is usually called the *luminance signal* in color receivers) may be delivered to the picture tube through capacitive-coupling circuits. In this event the same problem of dc restoration occurs as with the video signal in black-and-white receivers. The block diagram in Fig. 13-7 shows that the luminance signal passes through a

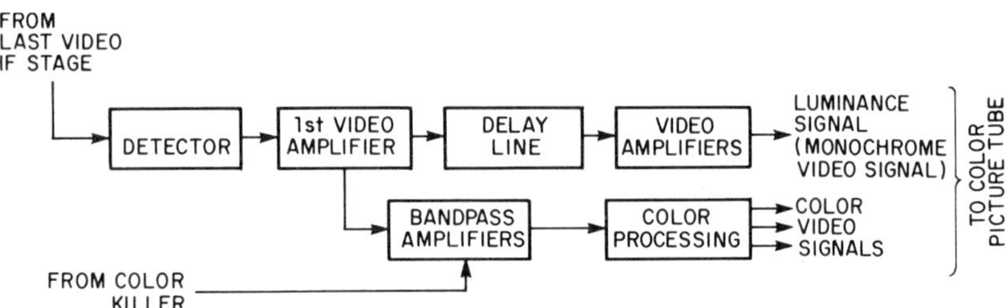

Fig. 13-7. Simplified block diagram of luminance and color-video sections.

video-amplifier stage, a delay line, and then through additional video-amplifier stages. The purpose of the delay line is to keep the luminance signal in phase with the color signals which must pass through different amplifier sections. If capacitive coupling is used with any of the video-amplifier stages, then dc restoration is needed. The circuitry is the same as that used in monochrome receivers.

In the block diagram of Fig. 13-7, the color signal is taken from the first video amplifier and fed through bandpass amplifier stages, then into color processing circuits. It should be pointed out that color receivers vary in their circuitry according to the manufacturer's preference, and this is only a typical color receiver. However, most receivers have a section called *the bandpass amplifiers.* These are simply tuned amplifier stages which pass only the color signals that are grouped around the 3.58 MHz color subcarrier.

Figure 13-8 shows the block diagram of an actual color-receiver (RCA CTC 40) chroma section. The color-killer control voltage prevents the color-bandpass amplifier stage from producing an output signal when there is no color signal being received. It is like a switch that automatically turns the amplifier on when a color transmission is being received. The color bandpass amplifiers deliver the color signals simultaneously to three demodulators—one for each of the primary

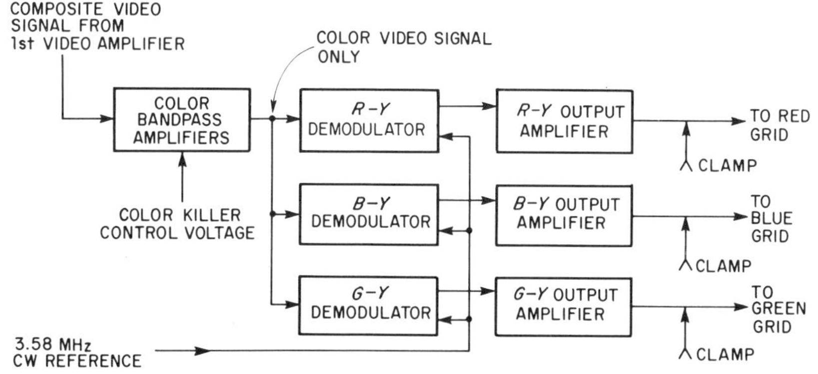

Fig. 13-8. Simplified block diagram of a color section showing clamping (dc reinsertion) voltage delivered to each amplifier output.

colors. At the same time, the 3.58 MHz reference signal is also fed to the demodulators. The output of each demodulator is then amplified before it is fed to the respective color grid. The amplifiers that follow the demodulators are capacitively coupled, and dc restoration must be used. Note that in the block diagram of Fig. 13-8, each color signal output is *clamped*. This is another way of saying that the dc level has been restored.

Since there are no sync or blanking signals present in the chroma-video signals to act as a reference, an artificial reference is established by the clamping amplifier circuit shown in Fig. 13-9. Note that while each color grid is clamped separately by the action of its own particular signal, the dc reference established by the clamping amplifier is used for all three color grids.

A negative pulse from the flyback transformer is fed to the emitter of clamping amplifier, Q_{101}, and causes conduction of Q_{101} only

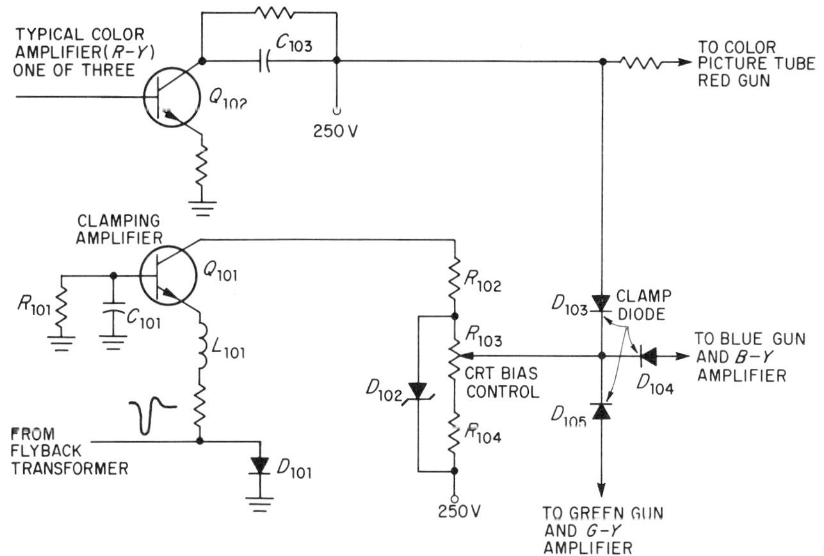

Fig. 13-9. Circuit for obtaining the clamping voltage. Only one color-amplifier circuit (*R-Y*) is shown in detail. The *B-Y* and *G-Y* amplifier circuits are connected in the identical manner as the *R-Y* amplifier circuit.

during the negative pulse time. Diode D_{101} removes any positive portion of the flyback pulse. Inductor L_{101} suppresses radiations from this circuitry. The negative, amplified pulse appearing across the CRT bias control and resistor R_{104} is clamped to a maximum value of 180 volts by the zener diode D_{102}. A portion of the negative pulse voltage, as determined by the setting of the CRT bias control, is applied to each of the three clamping (dc restoring) diodes, D_{103}, D_{104}, and D_{105}. The action of the clamping diodes causes the three grids of the color tube to be initially clamped to the bias pulse voltage, or about minus 80 volts. In Fig. 13-9, only one color amplifier is shown in detail, to simplify the diagram. However, the other two color amplifiers are connected to the clamping circuit in the identical manner.

When the chroma signals are passed through the coupling capacitors in the output of each chroma amplifier, such as C_{103} in the collector circuit of Q_{102}, the dc level is lost. It is necessary to provide dc restoration individually for each color grid, so that the proper color brightness values will be reproduced. This is accomplished by the same basic method as described previously for the composite video signal in a monochrome set. Each chroma signal causes conduction of its respective clamping diode, charging and discharging its respective coupling capacitor, (for example, C_{103}, for the *R-Y* amplifier), in accordance with the average dc level of the chroma signal from each color amplifier. In turn, this causes the dc voltage applied to each color-tube grid to vary independently in accordance with the average color brightness value of each color signal. In this manner, the color brightness value is caused to track with the values originally seen at the color-camera tubes.

13.7 TROUBLES IN D-C RESTORER CIRCUITS

Since dc restoration affects the bias voltage on the picture tube, it is reasonable to expect that a faulty dc restorer can result in a complete loss of picture and/or brightness. If the dc restorer circuit is faulty because of a change in the value or a defect in a component, the dc level may be lost. Also, since the bias voltage on the picture tube does not vary properly with changes in the average dc value of the video signal when the restorer is not functioning properly, it is possible for the picture background to become darker. Loss of contrast may be a symptom of dc restorer trouble. If the brightness control is then turned up, vertical retrace lines may become visible.

In color receivers it is possible to have a defective dc restorer in the luminance circuit as well as in the chroma circuitry. If the dc restorer in the luminance channel only becomes defective, the result may be poor picture contrast which will be noticeable on monochrome pictures. There may also be some loss of monochrome portions of the color picture. The result will be that poor color displays may be seen when a color picture is being viewed, and poor contrast may be seen when a monochrome picture is displayed.

If the clamping circuit in one of the color sections is faulty, it can result in a complete loss of one color, or in a washed-out color. DC voltage checks on the grid and the cathode circuits of the picture tube should be made if a dc restorer circuit is suspected. Remember that a dc restorer fault will change the dc bias on the affected gun of the color tube.

REVIEW QUESTIONS

1. Which types of coupling between the amplifier stages will result in a loss of the dc reference level?
2. Why is a reference signal obtained from the flyback circuit in the clamping circuit of a color-television receiver?
3. What is the dc component of a composite video signal?
4. What information is contained in the ac component of the composite video signal?
5. If transformer coupling is used between the stages immediately preceding the video detector, will the output signal from the detector have lost its dc reference?
6. Describe some of the symptoms of a faulty dc restorer circuit.

14 TV Picture Tubes

14.1 INTRODUCTION

A monochrome picture tube is a specialized form of cathode-ray tube. An electron gun in the tube directs a beam of electrons toward a fluorescent material on the screen, which glows when struck by the electrons. Between the gun and the screen are deflection coils which deflect the beam horizontally and vertically to form a raster. The brightness of the screen at any point depends upon the number (and velocity) of electrons striking that point. Therefore, the brightness of the picture may be controlled by varying the grid-bias voltage with respect to the cathode voltage. This bias can be changed by varying either the cathode voltage or the grid voltage.

A color-picture tube operates on the same basic principle, except that the screen is coated with different types of phosphors which produce colors when struck by electron beams. There are three basic or primary colors, which are used in combination to produce all the other desired colors. These primary colors are red, green, and blue.

In a three-gun color picture tube there is a separate gun for each of the color phosphors. Color phosphor dots (or strips) are located so close together on the screen of the picture tube that the eye cannot distinguish between them. When all three adjacent color phosphors are simultaneously illuminated by electron beams, the screen will radiate a white light. The various mixture colors are obtained by controlling the strength of the individual electron beams striking each color phosphor. The brightness (or saturation) of each color is controlled by varying the grid or cathode voltage of its respective electron gun. In order to insure that each electron beam strikes only one particular color phosphor dot or strip, a shadow mask, or aperture mask, is located near the screen. It is very important that each electron beam be carefully directed to pass through the desired hole in the mask and strike only the desired color phosphor.

This chapter covers the theory and operation of the picture tubes used in monochrome and color receivers. Also discussed are some of the more recent developments in color-picture tubes.

14.2 MONOCHROME TUBE THEORY

The formation of the electron beam starts naturally at the cathode. The emitting surface, composed of thoriated tungsten or barium and strontium oxides, is restricted to a small area in order that the emitted electrons progress only toward the fluorescent screen. They would serve no useful purpose in any other direction. The emitting material is thus deposited on the end of the nickel cathode cap that encloses the heater in the manner shown in Fig. 14-1 for a typical construction. The electrons, after emission, are drawn by the positive anode voltages into electric or magnetic lens systems. These form and focus the electrons into a sharp, narrow beam that finally impinges on the fluorescent screen in a small round point.

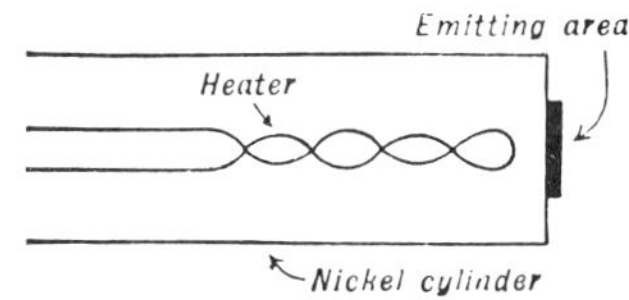

Fig. 14-1. Cathode and heater construction for a cathode-ray television tube.

The use of the word *lens* may puzzle the reader who thinks of this term only in connection with light rays, not electron beams. The purpose of a glass lens is to cause light rays either to diverge from or to converge to a point. The same results can be achieved electronically; hence the reason for the carry-over of the name.

The First-Lens System In the first lens we find the cathode, the control grid, and the first anode arranged in the manner shown in Fig. 14-2. The grid, it is noticed, is not the familiar mesh-wire arrangement found in ordinary tubes. For the present purpose it is a small hollow cylinder with only a small pinhole through which the electrons may pass. This restricts the araa of the cathode that is effective in providing electrons for the beam and aids in giving the beam sharpness. Following the grid cylinder is the first anode. Here, again, baffles permit only those electrons near the axis of the tube to pass through.

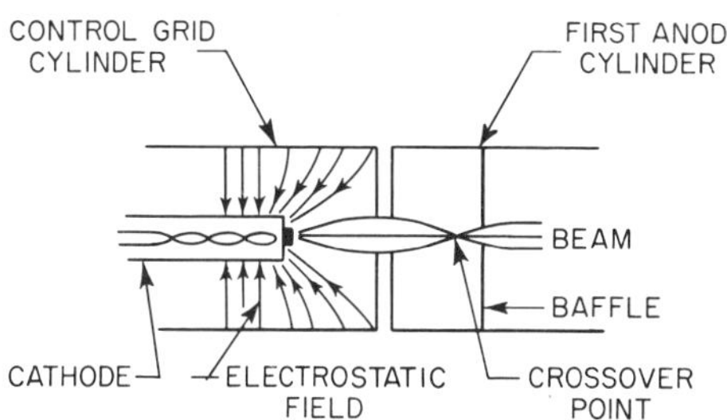

Fig. 14-2. The first lens system of a cathode-ray tube.

Because of the energy imparted to them by the heated cathode, the electrons leave the cathode surface with a small velocity. With no positive electric force (or field) to urge them forward, the electrons tend to congregate in the vacuum space just beyond the cathode and form a space charge. Eventually, just as many electrons will leave the heated cathode surface as are repelled by the negative space charge, and a state of equilibrium will exist. This condition can be overcome and a flow of electrons allowed to take place down the tube, if a high positive voltage is placed on the first anode.

The first anode, which is a hollow cylinder, does not have its electric field contained merely within itself; it also reaches into the surrounding regions. To be sure, the farther away from the anode, the weaker the strength of the field. With zero and low negative potentials on the control grid, the influence of the positive anode field extends through the baffle of the control grid right to the cathode surface. Electrons leaving this surface are urged on by the positive electric field and accelerated down the tube, with the baffle restricting their direction to very small angles with the axis of the tube.

Figure 14-2 illustrates the distribution of the electrostatic lines between the cathode and control grid. It is interesting to note that these lines are not straight, but tend to curve, the amount of curvature being influenced by the distance from the first anode and the control grid and by the voltages on these elements. Cathode-ray tube design engineers use such field distribution diagrams to determine the effect of each electrode on the electrons at the cathode and in the beam.

As a result of the bending of the electric field at the cathode, it can be proved by means of vectors that all electrons passing through the small hole in the control-grid baffle will come to a focus or converge toward a small area located just inside the first anode. This region is on the axis of the tube and is known as the *crossover point*. The effect of the electric field is such that electrons near the outer edges of the control-grid opening travel at an angle in order to get to the crossover point, whereas electrons on the axis of the lens move straight forward to this point. The direction of some of the electrons is shown in Fig. 14-2.

It is well to keep in mind that the shape of the electric field is determined by the placement of the electrodes and the voltages applied to them. The electrons are forced to converge toward the crossover point, because this point can more readily serve as the supply source of the beam electrons that the cathode from which they initially came. The area of the crossover point is more clearly defined than the relatively larger cathode surface, and it has been found that the electron beam is easier to focus if the crossover area is considered as the starting point rather that the cathode itself. The electrons that compose the final beam are then drawn from the crossover point while other electrons come from the cathode to take their place. The greater the number of electrons drawn from this point, the brighter the final image on the fluorescent screen.

For ordinary purposes, a negative bias is placed on the grid. In the larger cathode-ray tubes, the bias may rise as high as −60 volts. With a negative voltage on the control grid, the extent of the positive electric field is modified, and it no longer affects as large an area at the cathode surface as it did previously with zero grid volts. Now, only electrons located near the center of the cathode are subjected to the positive urging force, and the number of electrons arriving at the crossover point is correspondingly less. The intensity of the final electron beam likewise decreases. In the television receiver, the video signal is applied to the control grid, and the resulting variations in potential cause similar changes in electron-beam intensity.

For the beam arriving at the screen to remain in focus once the controls have been set, the position of the crossover point must remain fixed. With the normal variations of control-grid voltage, this condition is obtained. With large variations, however, the position of the crossover point tends to change, moving closer to the cathode as the grid becomes more negative. Thus, a certain amount of defocusing will take place. Proper design generally keeps this at a minimum, and for most of the voltage variations encountered in television work, defocusing is scarcely noticeable.

To summarize the purpose of the first-lens system: We see that electrons leaving the cathode surface are forced to converge to a small area near the anode. This offers a better point for the formation of the beam and its subsequent focusing.

The Second-Lens System The second-lens system draws electrons from the crossover point and brings them to a focus at the viewing screen. The system consists of the first and second anodes, as shown in Fig. 14-3. The second anode is operated at a higher potential than the first anode, is larger in diameter, and frequently overlaps the first anode to some extent. It is at the point of overlap of the two anodes that the second lens is effective, and it is here that the focusing action of the electron beam takes place. Electrons, when drawn from the crossover point established by the first lens system, are not all parallel to the axis of the tube. Some leave at various small angles. The beam thus tends to diverge, and it is due to the second lens that these diverging electrons alter their path and meet at another point on the axis. This second point is at the screen. Those electrons moving straight along the axis of the tube are not affected, in direction, by the focusing action of the second lens.

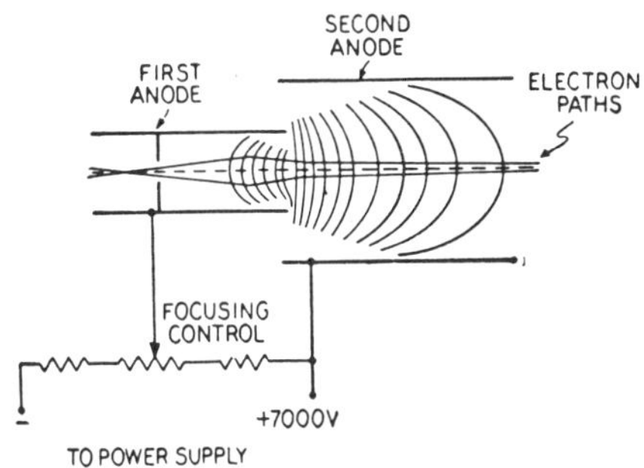

Fig. 14-3. The second lens system. The focusing of the electron beam at the viewing screen is accomplished by varying the voltage at the first anode.

The operation of the second lens depends upon the different potentials that are applied to the first and second anodes and the distribution of the resulting electric field. The equipotential lines for this lens are drawn in Fig. 14-3. It is to be noted that the curvature of these lines changes at the intersection of the two anodes. On the left-hand side, the electric-field lines are convex to the approaching electron beam, while to the right of the intersection the lines are concave. The effect of these oppositely shaped electric-field lines on the beam is likewise opposite. Since we have seen that some of the electrons tend to diverge after they leave the crossover point, the field distribution must be designed to overcome such a tendency. In action, the convex equipotential lines force the electrons to converge to a greater extent than the concave lines cause the electrons to diverge. Inasmuch as the convergence exceeds the divergence, the net result is a focusing of the electrons on the screen.

The ratio of the voltages, the size of the anode cylinders, and their relation to each other will determine the distribution and curvature of the electric lines of force; the latter, in turn, will determine the amount and the point at which the focusing takes place. In cathode-ray tubes, the ratio of the second to the second anode voltages ranges form 3 to 1 to more than 6 to 1.

In order that the electron beam leaving the crossover point shall not diverge too much, a baffle is placed just beyond this point, similar in construction to the baffle previously described for the control grid. The baffle again limits the width of the electron beam to the desired size. Practically, focusing control can be accomplished by varying the voltage on the first anode by an arrangement shown in Fig.14-3. This is one way of altering the voltage ratio between the first and second anodes and, with it, the distribution of the electric lines of force of the

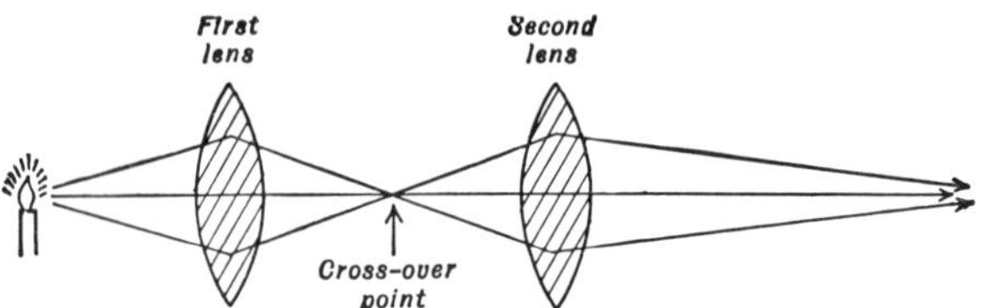

Fig. 14-4. The glass lenses used in focusing light rays illustrates the similarity between light-wave and electron-beam focusing.

lens system. An approximate optical analogy of the lens system is shown in Fig. 14-4 and may prove helpful in indicating the operation of the electric system.

Electrostatic Deflection Once past the second anode, the electron beam speeds toward the fluorescent screen. However, to present an image on the screen, a means of deflecting the beam is in order. This deflection may be achieved electrostatically by using two sets of plates, or electromagnetically by using two sets of coils. Currently, electromagnetic deflection is in use.

Electromagnetic Focusing It is well known that a wire carrying a current has a circular magnetic field set up around it, as shown in Fig. 14-5(A). Suppose the wire is placed in a magnetic field parallel to the magnetic lines of force (see Fig. 14-5(B)). There will be no interaction between the magnetic lines of the field and those set up by the wire. Why? Because the two *fields* are at right angles to each other.

For the opposite case, illustrated in Fig. 14-5(C), the current-carrying wire is placed at right angles to the field lines of magnetic force. Above the wire the lines of both fields add; underneath the wire they oppose and tend to cancel each other. Experiment indicates that a resulting force will act on the wire in such a way that it moves from the stronger part of the magnetic field to the weaker part. This is indicated in the figure. The illustration represents the two extreme angles that the wire and the field can make with each other. Intermediate positions (those between zero and 90 degrees) will cause intermediate values of force to act on the wire.

The transition from a wire carrying electrons to the electrons themselves without the wire is quite simply made. With only electrons moving through space, the same circular magnetic field is set up about their path. From the preceding discussion we know that electrons traveling parallel to the lines of force of an additional magnetic field experience no reaction from this field. On the other hand, if they enter the magnetic field at an angle to the flux lines, a force will be brought to bear on them and their path will be altered.

It is well to reiterate that for an electron to react with a magnetic field: (1) The electron must be moving, otherwise it does not generate a magnetic field; and (2) the moving electron must make an angle with the magnetic field in which it is traveling.

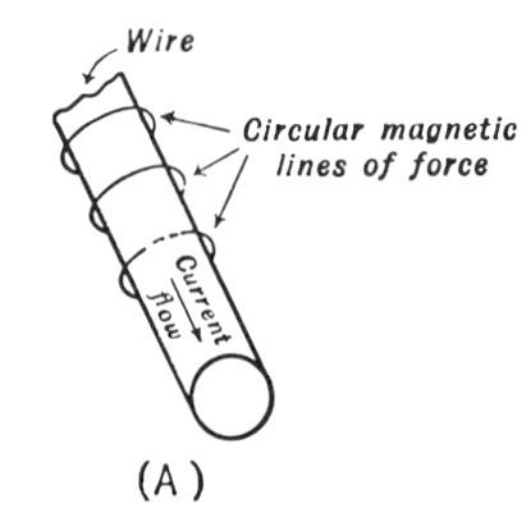

(A)

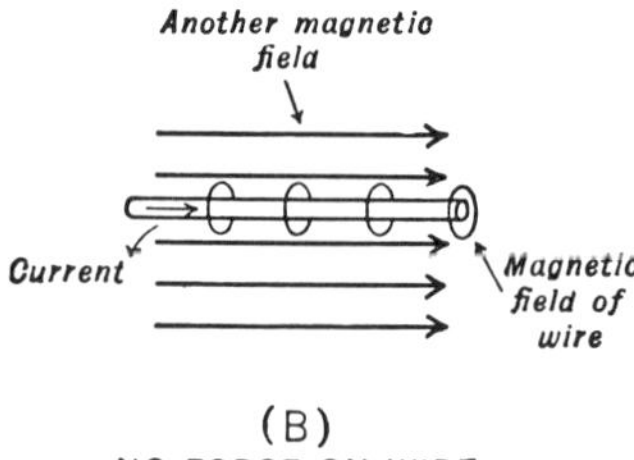

(B)
NO FORCE ON WIRE

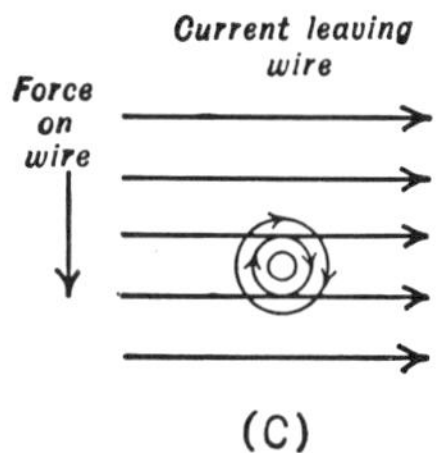

(C)

Fig. 14-5. The action of a wire-carrying current when placed in a magnetic field. No force is on the wire.

Now let us apply these considerations to magnetic focusing. The focusing coil is slipped over the neck of the cathode-ray tube and placed beyond the first anode. The first-lens system remains essentially as described; it still converges electrons to the crossover point. From this point, the electrons spread out and the focusing action of the coil begins to function. The focus coil, then, represents the second lens in electromagnetic tubes. An accelerating anode (anode number 2) is positioned close to anode number 1 in order to accelerate the beam down the tube and also to provide a means of removing ions from the beam. The accelerating anode is connected internally to the aquadag coating (Fig. 14-6) and receives its voltage from the coating. The latter, in turn, obtains its potential via a metal cavity or a ball-insert terminal located on the side of the glass bulb.

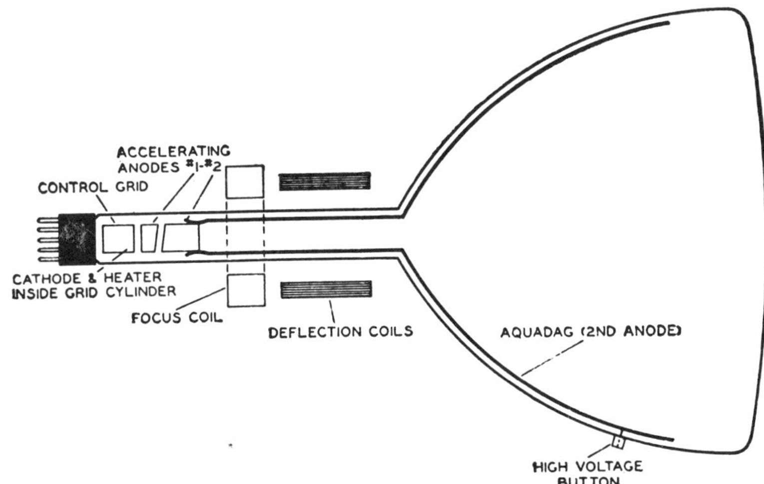

Fig. 14-6. The internal construction of an electromagnetic deflection-and focus-tube.

The field of the focusing coil is parallel to the axis of the tube and is generated by direct current flowing through the coil. So long as the electrons leave the crossover area and travel down the tube along the axis, the magnetic lines do not interfere with their motion. However, many electrons tend to spread out beyond the crossover region, and it is on these electrons that the magnetic force reacts because they are moving at a small angle to the magnetic flux lines.

The path taken by electrons that are acted on by a magnetic field can be more easily understood if it is recalled that the resulting force on the electrons is at right angles to both its motion and the magnetic field. This force, as seen in Fig. 14-7(A), causes the electrons to move in a circular path. In this way the force on the electrons, the electron motion, and the magnetic force are always at right angles to each other.

Apply these ideas to the action inside the cathode-ray tube. As the

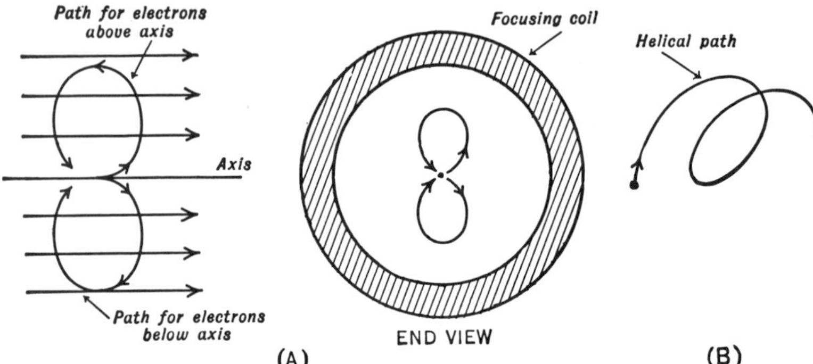

Fig. 14-7. The effect of magnetic lines of force and electrical force upon electrons.

electrons leave the crossover point at small angles to the magnetic field, they are subjected to a force that tends to make them turn in a circle. But at the same time that they are being forced to travel this circular path, they are also speeding forward. The resulting motion of the electrons is known as *helical* and is similar to the action of a screw being turned into a piece of wood. It rotates while moving forward. Figure 14-7(B) may aid the reader to visualize the motion.

The electrons that are acted on by the magnetic field all come from the crossover point that is situated on the axis of the tube. The instant they leave this point at an angle, the magnetic force starts to act, forcing them to move in a circular path back to the axis again. In the cathode-ray tube they are, at the same time, also moving forward; hence when the circular path is completed, the electrons will again be on the axis to the tube some distance away. The exact position down the tube where the electrons return to the axis is dependent upon the strength of the magnetic force and the forward velocity.

By suitable variation of the intensity of the magnetic field, it is possible to have the electrons return to the axis of the tube exactly at the screen. The beam is now focused. The greater the speed of the electrons, the stronger the magnetic field required. Thus, any changes that affect the velocity of the electrons, for example, varying the first anode voltage, will also require readjustment of the current through the focusing coil.

At other values of the magnetic field, defocusing occurs. As an exception to this statement, it should be mentioned that by continually increasing the strength of the magnetic field, the electrons can be made to do two(or more) complete revolutions before striking the screen. As each complete revolution brings the beam to the screen, a focused spot will appear. This process can be continued as long as the magnetic coil will carry current.

It would appear from the preceding discussion that the magnetic field must extend all along the tube in order that the electrons will always be under its influence. Their path will then be helical, as described. However, for practical applications, only a small iron-core coil is slipped over the neck of the tube. The electron beam is thus subjected to the magnetic force for only a short time. During this period, it is given enough of a twist so that it will move toward the axis; the forward motion then keeps it traveling along this path. The motion now is not truly helical, but the end result is satisfactory.

It will occur to the reader that magnetic fields need not necessarily be obtained from coils only. Permanent magnets (PM) are also suitable and these are extensively used. One type of PM focus magnet is shown in Fig. 14-8. The unit consists of three (sometimes four) small bar magnets which are placed along and around the axis of the tube neck. The magnets are equally spaced and held in position by two disc-like pole pieces made of a low-carbon steel. The magnets are mounted with similar poles at the same end. Thus, the flux lines joining the ends of the magnets pass through the tube along its axis. Focusing is accomplished by the interaction of these flux lines with the electron beam passing down through the tube.

Flux variation through the tube is achieved by a sliding steel collar controlled by a lead screw at the end of a flexible shaft. By rotating the knurled screw at the end of the shaft, either clockwise or counterclockwise, the steel collar can be moved back and forth. The end of the screw drive extends beyond the back cover of the receiver, permitting focus adjustments to be made without the necessity of removing the back cover.

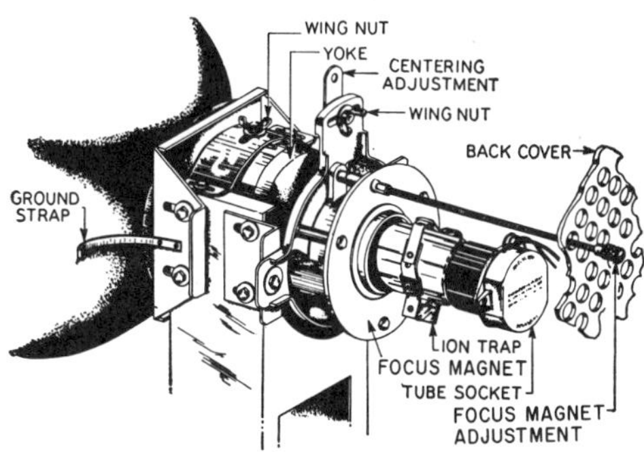

Fig. 14-8. A PM focusing unit using a movable collar for focus adjustment.

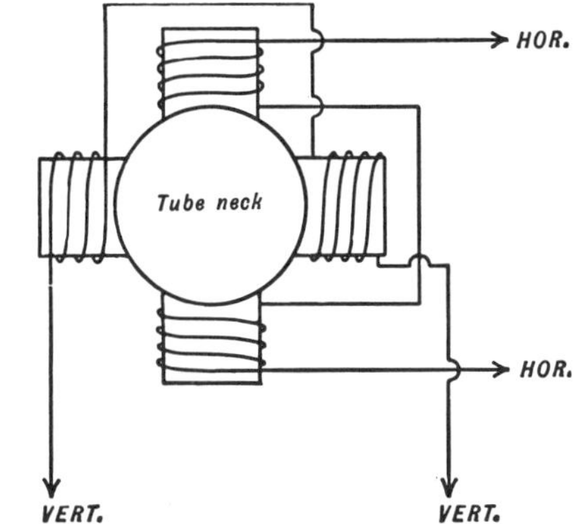

Fig. 14-9. Arrangement of coils for electromagnetic deflection.

Electromagnetic Deflection Actually, little new need now be added to understand the action of the deflection coils on the electron beam. Two sets of coils are placed at right angles to each other and mounted on the section of the tube neck where the electron beam leaves the focusing electrode and travels toward the screen. There are four coils in all (two in each set), with opposite coils comprising one set. These are connected in a series in order to obtain the proper polarity (see Fig. 14-9).

Figure 14-10 shows the actual physical placement of the deflection coils. For horizontal deflection, the coils are vertically placed, whereas, for vertical deflection, the coils are horizontally mounted. This reverse placement of the coils is due to the fact that the force on traveling electrons in a magnetic field is at right angles to both the direction of the motion and the lines of the field. After the coils have been oriented, sawtooth-shaped current variations are sent through them. The magnetic-field flux follows these current changes and causes the electron beam to move back and forth (or up and down) across the screen, sweeping out the desired pattern.

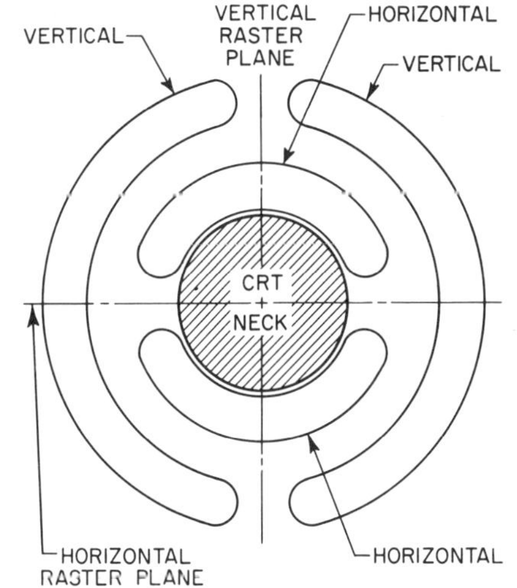

Fig. 14-10. The actual physical placement of the deflection coils about the neck of the picture tube.

(A)

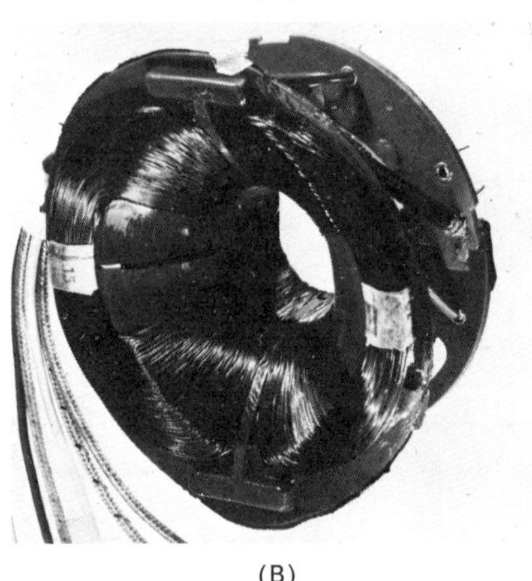

(B)

Fig. 14-11. Two typical deflection yokes. (A) A 70-degree deflection unit is shown. (B) A 110-degree unit is shown. Note how the 110-degree yoke windings flare out.

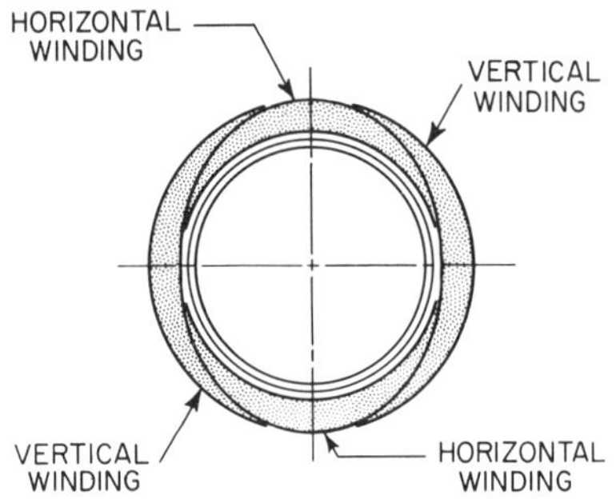

Fig. 14-12. The appearance of cosine windings in a deflection yoke. Note the uneven thickness of each winding.

The entire assembly of deflection coils is known as a "deflection yoke." Two typical commercial units are shown in Fig. 14-11. Note how the forward windings lap over the front edge of the yoke housing. The yoke is thus positioned right up against the flare of the tube in order to achieve the complete coverage of the full screen area. This is particularly important for wide-angle tubes (110 degrees or more).

The deflection windings in Fig. 14-10 are shown wound uniformly; that is, there is no variation in the winding thickness from end to end. This type of winding was characteristic of the yokes employed when narrow-angle picture tubes were prevalent. As the deflection angle increased, it was found that the magnetic field produced was not uniform, particularly when the beam was deflected toward the edges of the raster. Visually, an elongated spot was produced, tending to develop an out-of-focus condition.

A more uniform field is developed when a cosine-type winding is employed. In this arrangement (Fig. 14-12) the thickness of a deflection winding varies as the cosine of the angle from a central reference line varies. For horizontal windings, the reference line is the horizontal line through the center of the yoke. For vertical windings, it is the vertical line. Nearly all present-day yokes are wound in this manner, or in a cosine-squared fashion.

The reader should not become confused by the seemingly different actions of the focusing and deflection coils. At first glance it might appear that one coil (the focusing coil) twists the electron beam around so that it ends up at the screen in focus, while the other coils (the deflecting coils) only cause the beam to move either to the right or left or up and down. Actually the action of all the coils is the same; the only difference lies in the manner in which they affect the beam. At the focusing coil, the magnetic lines of flux are parallel to the axis of the tube, and the electrons that are moving away from the axis of the tube are subjected to a strong twisting force that turns them back to the axis. Their forward motion, given to them by the positive first anode and sometimes by an intensifier ring, keeps them moving toward the screen.

At the deflection coils, the magnetic fields are at *right angles* to the path of the beam. The beam, in moving through these fields, has a force applied which is at right angles to the forward motion of the electrons and the direction of the magnetic lines of force. Here the effect of the field is not as great as at the focusing coil, and the beam is merely deflected rather than bent all the way around into a circular path. The influence of the field ends when the electrons pass the yoke, but any sideward or up-and-down motion imparted to the electrons while in the field is retained. By varying the direction of the flow of current through the vertical and horizontal-deflection coils, it is possible to reach all points on the screen. This type of deflection is used with all present-day television picture tubes.

When a yoke is inserted over the neck of the picture tube, it is very easy to position it so that the image is not properly oriented. This is indicated in Fig. 14-13. In this case, correction may be accomplished by rotating the yoke until the image is again properly positioned.

Fig. 14-13. Image incorrectly positioned because of the improper placement of the deflection yoke.

Electromagnetic Deflection with Electrostatic Focus Electrostatic focusing is particularly suited to wide-angle tubes, because the beam produced by the gun is smaller in diameter and there is less defocusing when the beam is swung to the edges of the raster. Also, this form of focusing is useful to short guns by eliminating the external focus magnet.

A cross-sectional view of an electrostatic focus gun is shown in Fig. 14-14(A). The labeling of the various elements in the gun assembly as consecutive grids is done frequently to simplify identification. Thus, the control grid is grid number 1. The first anode following this is grid number 2. The second anode is grid number 3, etc. Note that this change in name does not in any way alter the function or construction of the electrodes. Grid number 3 serves the same purpose as the accelerating anode in Fig. 14-6. It contacts and operates at the high potential of the aquadag coating.

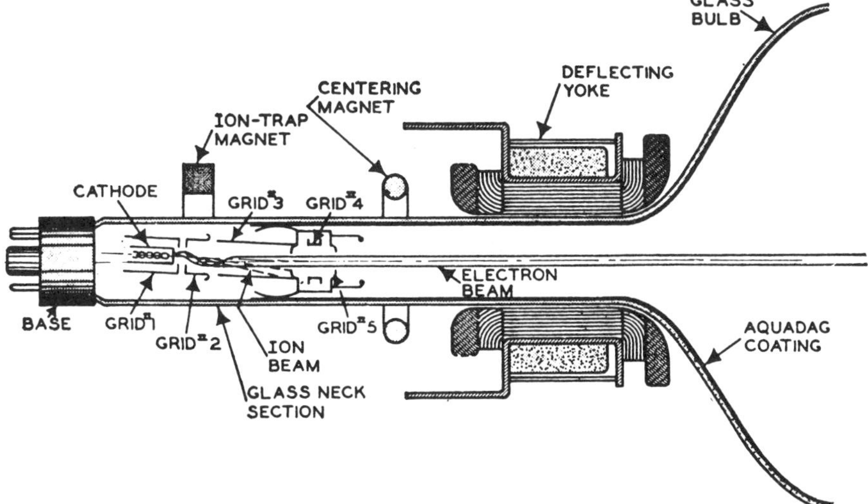

Fig. 14-14(A). Internal structure of an electrostatic focus tube. The deflection is accomplished magnetically.

The new elements, grids number 4 and number 5, provide the focusing field which directs the electron beam. The voltage applied to grid number 4 is lower than that which grids numbers 3 or 5 receive, and it is frequently made variable to permit the adjustment of the focus voltage to the proper value. Grid number 5 (which structurally surrounds grid number 4) is connected internally to grid number 3 and operates at the same potential as number 3.

The voltage applied to grid number 4 depends upon the manner in which grids numbers 3, 4, and 5 are constructed. The first electrostatic focus tubes manufactured required that the potential of grid number 4 be on the order of 20 percent of the accelerating (or second-anode) voltage. This meant that voltages between 2,000 and 3,000 volts had to be made available. A special potentiometer, inserted in this circuit, permitted the adjustment of this voltage for the sharpest picture focus.

In subsequent designs it was found that, by constructing grids numbers 3, 4, and 5 to closer tolerances, the necessary focusing action could be obtained with voltages on the order of 300 to 400 volts. These latter values can be obtained directly from the low-voltage or boost power supply, thereby avoiding the special circuit required when several thousand volts are needed.

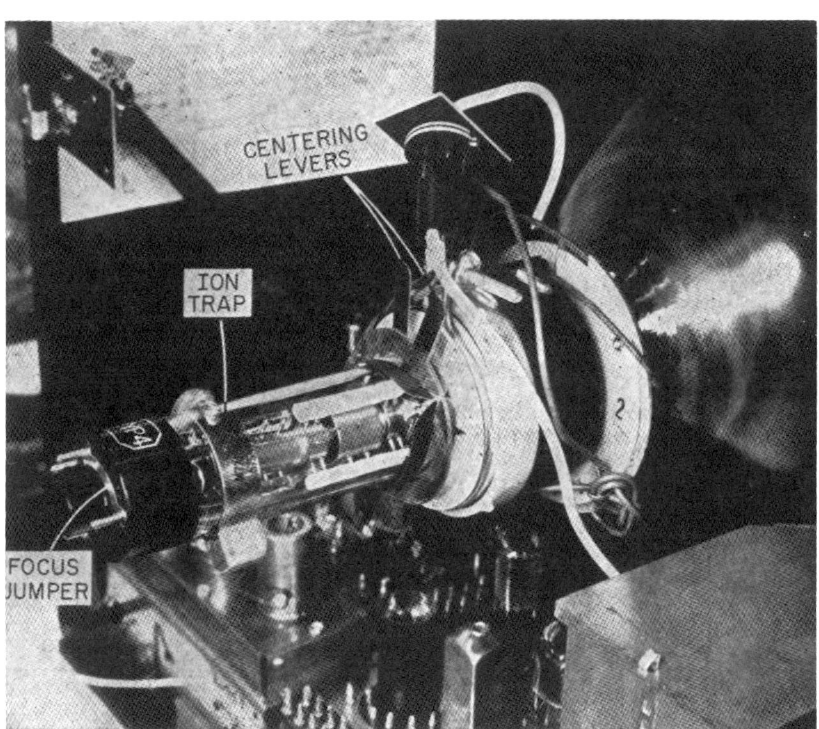

Fig. 14-14(B). Close-up view showing the centering device for electrostatic focus tubes. The ion trap is also evident at the left.

Since the focus magnet is dispensed with on electrostatic focus tubes, a new centering magnet is provided to center the picture on the screen (see Fig. 14-14(B)). The magnet assembly is in the form of two rings mounted on a nonmagnetic form which is placed around the neck of the picture tube and at a distance of about 3/8 inch back of the deflection yoke.

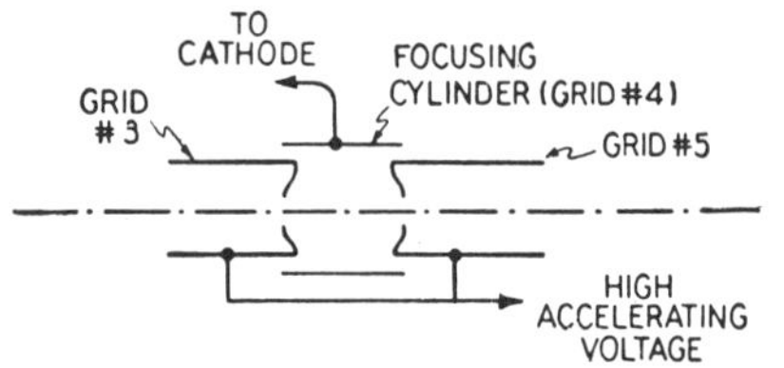

Fig. 14-15. The modified arrangements of grids numbers 3, 4, and 5, in the zero-focus tube.

It is also possible, by modifying the structure of grids numbers 3, 4, and 5 to the form shown in Fig. 14-15, to obtain the proper focusing action with a zero potential on grid number 4. Now, no external voltage or focusing potentiometer need by used.

14.3 PROBLEMS IN OBTAINING BRIGHTNESS AND CONTRAST

The principal objective in the design of a cathode-ray tube is the production of an image having good brightness and high contrast. When the electron beam strikes the back side of the fluorescent screen, the light which is emitted distributes itself in the following approximate manner:

50 percent of the light travels back into the tube.

20 percent of the light is lost in the glass of the tube by internal reflection.

30 percent of the light reaches the observer.

Thus, of all the light produced by the electron beam (and this, itself, is a highly inefficient process), only 30 percent reaches the observer.

Image contrast is impaired because of the interference caused by light which is returned to the screen after it has been reflected from some other points. Some of these sources of interference are given here in the order of their importance:

(1) Halation
(2) Reflections due to the curvature of the screen
(3) Reflections at the surface of the screen face
(4) Reflections from inside the tube

Halation If we take a cathode-ray tube and minutely examine the light pattern produced by a stationary electron beam, we find that the visible spot is surrounded by rings of light. These rings of light are due to a phenomenon known as *halation* (see Fig. 14-16). The light rays which leave the fluorescent crystals at the inner surface of the tube face travel into the glass and are refracted. Those rays which make an angle greater than θ do not leave the glass when they reach the outer surface, but instead they are totally reflected back into the glass. At each point where these reflected rays strike the fluorescent crystals they scatter, and it is this scattering of the rays that produces visible rings on the screen. These rings cause a hazy glow in the region surrounding the beam spot and reduce the maximum possible detail contrast. Contrast, it will be recalled, is the ratio of the brightness of two points, one of which is being bombarded by the electron beam, the other of which is under cutoff conditions. It is desirable to have this ratio as high as possible in order to achieve "rich-looking" or high-quality images. Due to the scattering of the light, however, areas which should be in total darkness receive some light, and the result is a reduction in the contrast ratio. A distinction is usually made between the detail-contrast ratio and the overall field contrast. The field-contrast ratio compares two sections of the screen which are widely removed from each other. Halation affects only the detail contrast.

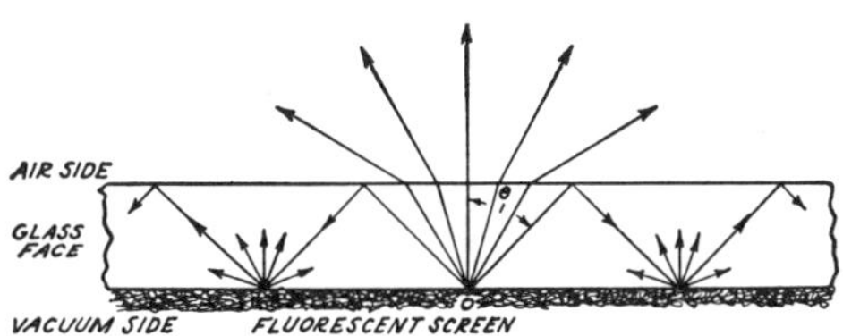

Fig. 14-16. Reflections between the two sides of the glass can cause halation.

Reflections due to the Curvature of the Screen Reflection arising from the curvature of the screen, as shown in Fig. 14-17, causes loss

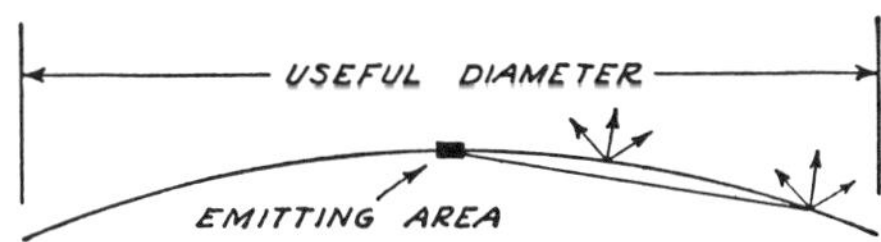

Fig. 14-17. Diffusion effects in non-flat screen.

in contrast. The remedy is the use of a flat screen. Much progress has been made in this direction, since the screen curvature greatly restricts the useful image area.

Reflections at the Surface of the Screen Face Light rays, when traveling from one medium to another, always lose a certain amount of energy at the intersection of the two media. At the cathode-ray tube screen, some light is reflected when it reaches the dividing surface between the air and the glass of the tube. The reflected light travels back to the inner surface and then back to the outer surface again. At each dividing surface, some of the light continues onward and some is reflected back into the glass. Absorption and dispersion quickly reduce the strength of these rebounding rays.

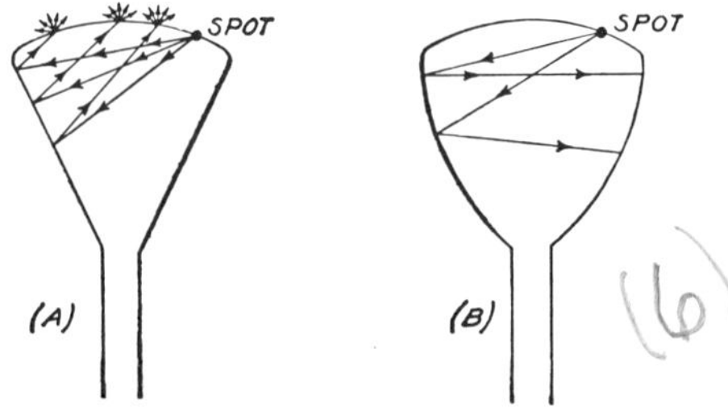

Fig. 14-18. The shape of the tube can reduce the internal reflections.

Reflections from inside the Tube In Fig. 14-18 we see how reflections from the inside surfaces of the tube can act to decrease the field contrast of the image. The loss of contrast from this source of interference can be made quite low by a special shaping of the walls of the bulb, as seen in Fig. 14-18, and the use of the black aquadag coating. The aquadag coating is also useful for electrical purposes; it acts as a shield and a path for the return of the secondary electrons emitted from the fluorescent screen. Secondary electrons must be emitted by the screen, otherwise the negative charge accumulation on the screen would soon become great enough to prevent the electron beam from reaching it.

One step taken toward improving screen brightness and contrast has been the addition of an extremely thin film of aluminium on the back of the fluorescent screen. The film is sufficiently thin to permit the electrons in the scanning beam to reach the fluorescent crystals. It will prevent, however, any of the light which is generated by the screen crystals from traveling back into the tube. This can be seen in Fig. 14-19.

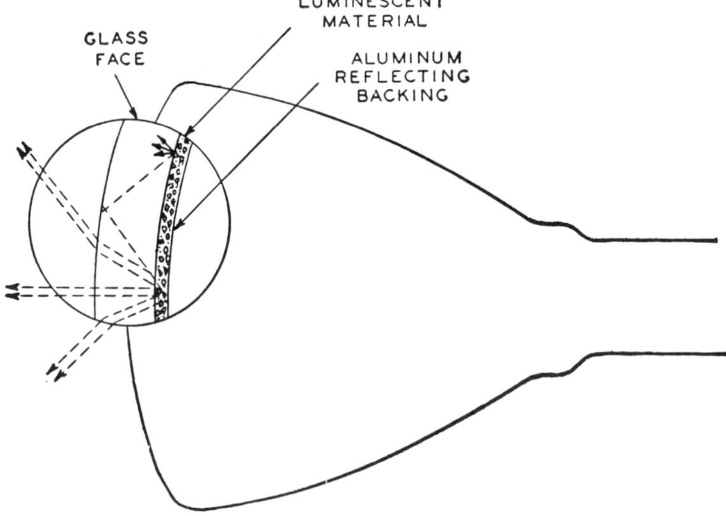

Fig. 14-19. An aluminum backing over the fluorescent screen prevents light from traveling back into the tube.

The light which previously went back into the tube is now reflected toward the observer. This is one improvement. In addition, the overall field contrast is improved as much as ten times. However, the detail contrast is not noticeably affected, since it is governed primarily by halation, and the addition of the aluminium layer does not affect this condition. An additional purpose which the aluminum film serves is to prevent undesirable effects due to poor secondary emission from the screen. It also greatly increases the range of substances which can be used as screen phosphors.

14.4 ION SPOTS

Another matter of considerable importance is the elimination of the ion spot in tubes using electromagnetic deflection. No matter how carefully a tube is degassed or how well a cathode-coating is applied, it will be found that ions are present in the electron beam. These ions are either gas molecules which have acquired an electron or else molecules of the outside coating material of the cathode. These ions possess the same charge as the electrons and are sensitive to the same accelerating voltages. In tubes employing electrostatic deflection, the ions and the electrons are similarly deflected and for all practical purposes may be considered as one. However, when electromagnetic deflection is employed, it will be found that these heavier ions are hardly deflected. As a result, they strike the center of the screen in a steady stream and, in time, deactivate the fluorescent material in this area. When the electrons in the scanning beam subsequently pass over this section of the screen, no light is emitted. To the observer this section appears as a dark patch.

Bent-Gun Ion Trap Several means are used to prevent the ions from reaching the screen. First there is the bent electron gun, shown in Fig. 14-20. The cathode is inclined at an angle to the rest of the gun structure, and both ions and electrons would, if permitted to travel in a straight line, impinge on the side of the electron gun and never reach the screen. If a strong magnetic field is placed in the path of the particles, it is possible to alter the paths of the electrons sufficiently so that they travel toward the screen. The heavier ions, however, are not deflected enough, and as a result they hit the side of the electron gun. The magnetic field which

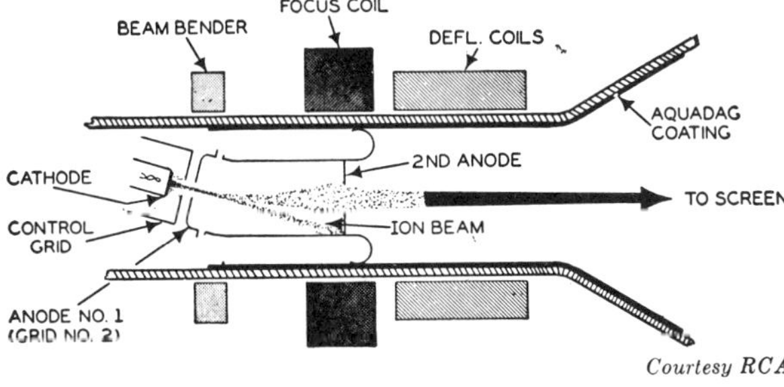

Fig. 14-20. A bent-gun ion trap.

causes this separation of ions and electrons is obtained from a small coil or a permanent magnet placed on the outside of the neck of the tube, above the cathode. The ion-trap magnet (or " beam bender," as it is sometimes called) is clamped onto the neck of the tube in the position shown in Fig. 14-20.

Diagonal-Cut Ion Trap Another approach to the prevention of ion spots is the diagonal-cut ion trap (see Fig. 14-21). The electrons and ions are emitted by the electron gun and are accelerated forward. The first and second anodes are so designed that the gap between then is oblique. The first anode has a low positive voltage; the second anode has a high positive voltage. The electrons, as they leave the cathode, are attracted forward by the first anode. However, the oblique gap between the first and second anode causes the electric field here to become warped, and the electrons and ions crossing the gap are bent in toward the second anode. With no other forces applied, the electrons and ions will strike the second anode and will be prevented from reaching the screen.

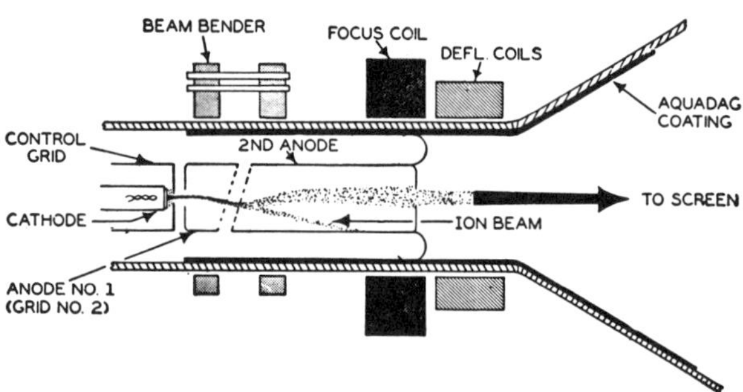

Fig. 14-21. A diagonal-cut (or slash-field) ion trap.

However, if a magnetic field is introduced at right angles to the electrode, the electrons receive a counterforce deflecting them upward and permitting them to continue through the gun. The ions, because of their greater mass and because the magnetic field scarcely deflects them, strike the second anode and are removed from the beam path.

Metal-Backed Screen Tubes The third method of preventing ions from reaching the screen uses the aluminium layer mentioned previously. The depth of penetration of any particle is governed by the relationship,

$$\textit{Depth of penetration} = \frac{K(V_e)}{m}, \text{ where} \qquad 14\text{-}1$$

K = Constant
V_e = Energy of particle
m = Mass of particle

Since an ion has considerably more mass than an electron, its depth of penetration is less. By properly proportioning the thickness of the metallic screen, the ions are exluded but the electrons in the beam are able to pass through.

The trend has been to aluminize all screens and, in most instances, eliminate the ion trap. This tendency has been accelerated particularly by the use of short guns.

Rectangular Screens It had been recognized for many years that a rectangular image on a circular screen is wasteful not only of screen area, but of cabinet space as well. The sensible solution was a rectangular screen dimensioned in the standard 4:3 ratio of the transmitted image. At first, manufacturing difficulties and obstacles presented by the rectangular shape of the tube prevented mass production, but in time these were overcome and rectangular tubes are now used almost exclusively.

A rectangular tube with a 21-inch screen is illustrated Fig. 14-22(A). This has a fairly long neck and a conventional socket base. A 110 degree picture tube possessing a shortened neck and a modified plastic base is pictured in Fig. 14-22(B). With the deflection yoke in place, as shown, not much room remains for the additional components.

Tube Shields In most TV receivers, a plate-glass shield is mounted close and in front of the picture tube as a safeguard to the viewers against possible serious injury should the tube implode. This shield, however, is not always easily removable, making it difficult to wipe dust or fog from either the tube face or the inner shield surface. There is also a loss of contrast stemming from light reflections between the shield and the tube.

The separate safety shield has been replaced by a flat sheet of glass possessing the same contour as the external surface of the tube screen and permanently bonded to this screen by an epoxy resin (see Fig. 14-23) The result is a virtually implosion-proof tube that is easy to keep clean, produces a brighter, sharper image, and permits a picture shaped more

(A)

(B)

Fig. 14-22. (A) A rectangular picture tube with a 21-inch screen. The deflection angle is 90-degrees. (B) A 110-degree picture tube possessing a shortened neck and a modified plastic base. Note the deflection yoke on the neck.

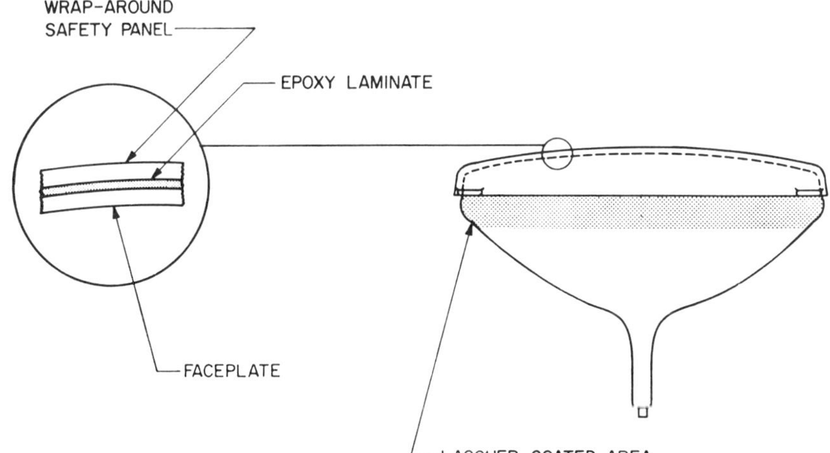

Fig. 14-23. Cross section of tube possessing a bonded shield. Lacquer band protects critical area of the bulb behind the wrap-around panel.

nearly rectangular. Finally, the use of this technique has made it possible to design a bulb having a flatter face, which reduces distortion at wide viewing angles.

14.5 TRI-GUN COLOR PICTURE TUBES AND ASSOCIATED CIRCUITS

Most of the color receivers that have been marketed have employed a tri-gun, tri-color picture tube. The tri-gun portion of the name indicates that the tube possesses three electron guns. The conventional black-and-white employs only one electron gun.

The second half of the name, tri-color, reveals that the screen of the tube possesses three different color-emitting phosphors. This, of course, is basic to the entire color television system which employs the three primary colors—red, green, and blue—to synthesize the wide range of hues and tints required for the satisfactory presentation of a color picture.

The tie-in between the three electron guns and the three different types of screen phosphors now becomes evident. Each gun is concerned with one type of phosphor. Thus, one of the electron guns develops an electron beam which strikes, say, the phosphor which emits red light. This gun is labeled the "red gun." A second electron gun directs its beam only at the green phosphor dots and it is the "green gun." The third gun is concerned in similar manner only with the blue dots. In each case, it is not the color of the phosphor referred to, but the light which this phosphor gives off when struck by an electron beam. The actual color of the substance and the color of the phosphorescent light it emits do not necessarily bear any relationship to each other.

The overall color that is seen on the screen is determined by two general factors: (1) the phosphors which are being bombarded by the three guns, and (2) the number of electrons which are contained in

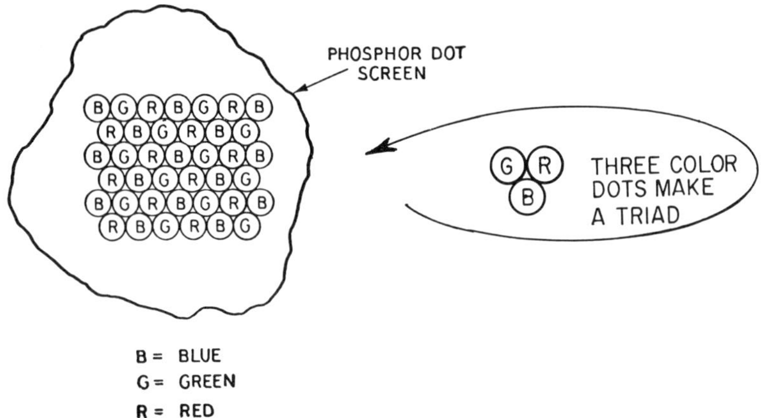

Fig. 14-24. The phosphor dot screen of a tri-gun, color-picture tube.

each beam. Thus, suppose you turn one beam off completely—say, the beam from the red gun. Then, only the blue and green dots will be emitting light, and what you see is a mixture of blue and green light which can range from a greenish blue to a bluish green. This color is called " cyan." The exact color is determined by which of the two beams is the stronger.

In the same way, we could cut off the green gun, leaving only the red and blue guns in operation. Now the screen color would fall somewhere in the purplish range and this color is known as "magenta." If the blue gun were stronger than the red gun, the color would appear closer to blue, say bluish purple. On the other hand, if the red beam were made more intense, the resultant color would be nearer a purplish red.

It is, of course, not necessary to turn any gun off. All three may be operating simultaneously, and when they do, you generally see the lighter or pastel shades on the screen. This is because red, green, and blue combine in some measure to form white, and although white may not be predominant, it will mix with whatever colors are present and serve to lighten, or *desaturate* them.

The phosphorescent dots which produce the colored light are arranged on the screen in any orderly array of small triangular groups, each group containing a green-emitting dot, a red-emitting dot, and a blue-emitting dot. The combination of the three dots is called a triad (see Fig. 14-24). The actual number of such dots, for a 21-inch screen, like the one in Fig. 14-25, is somewhere in the neighborhood of 1,071,000.

With a 1,071,000 dots on the screen, there are 357,000 triads. Each dot has a diameter of approximately 16 mils. If all three dots in a group are bombarded at the same time, the combined red-, green-, and blue-light output will present one mixed color to the observer's eyes.

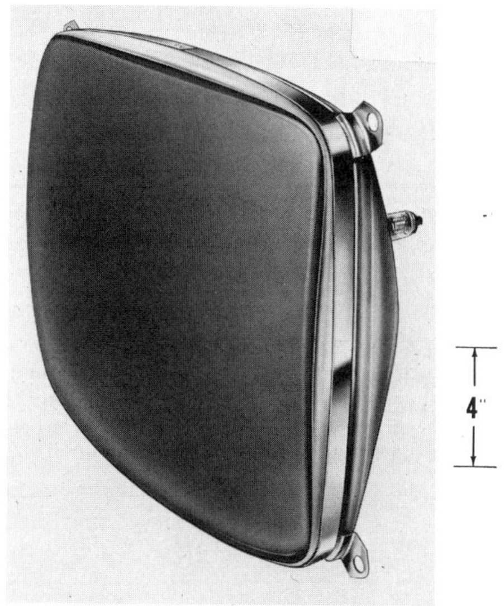

Fig. 14-25. A 23-inch, color-picture tube. (*Courtesy of RCA.*)

Electron-Gun Structure At the other end of the color picture tube there are three parallel, closely-spaced electron guns which produce three independent electron beams (see Fig. 14-26). Each gun consists of a heater, a cathode, a control grid (grid number 1), an accelerating (or screen) grid (grid number 2), a focusing electrode (grid number 3), and a converging electrode (grid number 4). The heaters of all three guns are in parallel and require only two external connections to the tube base. Each control grid has its own base pin, and the same is true of each screen electrode. The focusing electrodes (grid number 3) of all the guns are electrically connected, because one overall voltage variation will bring all three beams to a focus at the phosphor dot screen.

The final electrode in the gun structure is grid number 4, the converging grid. This is a cylinder of a small diameter, which is internally connected to (and operated at the same high potential as) the aquadag coating (20,000 to 30,000 volts). Also associated with each number 4 grid is a pair of pole pieces. These are mounted above each grid. External coils on the neck of the tube induce magnetic fields in each set of pole pieces,

Fig. 14-26. Electron-gun assembly of a 21-inch color-picture tube. Three electron-gun structures are employed, although only two are clearly visible.

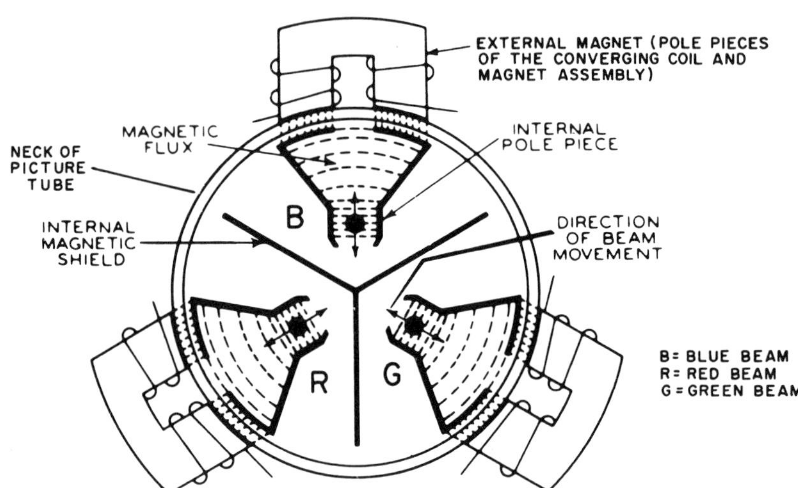

Fig. 14-27. External coils mounted on the neck of the picture tube induce magnetic fields in each set of pole pieces. These fields force the three beams to converge, so that each beam will strike the proper phosphor dot in each triad.

as shown in Fig. 14-27. These fields force the three beams to converge, so that each beam will strike the proper phosphor dot of a triad at any one instant of time; that is, one beam will strike the red dot, a second beam will strike the green dot, and the third beam will hit the blue dot, all three dots being in the same triad. The three dots are bunched so close together that the light they produce combines and appears to the eye as a single color. In the absence of the proper converging action, it is possible for the beams to hit phosphor dots at sufficiently separated points so that an observer sees three individual points of light. Under these conditions, proper mixing of colors to obtain different hues is not possible.

The Shadow Mask Proper beam convergence is an important aspect of tri-gun picture-tube operation. Thus, to insure that each beam strikes only one type of phosphor dot, a mask, called a *shadow* or *aperture mask*, is inserted between the electron guns and the phosphor dot screen (see Fig. 14-28). The mask is positioned in front of and parallel to the screen. It contains circular holes, equal in number to the dot triads. Each hole is so aligned with respect to its group that any one of the approaching beams can "see" and therefore strike only one phosphor dot. The remaining two dots of the triad are hidden by the mask; that is, the two other dots are in the "shadow" of the mask opening—hence the name of shadow mask.

What is true for one beam is true for the other two beams. Each can also see one phosphor dot. In this way, it is possible to minimize color contamination which occurs when a beam either hits the wrong dot or overlaps several dots at the same time.

Static and Dynamic Convergence In the foregoing discussion, beam convergence was covered in a general manner. Actually, there are two types of convergence: *static convergence* and *dynamic*

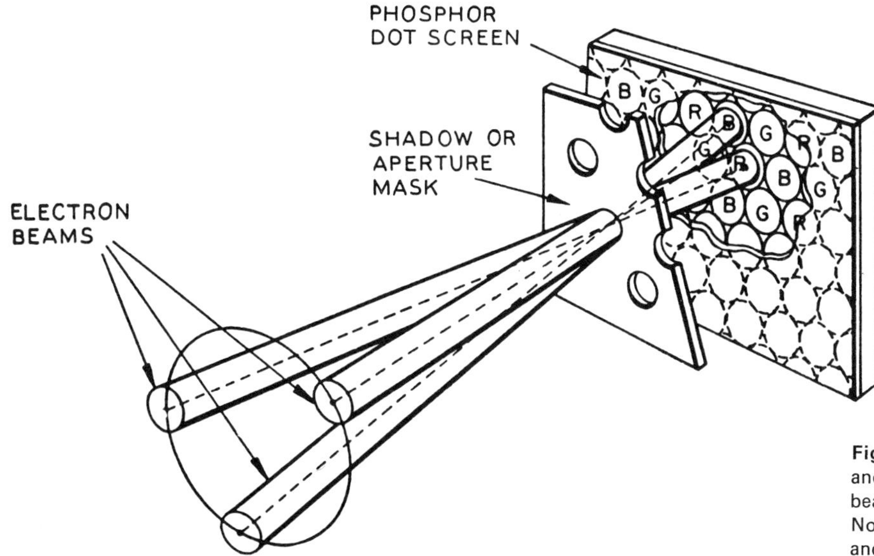

Fig. 14-28. Diagrammatic illustration of the mask and screen, showing the convergence of the three beams at a single hole in the shadow (aperture) mask. Note that the converged beams pass through the hole and strike their respective phosphor dots.

convergence. In static convergence the positions of the beams are adjusted by using either fixed dc voltages or fixed magnetic fields. As a further aid in this action, the electron guns are tilted inward slightly. If the adjustments are made carefully, the beams will converge properly over the central area of the screen.

To maintain this converged condition of the beams as they swing away from the center, it is also necessary to vary their relative angles slightly. This process of changing the beam angle so that it will be in step with the scanning is referred to as dynamic convergence. It is required because the distance traveled by the beams increases as they swing away from the center of the screen. The swing away from the center, in turn, occurs because the curvature of the screen is not perfectly spherical, and beams which are converged at the screen center will tend to converge in front of the shadow mask at points away from the center (see Fig. 14-29).

A moment's reflection will reveal that the extent of convergence changes, the farther the beams are from the center of the screen. Furthermore, there is a direct relationship between the convergence needed at any one point and the instantaneous horizontal-and vertical-deflection current values. Thus it is possible to obtain whatever correction currents are needed from the vertical and horizontal deflection systems. These additional currents are known as dynamic convergence currents to distinguish them from the dc or static convergence which is made over the central area. Where dynamic magnetic convergence means are employed, the static adjustment is usually made with permanent magnets. The dynamic convergence is then achieved by introducing varying magnetic fields via convergence coils mounted on the neck of the picture tube.

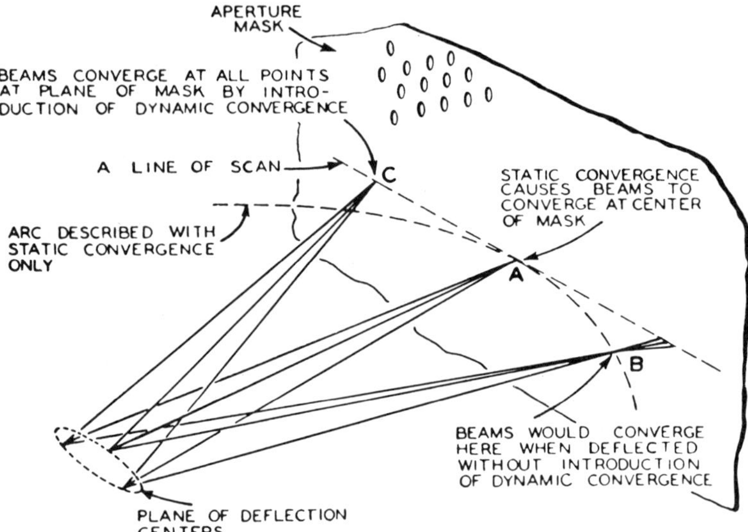

Fig. 14-29. Dynamic convergence voltages are required to cause the three beams to converge at the picture edges because the screen and the aperture (or shadow) mask are not perfectly spherical.

Fig. 14-30. The form of the dynamic convergence current is parabolic.

The basic form of the dynamic convergence correcting currents is parabolic, as pictured in Fig. 14-30. When the three beams are in the center of the screen, the correction current is zero. On either side of the center, however, the current varies, and the combined effect of the correction (that is, dynamic and static fields) is to keep the beams properly converged at every point of the screen.

On many late-model receivers the convergence circuitry and controls are contained on a separate, small panel which is easily removed and positioned for convenience in performing convergence adjustments. The connection to the main chassis is made via a cable-and-socket arrangement. Generally, there are six controls for adjusting the vertical convergence and six controls for adjusting the horizontal dynamic convergence.

External Picture Tube Components The components which are mounted on the neck of the tri-gun, tri-color picture tube will now be discussed (see Fig. 14-31). The deflection yoke is, to a considerable exent, similar to the deflection yoke used in a black-and-white receiver. However, its design is more complex because three beams must be deflected instead of one, and it is of the utmost importance that a symmetrical and uniform magnetic field be maintained throughout the deflection area.

Another component found on the neck of the color-picture tube is the purity-magnet assembly. This device adjusts the axis of each electron beam so that it approaches each hole in the shadow mask at the proper angle to strike the appropriate color phosphor dot. In other words, the purity magnet provides for the proper alignment of the three beams with respect to the phosphor-dot plate and the shadow mask. When this component is properly set, a uniform color field will be obtained for

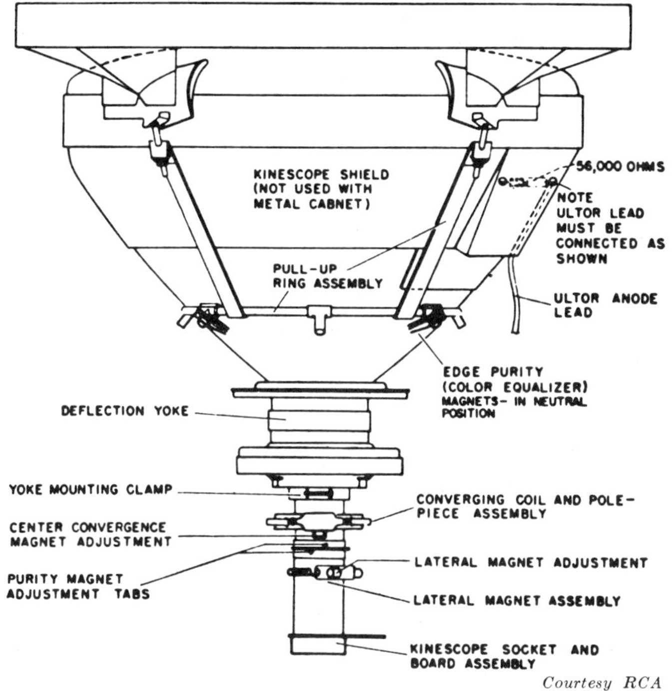

Courtesy RCA

Fig. 14-31. The location of the external components on the neck of the tri-color picture tube.

each gun. For example, with only the red gun in operation a uniform red raster should be observed. Any departure from pure red at any point on the screen indicates that the beam is striking phosphor dots other than red. Similarly, when only the green gun is in operation, a uniform green raster should be obtained, and when only the blue gun is active, a blue field should be visible.

The larger screen color tubes utilize magnetic convergence and, toward that end, they employ three sets of convergence coils, each positioned directly over the pole pieces which are internally associated with each number 4 grid. The magnetic fields set up by the coils are coupled through the glass neck of the tube to the internal pole pieces. The latter serve to shape and confine the fields, so as to affect only the particular electron beams to which the individual pole pieces correspond. For example, the change in the convergence angle of the red beam is a function only of the current through the external coil which couples to the internal set of pole pieces adjacent to the red beams. Similarly, the currents through the green and blue external magnets affect respectively only the green and blue beams (see also convergence discussions in Chapters 18, 19, 23 and 24).

Each external coil possesses two separate windings to provide for horizontal-and vertical-dynamic-convergence correction. For the static-convergence adjustment, each coil has a small permanent magnet associated with it. Its position can be varied to achieve center-screen static convergence.

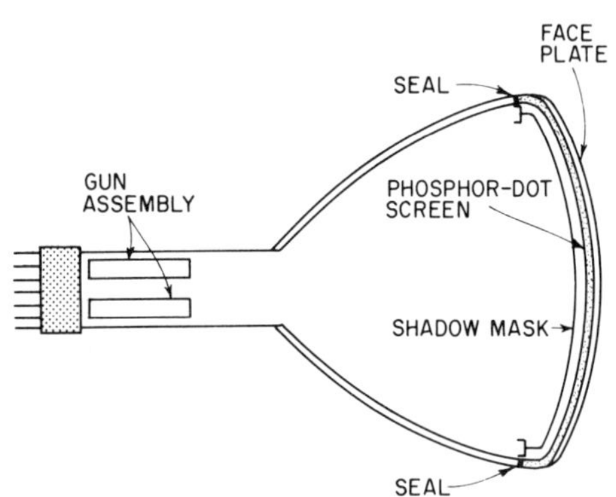

Fig. 14-32. The use of a curved-shadow mask, simplifies the problem of dynamic convergence.

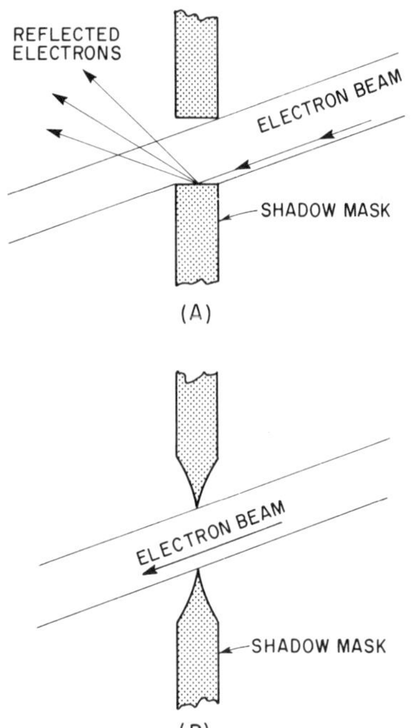

Fig. 14-33. The hole in the shadow mask affects the quality of the picture. (A) Electrons are reflected from the edge of the hole in the shadow mask. (B) By shaping the holes in the shadow mask as shown here, undesired electron reflections are practically eliminated.

A diagram of the individual dynamic convergence-control magnets is shown in Fig. 14-27. The heavy dots represent the individual electron beams as they pass through the gun on their way to the screen. The arrows at these beams indicate their direction of movement. Note that the red and green beams are confined to paths which make an angle of 60 degrees on either side of a perpendicular axis. The blue beam, on the other hand, can move only vertically.

Now it can readily happen that, although the color dots of the green and red beams fall within the triad of phosphors, those of the blue beam do not. This means that, while it is always possible to make the red and green beams (or color dots) converge, it may not be possible to have the blue beam meet the other two. Still another adjustment is required, that of being able to move the blue beam from side to side. To effect this, a special blue-beam lateral-positioning magnet is also found on the neck of the tube. This makes correct convergence of the three beams at the center of the screen always achievable (see Chapter 23 for the convergence procedure).

No ion traps are used with color-picture tubes, because the color screen is aluminized. The layer of aluminium presents a barrier to any oncoming ions and prevents them from reaching and damaging the screen. Electrons, having only 1/1800th of the mass of an ion, encounter little difficulty in passing through this aluminium layer.

Improvements in Color-Picture Tubes The size of the holes in the shadow mask is extremely small. As an example, in a typical 25 inch color tube the separation center-to-center between the holes is only 0.028 inch. The holes have a diameter of only 0.01 inch. The mask, which contains over 400,000 such holes, is manufactured with a photo etching process. The holes at the center of the mask are larger than those at the edge which gives the picture an overall brighter appearance. Once it is completed, it is exactly positioned into the tube to an accuracy within 0.00025 inch.

Two important improvements have been made in the design of the shadow mask. One of these simplifies the problem of dynamic onvergence and the other improves the quality of the color picture.

In early types of three-gun color tubes, the problem of dynamic convergence was aggravated by the fact that the shadow mask was flat and the screen of the tube was curved. One of the improvements in the design of shadow masks is a curved mask like the one which can be seen in Fig. 14-32. Use of the curved mask simplifies the problem of dynamic convergence.

The second improvement in the shadow-mask design is illustrated in Fig. 14-33. When the electron beam strikes the holes of a shadow mask at an angle as shown in Fig. 14-33(A), some of the electrons are reflected from the sides of the hole. These electrons strike phosphor dots in triads other than the ones they are supposed to strike. To get around this problem, the edge of the hole is shaped as shown in Fig. 14-33(B). Note that the edge of the hole is now very small, thus eliminating undesirable reflections.

The triads on the inside surface of the picture-tube screen emit light when struck by an electron beam. When there is a bright light striking the face of the tube, it is reflected. The reflected light combines with the light from the triads, and the overall effect is to give the television picture a washed-out appearance. This problem has been virtually eliminated by the use of the *black surround* which is simply an opaque, jet-black material between the dots in the triads. Since ambient light is not reflected from the front of the screen, the tube manufacturer does not need to use a filter-type glass screen that dims the picture. To the viewer the result is a much brighter color picture.

Brighter color pictures have also been achieved by the use of rare-earth phosphors instead of sulphide phosphors. Not only do the rare-earth phosphors radiate light much more efficiently, they also have the advantage that they reflect white light (since the phosphors are white) rather than the yellow light reflected by sulphide phosphors. While the reflected white light still dilutes the red light being emitted by the red rare-earth phosphor, this is still preferable to the red-yellow mixing that occurs with the sulphide phosphor.

The problem of convergence has been simplified in some receivers by the use of the color picture tubes having in-line cathodes. A comparison of the conventional delta arrangement and the in-line arrangement for the cathodes is pictured in Fig. 14-34. As shown in Fig. 14-34(A), the delta arrangement results in the use of two separate beam-control systems: one for converging the beam and the other for deflecting the beam. When the cathode and the triads are arranged horizontally as shown in Fig. 14-34(B), the vertical convergence remains constant with vertical deflection and thus, no vertical convergence correction is required. Horizontal convergence correction is still needed. General Electric first used color-picture tubes with in-line cathodes for their portable color-television receivers.

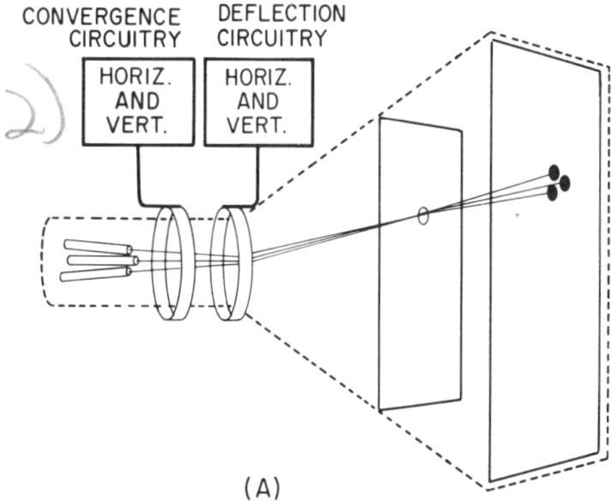

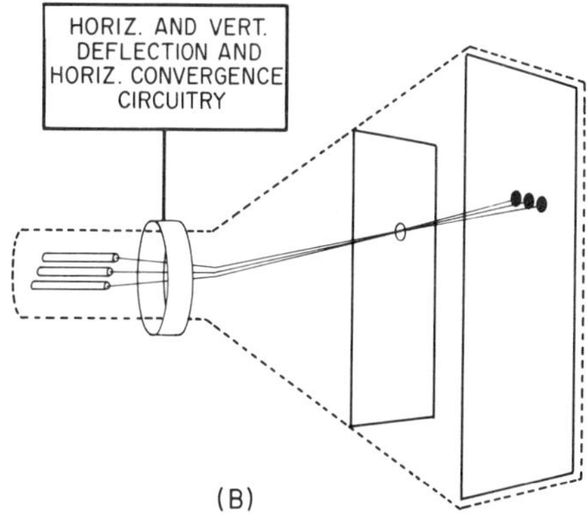

Fig. 14-34. Comparison of the delta and in-line cathode systems in the color-picture tubes. (A) A conventional tube with the cathodes in a delta formation. (B) The in-line tube. Note that the cathodes and the three color dots are in line horizontally.

14.6 TRINITRON PICTURE TUBES

The Trinitron color-picture tube employs three cathodes, but only a singly focusing lens. The color phosphors on the screen of a Trinitron are in alternating vertical strips, rather than in delta-shaped triads, and an aperture grill replaces the shadow mask in the conventional three-gun tube.

Figure 14-35 shows the arrangement of the electron gun, the aperture grid, and the vertically-striped color phosphors. Note that the three electron beams leave the cathodes and are focused in a single electron lens. Convergence is accomplished by four convergence control plates.

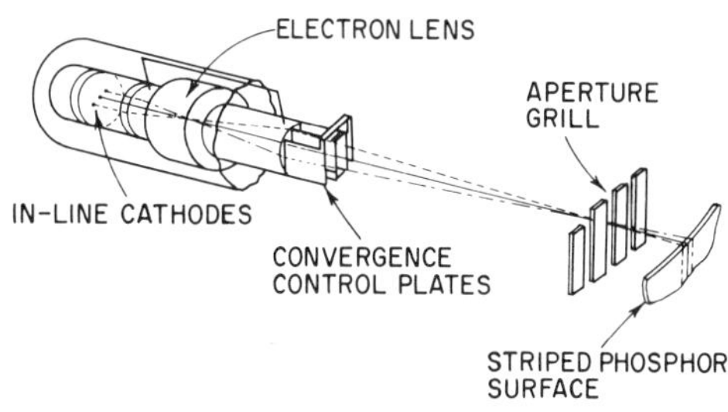

Fig. 14-35. The basic components of the Trinitron tube.

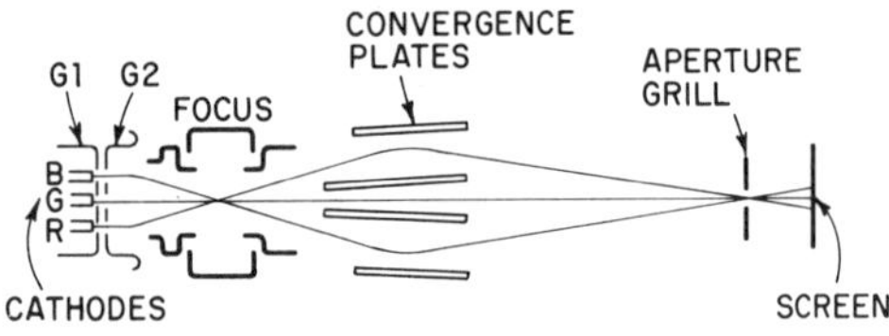

Fig. 14-36. Top view of the beam passing through the various sections of the Trinitron tube.

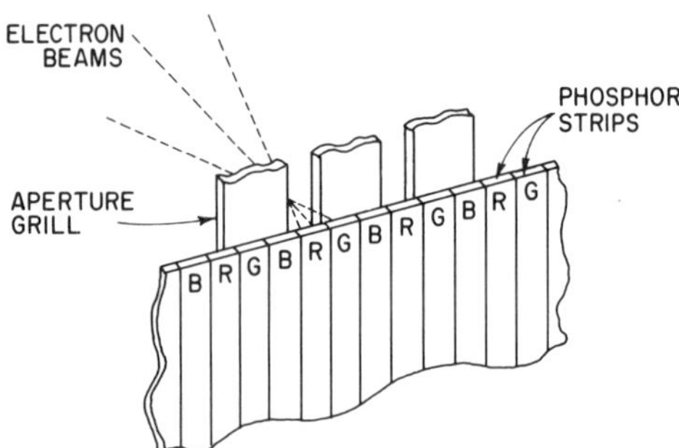

Fig. 14-37. This illustration shows that the phosphors of the Trinitron are in vertical strips rather than in triads. Note the relationship of the aperture grill behind the color strips.

The three beams converge at the aperture grill. Proper convergence is obtained by adjusting the dc voltage on the convergence control plates.

Figure 14-36 shows an illustration of a Trinitron gun, the aperture grill, and the screen. In this illustration, G_1 is the control grid, and G_2 is the accelerating grid which moves the electrons toward the screen. The focusing of the three beams is accomplished electrostatically in the single focus lens. Note the action of the convergence plates on the blue and red beams.

Figure 14-37 shows a clearer picture of the way the vertical color strips on the screen are related to the aperture grill. Note that the electron beams pass through the aperture grill and then diverge to light the color phosphors in a horizontal direction. The slots in the aperture grill are arranged so that each beam strikes only one phosphor, assuming that the electron beams in the tube are properly converged. This tube has the same advantage as that of the shadow-mask tube with the in-line cathodes, in that the triads are illuminated in a horizontal direction only. Therefore, it is not necessary to have vertical convergence circuits.

14.7 TROUBLES IN MONOCHROME AND COLOR-PICTURE TUBES

In this section some of the symptoms of defective monochrome and color-picture tubes are discussed.

Symptom: No Brightness There are a number of things in a television receiver that can cause a complete loss of brightness. For example, the high voltage section can be defective, or the picture-tube grid circuit can be defective. In the event that there is no brightness, these things should be checked in the normal troubleshooting procedure. One cause of loss of brightness is a burnout of the filament in the picture tube. As within the other vacuum tubes, failure of the filament is catastrophic. Fortunately, in modern television picture tubes, both monochrome and color, it is rare for the filament to burn out.

Symptom: Low Brightness, Cannot Set Contrast Control Properly If the picture is very slow to appear after the set is turned on, and if the picture is weak with very poor contrast (often with a silvery appearance), then the picture tube may be gassy, or its cathode emission may be low. If the problem is with low emission, some technicians prefer to install a *picture tube brightner* which is nothing more than a step-up voltage transformer for increasing the filament voltage. This makes the filament hotter and increases the emission from the cathode which, in turn, makes the tube useable for an additional amount of time.

Symptom: Excessive Brightness If the brightness control has no control over the amount of light coming from the picture tube, and the tube is at full brightness at all times, the problem may indicate a shorted control grid in the tube.

Symptom: Broad Dark Horizontal Bars in the Picture This type of problem can occur with both monochrome and color-picture tubes. It indicates a short between the picture-tube heater and cathode. It does not necessarily mean that the picture tube must be considered to be useless. Some technicians prefer to use an isolation transformer in the heater circuit. The transformer has a "floating secondary" which means that the heater and cathode voltage will be the same. This isolation transformer is especially useful for color picture tubes which are very expensive to replace.

Symptom: Large Blotches of Color (on a Color Tube Screen) that Cannot be Adjusted This is usually an indication that picture tubes or related mounting components have become magnetized. The magnetic field causes deflection on the beam in the color-picture tube which disturbs the proper beam landing needed for convergence. Hence, it is necessary to *degauss*—that is, demagnetize—the picture tube and mountings. A large coil connected to the ac power line is used for this purpose.

In many receivers the degaussing is accomplished automatically when the set is turned on.

Symptom: Only Two colors are Available in a Three-Gun Picture Tube This type of problem is normally due to one deflective gun in the color-picture tube, or to defective circuitry or voltages associated with the gun.

Symptom: Pincushion Raster In a color receiver special dynamic voltages are delivered to tube-sweep circuits in order to eliminate the pincushion type of raster. This circuitry is discussed in other parts of the book, and it will not be taken up here. A defect in this circuitry may result in the pincushion effect.

In some smaller monochrome picture tubes, permanent magnets for pincushion correction are mounted on the picture tube near the region of the aquadag coating. These permanent magnets must be exactly positioned or the sides of the raster will be bowed inward, giving a typical pincushion appearance.

REVIEW QUESTIONS

1. Which type of cathode system—directly heated or indirectly heated—is used in television picture tubes?
2. In which sections of a picture tube is focusing accomplished?
3. Why is degaussing necessary for color-picture tubes?
4. Explain why convergence is necessary in color-picture tubes?
5. What are the disadvantages of electrostatic deflection?
6. What is the function of the aquadag coating in a picture tube?

7. Describe two methods of preventing "ion spots"—that is, areas on the screen that have been damaged by ionic bombardment.
8. What is a bonded shield used for on picture tubes?
9. What is a triad?
10. What are the advantages of rectangular-picture tubes over round-picture tubes?
11. Why are three guns needed for a conventional color-picture tube?
12. What is meant by the expression "black surround"? How does it help to improve a color picture?
13. What is dynamic convergence as compared to static convergence?
14. How is dynamic convergence accomplished in a color receiver?
15. How does a curved shadow mask simplify the problem of convergence?
16. How does the shape of the hole in the shadow mask affect the operation of a color-picture tube?
17. What is an advantage of in-line cathodes in a color-picture tube?
18. How is the Trinitron different from a conventional three-gun color tube?

TV Power Supplies 15

15.1 REQUIREMENTS OF LOW-VOLTAGE, TV POWER SUPPLIES

The power supply requirements of a television receiver differ considerably from those of radio sets. The difference is due, in part, to the higher voltages required for the operation of the cathode-ray tube. In a television receiver, we are confronted with the task of supplying low voltages to tubes or transistors and also supplying up to 30,000 volts to the cathode-ray tube. It is possible to construct one supply for both or to employ two separate supplies. In the latter case, one supply would be used for the picture tube and the other for the remainder of the set. If one supply is decided upon, it must be capable of providing both the low voltages with the high currents and the high voltages with the low currents. The power supply would be bulky, expensive, and quite out of proportion with other sections of the receiver. As a result, TV receivers have both low and high voltage power supplies. This chapter will be concerned with low-voltage power supplies. The high-voltage supplies will be discussed in Chapter 18.

Voltages and Currents A tube-TV receiver may require voltages in the low-voltage power supply ranging from −100 to +350 volts. The currents for these voltages might range from a few microamps for bias voltages to hundreds of milliamps for plate voltages. If voltages higher than +350 volts, such as +400 or +500 volts, are required, then a boosted B+ supply is needed. This boosted power supply is actually a part of the high-voltage power supply and will be discussed in Chapter 18.

Transistorized TV receivers require low voltages at high currents. Voltages from −40 to +40 volts at currents up to several amperes are typical. However, some new transistors are designed to withstand several hundred volts. Receivers which have both tubes and transistors will require a large assortment of voltages.

15.2 TYPES OF POWER SUPPLIES

Power supplies can be classified in many ways. Perhaps the first thing noticed when looking at a power supply is the presence or absence of a power transformer. Some sets have large transformers, some have one or two small ones, and some sets have no power transformer. All will perform as required; however, those with transformers will probably operate more efficiently and have a longer life. Usually, the lack of a transformer means that more capacitors, rectifiers, and other components are needed.

Power supplies can also be classified by the number of diode rectifiers. Some receivers operate satisfactorily with two diodes, while others require as many as six. These diodes may be arranged as half-wave or full-wave rectifiers. Sometimes they are connected as bridge or multiplier circuits.

Most tube-type TV power supplies (except color sets) are unregulated. If the line voltage goes up or down, then the output voltages increase or decrease. Likewise, if the current from any output terminal increases significantly, that output voltage, and perhaps all the other output voltages, decrease. Under normal operating conditions, the TV-receiver performance will not be drastically affected, because the circuits are designed to operate properly over a specified range of supply voltages.

Many transistor-TV receivers contain regulated supplies. Some of their circuits must have supply voltages with a tighter tolerance to achieve "acceptable" performance. Since regulated supplies cost more to build, the manufacturer has to determine what an "acceptable" performance is.

The power supplies for tube receivers differ from those required for transistor set. Since transistor receivers require low voltages at high currents, the associated power supplies are optimized for such voltages and currents. Transformers, diodes, and resistors will have low resistance and high-current carrying capacity. Capacitors will have low-voltage ratings with very large capacitance values.

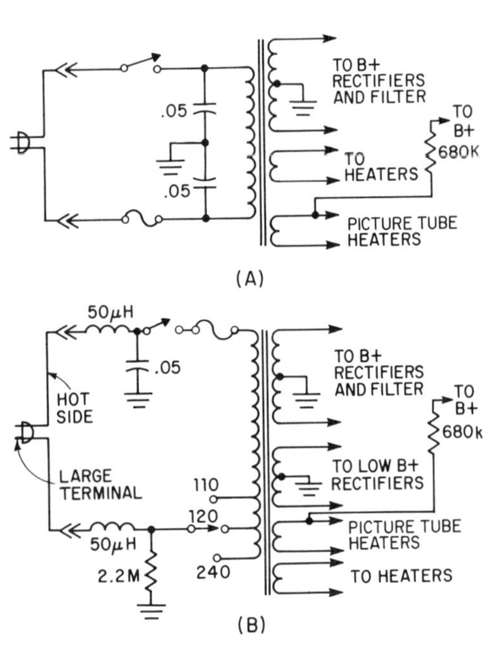

Fig. 15-1. The input portion of transformer power supplies: (A) The conventional transformer; (B) With tapped primary; 2, B+ windings, and polarized plug.

15.3 TRANSFORMER POWER SUPPLIES

Most better quality TV receivers use power transformers. A power transformer is more expensive than some of the components used to replace it in cheaper sets. Many advantages are possible by the proper use of a power transformer. They are:

1. Since the ac line is isolated from the B+ and filaments, there is no shock hazard to persons touching the chassis or metal cabinet.
2. B+ voltages are generated more efficiently. A longer receiver life and better regulation are the result.
3. The filament voltages for tubes which require a high dc bias is easily available on a separate winding.
4. The center-tapped secondaries allow full-wave rectifier circuits to use only two diodes instead of four.
5. Since tubes are connected in parallel, they need not all have the same current requirements.
6. The set is easily adjusted for a 110, 120, or 240 volt operation if a tapped transformer primary is used. The input section of several transformer power supplies are shown in Fig. 15-1. The supply pictured in Fig. 15-1(A) totally isolates the ac line from all circuits.

The circuit of Fig. 15-1(B) is intended for use with a polarized plug, though it can be used on a floating home ac line. The purpose of the 2.2M resistor is simply to act as a short-time constant filter, together with the .05 μf capacitor. This filter reduces the ac line transients. Its impedance at 60 Hz is sufficiently high to effectively isolate the chassis ground from the ac line.

15.4 TRANSFORMERLESS POWER SUPPLIES

There are several varieties of power supplies, including those that use a small transformer for part of the power. We will discuss each of these briefly before examining the fully transformerless TV receivers.

B+ Transformer Supplies Many receivers use a transformer only for the generation of B+ voltages. This is a reasonable approach to cost saving, since it is quite easy to connect all of the filaments in series across the ac line. However, many receivers require a dc bias of 100 to 200 volts on the picture tube and damper-tube heaters. This is applied so that the voltage between the heater and cathode will not become too large, otherwise a heater-cathode short may occur. A separate filament winding may be provided for this purpose.

Figure 15-2 shows a typical TV receiver power supply which has a B+ transformer. A small filament winding for the picture tube and damper tube filaments is also illustrated. This latter winding is biased to approximately 2/3 of the B+ voltage. Note that one side of the ac line is grounded; similarly, one side of the filament string must be grounded. The heaters on the hot side of the ac line see $\pm(1.4)(120) = \pm 148$ volts during the peaks of the sine wave. This is dangerously close to causing a heater-cathode short.

If the filament string were removed from Fig. 15-2, we would have a power supply resembling that required for a transistor receiver. The B+ voltage would probably be only 50 volts or less, however. Power supplies for transistor receivers are usually quite simple compared to those found in tube sets.

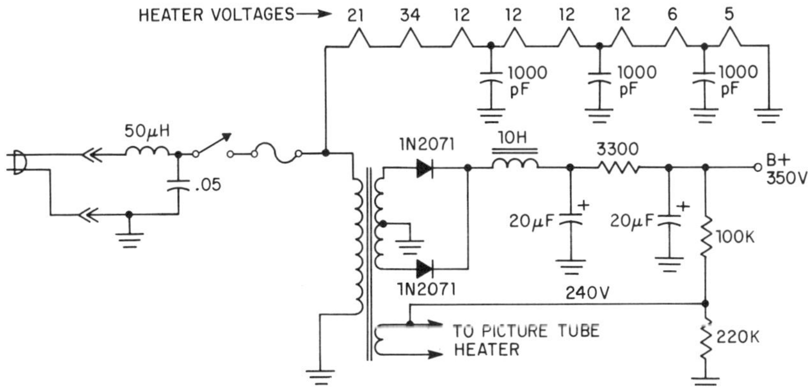

Fig. 15-2. TV receiver power supply which uses a transformer only for B+ and the picture tube filament.

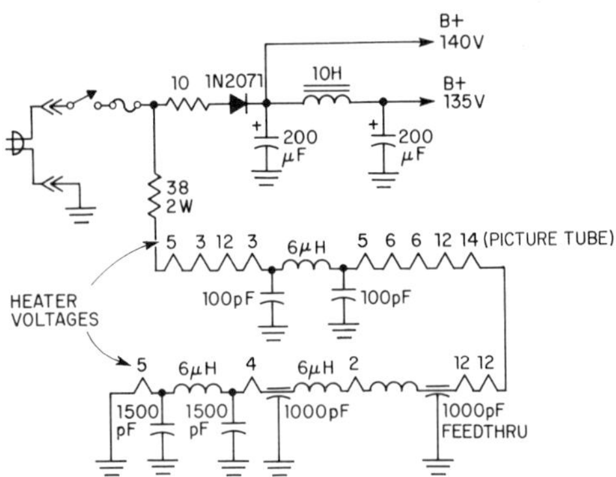

Fig. 15-3. A simple transformerless power supply for a 14-inch TV receiver.

Filament Transformer Receivers Since it is often required to bias the CRT and damper tubes to some dc voltage, transformers are the only solution. It is impossible to place a bias on the heater voltage if it is connected directly to the input ac line. Many receivers have a transformer just for this very requirement. Since this transformer only supplies current to one or two heaters, it is usually quite small.

Some receivers have two separate heater windings on the transformer. One is used for the picture tube and damper tube; the other winding is used for the remaining heaters.

Complete Transformerless Power Supplies Power supplies with no transformers are practical only for small screen portable sets. In small receivers, weight is very important. Since transformers are one of the heaviest components, they are the most obvious item to eliminate. Portable receivers are often switchable between line or battery operation. The power supply must, in these cases, be easily changed from ac to dc operation or vice versa. This is difficult if there are any transformers in the power supply.

A power-supply schematic for a 14-inch tube receiver is diagrammed in Fig. 15-3. The +140 volt output line has a large ripple and is used for an audio output stage. Either push-pull or negative feedback must be used in this audio amplifier to prevent a hum.

15.5 RECTIFIERS

The devices for changing ac to dc come in many sizes, shapes, and prices. Up until the early 1950's, nearly all TV power supplies utilized tubes for rectifiers. As selenium, germanium, and silicon rectifiers became more plentiful and cheaper, they gradually replaced tubes. Today, every manufacturer uses silicon rectifiers, and they will probably be the mainstay for many years to come.

Silicon Rectifiers Nearly all new transistors, diodes, and integrated circuits are made with silicon. It is silicon that enabled solid-state electronics to take over most of the functions of vacuum tubes. Silicon devices can withstand high temperatures. They are stable over long periods of time because of their resistance to the deleterious effects of air inside the device. The leakage current in the reverse direction is also low. In the silicon rectifier, a small wafer of silicon is securely fastened, generally by welding, to a copper plate which then is attached to the rectifier housing (see Fig. 15-4). This terminal is the cathode, and the copper serves to make an electrical connection to the silicon. By connecting the copper to the rectifier case, and then mounting the case to a chassis, an efficient heat dissipator is produced.

The other terminal of the rectifier is formed by alloying a small dot of gold antimony into the opposite face of the silicon wafer. Electrical

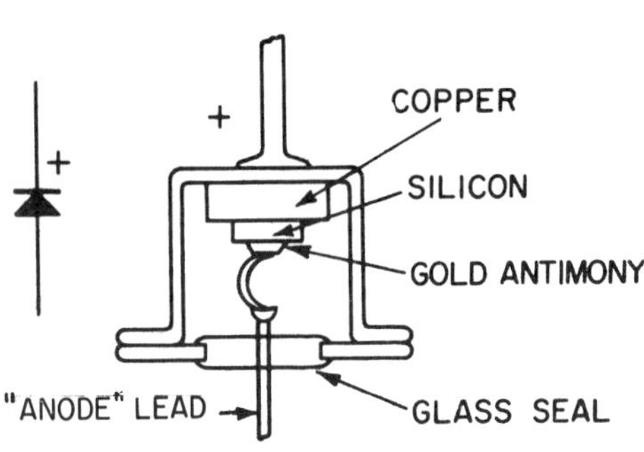

Fig. 15-4. Internal construction of a silicon or germanium rectifier diode.

connection is then made to this dot, and this lead, brought out of the case but insulated from it, becomes the anode of the rectifier.

One precaution must be observed when using silicon rectifiers. Their forward resistance is so low that when a set is turned on, a large current surge through them to the filter capacitors. In Fig. 15-3 the 10-ohm resistor is used to limit this surge current to a safe value. Let us determine what this current is with and without the 10-ohm resistor. The 1N2071 diode has a forward resistance of approximately one ohm at large currents. At the instant of turn on, the 200-microFarad capacitor looks like a short circuit. However, the peak value of the first half cycle of the voltage is equal to $120\ (1.414) = 170$ volts. Thus, a maximum current of $I = V/R = 170/1 = 170$ amps will flow during the first half cycle. This diode is only capable of withstanding a 25 amp surge. The 10-ohm resistor shown in Fig. 15-3 keeps this initial surge down to $I = 170/(10+1) = 15.5$ amps.

15.6 VOLTAGE MULTIPLIERS

Television receivers require a large variety of dc voltages for operation. Many of these voltages are larger than those available from the rectified ac line voltage. If a transformer power supply is utilized, then it is a simple matter to generate any required voltage. Transformerless receivers often use voltage multipliers composed of diodes and capacitors to generate larger voltages. We will briefly discuss several of these multipliers in this section. Another method of voltage multiplication makes use of energy salvaged during the horizontal retrace time. This is called the boosted B+ supply. Its description is contained in Chapter 18.

Voltage Doublers This is the most commonly seen voltage multiplier. It has been used by nearly every TV-receiver manufacturer for some of their models. It usually comes in two basic circuit arrangements, although other methods are possible. The first method, pictured in Fig. 15-5(A), is the most common circuit. It is called the half-wave voltage doubler. This voltage doubler is popular because it is a three-terminal device. This means that the input and output circuits use the same ground point. The major disadvantage of this doubler, compared to others, is that the output ripple is at 60 Hz. This requires large filter capacitors to smooth out the ripple.

The other common voltage doubler, shown in Fig. 15-5(B), is called the full-wave voltage doubler. This name is used because the output ripple is at 120 Hz, similar to a conventional full-wave rectifier circuit. This higher frequency ripple allows the use of small filter capacitors when compared to the half-wave multiplier. The regulation of this multiplier is also better than the half-wave type; however, both types have poor regulation when compared to the conventional rectifier circuits.

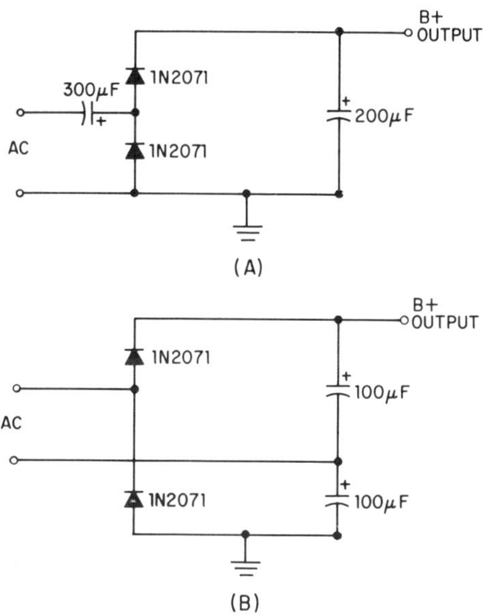

Fig. 15-5. Two common voltage doublers: (A) half wave; (B) full wave.

15.7 FILTERS

The dc output from the rectifiers alone always has a large amount of ripple. Some electronic devices can use this form of dc with no further processing. Television receivers, however, require that most of the ripple be removed. In this section we will examine a few of the many ways used to remove the ripple from rectified ac power. Only passive filters will be considered. These contain resistors, capacitors, and inductors. We will close this section by briefly discussing line filters and heater filters.

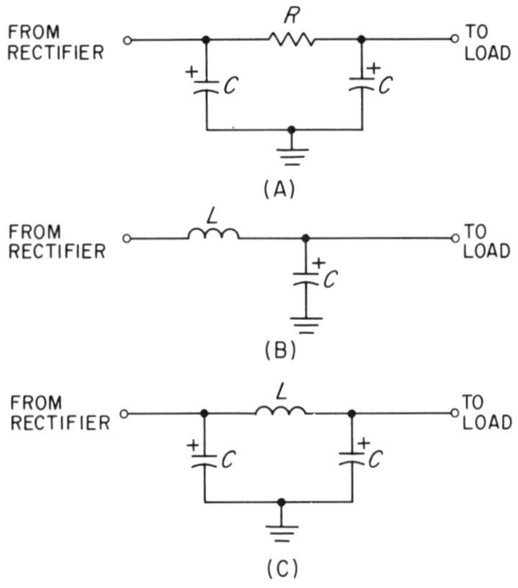

Fig. 15-6. A few of the filter circuits found in TV receivers: (A) *RC* π circuit; (B) choke-input *LC* circuit; (C) capacitor-input *LC* circuit.

RC Filters An *RC* filter commonly seen is the π filter shown in Fig. 15-6(A). For low-voltage power supplies, this type of filter is ideal. Two reasons make this true. First, the resistor is much smaller and lighter than the choke used in higher voltage power supplies. Second, high-capacitance, low-voltage capacitors are quite small. Some TV receivers use up to several thousand microFarads for these capacitors. The exact value of the *R*'s and *C*'s depends upon the ripple, current, and regulation requirements. For the same capacitor, a larger *R* will produce less ripple; however, this provides a lower output voltage and poorer regulation.

LC Filters Inductance and capacitance filters also come in several forms. Again, the exact circuit values depend on whether ripple or regulation is more important. The choke input *LC* filter is shown in Fig. 15-6(B). This is the most commonly seen *LC* filter. The choke is chosen so that it presents a high impedance to the 60- or 120-Hz ripple. A half-wave rectifier has a 60-Hz ripple, and a full-wave rectifier has a 120-Hz ripple. Since this impedance is in a series with the current to the output terminal, the output voltage is less than that achieved with the *RC* filters. The output voltage (if used with a full-wave rectifier) is usually between 60% and 70% of the peak sine-wave voltage. With an *RC* filter for a full-wave rectifier, the output voltage is 97% of the peak sine-wave voltage. However, the choke input *LC* filter has a better voltage regulation than any other filter which uses only passive components.

The π circuit *LC* filter is commonly called the capacitor input *LC* filter. It is shown in Fig. 15-6(C). The rectified current is fed directly into a capacitor before going to the choke. This results in an output voltage greater than 90% of the peak sine-wave voltage. However, like the π circuit *RC* filter, regulation is not too good.

15.8 HEATER POWER

Power for the tube heaters is obtained by several methods. We will briefly examine the most common methods in this section.

Series Heaters Small transformerless TV receivers use the simplest heater wiring scheme (see Fig. 15-3). All heaters, including the picture tube and the damper heaters, are connected in series across the ac line. Sometimes a low-resistance power resistor is placed in series with the heaters. In some receivers, this resistor is a power thermistor which has a high cold resistance and a low hot resistance. The thermistor will not allow the heater string to heat up too fast, as this may put excessive thermal stress on the tubes.

All heaters in a series string must have the same heater current requirements. Since they are in series, only one current value can possibly flow. If any tube requiring a larger current is plugged in by mistake, it will only pass the current required by the rest of the heater string. As a result, the voltage across that tube will be lower than its requirement.

Parallel Heaters With a transformer power supply, the heaters will likely be connected in parallel. Usually there are two heater windings; one for the picture tube and the damper tube, and the other winding for all the other tubes. A third winding is present if a vacuum-tube rectifier is utilized. Parallel connected heaters require that all tubes have the same heater voltage requirement. However, the currents in each tube can be different. If a tube with a lower voltage requirement is accidentally used, its current will be larger than that specified for that tube.

As shown in Fig. 15-2, some receivers use series heaters for most tubes and a transformer for the picture tube and the damper tube. This is done so that a dc bias can be placed on the heaters of the two latter tubes. Otherwise, because of the voltages normally present, a high voltage stress will exist between their heaters and cathodes.

15.9 VOLTAGE REGULATORS

Most circuits in a TV receiver are designed to operate with one or more power-supply voltages. These voltages must be kept within specified limits or the circuit will not deliver optimum performance.

Brute Force Regulation The most obvious way to provide voltage regulation is to build a power supply with much more current capacity than is required. However, this type of supply usually turns out to be expensive, heavy, and large. The modern tendency toward light, compact, and low-price receivers requires a more practical approach to voltage regulation.

Zener Regulation For low voltages at power levels under one watt, the zener diode is a good power-supply regulator. Typical positive voltage and negative voltage zener regulators are shown in Figs. 15-7(A) and (B), respectively. When reverse biased, a zener diode will not allow the voltage across its terminals to rise above its breakdown voltage. It does this by conducting current very heavily as soon as its reverse breakdown voltage has been reached. A zener diode is reverse biased when connected as shown in Fig. 15-7. The symbol used resembles that

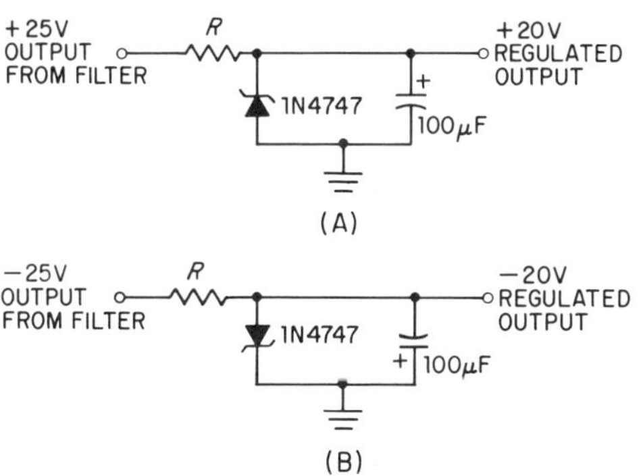

Fig. 15-7. Zener diode voltage regulators: (A) positive voltage, and (B) negative voltage.

of a regular diode, because it is quite similar to regular diodes. The only difference is in the tightly controlled reverse breakdown voltage required by zener diodes.

How well does the zener regulator stabilize the output voltage? As long as the input voltage is well above 20 volts, the output voltage will remain fairly stable. However, if the input voltage goes down to 20 or 21 volts, the output voltage will begin to drop. Likewise, the load cannot draw more than about 50 milliamperes before the output voltage begins to drop.

Active Regulators The zener regulator is a very simple device, and consequently suffers several limitations. The first problem is the power wasted in the diode. This wasted power is of the same magnitude as that used by the load. The second problem is the nonideal current-voltage characteristics of the zener diode. The voltage across the diode changes a few percent as the current through it changes. Both of these problems are nearly overcome by using a zener controlled active regulator. A simple version of such a circuit can be seen in Fig. 15-8. By placing a 1000-ohm resistor in series with the zener diode, the power wasted is reduced 90%. The zener diode maintains the base voltage of the 2N3738 power transistor at +20 volts. The emitter voltage must remain at approximately the same voltage. The transistor has a current gain of 50. Therefore, it can control up to 125 milliamperes with only 2.5 milliamperes going into the base. Thus, we have overcome both limitations of zener diodes by using a power transistor.

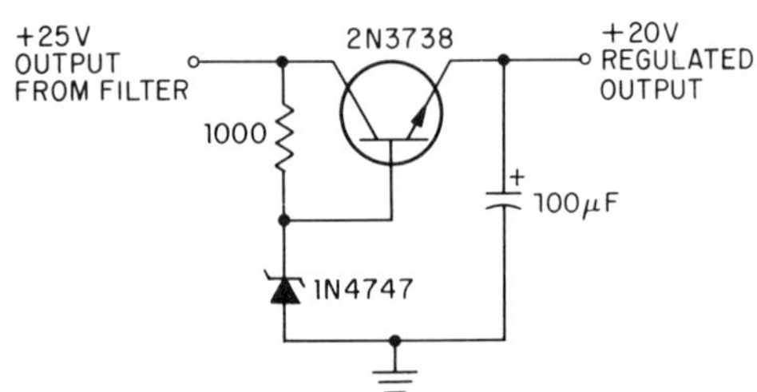

Fig. 15-8. A simple active regulator circuit.

15.10 TRANSISTOR TV POWER SUPPLIES

Solid-state receivers utilize lower dc voltages than do tube sets. Conversely, the current requirements are larger, because transistors are low-impedance devices. Some portable solid-state power supplies generate only 6 or 12 volts dc, which makes them easy to switch over to battery operation. Figure 15-9 shows a typical low-voltage power supply which may be operated from the ac line or from a battery.

The on-off switch has a third position for charging the batteries without operating the receiver. The batteries are also charged during receiver operation; in addition, they serve as voltage regulators during normal operation. They perform in a manner similar to the parallel combination of a zener diode and a filter capacitor. The 100-ohm resistor limits the battery charging current to a recommended value. At the same time, this resistor is part of the *RC* filter network. The 10-ohm resistor limits the initial capacitor charging current to a safe value for the 1N2070 diodes.

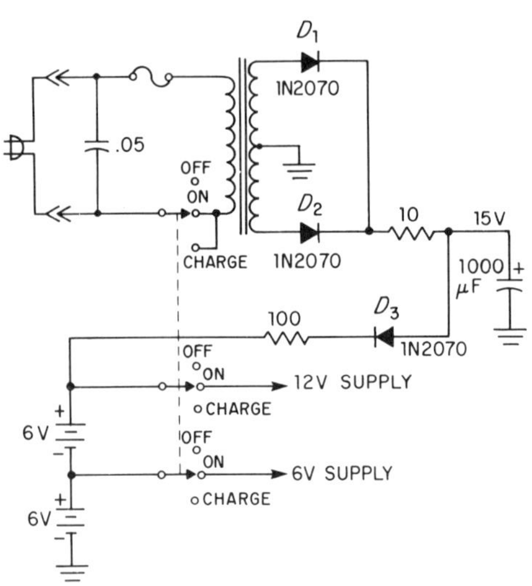

Fig. 15-9. The power supply for ac or a battery operated transistor TV receiver.

New solid-state TV receivers utilize both low-voltage and high-voltage transistors. This requires a large assortment of voltages. However, since transistors are low-impedance devices, they work best when driven by low output-impedance supply voltages. Some receivers have 3 or 4 rectifier-filter circuits for this reason. A large assortment of bleeder resistors would not provide this low impedance. A power supply of this type is shown in Fig. 15-10. The +24 and −20 output voltages drive most of the low-level stages in the set. A zener regulator is used for the +24 supply, and an active filter is required for for +20 volt supply. The +100 output voltage services the video-, vertical-, horizontal-, and audio-output stages. Note the wiring arrangement of the two full-wave rectifier circuits. Some receivers do this so that part of the diode can be physically connected to ground. The ground is then used as a heat sink, and no insulator is required between ground and the rectifier body.

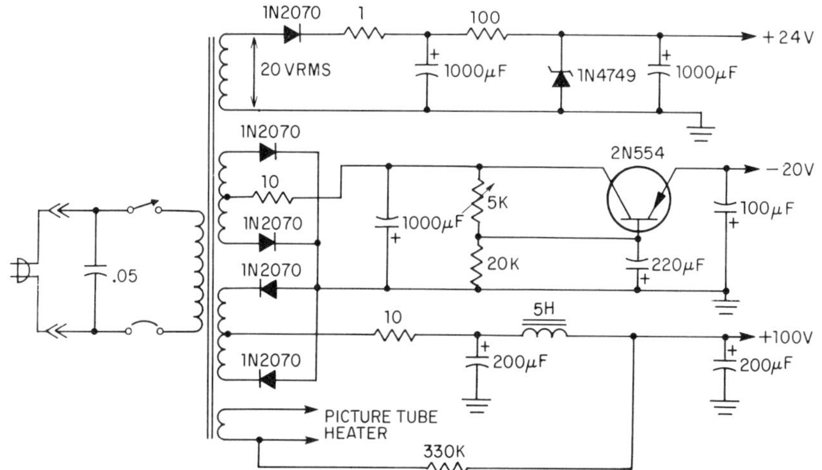

Fig. 15-10. Power supply for solid-state TV receiver which requires multiple low impedance-output voltages.

15.11 COLOR-TV POWER SUPPLIES

In general, color-TV power supplies are slightly larger and more complex than those in black and white sets. They must handle more power because of the many extra circuits in color receivers. The presence of automatic degaussers in the B+ line also makes the circuitry look more complicated. However, this degaussing circuit has nothing to do with the normal operation of a warmed-up power supply.

Figure 15-11 shows a power supply for a solid-state, large-screen color receiver. This set also has several tubes. Many TV sets fall in this catagory. Most of the transistors receive their power from the +20 volts developed by the active filter. The tuner also uses +20 volts with a little extra filtering. An adjustment R_1, is available for setting the exact voltage of the +20 supply. The other side of the transformer winding of the +20 supply is used for the tube heaters. The +290 volt B+ circuit has a degaussing circuit in series with it. A thermistor and varistor circuit control the current in the degaussing coil. As the circuit warms up, the thermistor decreases in resistance. At the same time, the varistor

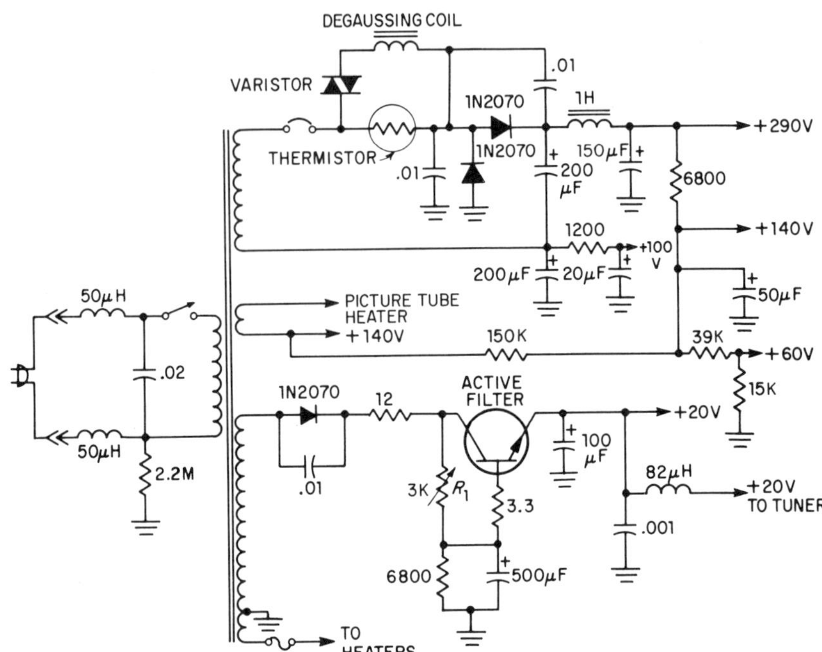

Fig. 15-11. Power supply for a transistorized color-TV receiver, which also has several tubes. (*Courtesy of Philco.*)

increases in resistance. The net result is a surge of current through the degaussing coil for a few seconds before the set warms up. Very little current flows in the degaussing coil during normal receiver operation.

Several other things are worth noting in Fig. 15-11. Notice that the picture tube heater is biased to +140 volts to prevent the cathode to heater shorts. The +290 volt supply uses a half-wave voltage doubler. The 0.01-microFarad capacitors across the diodes are for transient protection. Solid-state diodes in the rectifier circuits have a much longer life with these capacitors present. In some receivers, if the capacitor is removed, the diode will soon burn out.

15.12 TROUBLES IN POWER SUPPLIES

The dc voltages from a power supply will generally have only three types of troubles. These are: 1) total loss of one or more voltages; 2) reduction in the value of one or more voltages; and 3) an ac ripple superimposed on one or more dc voltages. The ac voltage used for the heaters will most likely be either operative or nonoperative.

No DC Voltage When this occurs on all dc voltages simultaneously, the receiver will have no picture or sound. If only one dc voltage is down, then many possible changes in performance can happen, depending on which circuits receive power from that one supply voltage.

Low DC Voltages TV receivers often have one or more dc voltages which decrease with time. This is usually evident on the CRT faceplate by a continual reduction of the horizontal and vertical sweep sizes. Sometimes the sync and contrast will be affected if the voltages sink too low. This voltage reduction is caused by the aging of electrolytic capacitors.

AC Ripple Ripple can be observed both on the CRT and from the speaker. On the CRT it usually appears as one or two horizontal black bars across the screen. One bar means a 60-Hz ripple, so a half-wave supply is at fault. A 120-Hz ripple means a full-wave supply is to blame. However, heater-cathode shorts in tubes may cause some of these same symptoms. A power-supply ripple may also be heard from the speaker as a hum. Aging electrolytic capacitors are usually the cause of ripple.

REVIEW QUESTIONS

1. Draw the circuit for a low-voltage power supply suitable for a television receiver.
2. Indicate how the same power supply can provide positive and negative voltages.
3. What advantages do transformer power supplies have over transformerless types?
4. In what ways are the voltage and current requirements different for tube-TV sets compared to solid-state sets?
5. Why are silicon rectifiers preferred by most manufacturers?
6. Why does a zener diode need a filter capacitor across it? Isn't the zener diode sufficient to keep the voltage stable?
7. What is the purpose of the 10-ohm resistor in the −20 volt supply of Fig. 15-10?
8. What is the function of D_3 in Fig. 15-9? Hint: it is used only when the set is turned off.
9. Which diode in Fig. 15-10 is likely to burn out first?

16 Synchronizing Circuit Fundamentals

16.1 AN OVERVIEW OF SYNCHRONIZING CIRCUITS

Up to this point we have studied the effect of the various stages on the video signal, so that it will finally be suitable for application to the grid of the picture tube. However, little has been said so far about the method of supplying the proper wave-forms to the deflection coils so that the image will be swept out in a suitable manner on the picture-tube screen. To accomplish this, we must obtain the synchronizing pulses from the video signal and apply them to other circuits that will eventually connect directly to the deflection system of the picture tube. Since each line has a separate synchronizing pulse, it becomes possible to lay the lines out on the screen in their proper position exactly as they were scanned in the camera tube. The synchronizing pulses responsible for the correct positioning of the various lines are referred to as the horizontal sync pulses. These pulses are diverted to amplifiers that control the action of the horizontal-deflection coils.

After the electron beam sweeps out the correct number of horizontal lines and arrives at the bottom of the picture, a vertical synchronizing pulse is applied to the vertical-deflection circuits, and the beam is rapidly brought back to the top of the screen again. This vertical pulse is transmitted together with the horizontal pulses in the video signal, separated from the horizontal pulses by filters and applied to a set of amplifiers that connect to the vertical-deflection coils.

Block Diagram of Sync Circuits The block diagram of Fig. 16-1 illustrates the general path of the sync pulses within a television receiver. This chapter will be concerned with the first three blocks in the figure: 1) The sync clipper, where the sync pulses are removed from the

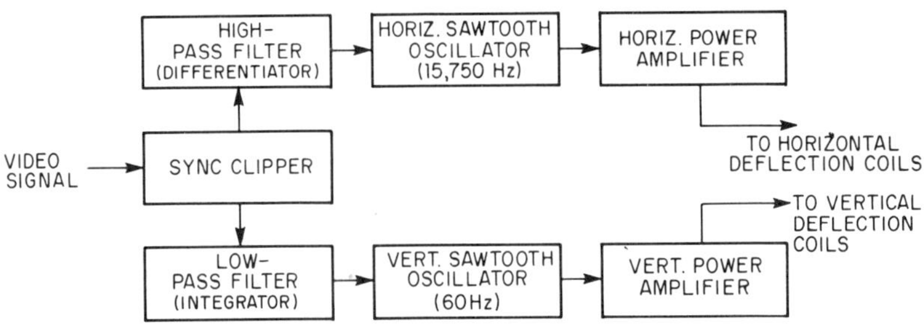

Fig. 16-1. The block diagram of the synchronizing section of a television receiver.

video signal, 2) The high pass filter (or differentiator), which separates the horizontal sync pulses from the combined sync signal, and 3) The low pass filter (or integrator), which separates the vertical sync pulses from the combined sync signal.

16.2 REVIEW OF SYNC AND BLANKING PULSES

Thus, far, only general terms have been used when discussing the synchronizing pulses of video waves. Their purpose has been stated in Chapter 2, but nothing definite has been stated about how their objective is accomplished. There is nothing in sound receivers that even closely resembles this action, and a detailed examination becomes necessary. Discussion of the pulses, separated from the rest of the wave will be held in abeyance while we explore in greater detail the form and functions of the horizontal and vertical pulses.

Horizontal It is already known that, as each horizontal line of video information arrives at the grid of the picture tube, the electron beam should be in the correct position, ready to sweep out the information contained in the signal. The position of the electron beam is controlled by sawtooth oscillators. In order for the oscillator to have the beam in the correct position, horizontal synchronizing pulses are inserted into the video signal. These could have been sent separately, but the present method is cheaper and simpler in operation. It should be noted and kept in mind that the function of the horizontal synchronizing pulses is to trigger an oscillator in order to bring the electron beam from the right-hand side of the screen to the left-hand side. Once the beam is at the left-hand side, the oscillator is no longer directly under the control of a pulse and goes about its normal function (with the horizontal amplifier) of sweeping the beam across the screen. Thus each horizontal pulse that precedes the line detail sets up the beam in readiness for the scanning of this information. The next pulse arrives when the beam is at the far right-hand side of the screen, at the end of a line.

There are 525 lines sent out every 1/30 sec. In one second, then, we have 525 times 30, or 15,750 lines. This means that the frequency of the horizontal pulses is 15,750 per sec, or one arrives every 1/15,750 sec. The time interval is quite small, being only 64 microseconds.

Vertical In a similar manner, the vertical pulses serve the purpose of bringing the electron beam back to the top of the screen for the beginning of each field. With interlaced scanning (described in Chapter 2), every other line is scanned, each field (1/2 frame) taking 1/60 sec. The beam next sweeps out the lines that were missed, this also in 1/60 sec. The total frame, with all lines, is completed in 1/60 plus 1/60 sec, or 1/30 sec. Thus we see that the vertical pulses must occur once every 1/60 sec, or 60 times in one second. This frequency is considerably less than that of the horizontal pulses, and it is because of this fact that they can be separated with comparative ease.

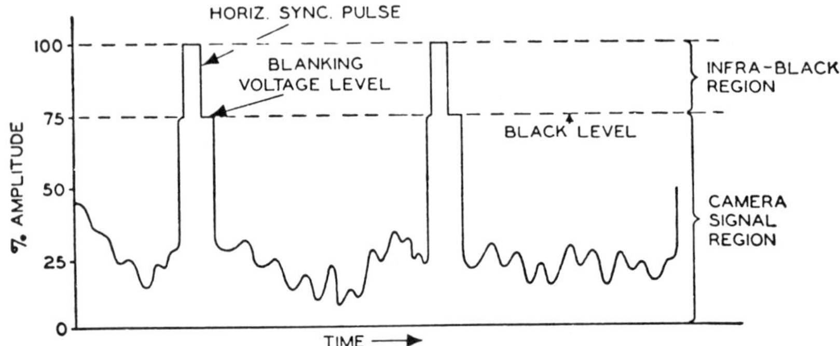

Fig. 16-2. A video signal with its synchronizing (horizontal) pulses.

Composite Sync and Video With the preceding ideas in mind, let us closely examine the construction of the video signal with its synchronizing pulses. In Fig. 16-2 several lines of an image are shown, complete with the detail information, blanking voltages, and horizontal synchronizing pulses. The blanking and synchronizing voltages occupy approximately 20 to 25 percent of the total signal amplitude. Note that the blanking voltage retains its control over the picture-tube grid for some time before and after each synchronizing pulse. This is done to make certain that no beam retrace is visible on the screen. As soon as the blanking voltage relinquishes control of the grid, the line detail becomes active again. All the lines of one field follow this form; the only difference from line to line is in the detail of the various sections of the image.

At the bottom horizontal line, it is necessary to insert a vertical pulse that will bring the beam back to the top of the screen again. During the period that the vertical pluse is active, it is imperative that the horizontal oscillator not be neglected. For, if this did occur, the horizontal generator would slip out of synchronization. To avoid this, the vertical pulses are arranged in serrated form and accomplish vertical and horizontal synchronization simultaneously. Figure 16-3 illustrates the composite video and sync waveform at the beginning of each field. These two wave-forms are slightly different so that interlacing can be performed. The first horizontal line of field 1 and the last horizontal line of field 2

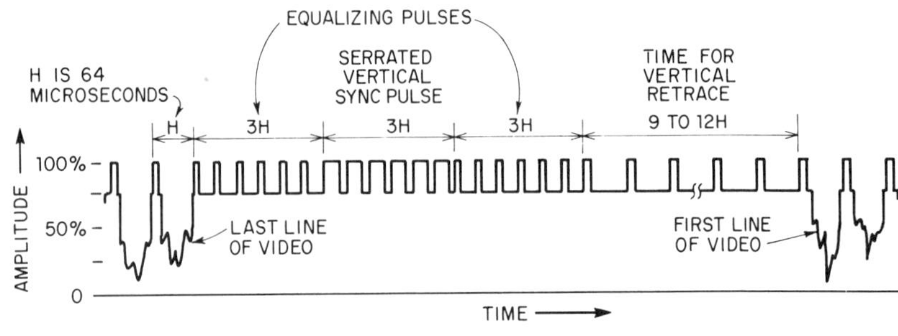

Fig. 16-3(A). Video and sync wave-forms at the beginning of the first field.

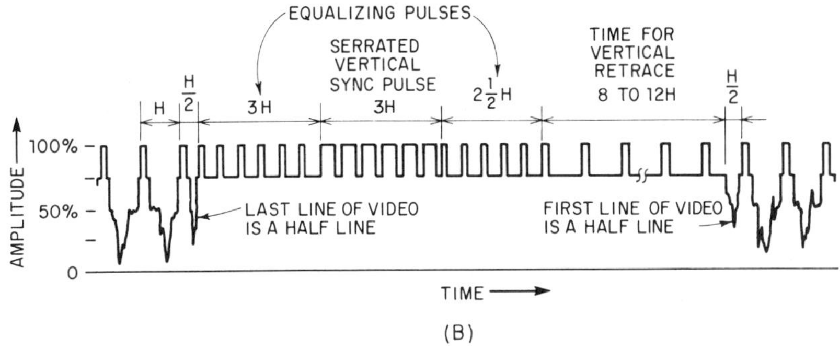

Fig. 16-3(B). Video and sync wave-forms at the beginning of the second field.

will be complete lines. The last horizontal line of field 1 and the first horizontal line of field 2 will be half lines. The serrated sync pulses will be discussed further in Section 16.7. The equalizing pulses will be explained in Section 16.9.

16.3 SYNC AND VIDEO SIGNAL SEPARATION

Before they can be used, the pulses of a video wave must first be separated from the other portions of the signal. The separation may occur anywhere from the video detector to the last video stage before the cathode-ray tube. In practice, receiver designers have generally chosen to obtain the input for the synchronizing stages from a point beyond the video detector, usually at the output of one of the video amplifiers. At such points, the signal has sufficient amplitude and is in proper form for controlling horizontal- and vertical-deflection oscillators with a minimum of additional stages. For example, circuit designers often do not apply the video signal to the sync separator until it has passed through the first video amplifier. In this way an extra pulse amplifier is eliminated.

Since it is necessary to obtain the synchronizing pulses from the incoming wave, it is first imperative that the signal be in its dc form. This should be evident by reference to the figures of Chapter 13, where the ac and dc forms of a video signal are illustrated. While the signal is always in its dc form at the output of the detector, it may not be so at the output of an amplifier that follows. In this case, dc restoration is necessary.

The circuit that separates the sync pulses from the rest of the video signal is called the "clipper" or "sync separator." Both horizontal and vertical synchronizing pulses are clipped by this circuit; the further separation of these two pulses then occurs at another point beyond this stage. The type of circuit that may be utilized for the sync pulse separation is not restricted. Practically every type is suitable since the action consists merely in biasing the circuit so that only the top portions of the video wave (where the pulses are found) cause current to flow.

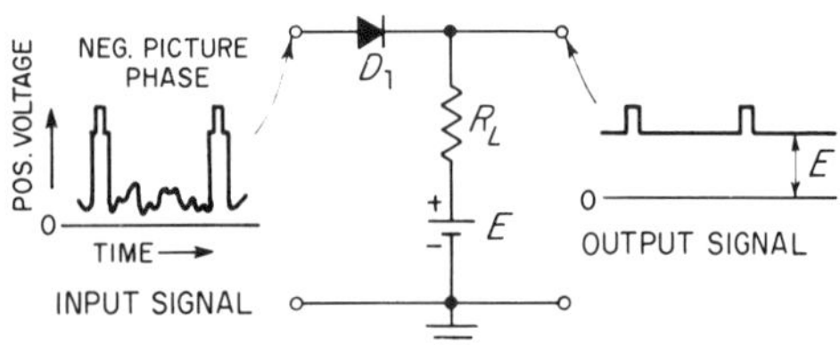

Fig. 16-4. A diode clipper operating with input-video signals having a negative picture phase.

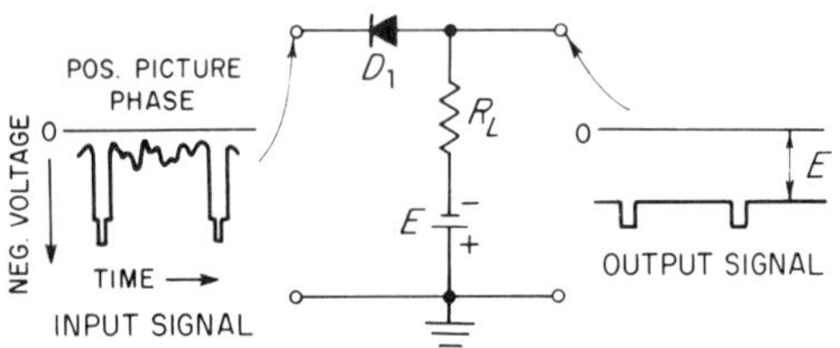

Fig. 16-5. An inverted diode clipper, suitable for input signals having a positive picture phase.

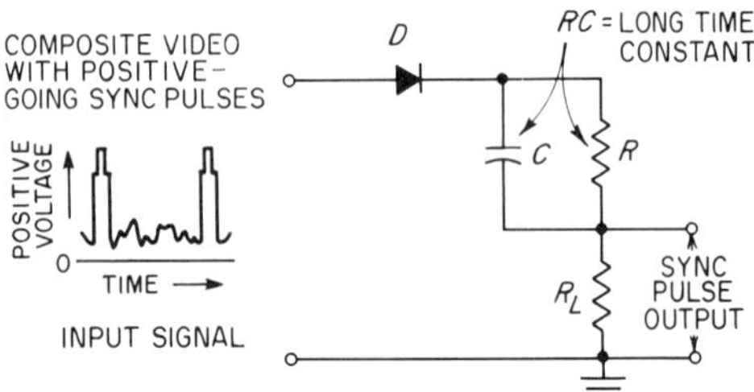

Fig. 16-6. A practical diode clipper with the battery removed.

16.4 SYNC SEPARATOR CIRCUITS

The separation of horizontal and vertical sync pulses from the video signal can be performed with many circuits. In this section we will begin with the simple diode sync separators and then advance to the more complicated tube and transistor circuits. The emphasis will be placed on the more recent transistor sync separators.

Diode Sync Separators A diode sync separator circuit is shown in Fig. 16-4. The video signal is applied between the anode and ground, while the output voltage is developed across the battery E and the load resistor R_L. The small battery is inserted with its negative end toward the anode, which prevents current from flowing until the video signal acting on the diode becomes sufficiently positive to counteract the negative biasing voltage. Current then flows. With the circuit constants properly chosen, current should flow only at the synchronizing pulses which are the most positive for a signal having a negative phase, and the output will consist only of these short pulses of current. The picture phase at the input of this diode must be negative, as in Fig. 16-4.

By inverting the diode, as in Fig. 16-5, it is possible to apply a positive-picture phase to the circuit and again obtain only the pulse tips across R_L. Note that the battery must also be inverted for this circuit. The dc biasing voltages necessary for these diodes may be taken from the low-voltage power supply.

It is generally not practical to use a bias battery or power-supply dc voltage for the diode-clipper. We require an arrangement that is completely automatic in its operation, altering its operating point as the amplitude of the received carrier varies. A simple, yet effective, circuit is shown in Fig. 16-6. The diode clipper uses the time constant of R and C to bias the circuit, so that all but the synchronizing pulses are eliminated. Capacitor C and resistor R form a low-pass filter with a comparatively long-time constant, equal to approximately ten horizontal lines. Therefore, the voltage developed across R (and C) will be determined by the highest voltage applied across the input terminals. This, of course, means the synchronizing pulses. Throughout the remainder of the line, although the video voltage is active, the anode is never driven sufficiently positive to overcome the positive cathode bias.

Triode Sync Separator Since a triode or a pentode can do anything a diode can, and provide amplification as well, it is natural to find these tubes used as sync separators. An application involving a triode sync separator is pictured in Fig. 16-7. The triode, V_1, is biased by grid-leak bias developed across C_1 and R_1. The pulses in the video signal fed to the sync separator triode V_1 possess the most positive polarity of the signal. Electrons will then flow in the grid circuit, charging C_1. Because of the high value of R_1, the charge on C_1 will leak off slowly, causing a fairly steady bias voltage to develop across the grid resistor. This bias voltage prevents plate current from flowing except

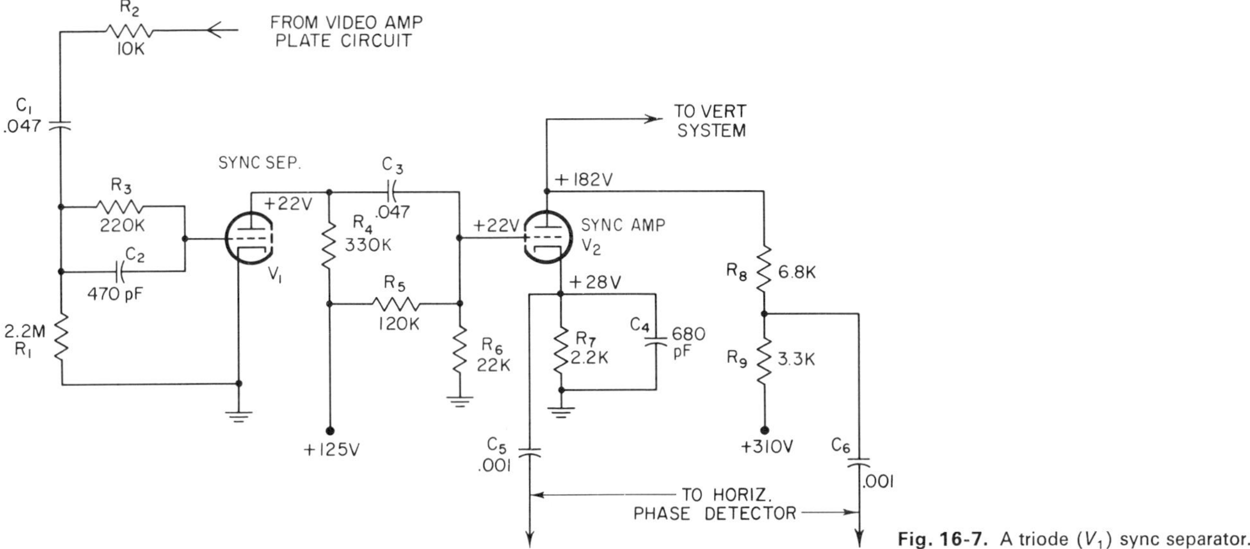

Fig. 16-7. A triode (V_1) sync separator.

for the most positive values of the incoming signal, which are the synchronizing pulses. A fairly low plate voltage causes V_1 to saturate readily, thereby tending to square off the sync pulses and to limit any noise pulses that may appear in the signal.

Resistor R_2 serves to isolate, or decouple, the sync separator tube from the video amplifier where the signal is obtained. In this way, the input circuit of V_1 and its capacity do not unduly load the video-amplifier plate circuit with subsequent deterioration of the image quality. R_3 and C_2 are inserted into the circuit to help minimize the effect of any noise pulses that may be present in the received signal. If a strong noise pulse (extending in the positive direction) should come along and R_3 and C_2 were not present, the resulting electron flow in the grid circuit of V_1 would charge C_1 to a fairly high negative voltage. Because R_1 is so large in value, the time required by C_1 to discharge enough through R_1 to enable the regular signal to again produce a current flow through V_1 might be so long that the set would lose synchronization. To avoid this, R_3 and C_2 are placed in the grid circuit. Now, when a noise pulse comes along, C_2 absorbs the additional current flow it produces. C_2 then discharges fairly quickly through R_3, which has a value one-tenth the value of R_1.

The video signal arriving at the grid of V_1 has an overall, or peak-to-peak, amplitude of 60 volts (see Fig. 16-8). However, the grid-leak bias developed by C_1 and R_1, permits V_1 to conduct only when the sync pulses are active. For the remainder of the video signal below the sync pulses, the tube is cut off by the bias. Also assisting in this action is the low-plate voltage of the tube.

Note that not all of the sync pulse produces a corresponding change in the plate circuit. Near the peak of the pulse, the grid is almost at a

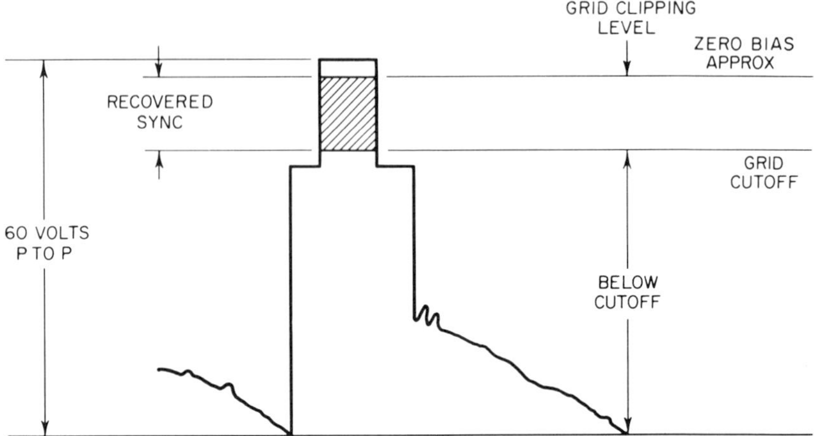

Fig. 16-8. The video signal at grid of V_1 in Fig. 16-7. Only the shaded portion is allowed to pass through V_1.

zero bias and the tube is passing as much current as it can with the low level of plate voltage. In other words, it is operating at saturation, and a further rise in input voltage produces little additional output voltage. This is the reason why the very tip of the sync pulse in Fig. 16-8 is shown unshaded. This small upper section does not develop much additional output voltage. In essence, this tends to square off the top of the output pulse.

From V_1, the sync pulses, which are now negative-going, travel to V_2, the sync amplifier. The grid of this tube is made 22 volts positive by R_5, which causes a considerable amount of current to flow in the grid and plate circuits, both currents returning to the cathode by way of R_7. The result is 28 volts at the cathode. This makes the grid negative with respect to the cathode by 6 volts. This voltage, together with the negative portion of the sync pulse from V_1, quickly drives V_2 into cutoff and helps to square off this end of the applied pulses. By the same token, since the grid already has a positive voltage (from B+), the positive portion of the pulses from V_1 cannot drive the grid far before plate-current saturation is reached. Thus, the positive portion of the signal is also clipped.

The sync amplifier effectively clips both the positive and negative extremes of the sync pulse fed to it from V_1. This amplifier section operates at considerably higher plate voltage, and the output is consequently greater than that of the sync-separator stage alone.

The sync pulses at the plate of V_2 are in the positive direction and are fed to the vertical-integrator network. At the same time, positive pulses from R_9 and negative pulses from the cathode are fed to the horizontal-phase detector.

Transistor Sync Separators The simple diode sync separator is easily converted into a transistor sync separator. This is possible because the transistor-input terminals (the base-emitter terminals) are actually a diode. The transistor merely adds gain to the basic diode sync separator. Transistors are ideally suited for sync separators because their input-output characteristics resemble those of a sharp cutoff pentode. Transistor sync separators will be examined in detail, so that the reader may appreciate why transistors fit this application so well.

Most newer TV-receiver designs use silicon transistors because of their superior temperature and reliability characteristics. A silicon transistor requires approximately 0.5 volt across its base-emitter terminals before it begins to conduct. As the base-emitter voltage varies from 0.5 to 0.7 volt, the transistor collector voltage will also vary. However, when the base-emitter voltage rises above 0.7 volt, the collector saturates at approximately 0.3 volt. These three situations are summarized by the following:

Input (base-emitter voltage)	Output at Collector (assume a 20 V collector supply)
below 0.5 volt	20 volts (cutoff)
varies from 0.5 to 0.7 volt	varies from 20 to 0.3 volts
above 0.7 volt	0.3 volt (saturation)

The above characteristics make it possible to perform sync separation, shaping, and gain in one transistor stage. Figure 16-9(A) shows the major part of the circuit required to perform these functions. We will show how the circuits of Figures 16-9(B), 16-9(C), and 16-9(D) are complete sync separators with the addition of only one more resistor, R, to the basic circuit of Fig. 16-9(A).

A video signal with negative picture phase is applied to the input of Fig. 16-9(A). Assume the peak-to-peak video voltage is 1 volt. This signal passes through C and is applied to the base-emitter terminals of Q_1. Before this signal rises to 0.5 volt, the collector voltage remains at +20 volts (cutoff). However, as the base voltage starts to rise above 0.5 volt, it begins to draw a current from C. At the same time, the collector voltage swings from +20 volts down towards 0.3 volt. When the input-video voltage is +1 volt peak, most of this voltage appears across the base-emitter terminals. However, when the video signal again drops down to zero, a 0.5 volt charge is left on the capacitor C. This charge remains, because after the transistor base stops drawing current (at 0.5 volt), the charge on C has no place to go. The polarity of this 0.5 volt charge is shown in Fig. 16-9(A).

When the other horizontal sync pulses arrive they find a new situation. They must overcome the 0.5 volt base-emitter cutoff voltage in series with the 0.5 volt charge across the capacitor C. If these pulses are the same voltage as the first pulse, namely +1 volt peak, then they will not cause the base to come out of cutoff. Now, how does the resistor R added to Fig. 16-9(B) make it a complete sync separator? If the sync pulse portion of the video wave-form is 0.25 volt in amplitude,

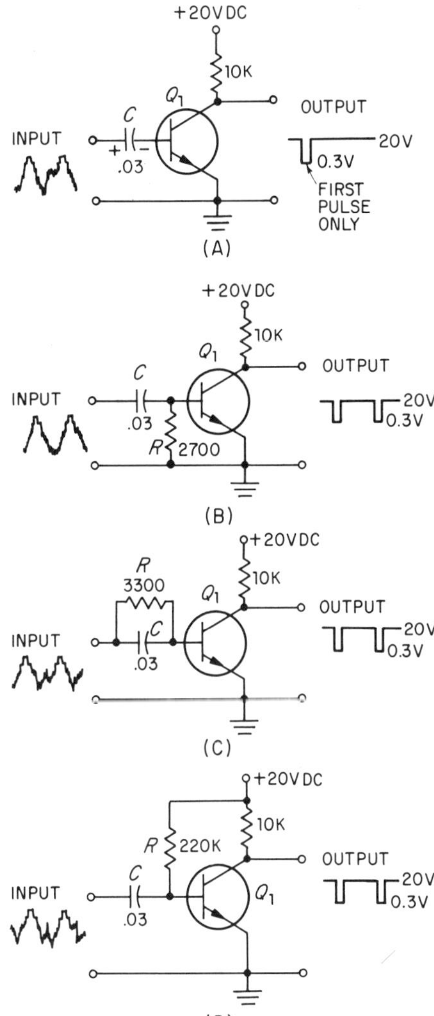

Fig. 16-9. The evolution of 3 types of transistor sync separators: (A) basic circuit; (B) discharge of C directly to ground; (C) discharge of C around itself; and (D) discharge of C to the power supply.

then we should lose exactly that much voltage from C during each horizontal scan period (63.5 microseconds). This will allow only the sync pulse to turn the transistor on. The rest of the video wave-form will be blocked by the remaining 0.25 volt on C and the 0.5 volt base-emitter cutoff voltage. Since $0.25 + 0.5 = 0.75$, the bottom 0.75 volt portion of the video signal will not pass through the sync separator.

Figure 16-9(B) partially discharges C to ground through a 2700-ohm resistor. Figure 16-9(C) partially discharges C from one of its plates to the other through a 3300-ohm resistor. Both of these circuits will have the same time constant, since R in Fig. 16-9(B) is not the total resistance around C. R must be added in a series to about 600 ohms from the previous stage to make 3300 ohms. Thus both figures have a time constant of 3300 ohms $\times$ 0.03 microFarads $=$ 100 microseconds. This is approximately the time of 1.5 horizontal scan periods (1.5 $\times$ 64 microseconds $=$ 96 microseconds). In one horizontal scan time C will discharge for 2/3 of a time constant. This is sufficient to remove 0.25 volt from C.

The circuit of Fig. 16-9(D) discharges C with a different method. A 220-kilohm resistor pulls electrons off the negative side of C to the positive supply voltage. Since this is a much more efficient (faster) way to discharge C, then R must have a large resistance, typically several hundred thousand ohms.

The resistor R in Fig. 16-9(B), 16-9(C), and 16-9(D) performs a critical function. It discharges the capacitor C by approximately the height of the sync pulse during each horizontal scan period. If R is too large, then C will not be sufficiently discharged. This will not allow the sync separator to utilize the full voltage of the sync pulse. If R is too small, then C will discharge too much. As a result, some of the blanking voltage or video signal will pass through the sync separator.

16.5 REDUCTION OF NOISE TO IMPROVE SYNCHRONIZATION

The appearance of noise in the sync pulses is much more serious than noise in the video or audio signals. The eye and ear tend to ignore a certain amount of video and audio noise. However, just one noise pulse at the right time may cause a receiver to momentarily lose vertical sync. A temporary loss of horizontal sync is also very annoying. For this reason, some receivers have a separate tube or transistor stage just to remove noise from the sync pulses. These stages are usually called noise cancellers or noise gates. We will now examine a transistor noise gate.

Transistor Noise Gate. The series method of noise cancellation uses a stage called the noise gate. A transistorized version of this device is shown in Fig. 16-10. The noise gate, Q_2, is in series with the emitter of

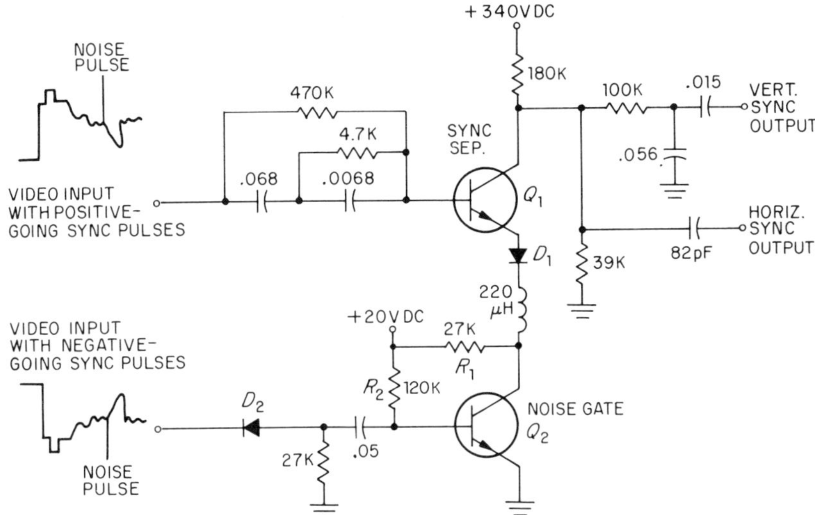

Fig. 16-10. A transistorized noise gate and sync separator. (*Courtesy of Sylvania.*)

the sync separator, Q_1. If the noise gate is fully on, as it normally is, then the voltage between its collector and emitter will be only 0.3 volt. The sync separator operates normally when the noise gate is on. If the noise gate is momentarily turned off then the sync separator-emitter rises to +20 volts. This is developed by the 20 volt supply connected to R_1. With a positive bias on the sync separator-emitter no sync separation can take place at this time.

What can cause the noise gate to turn off? Note that it is driven with video which has negative-going sync pulses. The sync separator is driven with a video of the opposite phase. The normal video signal applied to the base of the noise gate does not have sufficient amplitude to overcome the bias of R_2. However, if a noise spike with a negative amplitude beyond the sync tip appears, it will turn the noise gate off. Noise pulses which are not larger than the sync tip will not cause the noise gate to turn off.

16.6 COLOR TV SYNC SEPARATION

At this point one may ask how the sync separation process in a color receiver differs from that in a monochrome receiver. There is no important difference. However, since color receivers are usually higher priced instruments, we may find extra circuits in them such as noise cancellers. Better quality black and white receivers may also have these same circuits.

Chapter 21 mentions a modification of the horizontal sync pulse which is required for color TV. A burst of at least 8 cycles at 3.58 MHz is required on top of each blanking pulse just behind the sync pulse. This burst will not disturb the operation of the sync separator circuits, because its frequency is too high to be accepted by these circuits.

Another item to notice in color sets is the location of the sync pulse takeoff in the video amplifiers. These pulses are taken from the luminance (black and white) video amplifiers and not the chrominance (color) video amplifiers.

16.7 SERRATED VERTICAL SYNC PULSES

We will now discuss in further detail the various parts of the sync signals shown in Fig. 16-3. The serrated vertical sync pulses are a good place to begin. Consider first the basic form of the vertical sync pulse shown in Fig. 16-11. At the bottom of each field this basic vertical sync pulse is inserted into the signal. This controls the vertical synchronizing oscillator and forces the beam to be brought back to the top of the screen. No provision is made in the signal, in this preliminary form, for horizontal-oscillator control while the vertical pulse is acting. Such a condition is undesirable as it permits the horizontal oscillator to slip out of control. To prevent this, the vertical pulse is broken up into smaller intervals, then both actions can occur simultaneously. The vertical synchronizing pulse, in its modified form, is shown in Fig. 16-12 and is known as a "serrated vertical pulse."

While the vertical pulse is broken up to permit the horizontal synchronizing voltages to continue without interruption, the effect on the vertical pulse is substantially unchanged. It still remains above the

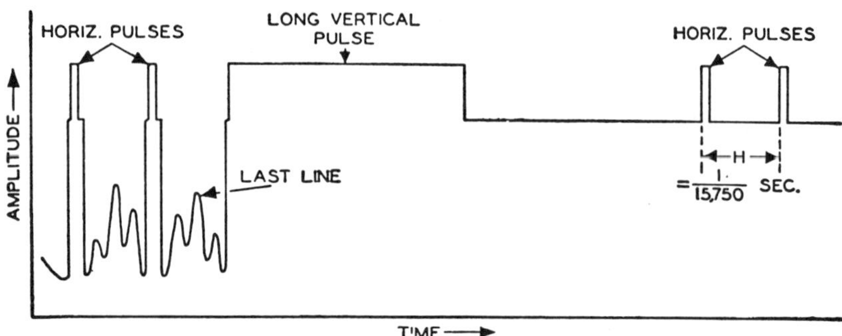

Fig. 16-11. The basic form of the vertical synchronizing pulse.

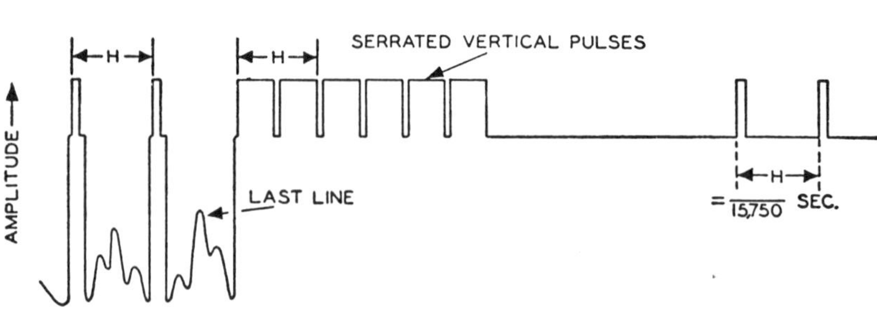

Fig. 16-12. The serrated vertical-synchronizing pulse.

blanking-voltage level practically all of the time it is acting. The interval is much longer than the preceding horizontal-pulse frequency. The two pulses are still capable of separation, however, because their waveforms are different, as is evident from Fig. 16-16.

Because an odd number of lines is used for scanning, the form of the signal just prior to the application of the serrated vertical pulse must be still further modified. With an odd number of lines, 525, each field contains $262\frac{1}{2}$ lines from the beginning of its field to the start of the next. This fact is important but has not been overly stressed before. Figure 16-13 indicates that the end of the visible portion of each field occurs at the bottom of the image. However, the actual end of that field is not reached until the beam has been brought back to the top of the screen again. At the end of the visible portion of the first field, the beam must be interrupted at point *D* and the vertical synchronizing pulse inserted. Point *D*, we see, occurs in the middle of a horizontal line. From *D*, the beam is brought up to point *E*, and the second field is begun. The visible portion of the latter field is completed at point *F*, the end of a complete horizontal line, and is returned to point *A* to repeat the entire sequence. These actions are mentioned here for review. The reasons for employing this particular method of scanning were explained in Chapter 2.

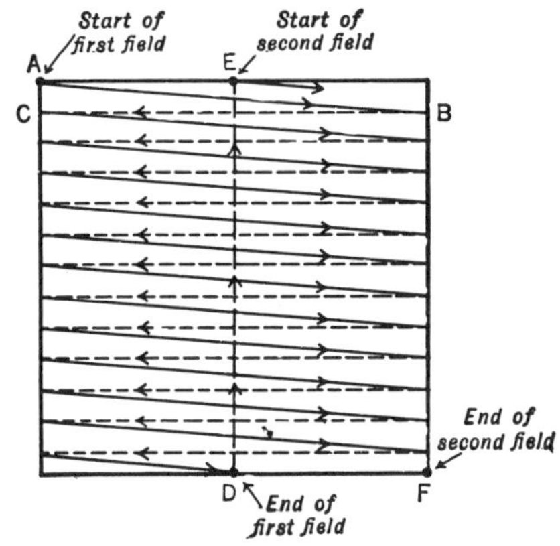

Fig. 16-13. The motion of the electron beam in interlaced scanning. For simplicity, the retrace from Point *D* to Point *E* has been shown as a straight line.

When the beam is blanked out at the bottom of an image and returned to the top, it does not move straight up; instead it moves from side to side during its upward swing. This movement is due to the rapidity with which a horizontal line is traced out as compared with the vertical retrace period. In fact, there are approximately 20 horizontal lines traced out while the vertical synchronizing pulses are bringing the beam back to the top of the picture. Thus, in each field, 20 horizontal lines are lost in the blanking interval between fields. Of the 525 lines which are sent out, only 525 − 2(20), or 485, are actually effective in forming the visible image.

The method of arriving at 20 horizontal lines is quite simple. The electron beam is blanked out for approximately 1,250 microseconds between fields while the beam is being shifted from the bottom to the top of the image. During this interval, the horizontal-sweep oscillator is also active. Thus, the beam, while it moves up under the influence of the vertical-deflection voltage, is also moving back and forth because of the horizontal-deflection oscillator. One horizontal line requires 1/15,750 sec, or 64 microseconds. Dividing this 64 into 1,250, we find that approximately 20 horizontal lines are traced out. In a frame, which contains two fields, 40 lines are thus lost. To see these retrace lines, turn up the brightness control on a television receiver when no station is being received and only the scanning raster is visible.

The fact brought out here, that the vertical pulse is once inserted into the video signal when a horizontal line is half completed and once inserted into the video signal at the end of a complete line, necessitates a further modification of the video signal just prior to the arrival of the

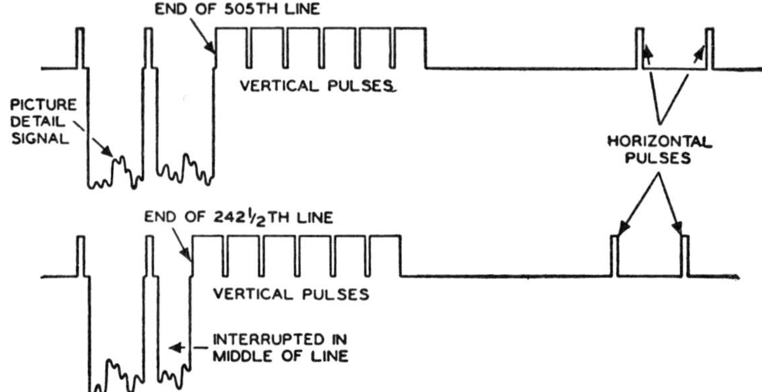

Fig. 16-14. The form of the video signal at the end of 242½ and 505 lines. The equalizing pulses are not shown.

vertical pulse. A serrated video signal for each case is illustrated in Fig. 16-14. The half-line difference between the two diagrams may not affect the horizontal-synchronizing-generator operation, but it can cause the vertical oscillator to slip out of control.

To have the vertical pulse oscillator receive the necessary triggering voltage at the same time after every field, a series of six equalizing pulses is inserted into the signal immediately before and after the vertical synchronizing pulses. These equalizing pulses, shown in Fig. 16-15, do not disturb the operation of either oscillator (as will be shown later), yet they do permit the vertical pulse to occur at the correct time after every field.

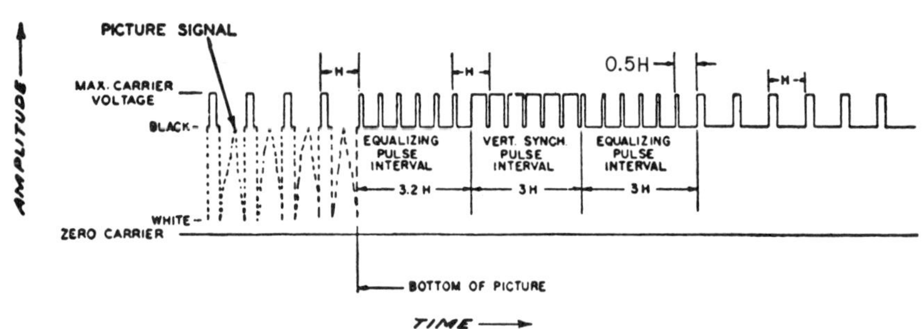

Fig. 16-15. The position of the equalizing pulses in the video signal.

Once the serrated vertical pulse is ended, the six equalizing pulses are again inserted into the signal, and the line detail assumes control while the next field is swept out. One vertical pulse occurs at the end of every 262½ lines, while a horizontal pulse appears at the end of each line.

16.8 VERTICAL AND HORIZONTAL PULSE SEPARATION

The separation of the vertical and horizontal pulses from each other is based on their frequency (or wave-form) difference and not on their amplitude since the latter is the same for both. The two pulses are shown

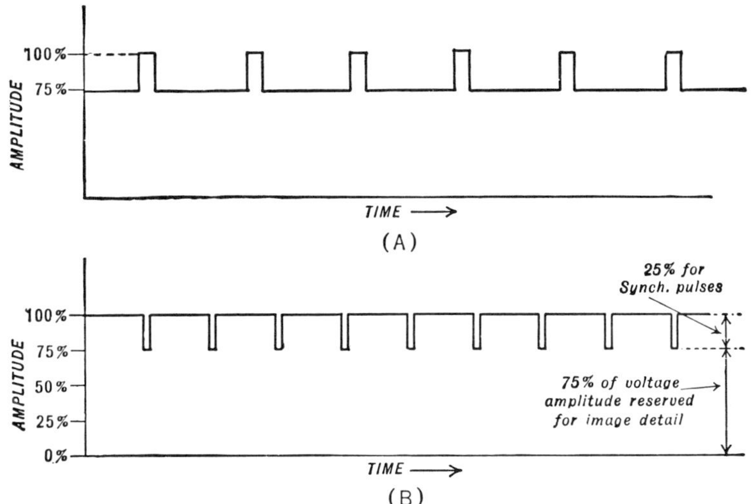

Fig. 16-16. The differences in wave-forms at (A) horizontal and (B) vertical pulses.

in Fig. 16-16. Note that the horizontal pulse is much shorter in duration than the vertical pulse, rising and falling in 5 microseconds, Essentially, then, a low-pass filter will develop the vertical-pulse voltage at its output, while a high-pass filter will have only the horizontal-pulse voltage at its output. These two distinct pulses can then be fed to their respective oscillators, and control them in accordance with the requirements of the signal being received.

Horizontal Sync Pulse Filters Let us consider a high-pass filter and its effect on the horizontal sync pulses (see Fig. 16-17). The filter in the diagram has a time constant of

$$\begin{aligned} T &= RC \\ &= 2000 \text{ ohms} \times 50 \text{ pF} \\ &= 2000 \text{ ohms} \times .00005 \text{ microFarads} \\ &= 0.1 \text{ microseconds.} \end{aligned}$$

A time of 0.1 microsecond is short compared with the 5 microseconds duration of the horizontal sync pulse.* At the application of the first

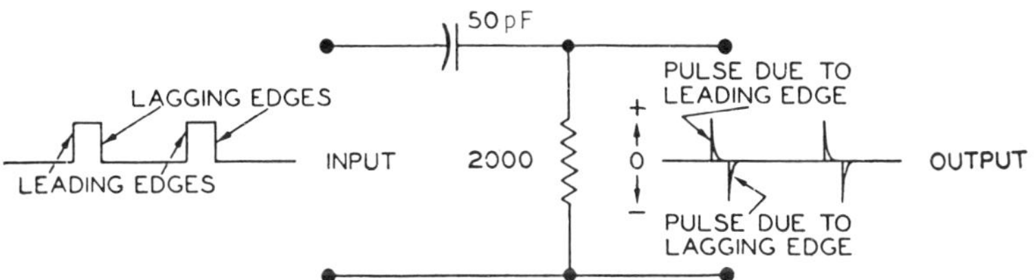

Fig. 16-17. A high-pass filter and its effect on the horizontal-synchronizing pulses.

* Any time constant one-fifth the duration of an applied pulse is said to be short with respect to that pulse. By the same token, any time constant five times longer than the duration of an applied pulse is said to be a long time constant.

edge of the horizontal pulse in Fig. 16-17, known as the "leading edge," a momentary flow of current takes place through the resistor to charge the capacitor to the full-pulse voltage. Once the capacitor has become fully charged, nothing further occurs all along the flat portion of the pulse, because a capacitor (and hence, a capacitor and a resistor in series) reacts only to changing or ac voltages, not to steady or dc voltages. At the lagging edge of the pulse, where the voltage drops suddenly, another short flow of current takes place, this time in the opposite direction, discharging the capacitor. The result of the application of the square-wave sync pulse to the input of the high-pass (or short-time constant) filter is the output wave indicated in Fig. 16-17.

Each incoming synchronizing pulse gives rise to two sharp pulses at the output of the filter, with one above and one below the reference line. This, of course, is due to the fact that one is obtained when the front edge of the incoming pulse acts on the filter, and one when the lagging edge arrives.

For control of the sweep oscillator, only one of these two output pulses is required. If the first pulse at the output of the filter is negative (below the line) and a positive pulse is required, the conversion is readily made. Merely apply these pulses to an amplifier, and the first pulse becomes positive. The amplifier introduces a phase shift of 180 degrees, which is equivalent to reversing every value in a wave. The oscillators that are used, respond to the first pulse, becoming insensitive immediately thereafter to other pulses that do not occur at the *proper point* in the oscillator period. When the next horizontal pulse arrives, it is again in position to control the oscillator action. In this manner, any pulse occurring at an intermediate interval is without effect. One or two exceptions will be noted later.

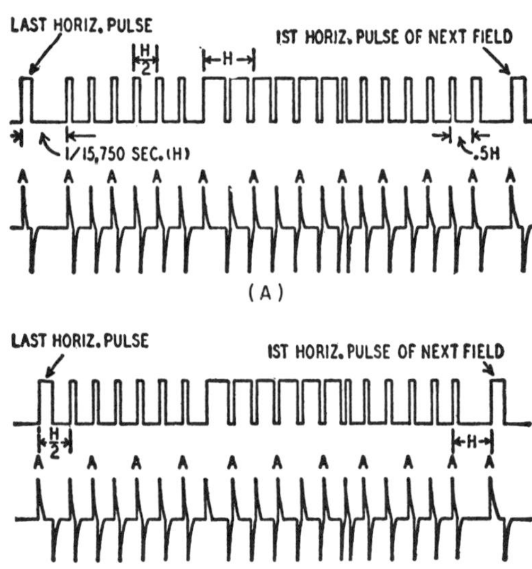

Fig. 16-18. The conditions during vertical pulses in the (A) First Field and (B) Second Field (see Fig. 16-3). The pips labeled *A* control the horizontal-sweep oscillator.

The foregoing action of a high-pass filter indicates how the serrations of the vertical pulse permit control of the horizontal synchronizing oscillator during the application of the vertical pulse. In Fig. 16-18 the input wave and the output pulses of a high-pass filter are shown. Of all those present, only the positive pulses that occur at the proper time (1/15,750 sec) affect the horizontal oscillator. These active pulses are indicated by *A* in the figure. Note that all active pulses are evenly spaced and differ by 1/15,750 second. The conditions illustrated in Fig. 16-18(A) occur only when the vertical pulses are inserted at the end of a full line. Figure 16-18(B) shows the situation when the field ends on a half line. Now the same equalizing and serrated pulse pips are not active in controlling the horizontal oscillator. Because of the difference in field ending, the control has shifted to those pips which were shown to be inactive in Fig. 16-18(A). However, the shift has in no way interfered with the timing in the control pips. This shift from field to field illustrates why all the equalizing and vertical pulses are designed to produce pips twice in each horizontal-line interval.

The long vertical pulses are without effect on this filter because of its short-time constant and, further, because the output is obtained from across the resistor. As soon as a vertical pulse is applied across the terminals of the filter, a short, sharp current fully charges the capacitor. With the capacitor charged to the full voltage value, no further current flows through the resistor until another change occurs. The output is taken from across the resistor and, with current flowing only a very short time, a short, sharp pulse of voltage is obtained. At the lagging edge of the input wave, another quick flow of current brings the capacitor voltage back to its previous value, and again a voltage pulse develops across the resistor. Hence, only *changes* in the input wave appear across the output resistor, because it is only at these times that a current flows in the filter, either to charge or to discharge the capacitor. The serrations inserted in the vertical pulse provide the changes that cause current to flow in the high-pass filter. Thus control can be maintained at the horizontal oscillator even when the vertical pulse is acting.

Vertical Sync Pulse Filters For vertical pulse separation, we use a low-pass or long-time constant filter of the type shown in Fig. 16-19. This appears identical with the high-pass filter, except that the positions of the capacitor and resistor have been interchanged and the output is obtained from the capacitor. Besides the difference of position, the time constant of the capacitor and resistor is much greater than that of the previous filter. A long-time constant means that the capacitor will charge and discharge slowly and will not respond as readily as a short-time constant filter to rapid changes in voltages. Hence, when a horizontal pulse arrives at the input of this filter, its leading edge starts a *slow* flow of current through the resistor, and the capacitor begins to charge. But this charging process is slow and, almost immediately afterward, the lagging edge of the wave reaches the filter and reverses the current flow, bringing the capacitor back to its previous value. Very little change has occurred during this short time interval. And the vertical-synchronizing oscillator is designed so that it does not respond to these small fluctuations.

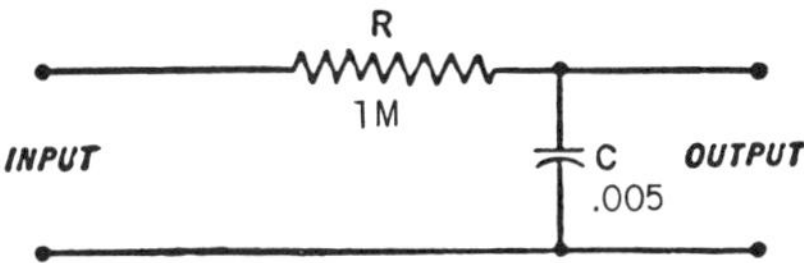

Fig. 16-19. A low-pass filter. The capacitor combines (or integrates) all the serrated vertical pulses until the output voltage rises to the level necessary for the vertical-sync oscillator to react.

What is true of the effect of the horizontal pulses on the vertical filter is even more true with respect to the equalizing pulses, which rise and fall much more rapidly. Essentially, then, we have eliminated the possibility of the higher-frequency pulses affecting the operation of the vertical-synchronizing generator. Figure 16-20 indicates the output voltage of the filter on the application of these higher frequency waves. Their voltage level is below the dotted line, which represents the point that the voltages must reach in order to affect the generator.

The building up of the voltage across the capacitor for the output begins when the serrated vertical pulses are reached. Even though the

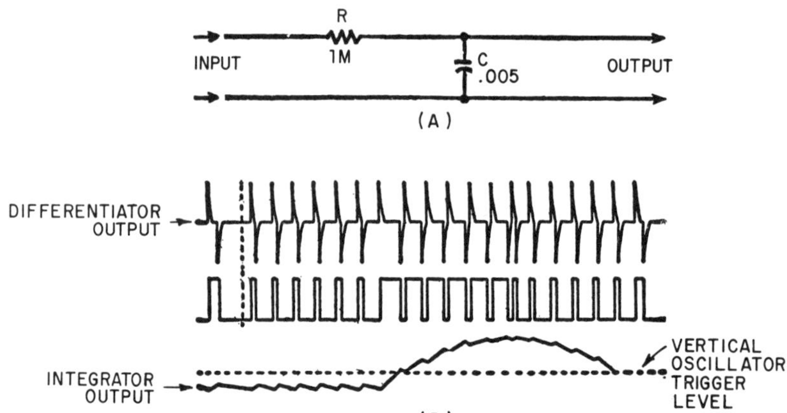

Fig. 16-20. (A) Low-pass filter for separating the vertical and horizontal pulses. (B) A wave-form of the rise in voltage across the capacitor due to vertical pulses.

pulse is serrated, it still remains above the reference line for a relatively long time. The capacitor charges slowly in the manner indicated in Fig. 16-20. The small notches in the wave are due to the serrations. At these points, for a fraction of a second, the voltage drops and then rises again. As previously noted, these changes affect the horizontal filter but leave the vertical filter output substantially unchanged because of their rapid disappearance.

16.9 EQUALIZING PULSES

We can pause for a moment here and determine more clearly the reason for the equalizing pulses. In Fig. 16-21 the build-up of a vertical deflection voltage across the output of the vertical filter is shown, once for the

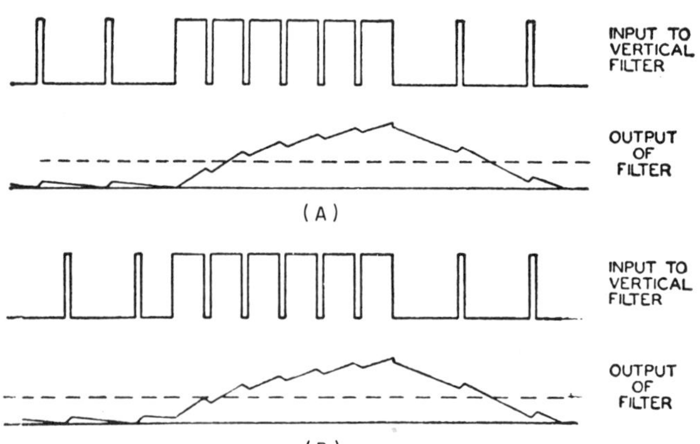

Fig. 16-21. The difference in voltage conditions before a vertical pulse when no equalizing pulses are used: (A) First Field and (B) Second Field (see Fig. 16-3).

vertical pulse that comes at the end of a line and once for the pulse that comes in the middle of a line. In Fig. 16-21(A), we see that each horizontal pulse causes a slight rise in voltage across the output of the vertical filter, but this is reduced to zero by the time the next pulse arrives. Hence there is no residual voltage across the vertical filter due to the horizontal pulses. Only when the long, serrated vertical pulse arrives is the desired voltage increase obtained.

However, the situation in Fig. 16-21(B) is slightly different. Here the last horizontal pulse is separated from the first vertical pulse by only half a line. Any horizontal voltage developed in the vertical filter will thus not have as much time to reach zero before the arrival of the first vertical pulse, which means that the vertical build-up does not start from zero, as in the top illustration, but from a low voltage value. As a result, the dotted line is reached sooner than it would be reached if the voltage rise had started from zero. Since the dotted line represents the firing point of the vertical oscillator, we see that the oscillator is triggered a fraction of a second too soon. The actual time involved is quite short, but it does prove sufficient to upset the precision interlacing of modern television images.

With the insertion of equalizing pulses before and after every vertical pulse, the voltage level established before the start of each vertical serrated pulse is essentially the same, and the vertical oscillator is triggered at the proper moment in each instance.

After the complete vertical pulse has passed through the filter, the charge on the capacitor output gradually returns to the small value it had previously, the voltage due to the horizontal pulses (Fig. 16-21). These pulses develop a very small voltage, far from sufficient to affect the vertical oscillator. Only the larger, longer vertical pulse 1/60 sec later accumulates enough voltage to trigger the oscillator.

From a comparison of the vertical- and horizontal-pulse forms shown in Fig. 16-20(B), we may get the impression that the vertical pulse is not very sharp. The reason is the vertical pulse is shown extended over quite a few horizontal pulses, and the comparison exaggerates the extent of the vertical pulse. If the vertical pulse were drawn to a larger interval, then it too would appear sharp. So far as the vertical synchronizing oscillator is concerned, this pulse occurs rapidly and represents a sudden change in voltage.

The polarity of the pulses, as obtained at the output of their respective. filters, may or may not be suitable for direct application to the controlled synchronizing oscillators. It all depends upon the type of oscillator to be controlled. For a blocking oscillator, the leading pulse must be positive. If a multivibrator type of oscillator is employed, either a positive or a negative pulse may be used, depending upon where it is introduced. This point will be more fully developed in Chapter 17.

16.10 VERTICAL AND HORIZONTAL BLANKING

We will now briefly look at the blanking portion of the video waveform. Blanking is required during both the vertical and horizontal retrace periods.

Horizontal Without horizontal blanking, the contrast of the picture would be impaired. The retrace lines would appear in between each normal scan line. If these retrace lines are bright enough, then they will give the picture a light colored background. This would be quite objectionable. However, the horizontal-blanking pulse does not allow this problem to occur. Under various signal levels, the dc restoration circuit (see Chapter 13) makes sure that the horizontal-blanking pulse is always effective.

Vertical Blanking the video signal during the entire vertical retrace period is a little more difficult than horizontal blanking. It is quite common to see all 20 retrace lines make their appearance on a TV receiver screen (see Section 16.7). These lines are caused from the-horizontal oscillator tracing out 20 scans during the vertical retrace period. Under varying signal levels, brightness settings, and contrast settings, these lines will occasionally appear. The existing blanking pulses in a video signal do not always handle these varying conditions. For this reason, many receivers use a vertical-blanking circuit. This circuit is usually a simple *RC* network between the vertical-output stage and the picture tube. It applies a large negative pulse to the picture tube cathode, which completely darkens the screen during vertical retrace. A typical vertical-blanking circuit is shown in Fig. 16-22. The resistor is required to attenuate the vertical-blanking pulse since its amplitude may be over 1000 volts. The capacitor is required for dc isolation.

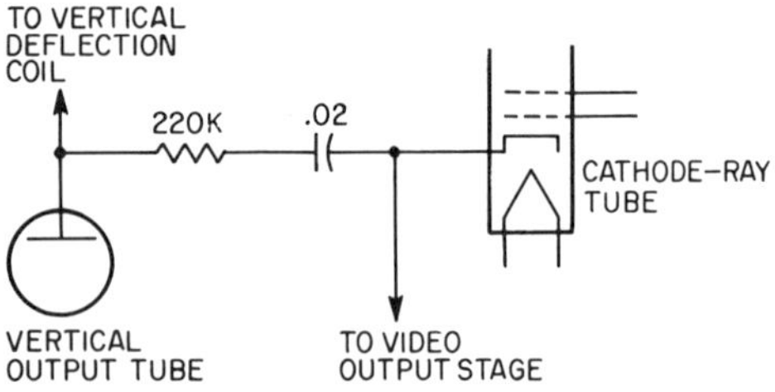

Fig. 16-22. A typical vertical blanking *RC* network.

16.11 TROUBLES IN SYNC CIRCUITS

When a problem arises in the sync amplifier or sync separator, it usually affects both the horizontal and vertical synchronization. A defect in the vertical sync filter, or beyond, will affect only the vertical sync. Likewise the horizontal sync filter, and following circuits, only affect the horizontal stability. Many other circuits in a TV receiver also have a strong influence on horizontal and vertical stability. Any defective stage preceding the sync stages may distort the video signal, thereby leaving a poor wave-form upon which the sync stages must work. The horizontal and vertical oscillators also can fail to utilize their sync pulses, thus causing loss of sync.

We can summarize the troubles caused by sync circuits (plus associated circuits) in the following table:

Circuit	Trouble Observed
Sync Amplifier and Sync Separator	1. Total loss of vertical and horizontal sync 2. Unstable vertical and horizontal sync 3. Horizontal pulling of picture 4. Vertical jitter
Any stage prior to the sync stages	Any or all of the above, usually combined with a poor picture
Vertical Pulse Filter or Vertical Oscillator	1. Total loss of vertical sync 2. Unstable vertical sync 3. Vertical jitter
Horizontal Pulse Filter or Horizontal Oscillator	1. Total loss of horizontal sync 2. Unstable horizontal sync 3. Horizontal picture pulling

REVIEW QUESTIONS

1. Do the horizontal and vertical synchronizing pulses ever reach the control grid of the cathode-ray tube? Explain.
2. Draw a diagram of a complete video signal, indicating where the horizontal sync pulses are located.
3. State specifically the action of the horizontal and vertical sync pulses in controlling the motion of the electron beam.
4. Explain what requirements must be present before the sync pulses can be separated from the rest of the video signal.
5. Why are transistor clippers more desirable than diode clippers?
6. Why do the blanking voltages last longer than either the vertical or the horizontal synchronizing pulses? What would happen if the blanking voltages were too short?
7. Explain why serrated vertical pulses are employed.
8. What is an active line? What is the approximate number of active lines per frame? Indicate how your figure was obtained.
9. Indicate the location and time duration of the equalizing pulses.
10. Why are equalizing pulses necessary? Illustrate your answer by means of diagrams.

17 Deflection Oscillators and Horizontal AFC

17.1 INTRODUCTION

In order to form a picture on the screen of the TV receiver that is synchronized with the one generated at the transmitter, it is necessary to first produce a synchronized raster. Having produced such a receiver raster, the video information automatically "paints" a copy of the transmitted picture.

While the actual movements of the electron beam (or beams) in the picture tube are controlled by magnetic fields produced in the horizontal and vertical yoke coils, the proper synchronized horizontal and vertical scanning rates must first be created by synchronized oscillators. This then, is the function of the horizontal and vertical deflection oscillators. These oscillators and their closely associated circuits must create suitable driving wave-forms at the correct synchronized frequencies. For vertical deflection, the frequency is 60 Hertz, while for horizontal deflection, the frequency is 15,750 Hertz. In color transmissions, these frequencies are slightly different. (The principles of scanning and synchronization were explained in Chapter 3.)

The driving waveforms created by the vertical and horizontal deflection oscillators are applied to power amplifiers, which are required to provide sufficient current to drive the yokes. As we will see, these driving waveforms may be either pure or modified sawtooth waves depending upon the deflection frequency and whether we are using vacuum tube or transistor circuits. The basic types of deflection oscillators are the same for vacuum-tube and transistor circuits, as well as for monochrome and color circuits. However, as will be shown in this chapter, some color sets may use circuits not normally found in monochrome sets.

In general, monochrome TV receivers employ blocking oscillators or a variation of the multivibrator oscillator, wherein the vertical output tube serves as part of the oscillator. Color-TV receivers may employ the same types for vertical deflection and, in addition, other circuits, such as the RCA "Miller Integrator" circuit will be found. All of these types of vertical deflection oscillators will be explained in this chapter.

In the case of the horizontal deflection oscillator, these are always accompanied by some type of automatic frequency control, to stabilize the oscillator frequency and render it immune to noise impulses. Both monochrome and color-TV receivers employ similar horizontal oscillator types. The most commonly used types are the blocking oscillator, the multivibrator, and the Hartley (or Colpitts) oscillator.

While the deflection oscillator circuits for both monochrome- and color-TV sets may be quite similar, there are important differences in the output amplifier circuits. These are explained in Chapters 18 and 19.

17.2 REQUIREMENT FOR SAWTOOTH CURRENT

The electron beam must be scanned across the screen at the horizontal rate of 15,750 Hz. At the same time, it is moved slowly down the screen at the vertical rate of 60 Hz. Its path, as explained in Chapter 2, is not straight across the screen, but tilted slightly downward. At the end of the line, it is brought rapidly back to the left-hand side of the screen. The type of current in the horizontal- and vertical-deflection windings that will accomplish this motion is the sawtooth wave shown in Fig. 17-1. This wave gradually rises linearly and then, when it reaches a certain height, returns rapidly to its starting value. The process then repeats itself, 15,750 times a second for the horizontal oscillator and 60 times a second for the vertical oscillator. Before sawtooth waves of current can be produced, it is first necessary to create sawtooth waves of voltage. The voltage wave is then used to drive a power amplifier, which suppliers the sawtooth current.

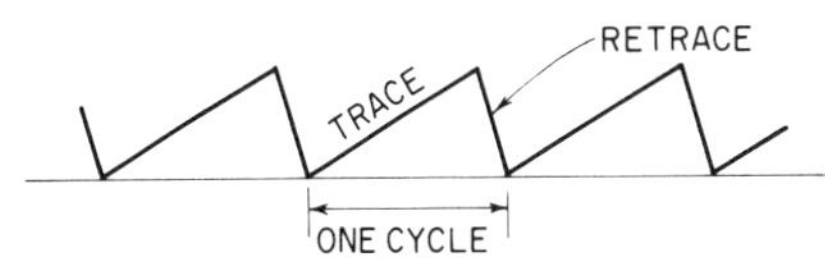

Fig. 17-1. A sawtooth-wave current. This type of current is required for the horizontal- and vertical-deflection coils.

Practically, the simplest way of intially obtaining the desired gradual rise followed by a sudden drop is by charging and discharging a capacitor. If a capacitor is placed in series with a resistor and a source of voltage, the flow of current through the circuit will cause the voltage across the capacitor to rise in the manner indicated by the curve of Fig. 17-2. This curve is not linear along its entire length, but approximation to linearity at the beginning section of the curve is close enough for most

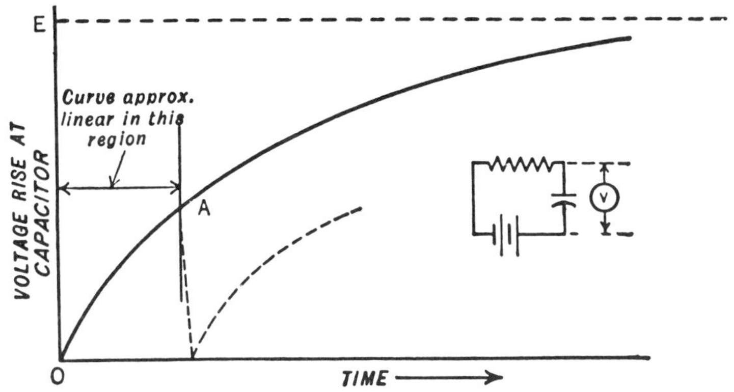

Fig. 17-2. The manner in which a voltage across a capacitor increases when a potential is applied through a series resistor.

practical purposes. Hence, if the capacitor is discharged just as it reaches point A on the curve, we will have a satisfactory sawtooth wave. The discharge of the capacitor should be as rapid as possible, since during the time the capacitor is discharging the electron beam is blanked out at the picture tube and no picture detail appears on the screen. The shorter the time spent in discharging the capacitor, the greater the interval during which the useful portion of the video signal may be acting at the screen.

Need for Trapezoidal Waveforms In electromagnetic deflection systems, the driving force in the picture tube is a magnetic field and, to develop such fields, current is required, a *SAWTOOTH DEFLECTION* current. However, in order to achieve a sawtooth-current flow through the deflection coils, we frequently must apply to these coils a voltage that possesses a modified sawtooth form.

The form of the voltage wave to be applied to the deflection coils is derived by analyzing the components of the coils and their action when subjected to voltages of various shapes. Each coil contains inductance plus a certain amount of resistance. So far as the resistance is concerned, a sawtooth voltage will result in a sawtooth current (see Fig. 17-3(A)).

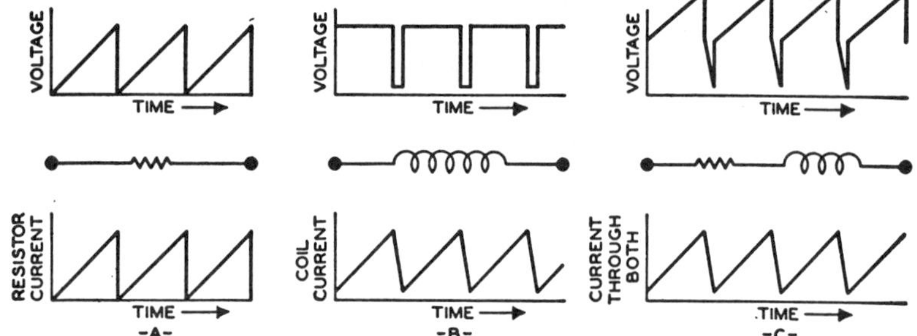

Fig. 17-3. By applying the voltage shown above the electrical component, the sawtooth waves shown below the component are obtained.

For the inductance, considering a pure inductance, a voltage having the form shown in Fig. 17-3(B) is needed for sawtooth-current flow. Combining both voltage waves, the result obtained varies in the manner shown in Fig. 17-3(C). A voltage of this type, when placed across the deflection coils, will give a sawtooth current, and the magnetic flux, varying in like manner, will force the electron beam to sweep across the screen properly. Note carefully that the resultant wave is not obtained by combining the two voltage waves in equal measure. If the deflection circuit contains more inductance than resistance, the resultant wave will be closer in form to Fig. 17-3(B). On the other hand, if the resistance predominates, then the resultant wave will more closely resemble Fig. 17-3(A). Hence, one may expect to find variations of this deflection wave ranging from almost a pure sawtooth wave to that shown in Fig. 17-3(C).

In vertical deflection circuits, which are driven by vacuum tubes, the ratio of the yoke inductive reactance to its resistance, is such that the resistance is an appreciable part of the ratio. In this case, it is generally found that a trapezoidal wave of voltage across the vertical yoke is required to produce a sawtooth of current through it. In vacuum-tube type horizontal deflection circuits, however, the situation is quite different. In this case, due to the much higher deflection frequency, the reactance of the horizontal yoke is very much higher than its resistance. Consequently, the horizontal yoke looks almost like a pure inductance and a square wave of voltage across the yoke will produce a sawtooth current through it. Note that the square wave of voltage across the yoke is produced by applying a sawtooth driving voltage to the input of the horizontal-output amplifier.

When transistors are used for the vertical and horizontal output amplifiers, you will find that the input wave to the output transistor is generally not trapezoidal, but is close to a pure sawtooth. The reason for this is that the transistor represents a very low driving impedance for the yokes, which now appear to be mainly inductive. When tubes are used for the output amplifiers, the series plate resistance must be considered as a part of the yoke resistance. This series plate resistance is much higher than the equivalent yoke driving resistance of a power transistor.

The next problem is to generate this trapezoidal voltage. It was found that this could be accomplished readily by obtaining the output from the charging capacitor and a series resistor in place of the capacitor alone. The circuit is shown in Fig. 17-4. This modified waveform is generated in the output circuit of the sweep oscillator. Resistors R_1, R_2, and capacitor C_1 form the normal sawtooth. A smaller value peaking resistor, R_3, connected to the side of C_1 which is normally grounded, provides the additional modification which results in a trapezoidal shape.

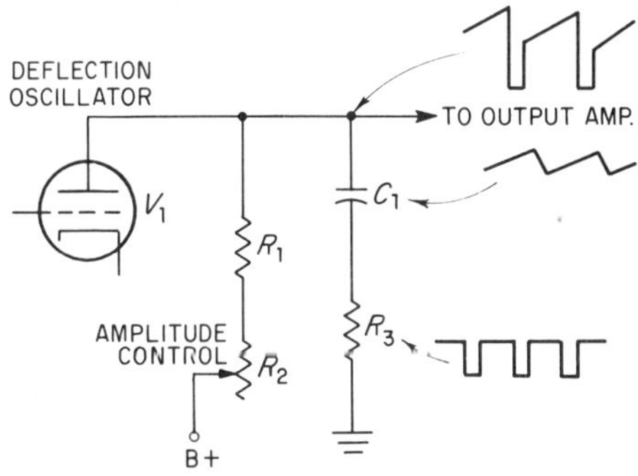

Fig. 17-4. Method of generating deflection voltages required to produce sawtooth currents in deflection coils.

With V_1 cutoff, C_1 charges through R_1, R_2, and R_3. R_3 is small in value compared to R_1 and R_2 and during this interval produces the positive portion of the rectangular wave across it. During the retrace period, with V_1 conducting, C_1 discharges through V_1 and R_3. The discharge current which now flows through R_3 and V_1 develops the negative segment of the rectangular wave across R_3. This rectangular wave and the sawtooth add to form the trapezoidal voltage waveform required by a tube-type, vertical-output amplifier.

17.3 THE VACUUM TUBE BLOCKING OSCILLATOR

The blocking oscillator is one of the popular deflection oscillators used in television receivers. In common with all oscillators, feedback of energy from the plate to the grid must occur and a transformer is employed for this purpose. Any change of current in the plate circuit will induce a voltage in the grid circuit which will aid this change.

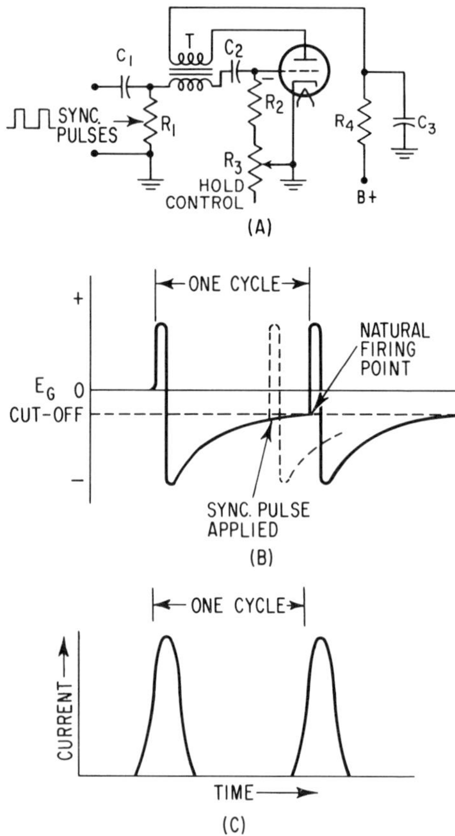

Fig. 17-5. (A) A blocking oscillator. (B) The grid-voltage variation. (C) The form of the plate current.

To examine the situation in detail, consider the operation of the oscillator when a disturbance occurs in the circuit which increases the plate current. To aid this increase, a positive voltage is induced in the grid through transformer *T* (Fig. 17-5(A). With the grid more positive than before, more plate current will flow, resulting in the grid becoming rapidly positive. A positive grid means that electrons will flow in the circuit, charging capacitor C_2. The electrons reaching the grid pile up on the right-hand plate of C_2. If resistors R_2 and R_3 were low in value, the charge on the capacitor would leak off rapidly and the action of the oscillator would continue. In practice however, R_2 and R_3 are made high, combining with C_2 to give a long time constant. The electrons on C_2 discharge slowly to the cathode, placing a negative voltage on the grid, as shown in Fig. 17-5(A).

Because of the slow discharge of C_2, electrons which have accumulated remain in sufficient numbers to give the grid a large negative bias, sufficient to block or stop the plate-current flow. Gradually the electrons accumulated on C_2 pass through R_2, R_3, and R_1 back to the other plate of C_2 and the negative bias on the grid slowly becomes less. When the discharge is such that the grid bias becomes less that cutoff, now the plate current starts up, quickly reaches its high value, drives the grid positive, and the process repeats itself. Thus, during every cycle there is a short, sharp, pulse of plate current, followed by a period during which the tube is blocked until the accumulated negative charge on C_2 leaks off agin. The frequency of these pulses is determined by C_2, R_1, R_2, and R_3.

The form of the voltage drop across R_2 and R_3 can be seen in Fig. 17-5(B). In Fig. 17-5(C), the plate-current pulse occurs once in every cycle. It is possible to control the frequency of this oscillator if a positive pulse is injected into the grid circuit at the time indicated in Fig. 17-5(B). To be effective, the frequency of the controlling pulse must be near and slightly higher than the free frequency of the oscillator. By free frequency is meant the natural frequency at which it will oscillate if permitted to function alone. This frequency is controlled by C_2, R_1, R_2, and R_3.

The point at which the synchronizing pulse should be applied to the grid of the oscillator is illustrated on the curve of Fig. 17-5(B). A positive pulse, applied to the oscillator grid when it is at this point of its cycle, will bring the tube sharply out of cutoff and cause a sharp pulse of plate current to flow. Then, at the application of the negative pulse of the horizontal synchronizing voltage which follows immediately, the oscillator is no longer in any position to respond. The grid has now become so negative that it is unaffected by the second negative synchronizing pulse. It is only when the grid capacitor C_2 is almost completely discharged that any pulse will effectively control the frequency of the oscillator. This accounts for the firm control of the synchronizing pulses. Equalizing pulses which occur at the halfway point in the oscillator cycle do not possess sufficient strength to bring the tube out of cutoff. This also explains why a positive synchronizing pulse is required.

In short, then, the synchronizing pulse controls the start of the oscillator cycle. If left alone, the oscillator would function at its natural period which, more often than not, would not coincide with the incoming signal. Through the intervening action of the synchronizing pulse, both oscillator and signal are brought together, in step. Naturally, for effective control, both the synchronizing pulse and oscillator frequency must be close enough together to permit the locking-in of the oscillator.

The resistor R_3 is made variable in order to provide adjustment of the oscillator frequency. It is commonly known as the "hold control," since it can be varied until the frequency of the blocking oscillator is held in synchronism with the incoming pulses.

A simple and inexpensive method of charging and discharging a capacitor to produce sawtooth waves is given in Fig. 17-6. The triode

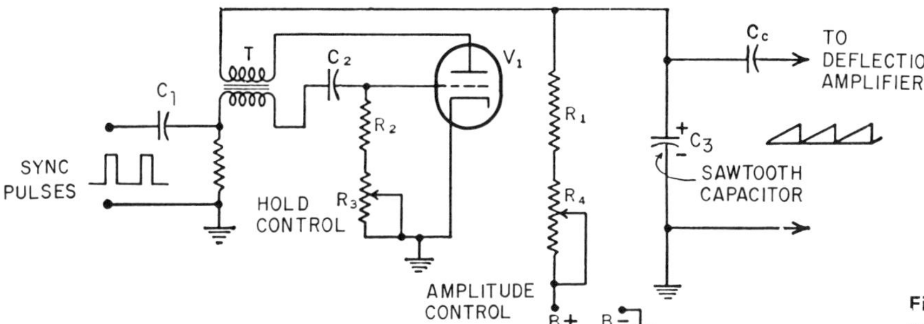

Fig. 17-6. A simple method of obtaining a sawtooth voltage from a capacitor using a blocking oscillator.

V_1 is connected as a blocking oscillator, and the sawtooth capacitor, C_3, is placed in the plate circuit. From the preceding discussion of the operation of these oscillators, we know that a short, sharp pulse of plate current flows once in every cycle. During the remainder of the time, the grid is negatively biased beyond cutoff and no current flows in the plate circuit.

During the time no plate current is flowing, C_3 is charging, because one side of this capacitor connects to the positive terminal of the power supply through resistors R_1 and R_4, and the opposite side is attached to ground. The charge that the capacitor absorbs assumes the polarity shown in Fig. 17-6.

When plate current starts to flow, it is for only a very short period, and during this time the resistance of the tube becomes very low. Capacitor C_3, which is actually in parallel with the tube, then quickly discharges through this low-resistance path. At the end of the short pulse of plate current, the grid has been driven very negative by the accumulation of electrons in C_2, and the tube becomes nonconducting again. C_3 no longer has this easy path for discharging and slowly starts to charge, as previously explained. The sawtooth variation in voltage across C_3 is transmitted to the next tube, an amplifier, through coupling capacitor C_c. The process repeats itself, either at the horizontal-scanning frequency or at the vertical frequency, depending upon the oscillator constants.

It will be noted from the foregoing action that the instant the synchronizing pulse arrives at the oscillator, it triggers the oscillator, the tube becomes conducting, and the capacitor developing the sawtooth voltage discharges. Hence, whenever a pulse arrives at the grid of the blocking oscillator, the capacitor discharges and the electron beam is brought back from the right-hand side of the screen to the left-hand side. This action is true in all such synchronizing oscillators.

Resistor R_4 may be made variable to permit adjustment of the amplitude of the sawtooth. As more of its resistance is placed in the circuit, the amount of charginng current reaching C_3 is lessened, with a subsequent decrease in the voltage developed across C_3 during its period of charging. A small voltage variation at C_3 means, in turn, a small voltage applied to the deflection amplifier. The length of the motion of the electron beam is consequently shortened. In a vertical-deflection oscillator circuit, this control will affect (and adjust) the height of the picture. It is labeled the "height control." Horizontal-deflection oscillators generally do not contain this control as the width is usually controlled in another manner, to be described later.

The blocking-oscillator circuit illustrated in Fig. 17-6 is used primarily for vertical deflection. Blocking oscillators are also employed to generate synchronized deflection sawtooth wave at 15,750 Hertz, for the horizontal system. The basic principle of these blocking oscillators is the same as for the ones described previously. However, for horizontal deflection, the oscillator is invariably controlled by an AFC system. Horizontal-deflection oscillators and AFC systems are described later in this chapter. (Horizontal- and vertical- deflection output amplifier systems are described in Chapters 18 and 19.)

17.4 THE TRANSISTOR BLOCKING OSCILLATOR

Transistor blocking oscillators are used for both vertical and horizontal deflection systems. While the principle of operation applies equally to both horizontal and deflection oscillators, our discussion at this point is directed mainly toward vertical oscillators. Horizontal deflection transistor oscillators and their associated AFC systems will be discussed later in this chapter.

The basic operating mode of tube and transistor blocking oscillators is the same, although there is a difference in the method which is used to turn the transistor oscillator on and off. This frequently involves the action of the sawtooth wave generated by the transistor circuit itself, rather than by a grid-leak bias action, as used in tube circuits. In transistor deflection oscillator circuits, we will frequently find one or more "driver" stages between the oscillator and the power amplifier, because in those instances the oscillator ouput is not adequate to fully drive the output stage. However, if a small screen is employed, and the oscillator transistor is powerful enough, then a direct connection between the oscillator and the output amplifier will produce the necessary power to adequately drive the deflection coils.

A Typical Vertical Transistor Blocking Oscillator, Sawtooth Generator A typical example of a transistor vertical-blocking oscillator, sawtooth generator is shown in Fig. 17-7. The tightly-coupled windings on transformer T_1 provide positive feedback from collector to base of transistor Q_1. The transistor Q_1 is operated Class C and is turned-on (conducts) for brief intervals.

This circuit operates similarly to its tube counterpart. An interesting feature to note is that one RC network, C_2-R_2, performs the dual role of determining the oscillator frequency and generating the sawtooth wave.

When the supply voltage is applied, the circuit operates as follows: A negative forward bias voltage appears at the base of Q_1 (PNP transistor) due to the voltage divider R_1-R_3. C_2 is not charged.

Transistor Q_1 conducts, and C_2 starts to charge in a negative direction. As the collector current increases, in the primary of T_1, a voltage is induced in the secondary winding (*i.e.*, the winding connected to the base of Q_1) with the polarity shown. Note that the polarizing dots show a phase reversal in T_1.

The base voltage now consists of two voltages in series to ground: the voltage across R_1 and the T_1 secondary voltage. Both voltages are series-aiding and force the base to a higher negative voltage (more forward bias). Q_1 thus conducts even more strongly and C_2 charges rapidly (and linearily) to produce the retrace portion of the sawtooth wave.

The regenerative cycle in which C_2 continues to charge to a higher negative voltage goes on until Q_1 saturates. Saturation causes the collector current in Q_1 to level off at the saturation level. T_1 now has a strong but steady magnetic field built up in the primary winding. No voltage is induced in the T_1 secondary winding at the moment the collector current saturates. (Voltage is generated in a transformer winding only so long as current is changing.) The base voltage on Q_1 drops to the voltage developed across R_1. At this time, the C_2 voltage (reverse bias) exceeds the base voltage (forward bias) and transistor Q_1 cuts off.

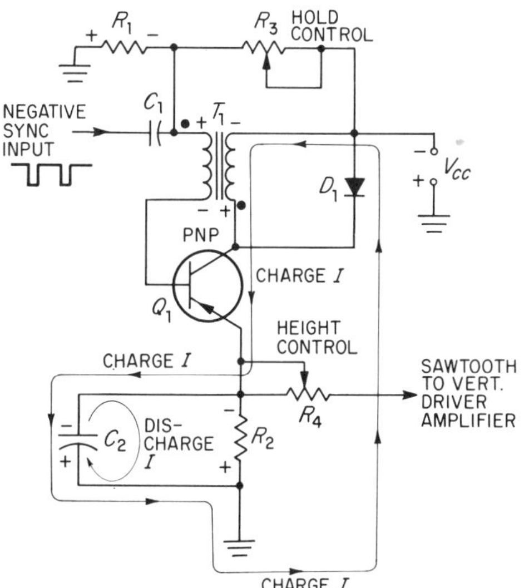

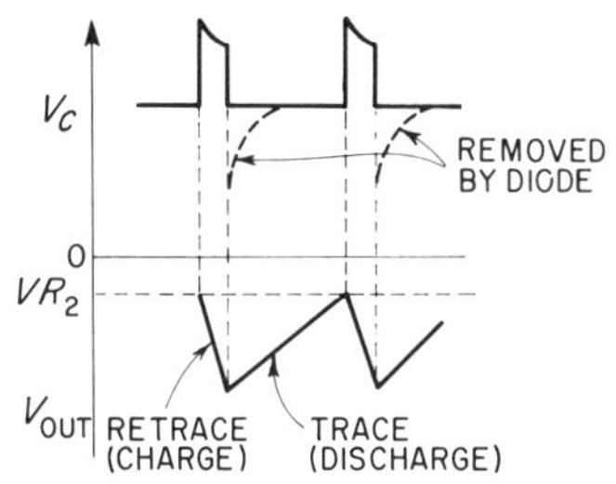

Fig. 17-7. A vertical deflection, transistor-blocking oscillator, sawtooth generator.

The steady magnetic field in T_1 now collapses and induces a voltage pulse in both windings opposite to the polarity shown in Fig. 17-7. The negative collector pulse is clipped due to the diode, D_1, becoming forward biased.

With Q_1 now cut off, C_2 begins to discharge thru R_2, producing the trace portion of the wave-form, in the positive-going direction. Q_1 remains cut off until the voltage across C_2-R_2 decreases to the same value as the forward-bias voltage across R_1. When this happens, the base-to-emitter junction again becomes forward biased, transistor Q_1 conducts, and the charging cycle begins again. The relatively slow discharge of C_2 through R_2 serves to develop the trace portion of the sawtooth and establish the trace-timing interval.

Lock-in or synchronization is accomplished by applying negative-going sync pulses to the base. As in tube-type blocking oscillators, the natural repetition frequency of the oscillator must be *slightly lower* than the sync-pulse frequency. Sync pulses force the transistor into conduction before C_2 completely discharges and causes transistor conduction.

The oscillator free-running frequency is controlled primarily by the R_2-C_2 network in the emitter circuit. Notice that during discharge, transistor Q_1 is cut off and the R_2-C_2 network is *isolated* from the remainder of the circuitry. This helps to provide excellent frequency stability. The natural frequency may be shifted over a narrow range by changing the dc forward bias (across R_1) applied to the base through the T_1 transformer secondary winding. Variable resistor R_3 is part of the bias network and performs this "hold control" function. An increase in negative voltage across R_1 (an increase in forward bias) allows the transistor Q_1 to break into conduction sooner. The natural frequency is then raised.

At the time Q_1 cuts off, the voltage transient pulses appearing across the T_1 secondary and primary might be troublesome because they may exceed the collector-to-base breakdown voltage rating of Q_1. Such transients are removed by connecting a diode across the primary winding, as indicated in Fig. 17-7. The diode conducts when the T_1 voltage polarities are opposite to those shown ,and the magnetic field energy is then dissipated harmlessly through the diode. The sawtooth output is applied to the vertical-driver stage.

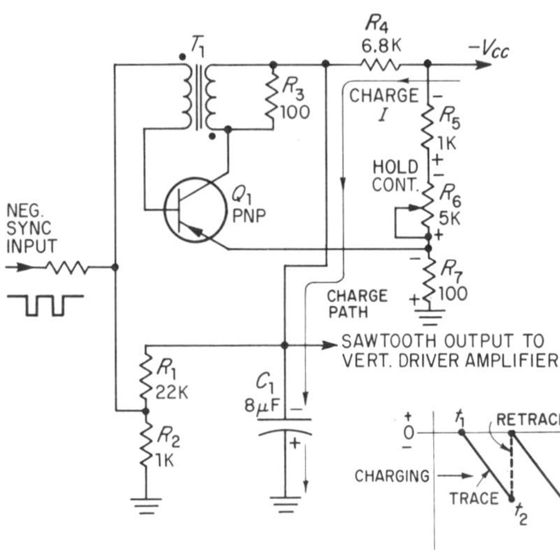

Fig. 17-8. Another vertical-deflection transistor blocking oscillator, sawtooth generator.

Another Transistor-Blocking Oscillator Sawtooth Generator

The circuit of Fig. 17-8 is similar in operation to a vacuum-tube blocking oscillator, in that the sawtooth forming capacitor, C_1, charges during the trace period and discharges during the retrace. Circuit action to form the required sweep sawtooth is as follows.

When the collector supply voltage ($-V_{cc}$) is first applied, the charge on the sawtooth forming capacitor C_1, is zero. The transistor Q_1 is cutoff by the reverse bias voltage across R_7 which is applied to the emitter. The trace starts at t_1 as C_1 charges through R_4 toward $-V_{cc}$. As the voltage across C_1 increases negatively, a portion of it is fed to the base of Q_1 through T_1 secondary, as a forward bias from the voltage divider R_1 and R_2. Time t_2 represents the end of the sweep as the forward bias across R_2 just exceeds the reverse bias across R_7. Q_1 conducts at this instant, and the collector current through transformer T_1 causes the base voltage to swing highly negative. The voltage across R_2 and the secondary voltage of T_1 are now series aiding to drive transistor Q_1 into saturation in order to begin the retrace portion of the sawtooth.

The transistor, primary of T_1, and R_7 form a very low resistance discharge path for C_1. At the end of this rapid discharge, C_1 has discharged to nearly zero and Q_1 is cut off again by the reverse bias across R_7. This completes the retrace.

Frequency control is accomplished by varying the amount of reverse bias to the Q_1 emitter. An increase in negative voltage from R_6 increases the time Q_1 is held in a cut off condition. This increases the time C_1 has to charge and effectivelv decreases the frequency of the oscillator cycle.

Resistor R_3 (100 ohms) performs the function of a damping resistor across the T_1 primary to reduce the retrace-pulse transient to safe limits. The Q_1 base-to-collector junction requires this, or voltage breakdown might occur and Q_1 would be destroyed.

Driver Amplifiers Sawtooth wave-forms are generated in tube circuits by allowing a capacitor to charge to a fraction (1/210) of the applied voltage. The voltage across the capacitor is then fed to the grid of a vacuum tube (output amplifier). The very high-input imepdance of the tube-grid circuit does not load the capacitor.

However, in transistor circuits, the sawtooth is fed to the base circuit of the following stage and the input impedance of this stage, representing a load on the sawtooth capacitor, may be only a few hundred ohms. This low resistance across the sawtooth capacitor draws appreciable current from the capacitor, lowers the sawtooth voltage amplitude, and destroys linearity.

A solution to the problem of loading the sawtooth-forming capacitor (and increasing the sawtooth current gain to drive the output amplifier) is to feed the sawtooth into a special driver amplifier. This amplifier must present a high-output impedance to the sawtooth capacitor so as not to load it and must also match the low-input impedance of the following amplifier. A common collector amplifier (emitter follower) does not severely load the sawtooth capacitor C_1 and yet has a low-output impedance to match the output-amplifier stage. Figure 17-9 shows this circuitry in a typical arrangement. Direct coupling is employed to the output amplifier and the requirement for an expensive and high-capacity coupling capacitor is eliminated.

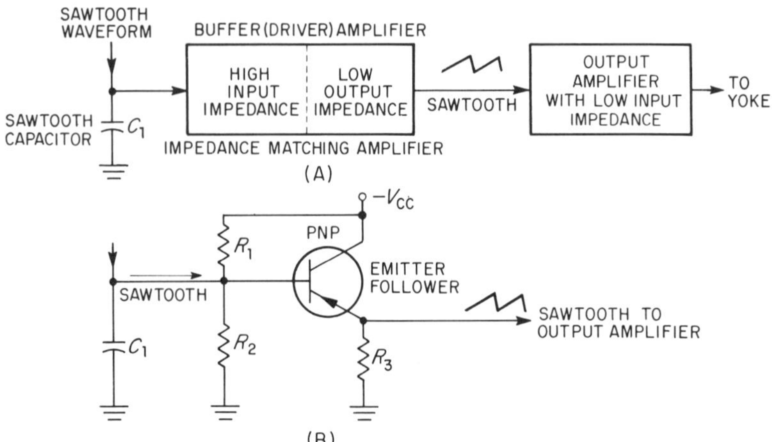

Fig. 17-9. A typical transistor-driver amplifier connected as an emitter follower.

Complete Vertical Deflection System The schematic diagram of a complete G.E. vertical deflection system is given in Fig. 17-10. This consists of a blocking oscillator (Q_{601}), a driver stage (Q_{602}), and a vertical output stage (Q_{603}). Explanations of additional vertical deflection systems will be given in Chapter 19; this circuit is presented at this time, mainly to show an example of a complete vertical deflection system. Note that the blocking oscillator, the sawtooth generator, and Q_{601} are of the same basic type as shown in Fig. 17-7, with the sawtooth capacitor (C_{604}) and its shunting resistor (R_{606}) located in the emitter circuit. However, note that Q_{601} is an NPN transistor, so that all of its voltages are reversed compared with the PNP circuit of Fig. 17-7. The driver, Q_{602}, is of the common emitter type. The sawtooth output of the driver is capacity-coupled to the vertical-output stage by capacitor C_{608}. The output of the power stage, Q_{603}, is choke coupled to the

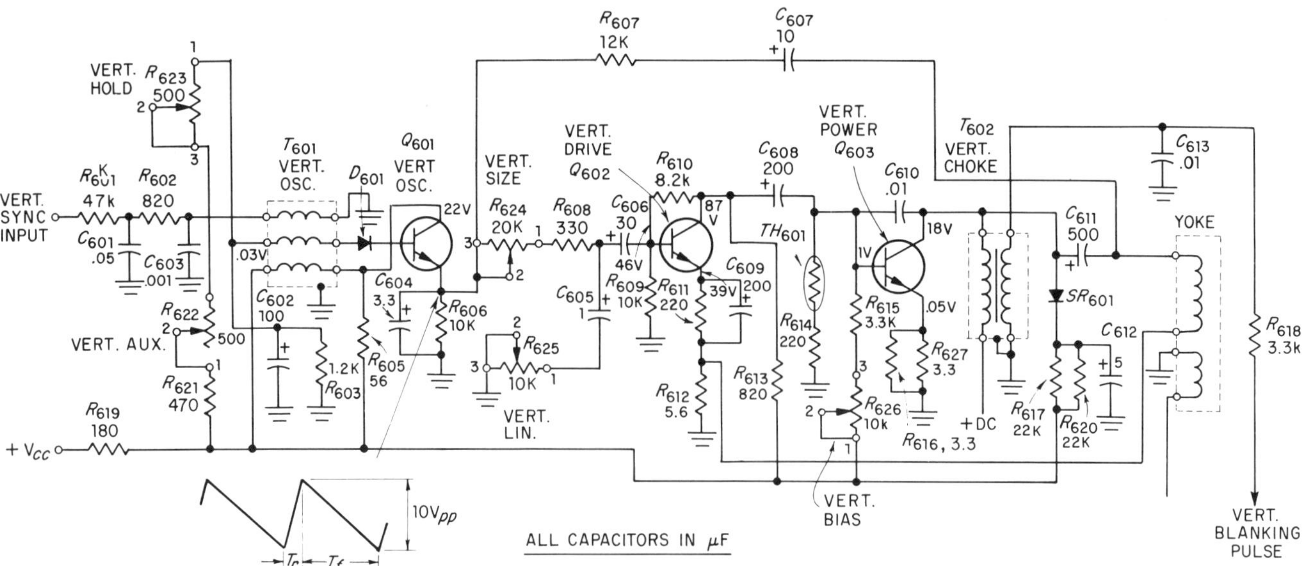

Fig. 17-10. Vertical-deflection system with an oscillator-driver (buffer) and output transistors. (*Courtesy of General Electric.*)

vertical yoke by the primary of the vertical choke-transformer, T_{602}. This provides the proper impedance match for both the output transistor and the yoke.

The vertical oscillator frequency is varied by changing the value of the bias applied through D601 to the base of the vertical oscillator (Q_{601}). Two controls, vertical hold (R_{623}) and vertical auxillary (R_{622}), have been provided for this purpose. R_{622} should be adjusted so that a vertical lock-in occurs when R_{623} is set at the mechanical center of its range. The output of the blocking oscillator circuit is a sawtooth wave that is amplified by the vertical driver (Q_{602}). The output of Q_{602} is further amplified by the vertical-output transistor (Q_{603}) and is then applied to the vertical coils of the deflection yoke. The function of C_{611} is to block out the dc component of the vertical output signal and to allow only the ac component to be applied to the vertical deflection coils.

The vertical linearity control (R_{625}) and C_{605} combine to form an integrating circuit that is used to regulate the shape of the sawtooth wave that is generated by the vertical blocking oscillator. By varying the resistance of R_{625}, the linearity at the start of the vertical scan (upper portion of the picture) can be controlled. R_{612}, R_{607}, and C_{607} serve to correct distortion that is introduced into the vertical circuit by Q_{602} and Q_{603}. R_{607} and C_{607} affect the linearity at the end (lower portion of the picture) of the vertical scan. The vertical-size control (R_{624}), located in the emitter circuit of the vertical oscillator (Q_{601}), is used to regulate the overall vertical size of the picture and also controls the linearity of the lower portion of the picture.

The vertical-output transistor (Q_{603}) must be operated as a linear amplifier. The vertical-bias control (R_{626}) is used to regulate the linear operation of this stage. Misadjustment of this control can produce a nonlinear condition in either the upper or lower portion of the picture.

A thermistor (TH_{601}) has been installed physically close to the vertical-output transistor (Q_{603}) to prevent excessive current flow through its collector circuit which could result from an increase in the ambient temperature. Thermal runaway, resulting in excessive collector current drain, could quickly destroy Q_{603}. Should the temperature of Q_{603} rise, the resistance of TH_{601} would decrease, lowering the drive to Q_{603} and hence counteracting a collector current rise due to the temperature. Q_{603} is further protected by a network consisting of the vertical-pulse limiter diode (SR_{601}), C_{612}, R_{617}, and R_{620}.

17.5 TUBE-TYPE MULTIVIBRATOR DEFLECTION OSCILLATOR CIRCUITS

Essentially, the multivibrator is a two-stage resistance-coupled amplifier, with the output of the second tube fed back to the input of the first stage. Oscillations are possible in a circuit of this type, because a voltage at the grid of the first tube will cause an amplified voltage to appear at the output of the second tube, which has the same phase as the voltage at the grid of the first tube. This is always the case with an even number of resistance-coupled amplifiers, but never with an odd number. The output of an odd number of such stages is always 180 degrees out of phase with the voltage applied at the input of the first tube. The two voltages thus oppose, rather than aid, each other. Two types of resistance-coupled multivibrators are used in TV receivers—plate-coupled and cathode-coupled. A third type commonly used, employs the vertical-output amplifier as one stage of the multivibrator. All three types will be discussed in this chapter.

The operation of a multivibrator is best understood if we trace the voltage and current changes through the various circuit elements. To start, assume that the power supply has just been connected across the circuit (see Fig. 17-11). Due perhaps to some slight disturbance in the circuit, the plate current of tube V_1 increases. This produces an increase in the voltage across R_1, with the plate end of the resistor becoming more negative. Capacitor C_1, which is connected to R_1, at this point, likewise attempts to become more negative, and the grid of V_2 also assumes the same potential. The net result is a lowering of the current through V_2 and R_2.

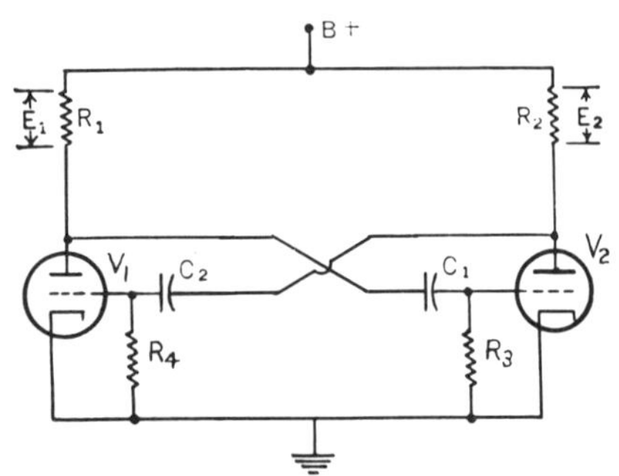

Fig. 17-11. The fundamental multivibrator circuit.

The lowered voltage across R_2 means that the plate end of this resistor becomes less negative, or relatively positive to its previous value. Capacitor C_2 transmits this positive increase to the grid of V_1 and, consequently, even more plate current flows through R_1. The process continues in this manner, with the grid of V_1 becoming more and more positive and driving the grid of V_2 increasingly negative by the large negative charge built up across R_3 and C_1. The plate current of V_2 is rapidly brought to zero by this action.

Tube V_2 remains inactive until the negative charge on C_1 discharges and removes some of the large negative potential at the grid of V_2. The path of discharge of C_1 is through the relatively low resistance r_p of tube V_1 and the relatively high resistance R_3. When C_1 has discharged sufficiently, plate current starts to flow through V_2, causing the plate end of resistor R_2 to become increasingly negative. This now places a negative charge on the grid of V_1, and the plate current through R_1 decreases. The lessening of the voltage drop at R_1 causes the plate end of the resistor to increase positiviely, and the grid of V_2 (through C_1) receives this positive voltage. The increased current through R_2 quickly raises the negative grid voltage on V_1 (through C_2) and drives this tube to cutoff. When the excess charge on C_2 leaks off, the process starts all over again. C_2 loses its accumulated negative charge by discharge through r_p of V_2 and R_4. Contrast this path with that of C_1.

The entire operation may be summed up by stating that first the plate current of one tube rapidly rises, driving the second tube to cutoff. This condition remains until the second tube is released from its cutoff state and commences to conduct. It is now the first tube which is cut off. When the first tube is again permitted to conduct, the second tube is driven into nonconduction. The switching continues in this manner, with the frequency largely determined by the grid resistors and capacitors, R_4, R_3, C_1, and C_2.

If a synchronizing pulse is applied to either of the grids, and if its frequency is close to the natural frequency of the oscillator, it is possible to control the period of the multivibrator effectively. Figure 17-12 illustrates how the multivibrator can control the charge and discharge of a capacitor, thereby developing the required sawtooth voltages. The same multivibrator is used, with the addition of the sawtooth capacitor C_3. When tube V_2 is not conducting, the power supply will slowly charge

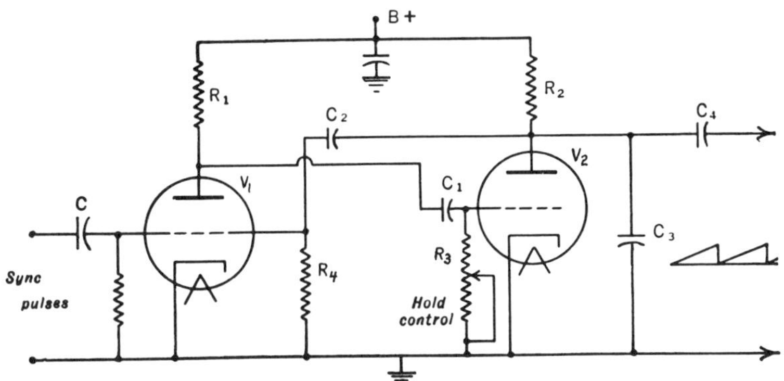

Fig. 17-12. How a multivibrator may be connected to control the charging and discharging of a capacitor to derive sawtooth waves.

C_3 through resistor R_2. The moment that the grid voltage of V_2 reaches the cutoff point of the tube, the tube starts to conduct and its internal resistance decreases. Capacitor C_3 then discharges rapidly through the tube. During the next cycle, V_2 is again nonconductive, and again C_3 slowly charges. C_4 transmits the voltage variations appearing across C_3 to the next amplifying tube. Resistor R_3 is made variable to permit adjustment of the multivibrator so that it can be locked-in with the synchronizing pulses. Hence R_3 is the hold control.

The desired form of the sawtooth output waveform is a slow rise in voltage, followed by a rapid decrease. Toward that end, C_1 and R_3 of Fig. 17-12 are designed to have a considerably longer time constant than C_2 and R_4. C_1 and R_3 will discharge slowly, maintaining V_2 in cutoff while C_3 slowly charges. During this interval, V_1 is conducting. Upon the application of a negative synchronizing pulse to the grid of V_1, this tube is forced into cutoff, while V_2 rises sharply out of cutoff and into conduction. C_3 now discharges rapidly. Because C_2 and R_4 have a small time constant, V_1 does not remain cut off very long and as soon as C_3 has discharged, V_1 begins to conduct, again cutting off the plate current of V_2. The ratio of the time constants of C_1, R_3 and C_2, R_4 is approximately 11 :1.

Typical Values of Components shown in Fig. 17-12

60 Hertz	15,750 Hertz
R_3—2.2 megohms	R_3—220,000 ohms
R_4—1.0 megohms	R_4—100,000 ohms
R_1—100,000 ohms	R_1—47,000 ohms
R_2—2.2 megohms	R_2—500,000 ohms
C_3—0.1 μf	C_3—500 pf
C_2—.01 μf	C_2—.001 μf
C_1—.05 μf	C_1—.005 μf

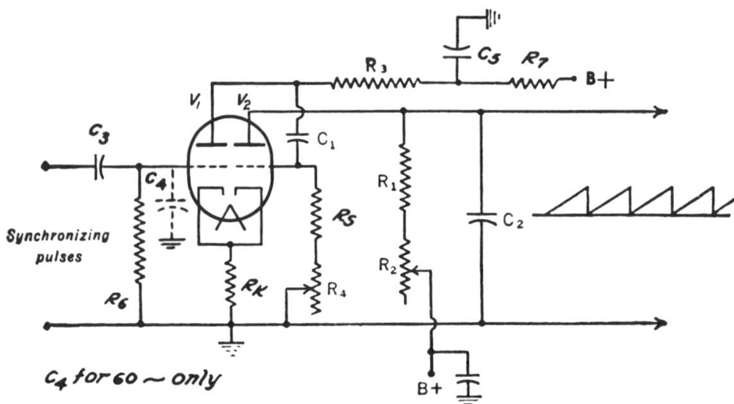

Fig. 17-13. Another widely used form of multivibrator, known as a cathode-coupled multivibrator.

In many commercial receivers, a slightly altered form of multivibrator circuit is used, although the basic operation remains the same. This oscillator is shown in Fig. 17-13. The feedback between tubes is accomplished in two ways: through the coupling capacitor C_1 and through the unbypassed cathode resistor, which is common to both tubes.

The sawtooth capacitor C_2 is placed in the plate circuit of the second triode. During the portion of the multivibrator cycle when V_2 is not conducting, C_2 is practically across the power supply and charges through resistors R_1 and R_2. When a sharp negative pulse of voltage is applied to triode V_1, the plate current of this tube decreases, causing the plate end of resistor R_3 to become increasingly positive. As the grid of V_2 is connected to this part of R_3, it too will become more positive. The plate current through V_2 will rise sharply, developing enough voltage across the common cathode resistor to bring V_1 to cutoff. V_2, however, continues to conduct because its grid has received sufficient positive voltage from the potential variation across R_3 to partly counteract this high negative cathode bias. V_1, not having this positive grid voltage, is forced into cutoff. During this period, when V_2 is conducting heavily, its internal resistance is low and C_2 discharges through it.

The high positive voltage on the grid of V_2, which resulted in a large plate current flow for an instant and permitted C_2 to discharge, makes the grid draw current. This immediately biases the grid to cutoff (similar to the blocking oscillator), brings V_1 out of cutoff, and permits C_2 to charge again. Resistor R_4 is made variable to permit adjustment of the frequency of the multivibrator. R_2 controls the amount of the charging current flowing into C_2, and this in turn regulates the extent of the electron beam sweep across the screen. R_2 is the width control (or height control).

Here, as before, the incoming synchronizing pulses serve to alter slightly the time at which a changeover from one tube to the other takes place. Without these pulses, V_1 would conduct for a shorter portion of the cycle, just as in the case of the previous multivibrator.

Typical Values of Components shown in Fig. 17-13

60 Hertz	15,750 Hertz
R_1—1.0 megohm	R_1—470,000 ohms
R_2—2.0 megohms	R_2—500,000 ohms
R_3—100,000 ohms	R_3— 47,000 ohms
R_4—1.2 megohms	R_4— 50,000 ohms
R_5—1.2 megohms	R_5— 33,000 ohms
R_6—2.2 megohms	R_6— 2,000 ohms
R_7—100,000 ohms	R_7—100,000 ohms
C_1— .01 μf	C_1—.001 μf
C_2— .1 μf	C_2—500 pf
C_3— .01 μf	C_3—50 pf
C_4— .001 μf	C_4—not necessary
C_5— .1 μf	C_5—.006 μf
R_k—470 ohms	R_k—470 ohms

While a negative synchronizing pulse at the grid of V_1 will cause C_2 to discharge, we may obtain the same effect if a positive synchronizing pulse were fed to the grid of V_2. The negative pulse, however, results in a more stable arrangement and is generally used. This fact explains the statement made several paragraphs before when it was pointed out that either a positive or a negative synchronizing pulse could be used to synchronize a multivibrator. For the blocking oscillator, it will be remembered that a positive pulse was required.

Synchronizing of the Multivibrator The phrase "synchronizing an oscillator" is quite frequently used in describing the operation of television circuits There are, however, many students who are not completely clear as to the exact mechanism of this synchronization. To clarify this point the following explanation is offered.

In a television receiver, the pulses of the incoming signal take control of the free-running sweep oscillators and lock them into synchronism with the pulse frequencies. We are referring, of course, to the horizontal and vertical synchronizing pulses. It is highly improbable that the first pulse, when it reaches the oscillator, arrives at such a time as to force the free-running oscillator exactly into line. Generally, this does not occur until several pulses of the incoming signal have reached the sweep oscillator. Let us examine the means whereby the receiver oscillator is gradually forced into synchronization with the incoming pulses.

In order to synchronize an oscillator, the pulses must be applied to the oscillator input. In Fig. 17-14 we have the grid-voltage wave-forms of a multivibrator and, beneath them, the triggering pulses as they are received from the preceding pulse-separator networks. Suppose the first pulse, *A*, arrives at a time when the grid is quite negative, and thus this pulse is unable to bring the tube out of cutoff. The second pulse, *B*, arrives when the tube is conducting. Thus, it drives the grid more positive and has very little effect on its operation. The conditions for the third pulse are similar to those for the second pulse. The fourth pulse, *D*, arrives at a time when the grid of the tube is negative. However, this pulse is able to drive the grid positive, thereby initiating a new cycle. Thereafter, each succeeding pulse arrives at a time when it will bring the tube out of cutoff and the sweep oscillator is securely locked in as long as the pulses are active. It is important that the pulses reach the grid of the oscillator when it can raise the tube above cutoff. Unless it can do this, it will be without power to lock in the oscillator.

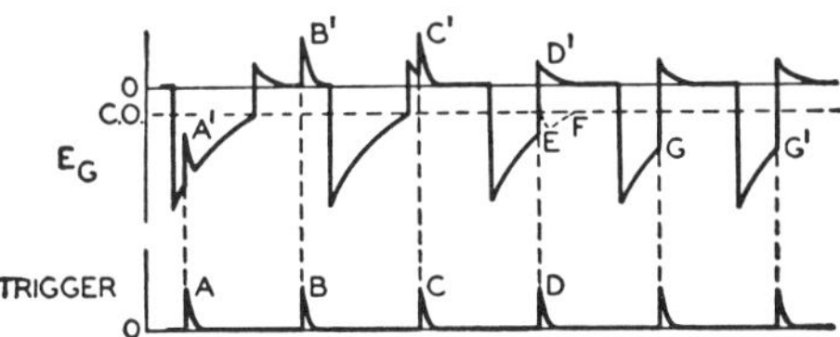

Fig. 17-14. How the sync pulses lock in the multivibrator oscillator.

One final word about the foregoing oscillators. As the grid voltage approaches the cutoff value, it becomes increasingly sensitive to noise pulses which may have become part of the signal. A sufficiently strong interference pulse, arriving slightly before the synchronizing pulse, could readily trigger the oscillator prematurely. When this occurs, the electron beam is returned to the left-hand side of the screen before it should be and the right-hand edge becomes uneven. Severe interference may cause sections of the image to become "torn." To prevent this form of image distortion, television-receiver manufacturers use synchronizing systems which respond only to long-period changes in the pulse frequency Since interference flashes seldom have regular patterns, they cannot affect these special systems Several such horizontal-deflection systems are analyzed later in this chapter. Vertical-deflection systems are relatively immune to impulse noise, since most of this is filtered out in the vertical-integrator circuit, preceding the vertical-deflection oscillator.

Combination Multivibrator and Output Stage Another vertical-deflection multivibrator system is shown in Fig. 17-15. Here two tubes serve as both the multivibrator and the output amplifier, an arrangement made possible by using the output tube to complete the multivibrator circuit. If the circuit is examined, it will be seen that there is a feedback

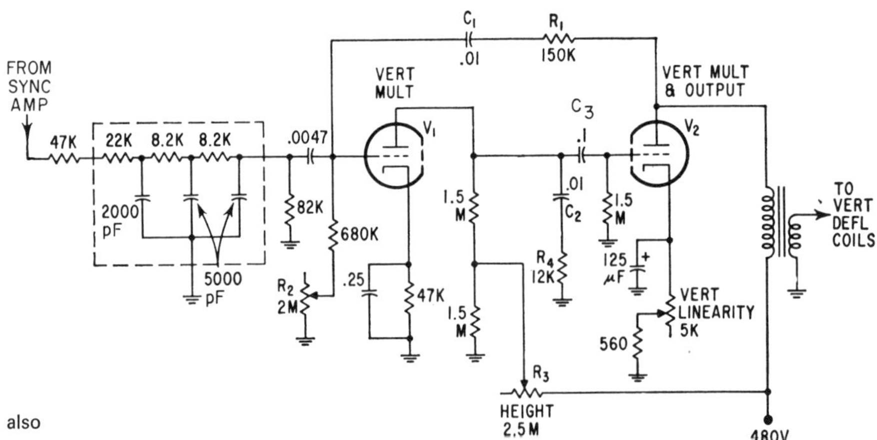

Fig. 17-15. The output tube in this system is also part of the multivibrator.

path from the plate of V_2 to the grid of V_1. This path is formed by R_1 and C_1, and by transferring back energy which arrives in phase at the grid of V_1, the oscillating action of a multivibrator is achieved. At the same time, a transformer in the plate circuit of V_2 feeds output signals to the vertical deflection coil. Thus, by utilizing the output tube as the second multivibrator tube, we accomplish both functions with only two triodes. In this case, since each triode is one-half of a dual-triode, the entire vertical In system contains only a single tube. This arrangement is very economical and for this reason has found widespread use.

The rest of the circuit is fairly conventional and follows the previous circuit quite closely. A potentiometer, R_2, in the grid circuit of V_1 serves as the vertical hold control. A second potentiometer, R_3, in the plate circuit of V_1 varies the amount of voltage fed to the capacitor C_2 and resistor R_4. R_3 is thus the height control. The peak wave-form which is developed across C_2 and R_4 is then transferred by C_3 to the grid of the output tube, where it is amplified and applied to the vertical-deflection coils. Part of this wave is fed back through R_1 and C_1 to the grid of V_1 to keep the multivibrator oscillating. Vertical linearity is accomplished conventionally by a potentiometer in the cathode leg of V_2. Despite the fact that V_2 is employed as the second half of the multivibrator, the vertical-linearity control functions in exactly the same fashion as it did in any of the previous circuits where the output tube is a straight amplifier.

It is possible to control vertical linearity by varying the grid bias of the output tube directly. Such a method is illustrated in Fig. 17-16. A

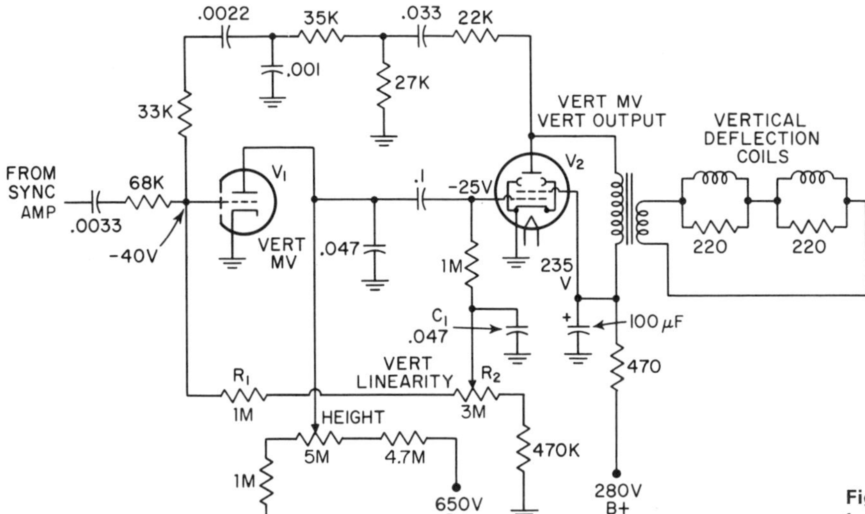

Fig. 17-16. The grid bias of V_2 is controlled directly for vertical linearity in this system.

negative voltage which is developed at the grid of the first multivibrator tube is fed through R_1 to the vertical-linearity control, R_2. The position of the arm on the vertical-linearity control determines the negative bias which the grid of V_2 receives. The resistor R_1, which precedes the vertical-linearity control, is used to isolate the control from the grid of V_1. Capacitor C_1, in the network leading to the grid of V_2, also assists in the filtering action so that the voltage fed to V_2 is essentially dc.

Figure 17-16 also demonstrates that the vertical multivibrator does not need to be formed by two tubes of similar type. In this case, V_1, is a triode, and V_2 is a beam-power tetrode. Just so long as there is sufficent feedback voltage from the output of V_2 to the input of V_1, the circuit will oscillate. It is not necessary that V_1 and V_2 be similar electrically.

Whether we are using a blocking oscillator or a multivibrator, the basic operation of the circuit remains unchanged. It is the function of the oscillator not only to provide the proper frequency, but also to develop a wave which will suitably deflect the beam from the top to the bottom of a picture tube. It is the purpose of the output amplifier to amplify the signal so that the complete screen will be covered by the image. A linearity control is generally incorporated in the output circuit to counteract any distortion that might arise either because of the nonlinear charging of the charge-discharge capacitor or because of the nonlinear characteristics of the output tube itself.

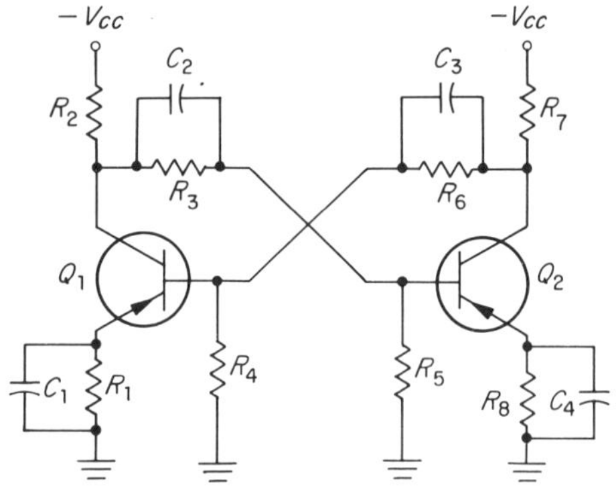

17.6 TRANSISTOR MULTIVIBRATOR OSCILLATORS

Figure 17-17 is a diagram of a transistor, collector-coupled multivibrator, whose operation is similar to the tube circuit of Fig. 17-11. These oscillators are used for both horizontal and vertical deflection systems. The circuit oscillates because of coupling from the collector of Q_1 to the base of Q_2 and feedback coupling from the collector of Q_2 to the base of Q_1. Both transistors shown are PNP.

PNP transistors are cutoff when the base is driven positive with respect to the emitter. The discharge of each coupling capacitor (C_2 or C_3) through the base bias resistors (R_4 or R_5) produces this cutoff voltage. The period of cutoff for Q_1 and Q_2 is determined by the RC time constant of their respective coupling capacitor and resistor (C_3, R_4, or C_2, R_5).

Fig. 17-17. Free-running, collector-coupled, transistor multivibrator.

The following properties of a transistor amplifier circuit should be considered in the analysis of multivibrator-oscillator circuits. These properties are applied to the common-emitter configuration which is used here because a 180 degree phase shift occurs (as in an electron tube amplifier) between the output and input signals.

(1) An increase in base current causes an increase in collector current through the transistor; conversely, a decrease in base current causes a decrease in collector current.

(2) An increase in collector current causes the collector voltage to decrease; a decrease in collector current causes the collector voltage to increase towards the value of the source voltage. With PNP-type transistors, an increase in collector current causes the collector voltage to become less negative. With NPN-type transistors, the collector voltage would become less positive.

(3) For normal functioning of a transistor amplifier, the base-emitter diode is forward biased and the collector-base diode is reversed biased. The polarity is determined by the type of transistor used (PNP or NPN).

(4) A transistor is saturated when a further increase of base current causes no further increase in collector current.

(5) A transistor is cut off when either the base or emitter is reverse biased.

Operation The free-running multivibrator is essentially a non-sinusoidal two-stage oscillator in which one stage conducts while the other is cut off, until a point is reached at which the stages reverse their conditions. That is, the stage which had been conducting cuts off, and the stage that had been cut off conducts. This oscillating process is normally used to produce a square-wave output. Most transistor multivibrator circuits are counterparts of those using electron tubes, whose operation has been previously described. For example, the collector-coupled transistor multivibrator operates on the same principles as the plate-coupled tube multivibrator of Fig. 17-11. In addition, the emitter-coupled multivibrator of Fig. 17-18 operates in the same manner as the cathode-coupled multivibrator of Fig. 17-13. Because of this similarity of operation, no individual discussion of the operation of the basic transistor multivibrators will be given here. The student should refer to the appropriate vacuum-tube multivibrator discussion for an explanation of the equivalent transistor multivibrator operation.

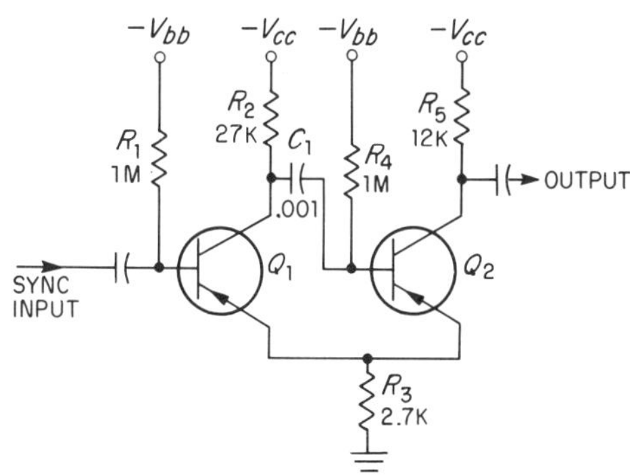

Fig. 17-18. Emitter-coupled multivibrator.

Transistor, Combination Multivibrator, and Output Stage Figure 17-19 shows the transistor equivalent of the tube circuit of Fig. 17-15. The circuit of Fig. 17-19 is a multivibrator circuit for vertical deflection, where the output stage forms a portion of the multivibrator. In the transistor circuit, however, note that a driver stage, Q_2, is located between the vertical-discharge transistor and the vertical-output transistor. The driver serves two important function. The first is to isolate the sawtooth-forming capacitors, C_2 and C_3, from the low-input impedance of the vertical-output transistor (as previously discussed). The second function of the driver transistor is to increase the sawtooth driving-current input to the output transistor to the required value. Note, however, that the driver is not an inherent part of the multivibrator action of the circuit. The driver is an emitter follower, which matches the relatively high impedance of the collector circuit of Q_1 to the low impedance of the base circuit of Q_3.

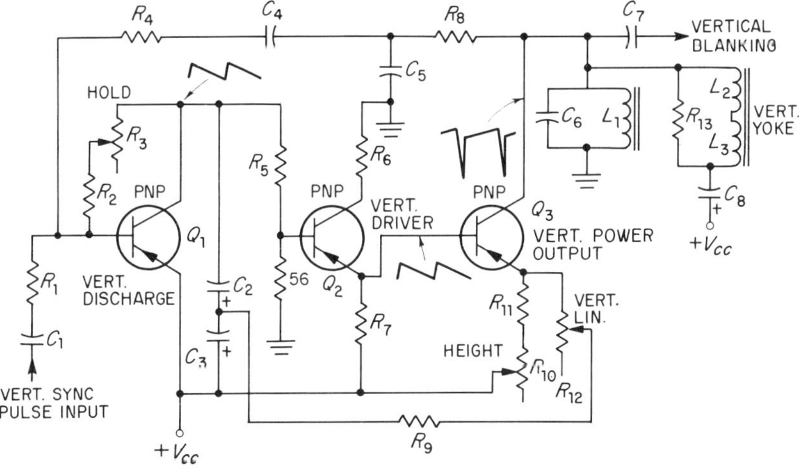

Fig. 17-19. A combination vertical multivibrator-output circuit, employing an intermediate driver stage. (*Courtesy of Motorola Consumer Products, Inc.*)

At the beginning of the vertical trace (top of picture), the sawtooth capacitors, C_2 and C_3, are completely discharged and the collector, base, and emitter of Q_1 are all at the same potential, preventing Q_1 from conducting. To begin the trace, the two capacitors start to charge through R_5 and R_6 toward $+V_{CC}$. As the capacitors begin to accumulate a charge, electrons flow into the top plates, making the polarity of these plates negative with respect to ground. This negative potential is dc coupled to the bases of both Q_2 and Q_3, representing forward bias for these two transistors. The increasing current through Q_3 flows through the vertical yoke coils, causing the picture-tube electron beam(s) to begin to be deflected downward.

At the same time, the voltage across Q_1 is increasing, because of the increasing charge across C_2 and C_3. This causes a negative voltage to appear at the collector of Q_1 and also at its base, through resistors R_2 and R_3. However, Q_1 will not conduct to any appreciable extent until the collector and base voltage have risen to a critical value. During this time, the sawtooth capacitors continue to charge and the electron beam(s) in the picture tube continues to be deflected toward the bottom of the tube. In practice, the sawtooth wave across the capacitors will reach a peak-to-peak amplitude of about three volts, before the Q_1 transistor conducts heavily to discharge them. Remember, that during this trace-forming time, transistor Q_1 is essentially non-conducting, while Q_2 is conducting more and more heavily as the trace approaches the bottom of the picture tube.

When the trace reaches the bottom of the picture tube, Q_1 begins to conduct heavily, and sawtooth capacitors C_2 and C_3 begin to discharge rapidly through the collector to emitter path of Q_1. This causes a positive-going (although still negative) voltage to be applied to the bases of Q_2 and Q_3. The positive-going voltage causes a negative pulse to appear at the collector of Q_3, as shown in Fig. 17-19. This negative pulse is coupled to the base of Q_1 through R_8, C_4, and R_4 and reinforces the current conduction of Q_1, which had initiated this part of the cycle. As a result of this regenerative feedback, Q_1 is driven into saturation and Q_3 is cut off. During the Q_1 saturation, the sawtooth capacitors are quickly discharged and return to approximately zero voltage condition to complete the retrace portion of the sawtooth wave. At this point, Q_1 again becomes cut off, because of the lack of voltage on its collector and base. The cycle then continues as previously described.

The height control, R_{10}, provides an adjustment of the amount of degeneration introduced into Q_3, and thus controls its ouput amplitude. A portion of the sawtooth wave in the emitter circuit of Q_3 is fed through the vertical-linearity control, R_{12} and resistor R_9, to capacitor C_3, where it is integrated to form a parabolic wave. This wave is used to modify the shape of the sawtooth wave fed to the base of Q_2 and Q_3 to adjust the linearity of the vertical sweep.

The hold control, R_3, adjusts the amount of forward-base bias and thus determines the point at which Q_1 will begin conducting on each cycle. In this manner, the hold control adjusts the free-running frequency of the system. Synchronization is accomplished by feeding negative vertical-sync pulses to the base of Q_1. Each pulse causes Q_1 to turn on before the free-running time and thus locks the oscillator to the sync-pulse rate.

17.7 MILLER INTEGRATOR VERTICAL DEFLECTION CIRCUIT

Some RCA color-television chassis utilize a novel type of vertical oscillator-deflection circuit, known as the "Miller integrator," or "Miller run-down circuit." This circuit differs from any of the vertical-deflection circuits discussed thus far and is an important type.

A simplified block-schematic diagram of this system is shown in Fig. 17-20. The basic operation of this system follows. A linear sawtooth

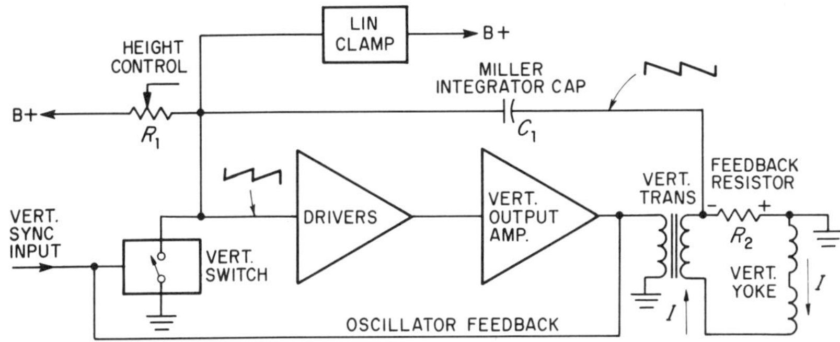

Fig. 17-20. Simplified diagram of the RCA Miller integrator vertical-sweep system. (*Courtesy of RCA.*)

wave is developed across the integrator capacitor C_1 by a rather unique method. To understand how this is accomplished, it is first necessary to briefly review some basic theory regarding the charging of capacitors. In order to develop a linear sawtooth voltage wave across a capacitor, it is necessary that a constant current be passed through the capacitor. In the usual sawtooth circuit, the wave is developed across the capacitor by the simple expedient of charging it from a power supply, through a resistor. If the charging is terminated when the capacitor has charged only a relatively small percentage of the power supply voltage, the sawtooth will be fairly linear, although it will possess some curvature toward its peak. This nonlinearity of the sawtooth requires that some means be incorporated in the vertical circuitry, not only to compensate for its effect on the vertical linearity, but also to compensate for other circuit factors which may affect the vertical-sweep linearity. In many circuits, this is accomplished by the adjustment of a vertical-linearity control. Because the circuit being described now produces a linear-vertical sweep by virtue of its inherent design, no vertical-linearity

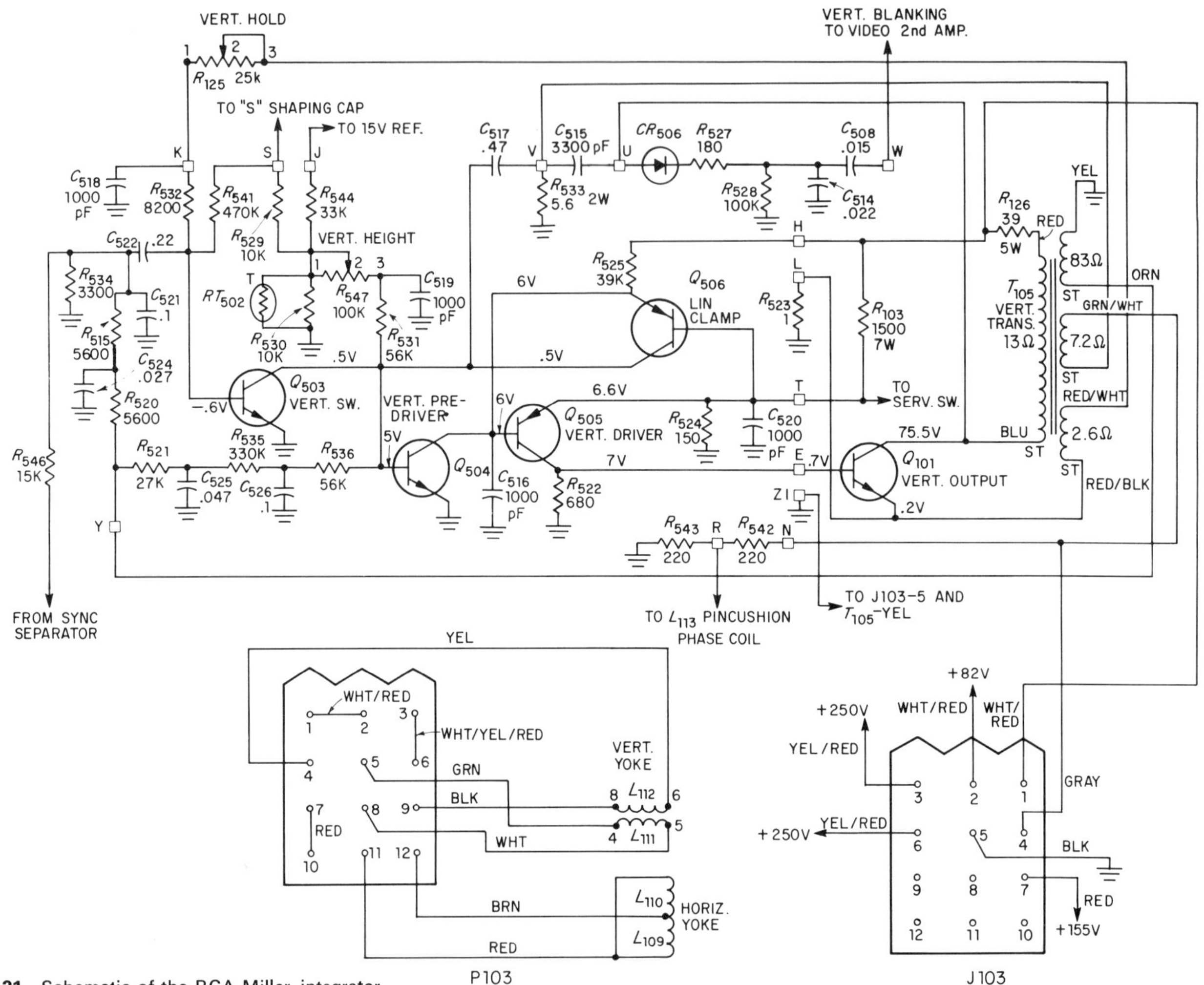

Fig. 17-21. Schematic of the RCA Miller integrator vertical-sweep system. (*Courtesy of RCA.*)

control is required. In the discussion which follows, refer also to the schematic diagram, shown in Fig. 17-21.

We said that to develop a linear sawtooth voltage wave, it was necessary to pass a constant current through our sawtooth-forming capacitor, in this case, C_1. One method of accomplishing this is to cause the usually grounded side of the sawtooth capacitor to vary its voltage in the opposite direction and in the same phase and amplitude, as it is varying its voltage on its charging side. That is to say, while the voltage of the left side of C_1, in Fig. 17-20 is rising in the positive direction, the voltage of the right side of this capacitor is decreasing at the same rate simultaneously. The sawtooth voltages on the left and right sides of C_1 have opposite polarities and thus are additive across the capacitor. In effect, the negative sawtooth at the right side of C_1

adds to the power supply voltage and provides C_1 with a charging voltage which is increasing at the same exact rate that the sawtooth is being formed. The net effect is to produce a constant current through C_1 rather than the exponentially decreasing capacitor current found in the usual simple capacitor charging circuit. Because of the constant current passing through C_1, during its normal charging time, a very linear sawtooth wave is formed across C_1 ,and this wave is used to drive the vertical yoke, after passing through several amplifier stages, and the vertical output transformer.

The negative sawtooth wave at the right side of C_1, is developed as follows: At the beginning of the vertical sweep (beam at the top of the screen), the left side of C_1 begins to charge (sawtooth fashion) toward the $B+$ supply through the height-control resistance, R_1. This positive-going voltage is applied to the predriver and the driver and arrives at the base of the vertical-output amplifier, again as a positive-going wave. The vertical-output stage is an NPN transistor, so the positive signal on its base causes the collector current to increase in proportion to the signal on the base. This increasing collector current (beginning of the sawtooth) is coupled through the vertical-output transformer and flows through the vertical-deflection yoke and through the series "feedback resistor," R_2. The direction of the yoke current is such that the voltage at the left side of the feedback resistor, R_2, is going in the negative direction at the same time, and at the same rate, as the voltage at the left side of capacitor C_1 is going positive. The value of resistor, R_2, is carefully chosen at the value of 5.6 ohms to insure that the amplitude of the negative-sawtooth wave is equal in amplitude to that of the positive-sawtooth wave applied to the input of the predriver. This type of linear sawtooth generation is also known as a "bootstrap" operation.

Vertical Switch Operation The free-running oscillator operation of this circuit is made possible by means of a vertical-switch transistor using feedback signals originating at the vertical output transformer. Vertical-sync pulses (positive polarity) are also applied to the input of the switch transistor and serve to lock the receiver-vertical sweep to the transmitted-vertical sweep. Simply stated, the vertical-switch transistor serves to discharge the sawtooth capacitor, C_1, quickly, at the end of the vertical trace. This initiates the vertical retrace, after which the next vertical trace begins.

The action of the vertical-switch transistor can be made clearer with the aid of the simplified diagram of Fig. 17-22. Note that there are two feedback paths, at the top and bottom of the diagram, from the vertical-output transformer to the base of the switch transistor. The path which makes it possible to sustain free-running oscillations is the one which includes R_{520}, R_{516}, and C_{522}. The wave fed back through this path becomes integrated by its network and appears at the base of the

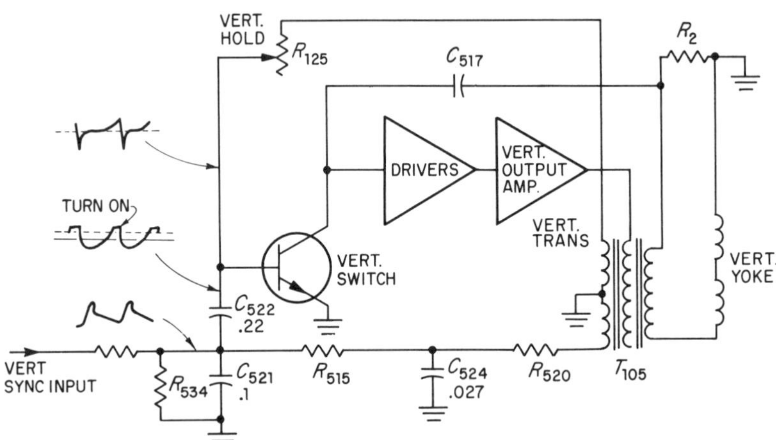

Fig. 17-22. Simplified schematic diagram of the RCA Miller integrator vertical-sweep system. (*Courtesy of RCA.*)

switch transistor as a modified parabola. When the positive peak of this wave appears above a threshhold voltage (represented by the dotted line), it forward biases the switch transistor, causing it to conduct heavily and to discharge the sawtooth capacitor, C_{517} (C_1 in Fig. 17-20). (The component designations in Fig. 17-22 are numbered to correspond to the schematic diagram of Fig. 17-21. Following the turn on period shown in Fig. 17-22, the switch transistor is turned off sharply by the negative-going portion of the parabola applied to the base. This initiates the start of the next vertical trace, as C_{517} once again begins to charge from B +.

The second feedback path from the vertical-output transformer is through the vertical hold control, R_{125}, to the base fo the switch transistor. The portion of this waveform (Fig. 17-22), above the dotted lines adds to the parabola turn-on voltage. In The free-running state, it is the combination of the two previously mentioned waveforms which are responsible for the rapid and stable turn-on action of the switch transistor. By using two waveforms, the turn-on action is made relatively immune from noise impulses which may be present on the sync-input line. The vertical-hold control may be varied to modify the shape of the second feedback wave. As a result, the total switch-transistor base turn-on voltage is changed and this in turn, changes the turn-on time and thus the free-running frequency of the system.

The third signal applied to the base of the switch-transistor base are the vertical-sync pulses, which are integrated by R_{546}, R_{534}, and C_{521}. Integration helps to remove noise pulses from the sync signal. The vertical-sync pulses are then added to the other two waveforms mentioned. This added sync-pulse voltage actually determines the turn-on time of the switch and thus locks-in the frequency of this vertical-deflection system to the incoming signal.

Lin Clamp Circuit In order to preserve the linearity of the sawtooth wave at the beginning of the sweep, it is necessary to momentarily connect the left side of C_{517} to a higher charging voltage than the +15 volts obtained from the height control. A simplified schematic indicating the operation of the lin-clamp circuit is shown in Fig. 17-23. The operation of this circuit is as described:

At the end of the vertical sweep (beam at bottom), the vertical-switch transistor is turned on, discharging sawtooth capacitor C_{517}. The negative-going sawtooth wave, at the input to the predriver, produces a positive-going wave at its collector circuit. Since this circuit is also connected to the emitter of the lin-clamp transistor (PNP), it provides forward bias for this transistor at the peak of the positive-going voltage. Thus, at the end of the sweep and at the beginning of the next vertical sweep, the lin-clamp transistor is turned on.

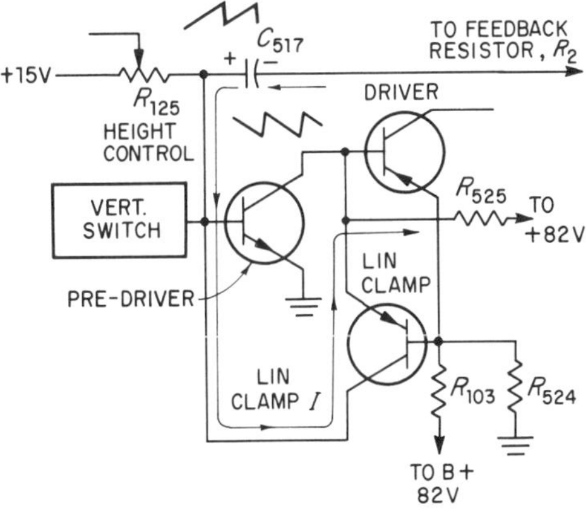

Fig. 17-23. Simplified diagram to illustrate RCA Miller integrator, vertical-linearity clamp operation. Arrows indicate momentary pin-clamp charging current to C_{517}. (*Courtesy of RCA.*)

You will note in Fig. 17-23 that there is now a charging-current path for C_{517} which passes through the lin-clamp transistor and R_{525}, to a source of +82 volts. This source voltage is considerably higher than the normal +15 volt source at the height-control input. The effect of connecting C_{517} to this higher source (momentarily) is to cause it to start charging more rapidly than if it were only connected to the +15 volt source at all times. The momentary higher charging voltage applied through the lin clamp is necessary because, as shown in Fig. 17-20, the sawtooth capacitor, C_{517}, must charge effectively, through the inductance of the yoke and thus requires a higher initial charging voltage to overcome the current opposition of the yoke inductance.

Once the yoke current has started to flow, C_{517} is returned to its normal charging source of +15 volts, through the height control. This happens as follows: When C_{111} begins to charge through the lin-clamp circuit, its left plate starts rising in the positive direction. This positive-going voltage is applied to the base of the predriver and produces a negative-going sawtooth voltage at the collector. Since the predriver collector is connected to the emitter of the lin-clamp transistor (PNP), the negative-going voltage cuts off the lin-clamp transistor and C_{517} can now charge only through the height control toward +15 volts, as previously explained.

17.8 HORIZONTAL DEFLECTION OSCILLATORS AND AFC

In general, the same basic types of deflection oscillators are used in the horizontal system as are found in the vertical-deflection system. These are the blocking oscillator and the multivibrator types, and they are, of course, found in both the tube and solid-state versions. The same general types are used for both monochrome and color sets, although, as with the vertical systems, some unique designs in some color sets may also be found.

The horizontal oscillators operate at 15,750 Hertz, compared to 60 Hertz for the vertical oscillators. Since the horizontal oscillators operate at over 200 times the frequency of the vertical oscillators, it must be expected that stability problems of the horizontal oscillators will be considerably greater than for vertical oscillators, and this is indeed the case. In addition the problem of false triggering of the horizontal oscillators by noise pulses, rather than by the horizontal-sync pulses, is far more serious than is the case with vertical oscillators. A major reason for this is the manner in which the vertical- and horizontal-sync pulses are separated before being fed to their respective deflection oscillators. In the case of the vertical-deflection oscillators, the vertical-sync pulses are selected by means of integrators, which are low-pass filters. Because the vertical-sync pulses are passed through low-pass filters, most of the noise pulses accompanying the composite-sync pulses are rejected by the integrator and never reach the vertical oscillator.

In the case of the horizontal-sync pulses, the situation is quite different. Horizontal-sync pulses are selected by the process of differentiation, and differentiators are high-pass filters. Thus, most noise pulses tend to be passed, rather than rejected, by the differentiators. This problem is particularly serious because of the much higher horizontal oscillator frequency and its attendant stability problems. If the horizontal-deflection oscillator were synchronized directly by the horizontal-sync pulses (a situation that was true in some of the very early TV receivers), poor stability would result because of the triggering by noise impulses. The result of this would be to produce horizontal "tearing" of the picture. In order to insure that the horizontal oscillator operates only on the correct frequency and is basically immune to noise pulses, all horizontal-deflection oscillators in today's TV receivers are controlled by an automatic-frequency-control circuit (AFC), which in turn is controlled by the horizontal-sync pulses as well as by the horizontal-deflection wave-forms. The result of using this indirect method of frequency control results in an excellent horizontal-oscillator stability and almost complete immunity from the effects of noise impulses.

Fundamentals of Horizontal AFC Figure 17-24 is a block diagram of a horizontal-deflection system. The horizontal-sync pulses from the sync separator (or sync amplifier or sync-phase inverter) are fed to the AFC circuit. With these pulses and others which will be discussed shortly, the AFC circuit determines whether or not the horizontal oscillator is on frequency. If it is not, then the AFC block develops a dc voltage which is applied to the horizontal oscillator and which serves to return the oscillator frequency to the correct value. The oscillator develops an appropriate deflection voltage which is then passed on to the output amplifier or amplifiers and, from there, to the horizontal-deflection coils.

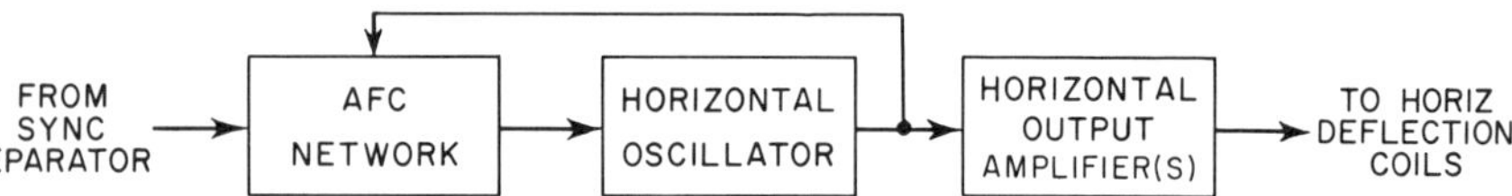

Fig. 17-24. Block diagram of a typical horizontal-deflection system.

This, then, is the overall action of a typical horizontal-deflection system. Since the AFC network is the first stage of this system, let us analyze more closely why it is required and how it functions.

The use of the incoming-sync pulses to trigger and control the vertical- and horizontal-sweep oscillators represents the simplest, most economical, and most direct method of controlling the motion of the electron beam in the image tube. Unfortunately, however, this method possesses limitations and disadvantages which outweigh its economy and simplicity. Perhaps the greatest disadvantage is its susceptibility to noise disturbances arising from electrical apparatus and equipment operating in the vicinity of the receiver. The noise pulses, combining with the video signal and extending usually in the same direction as the desired sync pulses, pass through the same stages as the pulses and arrive at the sweep oscillators. They do their greatest damage when they arrive during the interval between sync pulses. If the amplitude of the noise pulses is sufficently great, they will trigger the sweep oscillator, initiating a new cycle prior to its proper time. When the vertical oscillator is so triggered, the picture will move vertically either up or down, until the proper sync pulses in the signal can again assume control. If the horizontal oscillator is incorrectly triggered, a series of lines in a narrow band will be jumbled, giving the appearance of streaking or tearing across the image. When the interference is particularly heavy and persistent, the entire picture becomes jumbled and may even be thrown out of horizontal sync permanently, requiring manual adjustment of the horizontal-hold control.

Of the two sweep systems in a television receiver, interference is particularly destructive to the horizontal system. To understand why this is so, we must examine the nature of most interference voltages and their effect upon the vertical- and horizontal-sweep oscillators.

Whenever a blocking oscillator is triggered, for example by a sync pulse, its grid, after a short period of conduction, becomes highly negative as a result of an accumulation of electrons on the grid capacitor. This negative voltage is sufficient to keep the tube beyond cutoff until the charge on the grid capacitor has decreased to a value at which current is permitted to flow again through the tube. In most circuits now in use, the capacitor discharge occurs in the manner shown in Fig. 17-25. At the start, the discharge is fairly linear. However, as the amount of charge contained in the capacitor decreases, the discharging rate decreases exponentially. In Fig. 17-25 the region, usually called non-linear, extends from points A to B.

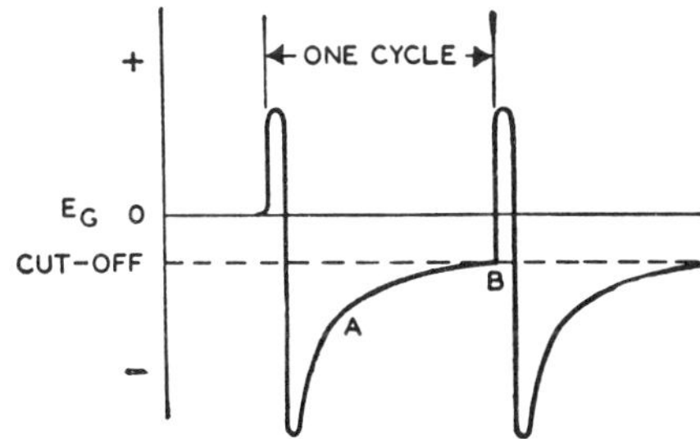

Fig. 17-25. The manner in which the grid voltage of a blocking or multivibrator oscillator varies.

Now, when the negative charge existing on the grid capacitor is large, the oscillator is relatively immune to incoming-positive pulses. With continued discharge, however, the immunity decreases. Experience has indicated that off-cycle triggering of the oscillator is generally concentrated in the last 15 percent of its discharge cycle. This is true regardless of the frequency at which the oscillator is operating. Hence, one would expect to experience equal difficulty with both deflection systems in the receiver. That this is not so is due to the nature of the noise pulses and the type of filters inserted before each sweep oscillator.

Noise pulses which are the most troublesome to television receivers possess a high amplitude, but are narrow in width or are of short duration. (The energy of the noise pulses is distributed over a wide range of frequencies. In order for a peak to occur, the phase relationship among the various frequencies must be such as to permit them to add, forming the high-amplitude pulse or peak. This condition, however, usually exists only for a brief interval, which explains the narrow width of these pulses.) When the pulses reach the path leading to the horizontal-sweep oscillator, they are readily passed because of the short time constant of the filter leading to the horizontal system. A short time constant filter is necessary, because the horizontal-sync pulses themselves have a duration of only 5 microseconds. On the other hand, the filters leading to the vertical system have a long time constant and automatically act to suppress the effects of all horizontal-sync pulses and noise pulses of short duration. The presence of this low-pass filter (the integrating network) is largely responsible for the greater immunity to noise pulses enjoyed by the vertical system. Of course, when a wide noise pulse is received, it contains enough energy to cause off-time firing of the vertical oscillator, but the annoyance caused to the viewer from this source is seldom great. To reduce the susceptibility of the horizontal-sweep system to noise pulses of any type, several automatic-frequency- (and phase-) control systems have been developed.

In each of these special control systems an oscillator is set to operate at 15,750 Hertz, and the output of the oscillator (through amplifiers) controls the horizontal motion of the electron beam across the screen of the image tube. The next step is to synchronize the frequency of this sweep oscillator with the incoming horizontal-sync pulses of the signal. This step is accomplished through an intermediate stage generally known by one of the following names: control stage, AFC phase detector, or horizontal-sync discriminator. Whatever the name, the function of this intermediate network is to compare the frequency of the incoming horizontal-sync pulses with the frequency of the receiver horizontal-sweep oscillator. If a difference exists, a dc voltage is developed, which when fed back to the horizontal-sweep oscillator, is used to change its frequency until it is exactly equal to that of the incoming pulses.

Note that the incoming-sync pulses are not applied directly to the sweep oscillator. They are merely compared (in frequency) with the output of the sweep oscillator and, if a frequency difference exists, then a dc voltage is developed which, when fed back to the sweep oscillator, forces its frequency back into line with that of the sync pulses.

Now, by having the dc control voltage pass through a long time-constant filter before it reaches the sweep oscillator, we can eliminate the effects of any noise impulses and permit only relatively slow changes in the frequency of the sync pulses (which may occur at the transmitter) to affect the sweep oscillator. Thus, a long time-constant filter somewhat similar to that present in the vertical-sweep system is incorporated in the horizontal AFC system. A long time-constant filter could not be used directly in the horizontal system because it would have prevented the desired horizontal-sync pulses (as well as the noise pulses) from reaching the horizontal-sweep oscillator. Hence the need for the indirect method outlined above.

17.9 VACUUM-TUBE PHASE-DETECTOR AFC SYSTEMS

Vacuum-tube phase-detector AFC systems, fall into two general categories. These are diode and triode systems, according to the type of tube used. Even within each category there exist several different methods of achieving the desired output-control voltage, although in most instances the general operation remains the same.

Diode Phase-Detector A common diode phase-detector circuit is shown in Fig. 17-26. Two diodes are connected so that they receive horizontal-sync pulses in phase opposition and a sawtooth wave in

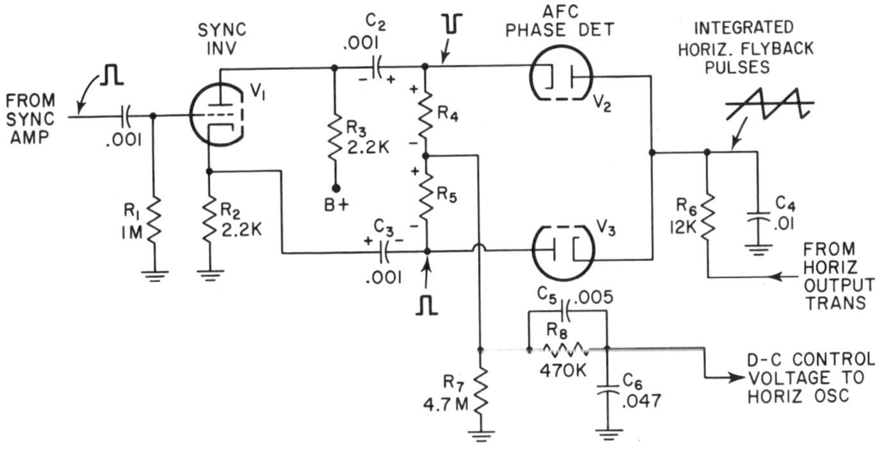

Fig. 17-26. A double-diode-phase detector.

common. From the interaction of these waveforms, a dc voltage is developed across R_7 that is governed by the relative-frequency difference between the incoming-sync pulses and the sawtooth wave.

In detail, the network functions as follows: When the sync pulses are received by V_1, positive and negative pulses of equal amplitude are applied to both diode sections of the phase detector. The cathode of V_2 receives a negative pulse at the same time that the plate of V_3 receives the positive pulse. These pulses cause both diode sections to conduct. The current flowing through V_2 charges C_2 to approximately the peak value of the applied pulse, while the current following through

V_3 charges C_3. The polarity of each voltage is indicated in Fig. 17-26. During the interval between the pulses, each capacitor discharges, the electrons moving from C_3 up through R_5, then down through R_7 to ground, and up through R_2 to the other plate of C_3. This current thus develops a negative voltage across R_5 and R_7. C_2 also discharges and its electrons travel down through R_3 to ground (by way of the power supply), then up through R_7 to the junction of R_4 and R_5, and then up through R_4 to the other plate of C_2. The result fo these two currents through R_7 is that the voltage produced by one current cancels the voltage produced by the other current, leaving a net potential of zero volts. This is desirable since these two pulses alone should produce no net-control voltage.

Because of the slow discharge of C_2 and C_3 through their respective networks, the voltages developed across R_4 and R_5 keep V_2 and V_3 from conducting until the arrival of the next pulse.

Fed to the phase detector is another voltage, a sawtooth wave which is developed across C_4 from pulses applied to it from the secondary of the horizontal-output transformer. This sawtooth voltage possesses the same frequency as the horizontal oscillator, since it is the oscillator which drives the horizontal-output amplifier. The sawtooth voltage is applied equally to each tube; thus, the plate of V_2 and the cathode of V_3 receives the same polarity voltage of the sawtooth wave at the same time. Hence, at the phase detector, we have both ingredients needed to check the oscillator frequency against the frequency of the incoming pulses.

Comparison of the two frequencies is possible only at the instant the sync pulses arrive, because it is only at this moment that V_2 and V_3 conduct and are in a position to respond to the sawtooth voltage. Three situations are possible.

First, if the sync pulses arrive at a time when the sawtooth voltage is passing through zero, current will flow through V_2 and V_3, replenishing any charge that C_2 and C_3 may have lost during the interval between pulses. No net voltage will appear across R_7, as indicated previously. This condition is the desired one because the frequency of the sweep oscillator and the sync pulses are in step with each other.

The second situation occurs when the sync pulses arrive. The sawtooth voltage is negative at this instant. (This occurs when the horizontal oscillator is running too slow.) Now V_3 will receive a positive pulse at the plate and a negative sawtooth voltage on the cathode and, hence, conduct more strongly than usual, producing a larger than normal voltage across R_7. At the same time, conduction through V_2 is reduced because the negative-sawtooth voltage at the plate partly offsets the negative-sync pulse at the cathode. The reduced current flow through V_2 cannot offset the voltage which the current of V_3 develops across R_7. Hence, a resultant negative voltage is developed which is fed to the horizontal oscillator, and its frequency is altered (in this case, speeded up).

In the third situation, the pulses arrive when the sawtooth voltage is positive. Now, V_2 conducts more strongly than V_3 and a resultant positive voltage is developed across R_7. This voltage, fed to the controlled horizontal oscillator, acts to slow it down or lower its frequency to bring it in line with the frequency of the incoming pulses.

Filters C_5, C_6, and R_8 respond only to slow changes in voltage level, preventing fast-acting noise pulses from affecting the operation of the horizontal oscillator. In this way, we tend to stabilize the circuit and avoid the false triggering that can happen when sync pulses are fed directly to the horizontal oscillator. In place of vacuum-tube diodes, solid-state diodes may be utilized with identical results.

DC Control of Oscillator Frequency The horizontal oscillator to which the dc control voltage developed in Fig. 17-26 is applied is shown in Fig. 17-27. This oscillator is a cathode-coupled multivibrator containing a special resonant-stabilizing circuit in the plate circuit of the first triode. More will be said on this point later.

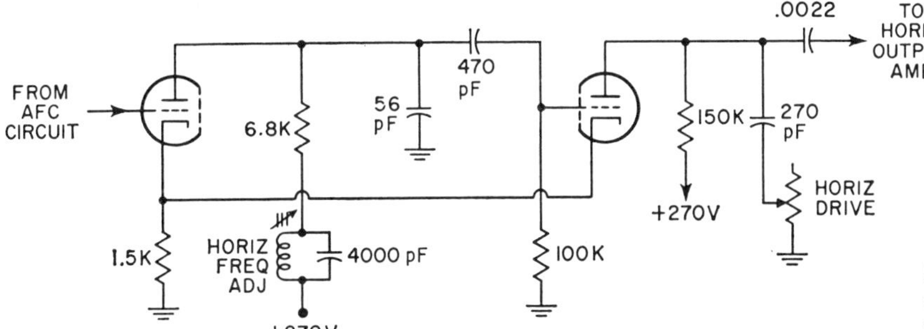

Fig. 17-27. A cathode-coupled multivibrator to which the AFC circuit of Fig. 17-26 would be connected.

To understand what happens when the dc control voltage is applied directly to an oscillator, consider the operation of a cathode-coupled multivibrator. In this oscillator, the first triode conducts during "trace" time. The second triode conducts only during the "retrace" time. Since the cathodes are tied to ground through a common resistor, the operating bias of the second triode is affected by the cathode voltage developed by the first triode.

The grid of the first triode is usually bypassed to ground and is not part of the feedback loop. This leaves the grid available as the controlling element of the system.

If the correcting voltage on the grid of the first triode is made positive (by the automatic-frequency-control network), current flow through the tube will increase and the cathode voltage will rise. This extends the cutoff time of the second triode. Since the time is lengthened before the retrace time occurs, the oscillator frequency is lowered.

Similarly, any negative voltage applied to the first-triode grid lowers the cathode potential and shortens the time of the RC discharge of the second triode grid circuit. This change increases the firing rate and raises the frequency of the system.

It is possible, by reversing the polarity of the sawtooth voltage which is fed to the phase detector to obtain control voltages of opposite polarity for the conditions of a fast or a slow oscillator. The oscillator is the controlling factor. For a cathode-coupled multivibrator, the required control voltages should possess the polarity indicated. For a blocking oscillator, an opposite set of polarity voltages would be needed.

The use of diodes in the phase detector just discussed requires horizontal-sync pulses of positive and negative polarity. A two-diode circuit in which only one set of sync pulses is required is shown in Fig. 17-28. The two cathodes of the diodes (in this case, germanium diodes) are connected and a negative-going sync pulse is applied at their junction. This arrangement applies the sync pulse equally across D_1 and D_2 because C_3 and C_4 are so much greater than C_1 that D_1 and D_2 are

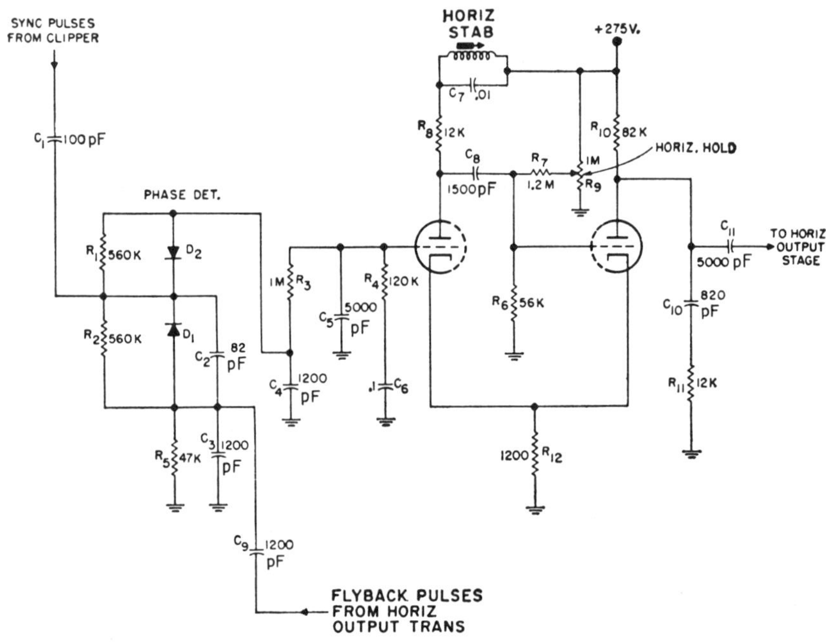

Fig. 17-28. A phase detector using germanium diodes and requiring only one set of input-sync pulses.

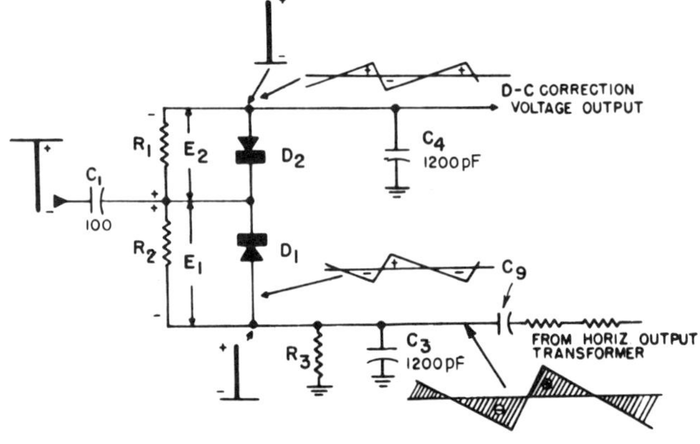

Fig. 17-29. A simplified diagram of the phase detector in Fig. 17-28, showing the waveforms in the circuit.

effectively connected in parallel. This being the case, current will flow in each diode, causing equal currents to flow in the load resistors R_1 and R_2. The currents, of course, flow in opposite directions, and the voltage drops across R_1 and R_2 will have opposing polarities and, therefore, will cancel out, producing a zero output voltage.

The sawtooth wave which is formed from the flyback pulses received at the horizontal-output stage is a sample of the horizontal-oscillator frequency. This voltage is applied across D_1 and D_2, effectively bringing one-half of the original sawtooth wave across each diode. It can be shown that the sawtooth wave across D_1 will be going positive when the voltage across D_2 is going negative, and vice versa (Fig. 17-29). The currents of both diodes will be equal but opposite in polarity, so equal and opposite voltages across R_1 and R_2 will produce a zero output voltage.

Thus, the incoming-sync pulses alone will not cause the phase detector to produce any voltage output. In like manner, the sawtooth wave alone will not cause the phase detector to produce any voltage output.

The sync pulse, possessing a much greater amplitude than the sawtooth wave, keeps the diodes biased so that they conduct only when the sync pulse is applied to them. Therefore, only that portion of the sawtooth wave that occurs at the instant of the sync pulse has any effect on the output of the phase detector.

Now, if the sync pulse occurs in the exact center of the sawtooth retrace (*i.e.*, the retrace passing through its ac axis), equal but opposite currents will flow and no output voltage will be developed (see Fig. 17-30(A).

If the oscillator is slow, the sync pulse willl arrive before the sawtooth retrace passes through its ac axis (see Fig 17-30(B)). On D_2, therefore, part of the sawtooth voltage will be added to the sync-pulse voltage, because the sawtooth voltage is on the positive half of its cycle when the sync pulse occurs. Part of the sawtooth voltage on D_1 will be subtracted from the sync-pulse voltage, because the sawtooth retrace here is still in the negative half of its cycle. The output voltage of the phase detector in this case will be a negative one, because the voltage drop across R_1 is greater than the drop across R_2.

If the oscillator is fast, the sawtooth retrace will pass through its ac axis before the sync pulse occurs (see Fig. 17-30(C)). On D_2, therefore, part of the sawtooth voltage will be subtracted from the sync pulse. On D_1, part of the sawtooth voltage will be added to the sync pulse, producing a higher voltage drop across R_2 than across R_1. This will produce a positive output voltage which slows down the horizontal oscillator.

Triode Phase Detector It is possible, through the action of a single triode, to achieve the same measure of control as with the double-diode circuits. A triode-phase detector circuit is shown in Fig. 17-31. Its chief distinguishing feature is its need for one source of sync pulses rather than two. The sync pulse is applied to the cathode, and its phase is negative.

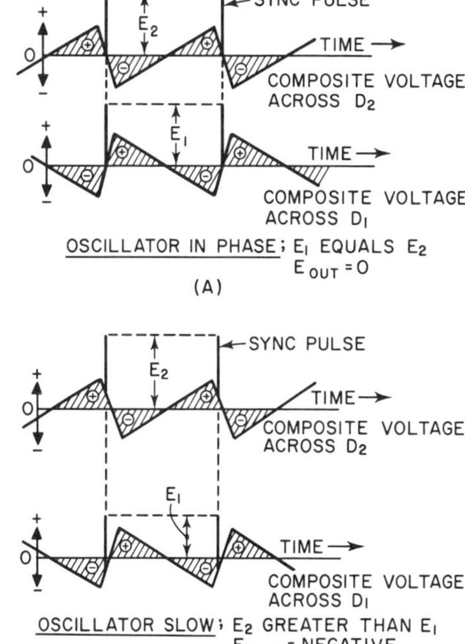

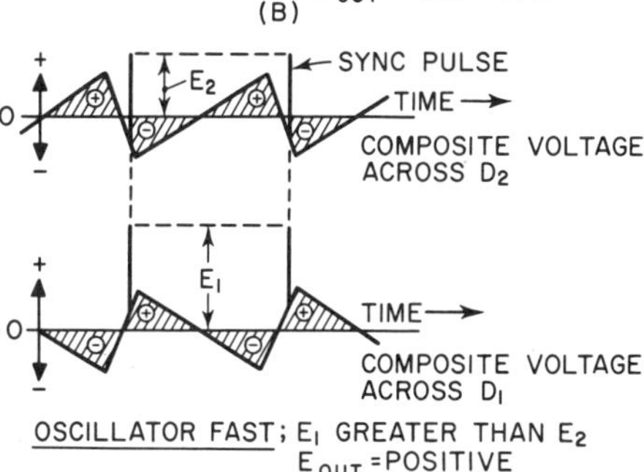

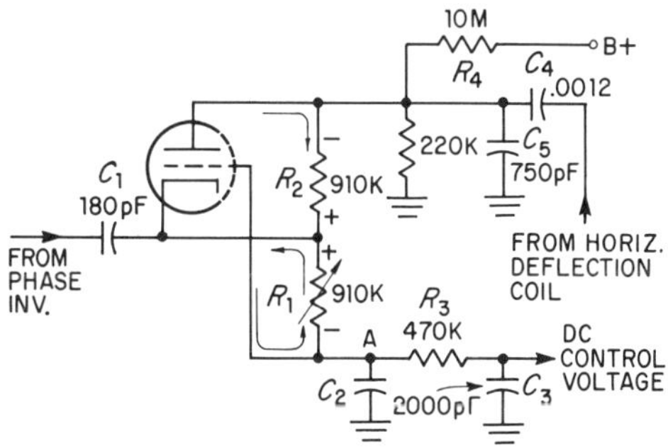

Fig. 17-30. The operation of the phase detector of Fig. 17-29. When the sync-pulse frequency and the oscillator frequency are (A) equal and (B and C) unequal.

Fig. 17-31. A triode phase-detector circuit. Here only one source of sync pulses is required.

The sawtooth wave is applied to the plate of the tube, while the dc control voltage is obtained from the grid. The over all circuit operation is as follows.

When a sync pulse arrives, it drives the cathode sharply negative or, what is the same thing, it makes the grid and plate positive with respect to the cathode. Current then flows in both circuits, through the paths indicated in Fig. 17-31. The voltages developed across R_1 and R_2 have the polarities indicated and, if the resistor values are correctly chosen, the two voltages will be equal and will cancel each other. Actually, C_2 and C_5 charge first when the current flows and then discharge between pulses. It is during the discharge interval that the voltages across R_1 and R_2 are developed.

When the oscillator frequency is too low, the sawtooth wave lags behind the sync pulses. Hence, when the pulses arrive, the sawtooth voltage at the tube plate is negative. This reduces the current flowing in the plate circuit, and in consequence, the voltage developed across R_2. The voltages across R_1 and R_2 no longer cancel, and a net negative voltage appears at point A and is transfered to the horizontal-sweep oscillator.

Conversely, if the oscillator frequency is higher than the sync-pulse frequency, the sawtooth wave is positive at the tube plate when the sync pulses arrive. Now, the voltage across R_2 rises above normal, and the dc control voltage is positive.

Circuit Adjustment In the foregoing AFC circuits there are practically no variable controls. Hence, there is actually nothing to adjust. Variable components, however, are generally found in the multivibrator, horizontal oscillator. A multivibrator is frequenctly used, because it provides a convenient input for applying the AFC-control voltage and because it operates with good stability, particularly when it has a stabilizing resonant circuit in the plate circuit of the first triode.

The cathode-coupled multivibrator employed with phase-detector-AFC systems may have one or two adjustments. If there are two adjustments, one is a horizontal hold control and the other is the movable core in the stabilizing coil. The multivibrator of Fig. 17-28 is representative of this group. The hold control is accessible on the front or rear panels of the receiver and may be adjusted from time to time, as required. In a number of sets, however, the hold control is dispensed with, and the only adjustment then is the movable core of the stabilizing coil. Figure 17-27 is an example of this approach. In both arrangements, the coil core should not require any attention once it is adjusted, unless some component changes value in the circuit and the stabilizing circuit is unable to keep the multivibrator in sync. The adjustment is quite simple and requires only that the core be rotated until the picture is properly synced in. If a hold control is present, it is set to the center of its range before the coil core is moved.

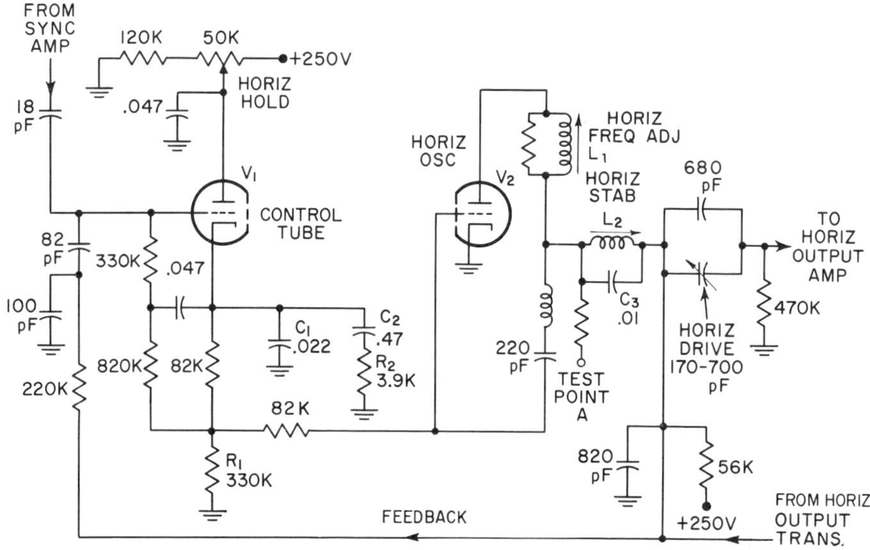

Fig. 17-32. The Synchroguide AFC System.

The Synchroguide AFC System* Another vacuum-tube, AFC circuit which has gained wide acceptance is that shown in Fig. 17-32. It consists of a single control tube, a long time-constant filter, and a blocking oscillator. Basically, the horizontal oscillator is a free-running oscillator and discharge circuit. It does not receive the incoming pulses directly, but should its frequency differ from that of the pulses, then the control tube, V_1, will alter the negative bias on the grid of the blocking oscillator and thereby change its frequency. It can do this because the cathode resistor, R_1, of the control tube is also common to the grid of the blocking oscillator. The incoming-sync pulses, positive in polarity, are applied to the grid of V_1 through an 18 pf capacitor. This grid also receives a wave from the horizontal-output transformer which possesses a shape which is a cross between a parabolic wave and a sawtooth wave. This wave combines with the incoming-sync pulses to control the blocking oscillator frequency.

The combined wave is specifically designed to have fairly steep sides (Fig. 17-33) in order that any difference in frequency between the blocking oscillator and the incoming-sync pulses will have a marked effect on the circuit. Here is how this occurs.

The control tube, V_1, is given enough negative bias to keep it cut off except when the incoming-sync pulse is high on the slope of the combined wave-form, as shown in Fig. 17-33(B). If the blocking oscillator changes phase so that the pulse arrives at a time when it is down along the slope, the amount of time during which V_1 conducts will decrease. This decrease is indicated in Fig. 17-33(C) by the narrow width of the wave-form extending above the cutoff point of the tube V_1.

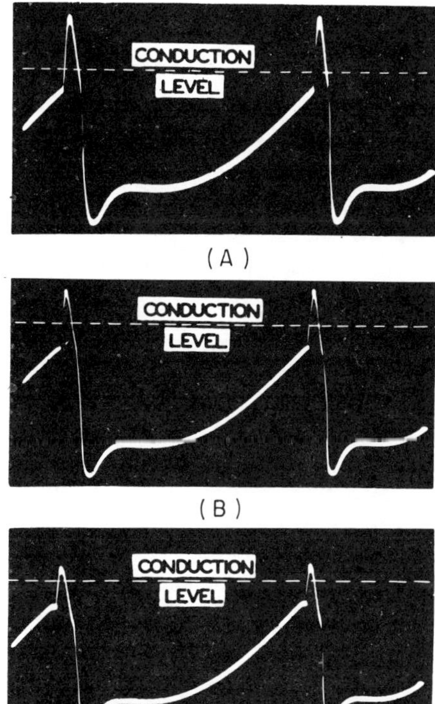

Fig. 17-33. The waveforms at the grid of V_1, Fig. 17-32, for different oscillator frequencies. (A) Oscillator slower than normal. (B) Oscillator in phase. (C) Oscillator faster than normal.

* Also known as the pulse-width system.

On the other hand, if the blocking-oscillator frequency changes so that the sync pulse arrives at a time when it is closer to the top of the combined wave (Fig. 17-33(A)), then the plate conduction time of V_1 will increase. When the control tube conducts, C_1 and C_2 in its cathode circuit will charge to a dc potential proportional to the length of time that current flows through the tube. This dc potential is applied as a bias to the grid of the blocking oscillator, altering its frequency and tending to bring it back into line. The components of the cathode circuit of V_1 form a long time-constant filter which averages out the plate current pulses.

Three controls are associated with this circuit: (1) The blocking oscillator transformer, L_1, is slug-tuned to permit coarse adjustments in oscillator frequency. (2) The horizontal hold control will affect the plate voltage of the control tube and, in this manner, affect the amount of voltage developed across R_1. This is the only front-panel control of the group. (3) The core of coil, L_2, is adjustable to permit tuning this resonant circuit to the frequency of the oscillator. This control is the most difficult to adjust.

L_2 and C_3 serve to stabilize the oscillator operation and thereby render it more immune to signal or circuit disturbances that may occur from time to time. The manner in which it accomplishes this is detailed in the circuit of Fig. 17-34 as follows.

The oscillator, being a blocking oscillator, is cut off during more than 90 percent of the cycle and conducts heavily 10 percent of the cycle or less. Let us consider the oscillator when it is conducting heavily. During this interval, the heavy current drawn through L_2 and C_3 prevents this circuit from oscillating. At the same time a voltage is developed across these components with the polarity indicated in Fig. 17-34(A). When the blocking oscillator drops into cutoff, the current flowing through the tube and the resonant circuit, L_2, C_3, ceases abruptly. This sudden stoppage of current excites L_2, C_3 into resonance, as the coil attempts to maintain the flow of the current. The polarity of the voltage developed across the coil required to do this is shown in Fig. 17-34(B).

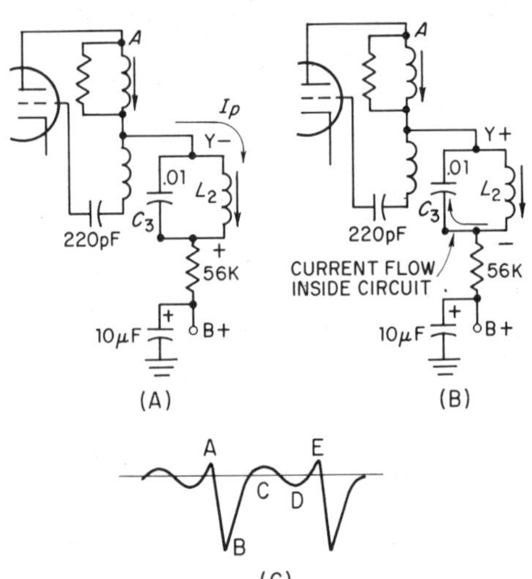

Fig. 17-34. Diagrams illustrating how L_2 and C_3 (of Fig. 17-32) help to stabilize the pulse-width AFC system.

Referring to the waveform developed in this circuit at point Y, at time A in Fig. 17-34(C), the blocking oscillator drops into cutoff, and at time B the resonant circuit reverses its voltage polarity in an attempt to maintain the flow of current. Thereafter, the current in the resonant circuit L_2 and C_3 flows first into the lower plate of C_3, then reverses and flows into the upper plate. The frequency of the circuit is close to 15,750 cps, and so it has time to complete one cycle before the blocking oscillator again conducts heavily at time E, damping out the oscillations in L_2 and C_3.

The ability of this additional resonant circuit to improve the noise immunity of the blocking oscillator is due to the fact that near the end of the discharge cycle, when the cutoff voltage on the oscillator grid is low, the voltage developed across L_2, and C_3 is negative. In Fig. 17-34(C) this is the region *C-D*. This negative voltage opposes the B+ voltage from the power supply and, in so doing, reduces the susceptibility of the blocking oscillator to any noise pulses that may appear at this time. Here is why this is so.

When the plate voltage of a triode decreases, the grid voltage required to cut off the plate current decreases. In other words, the cutoff point approaches closer to zero volts. With the B+ plate voltage going down (because of the opposition voltage across L_2 and C_3), we achieve the same effect as bringing the cutoff level of the tube closer to zero (see Fig. 17-35). With the cutoff level thus moving upward, it will require a stronger noise pulse to raise the grid voltage to this new cutoff level in order to have current flow through the tube again for the start of the next cycle.

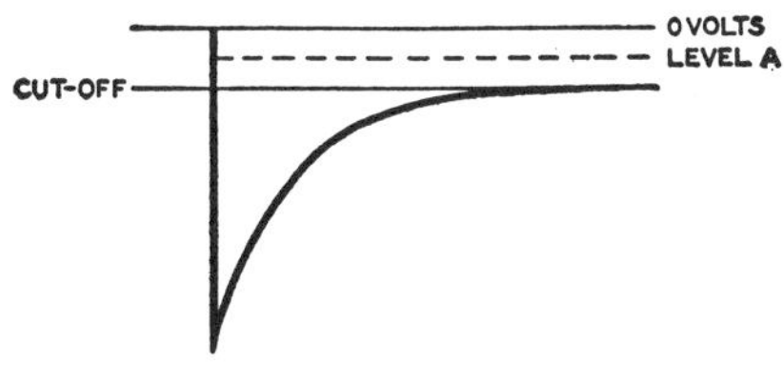

Fig. 17-35. See text for explanation.

On the other hand, in the region near *E* in Fig. 17-34(C), the voltage across L_2 and C_3 is going in the positive direction, aiding the B voltage and preparing the plate circuit for the arriving triggering pulse.

Multivibrator Stabilization Resonant-stabilizing circuits have been used in the multivibrator as well as in the blocking oscillator. One such circuit was shown in Fig. 17-27. The additional resonant coil and capacitor are placed in the plate circuit of the first triode and adjusted to 15,750 cps. The presence of this circuit alters the manner in which the grid of the second triode comes out of cutoff. The waveform of Fig. 17-36(A) appears at the plate of the first triode in the absence of the stabilizing circuit. Figure 17-36(B) shows the grid waveform of the second triode under the same condition. Now, when we insert the stabilizing circuit, its waveform, shown in Fig.17-36(C), will add to those existing in the circuit to produce the modified waveforms shown in

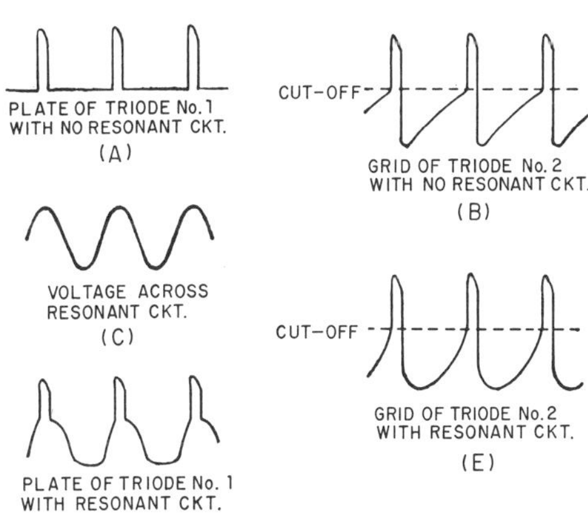

Fig. 17-36. The effect of a resonant stabilizing circuit on the operation of the multivibrator shown in Fig. 17-27.

Fig. 17-36(D) and (E). Of particular importance is the grid waveform of the second triode. Note that it now comes out of cutoff quite sharply. A considerably stronger noise pulse will be required to trigger this tube prematurely than would have been needed without the stabilizing circuit.

17.10 SOLID-STATE, HORIZONTAL AFC SYSTEMS

By studying the Motorola solid-state, AFC circuit in Fig. 17-37, you may note that it is similar to the AFC circuit employed in many vacuum-tube TV receivers (compare with Fig. 17-28). Also included in Fig. 17-37 is a set of wave-forms which indicate the phase relationship between the oscillator sawtooth and sync-pulse waveforms, for different frequencies of the oscillator.

The phase detector diodes, D_1 and D_2, are connected back to back and are encapsulated in a single plastic case. R_{501} and R_{502} form the output load across which the control voltage will be developed. R_{500} and C_{500} are used to couple the negative horizontal-sync pulses into the phase detector to establish a reference frequency against which

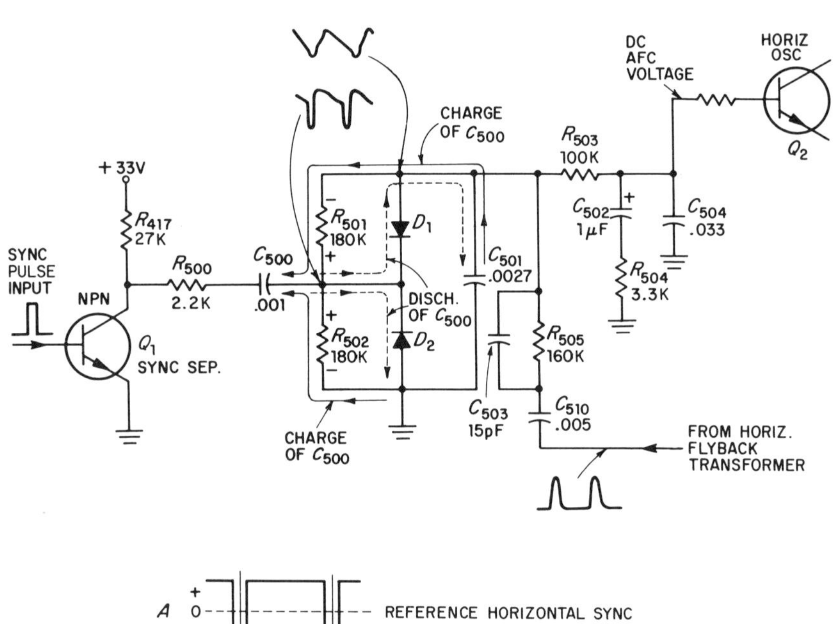

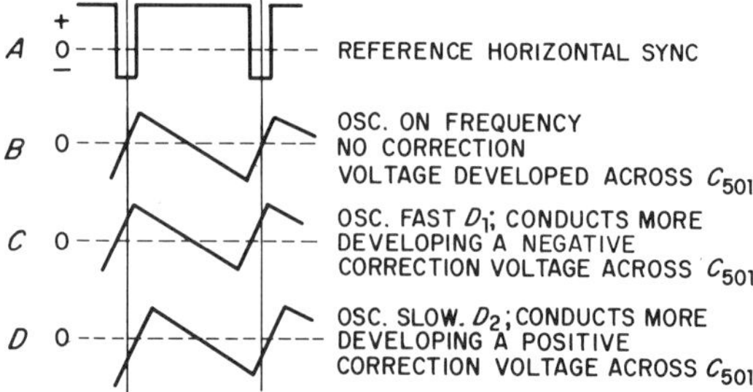

Fig. 17-37. Motorola solid-state AFC circuit. (*Courtesy of Motorola Consumer Products, Inc.*)

to compare the oscillator frequency. The capacity of C_{500} which differentiates the horizontal and the vertical serrated sync pulses is small, thus coupling only horizontal-sync pulses to the AFC network. C_{510}, C_{503}, R_{505}, and C_{501} couple integrated positive pulses from the flyback transformer (this pulse is repeated at the oscillator frequency) into the phase-detector network for frequency comparison with the horizontal-sync pulses. R_{503}, C_{502}, R_{504}, and C_{504} form the horizontal AFC antihunt network, thus preventing any ac variations from entering the horizontal-oscillator circuit and causing the oscillator to be unstable.

To explain the operation of the AFC circuit, let us begin with the sync-separator stage (Q_1). Prior to the arrival of the positive-sync pulses at its base, Q_1, is cut off and has a high-collector voltage. Since C_{500} is connected to the collector of Q_1, it will be charged to nearly the supply voltage. When an incoming positive-sync pulse turns on Q_1, its collector voltage decreases.

C_{500} will now discharge (dashed arrows), coupling the sync to the cathodes of D_1 and D_2. This discharging voltage will forward bias both diodes (D_1 and D_2) causing them to conduct simultaneously and equally. If the negative-going sawtooth at the anode of D_1 is going through zero at this time (this is indicative of the oscillator being in phase with the sync), both diodes will conduct equally since both the D_1 and D_2 anodes are now at zero volts. After the sync pulse ends, the collector voltage of Q_1 rises. C_{500} will now recharge (solid arrows), producing a voltage drop across the two load resistors (R_{501} and R_{502}) as indicated in Fig. 17-37. Since both diodes conducted equally, equal and opposite voltages will now appear across R_{501} and R_{502} when C_{500} recharges. This being the case, the net voltage across the two resistors will be zero, and no correction voltage will be developed for the oscillator across C_{501}.

If the sawtooth at the anode of D_1 is not going through zero when the sync arrives (indicative of the oscillator frequency being out of phase with the sync), the conduction of D_1 will be influenced by the sawtooth wave at its anode. When the sync pulse is completed and C_{500} recharges, the voltages developed across the load resistors are still opposite but are not now equal. The net voltage is no longer zero. A correction voltage is now developed across C_{501} and is applied to the oscillator to correct its frequency.

We can clarify the foregoing discussion by observing waveforms *A*, *B*, *C*, and *D* in Fig. 17-37. Waveforms *A* and *B* are impressed across D_1 when the oscillator frequency is in phase with the sync frequency.

Because the sawtooth (*B*) is going through zero at this time, it does not alter the conduction of D_1. Hence, both diodes will conduct equally since their anodes are referenced to zero. No correction voltage is developed under these conditions across C_{501}.

Waveforms *A* and *C* are impressed across D_1 when the oscillator is leading in phase (higher frequency) with respect to the sync. Now, when the sync pulse turns D_1 on, it will conduct more than D_2 since the waveform C caused its (D_1) anode to be positive at this time. As a result, the voltage drop across R_{501} will be larger than the voltage across R_{502}, and a negative correction voltage is developed across C_{501}. This voltage has the correct polarity to reduce the oscillator frequency and pull it back into phase lock with the sync-pulse frequency.

Waveforms A and D are impressed across D_1 when the oscillator is lagging in phase with respect to the sync. Now, when the sync pulse turns D_1 on, it will conduct less than D_2, since waveform D caused its anode to be negative at this time. As a result, the voltage drop across R_{501} is less than the voltage drop across R_{502}, and a net positive voltage is developed across C_{501}. This voltage now has the correct polarity to increase the oscillator frequency and pull it back into phase lock with the sync-pulse frequency.

Motorola Solid-State Horizontal Oscillator The horizontal oscillator circuit which is frequency stabilized by the foregoing AFC circuit is shown in Fig. 17-38. The horizontal oscillator acts somewhat like a

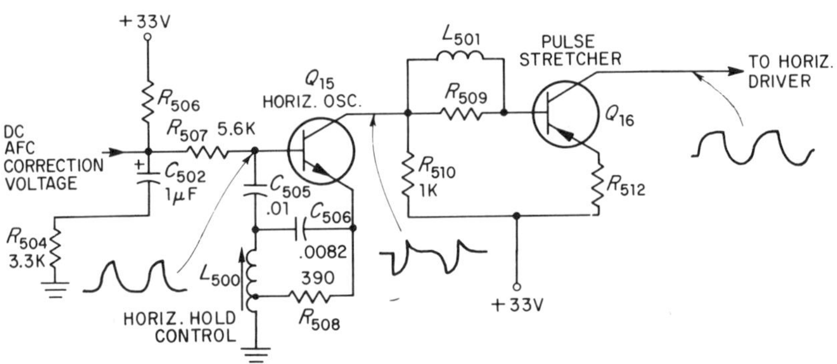

Fig. 17-38. Motorola solid-state, horizontal-oscillator circuit. (*Courtesy of Motorola Consumer Products, Inc.*)

switch, turning on and off at a predetermined rate (15,750 Hz). From its output, we obtain a driving signal which we can shape, amplify, and ultimately apply to the horizontal-deflection coils.

The horizontal oscillator, Q_{15}, receives its initial emitter/base forward bias for starting from R_{506} and R_{507}. The base of Q_{15} also receives a dc voltage from the AFC which acts to keep the oscillator locked in frequency with the transmitter sync pulses.

The horizontal oscillator is of the Hartley type, employing a tapped oscillator coil, L_{500}, in its emitter circuit to sustain oscillation. The main frequency determining components are L_{500}, C_{506}, and R_{508}. In the output of the oscillator, a waveform is produced which resembles a square wave. This waveform is produced by the abrupt switch on of Q_{15} to saturation, and the equally abrupt switch off of Q_{15}. This unsymmetrical square wave is not quite suitable for switching (or driving) the horizontal-output transistor, and must be modified.

L_{501} and R_{509} are employed to shape a more uniform square wave to be applied to the pulse shaper (Q_{16}) for further shaping and amplification. These two components form what is called a pulse stretcher. The name is very descriptive, because this is exactly what they do. The inductor, L_{501}, opposes a change in current similar to a choke in a power supply; thus, the spikes and abrupt voltage changes on the existing square wave are filtered out by the inductor.

RCA Solid-State AFC A solid-state RCA color-TV chassis, AFC circuit will now be explained. The following information was supplied by RCA and is reproduced by their permission.

Horizontal AFC The AFC circuitry used in any television receiver is designed to automatically hold the horizontal oscillator at the exact frequency and phase of the received horizontal-sync pulses. This function is especially critical in color-television circuitry because the color-sync (burst) amplifier is keyed from pulses derived from the horizontal circuitry. As a result, any slight frequency or phase discrepancy between the occurrence of the color burst and the horizontal-keying pulses for the burst amplifier may cause an incorrect color display.

The horizontal AFC circuitry employed in this RCA chassis will automatically hold the horizontal oscillator at the exact frequency of the received horizontal-sync pulses if the oscillator free-running frequency is within ± 300 Hz of the received sync-pulse frequency.

A block diagram of the horizontal AFC System is illustrated in Fig. 17-39. The circuit develops a dc oscillator-control voltage which is proportional to the difference between the frequency of the received sync pulses and the operating frequency of the horizontal oscillator. The AFC circuit is basically the same as the circuit shown in Fig 17-26. The functioning of the AFC circuit is also basically the same as explained for the dual-diode phase detector, at the beginning of Section 17.10. Therefore, only a brief description of the AFC circuit operation will be presented here. For a more detailed discussion of the operation of the AFC operation, refer to the beginning discussion of Section 17.10.

The horizontal AFC circuit senses the oscillator frequency through a feedback network connected between the horizontal-output transformer

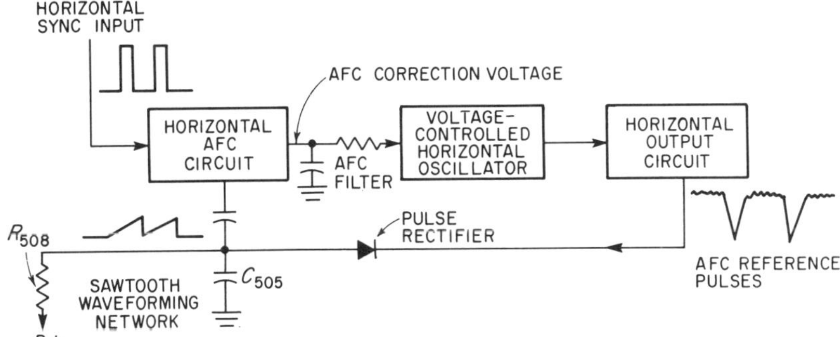

Fig. 17-39. The block diagram of an RCA, solid-state, horizontal AFC system. (*Courtesy of RCA.*)

and the AFC circuit. The feedback waveshape at the horizontal-output transformer is in the form of negative-going pulses, occurring at the operating frequency of the oscillator. A waveshaping network (C_{505} and R_{508}) transforms these pulses into a sawtooth waveshape which is suitable for application to the AFC circuit. The dc correction voltage developed by the AFC circuit is filtered and is then applied to the horizontal oscillator. The frequency of this oscillator may be varied by the AFC voltage, causing the oscillator to operate at the exact frequency and phase of the received horizontal-sync pulses.

A simplified diagram of the AFC circuit is illustrated in Fig. 17-40. Note the similarity to Fig. 17-26. A phase-splitter circuit is employed which supplies equal but opposite polarity sync pulses to a dual-diode

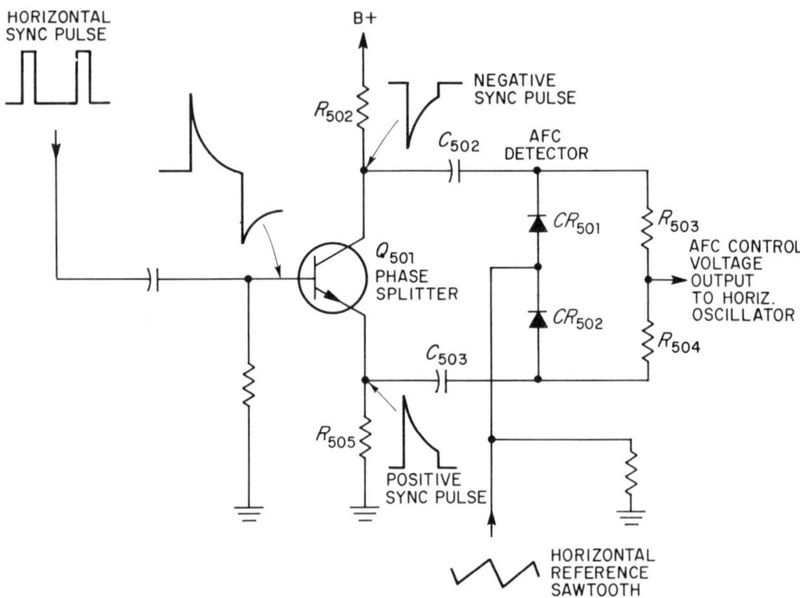

Fig. 17-40. RCA solid-state horizontal AFC phase-splitter and detector. (*Courtesy of RCA.*)

phase-detector network. The incoming horizontal-sync pulses are differentiated at the base of the phase-splitter transistor (Q_{501}). The phase-splitter output pulses are coupled to the AFC Detector circuit through capacitors C_{502} and C_{503}. The sawtooth reference voltage is applied to the common diode connection, as shown.

AFC Operation Figure 17-41(A) illustrates the action of the diode-detector circuit when no frequency difference exists between the applied sync pulses and the sawtooth-reference voltage. In this case, each diode is gated into conduction by the sync pulses as the reference voltage passes through *O*. The current through each diode is therefore equal, resulting in equal but opposite charges on capacitors C_1 and C_2. The discharging action of these capacitors develops equal but opposite polarity voltages across resistors R_1 and R_2. The voltage at the junction of these two resistors (Point A) is therefore OV (with respect to ground). This voltage is the AFC correction voltage. It is significant that the discharging action of the capacitors is sufficient to reverse bias the diodes between sync pulses. This action improves the noise immunity of the AFC system. Also, it should be noted that no potential will be formed at Point A if either the sync pulses or the reference voltage is absent.

The action of the AFC detector circuit when the oscillator is operating at a frequency less than that of the applied sync pulse is shown in Fig. 17-41(B). A decrease in the operating frequency of the oscillator is represented by a change in the relative position of the reference-voltage waveform during the application of the sync pulses. The sync pulses now gate on the diodes during the positive portion of the retrace slope causing diode D_1 to conduct more strongly than diode D_2. The charge on C_1 therefore becomes more positive, while the charge on C_2 becomes less negative. The resulting unbalance in the current flow through R_1 and R_2 (when the capacitors C_1 and C_2 discharge) causes the potential at Point A to become positive. This positive voltage will cause an increase in the oscillator frequency compensating for the discrepancy between the oscillator frequency and the horizontal-sync pulses. A similar but opposite action occurs when the oscillator frequency is higher than that of the applied sync pulses (Fig. 17-41(C). In this case the sync pulses will gate on the diodes during the negative portion of the retrace slope, causing diode D_2 to conduct more than D_1. The resulting C_1-C_2 capacitor discharging action through R_1 and R_2 will cause the potential at Point A to become negative. This negative voltage will soon cause an appropriate decrease in the oscillator frequency.

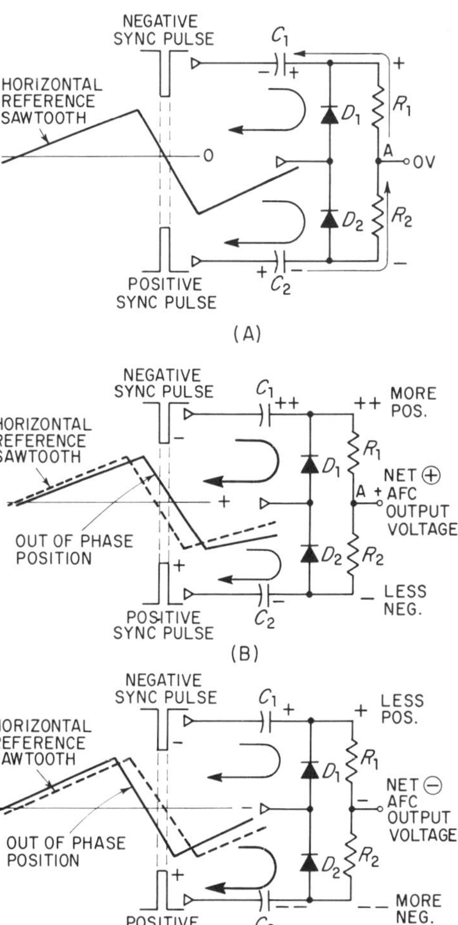

Fig. 17-41. Diagrams illustrating the operation of the RCA solid-state horizontal AFC system. (*Courtesy of RCA.*). (A) The action of the horizontal AFC phase detector with the oscillator in sync. (B) The action of the horizontal AFC phase detector with the oscillator frequency low. (C) The action of the horizontal AFC phase detector with the oscillator frequency high.

AFC Limiter and Filter Circuitry The dc voltage from the output of the AFC discriminator circuit is applied to a limiting and filtering network illustrated in Fig. 17-42. The limiting network consists of two silicon diodes, CR_{508} and CR_{507}, in a "front-to-back" configuration. The natural silicon-diode junction-barrier potential will result in diode conduction when a minimum of 0.5 V is applied across each diode. Therefore, diodes CR_{508} and CR_{507} will conduct when the AFC voltage reaches a value of +0.5 V or −0.5 V. This action (diode conduction)

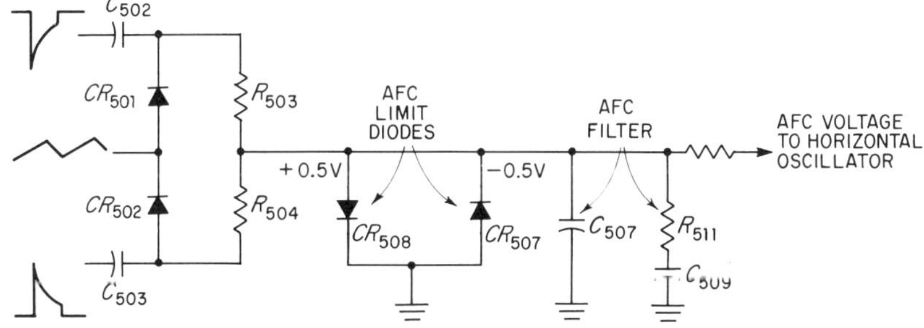

Fig. 17-42. AFC limiting and filter network. (*Courtesy of RCA.*)

will prevent the AFC voltage applied to the oscillator from swinging more positive or negative than ± 0.5 V, thus establishing the proper range of AFC control over the oscillator frequency (hold-in range).

The AFC filter network, also illustrated in Fig. 17-42, consists of R_{511}, C_{509}, C_{507}, and the distributed impedance of the detector-output circuitry. The purpose of the filter is to prevent the AFC voltage from varying at undesirable frequencies (antihunt action). These undesirable frequencies include 60 Hz variations, which would cause horizontal bending in the picture, and 15,750 Hz variations, which would create raster "edge ripple."

Reference Voltage Waveform Figure 17-43 illustrates the circuitry which forms the AFC reference-voltage waveshape. A negative-going horizontal pulse is "tapped" from the horizontal-output transformer, delayed slightly by inductance L_{501}, and applied to the pulse rectifier

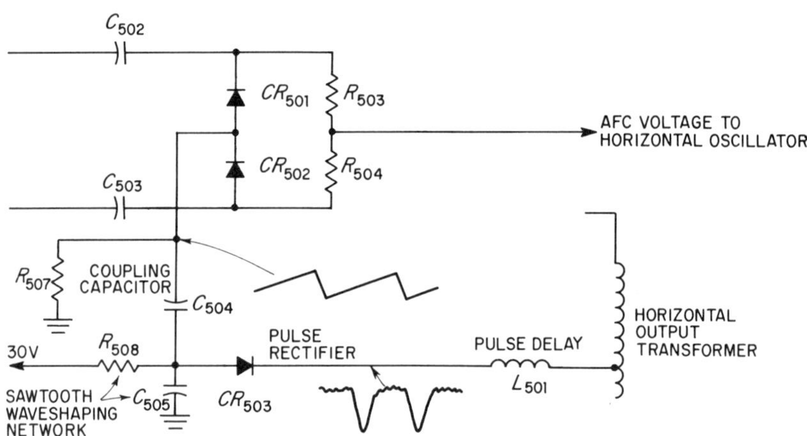

Fig. 17-43. Reference sawtooth generation. (*Courtesy of RCA.*)

diode CR_{503}. An *RC* network, consisting of capacitor C_{505} and resistance R_{508}, converts the pulses to a sawtooth waveshape. During scan time, capacitor C_{505} discharges through the resistance (R_{508}), creating the positive (trace) slope of the sawtooth. The appearance of the negative-going retrace pulse causes CR_{503} to conduct, resulting in the quick charging of C_{505}. This action creates the negative-going (retrace) slope of the sawtooth. Coupling capacitor C_{504} removes the dc component from the sawtooth. The resulting waveform is developed across R_{507} and is coupled to the AFC discriminator diodes.

The basic schematic of the RCA solid-state horizontal oscillator circuit is illustrated in Fig. 17-44. A blocking oscillator is employed, utilizing transformer T_{501} to maintain the regeneration and to couple the output pulses from the oscillator collector to the horizontal-output circuitry.

Basically, the oscillator functions as follows: Collector pulses which are coupled into the base circuit by T_{501} cause the oscillator transistor to be driven into cutoff. While the transistor is cut off, C_{510}

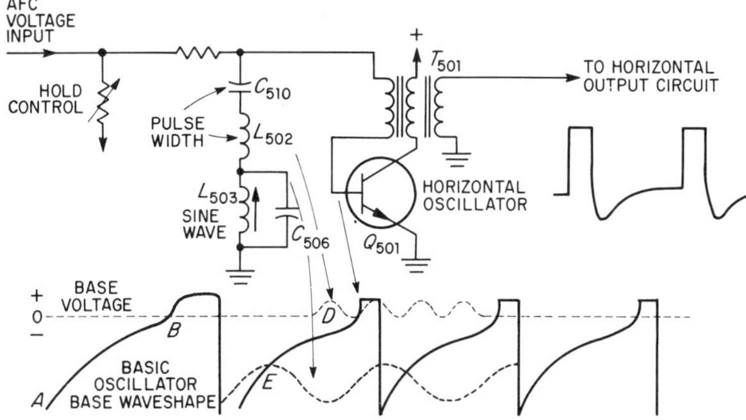

Fig. 17-44. Basic RCA solid-state horizontal oscillator. (*Courtesy of RCA.*)

discharges through the hold control and associated circuitry to a slightly positive "turn-on" potential of the transistor. This action is illustrated in Fig. 17-44 by the time interval from *A* to *B* of the basic oscillator waveshape. The resulting pulse of current appears in the collector circuit and is coupled back into the base circuit through T_{501}, driving the oscillator to saturation.

The hold control varies the discharging time of C_{510} and thus determines the instant at which the transistor will reach its "turn-on" voltage (Point *B*). The hold control, therefore, varies the frequency of oscillator operation. The AFC voltage is effectively added to the charge of C_{510} and thereby alters the instantaneous oscillator-base bias developed by the discharging action of this capacitor. In this manner, the AFC voltage dictates within a relatively narrow range the exact time the C_{510} discharging voltage reaches transistor "turn-on" (Point *B*). As a result the AFC voltage provides a means to control the oscillator frequency.

In order to provide the necessary oscillator-output pulse width and improve the stability of operation, two resonant circuits are employed in the oscillator circuit. The pulse width is determined by the series-resonant circuit consisting of C_{510} and L_{502}. The oscillatory action of this circuit effectively adds the waveform *D* to the basic oscillator waveshape. The period of the C_{510}-L_{502} circuit oscillation is approximately twice the desired pulse width. In this manner, the negative-going portion of one-half cycle of this oscillation will cause the transistor bias to instantaneously swing below the cutoff value. The resulting blocking-oscillator regenerative action will then complete the cutoff pulse.

Coil L_{503} and capacitor C_{506}, constitute a parallel-resonant circuit commonly called the "sine-wave" circuit. The action of this circuit effectively adds a sine wave (waveform *E*) to the basic oscillator waveshape. The frequency of the sine-wave oscillation is slightly higher than the operating frequency of the horizontal oscillator. The addition causes the potential at the oscillator base to quickly pass through the transistor turn-on voltage value. In this manner, the oscillator-operating frequency is made relatively immune from noise pulses and relatively independent of small changes in the supply voltage.

17.11 TROUBLES IN VERTICAL DEFLECTION OSCILLATOR SYSTEMS

The trouble in the vertical oscillator system can be grouped into four general categories. These are: (1) no oscillation; (2) incorrect frequency, or failure to synchronize; (3) height and/or linearity problems; and (4) interface problems.

No Oscillation If the vertical oscillator does not operate or if any stage following the oscillator is completely inoperative, there will be no vertical sweep, although there may be a normal horizontal sweep. The appearance of this problem on the picture tube is a single, bright horizontal line, indicating that only the horizotal-deflection circuits are operative. There are many possible causes for this symptom, including defective tubes or transistors, a defective vertical blocking-oscillator transformer, a loss of power supply voltage and, of course, defective parts. For more detailed troubleshooting precedures for this and other symptoms caused by problems in both the vertical- and horizontal-oscillator systems, see Chapter 24.

Incorrect Frequency or no Synchronization Incorrect vertical-sweep frequencies may be caused by an improper setting of the vertical-hold control, or by defective parts in the vertical-oscillator circuit. The appearance of this symptom is that a number of pictures appear vertically, instead of the desired single picture. You should remember that when a combination oscillator-vertical output system is employed, using two tubes or transistors (the transistor circuit may also include a driver stage), the components in both stages may affect the oscillation frequency.

Loss of vertical synchronization may be caused by a lack of the vertical-sync pulses being fed into the vertical oscillator. This should be checked first, since it indicates that the problem precedes and is not in the vertical-oscillator system. If sync pulses are present, they should be of the proper waveform and amplitude. The vertical-integrator circuit, at the input to the vertical oscillator, sometimes becomes defective and may cause this problem. Vertical-oscillator operation should also be checked to see that it is capable of being adjusted to the correct frequency. This may be determined by adjusting the hold control and trying to obtain a momentarily stationary, single picture on the screen. The effect, on the screen, of a lack of vertical synchronization, is shown in Fig. 17-45. In such a case, the picture may drift slowly up or down, depending on the setting of the vertical-hold control.

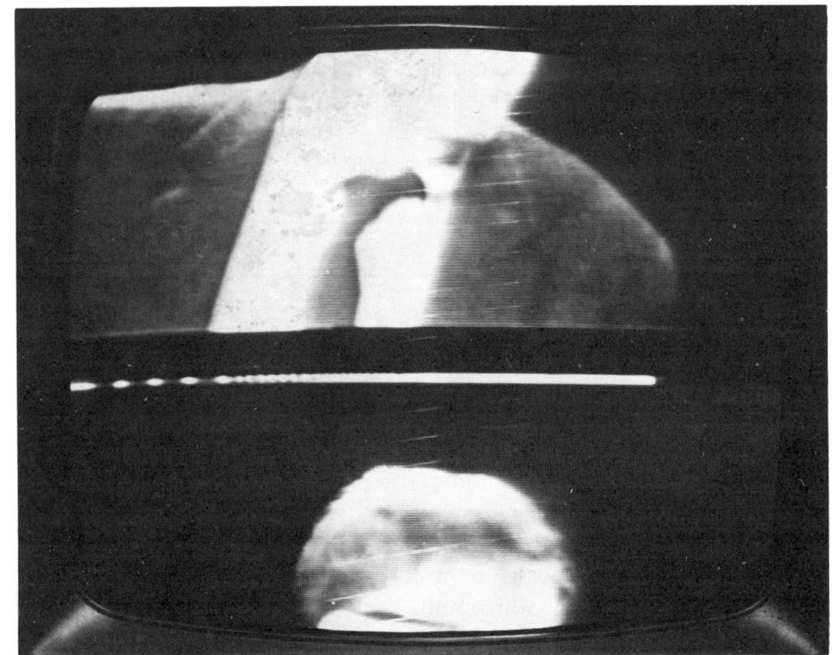

Fig. 17-45. This picture shows a lack of vertical synchronization.

Height/Linearity Problem Height and linearity problems are frequently related, but sometimes will occur separately. Height problems, caused by the vertical-oscillator system, are generally the result of a lower than normal amplitude of the generated vertical-deflection sawtooth. This may also affect the vertical linearity. The settings of the height and linearity controls should be checked before looking elsewhere for the source of trouble, as these may simply be misadjusted. Note that these controls interact and that both of them must be adjusted each time adjustment is required. These same vertical problems may also be caused by a defective tube(s) or transistor(s), by defective controls or other parts, or by improper power-supply voltages. Both height and linearity problems may also be caused by the vertical-output circuit (see Chapter 19), and thus the problem must first be isolated between the vertical-oscillator system and the vertical-output system. In some sets, the two functions are combined in a combination circuit, and this must be taken into account when troubleshooting these problems.

Interlace Problems As decribed in Chapter 2, a television picture is composed of two interlaced fields, which compose one frame, or one complete picture. If the two fields do not interlace properly, the picture will appear to lose detail and there may also be some slight vertical movements of the scanning lines if the problem is such as to allow variations of the interlace spacing between scanning lines. If alternate scanning lines "pair," visible spaces may actually appear between scanning lines. Poor interlace can be caused by a defect in the vertical-integrator circuit, or by the introduction of horizontal pulses into the vertical-deflection system. The horizontal pulses may be introduced either by a direct conducting path or by radiation.

17.12 TROUBLES IN HORIZONTAL OSCILLATOR AND AFC SYSTEMS

The troubles in horizontal oscillator and AFC systems can be generally grouped into three categories: (1) no oscillation; (2) incorrect frequency, sync, or phasing problems; and (3) width problems. Note that problems with this system may also affect the high-voltage generation, which depends upon correct horizontal-oscillator operation as well as correct horizontal-output circuit operation. You should remember that troubles in the horizontal AFC system will frequently cause changes in the oscillator frequency and may also produce improper sync and phasing but will rarely cause the oscillator to stop operating.

No Oscillation. If the horizontal oscillator is inoperative (or if any stage following the oscillator is completely inoperative), the screen will be blank. That is to say, no raster or any illumination will appear. This results because the high-voltage supply operation depends upon the operation of the entire horizontal-deflection system, and in this case, no high-voltage is being produced to operate the picture tube. In this situation, it is not immediately apparent which stage of the horizontal-deflection system is at fault and isolation procedures must be instituted (see Chapter 24).

Frequency, Sync, and Phasing Problems The problem of incorrect horizontal frequency may be caused by defects in either the AFC system or the oscillator system. While the oscillator is primarily responsible for generating the horizontal frequency, it is controlled to a degree by a dc control voltage from the AFC system. However, before assuming that there may be a defect in the system, you should check to see if the horizontal-oscillator adjustments and controls are correctly set. Incorrect settings may be the cause of incorrect horizontal frequency. Figure 17-46 illustrates the appearance on the screen of incorrect horizontal frequency which, however, is synchronized. To isolate between the AFC and the oscillator circuits, in a case of incorrect horizontal frequency, the AFC system should be disconnected from the oscillator, and the oscillator should be adjusted to see if a single picture can be momentarily obtained. If it can, then the fault is most likely in the AFC system.

Since the AFC system is responsible for synchronizing the oscillator to the horizontal-sync pulses, it is suspect whenever there is a condition of improper synchronization or phasing of the horizontal oscillator. In some horizontal systems, misadjustment or defect in the "stabilization-coil" circuit may be a cause of incorrect horizontal-oscillator phasing. If the phasing is incorrect, the picture will be shifted horizontally, as

Fig. 17-46. This picture is synchronized, but is displaying incorrect horizontal-sweep frequency.

shown in Fig. 17-47. If there is no horizontal synchronization, the appearance of the picture is one of multiple pictures in the horizontal direction. Remember that a loss of horizontal synchronization may also be caused by a lack of horizontal-sync pulse input to the AFC system and this should be checked first.

Fig. 17-47. In this picture, the horizontal phasing is incorrect.

Width Problems Most width problems generally stem from insufficient width, although excessive width may also occur at times. While insufficient width may be caused by defects in the horizontal-oscillator system, this system is not likely to be responsible for a condition of excessive width, which most likely originates in the horizontal-output system. An exception to this may be an improper setting of a horizontal oscillator "drive" control, and its setting should be checked first. Insufficient width may also be caused by an improper setting of the horizontal "drive" control.

In the oscillatory system, insufficient width may be caused by too low an amplitude of the sawtooth wave. This may be the fault of the oscillator circuit, or in the case of transistor systems, may be caused by a defective intermediate "driver" stage. Remember that frequently the cause of insufficent width may be a faulty horizontal-output system, rather than a faulty oscillator system, and this must be first determined before proceding with troubleshooting. However, here again, the setting of any controls affecting the picture width should be checked before assuming that there is an actual circuit defect. Insufficient width may not only be caused by defective tubes, transistors, or other parts, but may also be caused by insufficient supply voltage.

REVIEW QUESTIONS

1. Describe the main purpose of the deflection oscillators.
2. Draw a typical block diagram of a horizontal oscillator system (including AFC).
3. Blocking oscillators require one tube or transistor plus a special feedback ____________.
4. Draw a transistor collector-coupled multivibrator.
5. List the components responsible for a positive feedback in the transistor multivibrator drawn in question 4.
6. How is the feedback accomplished in a blocking oscillator?
7. Draw a tube version of a blocking oscillator.
8. Show how a sawtooth can be produced in the oscillator drawn in question 7 by the addition of two new components (capacitor and resistor).
9. Why must height and hold controls be utilized in vertical-deflection oscillator circuits?
10. Draw an emitter-follower amplifier modified for use as a driver amplifier. What is its function(s)?
11. What type of voltage waveform is required to pass a sawtooth current through the vertical yoke coils?
12. Which deflection oscillator, vertical or horizontal, would be more susceptible to noise interference? Why?
13. What might be the cause of vertical picture "rolling"?
14. Can there be an indication on a picture tube consisting only of a single, vertical line?

Horizontal Deflection and High Voltage 18

18.1 GENERAL REQUIREMENTS

Horizontal deflection and high-voltage circuits have passed through several stages of evolution in recent years. Originally these circuits used vacuum tubes exclusively, but now circuits will be found that use transistors and silicon-controlled rectifiers (SCR), in many cases, to the complete exclusion of tubes. All three types of circuits will be discussed in this chapter.

Electrostatic and Electromagnetic Deflection Most oscilloscopes use electrostatic deflection to sweep the electron beam across the faceplate of the cathode ray tube (CRT). Two sets of deflection plates inside the CRT utilize a sawtooth voltage waveform to perform this function. In this arrangement, the position of the beam is directly related to the voltage impressed across the deflection plates.

Deflection of the beam in a TV receiver utilizing magnetic deflection is not as straightforward as electrostatic deflection. A linear deflection in this case occurs only when the magnetic field changes in a linear fashion, and this in turn will occur if the current through two deflection coils varies linearly. The unit consisting of two deflection coils is called the deflection yoke, and the coils can be connected either in series or in parallel. The voltage across the deflection yoke is not related as directly to the magnetic field as the current. A sawtooth current waveform is utilized in the deflection yoke. However, the voltage across the coils is not a sawtooth; rather, it is a combination of a sawtooth and a pulse waveform. Why this particular form is required will be noted presently.

High-Voltage Generation The second major function of the horizontal deflection stage is high-voltage generation. We will see in this chapter that the high voltage is really a by-product of the deflection function. As the electron beam in the CRT quickly moves from the right side of the screen back to the starting position on the left side, a large amount of excess energy is produced. This excess energy, in the form of high-voltage pulses, is rectified to produce a high voltage for the CRT second anode. This portion of the deflection cycle is called the retrace or flyback period. The dc voltage produced may be as much as 30,000 volts for a color receiver and 20,000 volts for a monochrome set.

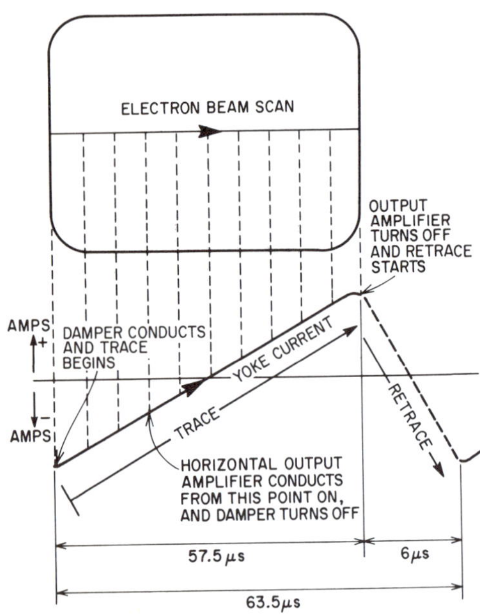

Fig. 18-1. Relationship between CRT beam position and the deflection yoke-current waveform.

Trapezoidal Waveforms We will now review the reason for utilizing a trapezoidal-voltage wave-form (combination sawtooth and pulse) on the deflection yoke. The horizontal position of the beam in the CRT depends on the magnitude of the current in the horizontal-deflection yoke. Figure 18-1 shows this one-to-one relationship between the beam location and the deflection-coil current. Note that if the sawtooth current is negative, the beam is left of screen center. Likewise, a positive current places the beam to the right of screen center. The negative coil current comes from a device called the damper. The positive coil current is obtained from the horizontal-output amplifier. In the following sections, we will discuss the manner in which these two currents mix in the deflection coils (yoke) to form a linear current sawtooth. When we say " linear," we are referring principally to the 57.5 microsecond sweep time. The 6 microsecond retrace time does not need to be linear since the screen is blanked out during that time.

The form of the voltage wave to be applied to the deflection coils is derived by analyzing the impedance components of the coils and the current waveforms produced when voltages of various shapes are applied. Each coil contains inductance plus a certain amount of resistance. So far as the resistance is concerned, a sawtooth voltage will result in a sawtooth current. This is shown in Fig. 18-2(A). For a pure inductance, as shown in Fig. 18-2(B), a rectangular (pulse) voltage waveform is required to produce a sawtooth-current waveform. If the voltage waveforms of Fig. 18-2(A) and (B) are combined, a result similar to the voltage waveform of Fig. 18-2(C) will be produced. If this waveform is applied to deflection coils containing both inductance and resistance, the sawtooth current waveform shown at the bottom of Fig. 18-2(C) will result. The magnetic flux, varying directly with the current, will force the electron beam to sweep across the screen properly. Note particularly that the resultant wave is not obtained by combining the two voltage waves in *equal* measure. If the deflection circuit contains more inductance than resistance, the resultant wave will be closer in form to Fig. 18-2(B). On the other hand, if the resistance predominates, the resultant wave will more closely resemble Fig. 18-2(A). Hence, one may expect to find variations of this horizontal-deflection wave

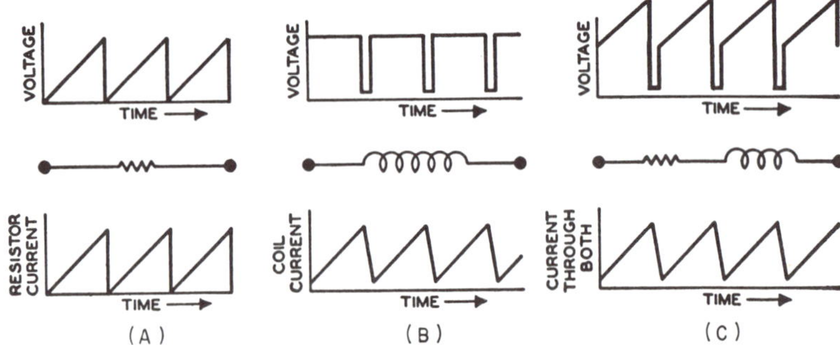

Fig. 18-2. Currents through and voltages across a resistor (A), an inductor (B), and an inductor with resistance (C).

ranging from the waveform of Fig. 18-2(B), which is found in transistor horizontal-output amplifier circuits, to some variation of Fig. 18-2(C), which is found in vacuum-tube circuits.

18.2 BLOCK DIAGRAM DESCRIPTION

The mode of operation of the horizontal deflection and high-voltage circuits can best be understood by referring to the block diagram shown in Fig. 18-3. As this chapter develops, we will see that this block diagram is generally accurate for most vacuum-tube, transistor, and SCR horizontal deflection systems.

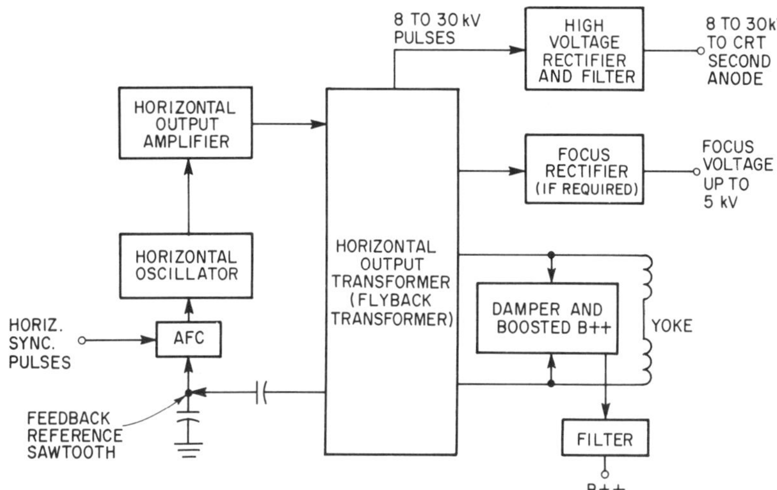

Fig. 18-3. Typical block diagram of the horizontal-deflection circuit.

Frequencies in Horizontal System The horizontal output amplifier must be driven by the correct voltage or current waveform in order to produce an accurate sawtooth current in the deflection coils. These waveforms are first generated in the horizontal oscillator. However, the voltage or current waveform in the horizontal oscillator may have a shape quite different from the yoke current. The reasons for this apparent paradox will become clear as we progress through the chapter.

We are all familiar with the audio-output amplifier which accepts low-power signals, amplifies them, and then drives a loudspeaker through an impedance-matching transformer. In many ways, this is analagous to the processes carried out in the horizontal-deflection system. The horizontal oscillator provides relatively low-power signals to the grid (for a vacuum tube), base (for a transistor), or gate (for an SCR) of the horizontal-output amplifier. In this latter stage, the power level of the signal is amplified and fed to the horizontal-output transformer. This transformer matches the impedance of the horizontal-output amplifier to (1) the deflection coils, (2) the high-voltage rectifier circuit, and (3) the damper and boosted B+ circuit. Other lower level waveforms are also obtained from the transformer for various functions.

It is difficult to draw much more of an analogy between an audio-power amplifier and a horizontal-output amplifier than we did in the above paragraph. The frequencies and waveforms involved in each case are quite different. The lowest frequency present in a horizontal deflection amplifier is the 15,750 Hertz scanning frequency. Since the flyback portion of the waveform is in the shape of a pulse, frequencies above 100 kHz are also required. The waveform is actually a mixture of 15,750 Hertz, 31,500 Hertz, 47,250 Hertz, and other harmonics which extend beyond 100 kHz. When these are all added together (with the proper phases and amplitudes), they result in the trapezoidal waveform. The output amplifier and the flyback transformer are designed to handle all of these frequencies at significant power levels. If the amplifier response ended at 50 kHz, for example, the rectangular pulse portion of the waveform would not be adequately transferred to the deflection coils. This would severely impair the retrace portion of the yoke-current waveform.

Deflection Yoke and Damper Circuits The deflection of the electron beam in the CRT is the primary function of the horizontal-output amplifier. This requires a large current through the deflection coils. Thus, the flyback transformer is connected as a voltage step-down, or current step-up device for the deflection-coils circuit. To preserve power, this secondary winding has an impedance close to that of the deflection coils.

As explained more fully later, the brief retrace period produces self-oscillations, and a special damper diode is used to absorb these oscillations. The energy from the first negative peak in these oscillations is passed from the damper diode to a boost filter where it is stored and used to provide additional B++ voltage to the horizontal-output amplifier. This voltage is usually several hundred volts above the regular B+ voltage. This voltage is utilized also in the vertical-output amplifier in some receivers. If the receiver uses solid-state horizontal- and vertical-output amplifiers, then this additional B++ voltage is not required.

High-Voltage Circuit A special high-*Q* secondary winding on the flyback transformer serves to generate high-voltage, low-current pulses during the flyback period. The pulses are rectified and filtered to produce up to 30 kilovolts dc for a color receiver or up to 20 kilovolts for a monochrome set. This high voltage is required for the CRT second anode. (See Chapter 14 for a detailed discussion of picture tubes.) In most receivers before 1970, the high-voltage rectifier was a vacuum tube, and a special winding on the flyback transformer was required to develop the heater voltage for this vacuum tube. In recent sets, silicon solid-state diodes are used for the high-voltage rectifier.

Other Functions of the Deflection Circuits A number of other functions also take place in the block diagram of Fig. 18-3. Foremost among these is the sawtooth waveform which goes from the flyback transformer to the AFC circuit. This sawtooth waveform is compared with the horizontal-sync pulses (from the video signal) to create a frequency correction signal for the horizontal oscillator.

One of the higher voltage taps on the flyback transformer is often used to develop a focus voltage. High-voltage pulses are rectified and filtered to provide up to 5 kilovolts dc in some color receivers.

Other taps on the flyback transformer are used for: (1) keyed AGC, (2) horizontal blanking, (3) burst-amplifier gating, and (4) chroma-amplifier gating. (1) is explained in Chapter 10, (2) is explained in Chapter 16, and (3) and (4) are covered in Chapter 22.

In the following sections we will explore in more detail the operation of each portion of the block diagram. We will be discussing horizontal output and high-voltage circuits for both color- and monochrome-TV receivers. In each case, both vacuum-tube and solid-state technology will be surveyed.

18.3 VACUUM TUBE MONOCHROME CIRCUITS

The trapezoidal waveform which drives the horizontal-output amplifier is produced in the output circuit of the horizontal oscillator. This oscillator is referred to sometimes as the horizontal-sweep oscillator or the horizontal multivibrator.

Coupling Circuits Careful attention is paid to the transfer of the complex trapezoidal waveform from the horizontal oscillator to the output amplifier. Time constants are chosen so that no part of the trapezoidal shape is degraded while driving the grid of the horizontal-output amplifier. The coupling network usually contains a variable capacitor, resistor, or inductor which controls both the amplitude and the waveshape of the trapezoidal waveform. This control is usually labeled "horizontal drive," or "horizontal waveform." It is often adjustable only with a screwdriver.

Two common coupling circuits between the horizontal oscillator and the horizontal-output amplifier are shown in Fig. 18-4(A) and (B). In Fig. 18-4(A), C_2 and C_3 form a voltage-divider network for the voltage developed across C_1. C_3 is the drive control and, as its capacitance is reduced, the impedance rises and more voltage is developed across it and passed on to the output amplifier. Figure 18-4(B) shows a series-connected drive control. In this case, more capacitance allows a larger voltage transfer to the output stage.

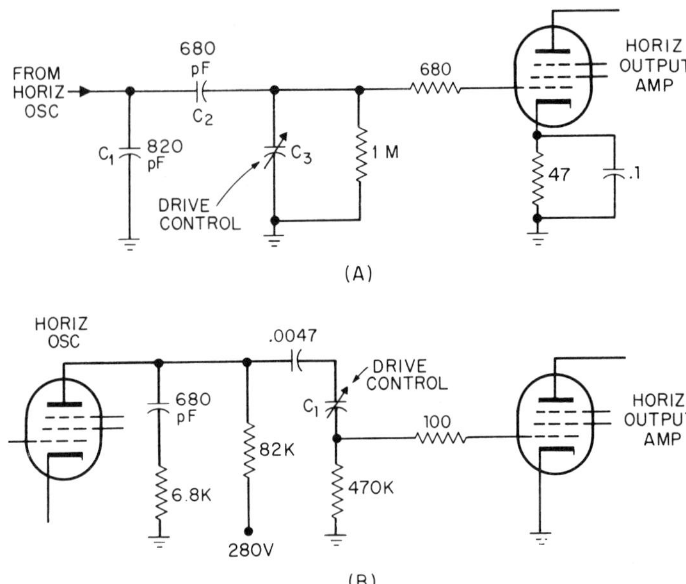

Fig. 18-4. Two different methods of controlling the drive voltage to the horizontal-output amplifier.

The Horizontal Output Amplifier After passing through the coupling network, the trapezoidal waveform is applied to the grid of the horizontal-output amplifier. A typical circuit is shown in Fig. 18-5, with the waveforms developed in this circuit shown in Fig. 18-6. E_1 is the voltage at the grid of V_1. The dotted line *CO* (Fig. 18-6) is the grid-cutoff voltage level. No plate current flows if E_1 is below the *CO* line. Current flows through V_1 from time 2 to time 3, from time 4 to time 5, etc. Thus, no plate current flows for approximately 30% of the time. The resultant plate current wave-form is shown in the *Ip* curve of Fig. 18-6.

Carefully note the timing relationship between E_1 and *Ip*. At time 1, V_1 is cutoff and no plate current flows. Likewise, no current flows

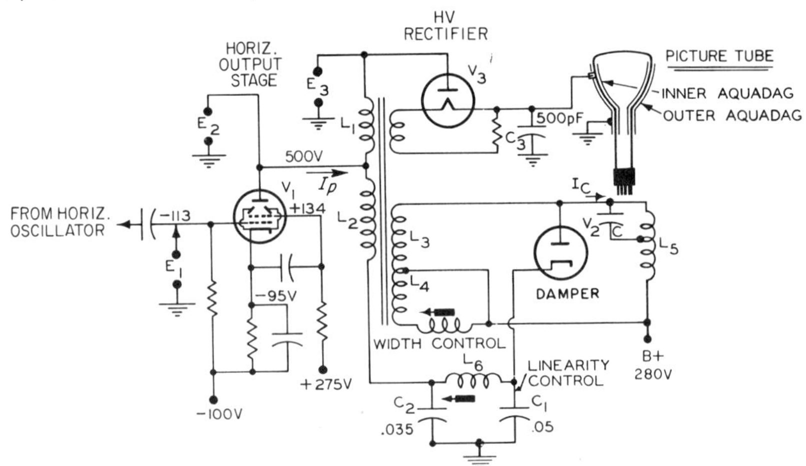

Fig. 18-5. A typical vacuum horizontal-output amplifier and high-voltage power supply circuit.

through L_2. At time 2, V_1 starts to conduct, current flows through L_2, and a field builds up around L_2. This expanding field cuts L_3, L_4, and L_1, inducing a voltage in each. At time 3, the tube is driven sharply into cutoff, plate current drops to zero, and the field built up around L_2 quickly collapses, inducing a high voltage in L_1, L_2, L_3, and L_4. The voltage induced across L_1 and L_2 is of the order of 15,000 volts or more, peak-to-peak (monochrome receivers). This sudden collapse of the magnetic field around L_2 does not occur in zero time. It decays in a finite time determined by the time constant of L_2 and all of its loads. Some loads are directly connected whereas others are reflected through the transformer. The collapsing field shock excites the circuit, composed of L_3, L_4, L_5, and C, into oscillation. The resonant frequency of this network (70 kHz) is such that the period of 1/2 cycle of an oscillation is about 5 to 7 microseconds, which is equal to the flyback time of the horizontal sweep. Thus, during the first half cycle of this oscillation, the beam, which is at the extreme right-hand side of the screen, is brought back to the left-hand side. The oscillations have now served their purpose and must be stopped. For this purpose we require a special damping circuit.

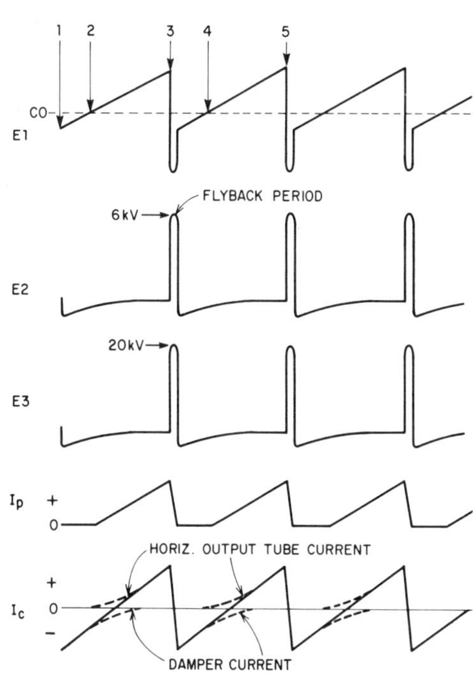

Fig. 18-6. Voltage and current waveforms from the horizontal-output circuits shown in Fig. 18-5. Note that the curves of the damper- and horizontal-output tube currents are of the opposite direction and produce a composite linear current sawtooth.

Horizontal Damping If the oscillations are not stopped after the first half cycle, they would continue into the next line and interfere with the proper motion of the beam. An expedient method of damping the oscillations quickly is accomplished by means of V_2. At about time 3, when V_1 is cut off and the field is collapsing about L_3, the top of L_3 and the plate of V_2 become negative, with the current flowing down through the deflection coil (L_5). When the current in the deflection coil reaches its negative peak, the voltage at the top of the coil begins to go positive, because the current lags the voltage by 90 degrees in an inductance. As soon as the voltage on the plate of V_2 becomes positive, the tube conducts and, in so doing, offers a low-resistance path for the deflection current into capacitor C_1, which thus becomes charged. The charging process is quite rapid at first, gradually slowing down as the voltage across C_1 rises. This slowing down is reflected in the gradual dying out of the current, as is indicated by the bottom dotted line in *Ic*. However, as the current approaches zero, V_1 comes out of cutoff and a magnetic field again builds up around L_2. This growing field induces a voltage in L_3, causing the current to build up to its maximum peak, time 5 in *Ic*. In this way, a linear sawtooth current is generated.

Both V_1 and V_2 in Fig. 18-5 should be regarded as switches. V_1 is turned on when the electron beam is about one-third of the way across the screen while the image is being traced out. It remains on until the beam reaches the far right-hand side of the screen, when the retrace starts. During all this time, V_2 is off. When the retrace starts, V_1 is also turned off, and the partial oscillation in the yoke brings the beam back to the left side of the screen. Now, V_2 is turned on and the gradual die out of energy (through absorption by C_1) brings the beam about one-third the way across the screen. V_2 now lapses into cutoff, and V_1 is turned on. V_2 can conduct only when its plate voltage is greater than its cathode voltage. This means that the plate voltage must exceed not only the B+ from the power supply, but also the deflection voltage across the yoke coils. This condition occurs only during the retrace interval and for the 1/3 of each scan period. For the rest of the cycle, the plate voltage of V_2 is lower than its cathode voltage and no conduction takes place.

As a review, a few more comments concerning the damping process are in order. This process is often misunderstood. Referring back to the E_2 and E_3 waveforms of Fig. 18-6, the flyback pulse tends to be followed by several oscillations. These oscillations must not reach the yoke current, or several bright vertical bars will appear on the left side of the CRT screen. By placing the damper diode directly across the yoke windings, this problem is overcome. The diode clips off the negative overshoot of the first oscillation. The oscillations are thus "damped" beginning at this point in time.

Boost B+ (or B++) Note that the voltage on the plate of V_2 (Fig. 18-5) is equal to the 280 volts (B+) from the power supply plus the deflection voltage which is developed across L_3, L_4. The greater the deflection voltage, the greater the voltage applied to the plate of V_2, and the greater the charge which C_1 receives.

The charge on C_1 produces a voltage across the capacitor in which the top plate is positive and the bottom plate is negative. This polarity stems from the fact that the electrons in V_2 flow from cathode to plate, through L_5 to B+, from here to ground in the power supply, and from ground to C_1. Thus, the voltage across C_1 will be greater than the B+ voltage, and the difference can be as much as several hundred volts. The name given this augmented voltage is boost B+, boosted B+, or simply *B++*. In Fig. 18-5, the B++ is applied to the plate of V_1 through L_2. In addition, B++ is frequently used to provide higher voltages for the other stages in the receiver besides the horizontal-output amplifier. (Usually the vertical-output amplifier uses the B++.)

V_2 conducts for about 30 percent of the sweep. While it is conducting, the voltage across C_1 and C_2 builds up. When V_2 ceases to conduct, the voltage across C_1 and C_2 falls. This rise and fall constitutes an ac ripple on the plate of V_1. By shifting the phase of this ripple voltage, it is possible to compensate for some of the nonlinearity of the current wave-form in the deflection coil. This change of phase is effected by the variable inductance (L_6). L_6 is called a "linearity control." C_1, C_2, and L_6 form a resonant circuit tuned to 15,750 Hz. The resonance of this circuit is indicated by a dip in the plate current of V_1. We will discuss linearity controls in more detail in Section 18.5.

The B++ is called "boost source" in some receivers, since it is used to raise the horizontal-output amplifier B+, yet the energy to create the higher voltage comes through the horizontal-output amplifier. The horizontal-output amplifier appears to be "picking itself up by its own bootstraps." The ultimate boost B+ may be +500 to +700 volts. The actual voltage of the B++ can be modified by the circuit designers in many ways. One common way is to move the damper tap on the fly-back transformer up or down with respect to the yoke tap. The yoke and damper diode would then not be exactly in parallel as assumed in the discussion thus far. Figure 18-7(A) shows a circuit where the damper tap is above the yoke tap. In this instance, the B++ would be higher than if there were a direct parallel connection between the damper and the yoke.

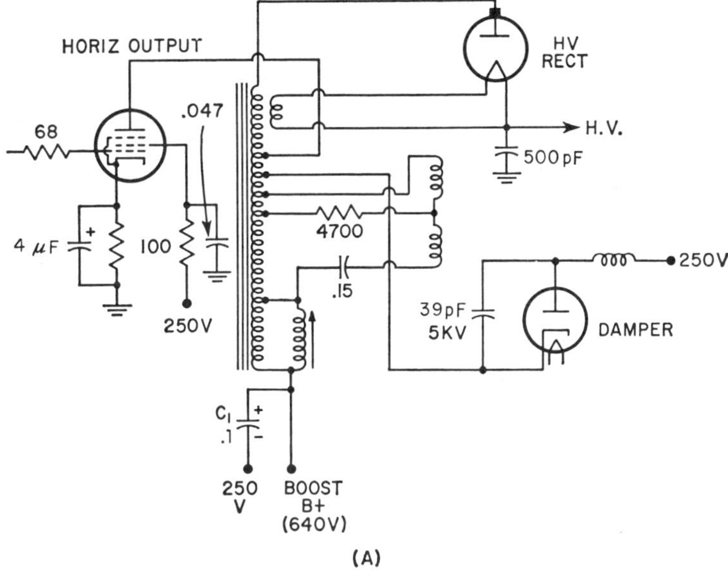

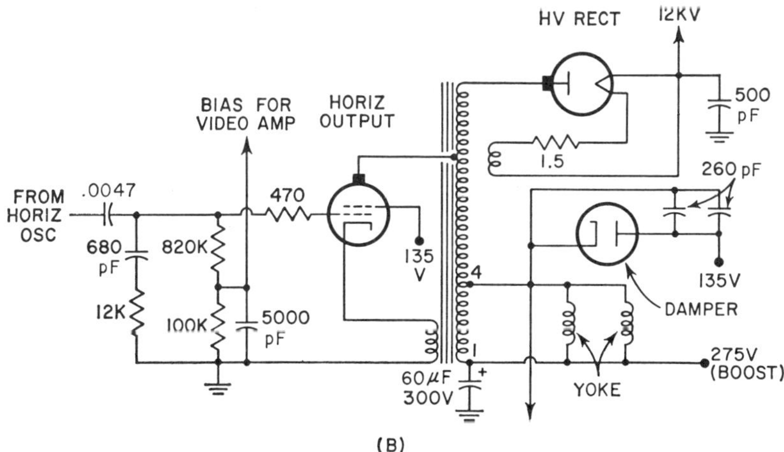

Fig. 18-7. Two horizontal-output circuits that have been used with 90 and 110 degree picture tubes.

Before leaving the subject of dampers and boosted B+, the subject of autotransformers should be addressed. Note in Fig. 18-5 that a separate winding (L_3 and L_4) on the flyback transformer is utilized for the damper and yoke circuit. L_3 and L_4 could also be connected in series with L_2 to form an autotransformer. Two circuits of this type are shown in Fig. 18-7. With this circuit configuration, the damper diode appears to be connected upside down. An analysis of the circuit reveals that the damper and B++ operation is not affected by this type of connection. We should carefully note that when the overshoot is negative at V_1, it is simultaneously positive at V_2. The polarities of all transformer connections are made so that this condition exists. The damper works while the horizontal-output tube rests, and vice versa.

Flyback High Voltage Referring back to Fig. 18-6, we note that a large voltage pulse (E_2) appears on the plate of the horizontal-output tube during retrace. This pulse has a peak-to-peak value of 5 to 6 KV. The damper tube is not allowed to kill this positive pulse, because it is useful for generating a high dc voltage for the CRT second anode. As shown in Fig. 18-5, a step-up autotransformer winding (L_1) is attached to the hot end of L_2. This winding increases the magnitude of E_2 a number of times. At the hot end of L_1, the peak-to-peak voltage may be as high as 20 KV for monochrome receivers and 30 KV for color sets.

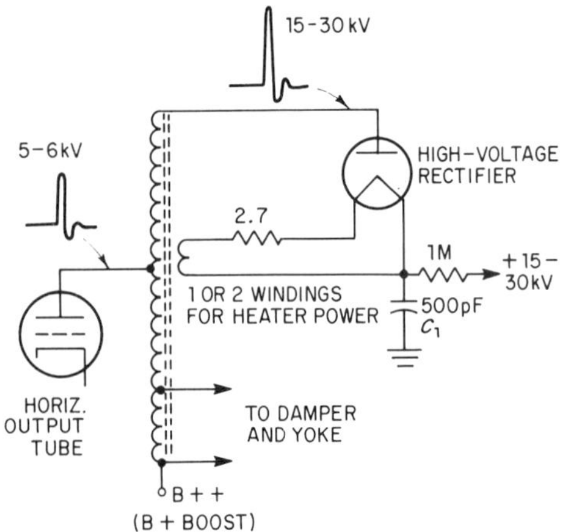

Fig. 18-8. The high-voltage portion of a horizontal-output stage.

Figure 18-8 is a simplified schematic showing the important circuit components in the **high-voltage** section. The 15-30 KV pulses are generated in a special high-voltage winding which is continuous with the primary, but made of finer wire. This part of the flyback transformer is nearly always an autotransformer connection. The high-voltage pulses are applied to the plate of a high-voltage, half-wave rectifier. The heater voltage for this tube, usually 1 to 3 volts, is obtained by merely wrapping one or two turns of high voltage wire around the flyback transformer core. One side of this heater becomes the dc output terminal for the high voltage. A high-voltage "doorknob" capacitor (C_1) filters the dc before it is sent to the CRT second anode. Sometimes a resistor is placed in series with this lead to provide an additional filtering. In some receivers, the filter capacitor is supplemented (or replaced) by the inner and outer aquadag coatings in the picture tube which form a capacitor of fairly large capacitance (this is shown schematically in Fig. 18-5).

Wide Angle Horizontal Deflection The first commercial television receivers produced in the late 1940's were designed principally for 53 degree picture tubes with approximately 9000 volts on the second anode. As the demand grew for picture tubes larger than 10 inches, it became necessary to increase also the deflection angle since the length of a large screen CRT with 53 degree deflection would be too long for public acceptance. Thus, the deflection angle has been gradually increased over the years to 110 degrees (somewhat greater in some designs).

Wide-angle deflection requires a greater high voltage and a greater deflection power than the 53 degree system. Some older receivers obtained this higher voltage by connecting two high-voltage rectifiers as a voltage doubler. This was relatively expensive. Similarly, the deflection power can be increased by paralleling horizontal-output tubes. This, too, is costly.

Recent vacuum-tube monochrome receivers use only one high-voltage rectifier and one horizontal-output tube. This is possible because of the development of highly efficient flyback transformers using ferrite cores and more powerful horizontal-output tubes. Two of these transformers are shown in Fig. 18-9.

Fig. 18-9. Two typical horizontal-output transformers. The single loop turn at the bottom of each unit provides the filament voltage for the high-voltage rectifier.

Two horizontal-output circuits for 90 and 110 degree picture tubes were shown in Fig. 18-7. In illustration A, the yoke windings are still connected in series across a portion of the horizontal-output transformer. However, a connection is made between the two half sections and a suitable tap on the transformer. The 4,700-ohm resistor in this lead is designed to minimize ringing effects in the yoke and also to help balance the two yoke sections. The 0.15 μf capacitor in series with the yoke is for dc blocking.

The B++ voltage is developed across C_1 with the polarity indicated. Note that the bottom end of this capacitor connects to the plate circuit of the damper tube since both attach to the 250-volt terminal in the power supply. The added voltage developed across this capacitor is equal to almost 400 volts, because the bottom end of C_1 has a potential of 250 volts and the top end provides a boost B+ of 640 volts.

In Fig. 18-7(B), the two horizontal yoke windings are connected in parallel. This arrangement eliminates the need for any balancing resistors or capacitors, but it does require more driving current (about 4 times as much). An interesting feature of the circuit is the horizontal-transformer winding that is connected to the cathode of the output tube. During retrace, a positive pulse of voltage is developed here, which helps to keep the output tube cut off during this interval.

The grid circuit of the output tube is also somewhat unusual in that it possesses a voltage divider consisting of an 820-kilohm and a 100-kilohm resistor. The voltage developed here is used, in part, as bias for the video-output tube. The 5000 pF capacitor at the tap-off point bypasses the 15,750 Hertz signal present here to prevent pickup by the video-output tube. Use of this point as a bias source is possible, because, under normal conditions, the negative voltage developed here is quite steady. It will change only when a defect occurs in the horizontal system, and this, of course, would tend to disrupt the normal functioning of the receiver anyway.

Combination Horizontal-Output and Damper Tubes Many new vacuum tubes for the horizontal-output and damper circuits have been introduced in recent years. Some of these are for series heaters, such as the 21GK5, 21GY5, 21HB5, 21JS6, 22JR6, 33GY7, 38HE7, and 38HK7 types. A few new types, such as the 38HE6, have both the damper and horizontal-output tubes enclosed in one glass envelope. Figure 18-10 shows a horizontal-output circuit using this type of vacuum tube. In addition to the vaccum tube, two other novel features in Fig. 18-10 are worthy of discussion. Inductor L_1 is placed in the plate lead of the damper to assist in damping the oscillations during retrace. Notice the factory-wired jumpers in the screen circuit of the horizontal-output amplifier. They provide a horizontal-size (width) adjustment capability and are to be changed only by technicians.

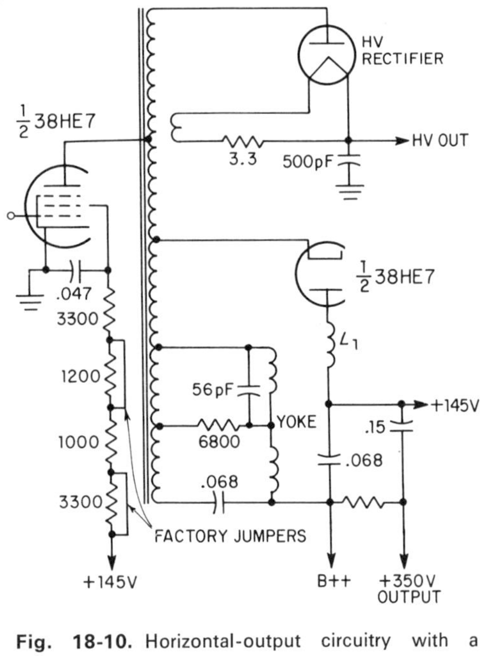

Fig. 18-10. Horizontal-output circuitry with a damper and a horizontal-output tube enclosed in the same glass envelope.

18.4 SOLID-STATE MONOCHROME CIRCUITS

In this section we will analyze horizontal-output and high-voltage circuits utilizing transistors and solid-state diodes. The discussion will be restricted to monochrome-TV receivers. We will not cover SCR circuits until the section on solid-state color receivers. Our first concern in this section will be the special requirements of solid-state horizontal circuits compared with their vacuum-tube predecessors.

Special Requirements for Solid-State Circuits When transistors are employed in the horizontal-output amplifier, four special requirements must be met:

(1) The transistors must be able to handle high currents and high voltages—both at moderately fast switching speeds.

(2) The transistors must be given overcurrent and overvoltage protection.

(3) The base drive waveform must be rectangular instead of trapezoidal.

(4) A bias system is required which will protect the output transistor in the event of drive failure.

Special requirement (1) was very difficult to satisfy for large screen receivers until the late 1960's. The transistors now used for the horizontal-output stage will switch a peak power of 2000 watts, while internally dissipating power on the order of only one watt. Silicon devices are extensively used for this application. They offer both high-voltage breakdown capability and a fast switching speed.

Despite the capabilities of the output transistor, it must be protected against overvoltage and overcurrent. One method of reducing the overvoltage transients at the output-amplifier collector is shown in Fig. 18-11(A). The reverse-peak voltage which appears on the collector during the retrace is reduced by adding to it a third harmonic of the pulse. This third harmonic is provided by a parallel-tuned circuit which is actually a part of the high-voltage system. Figure 18-11(B) illustrates how the fundamental pulse and its third harmonic are added to achieve a reduced pulse size.

A method used to protect the output transistor against overcurrent is shown in Fig. 18-12. Q_1 is the horizontal-output transistor and Q_2 is a current-limiting transistor, and both are in series between B+ (+30V) and ground. All current through Q_1 must pass through Q_2. However, the dc current (average current) through Q_2 is fixed by the bias resistor R_1. The average current through Q_1 will thus be fixed. The pulsed current through Q_1, caused by the horizontal oscillator, will be constrained to an average value determined by Q_2 and R_1.

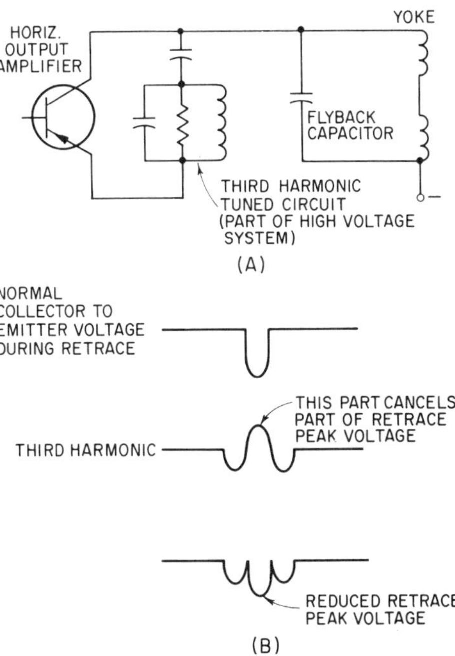

Fig. 18-11. Peak-pulse amplitude at the collector of the output transistor can be reduced by a tuned circuit: (A) horizontal-output circuit; (B) the mixing of waveforms on the collector to achieve reduced pulse voltage.

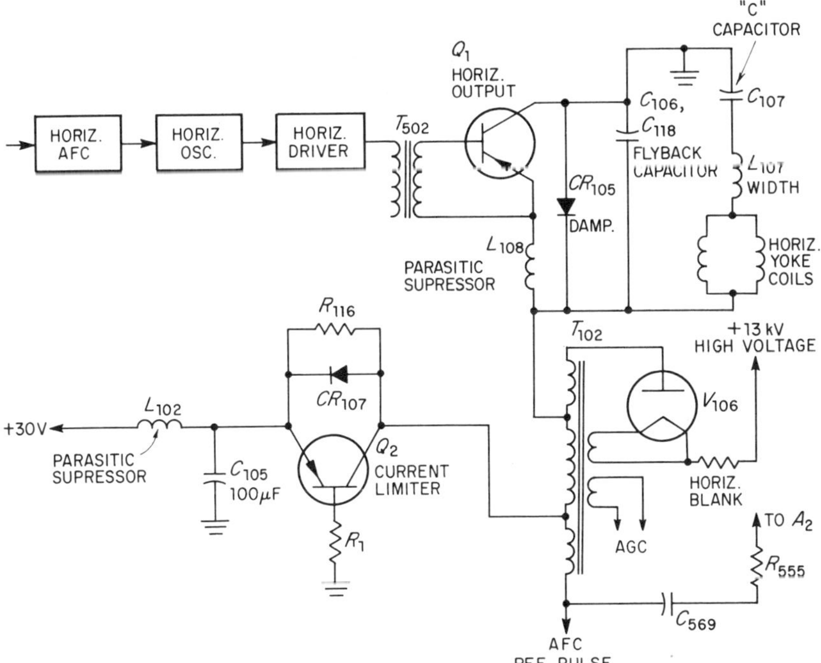

Fig. 18-12. Series current limiter protection of the horizontal-output transistor.

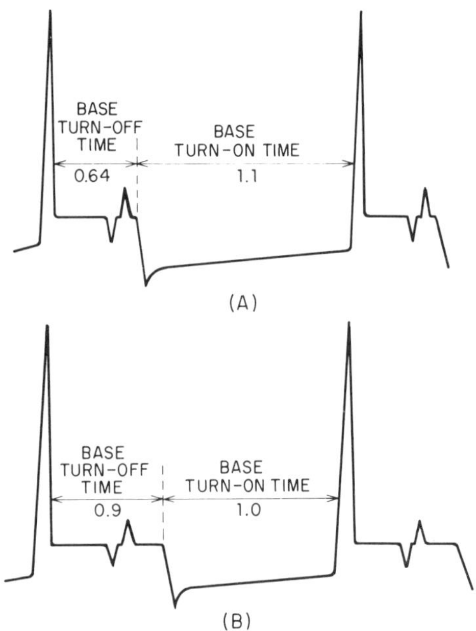

Fig. 18-13. Current limiting by control of the base-drive duty cycle (PNP horizontal-output transistor): (A) duty cycle = 1.7, and (B) duty cycle = 1.1.

Current limiting for the output transistor can also be accomplished by controlling its base drive duty cycle. The base-drive voltage waveform for a PNP output stage is shown in Fig. 18-13. The two waveforms in the figure illustrate the two types of base drive. Although the waveforms appear similar at first glance, they differ significantly. The ratio (duty cycle) of the turn-on drive time to the off time is (1.6/0.64) = 1.7 in Fig. 18-13(A). The same ratio is only 1 : 1 in Fig. 8-13(B). A circuit working normally only requires a 0.9 ratio. Thus, a circuit which is able to control this duty cycle from 0.9 to 1.7 is able to control the average current through the output transistor. This method effectively limits the average power that may be drawn from the power supply and therefore protects the output transistor.

The third special requirement of transistor horizontal-output amplifiers is the rectangular base-drive waveform. This is shown in Fig. 18-13. It is quite different in shape when compared to the trapezoidal waveform used on the grid of a vacuum-tube horizontal-output amplifier. The transistor output stage drives a low-voltage, high-current transformer and yoke. The yoke thus has high inductance compared to its resistance, hence, it requires a rectangular- (pulse) voltage drive wave-form. As noted in Section 18.1, this voltage waveform on a nearly pure inductance will produce a nearly linear sawtooth-current waveform.

The last special requirement for the output transistor is a means to protect this high-power device when drive from the horizontal oscillator fails. The output transistor is biased such that no current will flow unless a pulse waveform is applied to its base. If drive fails, no pulses turn the transistor on, and no power is dissipated in the device. It automatically shuts off completely when drive fails.

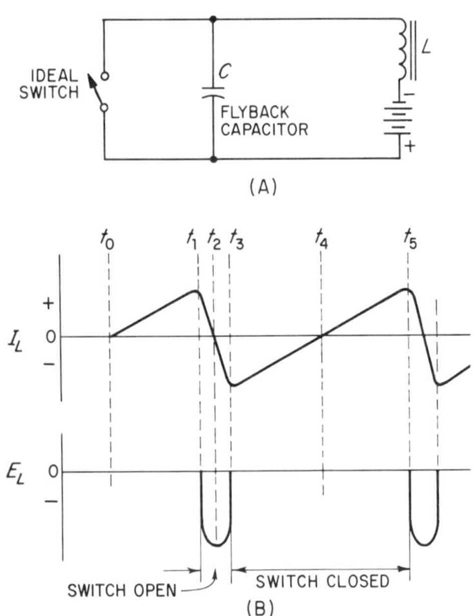

Fig. 18-14. (A) Simple deflection circuit utilizing an ideal switch. (B) Yoke current and voltage waveforms.

Transistor Horizontal Output Circuit The dc series resistance of a horizontal-deflection yoke in a transistorized receiver is less than one ohm. The yoke resistance is usually 5 to 30 ohms in a vacuum tube set. At the horizontal scanning rate of 15,750 Hertz, the yoke inductive reactance in a solid-state receiver is thus very large compared to its series resistance. In Fig. 18-2(B) we mentioned that a rectangular voltage waveform is required for this type of yoke. The circuit driving the yoke must also have a very low series resistance. An ideal switch, as shown in Fig. 18-14, would provide a low resistance and also a rectangular waveform. However, it would be difficult to open and close a mechanical switch 15,750 times per second. The transistor, driven between cutoff and saturation, comes closer to the characteristics of an ideal switch then does a vacuum tube. The series resistance of a high-power, high-speed transistor in the saturated state is less than one ohm; in the cutoff mode, the series resistance is usually above 50,000 ohms. This on-off resistance range makes the transistor a potentially good horizontal-output amplifier.

Before we examine a transistor horizontal-output circuit, it will be helpful to examine a sweep circuit utilizing an ideal switch. This is shown in Fig. 18-14(A). The yoke current and voltage waveforms appear in Fig. 18-14(B). In order to produce a linear sawtooth of current in a pure inductance, a constant voltage is required across the inductance. A linear sawtooth current is obtained by connecting a coil L to a constant source of voltage by means of an ideal switch. Current starts flowing in the coil by closing the switch at time t_0. If the circuit resistance is zero, the coil current will rise indefinitely in a linear fashion. In a practical circuit, the switch is held closed until one half (second half) of the trace is completed (time t_0 to t_1). When the switch opens, the coil current continues to flow in the same direction but flows now into capacitor C. The coil voltage variation follows a sine curve whose period is determined by $t = 2\sqrt{LC}$. When the coil voltage completes one fourth of a sine wave, the coil current is zero and the capacitor voltage is maximum. This is true because the coil and capacitor are now a resonant circuit which has completed one fourth of an oscillation. During the next one-fourth cycle, the capacitor fully discharges and the coil current is flowing in exactly the opposite direction to the one at the beginning of the retrace. At this instant (time t_3) the switch closes and the constant voltage causes the coil current to rise linearly. The low impedance of the battery and switch damps the oscillations. However, this time the current starts from a negative value. The linear rise continues until the switch again opens.

The switch is required to pass current in both directions—in one direction for the first half of the scan and in the other direction for the second half. This is not possible with a vacuum tube but is easily accomplished with a saturated transistor. Figure 18-15(A) shows a circuit with the switch replaced by a PNP transistor. As can be seen in Fig. 18-15(B), if the base-to-emitter voltage is +2 volts, no current flows from emitter to collector. The transistor appears to be a 50-kilohm resistor, or nearly an open circuit. During the scan period, from time t_3 to t_5 in Fig. 18-14(B), the base-to-emitter voltage is held at −1 volt. This drives the transistor into saturation conduction and makes it appear as a resistor whose vslue is less than one ohm—essentially a short circuit. In this saturated condition, current flows easily from emitter to collector or from collector to emitter. The circuit shown in Fig. 18-15(A) is thus capable of generating a linear-current sweep and retrace just as the ideal switch circuit did. Furthermore, the transistor circuit is capable of operating thousands of times faster than a mechanical switch.

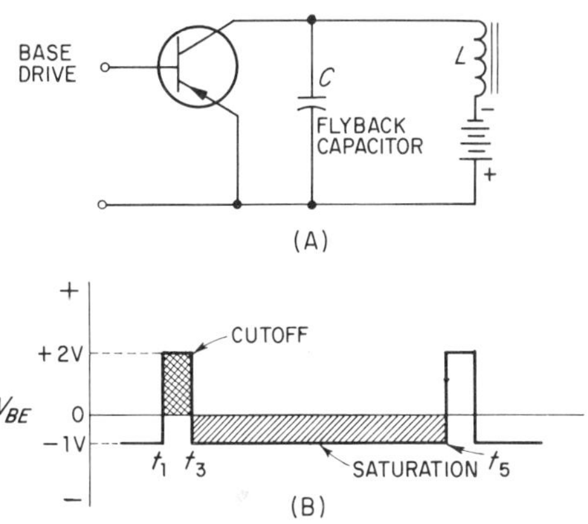

Fig. 18-15. Transistor functioning as an ideal switch: (A) transistor switch circuit; (B) base voltage waveforms.

The reader will recognise that the interval from t_3 to t_5 in Fig. 18-14(B) represents the trace interval and the period from t_1 to t_3 is the retrace interval in the scanning cycle.

Solid-State Damper Although a saturated transistor can carry current in either direction, it cannot reverse its direction of current fast enough at t_3 in Fig. 18-14(B). At this instant, the oscillations are stopped by the low-shunt resistance of the transistor and the damping diode, which are now effectively in parallel. The stored energy (current) in the yoke now begins to decay linearly toward zero, mostly through the diode. The

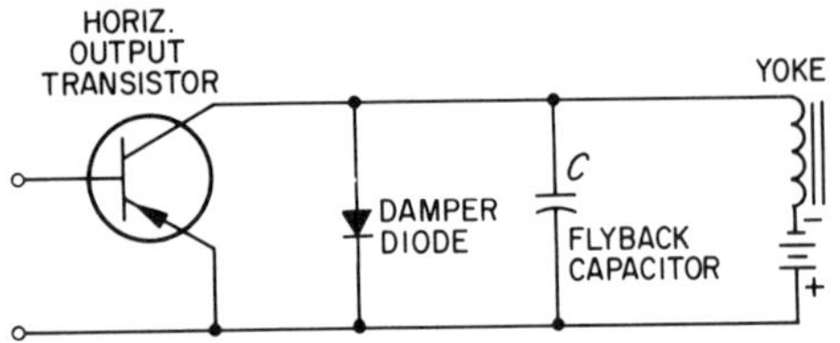

Fig. 18-16. Incorporation of the solid-state damper diode (see Fig. 18-17 for wave-forms).

transistor must be paralleled by a diode switch during most of this linear decay period to assure a very low resistance and linear-current decay. The diode, as seen in Fig. 18-16, is connected between the emitter and the collector of the transistor. Since this diode conducts current immediately after the retrace, thereby damping self oscillation in the circuit, it is called the damper diode. Most of the current through the yoke between times t_3 and t_4 is carried by the damper diode. The transistor will also carry part of the current, but has a higher starting resistance than the damper diode and does not conduct fully until sometime between t_3 and t_4, as shown in Fig. 18-17.

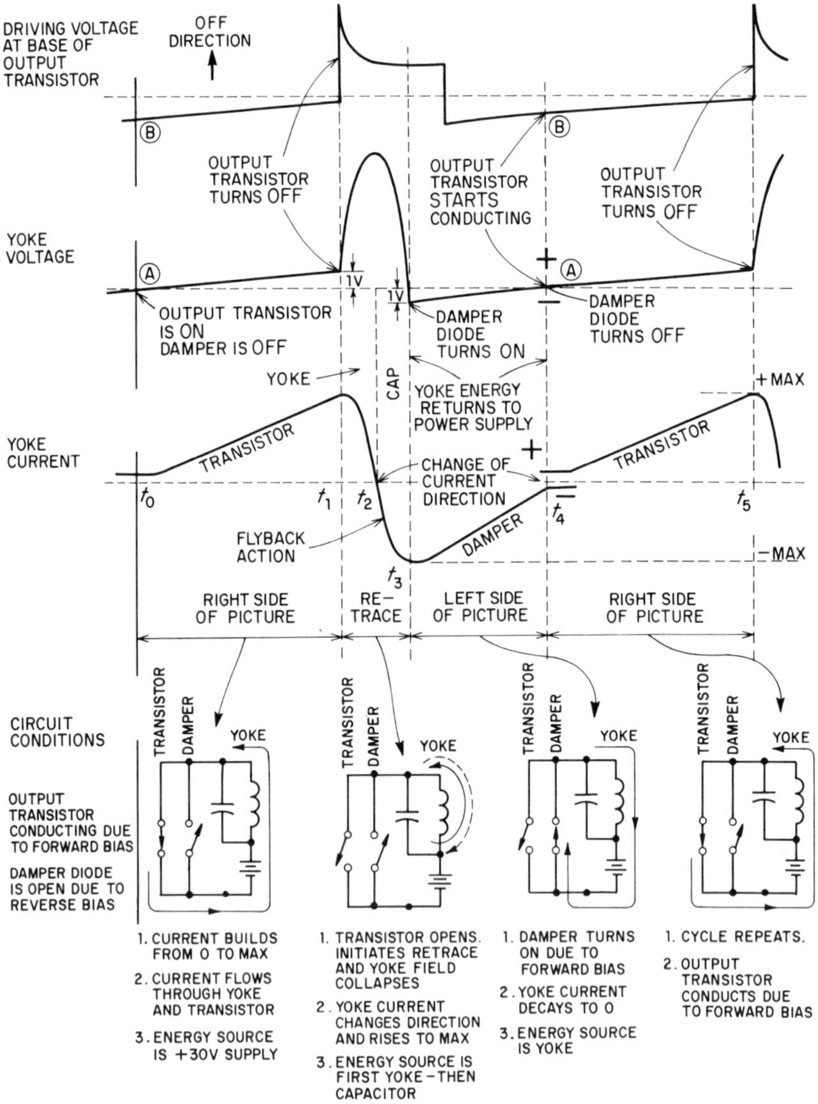

Fig. 18-17. Relative time relationship of the horizontal-output operational cycle (see Fig. 18-16 for circuit). (*Courtesy of RCA.*)

Figure 18-17 clearly illustrates the critical time relationship of the currents through the output transistor and the damper diode. The reader should study each step carefully shown at the bottom of the illustration and note the shape of each wave-form at that instant.

Unlike tube circuitry, the solid-state damper diode is not a source of boosted B+. It is employed only to carry the bulk of the yoke current when the CRT beam is at the left side of the screen. If boosted B+ is required, a separate diode is used.

The deflection yoke in a transistorized horizontal-output circuit has less than one ohm dc resistance. However, this small resistance is enough to create a slightly nonlinear sweep. It causes a left-hand stretch and a right-hand compression on the screen. To compensate for this nonlinearity, a capacitor is placed in series with the yoke as shown in Fig. 18-18. An "S" shaped voltage waveform is developed across the capacitor during scan. This compresses the left-hand side and stretches the right-hand side of the picture. The capacitor is appropriately named the "S" capacitor.

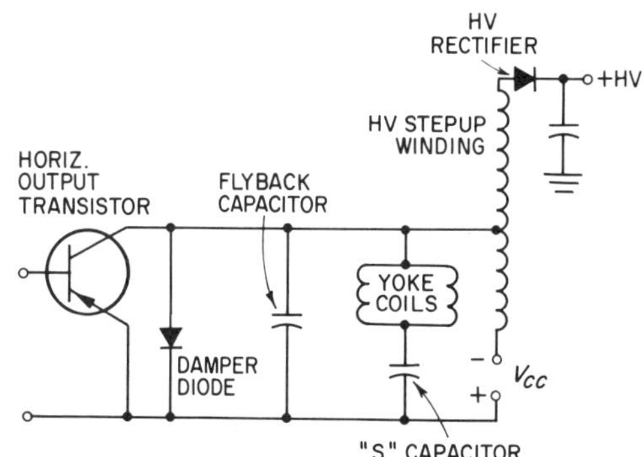

Fig. 18-18. Solid-state, horizontal-output stage using *S* capacitor.

Figure 18-18 shows how the flyback transformer is coupled to the yoke when an "S" capacitor is utilized. The capacitor also blocks dc current from flowing in the yoke. If dc current flows through the yoke, heat is generated, which in turn, increases the yoke dc resistance, thus causing more nonlinearity.

Note the simple high-voltage circuit in Fig. 18-18. The primary of the flyback transformer is placed in parallel with the damper, transistor, and yoke. The high-voltage winding provides pulses ranging from 10 KV to 30 KV which are rectified and applied to the CRT second anode. The most successful solid-state, high-voltage rectifiers are made from silicon. Several examples of these devices are shown in Fig. 18-19.

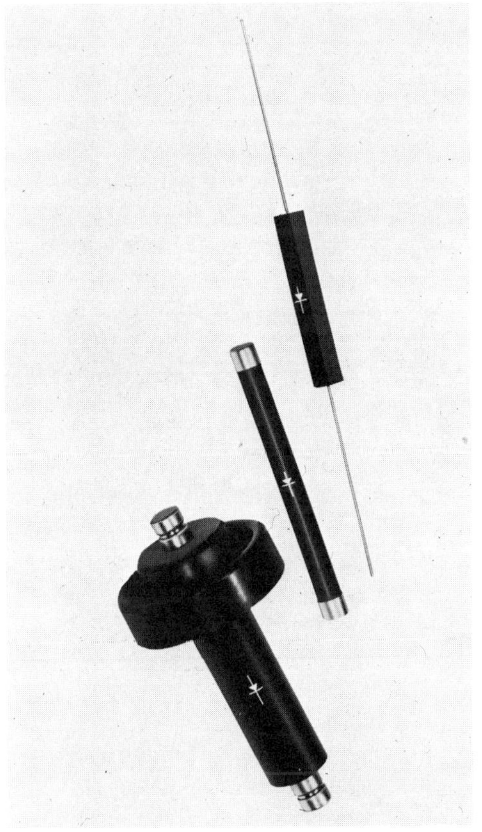

Fig. 18-19. Silicon high-voltage rectifier assemblies. (*Courtesy of Varo, Inc.*)

An NPN Horizontal Output Stage Figure 18-20 is an example of a typical NPN transistor monochrome horizontal-output circuit. The high-voltage rectifier circuit is similar to that used in vacuum-tube TV receivers. The remainder of the circuit is solid state.

The horizontal-yoke coils are connected in parallel. They are then connected in series with the "S" capacitor which is actually formed by two parallel capacitors, C_{16} and C_{17}. The yoke and the "S" capacitor are positioned directly across the primary of the horizontal-output transformer. However, the driving waveform from the transistor must pass through a filter before it can get to the yoke and the transformer. The filter is formed by C_{13}, C_{15}, C_{24}, and L_{603}. This filter, and the damper diode, serve to keep voltage transients from Q_1.

Q_1 is also protected against loss of drive since it is biased below cutoff. With no signal on the input-transformer primary, the base-to-emitter voltage is zero. At least +0.7 volts (for a silicon transistor) is needed on the base to bring the transistor out of cutoff. Thus, with no drive signal, a 0.7-volt safety margin exists, and no current will flow

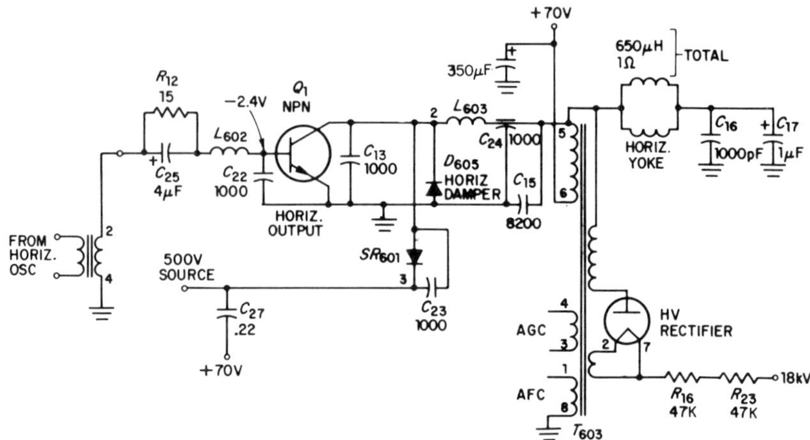

Fig. 18-20. NPN horizontal-output stage. (*Courtesy of Magnavox.*)

through the output transistor. When the drive signal is present, however, a −2.4-volt bias builds up on the base terminal. This is caused by the charging of C_{25} during the transistor "on" time, and its relatively slow discharge through R_{12} during the transistor "off" time.

The pulses on the transistor collector rise to approximately 500 volts amplitude during the retrace. These are rectified by diode SR601 and filtered with C_{23} and C_{27}. The resultant 500 volts dc is used for the CRT focus anode.

A completely solid-state horizontal output stage which utilizes a PNP transistor is shown in Fig. 18-21. The damper diode, D_2, is connected directly between the collector and emitter of the output transistor, as usual. Another diode, D_1, provides a +80-volt boost voltage since the regular B+ (a battery) is only 13 volts. The base and emitter are at the same dc potential, with only the low resistance of the coupling transformer secondary between them. This assures class "C" operation, so that the transistor is normally off until a pulse comes through the transformer.

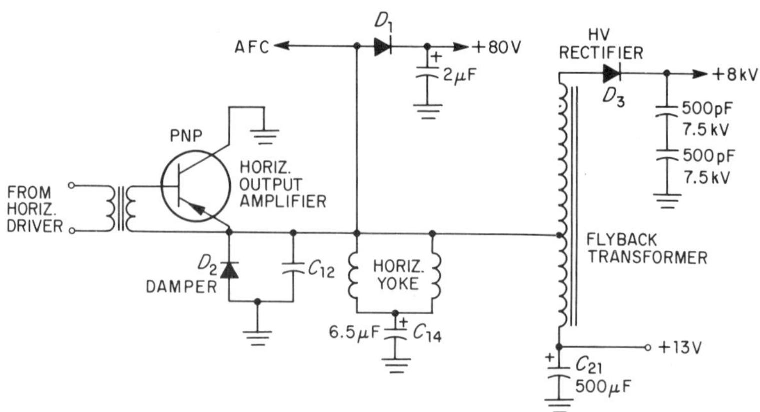

Fig. 18-21. PNP horizontal-output circuit.

High-voltage pulses from the flyback transformer are rectified with the solid-state diode, D_3. The resultant 8 KV dc potential is utilized for the second anode of the small screen CRT.

18.5 ADJUSTMENTS IN MONOCHROME RECEIVER CIRCUITS

The horizontal-output stages of a monochrome-TV receiver contain a number of controls and adjustments, each of which has a significant effect on the picture produced on the screen.

Drive Controls—Vacuum-Tube Sets Two common methods for controlling the grid drive were illustrated in Fig. 18-4. In Fig. 18-4(A), C_2 and C_3 form an ac voltage divider. C_3 is the drive control and, as its capacitance is reduced, its impedance rises and more voltage is developed across it and passed on to the output amplifier. The unit is usually a small trimmer capacitor adjustable by means of a screwdriver.

In the second circuit, Fig. 18-4(B), the drive capacitor C_1 is placed in series with the signal path between the oscillator and the output tube. More capacitance means more voltage; the variation is directly opposite to the variation in Fig. 18-4(A).

The drive control is basically utilized to control the amplitude of the grid-input waveform. This adjustment also has a significant effect on the linearity and width of the yoke-current waveform. In many receivers, the drive control is the only available width adjustment. These three controls: drive, width, and linearity are highly interactive. Thus, optimum width and linearity will be achieved only after making all three adjustments several times.

Drive Controls—Solid-State Receivers The horizontal-output transistor operates either completely on or completely off. As we noted in Section 18-4 (Fig. 18-13), we can control the dissipation of the output transistor by merely controlling the ratio of on to off time (that is, controlling the pulse width). This is effectively the same action as controlling the drive in a vacuum-tube receiver. Figure 18-22 shows this pulse-width control in the horizontal-oscillator stage of a General Electric receiver. If the horizontal pulse width is not properly adjusted, loss of horizontal sync and/or damage to the horizontal-output transistors can result. Normally, R_{739} should not require adjustment unless one of the transistors following the oscillator has been replaced. In the event any of these components has been replaced, it will be necessary to check the waveform at the collector of Q_{702}. Adjust R_{739} for an "on" time of 19 microseconds.

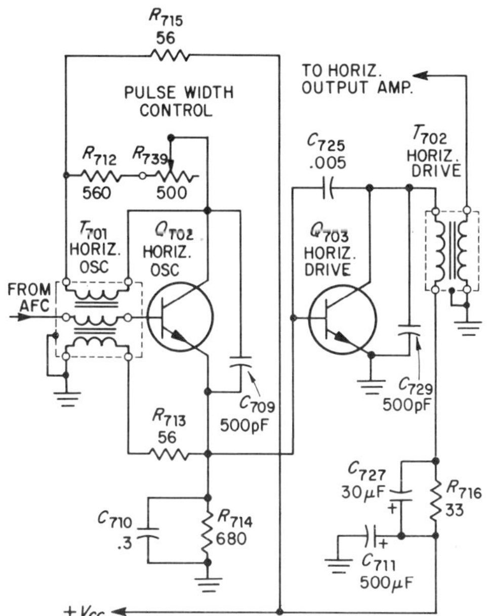

Fig. 18-22. Pulse width control (R_{739}) for adjusting drive to the horizontal-output transformer. (*Courtesy of General Electric.*)

Width Controls—Vacuum-Tube Sets A television picture is transmitted with an aspect ratio (ratio of width to height) of 4:3. Unless this ratio is maintained in the receiver, all the objects in the televised scene will be uncomfortably distorted; figures will appear taller and thinner, or shorter and broader than they should be. After the height of the picture is correctly set, the width of the picture must be adjusted. This adjustment is made by the width control or the horizontal-size control. In many receivers, the width control takes the form of a slug-adjusted coil connected across part of the horizontal-output transformer secondary winding. By changing the inductance of the width coil, the amount of deflection current flowing through the deflection coils is varied and the width of the raster changes accordingly. In this fashion, picture width can be varied by 1 1/2 inches or more. Figure 18-5 utilizes this type of width control.

Another method for picture-width variation is shown in Fig. 18-23. A potentiometer in the screen-grid circuit of the horizontal-output amplifier varies the gain of the stage and, with it, the deflection voltage (and current) fed to the yoke. In Fig. 18-10, the same method is employed, although here fixed resistors are brought into or cut out of the circuit by jumpers.

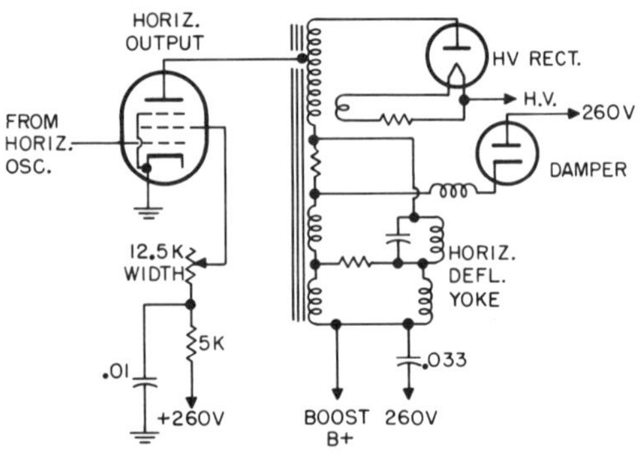

Fig. 18-23. Width adjustment by changing the gain of the horizontal-output amplifier

Still another approach to width control is finding favor among manufacturers. Rotation of a tapped shaft at the bottom of the flyback transformer varies an air gap in the ceramic core. As the air gap widens, the reluctance of the magnetic path increases, resulting in a reduction of the induced-signal voltage in the secondary winding. Thus, the width of the picture decreases. Although this method affects the high voltage as well as the picture width, the change in high voltage is not as noticeable as the change in width.

A simple and practical horizontal-width control can be fabricated using a thin metal sleeve that is slipped between the yoke and neck of the picture tube. This forms a "losser" type control (one-turn loop or short). The width is affected by the loop creating a magnetic field that reduces the sweep-magnetic fields. The sweep is of minimum length when the sleeve is completely under the yoke.

Width Controls—Solid-State Sets A capacitive type of width adjustment is shown in the transistor circuit of Fig. 18-24. A special capacitor, C_{514}, can be connected into the circuit by means of a special jumper accessible at the rear of the chassis. When C_{514} is in circuit, the width increases; when it is out of circuit, the width decreases. C_{514} has a loading effect when connected in the circuit, reducing the amplitude of the pulse applied to the high-voltage rectifier V_2. When V_2 conducts less, the high voltage decreases. The yoke magnetic field then has more effect on the beam and increases the picture width.

Linearity Controls The horizontal-output tube current sweeps the right section of the screen and the damper-tube current covers the left section of the screen. A transition region is apparent in Figs. 18-1 and 18-6. The final yoke current consists of a combination of currents from two tubes. They must "join up" and produce a linear wave-form. A simple joining of these two currents will *not* produce a linear wave-

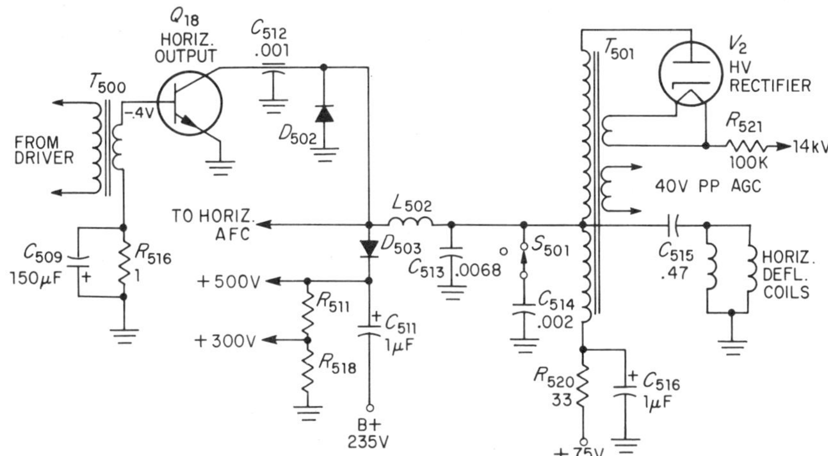

Fig. 18-24. Width-control circuit which utilizes a jumper connected capacitor. (*Courtesy of Motorola.*)

form. A waveform correction circuit is added to remove most of the nonlinearity caused by the addition or overlap of the two currents. In the circuit of Fig. 18-25, the plate-load impedance of the horizontal-output tube consists of two parts: the flyback transformer, and a π-type filter. The filter is a low-Q filter, resonant at 15,750 Hz. Resonance of this filter to the horizontal-sweep rate is achieved by varying L_1. At resonance, the impedance of the filter circuit is maximum and impedes the current flowing to provide the maximum output-tube efficiency. The linearity (π) filter shifts the phase of the ripple on the B+ boost voltage. This modifies the output-tube current waveform and permits center-screen, horizontal-linearity control.

Should a case arise where the horizontal-output tube current cannot be adjusted correctly, the judicious selection of a new value for C_1 may bring relief (this will change the resonant frequency of the tank circuit consisting of C_1, C_2, and L_1. Linearity controls are also sometimes called efficiency controls. It should also be pointed out that this control is not found very often in solid-state TV receivers.

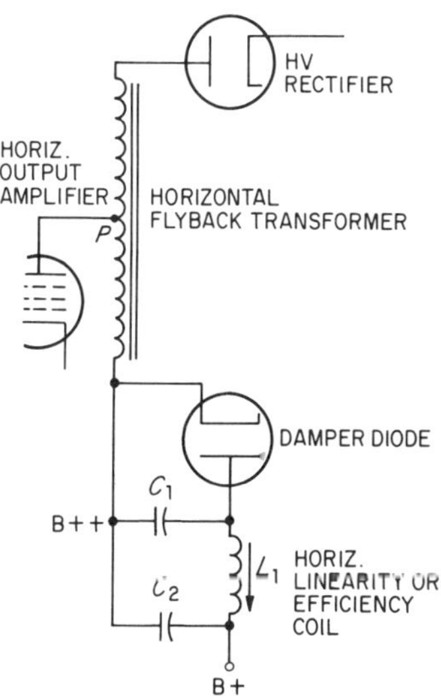

Fig. 18-25. Horizontal linearity or horizontal "efficiency" control.

18.5 VACUUM-TUBE, COLOR HORIZONTAL-OUTPUT CIRCUITS

In general, the circuits for horizontal deflection in color receivers are quite similar to their black-and-white counterparts. The principal point of departure is the power required in the deflection and high-voltage circuits; this is considerably higher in color receivers than in black-and-white receivers. The reasons for the increased power need are the larger kinescope-neck diameter (which requires a larger yoke current) and the greater high voltage. Horizontal-output tubes such as the 6DQ5, 6JE6A, 6LQ6, 23JS6A, 6LB6, and 26HU5 have been developed to meet these increased power requirements. Class "C" operation is utilized to keep the dissipation down in these tubes. This requires a grid-drive sawtooth with a peak-to-peak amplitude of several hundred volts.

To properly illuminate the CRT screen, the electron beams must be accelerated to a very high velocity. As discussed in Chapter 14, a high-voltage supply of 20-28 KV is required to accelerate the electrons to adequate velocities. The high-voltage supply must also provide up to 5 KV for the electrostatic focus electrode and 500 to 1000 volts for each screen grid in the electron guns.

The electrons in the three beams, after they strike the phosphor-dotted screen, are attracted to the internal aquadag coating inside the color kinescope. The aquadag, in turn, is connected to the high-voltage supply. For economy and safety reasons, only a limited current may be drawn from the high-voltage supply. Most of the high-voltage load current consists of the combined currents of the three electron beams. These beam currents will vary according to the relative brightness of their respective primary colors (red, green, and blue). When the entire screen is bright blue, for example, only one third of the maximum current will be drawn by the picture tube. If the entire screen is white, all three beams will draw maximum current. If the screen is just a dim green, very little beam current is drawn. Unless the high voltage is automatically regulated, the second anode potential will vary according to the amount of the current drawn. This would mean a change in focus and even in picture size, since the deflection system will produce a larger picture when the high voltage is lower. For these reasons, all of the color sets have a regulated high-voltage supply. The different circuits for providing such regulations are described later in this chapter.

In Chapter 15, when color picture tubes were discussed, we noted the need for static and dynamic convergence circuits. The horizontal-deflection section furnishes the basic dynamic convergence waveforms for this purpose. In order to compensate for the fact that the screen is flat and not spherical, the convergence field must be increased as the three electron beams move towards the edges of the screen. As shown in Fig. 18-26, a trapezoidal signal taken from a special winding on the horizontal-flyback transformer passes through a waveshaping circuit and produces a current waveform in the convergence coil which is roughly parabolic. The amplitude of the waveform can be controlled by potentiometer R_2 and the amount of relative curvature by the "dynamic phase adjustment" capacitor C_4. Since separate convergence correction coils are needed for each of the three electron beams, the horizontal-deflection system has to supply three separately adjustable dynamic convergence signals.

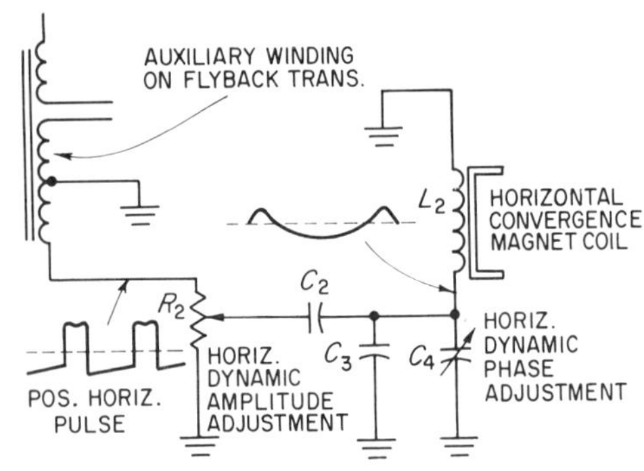

Fig. 18-26. A typical horizontal dynamic convergence circuit.

A Typical Circuit An example of a typical vacuum-tube color-TV horizontal-deflection circuit is shown in Fig. 18-27. The complexity of the drawing may obscure the fact that there are six sections contained within this overall circuit:

(1) Horizontal-output tube circuit.
(2) Yoke matching circuitry.

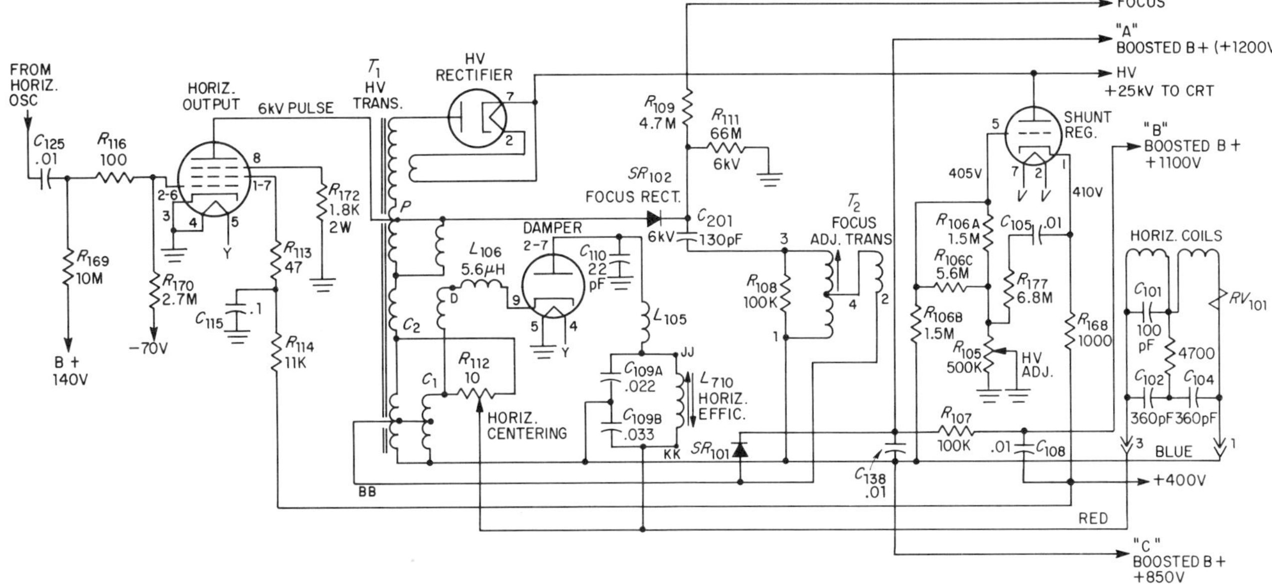

Fig. 18-27. Typical vacuum-tube, color-TV, horizontal-deflection circuit. (*Courtesy of RCA.*)

(3) Focus supply.

(4) Damper and boost circuitry.

(5) High-voltage supply.

(6) Regulator circuit.

(1) *The output-tube circuit.* Figure 18-28 shows the horizontal-output tube and its immediate circuit. C_{125} couples in the large amplitude sawtooth grid-drive voltage from the horizontal oscillator. V_1 is a class "C" amplifier and draws grid current on the peaks of this drive signal. During these peak intervals, C_{125} charges and then discharges through R_{169} to develop a -50 VDC bias at the control grid of the 6JE6A.

The plate circuit drives the flyback transformer which, with its associated components, make up a tuned circuit (70kHz). A dc milliampere meter M_1 is shown in the cathode circuit and proper adjustment of the tuned circuit is indicated by a dip in the reading on M_1. The "dip" adjustment is made with the horizontal efficiency coil (L_{710} in Fig. 18-27). This coil and its associated capacitors are designed to resonate to the horizontal frequency (15,750 Hz). In black-and-white television, this coil is called the horizontal-linearity coil. The tube should draw a maximum of 220 ma for long-life operation. In small-screen color sets, the output tube usually draws 190 to 200 ma; in larger screen sets (23″ to 24″), this tube will usually draw from 210 to 220 ma.

(2) *Yoke-matching circuitry.* Figure 18-29 presents a simplified drawing of the yoke-matching circuit. A 5 KV pulse during beam retrace, develops at point *P*. T_1 is an autotransformer, with the windings from *P* to *BB* functioning as the primary and the windings from C_2 or C_1 to *BB* as a step-down secondary. The pulse voltage is therefore stepped down with a resulting current step-up. This is necessary to match the horizontal-output tube impedance (relatively high *Z*) to the yoke windings (relatively low *Z*).

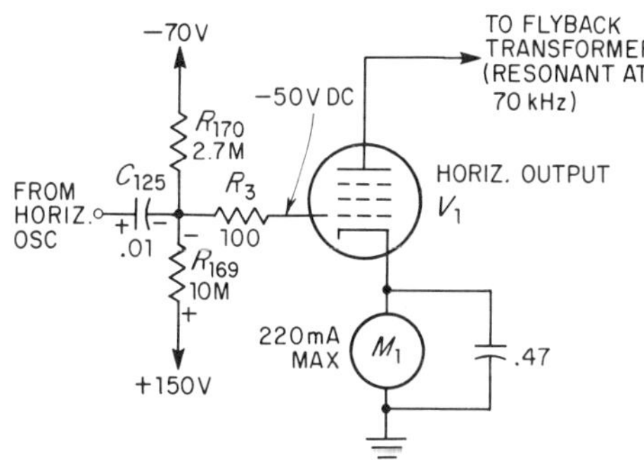

Fig. 18-28. Horizontal-output tube circuitry from Fig. 18-27.

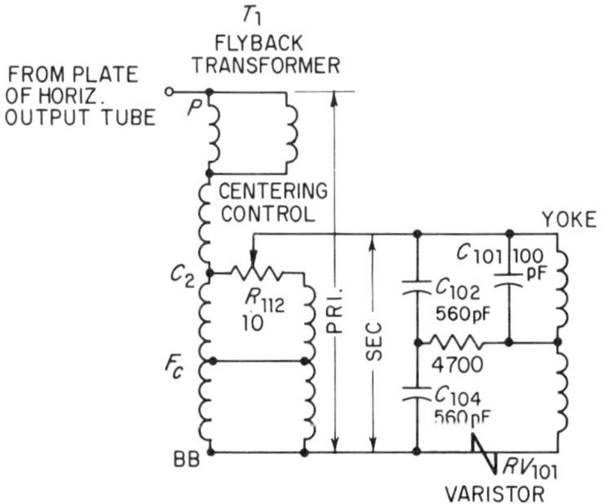

Fig. 18-29. Yoke matching circuit from Fig. 18-27.

Capacitor C_{101} is the yoke-balancing capacitor, and C_{102}-C_{104} are selected to resonate the horizontal circuit to 70 kHz to provide flyback timing (during retrace). RV_{101} is a varistor to hold the scan size constant. A varistor changes its resistance in opposition to the voltage drop across it. An increase in output voltage causes RV_{101} to decrease in value. This loads the output circuit and reduces the output voltage to its normal value. A decrease in output voltage works in the opposite manner—RV_{101} increases in value and loads the output circuit less. A typical value for RV_{101} is 175 ohms at 1 ma. R_{112} controls the centering by regulating the dc current through the yoke.

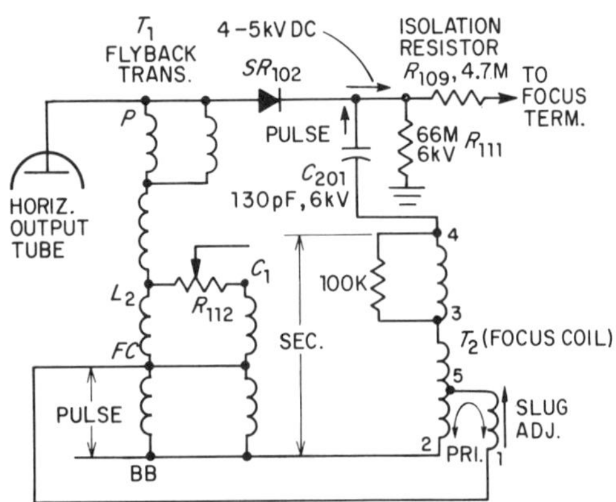

Fig. 18-30. The focus supply portion of Fig. 18-27.

(3) *Focus supply.* The focus supply section of Fig. 18-27 is shown in Fig. 18-30. If 25 KV is utilized on the CRT second anode, then 4 to 5 KV is required on the focus electrode. Since a 5 KV pulse is delivered to the flyback transformer from the output tube, the focus rectifier (SR_1) can be attached at that point.

Notice the special transformer which is connected between the 130 pF filter capacitor (C_{201}) and point *BB*. This is a specially wound buck-boost transformer (T_2) which permits adjustments in the focus voltage. The T_2 primary (pins 1 and 2) receives a pulse from point *FC* on the flyback transformer. Depending on the position of the slug adjustment, the secondary of T_2 (pins 4 to 2) generates either an in-phase or an out-of-phase pulse. A 100-kilohm resistor prevents the focus coil from breaking into oscillation during the retrace. The buck or boost pulse is coupled to the cathode of SR_{101} through the 130 pF capacitor. SR_{101} rectifies the difference between the two pulses (5 KV to the anode and buck or the boost pulse to the cathode). T_2 then allows a smooth adjustment of the focus voltage from 4 to 5 KV. The 4.7-megohm resistor in series with the lead to the focus anode provides isolation protection to the focus supply should a short occur on the focus anode lead or internally in the tri-color kinescope.

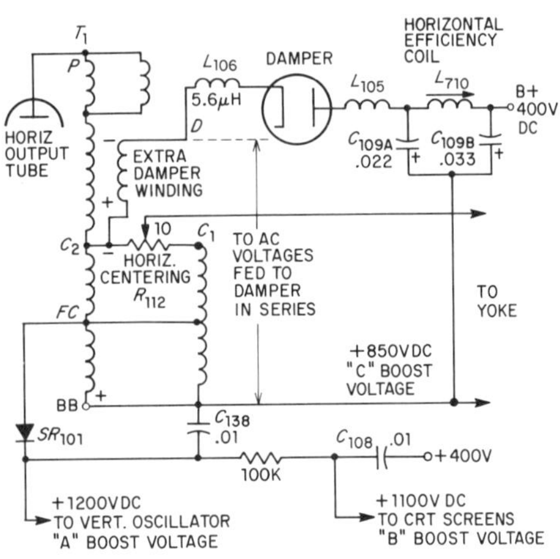

Fig. 18-31. The damper and boost circuitry.

(4) *Damper and B++ circuit.* The damper and B++ portion of Fig. 18-27 appears in Fig. 18-31. This particular receiver requires more boosted B+ than could be obtained directly from the yoke portion of the flyback transformer. Hence, an extra step-up section is added to the transformer so that an 850 volt B++ may be obtained. A rugged semi-high voltage rectifier, serves as the damper diode. The rectified pulses are filtered in the π filter composed of L_{710}, $C_{109}A$, and $C_{109}B$. Coils L_{106} and L_{105} prevent parasitic oscillations. This +850 volts dc is called the "C" boosted B+.

Two additional B++ voltages are provided by the circuit of Fig. 18-31. The diode SR_{101} has +850 Vdc plus a 350 volt pulse on its anode. It converts this to +1200 Vdc for use in the vertical oscillator. The same circuit also makes available +1100 Vdc for the CRT screen grids.

(5) *High-voltage rectifier.* The high-voltage transformer and rectifier tube circuit in a color receiver are quite similar to the same components

in a black-and-white receiver. The major difference is the larger voltage and current requirements of the color set. The high-voltage rectifier has a ruggedly-built cathode to help provide this extra capability. The high-voltage rectifiers in monochrome receivers usually do not have cathodes.

(6) *High-voltage regulator.* Figure 18-32 is a simplified drawing of the high-voltage rectifier and shunt-regulator portion of Fig. 18-27. Two basic facts regarding shunt-regulated power supplies should be emphasized. First, all power supplies have an internal impedance which causes the output voltage to vary with load changes. Second, if we desire a fixed output voltage, then the current through the internal impedance must be kept constant. This is performed with a shunt-regulator tube which functions as part of the high-voltage load. If the CRT draws more current, the shunt regulator draws less current, and vice versa.

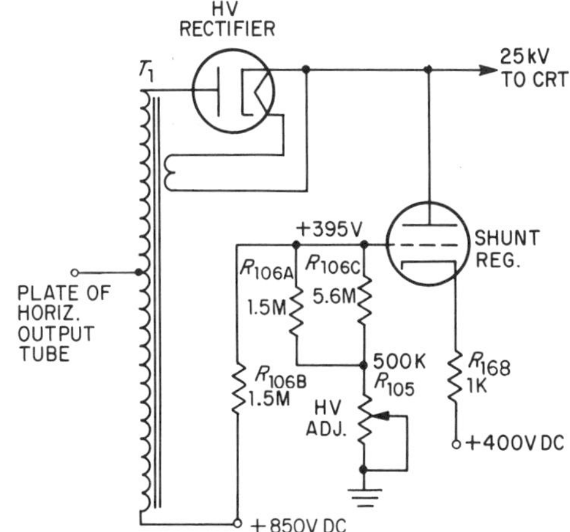

Fig. 18-32. Shunt regulator portion of Fig. 18-27.

It is difficult to continuously monitor the magnitude of the high voltage. Yet we must do this if we are to hold it constant. A good detection point is the boosted B+, because this varies directly with the high voltage. A sample of the boosted B+ voltage is fed to the grid of the shunt-regulator tube, as shown in Fig. 18-32. Resistors R_1-R_2-R_3-R_5 reduce the boosted B+ from +850 volts to approximately +395 volts at the grid of the tube. R_5 is a control that establishes the bias on the 6BK4 shunt regulator by changing the grid voltage. Note that the cathode is held at +400 volts from a separate power supply. The +400 volts does not change when the high voltage and the boosted B+ change. If these latter two voltages drop, then the +395 volts on the 6BK4 grid drops. This places more reverse bias on the 6BK4, which then draws less current, and permits the high voltage to rise back to its original value.

18.7 SOLID-STATE COLOR-TV HORIZONTAL-OUTPUT CIRCUIT

Transistorized small screen black-and-white television receivers have been commercially available for several years. It follows naturally that large-screen receivers, both black-and-white and color, are the culmination of this experience. Voltages and currents (and circuit stability) had to be extended appreciably to make the change from solid-state monochrome receivers to solid-state, large-screen color receivers.

As an example of a transistorized horizontal-output stage capable of handling a 25-inch color CRT, we will discuss the Motorola TS-915/919 chassis. This receiver has only 2 vacuum tubes, the CRT, and the high-voltage rectifier. The schematic for the horizontal output stage in this receiver is shown in Fig. 18-33.

The operation of this output stage is similar to the NPN output stage mentioned in Section 18.4. Two parallelled NPN transistors are now

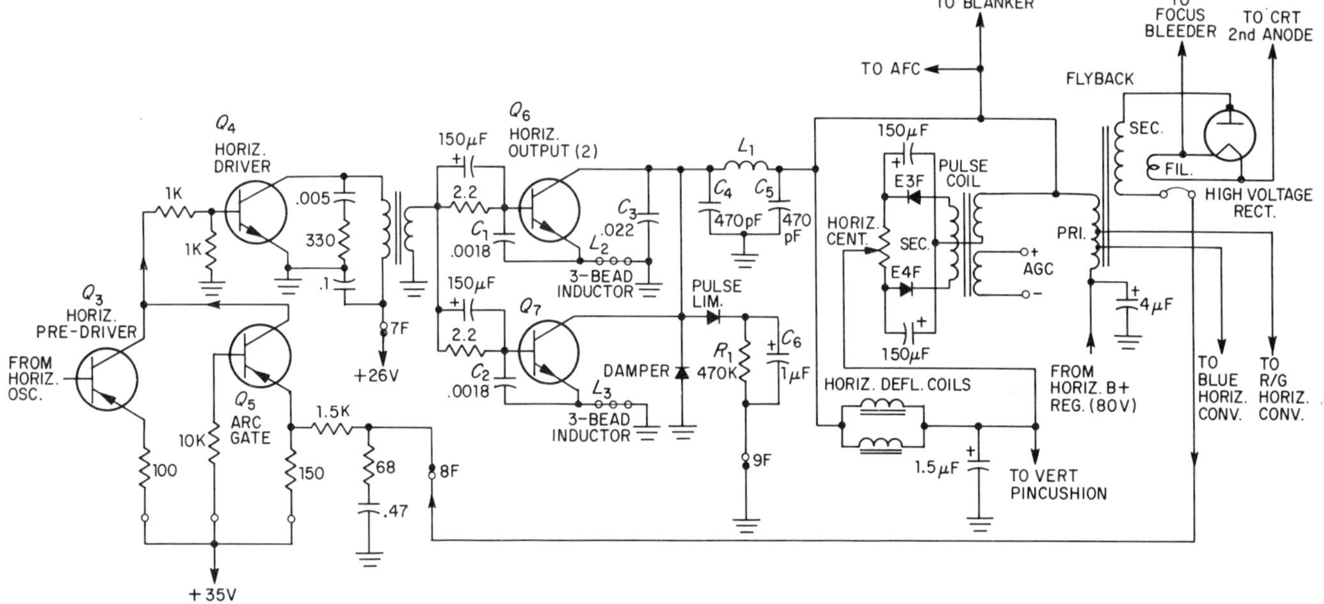

Fig. 18-33. Horizontal-sweep circuitry and regulator. Arc gate Q_5 protects output transistors from arc at high-voltage rectifier. B+ to the output pair is regulated. (*Courtesy of Motorola.*)

necessary to handle the higher power level in this color set. Two stages of amplification (Q_3 and Q_4) bring the oscillator power up to the level required by Q_6 and Q_7. All of these transistors (Q_3, Q_4, Q_6, and Q_7) operate as switches. That is, they are either completely on or completely off at different times during each cycle. As mentioned in Section 18.4, the output transistors must be on for the last half of each horizontal trace. Power flow during the first half of each trace is through the damper diode.

Quite a few parts are utilized to keep high-frequency transients at a safe level throughout the system. By-passing of these transients to ground is performed by capacitors C_1, C_2, C_3, and C_5. A π filter formed by C_4, C_5, and L_1 keeps the yoke induced transients from travelling back to the output transistors. The 3 bead inductors, L_2 and L_3, reduce the gain of the output transistors at high frequencies. This prevents fast transients from being amplified by Q_6 and Q_7.

Large-voltage transients at the collectors of Q_6 and Q_7 are shunted to ground through the pulse-limiting diode. A charge is built up on C_6 which is equal to the peak value of the flyback pulse (590 volts). R_1 allows this charge to leak off with a 470 kilohm $\times$ 1 μFarad = 0.47 second time constant. The time between the 590 volt pulses on the output transistor collectors is only 64 microseconds. Thus, this voltage stays quite constant between pulses. If any transients appear larger than 590 volts, they pass through the diode and are absorbed by C_6.

The output transistors are protected from current surges caused by high-voltage arcs. This is accomplished with the arc gate Q_5. This stage samples the current from the high-voltage winding of the flyback transformer; if an arc occurs, Q_5 is momentarily turned on, forcing

Q_6 and Q_7 off. Without this protection, the output transistors would be easily destroyed.

A shunt or feedback high-voltage regulator is not required in this Motorola circuit. The output impedance of the high-voltage circuit is sufficiently low, so that regulation of the +80 volt B+ keeps the high voltage regulated.

18.8 SILICON-CONTROLLED RECTIFIER HORIZONTAL-OUTPUT CIRCUIT

The basic objective of all horizontal-deflection circuits using electromagnetic deflection is the same; to cause an approximately linear current to flow in the yoke windings in such a manner as to deflect the picture-tube beam linearly across the screen. This current must be in synchronism with the video information provided in the received television signal. An RCA solid-state chassis generates the desired horizontal deflection current with circuitry using silicon-controlled rectifiers (SCR) and associated-circuit elements.*

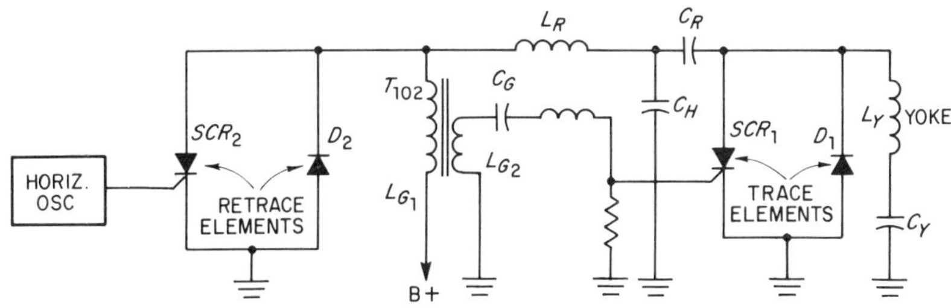

Fig. 18-34. The simplified schematic of an SCR horizontal-output circuit. (*Courtesy of RCA.*)

The essential components in the RCA solid-state horizontal-output circuit are shown in Fig. 18-34. The diode D_1 and SCR_1 provide the switching action which controls the current in the horizontal-yoke windings, L_Y, during the picture-tube beam-trace interval. The diode D_2 and SCR_2 control the yoke current during the retrace interval. The components L_R, C_R, C_H, and C_Y, supply the necessary energy storage and timing functions. Inductance L_{G1} supplies a charge path for C_R and C_H from B+, thereby providing a means to "recharge" the system from the power supply. Inductance L_{G2} provides a gating current for rectifier SCR_1 (L_{G1} and L_{G2} comprise transformer T_{102}). Capacitor C_H controls the retrace time, because it is charged up to the $B+$ voltage through L_R.

To assist in the explanation of the operation of the horizontal-output circuit, the illustrations in the following discussion have been greatly simplified. In Fig. 18-35 (and the illustrations following), SCR_1 and D_1 together comprise an S.P.S.T. switch labeled S_1, and SCR_2 and D_2

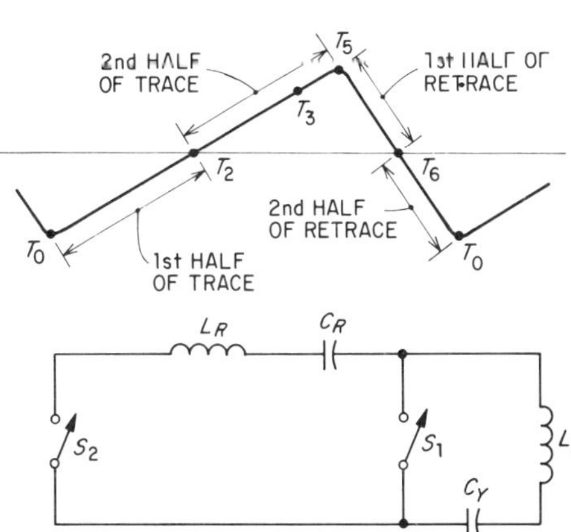

Fig. 18-35. Simplified SCR output circuitry. (*Courtesy of RCA.*)

* Information regarding the operation of this SCR circuit is provided by permission of RCA.

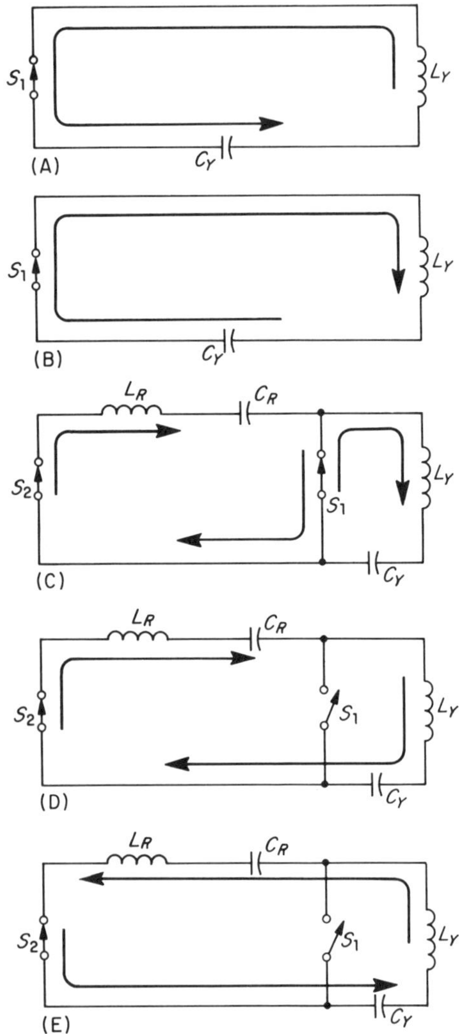

Fig. 18-36. Currents through portions of Fig. 18-35 at different times: (A) from T_0 to T_2, (B) from T_2 to T_5, (C) from T_3 to T_5, (D) from T_5 to T_6, and (E) from T_6 to T_0.

comprise another S.P.S.T. switch labeled S_2. The manner in which these switches are opened and closed is now discussed.

Trace Time Referring to Fig. 18-36(A), during the first half of the trace time (T_0 to T_2), switch S_1 is closed, causing a field previously produced about the yoke inductance, L_Y, to collapse and resulting in a current which charges the capacitor C_Y. This yoke current deflects the picture-tube beam to approximately the middle of the screen. The beam is at the center when this current decreases to zero at T_2. During the second half of the trace-time interval T_2 to T_5, the current in the yoke circuit reverses because the capacitor C_Y now discharges back into the yoke inductance L_Y as shown in Fig. 18-36(B). This current causes the picture-tube beam to complete its trace.

Retrace Initiation However, at time T_3 a pulse from the horizontal oscillator causes switch S_2 to close, releasing the previously stored charge on C_R. The resulting current flows in the circuit as illustrated in Fig. 18-36(C). Because of the natural resonance of L_R and C_R, this resonant current becomes equal in value to the yoke current, at T_5. At this time, S_1 opens, S_2 remains closed, and the retrace starts.

Retrace The simplified retrace circuit with S_2 closed and S_1 open is shown in Fig. 18-36(D). Basically, L_R, C_R, C_Y, and L_Y are connected in series. The natural resonant frequency of this circuit is much higher than that of the yoke circuit L_Y and C_Y, because the value of capacitor C_R is very much smaller than that of C_Y. As a result, the current change through the yoke windings L_Y during retrace time is much faster than that during trace time This, of course, causes the picture-tube beam to retrace (flyback) very rapidly.

During the first half of the retrace time, T_5 to T_6, the retrace-yoke current flows in the direction as shown in Fig 18-36(D). However, the retrace-circuit current soon reverses because of resonant circuit action, and the last half of the retrace action occurs from T_6 to t_0 as shown in Fig 18-36(E).

At time T_0, switch S_1 closes, and shortly thereafter S_2 opens. The field about the yoke inductance starts to collapse, and the resulting current again starts the trace interval. The trace-retrace cycle now has been completed.

SCR Color-TV High-Voltage System High voltage is generated in the RCA solid-state chassis in a manner silimar to most other color receivers. As shown in Fig. 18-37, a vacuum-tube, high-voltage rectifier provides a nominal 26.5 KV to the CRT second anode. This potential varies approximately 2.5 KV over the normal range of the picture-tube beam currents

Focus and screen voltages for the CRT are also derived from the high-voltage winding. A tap on this winding provides 1000-volt pulses which are rectified and filtered to produce 1000 Vdc. Another winding,

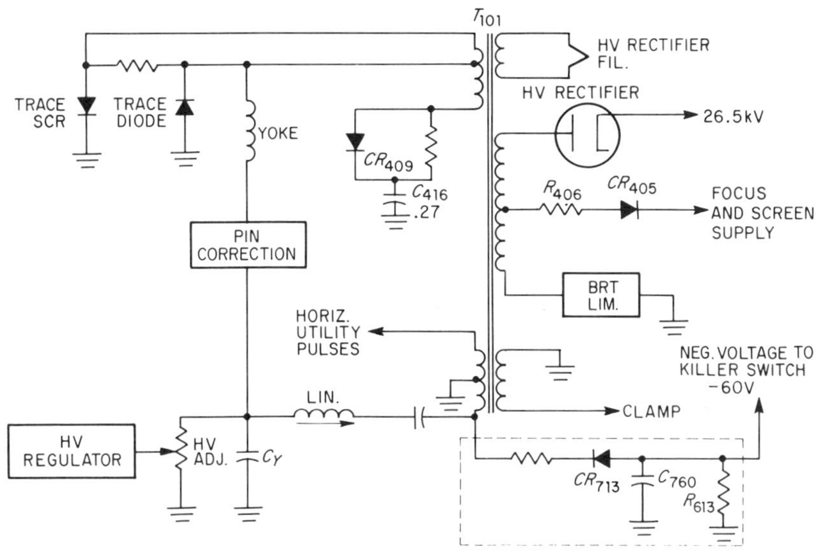

Fig. 18-37. High-voltage generation circuitry for SCR horizontal-output stage. (*Courtesy of RCA.*)

rectifier, and filter are used to provide −60 volts for operation of the color-killer switch.

SCR High-voltage Regulation The high voltage is regulated by controlling the amount of energy which is made available to the horizontal output circuitry. As previously stated, the output circuitry is supplied by the energy which is stored during trace time primarily on the commutating capacitors (C_R and the auxiliary capacitor C_H) as shown in Fig. 18-34. These capacitors are charged during trace time through inductance L_{G1}, which is part of transformer T_{102}. In order to provide some way to control the energy on the commutating capacitors, inductance L_{G1} is designed to resonate with these capacitors at a frequency whose period approaches twice the horizontal-scanning interval. The exact resonant frequency is made variable by the high-voltage regulator circuitry. Figure 18-38 illustrates the effect of this resonant action on the commutating capacitor charge It can be seen that the shape of the wave resulting from resonant action will determine the amount of charge developed in the capacitors. More charge means a larger pulse and a larger high voltage. Likewise, if the voltage at the initiation of the retrace is lower (as shown by the dotted lines of Fig 18-38) then the high voltage will be lower.

Figure 18-39 further illustrates the relationships that exist among the elements comprising the high-voltage regulating system. The resonance of L_{G1} and capacitors C_R, C_H is controlled by a saturable reactor, T_{103}, in parallel with L_{G1}. By changing the current in the reactor-control windings, the total inductance represented by L_{G1} and the reactor-load winding changes. The control current for the reactor is determined by the conduction of the high-voltage regulator transistor. In turn, the collector current of this transistor is controlled by the voltage across the

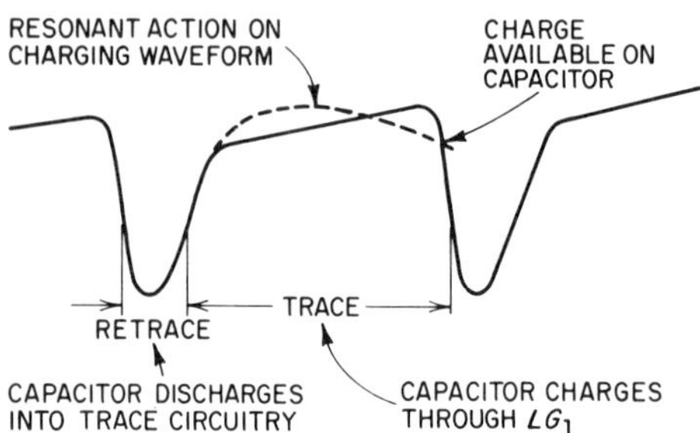

Fig. 18-38. Voltage on commutating capacitor. (*Courtesy of RCA.*)

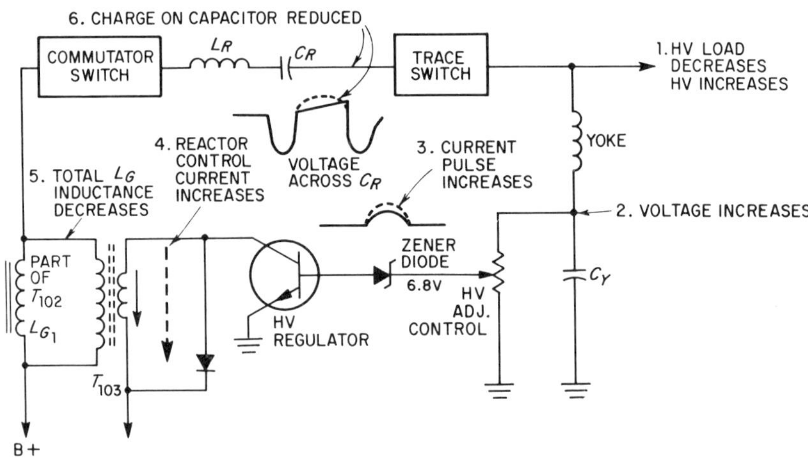

Fig. 18-39. The action of the regulator circuit. (*Courtesy of RCA.*)

yoke-return capacitor C_Y. This latter voltage, which reflects high-voltage changes, is sampled by the high-voltage adjustment control and compared to a reference voltage provided by a zener diode The resulting difference voltage reflects changes in the high voltage and controls the conduction of the high-voltage regulator transistor.

This regulating system is designed to maintain high voltage substantially constant for line-voltage variations ranging from 105 Vac to 130 Vac. The high voltage will drop only 2.5 KV from a nominal 26.5 KV with a picture-tube beam current increase from 0 to 1.5 ma.

SCR Protection Circuitry Two circuits are employed which protect the trace switch (both diode and *SCR*) from high currents resulting from high-voltage arcing (see Fig. 18-40).

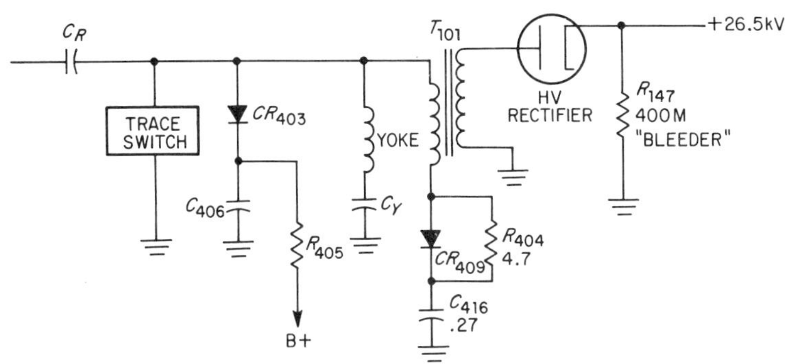

Fig. 18-40. Horizontal-deflection protection circuitry. (*Courtesy of RCA.*)

One protection circuit consists of resistor R_{404} and diode CR_{409} placed in series with the primary of the high-voltage transformer. These components act to dampen the high-ringing current which can occur under arcing conditions. This current is dissipated mainly in the resistor; the principal purpose of the diode is to allow normal initial flyback current to flow unimpeded, thereby preventing a reduction in high voltage.

The other protection circuit consists of diode CR_{403}, capacitor C_{406}, and resistor R_{405}. Diode CR_{403} conducts during the peak voltage of the retrace pulse, charging capacitor C_{406} to the peak voltage. Resistor R_{405} provides a high-resistance discharge path tor the capacitor, sufficiently long to keep the diode reverse biased during trace time. When high-voltage arcing produces a sharp voltage pulse, CR_{403} conducts, clamping the trace switch to the voltage on C_{406}, and preventing the arc-pulse voltage form exceeding the breakdown voltage of the trace-switch components.

A 400-megohm high-voltage bleeder resistor, R_{147}, located between the cathode of the high-voltage rectifier to ground, safely discharges the high voltage after the equipment is turned off.

18.9 ADJUSTMENTS IN COLOR-TV HORIZONTAL SYSTEMS

The adjustments and controls in color-TV sets are nearly the same as those found in black-and-white sets, as described in Section 18.5. The major exceptions are the adjustments related to the high-voltage regulating circuits and those of the high-voltage focus circuits. It must be remembered, however, that adjustment of the high voltage has an effect on the width and height of the picture. As the second anode voltage is increased, the picture will appear smaller; if the voltage is lowered, the picture will become larger. For this reason, adjustment of the width control may be necessary if the voltage at the second anode is changed substantially.

Adjustment of the linearity of "efficiency coils" has been described in Section 18.5. In this section, we will discuss the linearity control in an *SCR* horizontal-output stage.

High-voltage Controls Two basic high-voltage regulator circuits are used in color-TV sets: shunt and feedback. A typical shunt-regulator circuit was shown in Fig. 18-32. The high-voltage adjustment for that circuit is as follows:

(1) Place a voltmeter across the 1-kilohm resistor in the cathode leg of the regulator and connect a high-voltage probe and VTVM to the +25 KV terminal.
(2) Turn the brightness control down until the screen is *black.* No kinescope current will be flowing.
(3) Adjust the 500 kilohm high-voltage control for a reading of 25 KV on the VTVM. The voltmeter across the 1 K resistor should indicate a 1-volt minimum. One volt across 1 kilohm means one milliamp of regulator current is flowing.
(4) Turn the brightness control to its maximum brightness position. If the kinescope picture "blooms," do *not* adjust the high-voltage control. The picture-tube bias, screens, and drive controls must be adjusted, *not* the high-voltage control. Leave the latter control as set. Reset the picture-tube bias, screens, and drive controls.
(5) Notice that the voltmeter across the 1-kilohm resistor indicates zero volts (no regulator current) when the picture blooms. The *magic* ingredient to proper regulation is that at maximum brightness, the regulator never goes into cutoff. Cutoff means "blooming" and loss of regulation. Never allow the 6BK4 to be cut off by any adjustments or combination of adjustments.

Focus Adjustments A color picture tube requires up to 5 KV on its focus electrode. This must be adjustable at least ±10% to compensate for changes in the CRT characteristics or voltages on its various electrodes. One simple method used to obtain this voltage is to simply connect a bleeder between the 25 KV and ground. The 5 KV is tapped off with a potentiometer near the ground end. Other receivers utilize a separate winding on the flyback transformer. In this case, a separate rectifier and filter is required.

In Section 18.6, we showed an adjustable focus supply which used a buck-boost pulse transformer. Figure 18-30 showed a simplified diagram of that method. The reader should review Section 18.6 again since that method of focus adjustment is quite different from the normal potentiometer focus adjustment.

The procedure for the focus adjustment on a color set is quite similar to the procedure on a monochrome receiver. A strong station should be tuned in so that snow will not obscure the scan lines. The focus control is then adjusted until the scan lines are sharply visible over most of the screen. Brightness and contrast must be set at normal levels.

Linearity Adjustment (*SCR*) Horizontal-scanning current nonlinearity, which is caused by voltage drops across inherent resistances in the trace circuitry, is corrected by two means: (1) The voltage drop resulting from the resistive effects of the trace diode and *SCR* is minimized by placing the trace diode at a more negative voltage than the trace *SCR*, as shown in Fig. 18-37. This is accomplished by putting a resistor between the diode and the *SCR*, and also by connecting the diode and the *SCR* to different points on the flyback transformer. (2) The remaining nonlinearity is corrected by a damped series-resonant circuit illustrated in Fig. 18-41. This circuit, consisting of L_{402}, C_{413}, and R_{419}, produces a damped sine-wave current which effectively adds to or subtracts from the charge on the yoke-return capacitor (C_Y). The resulting alteration in yoke current corrects for the remaining trace current nonlinearities. L_{402} is adjustable and is set to obtain the best horizontal linearity.

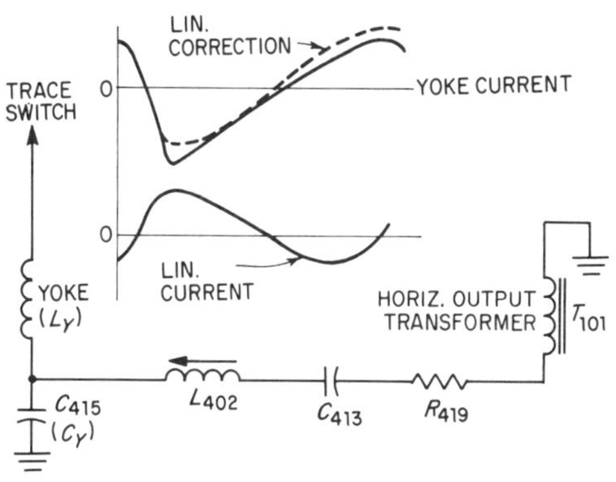

Fig. 18-41. Horizontal-linearity adjustment in SCR circuit. (*Courtesy of RCA.*)

18.10 HORIZONTAL-OUTPUT TRANSFORMERS, YOKES, AND SPECIAL ASSEMBLIES

In addition to its electrical circuitry, the horizontal-output system contains some major mechanical assemblies with which the reader should be familiar.

Output Transformers A photograph of a group of typical horizontal-output transformers shown in Fig. 18-42. All of these devices have at least the following terminals (or wires):

(1) Output tube plate connection.
(2) HV rectifier plate connection.

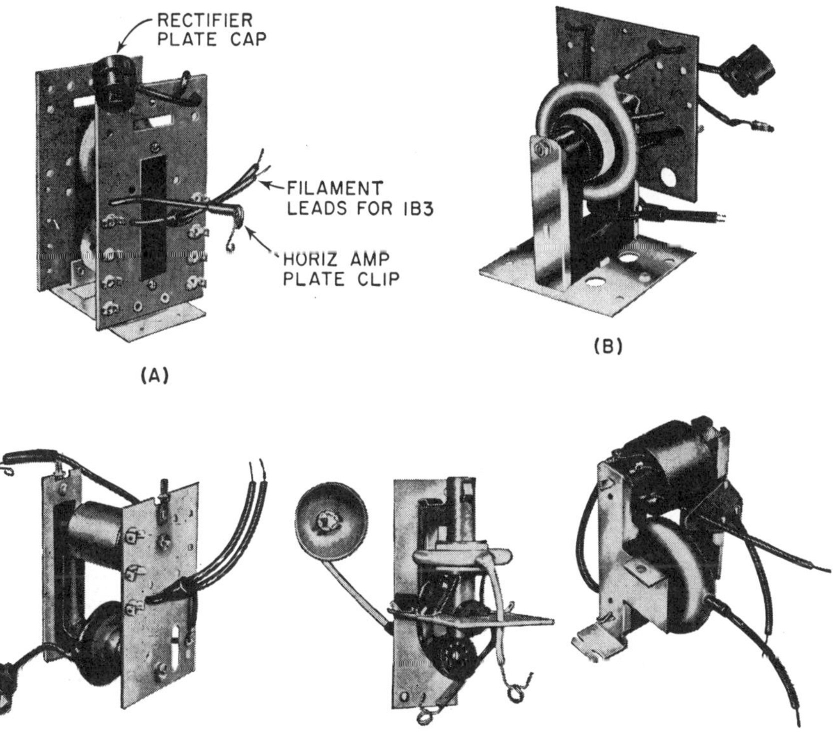

Fig. 18-42. (A) A typical horizontal-output transformer for the high-voltage flyback systems. (B) Another view of the same transformer. (C)-(E) Other horizontal-output transformers.

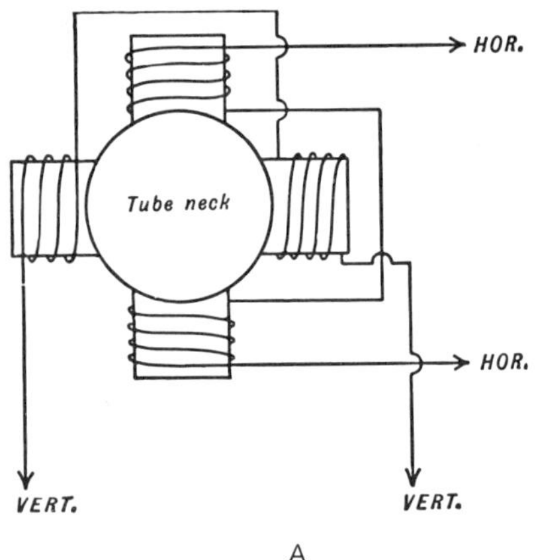

A

Fig. 18-43. The deflection yokes shown (A) schematically over the CRT neck and (B) pictorially.

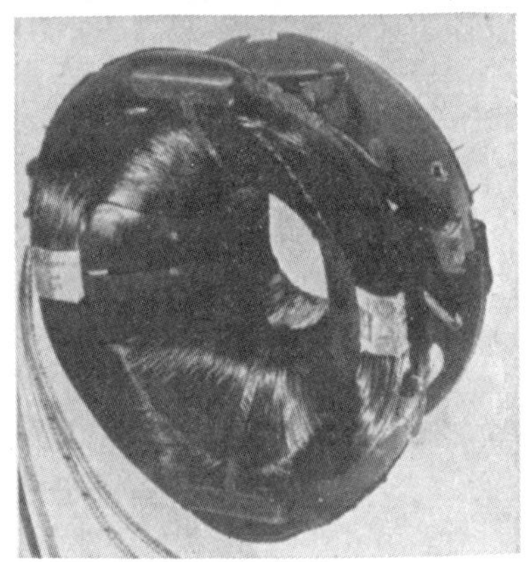

B

(3) Heater leads for HV rectifier.

(4) Terminals for yoke, B++, centering, linearity, width, etc.

Two distinct windings are visible. One winding, the primary carries the horizontal-output tube current. It is constructed of a comparatively large size wire, as the horizontal-output tube current is in the range of 150 to 250 ma. The other (secondary) winding is a "doughnut" type winding which serves only to generate the high voltage. A small wire size can be employed here since this winding generates from 12 to 30 KV pulses at less than 2 ma of current. A great number of turns are on this winding. A coating of special wax and/or a neoprene boot covers the winding to insulate and help the "doughnut" hold its shape.

Deflection Yokes The deflection yoke consists of two sets of coils positioned at right angles to each other and mounted on the section of the tube neck where the electron beam leaves the focusing electrode and travels toward the screen. Figure 18-43(A) shows schemetically how the horizontal- and vertical-deflection windings are placed over the CRT neck.

The entire assembly of deflection coils is known as a "deflection yoke." Two typical commercial units are shown in Fig. 18-43(B). Note how the forward windings lap over the front edge of the yoke housing. The yoke is thus positioned right up against the flare of the tube in order to be able to sweep the beam over the full screen area. This is particularly important for wide-angle tubes (110 degrees or more).

High-Voltage Assembly Figure 18-44 contains a closeup view of the high-voltage module introduced by Motorola. The solid-state, high-voltage rectifier and horizontal-output transformer are clearly visible. The high-voltage lead to the rectifier emerges from the top of the "doughnut" winding. A heavily insulated lead on the extreme right routes the high-voltage dc out to the CRT second anode terminal.

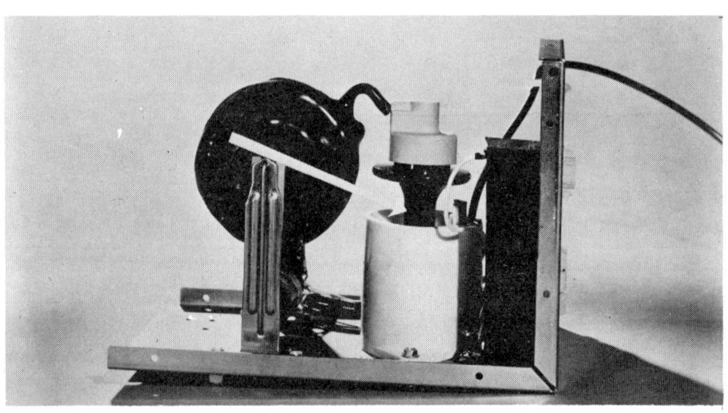

Fig. 18-44. Motorola high-voltage module. (*Courtesy of Motorola.*)

Solid-State HV Rectifier A photograph of a vacuum-tube HV rectifier and a new solid-state HV rectifier is shown in Fig. 18-45. Note the much simpler construction of the solid-state device. Since no filament power is needed, the flyback circuit can also operate more efficiently.

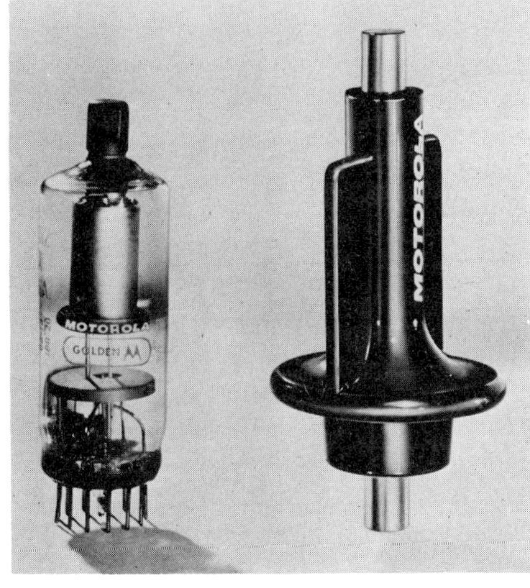

Fig. 18-45. Solid-state high-voltage rectifier compared to its vacuum-tube counterpart. (*Courtesy of Motorola.*)

18.11 TROUBLES IN HORIZONTAL-DEFLECTION CIRCUITS

Regardless of the type of horizontal-deflection system, most troubles originate in the amplifier or switching elements. Power tubes, output transistors, and *SCR* switches are handling great amounts of energy. These fail first, even if public relations departments have tried to convince us of the immortality of modern semiconductors. Tubes may be substituted and plate, screen, and grid voltages measured. Semiconductors should have element voltages measured before substitutions are made.

Failures of tubes and/or flyback transformers can sometimes be traced to incorrect adjustment of the horizontal linearity or efficiency coil. Refer to Section 18.9 for the correct adjustment procedures. Figure 18-46 shows the appearance on the CRT screen of a circuit defect (or adjustment) which causes poor horizontal linearity.

Replacing a horizontal oscillator tube while the set is operating will mean loss of drive to the output amplifier. In older units, without protection, this means possible destruction of the output tube. The circuit breaker may trip if the horizontal-efficiency coil is misadjusted.

In semiconductor systems, failure of the horizontal drive will cause the output transistor or *SCR* to turn safely off. This is just the opposite of the tube circuits. A bias check will spot this condition.

Remember that troubles on the left half of the screen are generally created by damper circuit difficulties. The right half of the screen is

Fig. 18-46. Horizontal-nonlinearity shown on picture tube.

controlled by the output amplifier or switch. Review the basic scanning principles regularly to keep your mind sharp. Stick to known factors, and eliminate the components systematically.

Width coils come in inductance values ranging from 0.05 mh up to 30 to 35 mH. Whatever the inductance, it must be matched to the transformer windings. Should any defect lead to a mismatch, such as a shorted turn in the transformer or in the width coil, it will be reflected in a picture which is either too narrow or distorted, or both. Even as little as one shorted turn can have a marked effect, particularly in low-inductance width coils.

18.12 HIGH-VOLTAGE TROUBLES

Probably the most troublesome section of the horizontal-deflection system, either in a black-and-white or color receiver, is the high-voltage section. Many of the defects that develop stem from a component that has failed in the high-voltage section itself. However, remember that a loss of high voltage may also be caused by a failure in some other section of the horizontal-deflection system. If the deflection wave is, for some reason, not reaching the output transformer, then no high voltage will develop.

The tube-type high-voltage rectifier is the cause of many high-voltage power supply failures. Low emission will cause picture "blooming." The filament circuit may incorporate a low ohm value resistor. This unit can open or change the value and also cause "blooming."

Corona discharge can cause a similar effect, as it bleeds off energy from the high-voltage circuit. Improper "lead dress" of the high-voltage lead may cause corona problems. Replace and "redress" any such lead.

In color receivers, the addition of a regulator circuit increases the possibility of part failures. Referring to Fig. 18-27, we can see how a shorted regulator tube will cause failure of the high voltage and perhaps the high-voltage rectifier and flyback transformer as well. Misadjustment of the high-voltage control can create problems from blooming (loss of regulation) to too high a value of high voltage.

The focus circuits in color-TV receivers are prone to fairly frequent failures wherever a selenium-focus rectifier has been used. The new replacement units of silicon are highly recommended. Focus coils handle high-pulse voltages and are subject to breakdown.

Whenever certain horizontal-circuit defects occur, the flyback may take excess currents drawn in the horizontal-output system and heat up the wax and/or neoprene insulation, and it may cause failure. Corona may create hot spots and melt the wax to again cause failure. Wax drippings in the bottom of the high-voltage cage mean that something is wrong. Be alert and perform the necessary adjustments to reduce excessive current flow in the flyback before the failure occurs.

Fig. 18-47. The appearance of the CRT screen if a portion of the horizontal yoke is shorted.

Figure 18-47 shows one possible way this problem may indicate itself on the picture-tube screen. In this case the original short circuit is in the yoke windings.

Horizontal-centering controls have been eliminated in many new sets, but in the older color sets this control carries yoke current. When the control is a carbon composition type, it is prone to pitting and failure. The usual defect shows up as loss of control or it works on "one end" only. In either instance, the picture on the kinescope will shift to the left or right.

Relationship of Horizontal and HV Problems Correct high-voltage generation is closely related to the proper operation of the rest of the horizontal system. A failure of the horizontal-output tube will stop all delivery of energy to the flyback transformer. No energy will be forwarded to the yoke, stored, or returned during the beam retrace. High voltage will not be generated.

Should the lower (primary) winding of the flyback transformer open, the B+ path to the output tube will become open. Again, there will be no high voltage. A short in the transformer will be most evident by the smoke and smell. A hot spot develops at the short and presents no difficulty to locate. High-voltage generation will be disabled.

The doughnut high-voltage winding may open, resulting in no high voltage. A short in the doughnut, if only one turn, can be found by its effect on the Q of the entire flyback transformer. Shorted turns absorb energy and reduce the Q of the flyback. This means a greatly reduced high voltage.

The output tube requires B++ in order to operate class C. A failure in the boost filter will decrease this voltage below normal. Insufficient energy will be delivered to the flyback and the yoke. The small amount of energy returned to the circuit during the retrace will generate an inadequate high voltage.

The damper tube charges the boost filter, and tube failure will cause either complete lack of high voltage or else produce a low high-voltage. A good maxim to follow is always to replace the damper tube.

The yoke connects across the flyback transformer and is matched to the system. An open yoke will mean no B++ voltage. Shorted turns in the yoke are indicated by reduced high voltage and horizontal distortion (see Fig. 18-47). Above all, check to see that the B+ from the TV receiver's power supply is satisfactory and low in ripple content.

Corona and Arcing Problems Corona and arcing are to be found in nearly all electrical high-voltage circuits. The corona discharge is like a very fine electrical spray into the air or to grounded surroundings. Areas with sharp bends or sharp points are most likely to form corona discharges.

The discharge has a very characteristic hissing or rushing sound. Placing the high-voltage section in complete darkness is an excellent way to see the faint blue glow around the discharge area. The discharge itself is not harmful, but the side effects can be disastrous to the flyback transformer. The corona discharge ionizes the air; once ionized, the air becomes conductive and a heavy arc can form which will burn through the insulation and wiring. At this point, a flame is nearly inevitable and the transformer is destroyed.

Arcing can occur if dust and small airborne debris are allowed to collect in the high-voltage cage or around the second anode of the picture tube. Cleanliness in these areas can prevent the arcing and its crackling sound, as it develops a path through the dust.

All wiring used in this area must be heavily insulated. Special wire with 25 to 30 KV insulation properties is employed to conduct the high voltage to the kinescope. The caps on the plate leads of the high-voltage rectifier and the horizontal-output tube are constructed to withstand such high voltages and prevent corona effects from occurring.

Special construction is used on the high-voltage rectifier tube socket to reduce corona effects. A corona ring is connected to the filament of the tube, as shown in Fig. 18-48. This ring causes a smooth, high-voltage, equipotential surface to form around any jagged surfaces on the tube socket. High-voltage points on the tube socket can cause arcing or corona only in relation to a nearby lower potential surface, and they cannot do so when surrounded by a smooth equal potential surface. The corona is thus minimized.

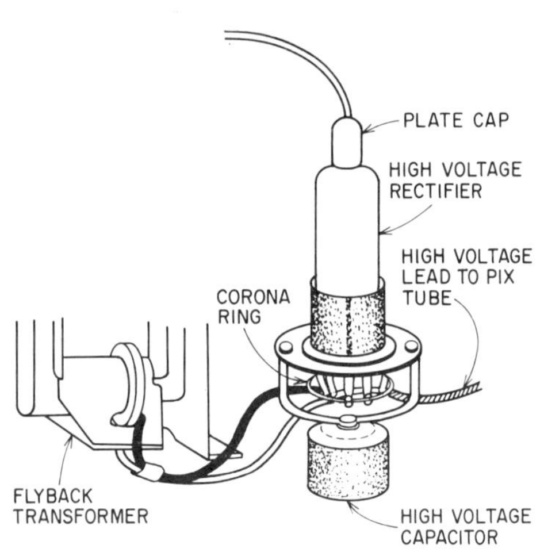

Fig. 18-48. A corona ring used to reduce corona discharge.

X-rays in Color-TV Receivers Much publicity has been given to reports that color-TV receivers radiate x-rays which can be harmful to the viewer. Before going into the detailed explanation of this phenomenon, a few facts should be understood. The type of x-rays produced by the high voltages used in color sets is called "soft" x-rays and cannot penetrate skin or tissue to any appreciable depth. The most likely damage might be to the cornea of the eye, but the level of radiation possible from color sets are so low that to be dangerous, it would be necessary to place one's eyes close to the source of the x-rays (the high-voltage rectifier and regulator tubes). While any TV x-ray exposure is not beneficial, the amount of radiation we may receive at the dentists or during a routine chest x-ray is generally far greater than we could get during years of close proximity to the high-voltage section of a color set.

X-ray radiation is cumulative in its effects, and any damage done to human tissue is irreparable. Even low levels are undesirable, and correctly operating TV sets do not emit any measurable amount of such radiation. It should be remembered that x-rays, like any radiation, decrease in intensity as the square root of the distance. A 16 milliroentgen level 4 inches from the source will be only 4 millirotentgen another four inches away. With normal viewing distances of several feet from the screen, the viewer is in no danger, even if the set is not operating correctly and emits some x-ray radiation.

X-rays are attenuated or reduced by any material placed in the x-ray field. Materials used as x-ray shields include metal, glass, ceramic, etc., all of which are used in television receivers. The degree of attenuation is determined by the type, density, and thickness of the material and the energy of the x-rays involved.

In color television receivers, x-rays may be emitted from three possible sources: (1) the high voltage regulator tube; (2) the high voltage rectifier tube; and (3) the picture tube.

Emission of x-rays from the regulator tube will vary with voltage and tube current. Maximum x-radiation from the regulator tube will occur when the picture tube is dark. The emission of x-rays from the rectifier tube occurs during the portion of the cycle when small reverse currents occur.

X-ray radiation from the picture tube depends upon beam current and voltage. Maximum emissions at rated tube voltage will occur when there is a bright picture on the screen. These emissions may be further increased if a condition exists where the picture does not completely fill the screen.

As noted previously, high voltage in excess of that specified creates even greater quantities of x-radiation. Conditions such as an excessive high-voltage setting, certain types of regulator tube failures or unusually high-line voltages can increase the potential of x-rays very significantly.

A color-television receiver presents no x-ray hazards if it is operated at the specified power-line voltage and has all the original factory installed shields and equipment in place, and is correctly adjusted. However, in servicing a color-television receiver there are certain precautions the electronic service technician should observe.

An x-ray is a form of radiation produced when a beam of electrons (typically in a vacuum tube) strikes some material at a relatively high velocity. Generally speaking, acceleration voltages above 10 to 15 KV are required before any significant quantity of x-rays is emitted. Due to the absorption of x-rays by glass envelopes, there is normally no significant escape of x-rays from the tubes until the voltages are in the range of 20 KV or higher.

When electron particles bombard material such as a metal target, some of the energy of these electrons is converted to x-rays. Unless shielded or otherwise absorbed, these x-rays are emitted from the target in all directions. The energy of an x-ray is proportional to the voltage which has accelerated the electron. The quantity of x-rays produced is very sensitive to the voltage, and a given increase in the voltage produces a much greater proportional increase in the x-rays. Therefore, it is important that the high voltage in color-TV receivers should not be set above the specified levels.

The roentgen (R) is the international unit used in measuring x-rays. One milliroentgen (mR) equals one thousandth (.001) roentgen. X-ray measurements in the area of a color-TV receiver would be measured in the milliroentgen range.

The Bureau of Radiological Health, the new arm of the United States Department of Health, Education, and Welfare, has set a schedule of standards for color-TV radiation. By June 1970, new color sets must have a means of limiting x-ray radiation to 0.5 mR/hr, even when badly misadjusted. By 1971, new designs must be such that radiation *cannot* exceed 0.5 mR/hr, no matter what type of trouble arises in the set.

REVIEW QUESTIONS

1. Draw a block diagram of a typical vacuum-tube, horizontal deflection circuit.
2. Describe the current flow during the period the horizontal-output amplifier is conducting. Explain the action of the electron beam during this time.
3. Describe the current flow during the retrace. Explain the movement of the electron beam during retrace.
4. Explain how the flyback oscillation is stopped after the retrace period.
5. How is the boosted B+ obtained?
6. Are high-voltage pulses developed in the flyback transformer during the trace or the retrace period?
7. List the requirements for a transistor horizontal-deflection circuit that differs from the tube version.
8. What happens to the horizontal-output transistor should the base drive fail?

9. Explain the reason for the automatic regulation of the high-voltage supply in a color-TV receiver.
10. Refer to Fig. 18-32 and give a detailed explanation of the circuit action if the high voltage should increase due to decreased load on the high-voltage line.
11. How are x-rays generated? Their unit of measurement is the ________

19 Vertical-Output Deflection Circuits

19.1 FUNCTIONS OF THE VERTICAL-OUTPUT DEFLECTION CIRCUITS

The vertical-output deflection circuits provide the current needed to drive the vertical windings of the CRT deflection yoke during the scanning process. These circuits amplify the output of the vertical-sweep oscillator to a level necessary to drive the coils in the deflection yoke, and also shape the waveform so that the current driving the yoke will produce a *linear* vertical sweep. The output voltage is sometimes fed back to the output amplifier driver circuits to control linearity of the sawtooth waveform.

In addition to driving the yoke, the vertical-output deflection circuits provide blanking pulses to "blank out" the CRT during vertical retrace after each scanned field. Some circuits also provide for linearity correction, and, in color-TV receivers, for convergence of the three scanning beams.

In the early days of television, vacuum tubes were the sole means for controlling the yoke-drive current. Later, the transistor came into widespread use in television, and today, the trend is toward all solid-state sets, which include transistors and integrated circuits. Transistors are particularly well suited for use as current amplifiers in the present-day magnetic deflection systems. There are, however, several different vertical-output deflection-circuit configurations in use. These configurations include all-vacuum tube, combination vacuum tube and transistor (hybrid), and all-transistor circuits.

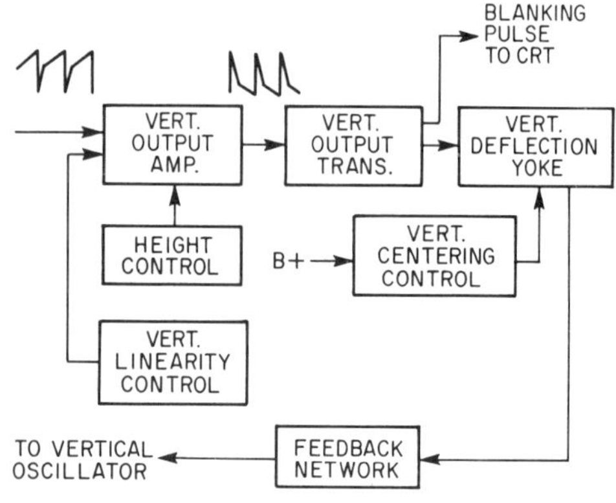

Fig. 19-1. A basic vertical-output circuit block diagram.

19.2 BLOCK-DIAGRAM OPERATION OF THE VERTICAL DEFLECTION CIRCUITS

Vacuum-Tube Circuits Figure 19-1 is a block diagram of a basic vertical-output circuit. This circuit uses a single vacuum tube to amplify the signals from the vertical-sweep oscillator and a transformer to couple the amplified signals to the magnetic-deflection yoke. Usually, the vertical-output tube is one section of a multisection vacuum tube, the other sections being used in the vertical-sweep circuits, as discussed previously.

As shown in the block diagram, the positive-going sawtooth waveform from the vertical-sweep oscillator is applied to the grid of the vertical-output tube. The operating point on the characteristic curve of this tube is established by a vertical-linearity control, and the output amplitude is set by a height control. These controls will be discussed later is this chapter. The output of the tube, an inverted replica of the input waveform, is impressed across the primary winding of the vertical-output transformer (VOT). The main function of this transformer is to match the low impedance of the yoke to the high impedance of the vertical-output tube. Also, since the vacuum tube is a high-voltage, low-current device, the VOT provides a means to step-down the voltage and thus step-up the current.

In some vertical-output circuits, a means for centering is provided. This portion of the circuit is simply a voltage divider that sends a dc centering current through the vertical-deflection windings of the yoke. Even though the centering current flows through the same windings of the yoke as does the vertical-driving sawtooth current, the two currents are completely separate. The centering current may flow in either direction, as may be required to move the raster up or down.

A feedback network, also shown in the block diagram, is employed is some vertical-output circuits. This circuit feeds a portion of the vertical-output signal back to the input of the output tube or to a prior stage. In this way, the rise of the sawtooth current is kept uniform, providing the linear magnetic-deflection field required in the yoke.

Vertical-blanking pulses are produced by a network in the output stage. This network generates a pulse that is applied to the CRT control grid or cathode, to cut off the scanning beam during vertical retrace.

Transistor Circuits Figure 19-2 is a block diagram of a typical transistor vertical-output circuit. The sawtooth waveform from the vertical-sweep oscillator is fed through a preamplifier, an amplifier, and a yoke-driver stage to provide the necessary power gain for driving the vertical windings of the deflection yoke. The amplifier stages also include networks for properly shaping the waveform. Linearity adjustments are made in the biasing networks of the preamplifier stage. The height control is shown in the input circuit of the preamplifier. This control merely establishes the amplitude of the input signal. Linearity is

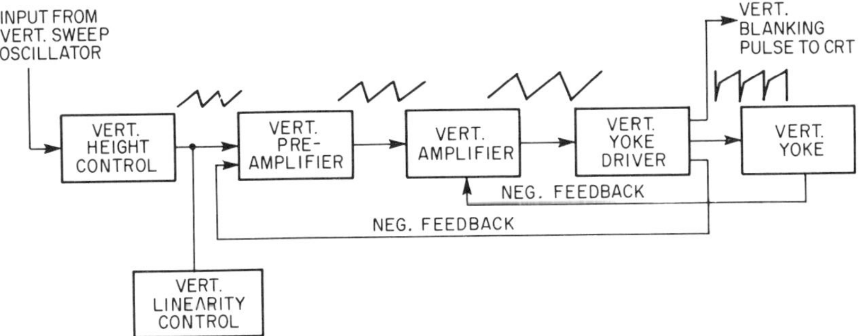

Fig. 19-2. A block diagram of a transistor vertical-output deflection circuit.

corrected by feedback from the yoke and yoke driver. The driver stage also provides vertical-retrace blanking pulses.

Vertical Color TV Circuits The block diagram of a vertical-output circuit suitable for a color-TV receiver is shown in Fig. 19-3. In this circuit, two main features, different from black-and-white circuits, are added: pincushion correction and dynamic convergence. The pincushion correction network is usually used only with some wide-deflection angle CRT's, but dynamic convergence is required in all color-TV receivers. The service switch shown in the diagram disables the output amplifier to facilitate gray scale adjustment.

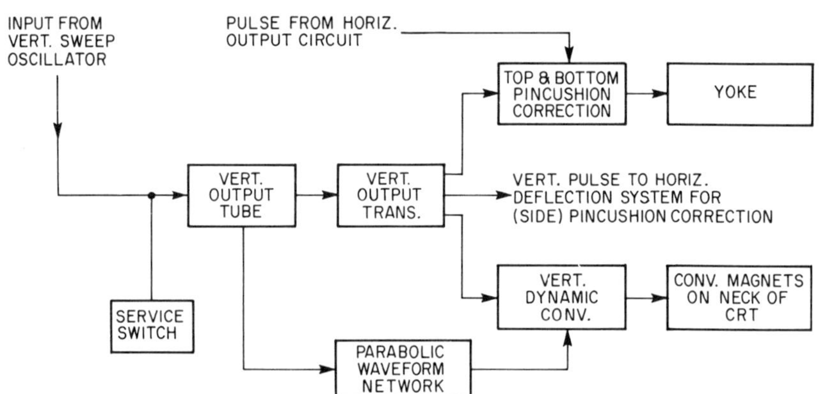

Fig. 19-3. A color vertical-output circuit block diagram.

Pincushion effects consist of the bowing of the horizontal lines in a vertical direction (see Fig. 19-6). To correct this condition, a parabolic-shaped voltage from the horizontal-deflection system is applied to the dynamic pincushion-correction circuit. Here, the parabolic-shaped voltage, amplitude modulates the vertical-output sawtooth that drives the vertical windings of the yoke. The vertical sawtooth, modulated by the parabolic waveform at the horizontal-scanning rate, increases vertical deflection at the center top and center bottom of the raster to compensate for the bow in the horizontal-scanning lines. Service controls are provided to permit precise adjustment of the correction waveform. The circuit (for top and bottom pincushion correction) is discussed in Section 19-4.

Provision is also made in some TV receivers for side pincushion correction, which is basically a function of the horizontal-deflection system. Vertical-sawtooth current is fed from the VOT into the side pincushion-correction circuit. Here, the current amplitude modulates the horizontal-scanning current at a vertical rate. This modulated current, when applied to the horizontal windings of the yoke, expands the horizontal width at the center of the raster, and does not expand the width at the top or the bottom. As in the top and bottom pincushion-correction circuit, controls are provided to adjust the phase and the amplitude of the side pincushion modulating current. In this way, the necessary amount of correction is attained.

The purpose of the dynamic convergence circuit is to cause the three electron gun beams to converge at the same spot on the face of the CRT. As shown in Fig. 19-3, a parabolic waveform is developed in a special network and is applied to the vertical-dynamic convergence circuit to achieve this. Convergence windings on the VOT (vertical-output transformer) provide sawtooth voltages to drive the circuits. The sawtooth from the VOT is modified by the parabolic waveform and is then applied to the convergence coils mounted on the yoke housing around the neck of the CRT. Controls are provided to adjust the amplitude and the phase of the sawtooth voltage, as well as the parabolic waveform. By properly setting these controls, the current flow through the convergence coils develops a magnetic field that converges the three beams at the edges of the raster. Dynamic convergence is discussed in greater detail in Section 19-4.

19.3 SPECIAL REQUIREMENTS FOR MONOCHROME TRANSISTOR VERTICAL-DEFLECTION CIRCUITS

Transistor vertical-output circuits follow generally the same configuration as their counterpart vacuum-tube circuits. The major difference between the two systems lies in component values, which are necessarilly different because the transistor is a low-voltage high-current device, whereas the vacuum tube is a high-voltage low-current device. Because of this high-current flow, the transistor circuits require some form of thermal stabilization. Also, transistor operating parameters are more critical than those of the vacuum tube. Consequently, additional variable controls are required to achieve proper operation of the output circuits, especially if transistors are replaced during maintenance.

To produce a linear vertical scan, the yoke current must rise at a uniform rate, with no more than a 10 percent deviation allowable. This requirement is accomplished by operating the output transistor as a class A* linear amplifier for the 60 Hz vertical-deflection sawtooth waveform. Phase shift in the amplifier is minimized by maintaining a frequency response that is flat down to 1 Hz. Wherever possible, direct coupling is employed, or else very large coupling and decoupling capacitors are used. Waveshaping and negative feedback are also used extensively throughout the vertical-deflection circuits. Another technique for obtaining a linear-sawtooth current is to employ a high-gain circuit with a rather limited low-frequency response and then using about 25 db of the output signal as negative feedback to linearize the waveform.

* Class A operation in this case means that neither the positive nor the negative peak of the output waveform is clipped.

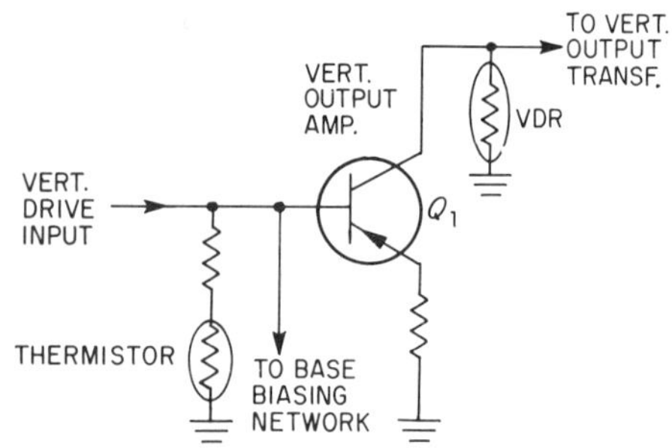

Fig. 19-4. A vertical-output transistor amplifier with thermal protection.

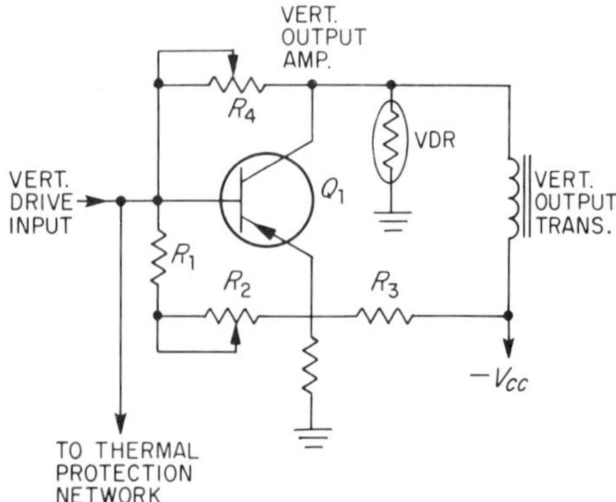

Fig. 19-5. A vertical-output transistor amplifier bias control.

Figure 19-4 is a simplified diagram of a vertical-output stage showing how thermal protection of the transistor is achieved. Since Q_1 is a power transistor, it operates with a substantial collector current. This high current causes appreciable heating of the collector junction. Even though heat sinks are used, the junction resistance tends to decrease, which further increases collector current flow and tends also to increase the base-emitter current. To prevent thermal runway of Q_1, the base-emitter bias current is stabilized by the thermistor shown in the base circuit of Q_1. The increasing collector current and temperature, causes the base-emitter current to increase. Thus, more current is drawn through the thermistor, causing its temperature to increase. Increasing the temperature of the thermistor decreases its resistance. The voltage divider action of the thermistor circuit then reduces the forward bias of Q_1. This, in turn, prevents the collector current from increasing and stabilizes the amplifier.

The thermistor alone, though, cannot provide complete stabilization of the forward bias. For this reason, a VDR is connected in parallel with the collector of Q_1. When the amplitude of the sawtooth waveform at the collector of Q_1 increases, the current flow through the VDR increases. The additional current flow through the VDR causes its resistance to decrease, thereby further increasing its current flow. This VDR action tends to stabilize the drive current applied to the yoke.

Figure 19-5 is a simplified diagram showing two manual controls in the vertical-output circuit. Unlike vacuum tubes, transistors used in vertical-output circuits require fairly critical biasing. To achieve this biasing, variable voltage dividers are included in the base circuit of the transistor. In the diagram, R_1, R_2, and R_3 are connected in series from the base of Q_1 to the $-$Vdc supply. Adjusting R_2 establishes the proper bias for the transistor. Variable resistor R_4 is connected from collector to base of Q_1 to provide the additional feedback and to make tolerances on replacement transistors less critical.

19.4 SPECIAL REQUIREMENTS FOR COLOR VERTICAL-OUTPUT CIRCUITS

Vertical deflection circuits for color-TV receivers are virtually identical to those for black-and-white receivers, except for pincushion correction and dynamic convergence. Also, since there are three electron beams in the color CRT, each beam must be set at a precise strength to produce a black-and-white picture. A service switch is provided to facilitate this setting. The circuits presented in this section operate equally well in vacuum-tube or solid-state TV receivers.

Top and Bottom Pincushion Correction Wide deflection-angle yokes used with rectangular CRT's produce a stretching effect at the four corners of the tube face. This stretching, which causes a bowing of the upper and lower horizontal scanning lines, is called "pincushion effect." Figure 19-6 shows the effects of pincushioning on a typical raster, reduced to twelve scanning lines.

To eliminate the pincushion distortion, and thus straighten the scanning lines, a maximum correction in a direction opposite to the pincushion bow must be applied at the top and bottom of the raster. As vertical-scanning progresses from the top of the screen, the pincushion effect decreases until, at the center of the screen, the horizontal lines are perfectly straight. Then slightly below the center, pincushioning again becomes apparent, but in the opposite direction, reaching a maximum at the bottom of the raster.

To correct, or bow upward, a pincushioned top-raster horizontal-scanning line, the portion of the vertical-deflection sawtooth corresponding to the period of time that the horizontal line is sweeping across the screen must curve upward in a parabolic form. Figure 19-7(A) shows

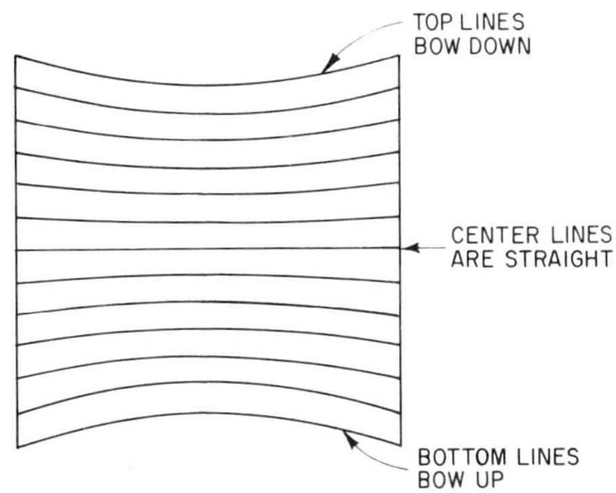

Fig. 19-6. Pincushion effects on a 12-line raster.

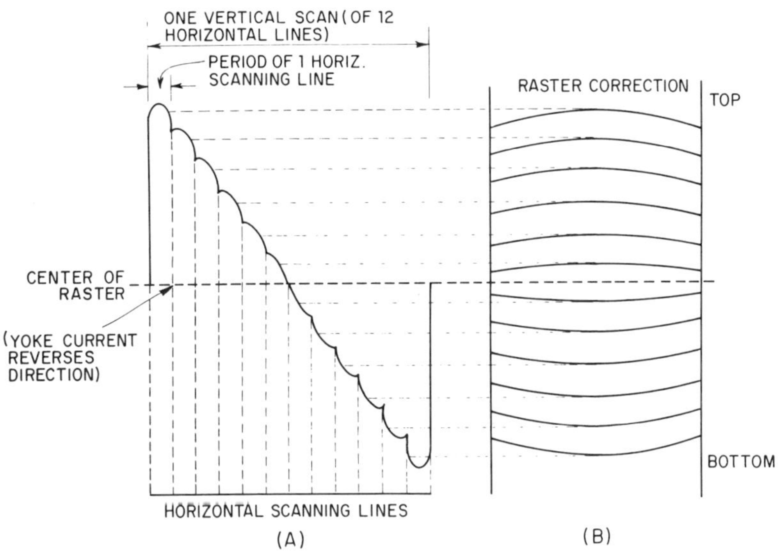

Fig. 19-7. A top and bottom pincushion correction.

that for each horizontal line, a parabolic curve is superimposed on the vertical sawtooth. These curves counteract the pincushion effect on the horizontal lines and produce a straight-line raster.

At the center of the raster, current flow in the vertical windings of the yoke passes through zero, as shown in Fig. 19-7. The horizontal-scanning lines at the center of the raster will therefore be perfectly straight, and correction will not be needed here. As scanning progresses in the lower half of the screen, the vertical sawtooth must provide a downward-curving correction for each horizontal line. During vertical retrace, this correction process repeats, but at a very rapid rate, so that the pincushion correction will begin at the proper place for the next vertical scan.

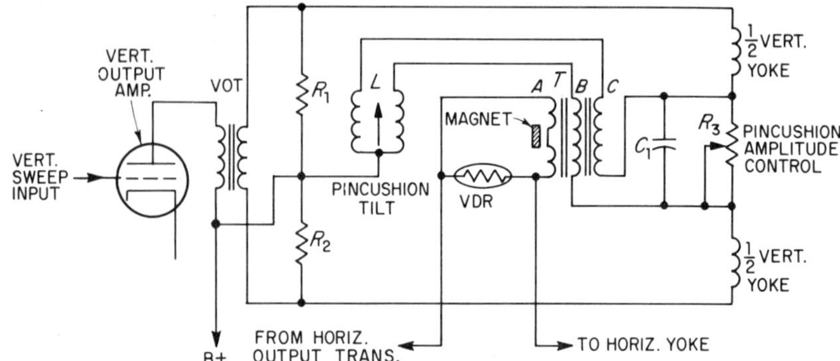

Fig. 19-8. A top and bottom pincushion correction circuit.

Developing the pincushion-correction waveform is a relatively simple process. Since each horizontal line is corrected by a parabolic curve superimposed on the vertical sawtooth, a parabolic waveform at the horizontal-scanning frequency is generated and used to modulate the vertical deflection-current waveform. Figure 19-8 shows the circuit arrangement for accomplishing this modulation.

In Fig. 19-8, notice that the only difference between this circuit and one without pincushion correction is the addition of L, the VDR, T, C_1, and R_3. Coil L and capacitor C_1 form a tuned circuit to shape the horizontal pulse into a parabolic waveform. Transformer T acts as a form of modulation transformer, and R_3 is the pincushion-correction amplitude control. When R_3 is set at its lowest amplitude position, the two vertical windings of the yoke are connected directly together and function as an ordinary yoke. Changing the setting of R_3 places a resistance between the two windings and causes a proportional current to flow through windings B and C of the transformer.

Winding A of transformer T, as shown in Fig. 19-8, is a two-coil balanced winding connected in series with the horizontal yoke. A small permanent magnet is mounted near the junction of the two coils. The VDR is connected across winding A to stabilize the total current flowing through the winding. Vertical-deflection current flowing through windings B and C, and the action of the magnet, unbalances winding A. When A is unbalanced, the horizontal pulse is induced into windings B and C to modulate the vertical sawtooth. Polarity of the induced pulse is determined by the direction of the unbalance across winding A.

The slug in coil L is used to change the phase of the parabolic-correction curve. Changing this phase shifts the curve from side-to-side across the CRT screen. Coil L and amplitude control R_3 can be adjusted to produce correction curves that exactly oppose the curved horizontal-scanning lines and thus cancel the pincushion effect.

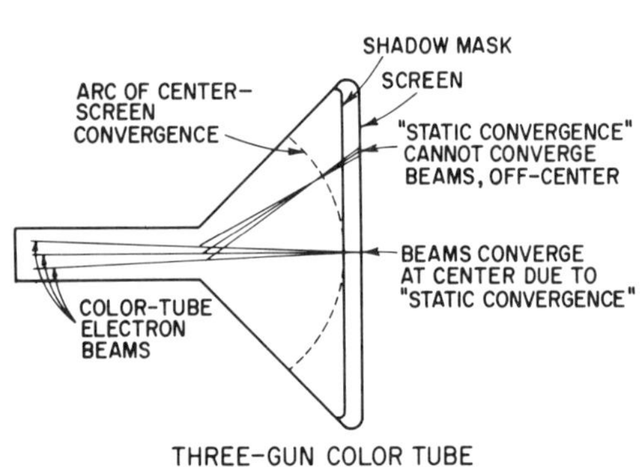

Fig. 19-9. Illustrating that "static convergence" is only effective at the center of the screen.

Dynamic Convergence In three-gun rectangular color CRT's using a shadow mask, converging the three beams at the center of the screen is not a particularly difficult task, as shown in Fig. 19-9. This center screen convergence is accomplished by permanent magnets mounted on the rear end of the deflection yoke housing and is called "static convergence." When the beams are deflected to the edge areas of the screen, though, a significant problem is encountered. Figure 19-9 shows that because the shadow mask and the CRT screen are not spherical, off center the convergence arc curves away from the mask and screen resulting in a loss of beam convergence. To correct this condition, "dynamic-convergence" correction currents are generated and applied to three convergence coils 120 degrees apart, located adjacent to the deflection yoke. Each of the convergence coils, one for each color, consists of two sets of windings, one set for vertical-convergence corrections and the other for horizontal corrections, wound on the same core. In this discussion, only the vertical-correction circuits will be explained.

Figure 19-10 shows what happens in the vertical center-edge areas when there is incorrect vertical dynamic-convergence correction. Note the lines shown are the vertical bar pattern of a crosshatch generator. Corrections to converge these three lines must be provided by the vertical dynamic-convergence circuits only. This condition is due to the fact that along the vertical centerline, all horizontal corrections are zero. Away from the vertical and horizontal centerlines, in the four quadrants of the CRT screen, dynamic convergence corrections require a combination of signals from both the vertical-and horizontal-correction circuits.

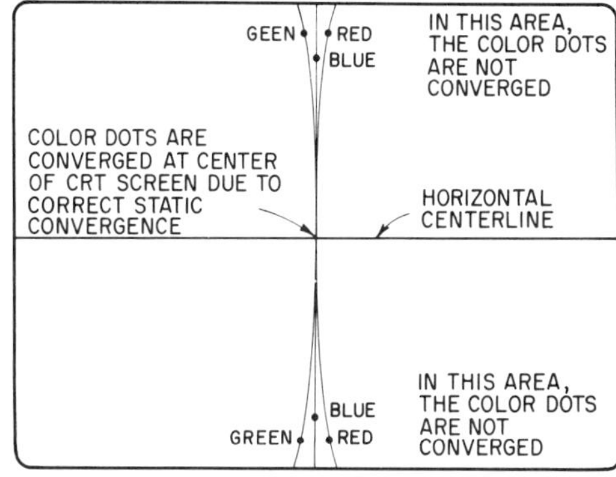

Fig. 19-10. Vertical lines from a vertical-bar generator are misconverged because of poor vertical dynamic convergence. When properly converged, the three lines will be superimposed.

The current necessary for dynamic-convergence correction is a parabolic waveform that can be set at any particular amplitude and can be tilted to either side. This waveform is derived from three separate signals taken from the vertical-output circuit. One of these signals, a parabolic wave, is coupled into the convergence circuits from the cathode of the vertical-output tube. The other two signals are 180-degree out-of-phase sawtooth waveforms developed across special windings no the VOT. Combining the parabolic waveform with a right- or left-hand sawtooth produces a parabolic waveform with a corresponding tilt, as shown in Fig. 19-11. The amount of tilt depends upon the amplitude of the sawtooth.

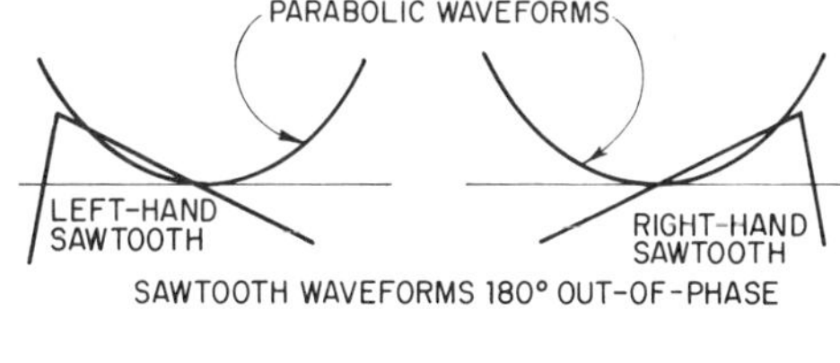

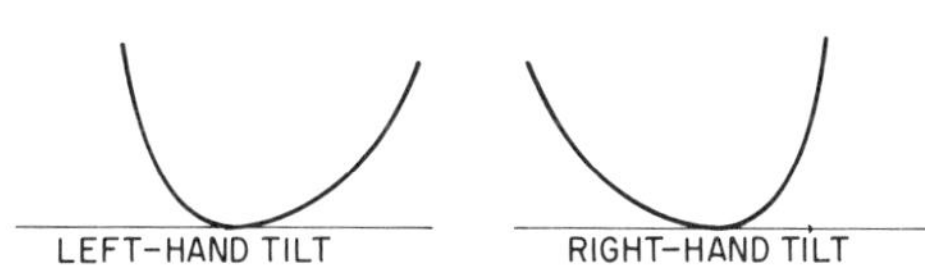

Fig. 19-11. Combining parabolic and sawtooth waveforms to provide tilt.

A typical vertical dynamic-convergence correction circuit is shown in Fig. 19-12. This circuit is divided into three functional sections: parabolic-shaping network, blue convergence, and red and green convergence. A simplified diagram of the parabolic-shaping network is presented in Fig. 19-13. Voltage from the cathode of the vertical-output tube is applied through C_1 to the integrator-phase-shifting network consisting of R_2 and C_2. Here, the sawtooth-shaped voltage is converted into a parabolic waveform. This voltage is then impressed across the voltage divider consisting of variable resistors R_1 and R_4. Pick-offs on the two resistors provide a means for varying the input voltages fed into the blue and the red and green convergence circuits. Resistor R_5 and diode D_1 clamp the convergence-correction reference at the

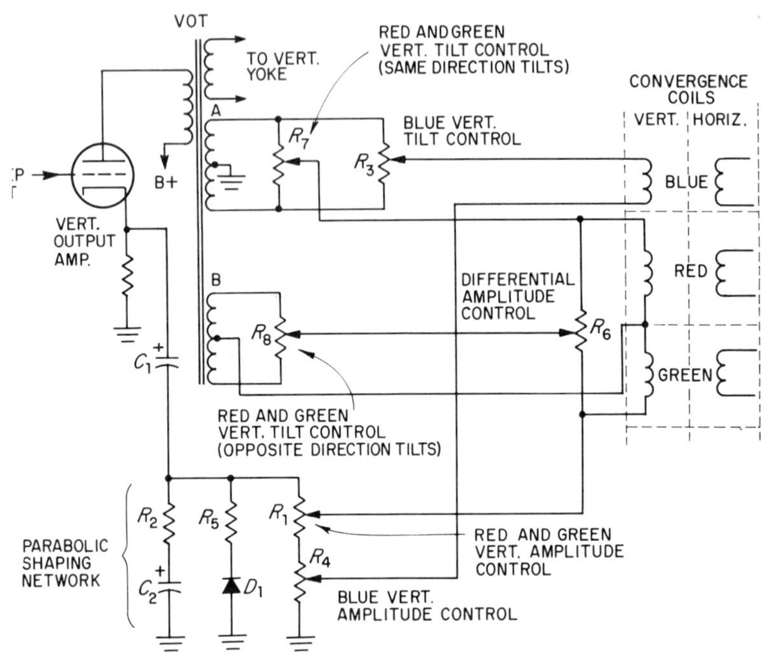

Fig. 19-12. A typical, vertical dynamic convergence correction circuit.

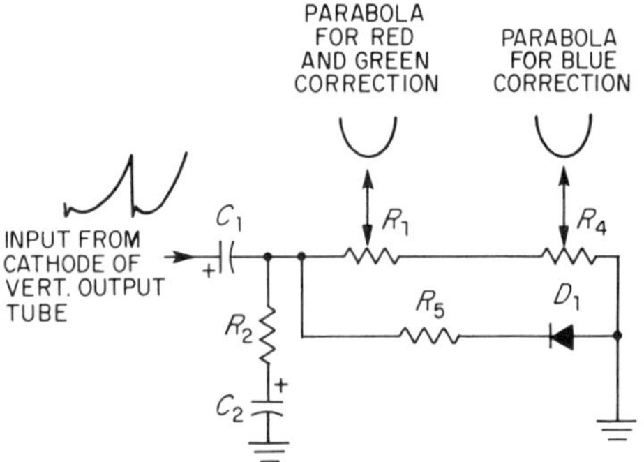

Fig. 19-13. Parabolic-shaping network (see also Fig. 19-12).

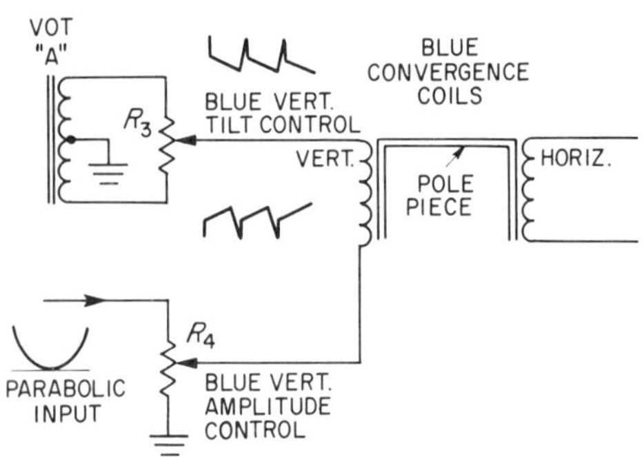

Fig. 19-14. Simplified, blue-convergence circuit.

center of the screen. This clamping is necessary to prevent any adjustment of the dynamic-convergence controls from changing the static convergence. Remember that static convergence is accomplished at the center of the CRT screen with three permanent magnets.

Sawtooth currents for the convergence coils are developed in two sets of center-tapped windings on the VOT, *A* and *B* in Fig. 19-12. In Fig. 19-14, R_3 is shown connected across winding *A* on the VOT. When the pick-off on R_3 is at the center of resistance, no sawtooth tilt-current will flow in the blue vertical coil. In this condition, only a symmetrical parabolic current will flow in the coil. The amplitude of the current is determined by the setting of R_4. This current in the blue convergence coil will cause an equal correction on both sides of the raster; the blue-beam line will remain in the center of the screen. Any magnetic-field developed by the blue horizontal coil, wound on the same pole-piece as the vertical coil, will combine with the vertical field and will move the beam accordingly.

Moving the pick-off on R_3 to either side will cause a sawtooth current to also flow through the coil. Direction of this sawtooth, right-hand or left-hand, is determined by the direction of movement from center of R_3. The sawtooth, when combined with the parabolic waveform from R_4, will tilt the correction-current waveform to the right or the left, depending on the direction of the sawtooth from R_3.

Figure 19-15 is a simplified diagram of the red and green vertical-convergence circuits. Here, R_1 provides the parabolic waveform for the red and green coils. Since current flows in the same direction through both coils, differential control R_6 is used to vary, inversely, the amplitude of the current in the two coils. In other words, changing the setting of R_6 increases the current amplitude in one coil and, at the same time, decreases the amplitude an equal amount in the other coil.

The sawtooth tilt-current for the two coils is provided by R_8, connected across winding B on the VOT. This control, similar to the action of the blue tilt control previously discussed, tilts the red and green waveforms to an equal degree, but in opposite directions; *i.e.*, if the red-coil waveform is tilted right, the green-coil waveform is tilted left. Resistor R_7, connected across winding A on the VOT, provides a sawtooth current that tilts both the red and green waveforms equal amounts and in the same direction, as determined by the direction of the setting of R_7. Again, any magnetic field developed by the horizontal-convergence circuits will combine with the vertical-magnetic field in the convergence coils. The action of both magnetic fields thus provide the means for converging the three beams at all points on the CRT screen.

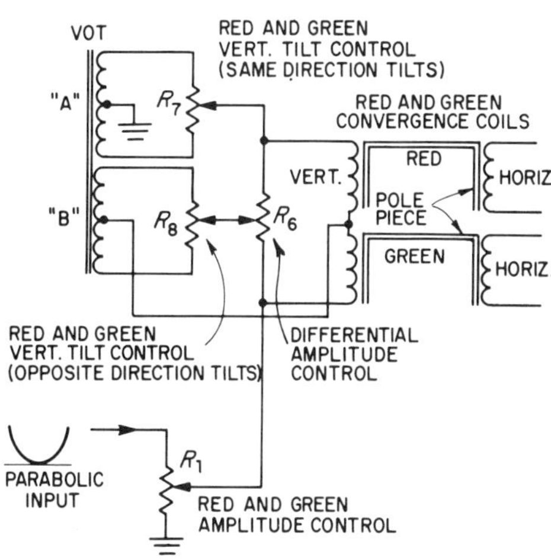

Fig. 19-15. Simplified red-and-green vertical-convergence circuit.

19.5 TYPICAL MONOCHROME VACUUM-TUBE VERTICAL-OUTPUT CIRCUIT OPERATION

Triode Amplifier Figure 19-16 shows a vertical-output circuit using a single vacuum tube, or one section of a multisection tube. This circuit is usually driven by a blocking oscillator; however, it can also be driven

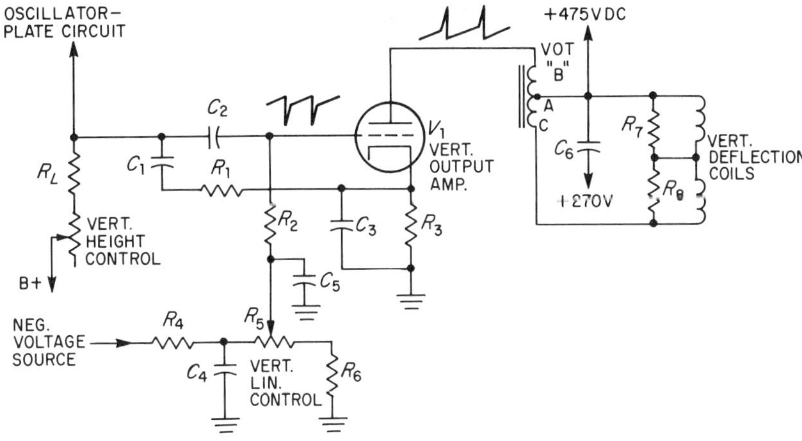

Fig. 19-16. Triode amplifier, vertical-output circuit for a monochrome-TV receiver.

by any pulse-type oscillator. Several TV-receiver manufacturers use this circuit in conjunction with a transistor-oscillator stage. The vacuum tube shown in the figure is a triode, which possesses a relatively low internal plate resistance. For this reason, the inductance of the output transformer dominates the circuit. Consequently, a peaked waveform is needed at the plate of the tube for linear deflection.

Vacuum tube V_1 is a high-current power triode connected in a conventional amplifier circuit. The operating point of the tube is determined by negative grid bias in conjunction with cathode bias developed across R_3. Capacitor C_3 by-passes most of the sawtooth waveform at the V_1 cathode to minimize degenerative losses. To provide the grid bias, voltage divider R_4, R_5, and R_6 is connected across a negative-voltage source, usually the horizontal-output tube grid circuit. Resistor R_5, vertical-linearity control, is a variable resistor permitting the adjustment of the V_1 grid bias and thus the establishment of the operating point of the tube on the most linear portion of its characteristic curve. Capacitors C_4 and C_5 are decoupling capacitors preventing the interaction between the V_1 grid circuit and the source of negative voltage.

Capacitor C_2 isolates the grid of V_1 from the dc plate supply of the oscillator. The height control and R_L are actually in the oscillator circuit but are shown in the figure to simplify the discussion of the C_1-R_1 action. The sawtooth voltage is developed across C_1 and is peaked by the resistor R_1, which is returned to the cathode of V_1 instead of ground. This cathode connection eliminates a small ac ripple that is produced by cathode by-pass capacitor C_3. The plate circuit of V_1 includes the vertical-output transformer and the vertical windings of the deflection yoke.

In the circuit of Fig. 19-16, the oscillator tube is cut off during the vertical-scan interval. In this condition, C_1 is connected to B + through R_L and the height control. Since the resistance of R_1 is very small in comparison with the resistance of RL and the height control, C_1 charges slowly, developing a sawtooth voltage at the grid of V_1. This voltage causes V_1 to conduct. At the end of the scan interval, the oscillator conducts heavily, effectively grounding the RL side of C_1. C_1 then discharges rapidly through R_1 to develop the negative spike shown in the waveform at the grid of V_1. This negative spike cuts off V_1. The oscillator conduction, or retrace interval, is too short for C_1 to discharge completely; therefore, the next cycle starts with some voltage still appearing across C_1, producing the peaked waveform shown in the figure.

By varying the V_1 grid bias with the linearity control, R_5, the operating point of the tube can be moved from one point on its characteristic curve to another where the curvature is less. In this way, it is possible to use the nonlinearity of the characteristic curve of V_1 to counteract any nonlinearity that may develop in the sawtooth section of the deflection wave. This can be done successfully, because the curvature of the tube characteristic is in a direction opposite the curvature which develops in the deflection wave. If such correction is not made, sections of the image will crowd together. It must be remembered though that overcompensation will lead to the opposite distortion in which sections of the image are stretched out.

The peaked sawtooth waveform applied to the grid of V_1 is amplified and applied to the vertical-output autotransformer. The vertical-deflection coils are connected across part of the transformer and carry the sawtooth current necessary to deflect the electron beam in the CRT. B+ is brought to the plate of V_1 from point *A*, a tap on the output-transformer winding. Insofar as V_1 is concerned, its plate load consists of the windings between points *A* and *B*. The bottom half of the transformer, between points *A* and *C*, is the output section of this unit, and it is across points *A* and *C* that the vertical windings of the deflection yoke are connected.

Point *A* on the output transformer is at ac ground potential because of the presence of electrolytic capacitor C_6. One end of C_6 connects to point *A*, while the other end connects to a +270-volt point in the power supply. This connection is used, rather than have C_6 return to ground directly, because the positive potential at point *A* is close to 475 volts. If C_6 were returned directly to ground, it would have to possess a voltage rating in excess of 475 volts, and an electrolytic capacitor with a voltage rating that high is fairly expensive. By returning C_6 to the 270-volt point, the capacitor must have a voltage rating of 475 minus 270 volts, or somewhat over 200 volts. Such electrolytic capacitors are considerably cheaper. Insofar as the grounding action of C_6 on point *A* is concerned, the same results are obtained as if C_6 had been tied directly to ground.

Beam-power Pentode Amplifier A vertical-deflection output amplifier, using a beam-power pentode instead of a triode, but basically the same circuit arrangement as that of Fig. 19-16, is shown in Fig. 19-17. Two interesting variations appear in this circuit. First, height control is achieved by resistor R_1, in which the center arm picks off whatever voltage is required by V_2 to attain the necessary picture height. This arrangement closely parallels that of volume control in a conventional radio receiver. With the arrangement of Fig. 19-17, the deflection voltage developed across capacitor C_1 remains constant. This voltage, which here

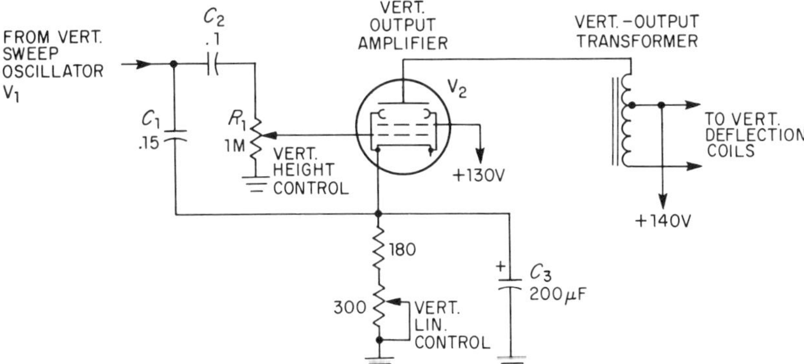

Fig. 19-17. Beam-power pentode, vertical-output circuit for a monochrome-TV receiver.

is practically sawtooth in form, is then passed on to R_1, which governs the peak-to-peak deflection signal fed into the vertical-output amplifier.

A fairly linear sawtooth-wave deflection voltage is developed across C_1. A peaking resistor (such as R_1 in Fig. 19-16) is not used in the pentode-amplifier circuit shown in Fig. 19-17. This arrangement is possible because V_2 has a high-internal resistance which completely dominates the output circuit. For a circuit that is essentially resistive, but contains a small amount of series inductance, a sawtooth voltage will produce a sawtooth-current flow. As indicated previously, the deflection voltage applied to the output circuit will vary from a pure sawtooth to one that is peaked, depending upon the proportion of resistance and inductance in the output circuit.

19.6 COMBINED VERTICAL MULTIVIBRATOR OUTPUT CIRCUIT

Many of the vertical-deflection systems found in TV receivers employ a multivibrator circuit in which the two transistors (or tubes) serve as both the vertical oscillator and the vertical-output stage. The theory of such circuits and typical examples of both transistor and tube systems were presented in detail in Chapter 17. It is advised that the reader refer to the discussions of these systems given in Chapter 17, before proceeding further.

An additional circuit, of this type, utilizing transistors is shown in Fig. 19-18.

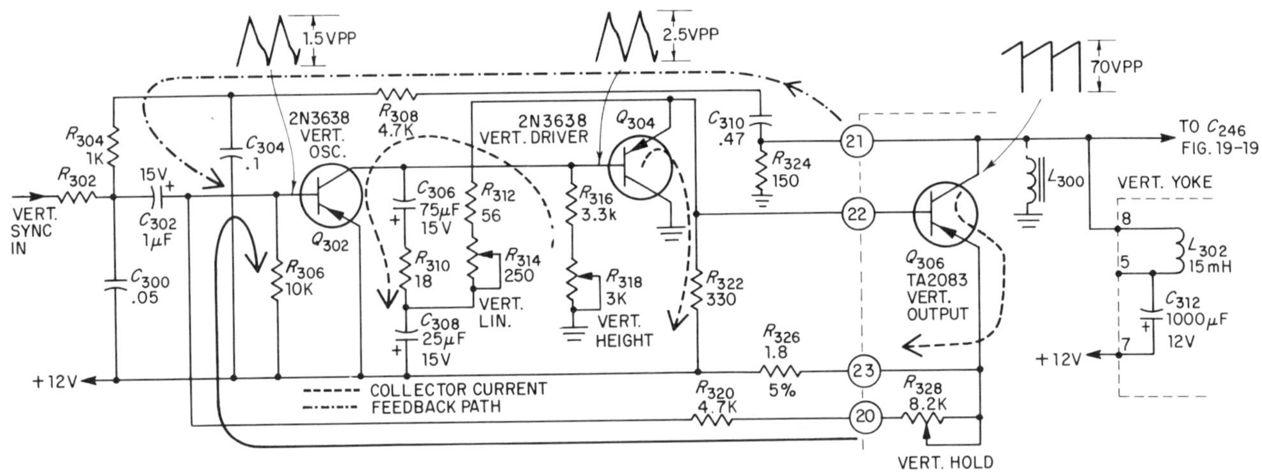

Fig. 19-18. A transistor vertical-deflection system (monochrome receiver). (*Courtesy of Sylvania Electric Products, Inc.*)

A Typical Monochrome Transistor Vertical-output Circuit A Sylvania all-transistor black-and-white chassis which uses the vertical-deflection system is shown in Fig. 19-18. In this three-stage system, Q_{302} and Q_{306} make up a multivibrator. Q_{304} is an emitter follower. Collector current flow and the feedback path of the PNP transistor circuits are also shown in Fig. 19-18.

In operation, vertical-sync pulses are integrated by R_{302} and C_{300} and are then applied to the base of Q_{302} through C_{302}. At the end of each vertical-scanning period, these positive-sync pulses trigger Q_{302} into conduction to recharge the sawtooth-forming capacitors, C_{306} and C_{308}. A portion of the Q_{306} emitter voltage is applied through R_{328}, the vertical-hold control, to the base of Q_{302}. This voltage establishes the average base bias of Q_{302} and thus determines the duration of each cycle. The positive feedback to sustain oscillations is applied to the base of Q_{302} through a shaping network consisting of R_{324}, C_{310}, R_{308}, C_{304}, and R_{304}. C_{304} also decouples horizontal-crosstalk frequencies from the feedback circuit.

During the vertical-scanning period, Q_{302} is cut off and the sawtooth-forming capacitors discharge through R_{316} and R_{318}, the vertical-height control. This discharge current develops a positive sawtooth at the base of Q_{304}, which drives the output transistor Q_{306}. The height control determines the base bias of Q_{304} and thereby establishes the amplitude of the output-transistor driving voltage. The linearity of the circuit is achieved by feedback from the collector of Q_{304} applied through R_{312} and R_{314}, the linearity control, to the junction between C_{306} and C_{308}. This voltage modifies the discharge slope of the sawtooth-forming capacitors to provide the uniformly rising current needed to drive the yoke.

The vertical windings of the yoke, L_{302}, are coupled to the collector of the output transistor, Q_{306}, across the choke L_{300}. To completely eliminate the dc component of the deflection current, the yoke is returned to +12 Vdc through the capacitor C_{312}. During retrace, Q_{306} is cut off, and the field across L_{300} collapses very rapidly, allowing C_{312} to discharge. This action returns the scanning beam to its starting position at the top of the CRT screen to begin the next field.

19.7 VERTICAL BLANKING

At the end of each scanning field, the vertical-deflection system returns the electron beam in the CRT to the top of the screen. During this return interval, the much faster horizontal-deflection system sweeps several lines across the face of the CRT. To prevent these lines from appearing on the screen, the vertical-output circuits produce blanking pulses. These pulses are then applied to either the cathode or to the control grid of the CRT to turn off the electron beam until it is at the top of the screen and in the proper position to begin the next field. The polarity of the blanking pulses is dependent upon the CRT element to which the pulses are applied: cathode blanking requires positive pulses, whereas control-grid blanking requires negative pulses.

Negative-Pulse Blanking During the vertical-retrace interval, the inductive kickback from the yoke inductance generates a peaking

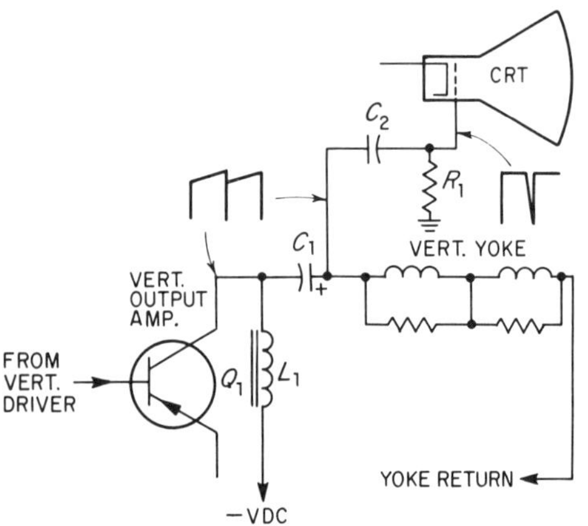

Fig. 19-19. Simplified circuit for negative-pulse blanking.

spike on the yoke-driving sawtooth waveform. This spike extends in either a positive or negative direction, depending upon the particular vertical output-circuit configuration. Figure 19-19 shows a simplified blanking circuit in which negative-peaked sawtooth waveforms are tapped from the vertical-yoke circuit. Even though these waveforms contain a sharp negative spike, they cannot be applied directly to the CRT control grid because they also contain components of the sawtooth waveform. These sawtooth components, if not eliminated from the blanking pulses, would cause raster shading. To provide properly shaped blanking pulses, the peaked-sawtooth waveforms are differentiated in the network comprising C_2 and R_1, as shown in the diagram. The output of the differentiating network, which is basically a high-pass filter, consists of narrow pulses built up from the higher frequencies of the sharp-peaked sawtooth waveforms. The lower frequencies that make up the ramp portion of the sawtooth waveform are attenuated and removed. In this way, only the narrow negative pulses are applied to the control grid of the CRT. The diagram pictured in Fig. 19-19 is typical of a large number of solid-state TV receivers.

19.8 YOKE-ANALOG, VERTICAL-OUTPUT CIRCUIT*

The yoke-analog circuit is a vertical-deflection system in which a voltage is developed that is a direct replica, or analog, of the voltage across the inductive component of the yoke. This analog voltage is compared to a dc reference, and then an error voltage that is proportional to any difference between the analog and the reference is developed. After amplification, the error voltage modifies the yoke current sufficiently to reduce the error voltage to zero. Figure 19-20 is a simplified diagram of the circuit.

The chief advantage of the yoke-analog, vertical-deflection circuit is that scanning linearity is dependent on only the time constants of the yoke and the analog network. Thus, there is no need for a linearity

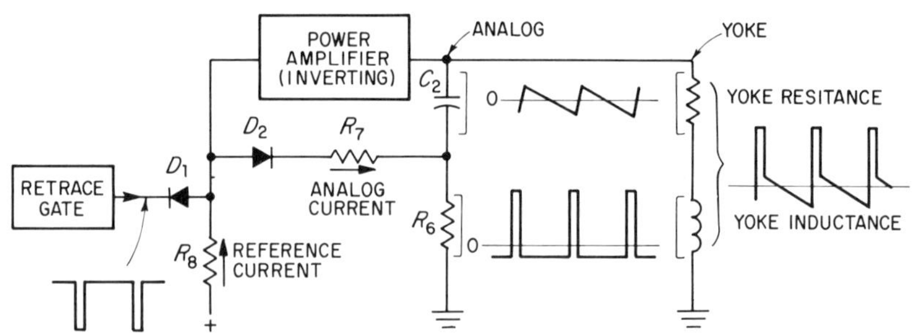

Fig. 19-20. Simplified diagram of the yoke analog circuit. (*Courtesy of The Institute of Electrical and Electronic Engineers, Inc.*)

* This circuit was presented in a paper by D. W. Taylor, Motorola Semiconductor Products, Inc., and published in the Institute of Electrical and Electronic Engineers, Inc., (IEEE) *Transactions on Broadcast and Television Receivers,* Volume BTR-11, Number 2, July 1965. The diagrams in Section 19.8 are reprinted with permission of IEEE.

control. Also, temperature shifts of the resistive components of the yoke are automatically compensated for by using a negative temperature coefficient thermistor for all or part of the analog-network resistance. Additional advantages are a simple vertical oscillator that is relatively free from horizontal crosstalk fed back from the yoke, and that the output transistors and other components need not have highly linear characteristics.

The vertical oscillator used in this circuit performs the single function of gating in and out the vertical-retrace interval. During the scanning period, the oscillator is completely isolated from the circuit by diode D_1. At the end of the vertical-scanning period, a sync pulse turns on the oscillator to develop the gate pulse shown in Fig. 19-20. This pulse forward biases D_1 and turns off the power amplifier that drives the yoke. With the power amplifier turned off, the yoke retraces very rapidly and develops a high-positive analog voltage across R_6. Diode D_2 prevents this high-voltage retrace pulse from turning on the amplifier before the end of the retrace interval.

As shown in Fig.19-20, the output of the power amplifier is fed to the parallel-connected analog circuit and yoke. The error voltage, which is developed by comparing the dc reference voltage across R_8 to the analgo voltage sensed by R_7, is applied to the input of the power amplifier. The error signal increases the yoke-current flow, which in turn increases the analog voltage across R_6 sufficiently so that there is no potential difference between the reference voltage across R_8 and the analog voltage across R_7. In this state, the scanning current driven through the yoke rises at a uniform rate to produce a truly linear scan.

Figure 19-21 is a schematic diagram of the complete yoke-analog vertical-deflection circuit. Q_1 is a unijunction relaxation oscillator. Q_2 is the error amplifier, and Q_3 the power amplifier. The only controls

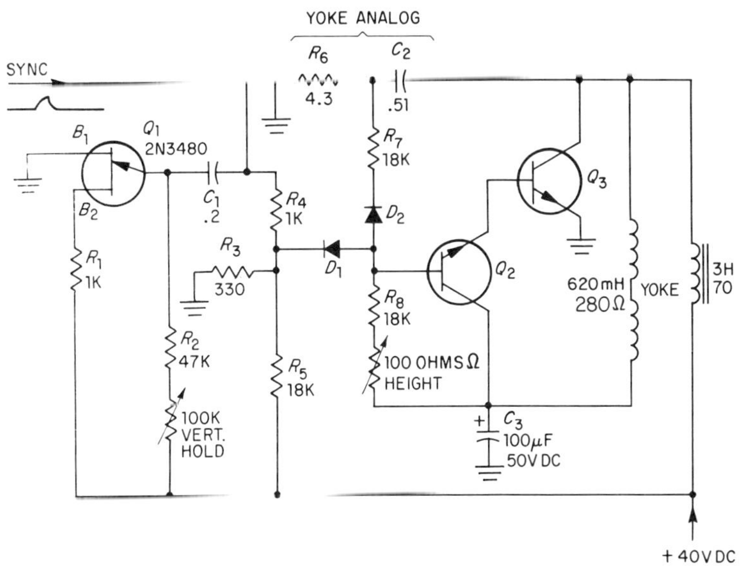

Fig. 19-21. Schematic diagram of the yoke analog circuit. (*Courtesy of The Institute of Electrical and Electronic Engineers, Inc.*)

in the circuit are the hold control, which establishes the oscillator-gating frequency, and the size control, which merely changes the dc reference level to increase or decrease the size of the picture. Operating power is applied to the circuit through a 3 henry choke.

The discharge time of the relaxation oscillator is established by the time constant of C_1, R_3, and R_4 to produce a 500 microsecond gating pulse. During the scanning period, a ramp voltage is developed across C_1 as the capacitor charges through R_5. This ramp voltage is fed through D_1 to the base of Q_3 via Q_2.

Even though the yoke-analog system offers a number of advantages over conventional systems, it has several disadvantages. Foremost among its disadvantages is its high production cost. An additional transistor-amplifier stage is required, along with considerable negative feedback to provide stability. Also, the oscillator must be a separate unit and cannot perform as a combined oscillator and driver stage, as in the multivibrator circuits.

19.9 TRANSISTOR COLOR-SET VERTICAL-OUTPUT CIRCUIT

Except for the convergence and pincushion correction circuits, vertical-deflection systems for color television are similar to those used in black-and-white sets. Figure 19-22 is a schematic diagram of the vertical-deflection circuit in the Sylvania all-transistor color chassis, E_{01}. In this circuit, the programable unijunction transistor Q_{307} is the vertical-sweep oscillator, and the driver is a Darlington-configuration amplifier consisting of Q_{308} and Q_{310}. Q_{312}, the vertical output stage, drives

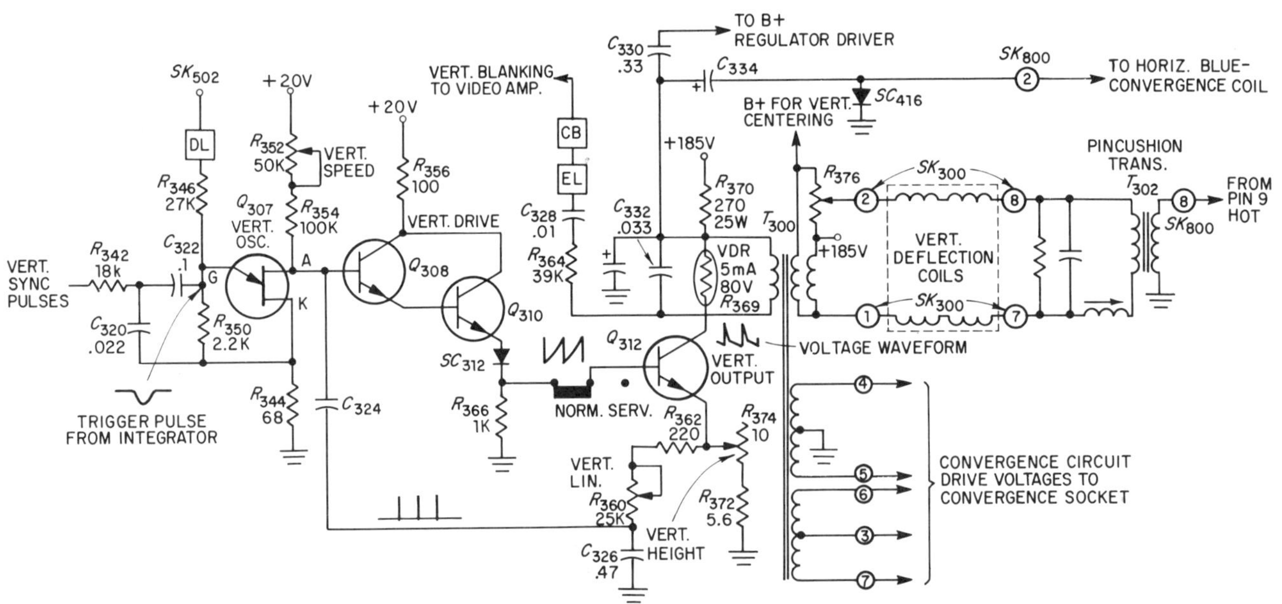

Fig. 19-22. A Sylvania color-TV chassis vertical-deflection system. (*Courtesy of Sylvania Electric Products, Inc.*)

transformer T_{300}, which couples the deflection voltage to the vertical windings of the yoke. The transformer also provides the drive voltages for dynamic convergence and pincushion correction. Aside from T_{300}, the service switch is the only component not found in the black-and-white chassis.

During the vertical-scanning period, C_{324} and C_{326} charge through the vertical-speed control R_{352} and R_{354} to provide the vertical-sawtooth sweep voltage. At the peak of the sawtooth, negative trigger pulses from the integrator network, R_{342} and C_{320}, drive Q_{307} into conduction. In this condition, the sawtooth capacitors C_{324} and C_{326} discharge through the vertical oscillator and initiate the retrace interval. The charging-time constant of the sawtooth capacitors is established by the vertical-speed control R_{352}. Adjusting this control raises or lowers the gate voltage, hence the firing voltage of Q_{307}. Adding more resistance into the circuit lowers the firing point, increasing the frequency of the oscillator, and deceasing the resistance raises the firing point to slow the oscillator.

The driver stage consists of Q_{308} and Q_{310} connected in a Darlington configuration. In this configuration, the collectors of the two transistors are common and the emitter of Q_{308} is connected directly to the base of Q_{310}. This cascaded amplifier operates as an emitter follower and develops the output-sawtooth waveform across R_{366}. SC_{312} is a vertical-drive gate that is forward biased during the vertical-scanning period to couple the positive-going sawtooth waveform through the service switch to the output transistor Q_{312}. The feedback voltage from the emitter of Q_{312} is applied through R_{362} and the vertical linearity control R_{360} to the junction between C_{324} and C_{326}. By combining this shaped-feedback voltage with the sawtooth voltage developed across C_{324}, the linearity of the sawtooth fed to Q_{308} is controlled and likewise the current waveform from Q_{312} to the vertical-output transformer T_{300}. Placing the service switch in the service position merely opens the base circuit of Q_{312} to cut off the vertical-deflection drive voltage.

Vertical-output transistor Q_{312} drives transformer T_{300}, which in turn drives the vertical windings of the deflection yoke and pincushion-correction transformer T_{302}. Two center-tapped secondaries on T_{300} supply the sawtooth voltages for the dynamic-convergence circuits. The height control, R_{374}, adjusts the emitter voltage and establishes the base-to-emitter bias of Q_{312}. Adjusting R_{374} for less resistance in the emitter circuit increases the current gain. This increased gain then increases the vertical-deflection current and the magnetic field in the yoke windings to expand the vertical deflection.

Raster centering is accomplished by applying a separate, regulated dc voltage to one end of centering control R_{376}. Adjusting this control allows the raster to be moved up or down on the CRT screen.

During the retrace interval, vertical blanking is achieved by applying peaked pulses, developed across the T_{300} primary, through R_{364} and C_{328} to the video amplifier. These pulses cut off the amplifier and blank out the raster until the beginning of the next field.

Also shown in Fig. 19-22 is part of the horizontal blue convergence circuit. Vertical-output voltage is applied through C_{334} to the blue lateral coil in the horizontal convergence coils. SC_{416} provides clamping for the circuit. This portion of the circuit is not a part of the vertical-deflection system but is shown here because SC_{416} and C_{334} are mounted on the vertical-deflection panel in the E_{01} chassis.

19.10 VERTICAL OUTPUT TRANSFORMERS

If the vertical-output circuits of all the present-day TV receivers were examined, it would be found that a large number of output-transformer configurations are in general use. These many configurations, however, perform the same basic function; *i.e.*, to couple the peaked-sawtooth voltage from the vertical-output stage to the vertical windings of the yoke. Also, some of these transformers include separate windings to provide convergence-circuit driving voltages, as well as feedback to linearize the output stage.

To understand fully why transformers are used to drive the vertical windings of the yoke, consider first the input and output conditions under which the transformers operate. Vacuum-tube output circuits operate at a high impedance, whereas the yoke presents a relatively low impedance. This means that the vacuum-tube output signal consists of large voltage variations but small current variations. The opposite condition is true at the yoke. Yoke-current variations must be large and the voltage variations small. Current flowing through the yoke must also be alternating in nature. A dc component would displace the scan toward the top or the bottom of the CRT screen. Thus, to prevent decentering of the raster, the dc component of the yoke-driving signal must be removed.

Transformers are ideally suited for matching the high output-circuit impedance to the low yoke impedance. This step-down transformation also provides the high-current low-voltage sawtooth that is necessary to drive the yoke. Remember that if efficiency is neglected, the power developed across the primary winding of a transformer is also reflected across the secondary winding; *i.e.*, $W_p = W_s$, where W is in watts. From this it is easy to see that, by the equation $W = EI$, a high value for E and a low value for I across the primary must be replaced in the secondary by a high value for I when E is stepped to a low value by the primary-to-secondary turns ratio. This current step-up is necessary to maintain the relationship $W_p = W_s$. Typical VOT turns ratios vary from 5 to 1 to as great as 15 to 1.

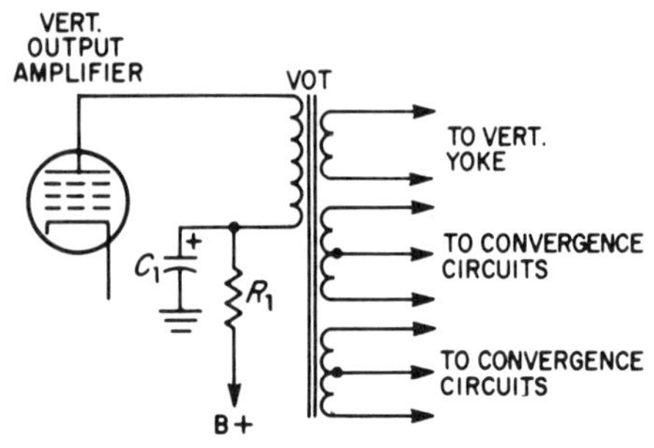

Fig. 19-23. Isolation transformer used in a vertical-output circuit of a tube color-TV receiver.

Electrically, the vertical-output transformers can be classed as either isolation or autotransformer types. Figure 19-23 is a simplified diagram of a typical isolation-type transformer circuit. In the diagram, C_1 and R_1 decouple the vertical-output circuits from the power supply. The main disadvantage of this circuit is frequency distortion, inherent in iron-core transformers. Also, the inductive kickback from the yoke during retrace develops a high-voltage pulse across the primary. Transformation at the step-down ratio reduces the voltage applied to the yoke. Voltages induced across the yoke by the inductive kickback, though, are applied back across the secondary of the transformer and are therefore stepped up by the turns ratio of the transformer. For this reason, the transformer primary must be insulated to withstand the kickback voltage peaks. The convergence-circuit secondaries shown in the diagram are not used on VOT's designed for black-and-white TV receivers.

A simplified autotransformer vertical-output circuit is shown in Fig. 19-24. This transformer provides the necessary step-down ratio, but the dc potential is not isolated from the secondary. In using a circuit of this type, the yoke is usually connected to the transformer through a capacitor. C_1 in the diagram improves the frequency response of the transformer by speeding-up the collapse of the inductive field during retrace. Decoupling from the power supply is accomplished by R_1 and C_2. An important advantage of the autotransformer over the isolation type is in its lower cost.

Vertical-output transformers are usually replaced by a choke in solid-state vertical-output circuits. In these circuits, the output transistor provides a signal at a low impedance that can be coupled through a capacitor directly to the yoke. The choke isolates the collector signal path from the power supply. There are, however, some solid-state vertical-output circuits that make use of the isolation-transformer. These transformers possess a turns ratio of 1 : 1 and are used mainly to provide an ac driving-current for the yoke. For color-TV receivers, convergence-circuit secondaries are provided on the transformers. Also, some transformers, as well as some chokes, include separate windings for the feedback to various stages in the vertical-deflection system.

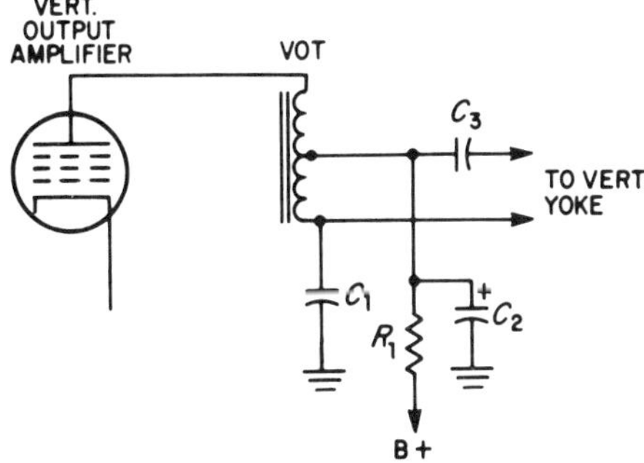

Fig. 19-24. Autotransformer used in a vertical-output circuit of a tube monochrome-TV receiver.

19.11 VERTICAL YOKE WINDINGS

Vertical scanning in the CRT is achieved by a magnetic field, whose lines of force cut horizontally through the neck of the tube. This field is developed by two coils mounted 180 degrees apart on the neck of the CRT. Physically, the coils are part of a yoke assembly that also includes the horizontal-deflection windings. During the vertical-scanning period, the magnetic flux moves the electron beam above and below the center of the screen, depending upon the magnitude and direction of current flow in the coils. As discussed previously, a sawtooth-current waveform drives the yoke.

During the vertical-scanning period, the sawtooth current decreases from a maximum value, in one direction, to zero, and then increases to the same maximum value but in the opposite direction. When the current flow in the vertical coils reaches this last peak value, the electron-scanning beam is at the right-hand end of the last horizontal line at the bottom of the screen. At this time, the retrace interval begins. Current flow in the coils suddenly decreases to zero and then increases to maximum, once again, in the opposite direction. During this retrace interval, though, the current changes very rapidly to reposition the beam at its top-screen position to begin the next field.

The VOT, during the vertical-scanning interval, delivers to the yoke windings a sawtooth current that achieves full-screen deflection of the electron beam. To accomplish this deflection efficiently, yoke inductance and resistance are kept as small as practical, for it is these two values that determine the peak-to-peak value of current required for full-screen deflection. In other words, the deflection sensitivity of the yoke depends on the peak-to-peak value of current required to drive the beam through the full deflection angle of the CRT. In this light, yokes are wound with the shortest length wire having the largest diameter that is practical. The diameter of the neck, the anode voltages, and the length of the CRT are often varied to increase the effectiveness of the yoke.

In present-day TV receivers, the vertical yoke inductances range from 3 to approximately 50 MH. Most of the 3 mH yokes are used on CRT's that have a narrow deflection angle, such as 70 degrees. The wider deflection angles generally require a higher, deflection-sensitive yoke. Recently, wide-angle CRT's have been developed to operate satisfactorily with yokes ranging from 10 to 15 mH. This has been accomplished by decreasing the length of the CRT and reducing the neck diameter from 1-1/8 to 3/4 inch. Typically, a 17-inch CRT requires a 48 mH yoke and a driving current of 400 ma, peak-to-peak.

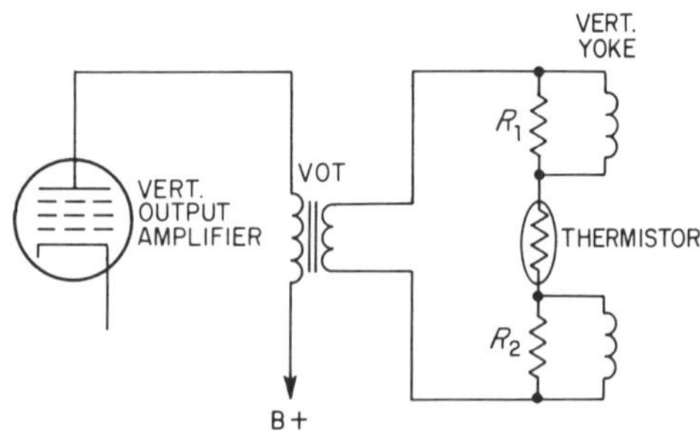

Fig. 19-25. A typical vacuum-tube vertical-yoke circuit which employs thermistor compensation.

A simplified diagram of a typical vertical-yoke circuit is shown in Fig. 19-25. Resistors R_1 and R_2 are connected across the two yoke coils to damp out the shock-induced oscillations caused by the sudden current change during the retrace interval. The thermistor is added between the coils to equalize the yoke resistance under high-heating conditions. When the yoke temperature increases due to the heavy current flow, the resistance of the yoke increases by about 8 ohms. During this temperature rise, the thermistor also is heated and its resistance decreases sufficiently to compensate for the increase in the yoke resistance. This maintains a constant yoke current to prevent height changes.

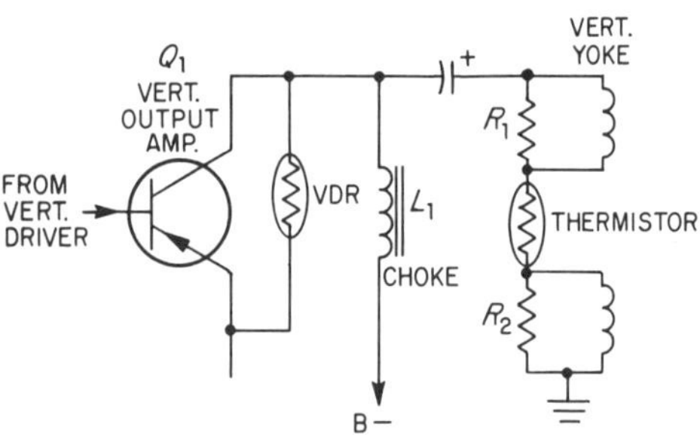

Fig. 19-26. Simplified, solid-state vertical-yoke circuit.

Figure 19-26 shows a simplified diagram of a typical solid-state yoke circuit. Since a choke is used in the collector circuit, instead of an isolation-type transformer, the coupling capacitor is necessary to block dc from the yoke. Also, a VDR is connected in the output circuit to protect the transistor from the high-kickback pulse developed across the yoke during the retrace interval.

19.12 TROUBLES IN VERTICAL-OUTPUT CIRCUITS

Troubles in the vertical-output circuits, in most cases, are clearly discernible on the CRT screen. For this reason, the affected picture is by far the best place to begin a logical analysis of output-circuit troubles. By carefully studying the CRT display, the troubles can even be localized to particular circuits within the vertical-deflection system. Once this localization has been accomplished, faulty components can be quickly isolated and replaced. In dealing with any vertical-output circuit troubles, bear in mind that the primary function of the vertical-deflection system is to provide a linear-scan current at a sufficient amplitude to fully deflect the beam and to provide for vertical convergence. With this function in mind, it is apparent that all troubles can be categorized in accordance with their effect on the picture: height, linearity, convergence, and retrace blanking. Except for convergence, the troubles are the same for either black-and-white or color receivers.

Height The most obvious of all vertical-output circuit problems is the complete lack of vertical deflection. This problem is manifest on the screen by a bright horizontal line across the center of the screen. This line is similar to the condition that exists when the service switch, in color-output circuits, is placed in the service position: vertical deflection is disabled. If this line is left on the screen too long, it may burn the phosphor coating inside the CRT. This damage cannot occur when using the service switch because the intensity of the beam is reduced.

Complete loss of vertical output may be caused by any of the following problems:

(a) No sweep input from vertical oscillator.
(b) Loss of operating voltages in output circuits.
(c) Defective output tube or transistor.
(d) Defective vertical-output transformer.
(e) Open coupling capacitor.
(f) Open output-tube cathode circuit.
(g) Open circuit in deflection coils.
(h) Both deflection coils or damping resistors shorted.

Other height problems include insufficient vertical deflection and the keystone effect. Both of these problems produce a picture on the screen that cannot be expanded to completely fill the mask vertically. In addition, the keystone effect tapers one side of the diminished-height pattern to give a keystone shape. Even though both of these problems affect the picture height, they are completely unrelated.

Insufficient height is due mainly to increased resistance values in plate-feed and other dc networks. This increase in resistance decreases the amplification of either the output or one of the driver stages. Also, weak or defective tubes and transistors fail to provide sufficient amplification. Low values of operating voltage are eliminated rapidly as a suspect if other receiver circuits appear to operate normally. The keystone effect is caused by a shorted circuit across one-half of the deflection coils. The remaining coil cannot develop a strong enough magnetic field to fully deflect the beam. Shorts of this nature are usually found in the damping resistors across the coils. Shorts also develop between windings of the coils and require yoke replacement.

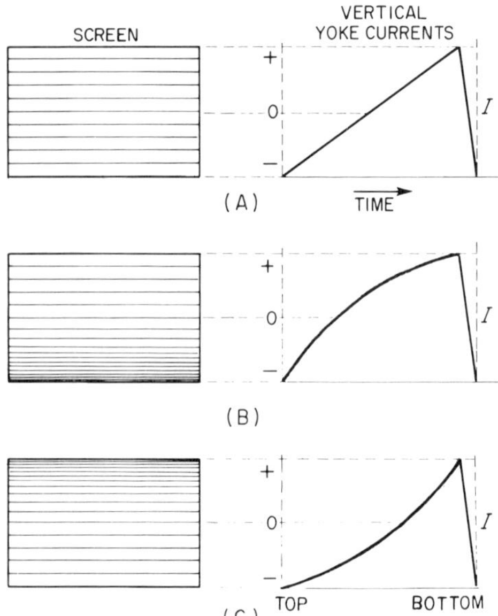

Fig. 19-27. Illustrating linear and nonlinear vertical sweep and the sawtooth currents producing them.

Linearity Figure 19-27 shows the effect of a linear and a nonlinear scan on the picture. In *A* of the figure, the linear scan allows equal spacing between each horizontal line. In other words, horizontal scanning progresses at a uniform rate down the screen. At *B*, scanning progresses normally until the beam is in the lower half of the screen. Here, the magnetic field increases at a reduced rate because of the nonlinearity in the positive portion of the current waveform. This reduced vertical-scanning rate allows the horizontal lines to "stack up" or crowd together at the bottom. The opposite curvature at *C*, in the figure, causes crowding at the top of the screen.

The following problems are the chief causes of nonlinearity in the vertical-output circuits:

Resistance and capacitance value changes in the sawtooth-forming network;
Defects in feedback loops;
Vacuum-tube or transistor-bias changes;
Open decoupling capacitors;
Poor filtering of horizontal "cross talk."

In the sweep system, the sawtooth is formed by an *RC* network at a rate determined by the total time constant of the network. All resistance or capacitance value changes vary this time constant and thus change the charge-discharge rate of the circuit. Also, grid- or cathode-bias changes in vacuum-tube circuits shift the tube operation to a nonlinear point on its characteristic curve. In a similar manner, the changes in transistor biases shift the operating point of the transistor. Since many present-day circuits make extensive use of feedback for linearity correction, circuit faults could disable or distort the feedback voltage and introduce nonlinearity. These faults are usually found in leaky or shorted coupling capacitors.

One particularly bothersome feedback problem is unwanted horizontal-sweep frequency feedback, or horizontal cross talk, which often causes loss of interlace. In this case, a decoupling capacitor in the output stage may be open. The same problem occurs in multivibrator circuits when the integrator network in the plate-to-grid circuit from the output to the input tube fails to filter out the horizontal frequency. Defects in this circuit may also affect the hold control or lock-in range of the vertical-sweep oscillator.

Another result of poor linearity is vertical foldover, a condition that exists when the retrace interval is too long or blanking is insufficient. This condition causes a bright horizontal bar, or haze, to extend up from the bottom or down from the top of the screen for about 1/2 inch. A leaky coupling capacitor in the output stage usually causes this trouble; however, it can also be caused by most of the linearity defects previously discussed.

Convergence The convergence circuits in the color-TV receiver are perhaps the least trouble prone of all the circuits in the set. Most of the circuit resistance is variable and can be readjusted to compensate for the normal changes due to aging and heating. In Fig. 19-12, C_1 couples a portion of the cathode voltage into the convergence circuits. The opening of C_1 results in a complete loss of the parabolic voltage. On the other hand, the shorting of C_1 distorts the parabolic waveform and changes the cathode bias of the output tube. This bias change then contributes to output-stage nonlinearity. Other changes in the wave shaping network cause distortion or the decreased amplitude of the parabolic waveform. Defects in the VOT affect the sawtooth voltage applied to the convergence circuits.

Retrace Blanking To provide vertical-retrace blanking, the peaked sawtooth waveform from the vertical-output stage is differentiated and applied to a grid on the CRT. Any portion of the sawtooth waveform that remains on the blanking pulse will cause picture shading. This condition is usually caused by component value changes in the blanking circuit. Complete loss of retrace blanking is caused by an open-coupling capacitor. Also, weak video output often appears as defective vertical-retrace blanking in some TV receivers.

Trouble Isolation To isolate troubles in the vertical-deflection circuits, first check the output tube or transistors. Second, make voltage and resistance measurements. To measure resistance in the yoke circuits, disconnect the yoke from the VOT and then disconnect both damping resistors. The third step is to substitute a new VOT or yoke. In substituting the VOT, do not disconnect and remove the entire unit. In some receivers this is a difficult task. The simplest method for testing is to disconnect only the primary and the yoke-driving secondary windings, and then, holding the new VOT is position, reconnect only these windings. A quick operational check will then show whether or not the trouble has been corrected.

REVIEW QUESTIONS

1. What is the purpose of the vertical output transformer?
2. Where is the sync signal applied in the combined vertical multivibrator-output circuit?
3. How many transistor stages are used in a typical black-and-white vertical-output circuit? Name them.

4. What protection is provided for the output transistor? Draw a simple diagram of a PNP output-transistor circuit, showing its protective devices.
5. What is the purpose of the thermistor between the coils of the vertical yoke?
6. Draw a diagram of a typical vacuum-tube, vertical-output circuit.
7. Describe the vertical-retrace blanking pulse. When is this pulse formed?
8. What is the difference between black-and-white and color vertical-output circuits?
9. How many controls are used for vertical convergence when all three beams are converged by the vertical system? Name them.

The FM Sound System 20

20.1 FUNDAMENTALS OF FM*

The audio portion of all television programs is transmitted by frequency modulation. This choice was the result of several important factors. Of the two broadcasting systems in use today, AM and FM, the latter has proven capable of better reception under adverse conditions. First, FM reception is almost noise free, as compared to AM reception. Second, the FM system is capable of providing much better rejection of undesired signals on the same or adjacent channels. Third, for a given area coverage, the FM system is more economical and efficient, insofar as the transmitter is concerned.

Amplitude-modulated Waves The ordinary amplitude-modulated radio signal, as it might appear for 100 percent modulation, is shown in Fig. 20-1(A). The audio-modulating signal adds to or subtracts from the amplitude of the carrier. When the modulating signal becomes too strong, overmodulation occurs and the carrier is driven to zero for a short time. This is illustrated in Fig. 20-1(B). Note that, whereas the amplitude of the wave may increase as much as possible, it can only decrease to zero. Whenever over-modulation occurs, the waveform becomes distorted and the greater number of frequencies that are generated by this process causes the bandwidth of the station to increase. Hence all commercial broadcasting stations are careful to see that their output never reaches 100 percent modulation.

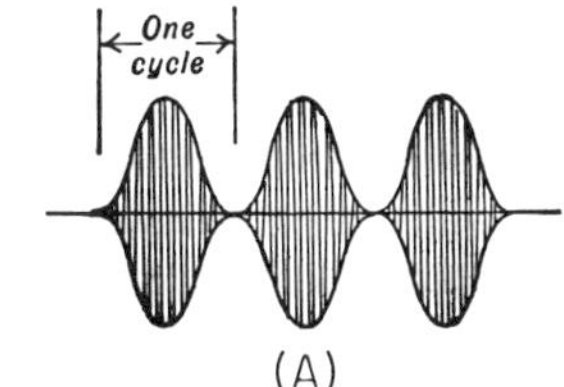

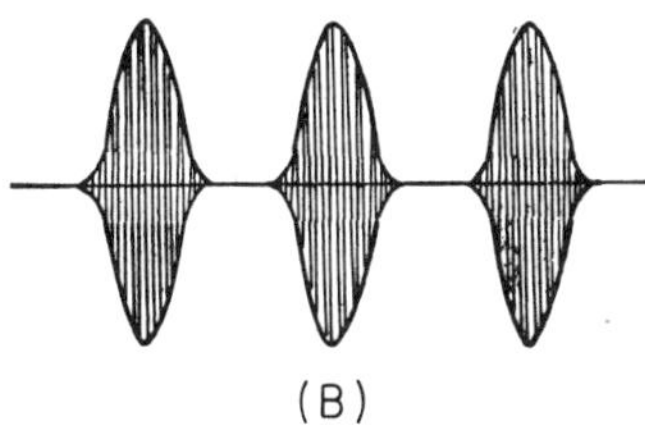

Fig. 20-1. Amplitude-modulated waves: (A) 100 percent modulation, (B) overmodulation.

Frequency-modulated Waves A frequency-modulated wave is constant in amplitude, but varies in frequency. It would appear as pictured in Fig. 20-2. The property of constant amplitude makes the frequency-modulated wave so important. Most man-made and natural interference has been found to affect the amplitude of a wave much more than its frequency. For the AM signal, the interference distorts the waveform and, with this, the intelligence contained therein. FM, on the other hand, contains its intelligence in its changing frequencies. At the FM

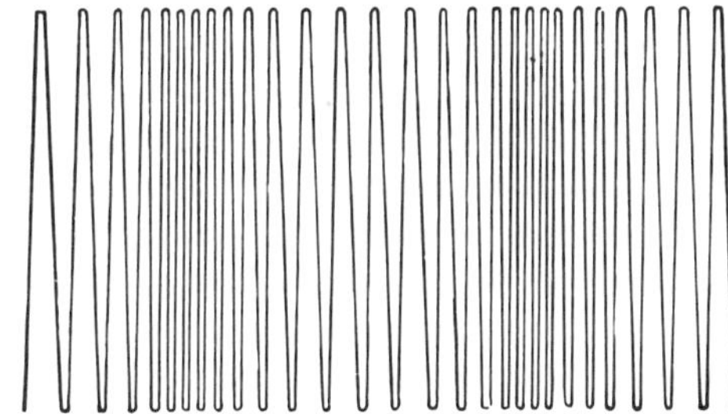

Fig. 20-2. Frequency-modulated wave. The changing pattern between cycles indicates different modulating frequency intensities.

* Due to space limitation, only the most important aspects of FM can be considered here. For a more detailed analysis, the reader is directed to M. S. Kiver, *F-M Simplified,* 3rd Edition, published by Van Nostrand Co., Inc., 1960.

receiver, one of the IF stages, acting as a limiter smooths any irregularities in the amplitude of the incoming signal and by this process eliminates the interference.

The frequency bandwidth of an FM wave depends upon the strength of the impressed audio voltage. At the transmitter, the carrier frequency is fixed by a self-excited oscillator. This frequency is the mean, or center, frequency of the broadcast station. When the sounds that are to be transmitted are fed into the microphone, the mean frequency of the transmitter is varied. The louder the audio signal, the greater the deviation. For example, a frequency deviation (or change) of 25 kHz in the output might occur for a strong audio voltage, whereas only a 1-kHz change would occur if the audio voltage were weak. In the AM case, the amplitude, and not the frequency of the wave, changes for different audio sound levels.

The rapidity with which the FM transmitter frequency moves from one point to another is determined by the frequency of the modulating sound. A high-pitched sound will cause the frequency of the FM transmitter to change more rapidly than when 60 or 100 Hertz are used.

Differences between AM and FM AM and FM differ in many respects and this is perhaps best revealed by the following table:

Factor	FM	AM
Amplitude of signal	Remains constant	Varies with percentage of modulation
Audio voltage	The frequency spread of the signal is determined by the strength of the audio voltage	Determines the amplitude of the wave
Audio frequency	The frequency of the audio-modulating voltage will determine how rapidly the FM wave will change from one frequency to another	The audio frequency controls the speed with which the amplitude of the wave changes
Signal spread	The number of sidebands depends upon the amplitude of the modulating signal. In television, this spread is restricted to 25 kHz on either side of the carrier	Limited to 5 kHz on either side of the carrier frequency. It is determined by the frequency of the audio-modulating wave

20.2 ADVANTAGES AND DISADVANTAGES OF FM

Elimination of Noise Of the two broadcasting systems, AM and FM, the latter is capable of better reception under adverse conditions. It is easier to minimize interference from the other nearby stations operating on the same frequency with frequency modulation than with amplitude modulation.

Cost This is a factor especially applicable to transmitters. Because of the arrangement of the circuits in a frequency-modulated transmitter, a given wattage signal can be developed more economically with this equipment than with amplitude-modulation equipment. Specifically, the large difference in cost between the two systems lies in the audio power required to produce a certain strength signal. With AM, the audio power is generally 50 percent of the carrier power, and this may entail many thousands of watts for a powerful station. In FM, on the other hand, the audio required represents only a fraction of the output power and is more easily generated.

The power relationship that exists in an amplitude-modulated wave between the sidebands and the carrier is in the ratio of 1 : 2 for 100 percent modulation. This is only the average power, and when the equipment is designed, it must be capable of handling the much higher peak (or surge) power. Naturally, this requirement materially increases the cost of the station. In FM transmission, the power output does not increase with modulation and no additional provision for handling excess power need be made.

Fidelity The matter of fidelity is not stressed because, contrary to popular opinion, just as much fidelity is available with AM as with FM. It is only on the present crowded broadcast band (500 to 1,500 kHz) that space is not available to permit the full 15,000 Hertz audio bandwidth to be reproduced. Given sufficient spectrum space, both systems may have equal fidelity.

Stations on the Same Frequency One definite advantage obtained with frequency modulation is due to the observed (and calculated) fact that, if two signals are being received simultaneously, the effect of the weaker signal will be eliminated almost entirely if it possesses less than half the amplitude of the other stronger signal. This means that for one signal to completely override another at the receiver, their amplitudes need to be in the ratio of 2 : 1, or more. With a good antenna, it is frequently easy to tune in one station in sufficient strength so that the other interfering station or stations are eliminated entirely. No such situation exists with AM signals, where interfering stations can be heard when even a 100 : 1 relationship exists between the different carrier amplitudes.

Summary of Advantages and Disadvantages The highlights of this section are briefly shown in the following table:

ADVANTAGES	
FM	AM
(1) Better reception under adverse conditions (2) Easier to eliminate weaker stations on the same frequency as the desired station (3) Transmitter cheaper to build because less audio power is needed (4) Easier to separate audio and video signals	(1) Simpler circuitry at the receiver (2) Alignment very easily performed
DISADVANTAGES	
FM	AM
(1) Receiver more complicated (2) Alignment more difficult than AM receivers	(1) Overmodulation causes distortion (2) Most noise adds to the AM signal and is heard at the speaker (3) Difficult to eliminate weak station on the same frequency as the desired strong station (4) Much power needed in audio portions of the transmitter

20.3 COMPARISON OF FM AND AM RECEIVERS

Although it is difficult to draw a direct comparison between AM and FM transmitters, it is possible to show the similarity between the respective receivers. The FM receiver is a superheterodyne in all instances.

Block Diagrams The block diagrams in Fig. 20-3 illustrate the differences between AM and FM superheterodynes. Besides the limiter and discriminator stages in the FM receiver, both sets appear to be exactly alike, and indeed might easily be taken for each other in an ordinary schematic. Up to the limiter stage, the primary difference between the two types of receivers resides almost wholly in the tuning circuits that connect each stage. In FM, these circuits must be capable of receiving higher frequencies and of passing the wider band of side frequencies associated with the FM carrier. In the ordinary FM receiver, designed for use between 88 and 108 MHz, each station is allowed sidebands ranging up to 75 kHz on both sides of the carrier. For television audio, 25 kHz is used, the narrower bandwidth simplifying somewhat

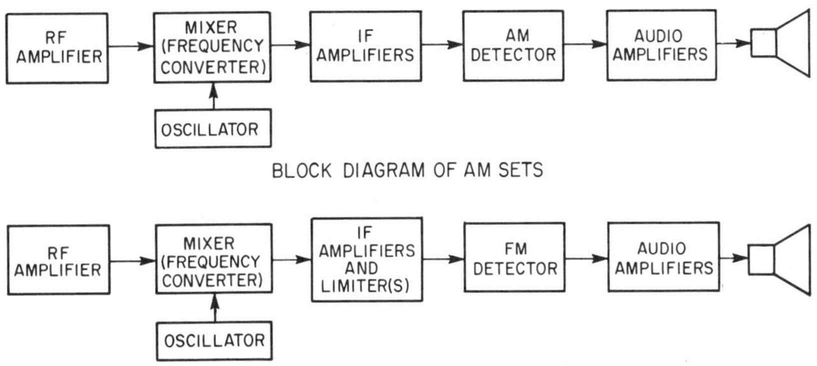

Fig. 20-3. A comparison of the block diagrams of AM and FM receivers.

the problem of receiving both the video and audio carriers simultaneously.

High-quality Circuits AM receivers are usually cheaper, and made much simpler, than most FM receivers. The AM oscillator and mixer functions are quite often carried out in one tube or transistor. One feature common to most FM receivers is the use of a separate oscillator that feeds the mixer tube and helps generate the IF. At the higher frequencies employed for the television audio, the drifting of the oscillator produces more marked effects than it does in the lower-frequency broadcast receivers. At 60 MHz, a 2 percent frequency drift would shift the signal beyond the bandpass of the audio IF circuits; at 1 MHz, the same percentage shift is only 20 kHz and would not shift a signal beyond the bandpass of the circuits designed for a $\pm$75 kHz signal spread. The separate oscillator arrangement results in a greater stability, with the drift reduced to a smaller fraction than would be present in designs using the same device for mixing and generating the oscillator voltage. Often, such additional devices as compensating ceramic capacitors are placed in the oscillator tank circuit in order to counteract tendencies on the part of the other frequency-determining components to change with the operating conditions.

20.4 MINIATURIZED FM-SOUND CIRCUITS

Miniaturization is a trend common to all electronic equipment. The FM-sound circuits (and others) of television receivers are no exception. In the following section of this chapter, we will point out methods by which the physical dimensions of the various circuits are being reduced.

Tube Circuits One common way to miniaturize tube FM circuits has been to combine several tubes into one envelope. As a result, many glass envelopes containing two or three tubes are available.

Transistor Circuits The conversion from tube to transistor circuitry in the FM-sound system offers many advantages. For small-screen TV, where space is a premium, a 10:1 savings in space for some of these circuits is possible. High reliability and cool operation are also important results of this conversion. However, many manufacturers are still building tube type sets because transistor circuits have not generally been cheaper. Recent developments in plastic transistors are reversing this situation.

Integrated Circuits The recent availability of low-cost plastic IC's is expected to revolutionize the TV industry. The first usage of these devices in TV sets is usually in the FM-audio section. Early applications merely used a simple IC in place of each tube or transistor. More advanced versions have up to four separate stages in each package. Thus, one package may have an IF amplifier, limiter, discriminator, and a low-level audio amplifier. A small amount of external circuitry such as a power transistor, several coils, a volume control, several small resistors, and several small capacitors are also needed. Some manufacturers are even trying to eliminate the tuned circuits with ceramic resonators. These devices are similar to the familiar quartz crystals and can be packaged in very small containers.

20.5 THE FM INTERMEDIATE-FREQUENCY AMPLIFIER

The FM, IF amplifier looks very similar to an AM broadcast IF amplifier. The major difference is in the bandwidth and the operating frequency.

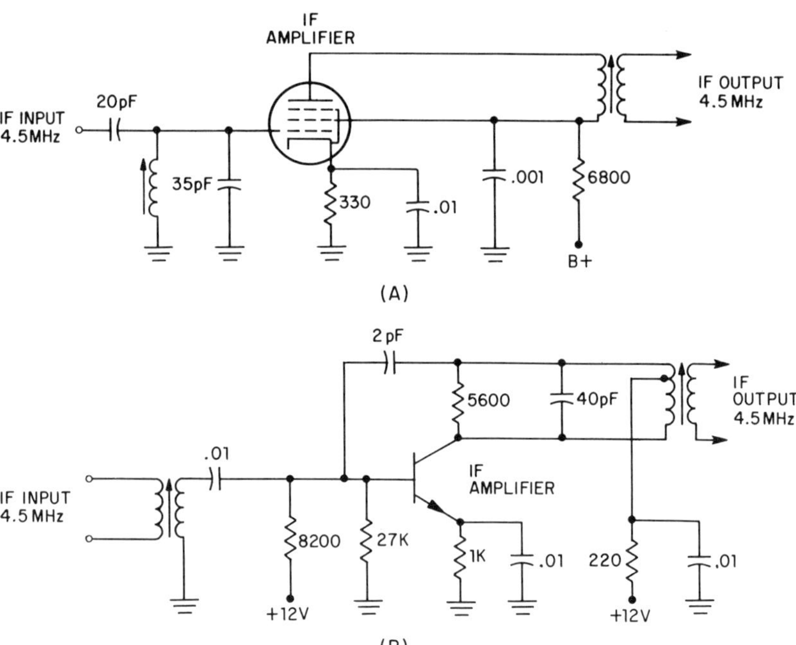

Fig. 20-4. Typical FM IF amplifiers for TV receivers: (A) tube IF amplifier, and (B) transistor IF amplifier.

IF amplifiers for TV audio require a capability to amplify signals with a total bandwidth of 50 kHz. This must be done at a center frequency of 4.5 MHz for an intercarrier set.

As noted in Chapter 5, a wide-band amplifier has less gain than a narrow band amplifier. For this reason, two or more IF amplifiers are required in most FM receivers and TV sound sections. The last IF amplifier is sometimes also used as a limiter (to be discussed in Section 20.7).

The wider bandwidth is required for several reasons. First, the maximum deviation of the TV FM sound is ±25 kHz. AM radio has a bandwidth of only ±5 kHz. Second, the wider bandwidth will allow a small amount of oscillator drift before the sound-output level drops or distorts noticeably.

Figure 20-4(A) shows a typical tube FM, IF amplifier. Transformer coupling is used on the output while input tuning is accomplished with a simple parallel tank. Other types of *L-C* networks can also be used for the input and output coupling.

A transistor version of an FM, IF amplifier is shown in Fig. 20-4(B). Two transformers are used for coupling. The output transformer has several extra turns on the primary winding to provide the neutralization through a 2 pF capacitor. This provides feedback which effectively cancels the internal base to the collector capacitance of the transistor. Without this neutralization, the stage would tend to oscillate and most likely cause undesirable high-pitched sounds from the speaker.

20.6 THE FM LIMITER

The first significant difference between the AM and FM superheterodynes is noted at the limiter stage or stages. Essentially, the purpose of a limiter is to eliminate the effects of the amplitude variations in the FM signal. While it may have been true that the frequency-modulated signal left the transmitter with absolutely no amplitude variations, this is almost never true by the time the signal reaches the limiter.

To digress for a moment, let us see where, in the receiver itself, various parts of the FM signal could have received more amplification than other parts of the signal. An ideal response curve for a tuned circuit is shown in Fig. 20-5(A). With such a characteristic, each frequency within the signal receives uniform amplification. Such a happy situation, however, is seldom encountered in practice. The more usual state of affairs is illustrated by the curve in Fig. 20-5(B). Here it is apparent that the center frequencies receive more amplification than those located farther away. Hence, even if the incoming signal is perfectly uniform, by the time it arrives at the limiter, amplitude variations would be present. The result is distortion if this wave is allowed to reach the speaker. It is for the limiter to remove the amplitude variation.

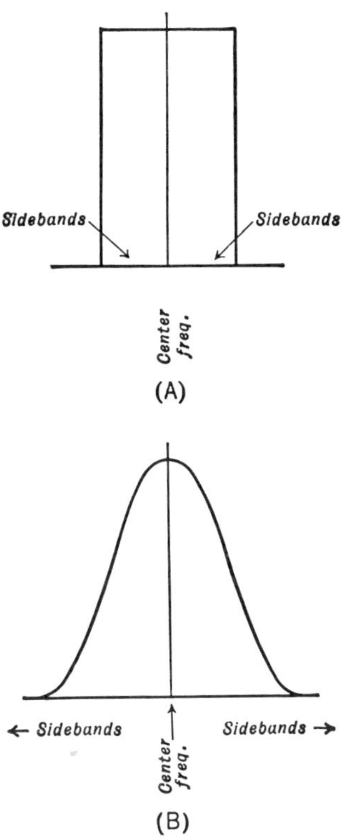

Fig. 20-5. Receiver response curves: (A) the ideal curve, (B) a typical practical result.

A Vacuum-tube Limiter A typical limiter stage is shown in Fig. 20-6. Except for the grid-leak method of achieving bias, its appearance is almost identical to an IF stage. Further inspection reveals that low-plate and low-screen voltages are used. The low electrode voltages cause the tube to reach current saturation with moderate signals at the grid. The use of the grid-leak bias aids in keeping the output-plate current (and hence the output signal) constant for different input-voltage levels. It is readily apparent that, with FM signals of different amplitudes arriving at the grid of the limiter, a constant output for each would mean the elimination of any amplitude distortion, which is exactly what is desired. With the limiter so designed that it will easily saturate, amplitude variations can be eliminated and, with them, most

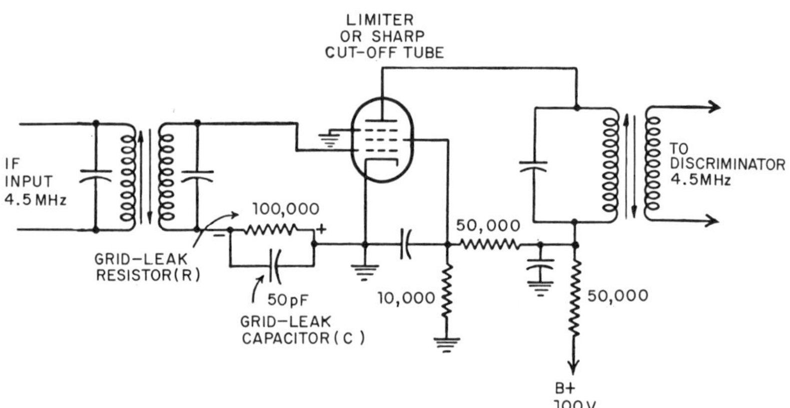

Fig. 20-6 A vacuum-tube limiter.

disturbing noises. This is all possible because of one fact, namely, that much man-made or natural interference affects the amplitude of the radio signal more than it does its frequency. By the simple technique of smoothing out the amplitude differences of the incoming waves (without affecting their fundamental frequency), we eliminate the noise or interference. This constitutes one reason for the extensive use of FM.

It is possible to design limiters on the basis of low-plate and low-screen voltages alone, but better results and more amplification are obtained if the grid-leak bias is added to this combination (see Fig. 20-6). With the insertion of grid-leak bias, it is possible to raise the electrode voltages, somewhat increasing the gain. The tube initially has a zero bias with no signal at the grid. As soon as a signal acts, the grid is driven slightly positive, attracts electrons, and charges the capacitor *C*. This capacitor attempts to discharge through *R* but, due to the relatively long time constant of *R* and *C*, the discharge occurs slowly. Because of current flow through *R*, a voltage is developed, with the end nearest the grid becoming negative. This voltage will act as a bias, varying in value as the incoming signal varies and in this way tending to keep the plate current steady within rather wide limits of input voltage. A strong signal causes the grid to become more positive, resulting in greater current flow through *R*. A larger bias is then developed. A weaker signal will cause less voltage, resulting in essentially the same amount of plate current. The usual values of *C* range from 30 to 60 pF and, for *R*, between 50,000 and 200,000 ohms.

Because the voltage across *R*, the grid-leak resistor, will vary with the amplitude of the incoming signal, this point of the limiter is generally used for aligning the preceding IF amplifiers.

Graphical Analysis of a Limiter A limiter characteristic curve is shown in Fig. 20-7. Notice that the output signal of the tube increases with the input signal until a certain voltage is reached. Beyond this point, known as the "knee" of the curve, point *A*, the plate current of the limiter remains substantially constant for all stronger input voltages. Since complete limiting begins at this point, the signals at the antenna of the receiver must receive sufficient amplification to force the limiter tube to operate beyond point *A*. From this point, the output of the limiter will remain constant. Any signal which is so weak that it is unable to operate the tube beyond *OA* will have its noise appear in the limiter output.

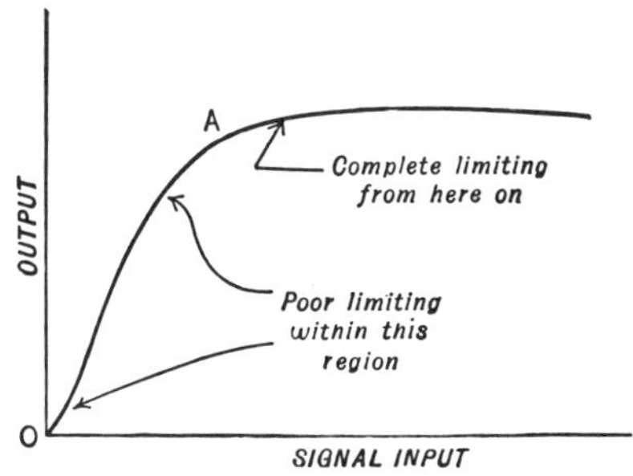

Fig. 20-7. The characteristic curve for a limiter.

The situations for weak and strong signals can be shown graphically. The curve *OAB* in Fig. 20-8 shows the relationship between the input-grid voltage, or signal, and the resulting plate current in the output of the limiter. With the tube biased to point *C*, the input-signal voltage will vary about this point. Consider the first small signal coming in. As it varies the grid bias, corresponding changes take place in the plate

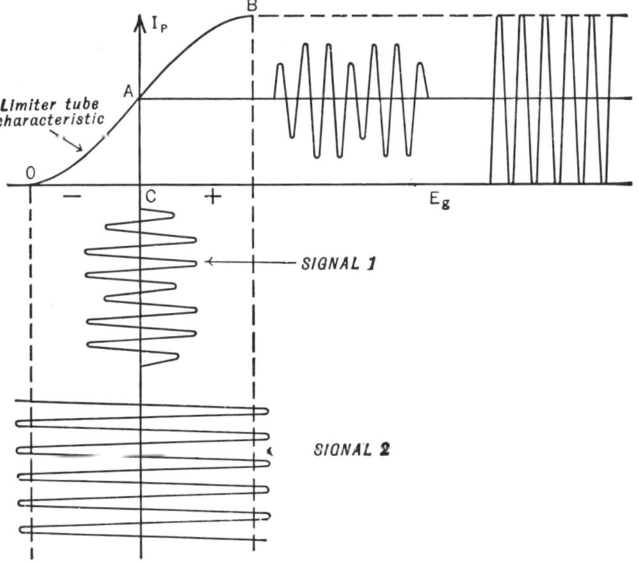

Fig. 20-8. These curves illustrate that the incoming signal must reach a certain amplitude before the limiter stage will saturate.

circuit, and at no time will the plate current be forced to its saturation value. This means that any noise and amplitude distortion contained in this signal will be amplified and reproduced in the plate circuit and, from here, go to the discriminator.

Now consider the voltage of the second signal. At all peak points of the signal, plate-current saturation is reached on the positive peaks, while current cutoff is responsible for smoothing out the negative peaks. In the output circuit, all amplitude distortion has been clipped or eliminated. When this signal is fed to the discriminator, it should give noise-free operation. Thus, while a limiter provides FM with its greater advantages, care must be taken to see that it is properly operated; otherwise its usefulness is lost. The FM receiver must be so designed that all desired signals to the input receive sufficient amplification. When this is done, the plate current of the limiter will give a constant output.

Dual Limiters Although one limiter stage serves satisfactorily, better results can be obtained with two stages, one following directly after the other. These limiters can be identical to each other or completely different. A transistor and diode dual limiter is shown in Fig. 20-9. The

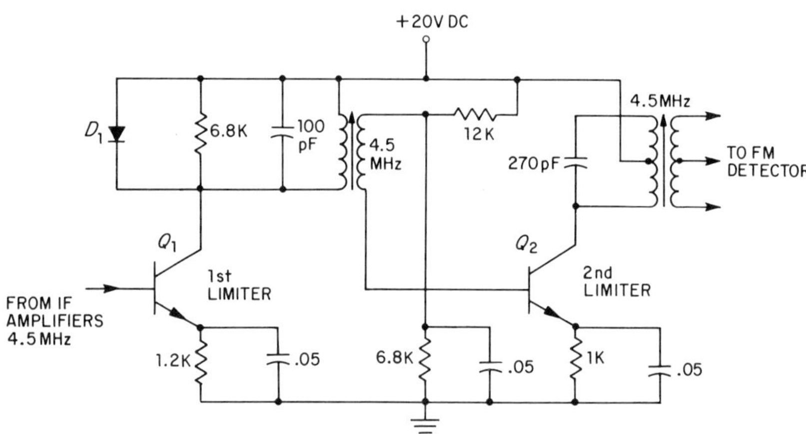

Fig. 20-9. A solid-state dual limiter.

first limiting action is achieved by placing a diode across the primary winding of the interstage transformer. The abrupt knee in the forward characteristics of the diode causes this limiting action. The last IF stage is biased so that its transfer characteristics look like Fig. 20-8. With two limiters, the knee of the resulting curve becomes sharper and provides better limiting action.

20.7 THE FM DISCRIMINATOR

The purpose of the second detector in an amplitude-modulated set is to obtain the audio variations from the incoming modulated signal. The same stage in a frequency-modulated receiver must derive the audio variations from the different incoming frequencies. Thus, although the end product in both cases is the same, the methods used are quite different. We know that with FM a large frequency deviation from the carrier means a loud audio signal, whereas a small frequency deviation means a weak audio note. Hence, some circuit must be devised that will develop voltages proportional to the deviation of the various incoming frequencies about the carrier.

A simple circuit that discriminates against the various frequencies is the elementary parallel- (or series-) resonant circuit. As is well known, this circuit will develop maximum voltage at the resonant frequency, with the response falling off as the frequency separation increases on either side of the central or resonant point.

One of the first discriminators used in FM receivers contained two resonant circuits in an arrangement as shown in Fig. 20-10. The primary coil, L_1, is inductively coupled to L_2 and L_3, each of which is connected to a diode. Each diode has its own load resistor, but the output of the discriminator is obtained from the resultant voltage across both resistors.

In order to determine the frequencies to which L_2 and L_3 must be tuned, it should be recalled that when an audio-modulating signal alters the frequency of an FM transmitter, it varies this frequency above and below one central, or carrier, value. Thus, for a sine wave, the maximum positive portion would increase the frequency, say by 40 kHz, while the maximum negative section would decrease the carrier frequency by the same amount. At intermediate points, less voltage would cause correspondingly less frequency deviation.

To have the discriminator function in a similar manner over the same range, L_2 and L_3 are each peaked to one of the two end points of the IF band. For example, if the IF band-spread extends from 4.25 MHz to 4.75 MHz (with 4.50 as the mean, or carrier, frequency), L_2 could be peaked to 4.25 MHz, and L_3 to 4.75 MHz. The response curves would then appear as in Fig. 20-11.

The two curves are positioned in the manner shown because of the way the load resistors and diodes are connected in the circuit. According to the arrangement, the voltages developed across the resistors tend to oppose each other, as indicated by the polarities shown in Fig. 20-10.

At the center frequency, point A of Fig. 20-11, the two voltages developed across the load resistors cancel each other and the resultant voltage is zero. Similarly, by adding the voltages at other points about the carrier, we obtain the overall resultant curve shown in Fig. 20-12. This is the familiar S-shaped curve of all frequency discriminators which shows how the output voltage of the second detector will vary as the incoming frequencies change. Specifically, suppose the signal acting at the input to the discriminator at any one instant has a frequency of 4.65 MHz. The amount of voltage developed at the output is given by point A on the vertical axis. Then, at the next instant, if the frequency

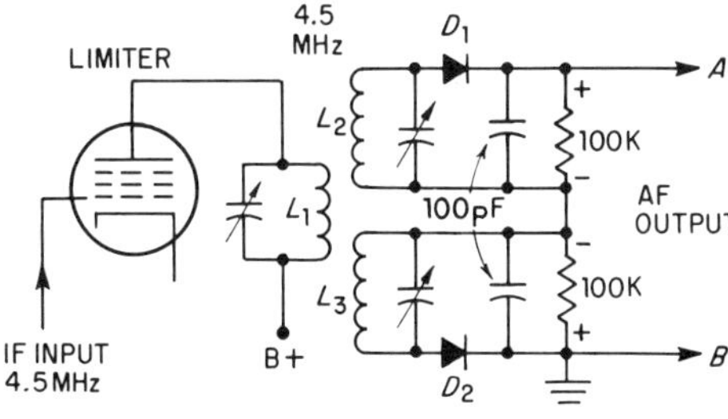

Fig. 20-10. A simple discriminator circuit.

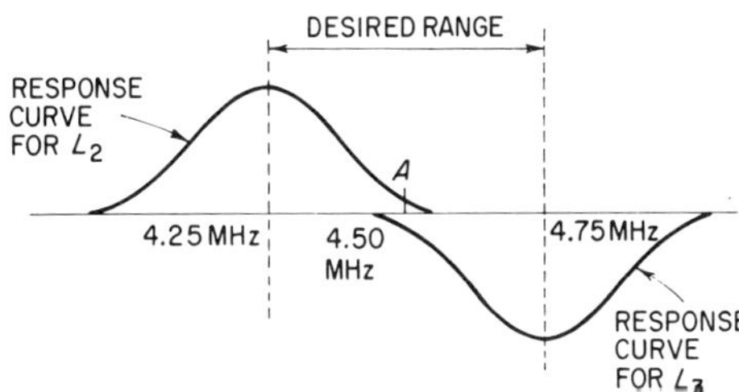

Fig. 20-11. The overall response curve for the discriminator of Fig. 20-10.

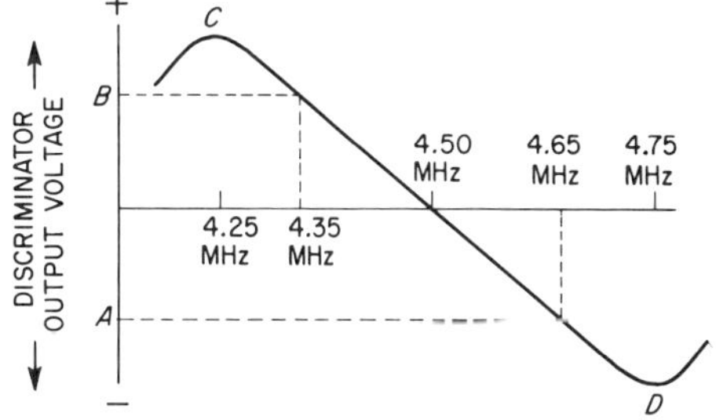

Fig. 20-12. The resultant S-shaped discriminator characteristic curve obtained by adding the two separate curves of Fig. 20-11.

should change to 4.35 MHz, the output voltage is indicated by point B. Notice that all frequencies below 4.50 MHz result in positive output voltages, whereas all those above 4.50 MHz give rise to negative output voltages. In this way, the audio voltages that modulated the carrier frequency at the transmitter are extracted in the receiver.

The useful segment of this characteristic curve of the discriminator is the linear portion included between the two maximum points, C and D. Any nonlinearity along this section of the curve would produce amplitude distortion in the output audio signal. When discriminators are designed, the maximum points C and D are generally set much farther apart than is required in the particular receiver. This insures a linear curve at those frequencies that are actually used, since the response characteristic has a tendency to curve near the maximum peaks. By utilizing a smaller range, amplitude distortion in the output signal is kept to a minimum. The sections of the curve of Fig. 20-12 beyond points C and D are completely disregarded.

The frequency of the output voltages is determined by how rapidly the frequency of the incoming IF signal varies. A large frequency deviation in the input signal gives rise to a strong output wave, and the rapidity with which this incoming frequency changes determines whether the strong output will be pitched high or low.

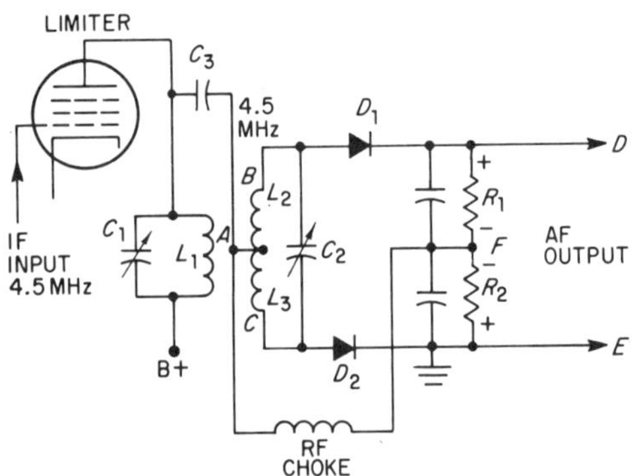

Fig. 20-13. A Foster-Seeley discriminator.

20.8 THE FOSTER-SEELEY DISCRIMINATOR

One may wonder why the preceding circuit was described in such detail if it is not used in modern receivers. The reason lies simply in the ease with which this circuit brings out the fundamental conversion process at the second detector of an FM receiver, and because it is basically the same as the present-day discriminator of Fig. 20-13.

Vector-diagram Analysis Instead of employing two separate tuning capacitors for the secondary circuit, only one is used. R_1 and R_2 are the load resistors, one for each diode, and the resultant output audio voltage is obtained across points D and E. The use of one capacitor, instead of two, results in greater ease in aligning the circuits and economy in construction. The tap divides the secondary coil into two identical coils, L_2 and L_3. Vector diagrams are required to understand how the circuit responds to frequency modulation. The capacitor C_3 couples a voltage V_A at point A. We will consider this voltage as our reference at zero degrees (see Fig. 20-14(A)). This is a voltage vector coming from the L_1-C_1 tank circuit. The other two vectors we require for our vector diagram come from the transformer action across L_2 and L_3. We shall call these voltages V_{BA} and V_{CA}. These voltage vectors are ± 90 degrees out of phase with V_A when the incoming signal is at the IF center frequency. Figure 20-14(A) shows the vector relationship of these three voltages. The voltage V_B is simply the vector sum of V_A and V_{BA}. Similarly, $V_C = V_A + V_{CA}$ (vector addition again). These new

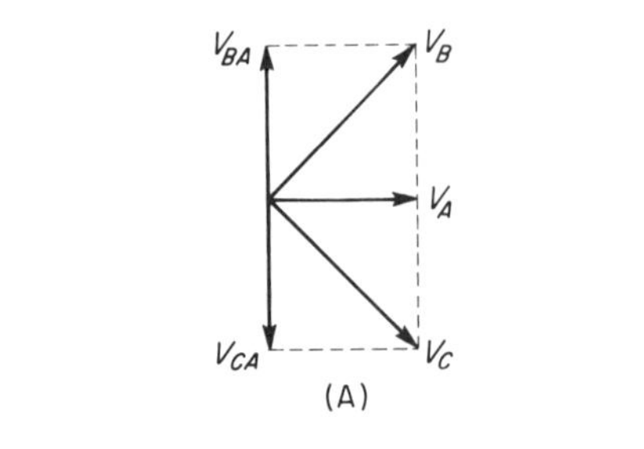

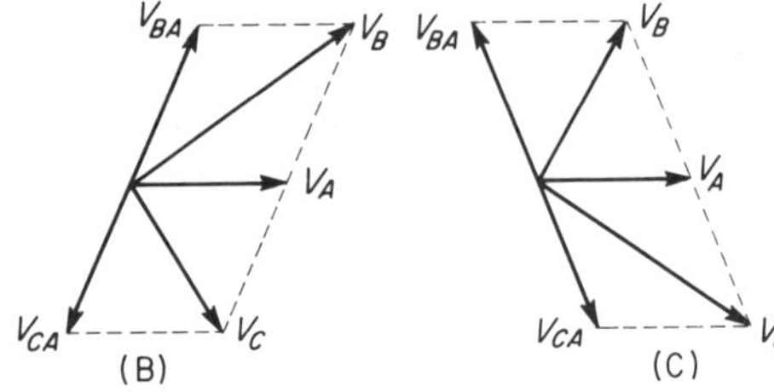

Fig. 20-14. Vector diagrams for the Foster-Seeley discriminator: (A) at resonance; (B) above resonance; and (C) below resonance.

vectors are shown by graphical construction in Fig. 20-14(A). The detected voltages across $R_1(V_{DF})$ and $R_2(V_{EF})$ are equal in magnitude to the peak RF voltages at points B and C, respectively. V_{DF} and V_{EF} are equal and opposite at resonance so the voltage V_{DE} is zero.

What happens above and below resonance? Consider the loop formed by L_2, L_3, and C_2. Above resonance the current in this loop lags and hence the voltages V_{BA} and V_{CA} lag from their mid-frequency phase. This results in the vector relationship shown in Fig. 20-14(B). V_B is now larger than V_C, so the detected output, V_{DE}, is a positive voltage. Below resonance the secondary loop current leads its mid-frequency phase. This makes V_C larger than V_B, and a negative voltage appears at V_{DE}. A characteristic curve similar to the S-shaped one in Fig. 20-12 is obtained for this discriminator.

The need for limiter stages arises because the discriminators noted in the preceding paragraphs are sensitive to the amplitude of the incoming signal. In other words, these discriminators are not pure FM detectors. There are other FM detectors, however, which are sufficiently immune to amplitude variations to enable us to dispense with the limiter. These detectors are discussed by the following.

20.9 THE RATIO DETECTOR

To understand why a ratio detector enjoys greater immunity from AM distortion in the incoming FM signal, let us compare its operation with that of the ordinary discriminator.

In the discriminator circuit of Fig. 20-13, let the signal coming in develop equal voltages across R_1 and R_2. This would occur, of course, when the incoming-signal frequency is at the IF center value. Suppose that each voltage across R_1 and R_2 is 4 volts. When modulation is applied, the voltage across each resistor changes, resulting in a net-output voltage. Say that the voltage across R_1 increases to 6 volts and the voltage across R_2 decreases to 2 volts. The output voltage would thon be equal to the difference between these two values, or 4 volts.

However, let us increase the strength of our carrier until we have 8 volts each, across R_1 and R_2, at mid-frequency. With the same frequency shift as previously, but with this stronger carrier, the voltage across R_1 would rise to 12 volts and that across R_2 would decrease to 4 volts. Their difference, or 8 volts, would now be obtained at the output of the discriminator in place of the previous 4 volts. Thus the discriminator responds to both FM and AM. It is for this reason that limiters are used. The limiter clips all amplitude modulation off the incoming signal, and an FM signal of constant amplitude is applied to the discriminator.

When unmodulated, the carrier produced equal voltages across R_1 and R_2. Let us call these voltages E_1 and E_2 respectively. With the weaker carrier on modulation, the ratio of F_1 to E_2 was 3:1, since E_1 became 6 volts and E_2 dropped to 2 volts. With the stronger carrier on modulation, E_1 became 12 volts and E_2 dropped to 4 volts. Their ratio was again 3:1, the same as with the previous weaker carrier. Thus, while the difference voltage varied in each case, the ratio remained fixed. This example demonstrates, in a very elementary manner, why a ratio detector could be unresponsive to carrier changes.

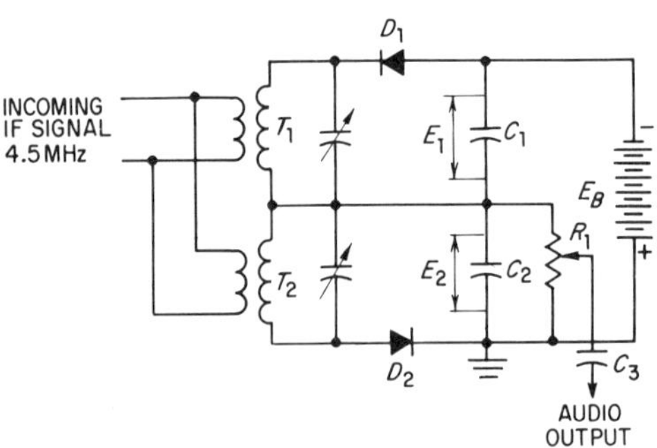

Fig. 20-15. Preliminary form of the ratio detector.

The Elementary Ratio Detector An elementary circuit of a ratio detector is shown in Fig. 20-15. In this form, the detector is similar to the detector in Fig. 20-10, where each tube has a completely separate resonant circuit. One circuit is peaked slightly above the center IF value (say T_1); the other is peaked to a frequency below the center (say T_2). The output voltage for D_1 will appear across C_1, and the output voltage for D_2 will be present across C_2. The battery, E_B, represents a fixed voltage. Since C_1 and C_2 are in series directly across the battery, the sum of their voltages must always equal E_B. Also, due to the manner in which the battery is connected to D_1 and D_2, no current can flow around the circuit until a signal is applied. Now, while $E_1 + E_2$ can never exceed E_B, E_1 does not have to equal E_2. In other words, the ratio of E_1 to E_2 may vary. The output voltage is obtained from a resistor connected across C_2.

When the incoming signal is at the IF center value, E_1 and E_2 will be equal. This is similar to the situation in the previous discriminator. However, when the incoming signal rises in frequency, it approaches the resonant point of T_1 and the voltage across C_1 likewise rises.

For the same frequency, the response of T_2 produces a lower voltage. As a consequence, the voltage across C_2 decreases. However, $E_1 + E_2$ is still equal to E_B. In other words, a change in frequency does not alter the total voltage, but merely the ratio of E_1 to E_2. When the signal frequency drops below the IF center point, E_2 exceeds E_1. The sum, however, of $E_1 + E_2$ must equal E_B. The audio variations are obtained from the change of voltages across C_2. The capacitor C_3 prevents the rectified dc voltage in the detector from reaching the grid (or base) of the audio amplifier. Only the audio variations are desired.

The purpose of E_B in this elementary explanatory circuit is to maintain an audio-output voltage which is purely a result of the FM signal. E_B keeps the total voltage ($E_1 + E_2$) constant, while it permits the ratio of E_1 to E_2 to vary. So long as this condition is maintained, we have seen that all amplitude variations in the input signal will be without effect.

Replacing the Battery with a Capacitor The problem of selecting a value for E_B is an important one. Consider, for example, that a weak signal is being received. If E_B is high, the weak signal is lost because it cannot possess sufficient strength to overcome the negative polarity placed by E_B on the diodes D_1 and D_2. The diodes, with a weak-input voltage, could not pass current. If the value of E_B is lowered, then powerful stations are limited in the amount of audio-voltage output from the ratio detector. This is due to the fact that the voltage across either capacitor, C_1 or C_2, cannot exceed E_B. If E_B is small, only small audio-output voltages are obtainable. To get around this restriction, it was decided to let the average value of each incoming carrier determine E_B. Momentary increases could be prevented from affecting E_B by a circuit with a relatively long time constant.

The practical form of the ratio detector is shown in Fig. 20-16. The detector uses the phase-shifting properties of the discriminator of Fig. 20-13. R and C_3 take the place of E_B, and the voltages developed across R will be dependent upon the strength of the incoming carrier. Note that D_1 and D_2 form a series circuit with R (and C_3), and any current flowing through these diodes must flow through R. However, by shunting the 8-μf electrolytic capacitor across R, we maintain a fairly constant voltage. Thus, the momentary changes in the carrier amplitude are merely absorbed by the capacitor. It is only when the *average* value of the carrier is altered that the voltage across R is changed. The output audio-frequency voltage is still taken from across C_2 by means of the volume control.

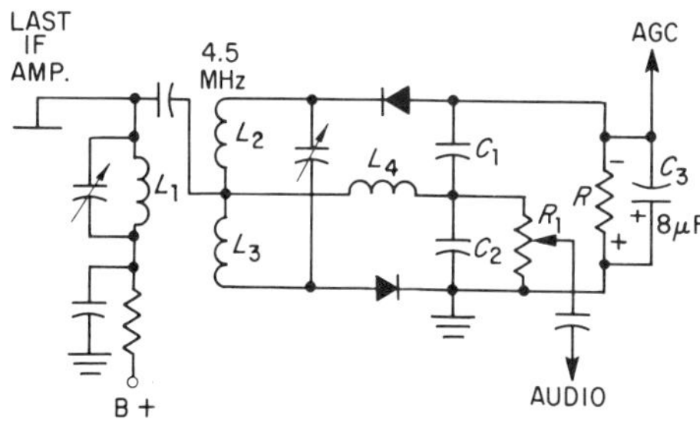

Fig. 20-16. Practical form of the ratio detector.

Since the voltage across R is directly dependent upon the carrier strength, it may also be used for AGC voltage. The polarity of the voltage is indicated in Fig. 20-16.

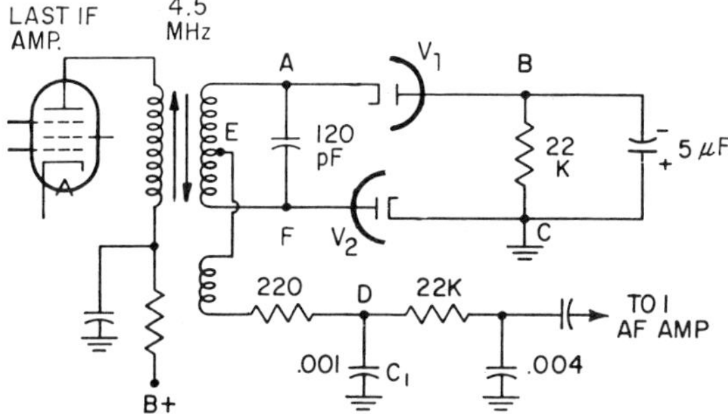

Fig. 20-17. An economical ratio detector that is widely used.

Economical Ratio Detector The urge to simplify these circuits and thereby reduce the cost is ever present among designers of television sets. Such simplification is possible with the ratio detector, as revealed by the design shown in Fig. 20-17. This arrangement lacks the C_1-C_2 divider shown in Fig. 20-16. In spite of the reduction, the circuit still functions satisfactorily. However, with fewer capacitors, the reader may fail to see how the difference voltage is established to provide the necessary audio-output signal.

To understand how the circuit in Fig. 20-17 operates, it has been drawn with lettered identification points. Current that flows through V_1 can take one of two paths. In one path, the current flows from the cathode of V_1 to the plate, to points B, C, F, E, A, and then back to the cathode again. The second path is: cathode to plate, to points B, C, then to ground, and up through C_1 to point D, then to point E, A, and finally back to the cathode again.

Now let us consider V_2. One path for its current is: cathode to plate, to F, E, A, through V_1 to B, C, and back to the cathode again. The second path is from cathode to plate to F, E, and down to point D, through capacitor C_1 to ground, then to point C and back to the cathode again. Note then that part of the current of V_1 flows up through C_1 while part of the current of V_2 travels down through the same capacitor. It is from these two opposing currents that the difference is established, this difference representing the audio-output voltage of the detector.

Balanced Ratio Detectors The preceding ratio detector shown in Figs. 20-15 and 20-16 is an "unbalanced" circuit, so-called because D_1 and D_2 are not equally balanced against ground. We can transform these circuits into a balanced ratio detector by moving the position of the ground connection, as shown in Fig. 20-18. In place of the resistor, R, we now have two. Their function, however, remains the same.

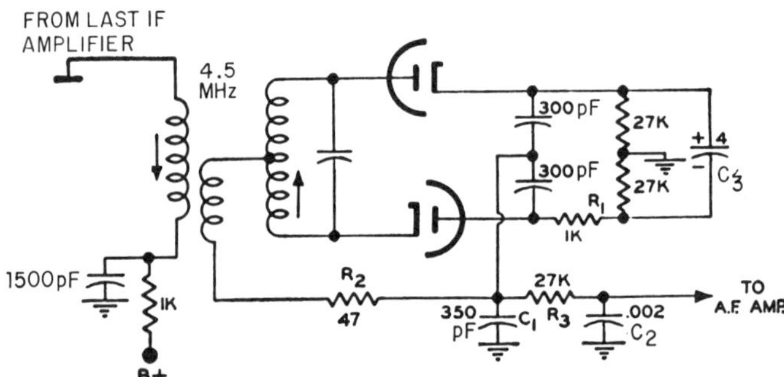

Fig. 20-18. A common form of the balanced ratio detector.

R_1 is inserted to provide a better balance between both halves of the circuit, and R_2 limits the plate current drawn by each diode. C_1 shunts IF voltages away from the audio output, while R_3, C_2 is a de-emphasis filter to equalize the audio signal back to its original form. More will be said on this point later in the chapter. An ACG voltage can be obtained from the negative side of the 4-μf stabilizing capacitor, C_3.

20.10 QUADRATURE DETECTORS

Another approach to FM detection is provided by the quadrature detector. This circuit acts as both a discriminator and limiter in one stage. The original method required a special tube such as the 6BN6. The technique has now been expanded to use more ordinary appearing tubes such as the 6DT6. Solid-state circuits are also used to provide quadrature detection.

The 6BN6 Detector The 6BN6 gated-beam tube was invented by Dr. Robert Adler of the Zenith Radio Corporation. It possesses a characteristic such that when the grid voltage changes from negative to positive values, the plate current rises rapidly from zero to a sharply defined maximum level. This same maximum value of plate current remains no matter how positive the grid voltage is made. The current cutoff is achieved when the grid voltage is about 2 volts negative.

The reason for this particular behavior of the tube stems from its construction (see Fig. 20-19). The focus-electrode and first accelerator slot together form an electron gun which projects a thin-sheet electron stream upon grid no. 1. The curved screen grid, together with the grounded-lens slot, and aided by the slight curvature of grid no. 1, refocuses the beam and projects it through the second accelerator slot upon the second control grid. This grid and the anode which follows are enclosed in a shield box. Internally, the focus, the lens, and the shield electrodes are connected to the cathode. The accelerator and the screen grid receive the same positive voltage because both are connected internally.

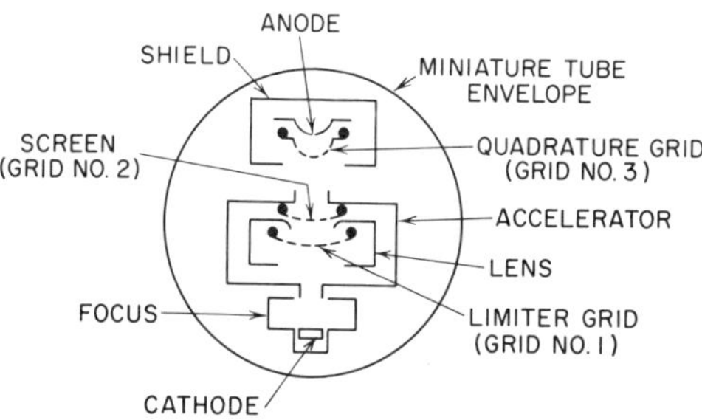

Fig. 20-19. The internal construction of the gated-beam tube 6BN6.

The foregoing design is such that electrons approaching the first grid do so head-on. Hence, when grid no. 1 is at zero potential or slightly positive, all approaching electrons pass through the grid. Making the grid more positive, therefore, cannot further increase the plate current. When, however, grid no. 1 is made negative, those electrons that are stopped and repelled toward the cathode do so along the same path taken in their approach to the grid. Because of the narrowness of the electron beam and its path of travel, electrons repelled by the grid form a sufficiently large space charge directly in the path of other approaching electrons, thus causing an immediate cessation of the current flow throughout the tube. In conventionally constructed tubes, the spread of the electron beam traveling from the cathode to the grid is so wide that those electrons repelled by the grid return to the cathode without exerting much influence on other electrons which possess greater energy and therefore are able to overcome the negative-grid voltage. It is only when the control-grid voltage is made so negative that no emitted electrons possess sufficient energy to overcome it that current through the tube ceases. These differences between tubes can be compared to the difference between the flow of traffic along narrow and wide roads. On narrow roads, the failure of one car to move ahead can slow down traffic considerably; along wide roads, where there is more room, the breakdown of one car has less effect.

The electron beam in the form of a thin sheet leaves the second slot of the accelerator and approaches grid no. 3. Thus, this section of the tube can also serve as a gated-beam system. If this second grid is made strongly negative, the plate current of the tube is cut off no matter how positive grid no. 1 may be. Over a narrow range of potential in the vicinity of zero, the third grid can control the maximum amount of current flowing through the tube. However, if the third grid is made strongly positive, it also loses control over the plate current, which can never rise beyond a predetermined maximum level.

Now let us see how this tube can be made to function as a limiter-discriminator. A typical circuit is shown in Fig. 20-20.

It has been noted that when FM signals reach the discriminator they contain amplitude variations. When the 6BN6 gated-beam tube is used, these signals are applied to control grid no. 1. If the signal receives sufficient prior amplification, it will have a peak-to-peak value of several volts. Upon application to grid 1, current through the tube will flow only during the positive part of the cycle and will remain essentially constant no matter how positive the signal may become, or what amplitude variations it may contain. Thus, signal-limiting is achieved in this section of the tube; the electron beam is passed during the positive

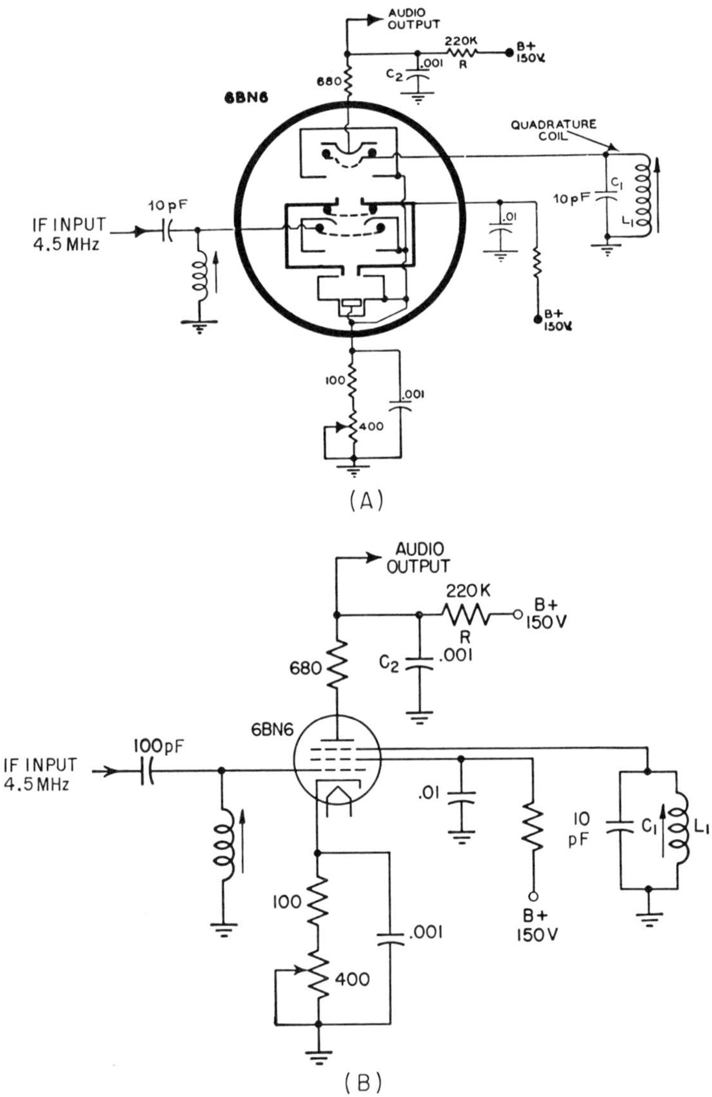

Fig. 20-20. The beam-gated tube connected as a limiter-discriminator: (A) tube shown in pictorial form; and (B) tube drawn schematically.

half-periods of the applied signal and cutoff occurs during the negative half-periods. The groups of electrons that are passed then travel through the second accelerator slot and form a periodically varying space charge in front of grid no. 3. By electrostatic induction, currents are made to flow in the grid wires. A resonant circuit is connected between grid no. 3 and ground, and a voltage of approximately 5 volts is developed in grid no. 3. The phase of this voltage is such that it will slow down the input voltage in grid no. 1 by 90 degrees, assuming that the resonant circuit is tuned to the intermediate frequency. (Because of this 90 degree difference between the grid voltages, grid no. 3 is often referred to as the "quadrature grid.")

Electrostatic induction, referred to previously, may be new to the reader. Whenever a group of electrons approach an element in a tube, electrons at that element will be repelled, resulting in a minute flow of current. By the same token, electrons receding from an element will permit the displaced electrons to return to their previous position. Again a minute flow of current results, this time in a direction opposite that of the first flow. If a sufficient charge periodically approaches and recedes from an element, the induced current can achieve substantial amplitudes. This is precisely what occurs at grid no. 3 in the 6BN6.

In the gated-beam tube, grids no. 1 and no. 3 represent electron gates. When both are open, current passes through the tube. When either one is closed, there is no current flow. In the present instance, the second gate lags behind the first. The plate-current flow starts with a delayed opening of the second gate and ends with the closing of the first gate. Now, when the incoming signal is unmodulated and L_1 and C_1 of Fig. 20-20 are resonated at the intermediate frequency, the voltage on grid no. 3 will lag the voltage on grid no. 1 by 90 degrees. However, when the incoming signal is varying in frequency, the phase lag between the two grid voltages will likewise vary. This, in turn, varies the length of time during which plate current can flow (see Fig. 20-21 (A)). Thus, plate current varies with frequency. The circuit is designed so that the current varies in a linear manner. By placing the resistor in the plate lead, R of Fig. 20-20, we can obtain an audio voltage to drive the audio amplifier that follows.

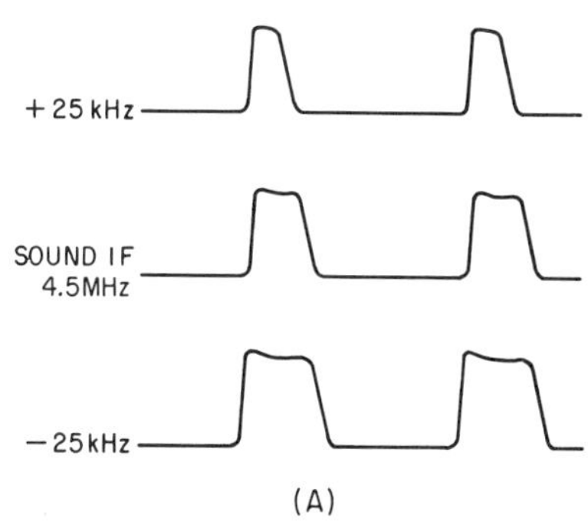

Fig. 20-21 (A). The variation in current pulse duration with different incoming frequencies in a 6BN6.

A typical response for a 6BN6 FM detector is shown in Fig. 20-21 (B). Note that this curve does not possess any sharp bends at frequencies beyond the range of the normal signal deviation. This makes the receiver easier to tune.

In the circuit of Fig. 20-20, a 680-ohm resistor is inserted between the load R and the plate of the tube. The by-passing of the IF voltage is accomplished by C_2, but since this capacitor is placed beyond the 680-ohm resistor, a small IF voltage appears at the anode of the tube. Through the interelectrode capacitance that exists between the anode and grid no. 3, the IF voltage developed across the 680-ohm resistor is coupled into L_1 and C_1. The phase relations existing in

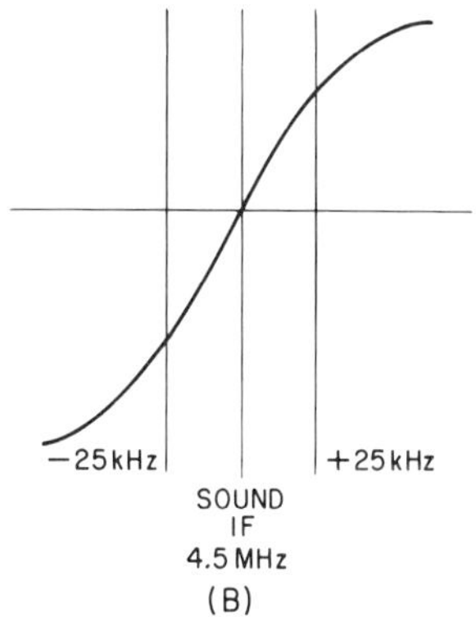

Fig. 20-21 (B). Discriminator response of the 6BN6 when connected as shown in Fig. 20-19.

this circuit are such that this feedback voltage aids in driving the tuned circuit.

The bias for grids no. 1 and no. 3 is obtained by placing a resistor in the cathode leg of the tube. Since amplitude rejection, especially at low-input signals near the limiting level, is a function of the correct cathode bias, the cathode resistor is made variable. This permits adjustments to be made in order to compensate for tube or other component changes.

The 6DT6 Detector The 6BN6 tube is, as we have seen, of special construction. Recently another tube, a 6DT6, has been similarly employed, although its internal structure is more like that of an ordinary pentode. However, in the new tube the control and suppressor grids are both capable of sharply cutting off the plate current. For this reason, they resemble grids no. 1 and no. 3 of the 6BN6. The circuit of an FM detector using a 6DT6 (or a 3DT6) is similar to the 6BN6 circuit (see Fig. 20-22). So long as the incoming signal is moderately strong, quadrature-grid detection takes place essentially as it does in the 6BN6 arrangement.

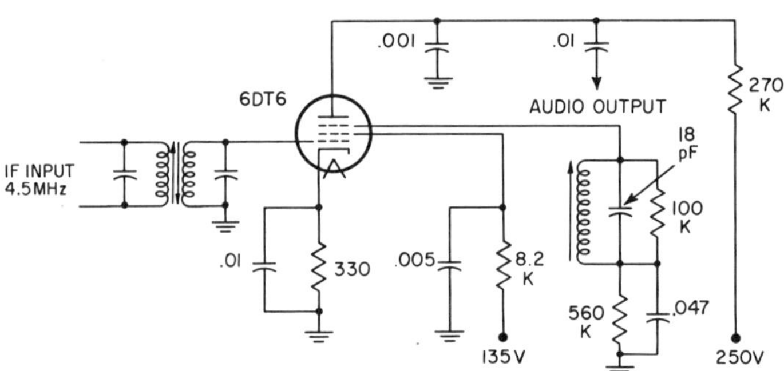

Fig. 20-22. FM-detector circuit using a 6DT6.

On weak signals, the 6DT6 circuit has a tendency to break into oscillation at the IF value. This serves to maintain the detected output-signal constant in spite of the fact that weak signals tend to vary considerably in amplitude because of noise and fading. The oscillations arise because of the feedback which takes place between the suppressor grid and the control grid within the tube. The incoming signal at grid no. 1 locks in with these oscillations and actually causes them to shift in frequency as the modulation moves the signal frequency back and forth. Normal quadrature-grid detection takes place in the oscillating detector. This oscillation boosts the sensitivity of the circuit to weak signals, causing it to deliver a clearer output under adverse conditions. However, if the applied signal becomes extremely weak, the oscillator will become unlocked, resulting in a loss of detection. Locking will occur only over a limited range of weak signal strength.

When moderate or strong signals are received, the control grid draws grid current and this loads down the circuit of the input tube. This loading not only kills any tendency to oscillate, but it also broadens the tuning response, all of which tend to limit these signals, thereby providing a certain amount of limiter action. The strong signal tends to drive the tube from plate-current cutoff to plate-current saturation. Thus, the current flow in the plate circuit will be essentially the same as that of a square-wave, and limiting action is produced. In the 6BN6, limiting is achieved by the characteristics of the tube itself.

Transistor-quadrature Detector The vacuum-tube quadrature detector is easily duplicated with only 3 transistors. These transistors must be identical to each other, so it is desirable to include them all in one integrated circuit (IC). The general circuit in use by a number of manufacturers is shown in Fig. 20-23. The portion inside the dotted lines is the part actually inside the IC. Usually the IC contains other transistors for such things as IF amplifiers or low-level audio amplifiers.

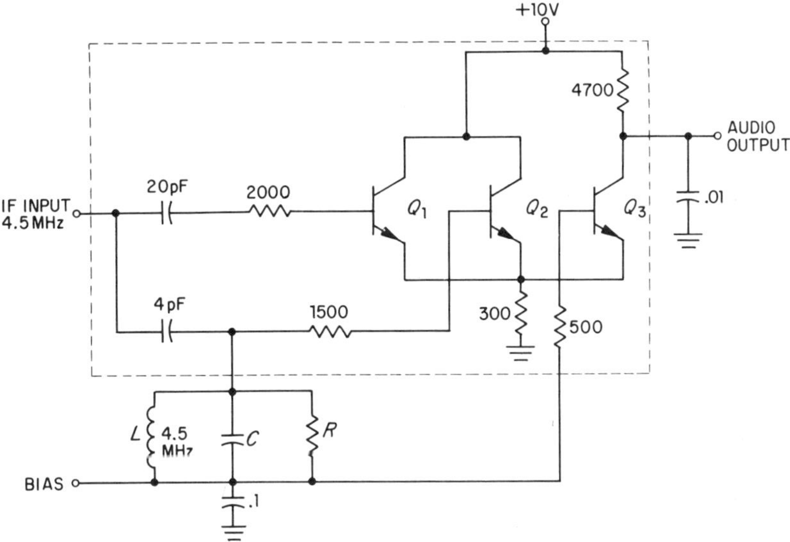

Fig. 20-23. A three transistor quadrature detector.

The principle of operation of the transistor-quadrature detector is no more difficult to understand than the gated-beam type. The IF signal has received sufficient gain before this stage so that Q_1 and Q_2 are driven very hard. The outputs on the emitters therefore have the appearance of a rectangular wave as shown in Fig. 20-21 (A). These two transistors perform the same function as grids 1 and 3 in the 6BN6. The 90-degree phase shift is accomplished in Q_2 by driving it through a small capacitor. The tuned circuit formed by L and C does not produce a phase shift at its resonant frequency (the IF center frequency). However, it does shift the phase of the signal linearly for small frequency changes from resonance. The net result of Q_1 and Q_2 being driven in quadrature is identical with that shown in Fig. 20-21(A). The only difference is that the combining of pulses is done in parallel at the emitters of Q_1 and Q_2. The gated-beam tube combines the pulses in series as they progress through the tube. The transistor version therefore produces wider pulses above resonance and narrower pulses below resonance. As seen in Fig. 20-21(A), the gated-beam tube does just the opposite. The audio output is identical in both methods. The final transistor Q_3 merely amplifies the pulses and, with the help of the 0.01 microFarad output capacitor, removes the IF components from the audio signal.

20.11 PREEMPHASIS AND DE-EMPHASIS

There is one other circuit found in FM receivers that is not used in AM sets—the so-called "de-emphasizing filter." It was required because the greatest amount of audio-frequency noise is generated in the transmitter at the higher frequencies—from 5 kHz up. To reduce the effect of the noise, a preemphasis network is inserted in the audio system of the transmitter. The function of the circuit is to favor the frequencies above 1,500 Hz. It accomplishes this by proportionately attenuating the lower frequencies more than the higher frequencies of the signals passing through the network. A typical accentuator filter is shown in Fig. 20-24(A). The higher frequencies, in passing through the network, lose less voltage than the accompanying low frequencies.

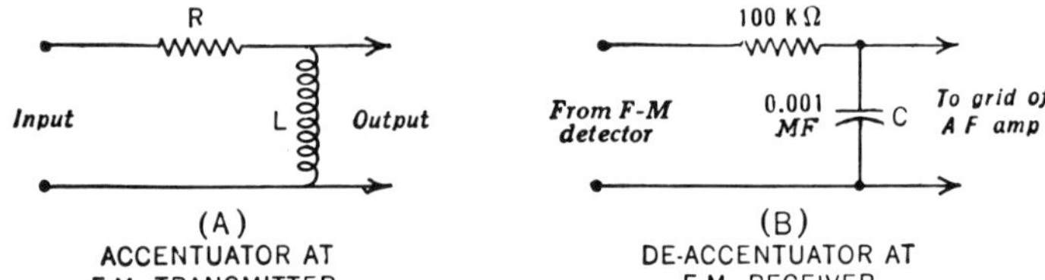

Fig. 20-24. Two circuits that help maintain the fidelity of the higher audio frequencies in FM systems.

Preemphasis is applied to the audio signals at the first audio amplifier. Beyond this network, the audio voltages combine in the usual manner with whatever noise is present in the system. At the receiver there is a de-emphasis circuit shown in Fig. 20-24(B), with the reverse properties of the preemphasis circuit. The frequencies above 1,500 Hz are reduced to their original values. At the same time a similar reduction in noise occurs. The overall effect is a return of the signal to its proper relative proportions, but with a considerable reduction in noise.

20.12 THE VACUUM-TUBE AUDIO-TV SYSTEM

A sound system which has been extensively employed in intercarrier tube sets is shown in Fig. 20-25. One stage of IF amplification (at 4.5

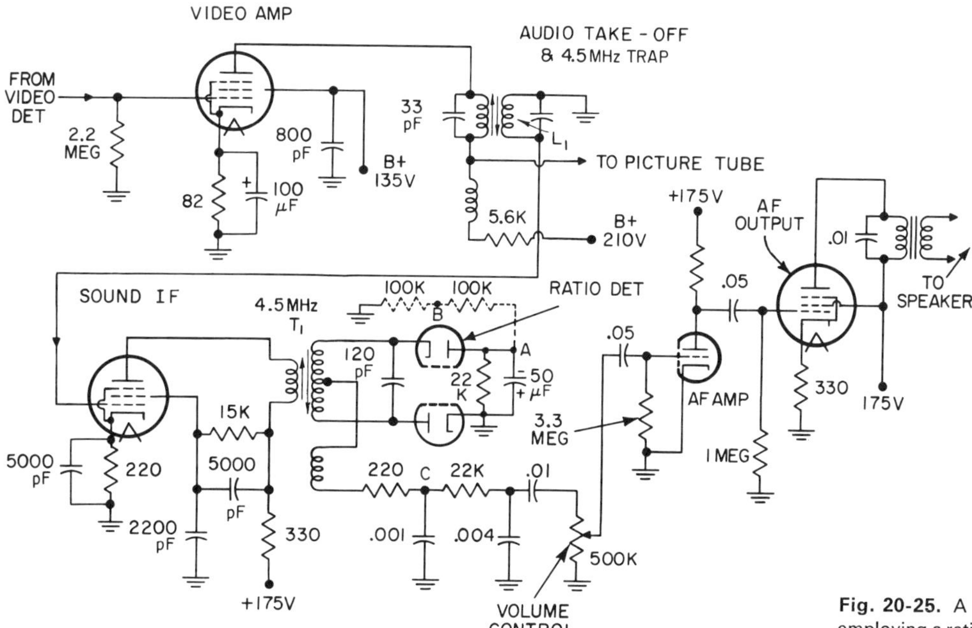

Fig. 20-25. A sound system of a television receiver employing a ratio detector (intercarrier sound system).

MHz) precedes an unbalanced ratio detector. The detected audio signal then is amplified by a triode amplifier and an audio power amplifier, and fed to a loudspeaker.

20.13 SOLID-STATE TV AUDIO SYSTEMS

Many of the early solid-state FM circuits in TV receivers were almost copies of tube sets with transistors substituted for tubes. Biasing and impedance matching also had to be changed. Recently the emphasis has been to fully exploit the unique capabilities of solid-state devices.

All-transistor Sets Figure 20-26 shows an all-transistor FM section of a color-television set. It has several novel features worth mentioning. Note that there is only one transformer operating at the IF frequency. This is the simple 2-winding transformer driving the unbalanced ratio detector.

L_1 and C_1 is a high-pass filter which passes only frequencies above 40 MHz. The sound IF frequency used is 41.25 MHz. The sound detector diode then removes the frequencies above 4.5 MHz. The remaining RF at 4.5 MHz is allowed to pass through the bandpass filter L_2, L_3, and C_2. An impedance-matching circuit, C_3 and C_4, converts this signal to the proper impedance for driving the base of the first IF transistor Q_2.

Q_2 and Q_3 form a dc coupled two-stage IF amplifier. The stabilization of the dc operating point and the frequency-response improvement is made possible by a negative feedback through R_1. The second IF

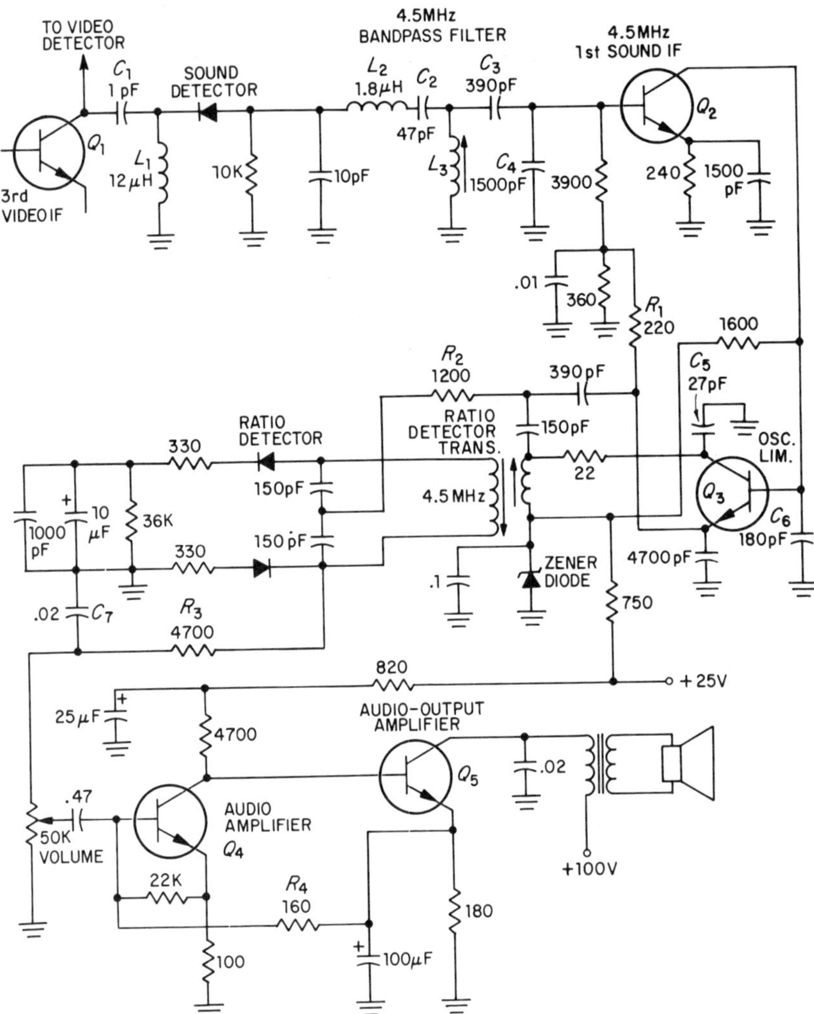

Fig. 20-26. An all-transistor FM circuit in a color-TV receiver. (*Courtesy of Admiral Radio Corp.*)

transistor, Q_3, also behaves as an oscillating limiter which possesses the same advantages as the oscillating 6DT6 quadrature detector. Weak signals are kept very constant in amplitude, and the resulting audio has a much better quality. The oscillation is sustained through the action of C_5, C_6, and the primary of the ratio-detector transformer. These components make the circuit a version of the common Colpitts oscillator. A zener diode keeps the voltages stabilized in this oscillator, so that its performance will remain constant.

The ratio-detector transformer receives its 90 degree out-of-phase signal from R_2. This out-of-phase signal combines with the transformer-output voltage similar to the Foster-Seeley discriminator (see Fig. 20-14). The two capacitors on the transformer secondary provide the same results as a center tap.

R_3 and C_7 provide the de-emphasis of high-audio frequencies. Audio amplification is achieved with a two-stage feedback amplifier. R_4 develops this feedback in a manner similar to R_1 in the IF amplifiers. This feedback broadens the bandwidth of the audio amplifier and also stabilizes the dc operating point of both stages.

Integrated-circuit Sets An ideal IC for a TV sound section would have 5 terminals: one for ground, one for power, a 4.5 MHz input, a lead for the volume control, and a connection which goes directly to the speaker. Figure 20-27 shows a TV-FM sound circuit which contains an IC and 13 other components. The two coils are single-slug tuned and simple to align.

The power transistor and output transformer are required for obvious reasons. IC's are usually small devices with low-power dissipation. Some additional gain is therefore required to change to the power and impedance requirements of a 4-ohm, 2-watt speaker.

The two IF amplifiers are not transformer-coupled or resonant-circuit coupled. Instead, adequate selectivity is obtained with just the resonance of L_1 and C_1. A gain of 2,000 at 4.5 MHz is typical. This large gain saturates the last IF amplifier and produces the square-wave signal required for the quadrature detector.

The quadrature detector is almost identical to the one shown in Fig. 20-23. This type of FM detector is easily produced as part of an integrated circuit. Tuning merely requires making L_2 and C_2 resonant at 4.5 MHz.

The audio-voltage amplifier is a 3 stage, 4 transistor, dc coupled amplifier. Direct coupling allows a good bass response and also avoids

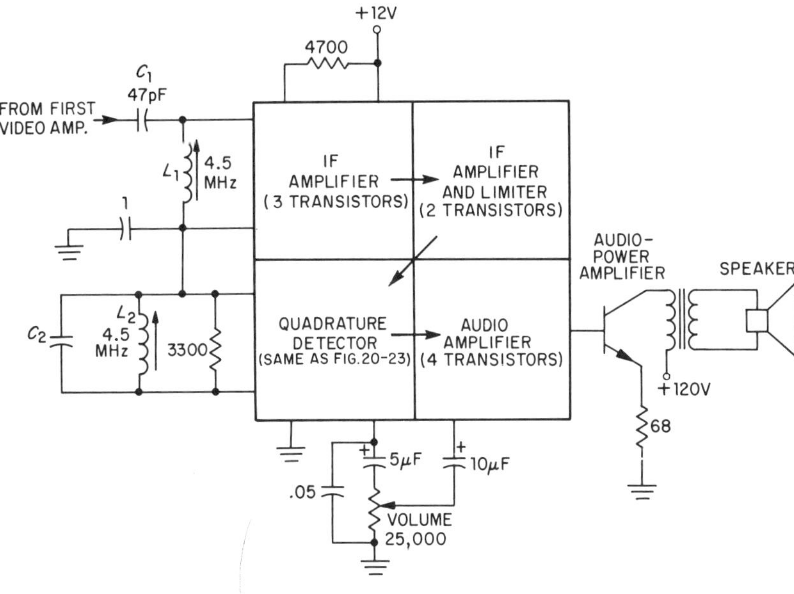

Fig. 20-27. A TV-sound section which uses one IC and a transistor as the only active elements. (*Courtesy of Fairchild Semiconductor Corp.*)

the use of capacitors. Note that, except for the speaker transformer and the two capacitors associated with the volume control, this system would have a bass response down to dc.

20.14 TROUBLES IN FM-SOUND CIRCUITS

Troubles in TV sound systems are usually in one of two categories—reduced sound volume or noisy sound.

Loss of Sound Volume A loss of gain in any transistor or tube in the FM section, except for the limiter and FM detector, will cause a proportionate loss of speaker volume. Tube (or transistor) testing or signal tracing with an oscilloscope usually isolates this problem.

Noisy Sound. This problem can come in the form of a 60-Hz hum (or buzz), random noise, or distortion of the sound. Hum can come from any stage after the limiter. The limiter usually eliminates the problem if the hum occurs in preceding stages.

Buzzing usually is caused by vertical-sync pulses getting through the limiter. The alignment of the first IF stage and its coupling to the previous stage are also suspect when buzzing occurs.

Random noise can come from any stage in the TV set. Often a weak stage causes other stages to work harder, thus generating noise. The volume control is helpful in determining if any of the noise sources originate before or after this point. If the noise goes up and down with the volume-control setting, then the source of the noise is before the volume control.

The distortion of the sound usually occurs when the gain of an audio stage becomes nonlinear. Bias drift can cause this to happen. Distortion can also be caused if the FM detector is grossly misaligned.

REVIEW QUESTIONS

1. In an AM wave, where is the intelligence contained? How does this differ from the conditions prevailing in FM?
2. A 100% modulated AM radio signal is the maximum modulation achievable before the distortion of the audio occurs. Does FM have this same problem? What problems would the overmodulation of FM cause?
3. What is a discriminator?
4. What purpose do the IF amplifiers serve in an FM superheterodyne?
5. Explain the operation of an IF limiter-amplifier.
6. Draw the schematic diagram of an early (basic) type of discriminator which employs two secondary windings.
7. Explain the operation of the circuit drawn for Question 6.

8. Would the foregoing discriminator function if one of the diodes became inoperative? Give the reason for your answer.
9. Draw the circuit of a ratio detector.
10. Explain briefly the operation of a ratio detector.
11. What properties of a 6BN6 enable it to be used as a limiter?
12. What component causes the 90 degree phase shift in Fig. 20-23.
13. Why does the ratio detector of Fig. 20-26 have a 1000-pF capacitor across the 10-μf capacitor?

21 Principles of Color TV

21.1 INTRODUCTION

Emphasis throughout the preceding chapters has been directed largely to the underlying principles of transmission and reception of black-and-white pictures. In such a system, only black, white, or intermediate shades appear on the receiver viewing screen. The result is similar in all respects to the ordinary motion picture. Although the reproduced image is certainly far from being an exact duplicate of the full-colored scene originally televised at the studio, it imparts sufficient information to prove entertaining. The public has long been accustomed to black-and-white pictures in the motion-picture theater and accepts with little or no objection the same type of image in a television receiver.

The appeal of color television lies in its greater naturalness. We live in an environment that contains many shades of color, and to desire the same lifelike qualities in television is quite understandable. Color in an image heightens the contrast between elements, brightens the highlights, deepens the shadows, and appears to add a third dimension to a flat reproduction. More detail appears to be present in colored images containing fewer lines than corresponding black-and-white pictures. Perhaps many readers have noticed the remarkable differences between color motion films and ordinary motion pictures. Similar differences are observed with television.

21.2 ELEMENTS OF COLOR

Color, physicists tell us, is a property of light. If we take sunlight and pass it through a glass prism, a variety of colors are produced. White sunlight contains all colors, but, owing to the limitations of the human eye and the fact that the colors produced by a prism blend into one another, we can count only six fairly distinct colors (red, orange, yellow, green, blue, and violet). Upon closer inspection of this color distribution, numerous fine gradations can be distinguished, both between different colors and within any one color itself. For example, red, when it first becomes definitely distinguishable from its neighbor, orange, possesses a different shade than that at the other end of the red band where infrared wavelengths are approached.

It is a common experience with all persons who are not color-blind to find that objects which possess one color under an electric light may assume a considerably different color when examined in the sunlight. The difference is due to the fact that the color of an object is a function of the wavelengths of the light which the object does not absorb.

Thus, if we shine white light on a body and none of it is absorbed, we see a white body. However, if under the same white light the object appeared blue, then the object would be absorbing all the other components of white light and reflecting blue.

To see the true color of an object, we must examine it under a light which contains all the wavelengths of the visible spectrum. Thus, a blue object appears much darker under an ordinary incandescent lamp than it does in sunlight. This is because the lamp has an excess of red light and a deficiency of blue. Since a blue object will reflect only blue rays, it will reflect less light under an incandescent lamp and give a darker appearance. In sunlight, blue and red are present to the same extent, and the object assumes its proper color.

With objects that are transparent, the color is determined by the light which is transmitted through the object. Thus, in a green piece of glass, green is permitted to pass through, whereas the other colors are absorbed.

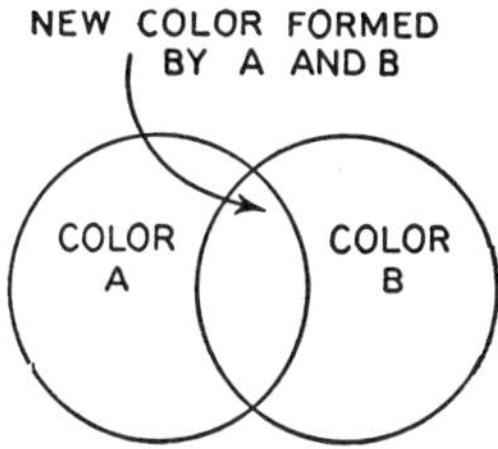

Fig. 21-1. Two circles of light, *A* and *B*. Where they overlap, they form a third color different from either *A* or *B*.

Color Primaries Anyone who has ever experimented with projector lamps has discovered that when differently colored lights from several projectors are combined, the resultant color seen by an observer will differ in hue from any of the projected lights. Thus, for example, yellow can be formed by combining red and green light; white light can be produced by combining red, green, and blue. The color of the mixed light will appear to the eye as a complete color, and the eye will be unable to distinguish the various components of the mixture that produced the color.

This method of color formation is illustrated in Fig. 21-1. Two circles of colored light are projected onto a screen and positioned so that they overlap to some extent. Within the overlapping region, a new color will be produced by the addition of color *A* and color *B*. Where the circles of light do not overlap, each light will retain its original color. If a third circle of light is added, as shown in Fig. 21-2, then additional colors can be obtained. These colors are

Color *A* (blue)	Color *D* (formed from *A* and *B*)—Cyan
Color *B* (green)	Color *E* (formed from *A* and *C*)—Magenta
Color *C* (red)	Color *F* (formed from *B* to *C*)—Yellow
	Color *G* (formed from *A*, *B*, and *C*)—White

and each differs from the others. In the areas where the circles of light overlap, the eye is not able to distinguish each of the colors forming the mixture, but instead the eye sees only a new color. Fig. 21-2(B) shows that the brightness of the colors produced at the junction of any two of the primaries is the result of the direct addition of the brightness of the two primaries. Yellow, for example, appears to be 89 percent as bright as the reference white, reflecting the addition of the 59 percent brightness of green and the 30 percent brightness of red.

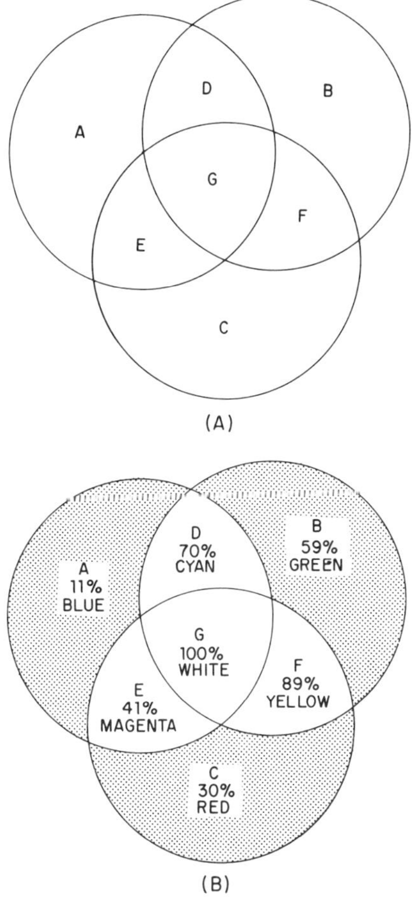

Fig. 21-2. The mixing of three colors—*A*, *B*, and *C*—results in four new ones: *D*, *E*, *F*, and *G*.

The number of different colors that can be formed by the use of three colored lights depends upon the colors chosen. Experience has indicated that the colors red, blue, and green, when combined with each other in various proportions, will produce a wider range (or gamut) of colors than any other combination of three colors. Note, however, that, if we use four different colors in our mixing process, we can produce an even wider range of different colors With the addition of more and more colors to our mixing scheme, the reproducible range will increase somewhat. Obviously, however, a line must be drawn, and the use of three colors has become standard. The three colors chosen—red, green, and blue—are thus referred to as the "primary colors."

Chromaticity Chart A diagram which is convenient to use for color mixing is the tongue-shaped (or horseshoe-shaped) curve shown in Fig. 21-3. This is known as a "chromaticity chart." The positions of the various spectrum colors from blue at one end to red at the other are indicated around the curve Any point not actually on the solid-line curve but within the area enclosed by the curve represents not a pure spectrum color but a mixture of spectrum colors Since white is such a mixture, it, too, lies in this area, specifically at point *C*. This particular point was chosen at an international convention in England and is generally referred to as "illuminant *C*." Actually, of course, there is no specific white light; sunlight, skylight, and daylight are all forms of white light, and yet the components of each differ considerably The color quality of a conventional black-and-white television receiver tube is represented by a point in the central region of the diagram about point *C*.

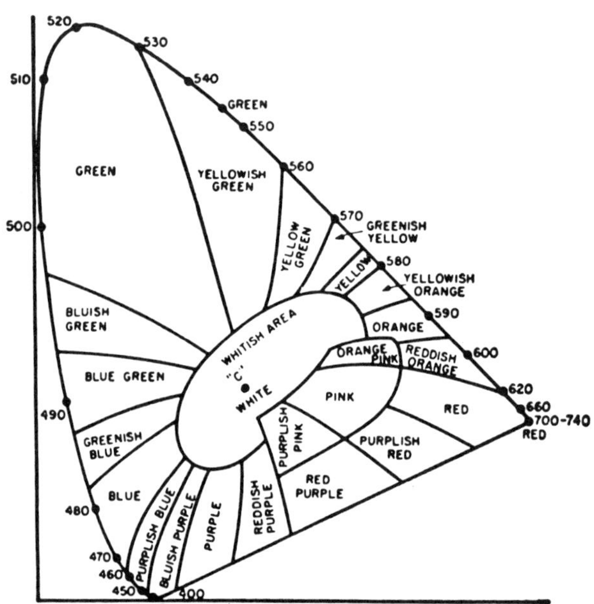

Fig. 21-3. A chromaticity diagram. The numbers listed around the perimeter of the chart represent the wavelengths of the various colors in millimicrons.

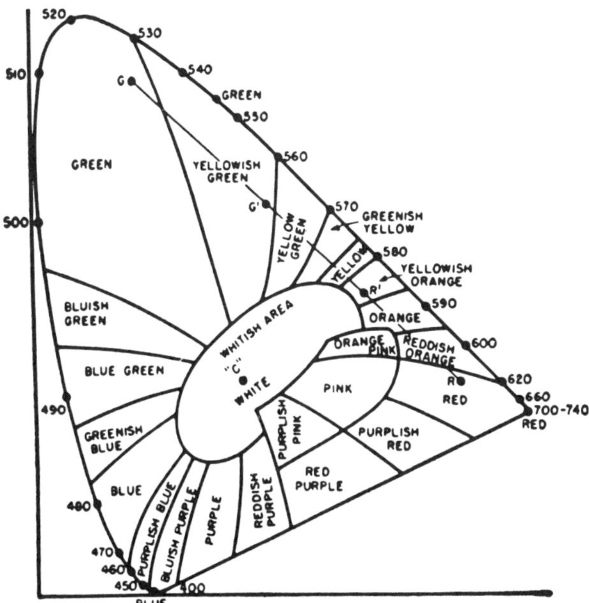

Fig. 21-4. The line drawn between points *R* and *G* passes through all the colors that can be obtained by mixing these two shades of the red and green hues.

If Fig. 21-4 were reproduced in full color, it would be seen that the color changes gradually from point to point. The deepest and most intense colors are obtained at the outer edge of the diagram. Here we find the real deep red, deep blue, and deep green shades which we actually see very seldom in everyday life. More familiar are the lighter colors, appearing as we move in toward the center. These are the pastels such as pink, light green, and pale blue. Finally at the center come the white with point *C* as the reference white, or, for our purposes here, the "whitest" white. Actually this is a rather nebulous shade, entirely arbitrary in value and simply chosen for certain conveniences.

The chromaticity chart lends itself readily to color-mixing because a straight line joining any two points on the curve will indicate all the color variations that can be obtained by combining these two colors in varying amounts. Thus, in Fig. 21-4 consider the line connecting points *R* and *G* as representing certain shades of red and green respectively. If there is more red light than green light, the exact point representing the new color will lie on the line, but be closer to *R* than to *G*. Point *R*′ might be such a color. On the other hand, if a greater percentage of green light is used, the new color will still lie on the line connecting *R* and *G*, but will now be closer to *G* than to *R*. Point *G*′ might be such a color. This method of combining colors can be carried out for any two colors on the chart.

It is possible to specify the purity of a color by its distance from point *C*. Consider point *B* in Fig. 21-5. This is halfway along the line between point *C* (white) and point *A* (green). Hence, point *B* represents a mixture of green diluted 50 percent with white light, and we can say

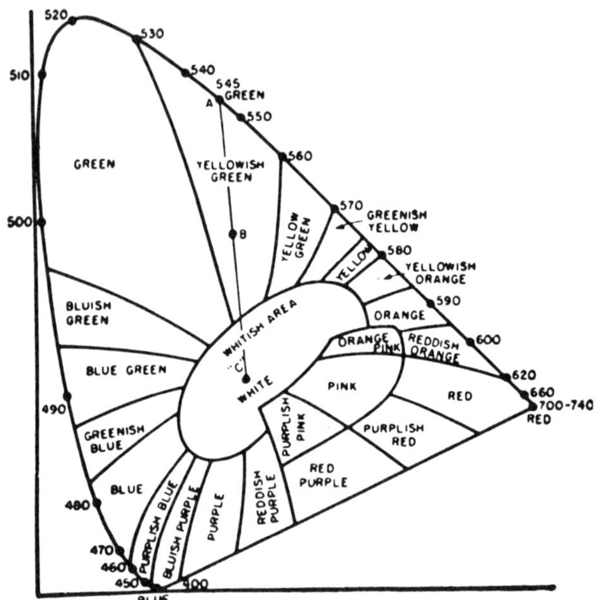

Fig. 21-5. When moving from *A* to *C*, the green becomes less and less saturated, or lighter in intensity.

that the purity of this color is 50 percent. Had the distance between point *C* and point *B* been 75 percent of the total distance between point *C* and point *A*, we would have stated that the purity of the color at point *B* was 75 percent. As point *B* moves closer and closer to the spectrum curve, the purity of the color it represents increases until it becomes 100 percent at the curve—point *A*. As point *B* moves closer to point *C*, its purity decreases. At point *C*, the purity is said to be zero.

In place of purity, the word "saturation" is frequently used. Any point located on the tongue-shaped curve is said to be completely saturated. As we leave the curve and approach closer to point *C*, more and more white light is added to the color and it becomes lesss aturated or, what is the same thing, more desaturated. And, at point *C*, the saturation is zero.

In connection with saturation, the word "hue" is frequently heard. Hue represents color, such as red, green, and orange. The term is associated with color wavelength, and when we call a certain color green, or orange, or red, we are specifying its hue. Thus, hue refers to the "basic" color as it appears to us, while saturation tells us how "deep" the color is. If the color is highly saturated, we say that it is a deep color, such as deep red or deep green. If it contains a considerable amount of white light, we say it appears "faded," as a faded red or a faded green.

21.3 THE NTSC* COLOR TELEVISION SYSTEM

We are now ready to study the NTSC color-television system, the system officially adopted by the FCC in 1953. This system has been so designed

* National Television System Committee.

that its signal occupies no more than 6 MHz (video and sound), and it carries not only the full black-and-white (or monochrome) signal but, in addition, the color information.

The question is: How is all this information compressed into a 6MHz bandspread? The answer is to be found in the nature of a television signal. It was discovered as far back as 1929 that a monochrome 4 MHz video signal does not occupy every cycle of the 4 MHz assigned to it. Rather, this signal appears in the form of "clusters" of energy located about the harmonics of the 15,750 Hz line-scanning frequency. The monochrome signal energy is grouped around these points, with relatively wide gaps between them (see Fig. 21-6). Since these empty spaces are not being used, they can be employed for the trasnmission of additional information, and here is specificially where the color information of the NTSC color-television signal is placed (see Fig. 21-7). The practice of placing the information of one signal between the clusters of energy of another signal is known as "interleaving."

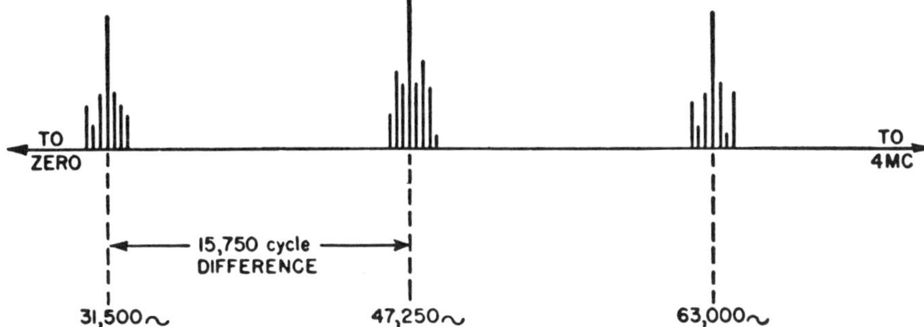

Fig. 21-6. Partial spectrum distribution of a monochrome signal.

Thus, a total color signal consists of two components—a monochrome signal and the signal which carries information concerning color. Let us examine each component separately.

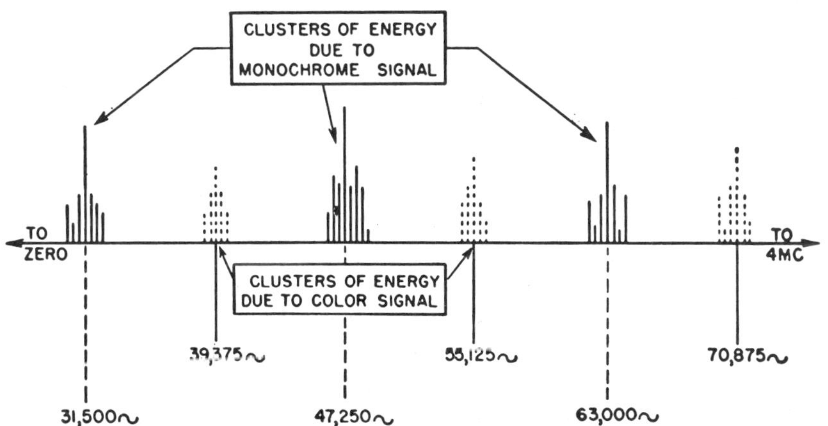

Fig. 21-7. The information of the color signal is inserted in gaps between the clusters of energy of the monochrome signal.

The Monochrome Signal The black-and-white, or monochrome, portion of the total color signal is equivalent in all respects to the present black-and-white signals. It is formed by combining the red, green, and blue signals from their respective color cameras in these proportions (see Fig. 21-2):

$$Y = 0.59G + 0.30R + 0.11B$$

where

Y = a mathematical symbol representing the monochrome signal
G = green signal
R = red signal
B = blue signal

This particular proportion was chosen because it closely follows the color sensitivity of the human eye. That is, if you take an equal amount of green light, an equal amount of red light, and an equal amount of blue light and superimpose the rays from these lights on a screen, you will see white. However, if you then look at each light separately, the green will appear to be twice as bright as the red, and from 6 to 10 times as bright as the blue. This is because the eye is more sensitive to green than to red and more sensitive to red than to blue. It is in recognition of this fact that the proportions given above were chosen.

Thus, the monochrome signal is composed of 59 percent green signal (that is, 59 percent of the output of the green camera), 30 percent red, and 11 percent blue, and contains frequencies from 0 to 4 MHz. (The use of the letter Y to denote the monochrome portion of the color signal is a common practice and should become familiar to the reader.)

Other names for this monochrome signal are luminance signal and brightness signal. These terms were chosen, because they indicate more clearly the action of this signal. Every monochrome-video signal contains nothing but the variations in amplitude of the picture signal, and these amplitude variations, at the picture tube, produce changes in light intensity at the screen.

The Color Signal The second component of the television signal is the color signal itself. This, as we have just seen, is interleaved with the black-and-white signal. To determine what information this portion of the total signal must carry, let us first see how the eye reacts to color, since it is the eye, after all, for which the color image is formed.

A number of men have investigated the color discerning characteristics of the human eye, and, briefly, here is what they found. The typical human eye sees a full color range only when the area or object is relatively large. When the size of the area or object decreases, it becomes more difficult for the eye to distinguish between colors. Thus, where the eye required three primary colors, now it finds that it can get along very well with only two. That is, these two colors will, in different combinations with each other, provide the limited range of colors that the eye needs or can see in these medium-sized areas.

Finally, when the detail becomes very small, all that the eye needs or can discern are changes in brightness. Colors cannot be distinguished from gray, and, in effect, the eye is color-blind.

These properties of the eye are put to use in the NTSC color system. First, only the large- and medium-sized areas are colored; the fine detail is rendered in black and white. Second, as we shall see later, even the color information is regulated according to bandwidth. That is, the larger objects receive more of the green, red, and blue than the medium-sized objects.

The color signal takes the form of a subcarrier and an associated set of sidebands. The subcarrier frequency is approximately 3.58 MHz. This represents a figure which is the product (approximately) of 7,875 Hz multiplied by 455. The value 7,875 is one-half of 15,750, and if we use an odd multiple (1, 3, 5, etc.) of 7,875 as a carrier, then the frequency will fall midway between the harmonics of 15,750 Hz. If we used even multiples of 7,875, we would end up with 15,750 Hz or one of its harmonics, and this would place the color signal at the same points (throughout the band) as those occupied by the black-and-white signal. Refer back to Fig. 21-7. By taking an odd multiple of 7,875, we cause the second signal to fall in between the bundles of energy produced by the first signal, and the two do not interfere.

Now that we have a color carrier (or "subcarrier," as it is known), the next step is to provide it with a properly modulated signal to enable the receiver to develop a color picture. Ordinarily, the information required would consist of *R*, *G*, and *B*, since these are the three primary colors from which all other colors are derived. This means modulating the color subcarrier with three different voltages. Actually, however, we can do the same job by using only two quantities if we resort to the following modification. Take the *R*, *G*, and *B* voltages and combine each with a portion of the monochrome signal after the latter has been inverted 180 degrees. This produces $R-Y$, $G-Y$, and $B-Y$ signals. We can do this by taking a portion of the brightness signal (*Y* signal) and passing it first through a low-pass filter (see Fig. 21-8).

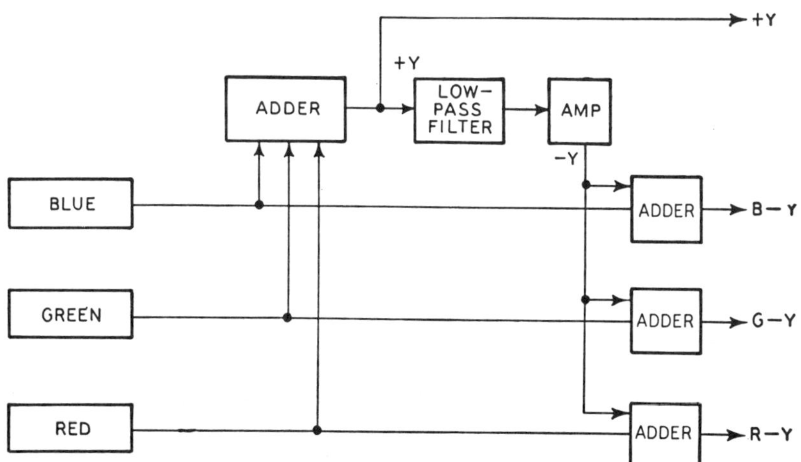

Fig. 21-8. Block diagram illustrating how color signals minus brightness signals are formed.

This permits only the lower frequency components (0–1.5 MHz) to get through, which is satisfactory since the color signals are concerned only with the lower frequencies. Then the brightness signal is passed through an amplifier where it is inverted. This gives us the desired $-Y$. This is then added to each of the three color signals or voltages to produce a $G-Y$, and $R-Y$, and a $B-Y$ signal.

At the receiver, the original R, G, and B can be reobtained by adding Y to $G-Y$ to obtain G, by adding Y to $R-Y$ to get R, and by combining Y with $B-Y$ to get B.

Thus far, it would seem that we have only exchanged R, G, and B for $R-Y$, $G-Y$, and $B-Y$. However, once this is done, it turns out that, instead of requiring the three color-difference signals, all we really need are two, say $R-Y$ and $B-Y$. This is so because the G information is already present in the Y, or brightness, signal, since the latter contains voltages from all three colors ($Y = 0.59G + 0.30R + 0.11B$). Hence, if we send along only $R-Y$ and $B-Y$ in the color signal to the receiver, we can use these to obtain the $G-Y$ information that we need.

We now have only two pieces of color information to send, and somehow the 3.58 MHz color subcarrier frequency must be modulated by $R-Y$ and $B-Y$ voltages without conflict with each other. The best solution to this problem, designers have found, is to take the $B-Y$ and $R-Y$ signals and apply each to a separate modulator. At the same time, 3.58 MHz carriers are also applied to each modulator, but with this difference. Their frequencies are the same, but one carrier is 90 degrees out-of-phase with the other. After the carriers are amplitude-modulated, they are then combined to form a resultant carrier. This is best illustrated by means of vectors.

In Fig. 21-9(A), the $B-Y$ vector represents the $B-Y$ modulated carrier; the $R-Y$ vector represents the carrier modulated by the $R-Y$ voltage. When these voltages, or signals, are combined, a resultant is formed. If the $R-Y$ and $B-Y$ signals are equally strong, the resultant will occupy the position shown in Fig. 21-9(B). If the $B-Y$ signal is predominant, the resultant will be drawn closer to it (see Fig. 21-9(C)). On the other hand, if the $R-Y$ signal is the stronger, the position of the resultant vector will shift toward it (see Fig. 21-9(D)). Thus, we can see that the phase angle of the resultant will be governed by the coloring or hue of the picture, whereas the amplitude (or length) of the vector will determine the saturation of the colors.

This particular fact is of great importance in the receiver. If we should change the phase of the resultant with respect to $B-Y$ or $R-Y$, then the colors reproduced on the screen will be incorrect. Hence, present circuit designs incorporate a special phasing control which enables us to compensate for any phase shift that may occur. The position of this control in the circuit will be discussed presently.

Note that the $B-Y$ and $R-Y$ signals amplitude-modulate their separate carriers prior to the addition, and that therefore each modulated signal possesses a 3.58 MHz carrier and a series of sidebands (like every AM signal). When the resultant is formed, the sidebands are brought along with it.

If we pause and reconstruct our total color signal, here is what we find. First, there is the Y, or monochrome, signal and it extends over the entire video-frequency range from 0 to 4.0 MHz. Second, there is a color subcarrier, with a frequency of 3.58 MHz. This carrier is modulated by the $R - Y$ and $B - Y$ signals, and the modulation intelligence is contained in a series of sidebands that extend above and below 3.58 MHz. Just how far above and below depends on the band of frequencies

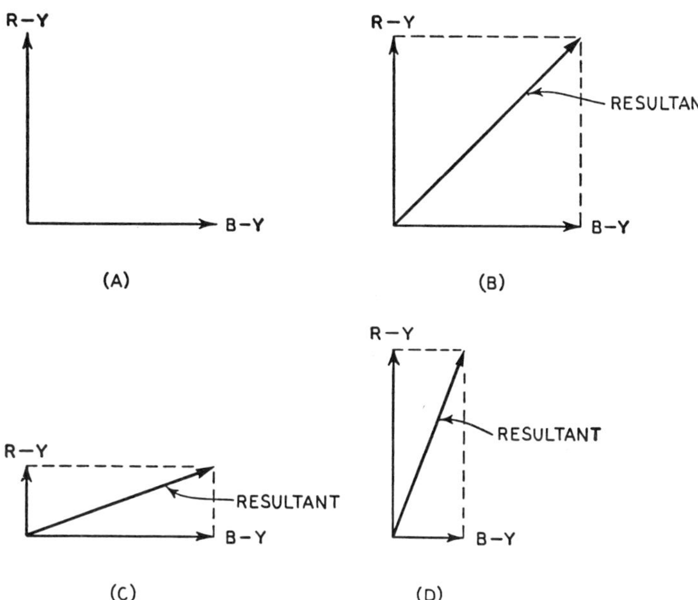

Fig. 21-9. The angular position and amplitude of the resultant carrier for various amplitudes of B-Y and R-Y: (A) The R-Y and B-Y vectors; (B) The resultant when R-Y and B-Y are equal; (C) The resultant when B-Y is stronger than R-Y; and (D) The resultant when R-Y is stronger than B-Y.

contained in the $R - Y$ and $B - Y$ modulating voltages. It was discovered that the eye is satisfied by the color image produced if we include color information only up to 1.5 MHz, while the portion of the image from 1.5 MHz to 4.0 MHz is rendered in black and white. Hence the sideband frequencies of the color-modulating voltages (so far, $R - Y$ and $B - Y$) need extend only from 0 to 1.5 MHz. Furthermore, we can even modify this set of conditions somewhat because the three primary colors are required only for large objects or areas, say, those produced by video frequencies up to 0.5 MHz. For medium-sized objects, say, those produced by video frequencies from 0.5 to 1.5 MHz, only two primary colors need be employed.

In other words, to take advantage of this situation, we need two color signals: one which has a bandpass only up to 0.5 MHz, and one which has a bandpass from 0 to 1.5 MHz. The next problem, then, is to determine the composition of these two color signals.

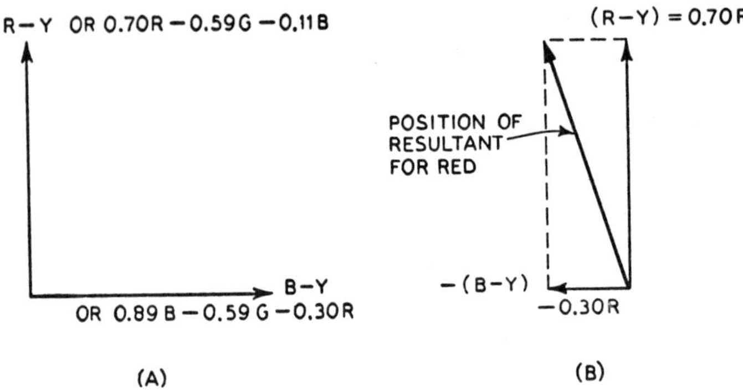

Fig. 21-10. How color determines the position of a resultant vector. (A) Equations showing the composition of *B-Y* and *R-Y* in terms of *R*, *G*, and *B*; (B) Position of the signal vector when the red field is being scanned.

To understand the answer to this problem, let us return to the vector diagram (Fig. 21-9) which shows the $R-Y$ and $B-Y$ signals. This diagram is redrawn in Fig. 21-10(A), and we have added the equivalent equation for Y.

$$Y = 0.59G + 0.30R + 0.11B$$

For $R-Y$, then, we have

$$R-Y = R - 0.59G - 0.30R - 0.11B$$

or

$$R-Y = 0.70R - 0.59G - 0.11B$$

and, for $B-Y$, we obtain

$$B-Y = B - 0.59G - 0.30R - 0.11B$$

or

$$B-Y = 0.89B - 0.59G - 0.30R$$

This means that the $R-Y$ and $B-Y$ vectors contain R, G, and B voltages in the proportions shown.

Now, let us suppose that the color camera is scanning a scene containing only red. Then, no green or blue voltages will be present and the $R-Y$ signal becomes simply $0.70R$, while the $B-Y$ signal is reduced to $-0.30R$. This set of conditions is shown in Fig. 21-10(B), with the position, too, of the resultant vector. In other words, this is the position the resultant vector will occupy when red only is being sent.

By following the same process, we can obtain the position that the resultant vector occupies when only green is being sent, or blue, or any other color formed by combining the three primary colors in any combination. A number of colors are shown in Fig. 21-11, and we see,

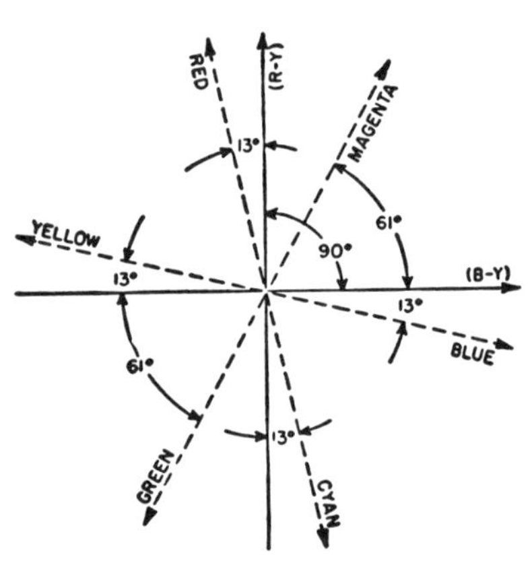

Fig. 21-11. The phase of the color subcarrier depends upon the color to be transmitted.

perhaps more clearly than before, how the phrase of the color subcarrier changes as the color to be transmitted varies. To reiterate: The phase angle of the resultant is governed by the coloring of the picture, whereas the amplitude (or length) of the vector determines the saturation of the colors.

The designers of the NTSC system found that, while they could use $R - Y$ and $B - Y$ for the color signals, a better system operation would result if they chose two other signals situated not far from the $R - Y$ and $B - Y$ signal. These two other signals were labeled I and Q, and their position with respect to $R - Y$ and $B - Y$ is shown in Fig. 21-12.

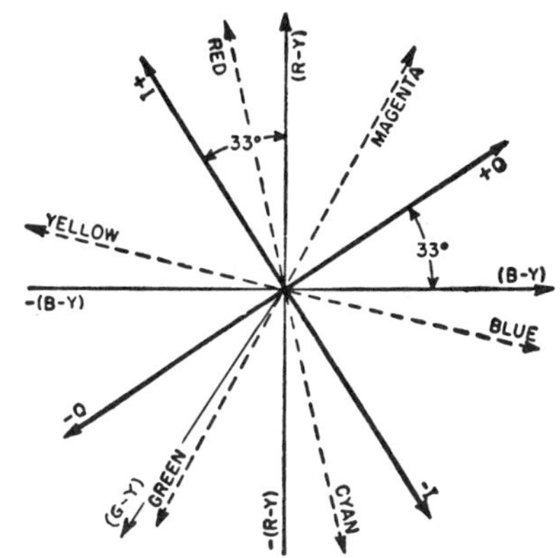

Fig. 21-12. The positions of the I and Q signals with respect to R-Y and B-Y.

Thus, where before we had $R - Y$ and $B - Y$ voltages modulating the the 3.58 MHz color subcarrier, we now substitute I and Q signals. Furthermore, the Q signal possesses frequencies up to 0.5 MHz, whereas the I signal is permitted to have sidebands up to 1.5 MHz.

Now, what do we gain from this? For all color-signal frequencies up to 0.5 MHz, both I and Q are active, and since they are 90 degrees apart, as were $R - Y$ and $B - Y$, they will act just the way $R - Y$ and $B - Y$ acted. That is, they will produce, in combination with each other, all of the colors shown in Fig. 21-12. Hence, whether we use I and Q or $R - Y$ and $B - Y$ as our modulating voltages for color-signal frequencies up to 0.5 MHz, we obtain precisely the same results.

Consider, however, the situation for color-signal frequencies from 0.5 MHz to 1.5 MHz. The Q signal drops out, and only the I signal remains to produce color on the picture-tube screen. From Fig. 21-12, we see that positive values of the I signal will produce colors between yellow and red, or actually a reddish orange. On the other hand, negative values of I will produce colors between blue and cyan, or, in general, in the bluish-green range. Hence, when only the I signal is active, the colors produced on the screen will run the gamut from reddish-orange to bluish-green.

But why do we want this arrangement? If you go back to an earlier paragraph, you will recall that, for medium-sized objects (say, those produced by video signals from 0.5 MHz to 1.5 MHz), the sensitivity of the eye is reduced. Actually, for medium-sized objects, it was found that the eye is sensitive principally to the bluish-greens or the reddish-oranges. The NTSC signal (via its I component) is fashioned to take advantage of this fact by producing only blue-greens or reddish-oranges for medium-sized objects.

We are now in a position to consider the color signal in all its aspects:

(1) There is a monochrome signal with components that extend from 0 to 4 MHz. This is the Y signal.

(2) The color subcarrier frequency is set at 3.58 MHz (actually it is 3.579545 MHz).

(3) This color subcarrier is modulated by two color signals called the I and Q signals with $I = 0.60R - 0.28G - 0.32B$ and $Q = 0.21R - 0.52G + 0.31B$.

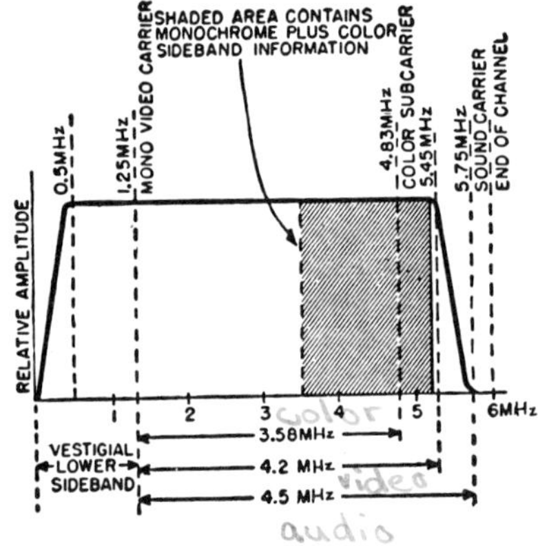

Fig. 21-13. The distribution of the full color signal within its allotted band.

(4) The *Q* signal has color frequencies that extend from 0 to 500 kHz, or 0.5 MHz. This means that the upper *Q* sidebands extends from 3.58 MHz up to 3.58 + 0.5, or 4.08 MHz. The lower *Q* sideband goes from 3.58 MHz down to 3.58 − 0.5, or 3.08 MHz.

(5) The *I* signal has color frequencies that extend from 0 to 1.5 MHz. When this modulates the color subcarrier, upper and lower sidebands are formed. The lower sideband extends from 3.58 MHz down to 3.58 − 1.5, or 2.08 MHz. If the full upper sideband were permitted to exist, it would extend all the way up to 3.58 + 1.5, or 5.08 MHz. Obviously this would prevent the use of a 6.0 MHz overall band for the television signal (video and sound). To avoid this spilling over beyond the limits of the already established channels, the upper sideband of the *I* signal is limited to about 0.6 MHz. This brings the upper sideband of the *I* signal to 4.2 MHz. The video pass band then ends rather sharply at 4.5 MHz (see Fig. 21-13).

There is one further fact of importance in the make-up of a color television signal, and this concerns the color subcarrier. We know that the 3.58 MHz carrier is modulated by the *I* and *Q* color signals. Now, in conventional modulation methods, both the carrier and the sidebands are present when the signal is finally sent out over the air. The intelligence (or modulation) is contained in the sidebands and actually is all that interests us. However, the carrier is sent along because it is required in the receiver in order to reverse the modulation process and recreate the original modulating voltages.

In the NTSC color system, the color subcarrier is not sent along with its sidebands (after the latter have been formed). Instead, it is suppressed by means of a balanced modulator. This particular practice is followed for two reasons. First, by suppressing the color subcarrier, we reduce the formation of a 920 kHz beat note between it and the 4.5 MHz sound carrier, which is also a part of every television broadcast. This 920 kHz beat note would appear as a series of interference lines on the face of the picture tube. Now, it is true that the color sidebands are present and that they can (and do) beat with the 4.5 MHz sound carrier to produce similar low-frequency beat notes. However, in any signal, the carrier usually contains far more energy than any of its sidebands; hence, when we suppress the carrier, in effect we are suppressing the chief source of this interference. Whatever other interference may be produced by some of the stronger sidebands near 3.58 MHz can be more easily dealt with by using traps in the IF system. This will be seen when we examine the circuitry of a receiver.

The second reason for using the suppressed-carrier method is that it leads to an automatic removal of the entire color signal when the televised scene is to be sent wholly as a black-and-white signal. When this occurs, *I* and *Q* become zero, and since the balanced modulators suppress the carrier, no color signal at all is developed.

With these advantages of carrier suppression comes one disadvantage. When the color sidebands reach the color section of the receiver, a carrier must be reinserted in order to permit detection to take place. Offhand, one might suppose that an oscillator operating at 3.58 MHz will be needed. This is one requirement. A second and vitally important consideration is the phase of this reinserted carrier. Remember that, at the transmitter, attention was given to the phase of *I* and *Q* as they were introduced into the modulator. If the same relative phase is not maintained in the reinserted carrier, the colors obtained at the output of the color circuits will not possess the proper hue.

Figure 21-14 shows the phase and color relationship as seen on a kinescope screen. Compare this figure to Fig. 21-12.

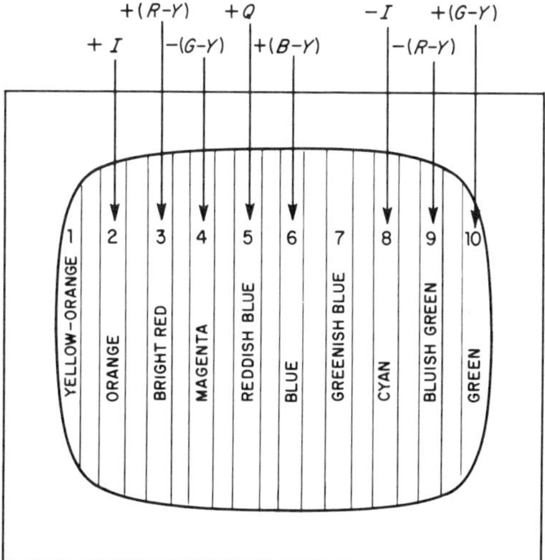

Fig. 21-14. The bar pattern developed by a color-bar generator. See Fig. 21-12 for corresponding phase angles.

21.4 THE COLOR-BURST SIGNAL

The proper detection of the color-subcarrier sidebands is not possible if the correct phase relationship is not maintained between the 3.579545 MHz subcarrier oscillators located at the transmitter and receiver. Since the 3.579545 MHz signal is suppressed at the transmitter, a short, separate "BURST" sample of the 3.579545 MHz transmitter oscillator is sent (see Fig. 21-15).

To provide information concerning the frequency and phase of the missing color subcarrier, a color burst is sent along with the signal. This burst follows each horizontal pulse and is located on the back porch of each blanking pedestal. It contains a minimum of 8 cycles of the subcarrier and is phased in step with the color subcarrier used at the station. In the receiver, this burst is used to lock in the frequency and phase of a 3.58 MHz oscillator, and thus we are assured at all times that the reinserted carrier will do its job correctly when it recombines with the color sidebands.

The color burst does not interfere with the horizontal sync because it is lower in amplitude, and follows the sync pulse. The burst performs in a similar manner as the horizontal-sync signal and keeps the receiver 3.579545 MHz oscillator at the proper phase.

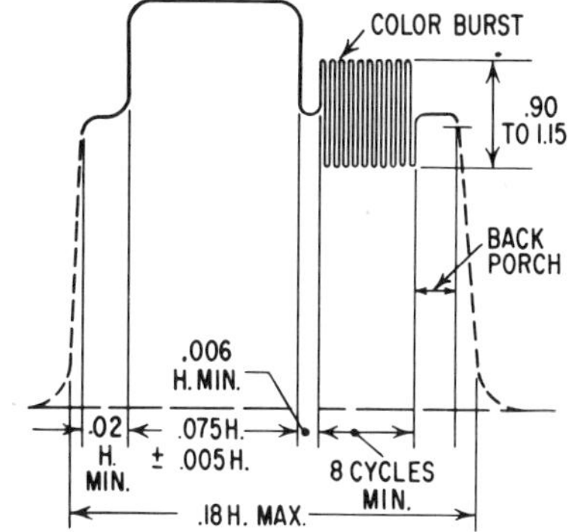

Fig. 21-15. The position of the color burst for the subcarrier oscillator sync; on the back porch of a horizontal-sync pulse.

21.5 DERIVATION OF THE COLOR-SUBCARRIER FREQUENCY

It has been mentioned in preceding text pages that the color subcarrier frequency is selected as approximately 3.58 MHz. This frequency was selected so that the chrominance video information will average out to nearly zero when a color program is being received on a black-and-white receiver. The color-subcarrier frequency was chosen to be an odd multiple of one-half the horizontal-scan frequency. More detail will follow on this choice.

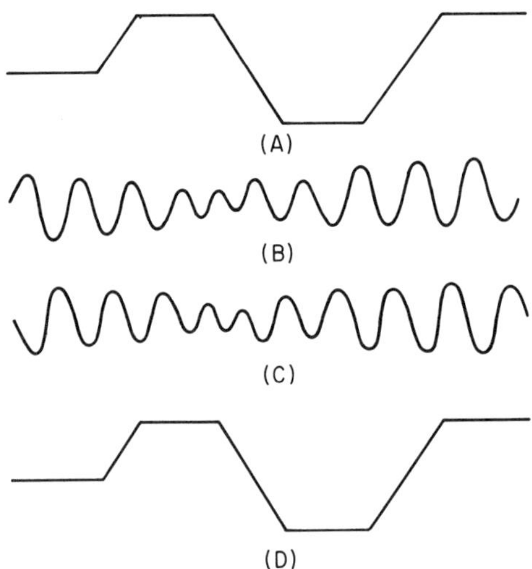

Fig. 21-16. Chroma interference cancellation in frequency interleaving.
(A) Black and white video during one scan line;
(B) Chroma video signal during one scan line;
(C) Same scan line, chroma video signal, exactly one frame later; (Note signal is 180° out of phase with part B.)
(D) Visual addition of waveforms A, B and C. (Note elimination of chroma video signals.)

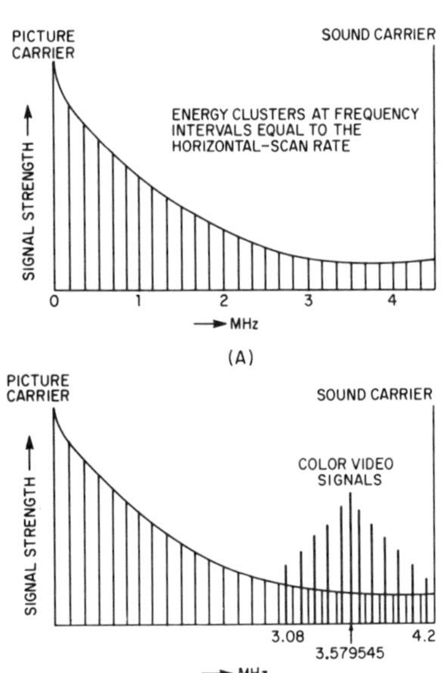

Fig. 21-17. Black-and-white video frequency energy distribution.

Color signals do not follow the brightness signal and would appear on a black-and-white screen as a spurious, unrelated dot pattern if not minimized. The choice of a color subcarrier frequency as an odd multiple of one half the horizontal-scan rate causes the color-video information on any scan line to be 180° out-of-phase with the next successive scan of that line on the next frame.

The chroma signal which varies about the brightness signal and modulates each horizontal-scan line, is inverted 180° each frame. This prevents the chroma video from producing an objectionable black-and-white interference pattern on the picture tube screen. The 180° inversion of the chroma video signal occurs because it passes through a whole number plus one-half cycles each frame.

Figure 21-16(A) is an illustration of a black and white video signal for a portion of one horizontal scan line. A portion of one modulated horizontal-scan line is shown in Fig. 21-16(B). Later, 526 lines or one frame later, the modulation is inverted 180° as in Fig. 21-16°C). As the two signals are nearly equal and opposite in signal polarity, they cancel to nearly zero, Fig. 21-16(D). The apparent cancellation is due to the human eye's persistence of vision; the eye retains the brightness of the horizontal line from frame to frame.

It is highly desirable to have the color-subcarrier frequency as high above the picture carrier as possible in order to minimize the interference with the black-and-white video information (see Fig. 21-17(A)). Research has shown that the average energy level of the black-and-white video falls very rapidly with increasing video frequencies. By placing the color subcarrier as high as possible as in Fig. 21-17(B), the difference in energy levels minimizes possible interference. Practical limits set the upper limit to 3.6 MHz.

There is a very objectionable 0.92 MHz signal generated from the beat between the sound carrier and the color subcarrier $(4.5 - 3.58 = 0.92)$. Experimental evidence has found that the beat is much less objectionable if the sound-carrier frequency is a multiple of the horizontal-scan frequency away from the video-carrier frequency. In standard black-and-white TV, the 285th and 286th harmonic of 15,750 Hz are at 4.48875 MHz and 4.5045 MHz.

The line frequency whose 286th harmonic is 4.5 MHz is 15,734.26Hz (F horizontal $= (4.5$ MHz$/286) = 15{,}734.26$ MHz). This frequency is within the deviation limit set by the NTSC monochrome standards. With the horizontal-scan frequency changed, the color subcarrier must be chosen to interleave. It must be an odd multiple of one-half the horizontal-scan rate. With 3.6 MHz as an upper limit, it was found that the 455th harmonic of half the horizontal scan rate becomes;

$$F \text{ color subcarrier} = 455 \times \frac{15{,}734.26}{2} = 3.579545 \text{ MHz}$$

Since there are 525 lines and using a 2:1 interlace, the new vertical-scan frequency becomes:

$$F \text{ vertical} = \frac{2}{525} \times \frac{15{,}734.26}{1} = 59.94 \text{ Hz}$$

21.6 COMPOSITE COLORPLEXED-VIDEO WAVEFORMS

The generation of a complete video signal, luminance and chrominance signals combined, is shown in Fig. 21-18. One horizontal-scan line crosses the test pattern in Fig. 21-18(A) to produce the video signals shown in Fig. 21-18(B) through 21-18(I). The pattern consists of the three primary colors, red, green, and blue; the three-two-color combinations of the primary colors, cyan, yellow, and magenta, and the one three-color combination of the three primary colors, white.

The waveforms *B*, *C*, and *D* give the primary color waveforms with an amplitude of one or 100% related to a fully saturated color. The luminance, the *Y* signal, in *E* shows the relative brightness for each color bar in the pattern (compare to Fig. 21-2). The *I* and *Q* signals are illustrated in *F* and *G*. Note that the *I* and *Q* signals can have positive or negative values. Figure 21-19 is another representation of Fig. 21-18. The *Y*, *I*, and *Q* video signals are formed from the basic *R*, *G*, and *B* camera outputs.

The *I* and *Q* video signals are passed to the modulators to generate the 3.58 MHz sidebands as shown in *H*. The amplitude of the 3.58 MHz signal in *H* is formed by the vectorial addition of the quadrature *I* and *Q* signals. For an example, the blue color bar *I* is (−0.32) and *Q* is (0.32), and the vectorial addition in *H* for blue is $\sqrt{(-0.32)^2 + (0.32)^2} = \sqrt{0.1985} = 0.44$. The complementary colors such as yellow and blue have the same peak amplitude but are 180° out-of-phase in the 3.58 MHz subcarrier signals. Refer to Fig. 21-12 for verification and comparison.

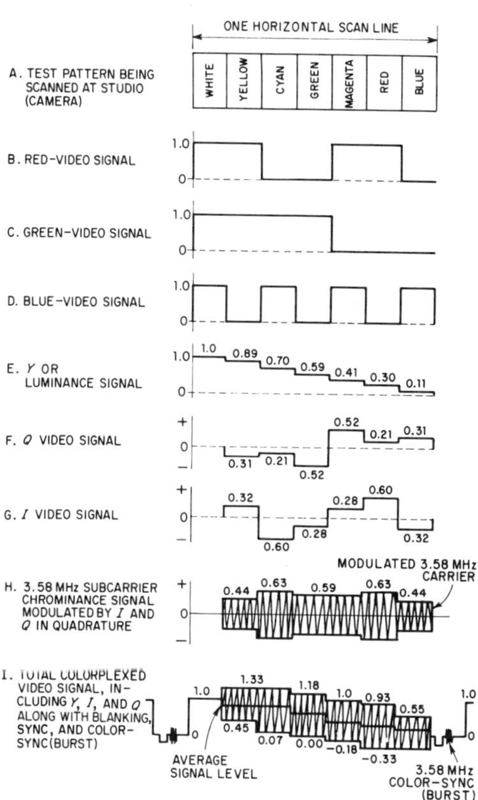

Fig. 21-18. Colorplexed composite video signal.

At the transmitter, the 3.58 MHz chrominance information is combined with the luminance signal, color sync, sync, and blanking to produce the composite colorplexed signal as shown in *I*. Adding the *Y* luminance signal in *E* causes a shifting of the 3.58 MHz chrominance information from a zero reference to a new level equal to the luminance value of the corresponding colors. This average dc level must be recovered at the receiver for the correct color renditions.

The preceding sections have introduced a number of diagrams and waveforms. In daily service activities only certain waveforms are of practical value.

Two waveforms of particular interest can be easily viewed on a wide-band service oscilloscope. The first is transmitted from the TV studio

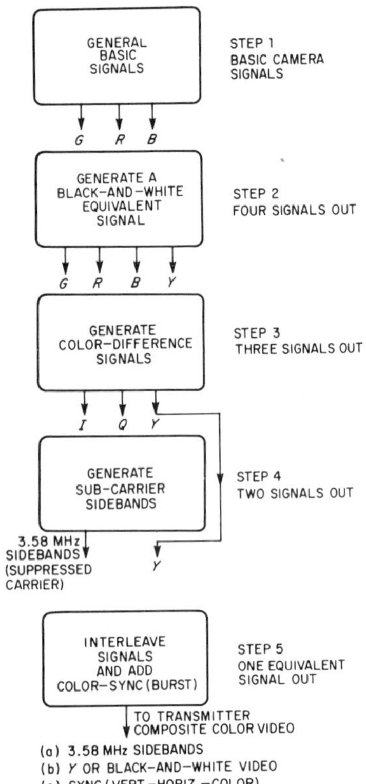

Fig. 21-19. Summary of color processing prior to the actual transmission.

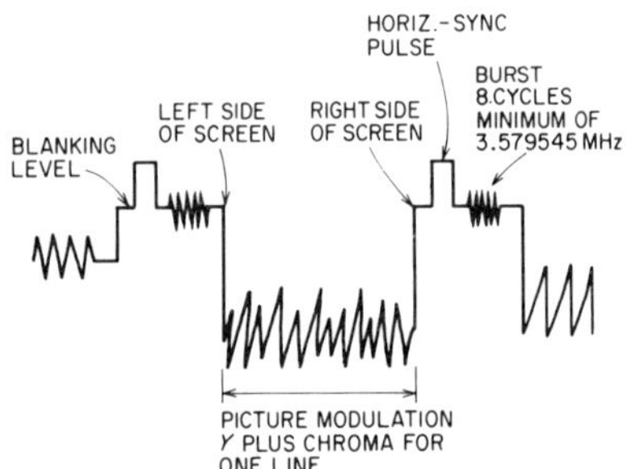

Fig. 21-20. Composite video signal waveform.

whenever color is being broadcast. The color sidebands are varying in nature and present a problem in that they are not repetitious. A service oscilloscope must have a repeating waveform to give a stationary pattern. The horizontal-sync pulses with the color-sync burst on the blanking pedestal back porch are repetitious (see Fig. 21-20). Note that the oscilloscope sweep is set to view two horizontal-sync pulses. A close observation will show several distinct signals. The blanking pulses, horizontal-sync pulses, color sync (burst), and video modulation (*Y* and chroma) are all illustrated. These signals can be traced through the *Y* or video section of color-TV receivers up to the tricolor kine.

Another important waveform can be observed if a color-bar generator is on hand. A group of color burst signals are produced in the color-bar generator and will produce color bars on a receiver. Figure 21-21 is a typical example of the output from such a generator. Each burst block contains roughly 8 cycles of subcarrier signal. The burst block immediately following the horizontal-sync pulse is used by the receiver for the color sync. The remaining burst blocks (10) generate a series of color bars as shown in Fig. 21-14. An oscilloscope can be used to trace this waveform through both the *Y* and the color sections of color-TV receivers. A great number of color-bar generators have peculiar shaped horizontal-sync pulses in the output signal, but the burst blocks are the key signals to observe. Use the manufacturer's service data on each chassis for the gain in each section of the receiver and compare the data with your particular wave-forms for future reference.

21.7 HOW COMPATIBILITY IS ACHIEVED

In order to properly reproduce color broadcasts in black and white, on monochrome-TV receivers, the color-broadcast signals must be specially constituted to produce a *Y*, or "luminance" signal. This signal is used to produce the black-and-white picture. The process of transmitting the color-broadcast signals in such a manner as to be normally useable on a monochrome-TV receiver, is called, "compatibility." (Technically, compatibility also includes the fact that a color broadcast includes the normal monochrome sync-and-blanking signals.)

Compatibility is illustrated in the diagram of Fig. 21-22. Note that the red, green, and blue outputs of the color camera are proportioned as 30%, 59%, and 11%, respectively, to produce the *Y* signal. (This composition of the *Y* signal is explained in Section 21.3 of this chapter.) The additional color-broadcast signals which are required to produce

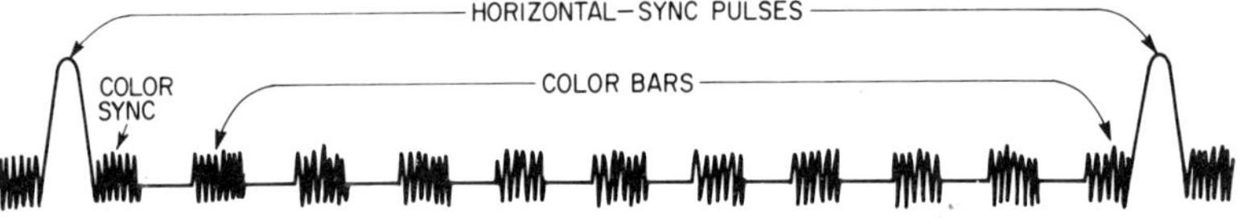

Fig. 21-21. An example of a color-bar generator signal.

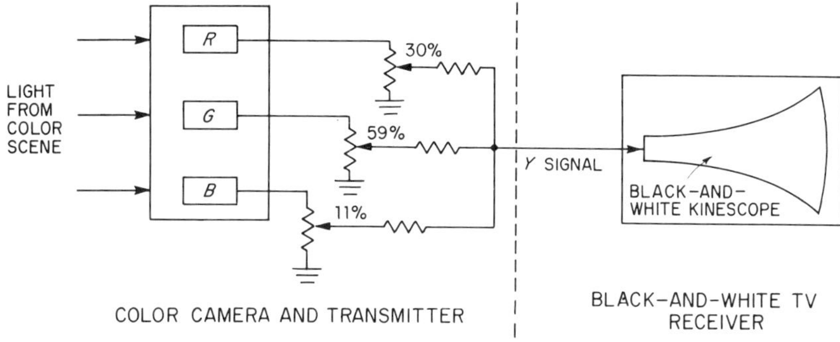

Fig. 21-22. Compatibility in the NTSC system.

color pictures on a color-TV set are accepted by the tuner of a monochrome set. However, few monochrome sets will pass the 3.58 MHz color subcarrier and its adjacent sidebands through the video IF's In addition, the monochrome set, of course, does not possess the various color processing circuits of a color set, and the special color signals simply do not affect the monochrome set.

Compatibility, not only means that a color transmission must be reproduced in black and white on a monochrome set, but also, that a black-and-white broadcast must be reproduced in monochrome on a color set. This is illustrated in Fig. 21-23. In order that the color picture tube will reproduce either a correct color picture, or a correct monochrome picture, it is necessary that certain static adjustments first be made to the picture-tube circuitry. Among these are the so-called "screen" adjustments. The color picture tube includes a red, green, and blue screen grid as shown in Chapter 14. Adjustments are made to the screen voltages (generally without a picture), so that the resulting light background on the color picture tube will be white. Since black is simply the absence of any light, the resultant picture on a color tube from a black-and-white broadcast will then be black and white only.

With regard to the operation of the color circuits during a black-and-white broadcast, these are disabled by a circuit called the "color killer," which is described in Chapter 22. If they were not so disabled, color snow or "confetti" might appear on the screen during black-and-white reception.

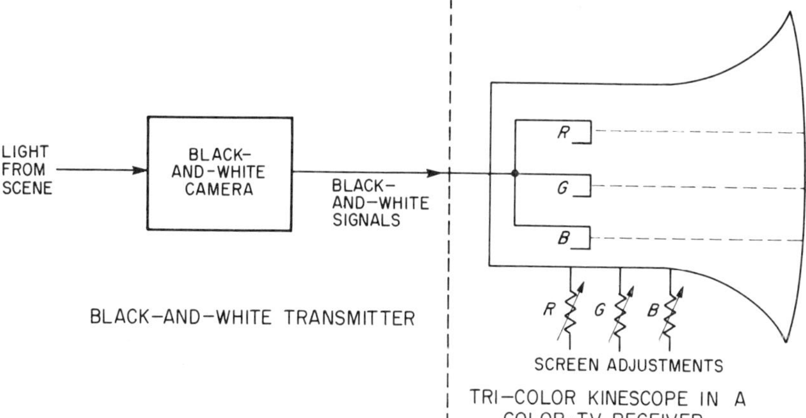

Fig. 21-23. Black-and-white broadcast TV signal compatibility on a color-TV receiver.

REVIEW QUESTIONS

1. What do the letters NTSC mean?
2. Most color receivers apply the black-and-white video signals to what elements in a tricolor kinescope?
3. Final color signal processing is accomplished in the color picture tube by modulating the three beams of ______________________.
4. Compared to white as 100%, how bright do red, green, and blue colors appear to the human eye?
5. Name the four video signals generated in a color camera.
6. Describe the method used to generate one set of sidebands from two modulators. (At transmitter.)
7. Give another name for the standard black-and-white signal.
8. How does the human eye see very small objects?
9. Is color sync transmitted during the vertical-blanking period?
10. Why were red, green, and blue selected as NTSC color primaries?
11. How are hue and saturation shown on a chromaticity chart?
12. Name the two signals that may beat in a color receiver to produce a 920-kHz interference pattern.

Color-TV Receiver Circuits 22

22.1 BLOCK DIAGRAM OF A COLOR-TV RECEIVER

The color-video signal requires special processing by several circuits which have not been described in any of the previous chapters. These new circuits are contained in the two upper blocks of Fig. 22-1, which is a block diagram of a color-TV receiver. This chapter will be devoted mostly to the analysis of the circuits in these two blocks.

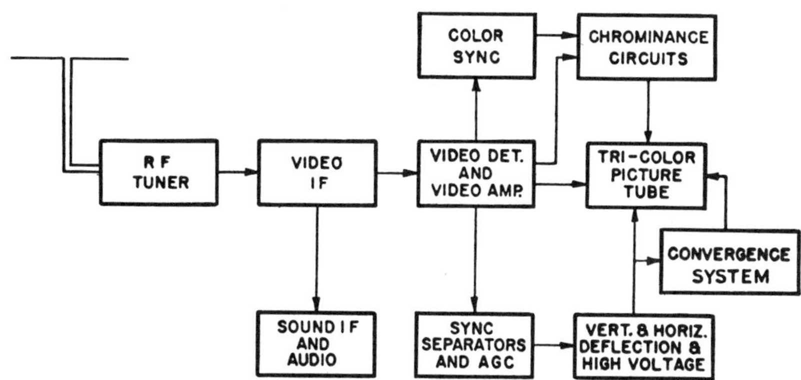

Fig. 22-1. Master block diagram of a color-television receiver.

Many of the other circuits in a color-TV receiver are slightly different from the black-and-white sets. We will begin this chapter by briefly discussing the entire color receiver. Then we will closely examine the content of the color sync and chrominance sections.

As indicated in Fig. 22-1, the major divisions of a color-TV receiver are:

(1) RF tuner.
(2) Video-IF system.
(3) Sound IF, FM detector, and audio system.
(4) Video detector and video amplifiers.
(5) Chrominance section.
(6) Color-sync section.
(7) Sync separators and AGC.
(8) Vertical and horizontal-deflection systems.
(9) High-voltage circuits.
(10) Tri-gun color picture tube and associated convergence circuits.

This chapter will concentrate on items 5 and 6 in this list. The subjects numbered 1, 2, 3, 4, 7, 8, 9, and 10 should be quite familiar to anyone who has studied the preceding chapters of this text. That this is so will become evident as we proceed through a detailed description of a color-television receiver.

RF Tuner The RF tuner consists of an RF amplifier, an oscillator, and a mixer stage. This section of the receiver is similar to that employed in the black-and-white receivers, since the RF requirements of both types of sets are alike. However, the allowable tolerance in the RF frequency response is more critical. A dip of 30% in the center part of the response curve might be satisfactory for monochrome reception, but for color-TV reception it would cause a degradation in the picture quality due to the unequal amplification. As shown in Fig. 22-2, the RF response curve of a tuner for color reception must have a more uniform characteristic within the signal portion of the band.

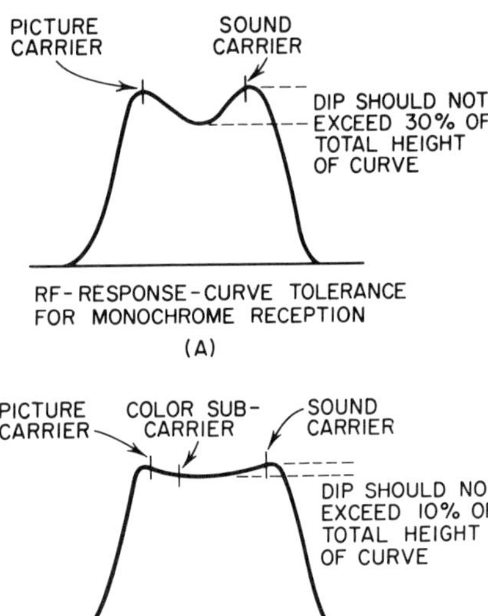

Fig. 22-2. RF-response curves: (A) Suitable for monochrome reception; (B) Required for color reception.

If a tuner possessing the response of Fig. 22-2(A) is employed for color reception, the color subcarrier will be attenuated as much as 20 to 30 percent compared with the picture and sound carrier. The response of Fig. 22-2(B) is required on all channels, since any station now sending monochrome signals is permitted by the FCC to send color broadcasts, provided the necessary additional equipment is installed at the studio and at the transmitter.

Video-IF System The video-IF system in color-television receivers, in general, contains three or four separate amplifiers (see Fig. 22-3). In form these closely resemble the IF section of a monochrome receiver, because the change to color reception in no way has altered the basic function of this section of the receiver, namely, that of establishing the overall bandpass and sensitivity of the receiver.

The stages in the video-IF system are stagger-tuned, generally in the 41–46 MHz range, with suitable traps for the accompanying sound (41.25 MHz), for the sound carrier of the adjacent lower channel (47.25 MHz), and for the video carrier of the adjacent higher channel (39.75 MHz). The adjacent channel traps generally have an attenuation of 55–60 decibels. The sound carrier of the same channel (41.25 MHz) may have more than one trap to insure that this signal is kept down at the proper level. Failure to observe this precaution will tend to produce a noticeable 920-kHz beat on the picture-tube screen, especially in those

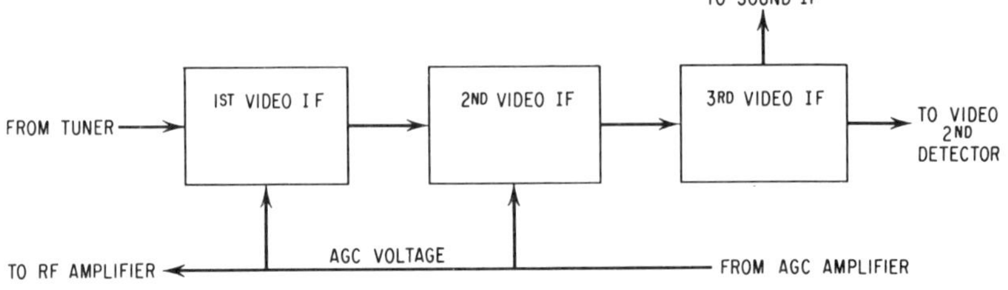

Fig. 22-3. A typical video-IF system. Note the similarity to the video system of monochrome receivers. In some color receivers one or two additional IF stages are used.

sections where the colors are highly saturated. This 920 kHz is the difference between the color subcarrier (3.58 MHz) and the sound subcarrier (4.5 MHz).

Typical IF-response curves for color reception are shown in Fig. 22-4. These curves differ mainly in the location of the color subcarrier. In Fig. 22-4(A), the color subcarrier is shown near the top of the curve with 41.6 MHz at the "knee," or limit of the response. This curve is typical of many early receivers.

The overall IF-response curve extends to approximately 4.2 MHz (at the 50-percent points) in order to include the color subcarrier and all its sidebands. It will be remembered from Chapter 21 that the upper sidebands of the color subcarrier extend about 0.6 MHz above 3.58 MHz or up to 4.2 MHz. Hence, to reproduce the picture in its full and true color, it is necessary that the upper color sidebands be permitted to pass. Beyond 4.2 MHz, the IF response drops sharply to the level of the sound carrier.

The IF response shown in Fig. 22-4(B) is of more recent design and is typical of later model receivers of many manufacturers. The same general shape of the response curve is maintained except for a slight decrease in the sharpness of the slope on the color-subcarrier side. Also, the color-subcarrier signal is carried through the IF stages at an amplification level equal to the picture IF (50 percent or 6 db). This design further reduces the tendency toward a 920-kHz beat pattern on the picture-tube screen. However, other effects created by this approach are compensated for in the video-amplifier and chroma-bandpass stages discussed later.

AGC voltage is applied to all the IF amplifiers, except the last, in order to control their gain in accordance with the level of the received signal. If the last IF stage is AGC-fed, excessive attenuation will result. A portion of the same voltage is also fed to the RF amplifier for similar control.

From the video-IF system, the signal is generally applied to two points: the video-second detector and the sound system. Let us consider the latter first.

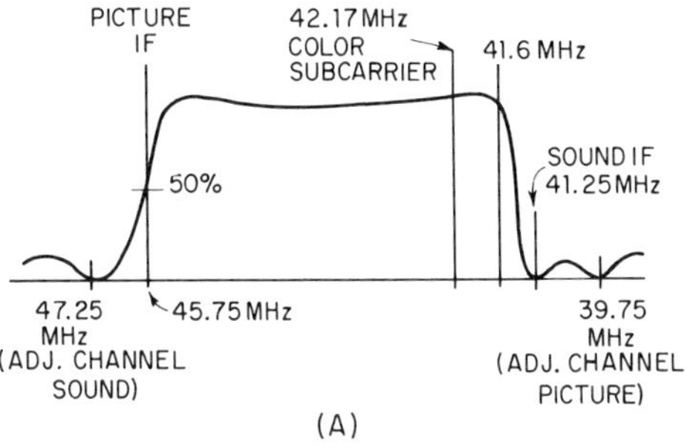

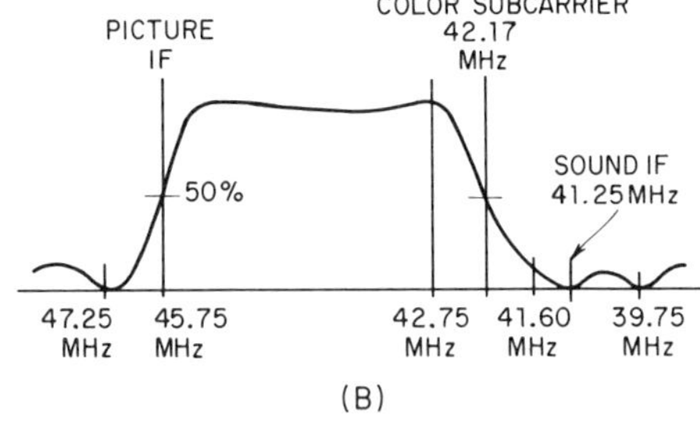

Fig. 22-4. Typical IF-response curves for color receivers: (A) Early receivers; (B) Recent receivers.

Sound IF, FM Detector, and the Audio System It is the practice in color-television receivers to separate the sound and video carriers just as soon as it becomes feasible to do so. This generally occurs at the last IF stage. The purpose, of course, is to keep the amount of 920-kHz beat interference voltage as small as possible so that its effect on the screen will be negligible.

The practice generally followed is to employ a separate detector (usually a solid-state diode) in which the sound and video carriers are mixed to produce the 4.5 MHz sound-IF signal. This signal is then amplified by one or two IF-amplifier stages and applied to a sound detector stage. From the sound detector, the original audio intelligence is recovered and fed to one or two audio-amplifier stages and finally to the speaker. The system is identical in all respects to the sound section of many present-day monochrome receivers.

Video Detector and Amplifiers Returning to the video system, we find that the signal enters the video-second detector after leaving the IF stages. Here the signal is demodulated, giving back the 0–4 MHz monochrome or luminance signal, and the color sidebands. Present, too, are the various synchronizing pulses plus the color burst. The latter, it will be recalled, is needed to reestablish the proper frequency and phase of the missing color subcarrier.

At the video-detector output, a number of things must occur. First, the brightness portion of the total signal must be fed to a separate amplifier. Second, the color sidebands must be separated from the full signal and transferred to a separate chrominance section. Third, the color burst must be made available to the color-sync circuits. And, finally, there are the sync separator and AGC system that must also be tied into the signal path.

There are a number of different ways in which all the foregoing functions can be carried out. The block diagram in Fig. 22-5 illustrates

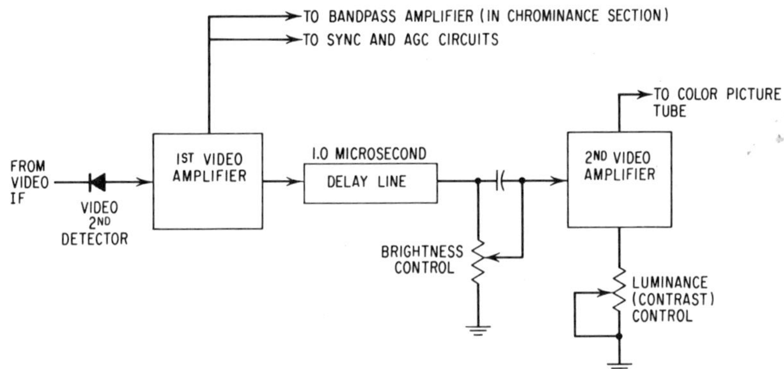

Fig. 22-5. Block diagram of video-second detector and video amplifier (luminance channel). Color receivers may use two to four video amplifiers.

one method. The signal from the video-second detector is fed to a video amplifier. Here both the chroma and monochrome signals are amplified. The monochrome signal is then transferred to a video-second amplifier and from this stage either to a matrix network or to the cathodes of the picture tube. The chroma signal is taken from the first-video amplifier and coupled to a bandpass amplifier in the chrominance section. A color receiver may employ one or two bandpass amplifiers. The chroma signal is also fed to a burst amplifier, which, by means of accurate gating, separates the burst signal from the chroma signal. The brightness and contrast controls are also associated with the video-output stage just as in monochrome receivers. A time-delay line (approximately 1.0 microsecond) is inserted between the first- and second-video amplifiers. The need for the delay line arises from the fact that the color signal passes through a rather narrow bandpass filter in its system which acts to slow down its passage. To insure the simultaneous arrival of the *Y* (or brightness) signal with the color signal at the matrix (or picture-tube grids), an artificial delay line having a delay of from 0.6 to 1.0 microsecond (depending on the receiver design) is inserted in the *Y* channel.

It is interesting to note that the bandpass of the *Y* channel beyond the color takeoff point is usually reduced below 3.58 MHz. This is done to minimize further any visual dot pattern which the 3.58 MHz color subcarrier signal may develop on the face of the picture-tube screen.

Chrominance Section We come now, for the first time, to a section of the receiver which has no counterpart in any monochrome receiver—one which is completely devoted to color. This section is usually known as the "chrominance section" and covers several stages. The number of stages used in this section and the type of circuitry employed to accomplish its function vary considerably among TV sets. However, regardless of the type of circuitry employed, the principal function remains the same: the section must demodulate the color signal in such a way that the original red, green, and blue components of the chroma signal are reobtained faithfully as originally seen by the color-camera tube. Essentially, the chrominance section amplifies the chrominance signal, demodulates (detects) the individual red-, green-, and blue-signal components, and by a matrixing, or signal-adding, network, couples the correct portions of these signal components to the tricolor picture tube. This, then, is the function of the chrominance section.

The chrominance section is one of the two major subjects discussed in this chapter. It will be further analyzed in Sections 22.2, 22.3, 22.4, and 22.5.

Color Sync Section The stability of the 3.58 MHz carrier signal reinserted into the chroma demodulators is an important factor in the reproduction of the original red, green, and blue voltages. It is the function of the color-sync section to develop a stable 3.58 MHz signal, and to make certain it possesses the proper frequency and, what is equally important, the proper phase. If the phase is wrong, the reproduced color will likewise be wrong, and in a color-television system this is a very noticeable form of distortion.

The color-sync section is the second of the major topics to be discussed in this chapter. It will be covered in detail in Sections 22.7, 22.8, 22.9, 22.10, 22.11, and 22.12.

Sync and AGC The sync separators and the AGC section of a color-television receiver do not differ in any important aspects from the same stages in monochrome receivers. Thus, the sync separators have, as their function, the separation of the horizontal- and vertical-sync pulses from the rest of the video signal. Once this is accomplished, the pulses are applied to the horizontal- and vertical-sweep systems through appropriate integrating and differentiating networks. Linked with the sync separator is a noise inverter which serves to prevent noise pulses from affecting the vertical- and horizontal-sweep systems.

For the AGC section, any method which has been employed in monochrome sets may be utilized in color receivers. At the present time, keyed AGC is favored, but this preference stems not from the fact that color television is being used, but from the inherent characteristics of keyed AGC itself.

The stages controlled by the AGC voltage include the RF amplifier and one or more video-IF stages. Clamping of the AGC voltage fed to the RF amplifier may also be employed.

Vertical- and Horizontal-deflection Systems In the horizontal- and vertical-deflection circuits we again encounter circuits similar to those found in monochrome receivers (see Fig. 22-6 and also Chapters

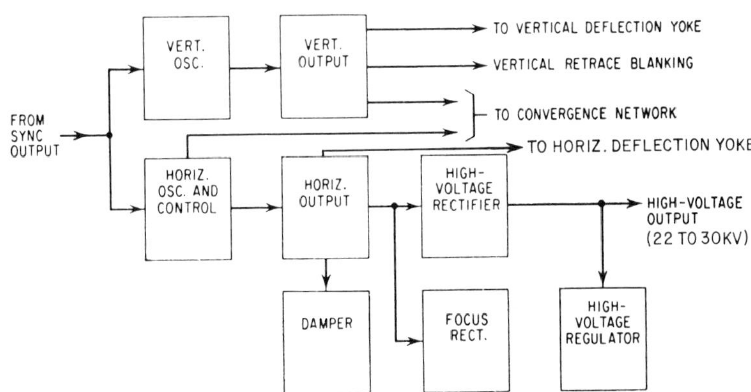

Fig. 22-6. Block diagram of the vertical and horizontal-deflection systems of a typical color receiver.

17, 18, and 19). The vertical system usually consists of a multivibrator with an output amplifier. In the horizontal section, the oscillator is preceded by an automatic frequency-control system. Beyond the oscillator, we have the output amplifier and the horizontal-output transformer. A damper diode is connected across the horizontal windings of the deflection yoke to eliminate oscillations which may occur during beam retrace. The energy absorbed by the damper in this process is converted into an additional voltage which when added to the normal B+ voltage results in a boosted B+.

A horizontal-retrace blanking circuit is included in color receivers to prevent the 3.58 MHz burst from reaching the grids of the color-picture tube. This burst, we have seen, follows the horizontal-sync pulse, and if it were permitted to reach the picture tube, it would develop a yellow strip during the horizontal-blanking interval.

High-voltage Circuits The high-voltage requirements of the three-gun, shadow-mask picture tube are considerably more critical than they are for a conventional black-and-white picture tube. The color tube requires up to 30,000 volts at a maximum current drain of 1,500 microamperes. There also must be available to the tube a focus voltage, variable between 5 to 8 kV (see Chapter 18).

The heavy requirements of the picture tube in regard to beam current result in some serious problems in the designing of a combination deflection and high-voltage system. Since the power used by the high-voltage circuit is an appreciable portion of the total, changes in the beam current due to changes in picture brightness can cause variations in scanning linearity and in the various operating potentials of the tube itself. To avoid such variation, it is necessary to maintain the high-voltage load constant whether the picture is bright or dim. This requires the use of a special high-voltage regulator.

Color Picture Tube and Convergence Circuits Most color receivers employ a tri-gun, tricolor picture tube (see Chapter 14). The tri-gun portion of the name indicates that the tube possesses three electron guns. The conventional black-and-white tube employs only one electron gun.

The second half of the name, tricolor, reveals that the screen of the tube possesses three different color-emitting phosphors. This, of course, is basic to the entire color-television system, since we employ the three primary colors—red, green, and blue—to synthesize the wide range of hues and tints required for the satisfactory presentation of a color picture. These color emitting phosphors are arranged in a dot pattern on the picture-tube screen.

Convergence circuits and convergence magnets are required to insure that the beams from each gun strike the proper dots on the screen. The magnets take care of any permanent misalignment of the guns or the offset caused by the earth's field. The convergence circuits, commonly called the dynamic convergence circuits, modify the instantaneous alignment of the beams to make further corrections to the convergence.

22.2 CHROMINANCE SECTION

Several ways of processing the color sidebands are possible. We will first briefly discuss the original system which utilizes the complete *I* and *Q* signals. We will then discuss the 3 systems which are in common use today.

***I* and *Q* System** Many early color receivers utilize the *I*, *Q* system shown in Fig. 22-7. Although this system is *functionally* identical to those used in recent receivers, it is similar in only certain aspects to present circuit design. This similarity will be discussed later.

The full color signal (Fig. 22-7) is obtained from the video system and fed to a bandpass amplifier. Also applied to the bandpass amplifier is a

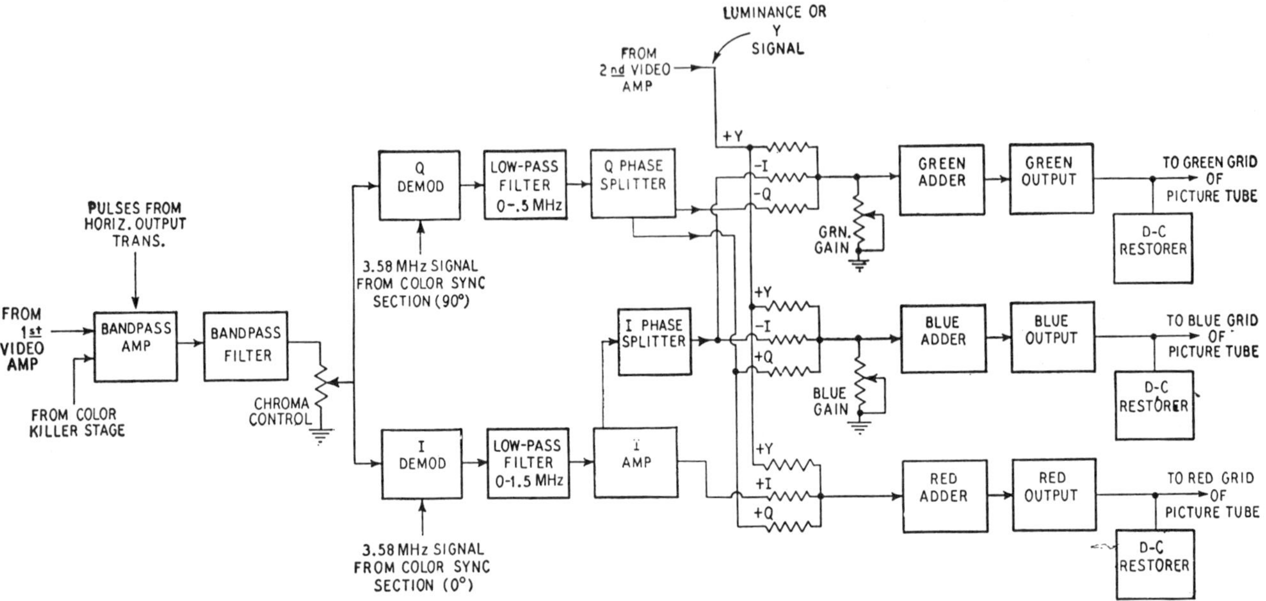

Fig. 22-7. The chrominance section of a color receiver employing *I*, *Q* demodulators. No special delay line is shown for the *I* stages, because the required delay is in distributed form.

gating pulse which keys off the amplifier by applying a pulse derived from the horizontal-deflection output transformer. The pulse arrives only during the horizontal-retrace interval when the color burst is passing through the system. By gating out the burst from the bandpass amplifier, we prevent color background unbalance in the picture tube. This unbalance may arise if the dc restorers in the chrominance channel clamp onto the color bursts rather than the normal chrominance signal.

The bandpass amplifier also receives a dc biasing voltage from a stage known as a "color killer." This stage is located in the color-sync section of the receiver, and its purpose is to bias the bandpass amplifier to cutoff in the absence of a color signal (that is, when black-and-white signals only are being received). This precaution is taken to insure that no random color appears on the picture-tube screen during a monochrome transmission. Such random color would be produced by noise or monochrome signals reaching the *I* and *Q* demodulators.

In the output circuit of the bandpass amplifier there is a bandpass filter which permits signals from 2.1 to 4.1 MHz to pass, but strongly attenuates all others. This filter thus serves to separate that portion of the signal containing the color sidebands from the section of the signal containing only monochrome information.

The bandpass filter is terminated in a color-intensity control (called the "chroma control") from which point we can take as much of the color signal as we feel is required and feed it to the following demodulators. This control actually determines how saturated (deep) the colors appear on the screen. It is a front-panel control which the user of the set can adjust as he sees fit. (It may also be labeled "chroma," "color intensity," "color saturation," or "color." Regardless of the name used, its function is the same in any color receiver.)

After this control, the color signal is fed in equal measure to the *I* and *Q* color demodulators. Also arriving at these demodulator stages is a 3.58 MHz signal. This signal represents the missing color subcarrier, and it must be recombined with the color signal of the demodulators so that the original *I* and *Q* signals can be detected. Both *I* and *Q* stages receive a 3.58 MHz voltage, the only difference being that one 3.58 MHz voltage lags behind the other 3.58 MHz voltage by 90 degrees. This particular phase relationship is required, because the color signal was modulated this way at the transmitter and demodulation is the reverse process.

With a color signal being received, the entire chrominance channel is operative, and the output of the two demodulators represents the original *I* and *Q* color signals that were originally developed at the transmitter. The *I* signal, then, is passed through a 0 to 1.5 MHz bandpass filter and a special 0.5-microsecond delay line. Then it may receive additional amplification before being made available to the adding, or matrix, network in positive and negative polarity. The double-polarity *I* signals are required in the final mixing process from which red, green, and blue voltages are recreated. A single phase splitter provides the positive and negative *I* signals.

The use of a 0.5-microsecond delay network in the *I* channel again stems from the narrow 0–0.5 MHz bandpass filter through which the *Q* signal is sent. The *Y* signal, it will be remembered, had to be delayed 1 microsecond for the same reason. The difference in delay between the *Y* and *I* signals arises from the different characteristics of their respective networks. In the *Y* channel, the bandpass of the circuits extends from 0–3.5 or 4.0 MHz. In the *I* channel, the bandpass extends only from 0–1.5 MHz. The narrower bandpass introduces some delay, requiring less additional delay in order to slow the *I* signal down to the *Q* signal.

In the *Q* channel, the demodulated *Q* signal passes through a 0–0.5 MHz bandpass filter and reaches a phase splitter from which positive and negative *Q* signals are made available to the matrix.

We now have at the matrix the *I*, *Q*, and *Y* signals and, by properly combining them, we can reobtain the red, green, and blue voltages that were originally combined to form the *I*, *Q*, and *Y* signals. The addition is carried out in rather simple fashion by using a series of resistors connected as shown in Fig. 22-7. At the output of the matrix section, each of the three color voltages is separately amplified and then transferred via separate dc restorers to the appropriate control grid of the tri-gun picture tube.

The demodulators in color receivers which reproduce the original *I* and *Q* components from the chroma signal are usually low level demodulators. Thus, demodulation is performed at a low color-signal level and, by necessity, the signals must be further amplified following demodulation.

At this point, it may be desirable to say a few words about color demodulators. The *I*, *Q* system just described, when properly designed, provides excellent color reproduction. However, price is an important aspect of receiver sales, and if it is possible to achieve acceptable results at lower cost, most circuit designers will use the more economical system. That is why the *I*, *Q* method is no longer being used; instead, several other systems have been developed. These other methods are the *R*-*Y* and *B*-*Y* systems; the *X* and *Z* system; and the *R*-*Y*, *B*-*Y*, *G*-*Y* system.

The *R*-*Y* and *B*-*Y* System In this commonly used system both channels possess the same bandwidth, generally 0 to 0.5 MHz. While this method does not color as much of the picture as the *I*, *Q* system, the visual results are acceptable. In this narrower system, only the larger objects are in color; medium and small detail are rendered in black and white only. By this modification, no time-delay networks are needed in the chrominance section, although the time-delay filter in the *Y* section is still retained. The change also permits other simplifications which are economically advantangeous.

R-*Y* and *B*-*Y* systems have been used with low-level and high-level demodulators. In high-level demodulation, the *G*-*Y* signal is often formed in a cathode circuit which is common to both *R*-*Y* and *B*-*Y* demodulators (see Fig. 22-8). The detected *R*-*Y*, *B*-*Y*, and *G*-*Y* color signals are then fed directly to the grids of the picture tube. This is what we mean by high-level demodulation; no additional amplifiers are needed beyond the demodulators. In such a system, the 3.58 MHz carrier that is reinserted with the two color sidebands does

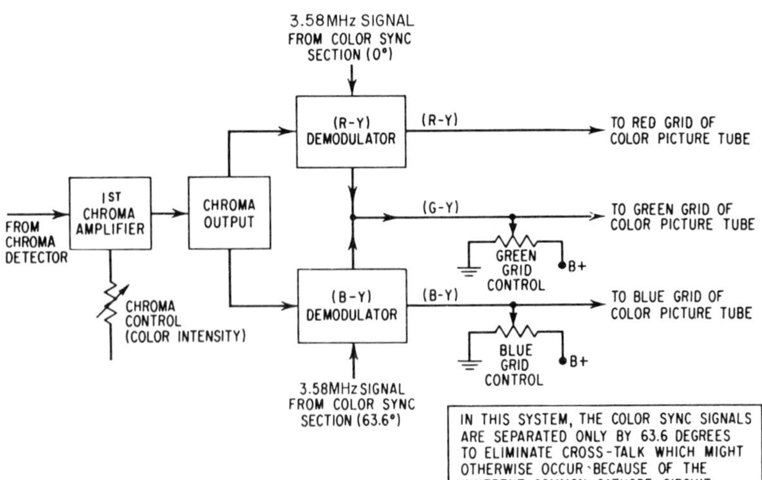

Fig. 22-8. The chrominance section of a color receiver employing the *R*-*Y* and *B*-*Y* demodulators. Note that bandwidths for the two demodulators are equal and that no delay line is required.

not possess the same 90-degree relationship previously indicated for the *I*, *Q* system, even when low-level *R-Y*, *B-Y* detection is employed. This stems from the fact that a common cathode is being utilized at the demodulators to provide the *G-Y* signal. If a 90-degree phase separation between the 3.58 MHz carriers being applied to each demodulator is maintained, cross talk will occur and cause improper coloring of the image. That is, the colors sent to each grid of the picture tube will not be as pure as they could be.

When low-level *R-Y* and *B-Y* demodulation is carried out and the *G-Y* signal is formed in a separate circuit, the reinserted 3.58 MHz carriers more nearly possess a 90-degree phase relationship.

In some receivers, *R-Y* and *G-Y* are demodulated and a portion of each is added to obtain *B-Y*. This is feasible because if we can add *R-Y* and *B-Y* to obtain *G-Y*, we can also use *R-Y* and *G-Y* to derive *B-Y*. It should be noted, however, that if we employ this method, a different phase relationship is required between the 3.58 MHz signals sent to the demodulators. This can be seen from Fig. 22-9.

The three color-difference signals are then applied to the grids of the tricolor picture tube at the same time that the luminance (or *Y*) signal is fed to the three cathodes of the picture tube. Thus, matrixing is actually performed within the tricolor tube itself. When the *Y* signal (appearing on the "red" cathode) is added to the *R-Y* signal appearing on the "red" grid, a voltage (bias) results that represents *R* alone; the *B-Y* and *G-Y* signals are added in the same manner to the *Y* signal, resulting in the correct voltage and beam current for *B* and *G*. Traps in the output circuits of the demodulators are resonant to 3.58 MHz, preventing this frequency from reaching the picture-tube grids.

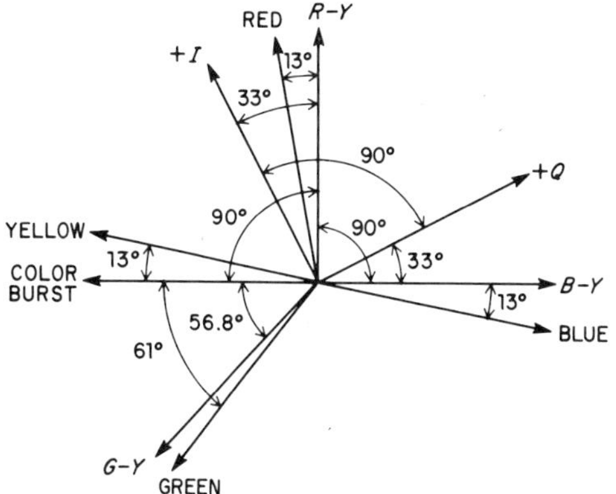

Fig. 22-9. The phase relationships between the various color signals, including *R-Y*, *B-Y*, and *G-Y*.

X and Z System Still another low-level modulation system uses what are called *X* and *Z* demodulators (see Fig. 22-10). This is basically an

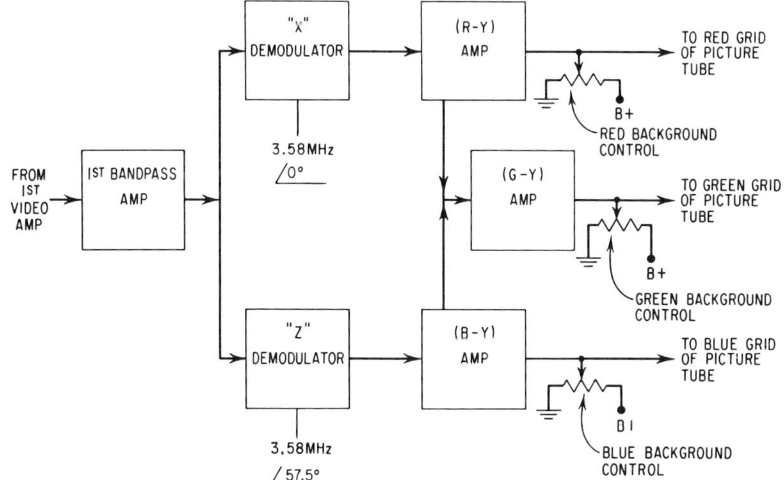

Fig. 22-10. The chrominance section of a color receiver which employs *X* and *Z* demodulators.

R-Y, *B-Y* system, but the angular separation of the reinserted 3.58 MHz carrier signals is 57.5 degrees and not 90 degrees. The output from the *X* demodulator is *R-Y*, and the output from the *Z* demodulator is *B-Y*. From the common emitters or the common cathodes of the following amplifier, the *G-Y* signal is obtained, amplified, and then transferred to the appropriate grid of the color picture tube. Notice that the terms *X* and *Z* have been selected arbitrarily to identify the demodulators; they have no other significance.

This matter of color demodulation in which phase angles other than 90 degrees are employed in the reinserted 3.58 MHz carrier often puzzles the service man working with color circuits. The precise angle selected depends upon the way the circuits have been designed to handle the color signals. Demodulation can be performed in a variety of ways, each with its own phase requirements. The point to remember in all these systems is that pure *R-Y*, *B-Y*, and *G-Y* signals are provided at their respective grids at the color picture tube.

The *R-Y*, *B-Y*, and *G-Y* System Demodulators can be designed to obtain all three color signals directly. This is evident from Fig. 22-9 which shows that the *G-Y* color signal is only 57 degrees ahead of the 3.58 MHz carrier. One method of demodulation accomplishes this by delaying the color signal 57 degrees. The 3.58 MHz carrier is then used to directly obtain the *G-Y* signal (see Fig. 22-11). The *B-Y* signal is

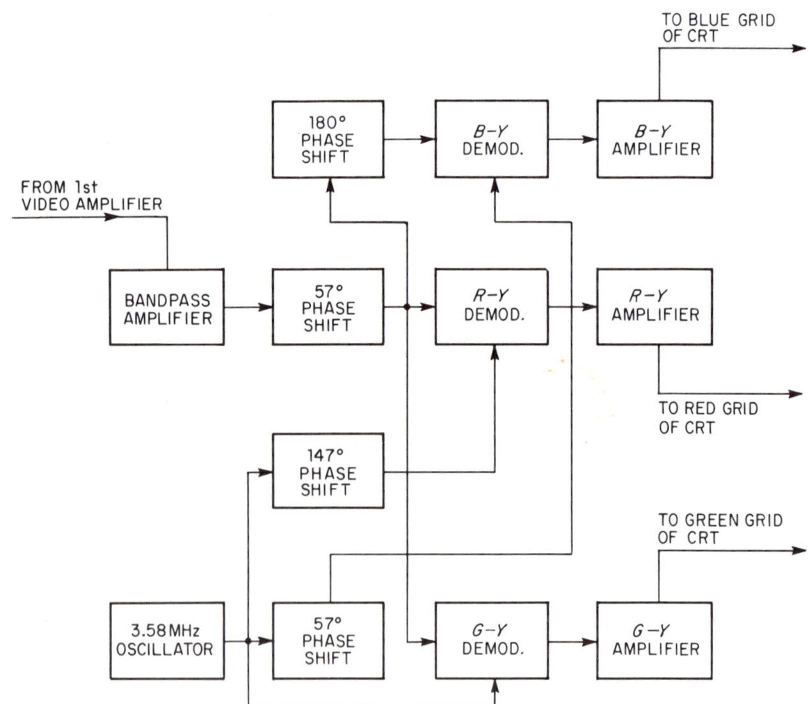

Fig. 22-11. The chrominance section of a receiver which uses *R-Y*, *B-Y*, and *G-Y* demodulators. (*Courtesy of RCA.*)

obtained by shifting the 3.58 MHz carrier 57 degrees and the color signal 180 degrees. The *R-Y* demodulator uses the same color signal as the *G-Y* demodulator. However, the 3.58 MHz carrier must be shifted 147 degrees for proper demodulation.

22.3 BANDPASS (CHROMA) AMPLIFIERS

We will now examine some typical circuits of bandpass amplifiers. The basic operation of these circuits was discussed in Section 22.2 for the *I, Q* system. In the newer systems, the bandpass amplifier has an identical usage. However, its bandwidth is only 1 MHz compared to 2 MHz for the *I, Q* system. The center frequency for the bandpass amplifier is 3.58 MHz in all cases. The *I, Q* system does not have symmetrical sidebands about this center frequency. It goes from approximately 2.1 MHz to 4.1 MHz. This is 1.5 MHz below and 0.5 MHz above the 3.58 MHz carrier. The newer systems use a symmetrical bandpass of 3.58 ± 0.5 MHz, or from approximately 3.1 to 4.1 MHz.

Typical Bandpass Amplifier Figure 22-12 is a typical two-stage solid-state bandpass amplifier. Q_1 and Q_2 are the first and second stages, respectively. Both stages use tuned input and output circuits. The input to Q_1 is series tuned. Parallel tuned transformers are used for the interstage and output couplings. In order to achieve a wide bandwidth of

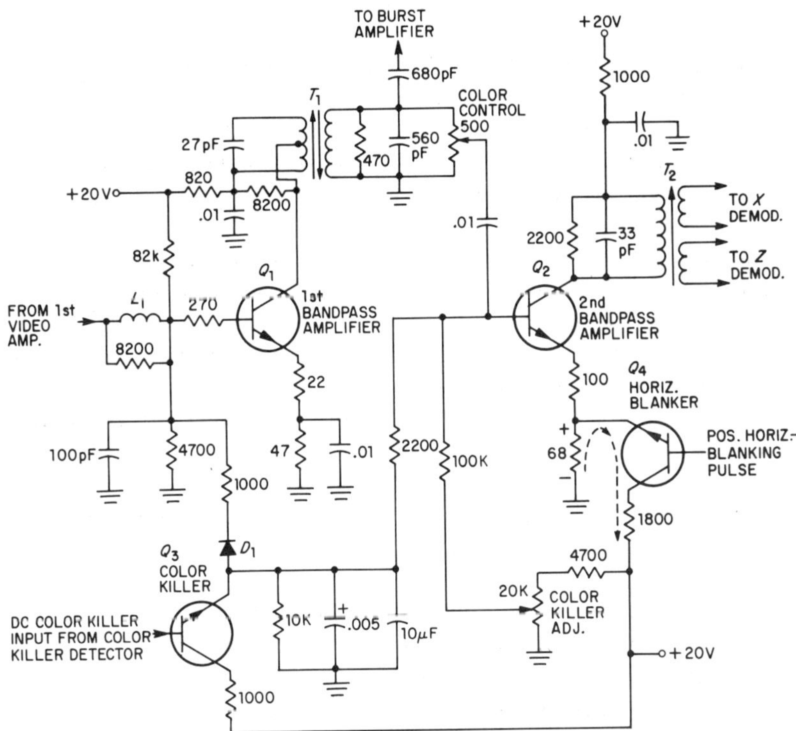

Fig. 22-12. Typical two-stage bandpass (chroma) amplifier. (*Courtesy of Philco.*)

1 MHz, an 8200-ohm resistor is placed across L_1 in the input-tuned circuit. The same scheme is used in the primary and secondary of T_1, and the primary of T_2, all to help widen the bandwidth.

The gain of the bandpass amplifier is adjustable with a potentiometer across the secondary of T_1. This is the color control on the front of the TV receiver. It simultaneously controls the intensity of all three color signals on the CRT.

Both Q_1 and Q_2 can be completely disabled by the color killer transistor, Q_3. During black-and-white reception Q_3 is turned completely on and, as will be presently explained, both bandpass amplifiers are disabled. The input to the color-killer transistor, Q_3, is a dc voltage which is obtained from a color-killer detector (not shown in Fig. 22-12). The output of the color-killer detector depends upon whether or not a color-burst signal is fed into it. During black-and-white reception, there is, of course, no color burst present at the color-killer detector. At this time, a positive dc voltage is applied to the base of Q_3, turning it fully on. This places a relatively high-positive bias on the bases of Q_1 and Q_2, thus saturating these two transistors. Since no ac signal can pass through a saturated transistor, these two stages become inoperative as far as signals are concerned. In this condition, no color snow, or "confetti" can pass through to the picture tube.

During color reception, the color burst is applied to the color-killer detector, which now has a negative dc voltage output. This is applied to the base of Q_3 (NPN) and cuts it off. The bases of Q_1 and Q_2 now are returned to their normal forward biases and will operate normally.

In order to insure the correct color-killer action, both in the presence of black-and-white and color signals, a color-killer threshold control is supplied, as shown in Fig. 22-12. This adjustment determines the amount of positive-base bias developed during the reception of black-and-white transmissions, for both Q_1 and Q_2. This is required because the base bias point of Q_1 and Q_2 is different for every receiver and for different transistors of the same type. If this control is incorrectly adjusted, the bandpass amplifiers will not be saturated on black-and-white signals, and in this case color noise signals will pass through them and will appear on the picture tube.

It is necessary to turn off the bandpass amplifier (Q_2) during the period of the 3.58 MHz color burst. If this is not done, the color burst will be demodulated and its effect on the color tube will be to produce a color background on the screen. In Fig. 22-12, a horizontal-blanker transistor (Q_4) reverse base biases Q_2 during the color burst (horizontal-retrace time) and prevents the burst signal from being applied to the color demodulators. This operates as follows: A positive pulse from the horizontal flyback transformer is applied to the base of Q_4. This pulse occurs only during the horizontal-flyback time. Since Q_4 is an NPN, the pulse turns it on hard. As illustrated in Fig. 22-12, current now flows through the 68-ohm resistor in the emitter circuit of Q_4, such that the top of the resistor becomes positive with respect to ground. This positive voltage is applied to the emitter of Q_2 and cuts it off. Since Q_2 is cut off during the horizontal-retrace period (and burst period), no burst signal can pass through Q_2 to affect the color demodulators.

Variable-phase Bandpass Amplifier In Section 22.2, we mentioned the requirement for a 57-degree phase shift in the color signal for the *R-Y, B-Y, G-Y* system. This can be accomplished with a circuit like Fig. 22-13. For simplicity, we have only shown the last two stages

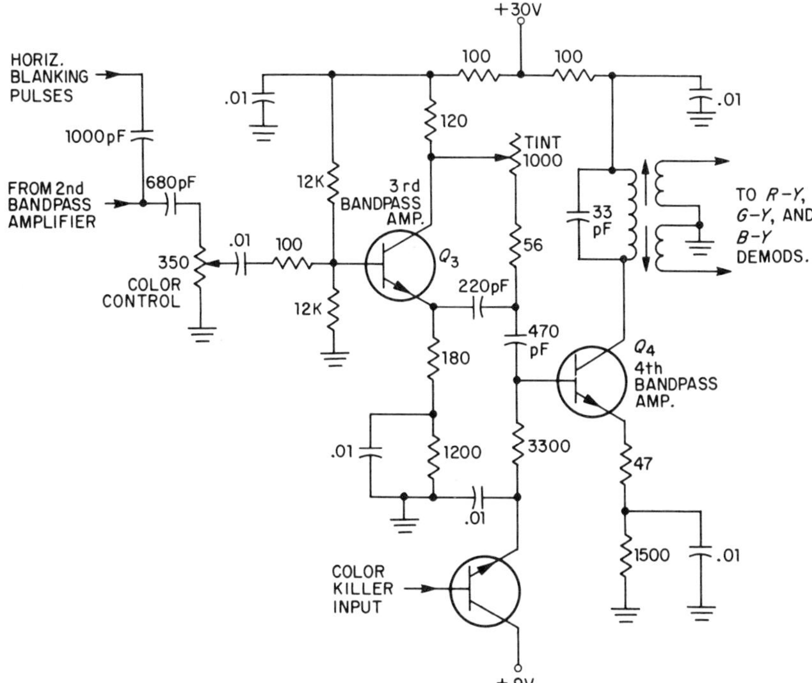

Fig. 22-13. A variable phase shift, bandpass amplifier. (*Courtesy of RCA.*)

of this four-stage bandpass amplifier. The third bandpass amplifier (Q_3) is where the 57 degree phase shift is performed. The collector of Q_3 shifts the output signal 180 degrees with respect to the input. The emitter signal does not have a phase shift with respect to the input. By mixing the collector and emitter signals in appropriate amounts, any phase shift between zero and 180 degrees can be obtained. A 1000-ohm potentiometer is placed in this mixing circuit to make the phase shift adjustable. A phase shift adjustment in this location produces the same effect on the screen as a phase shift in the 3.58 MHz oscillator circuits. Consequently, this control is used for the tint control on the front panel of the set.

The 350-ohm color control adjusts the gain of the bandpass amplifier. Also, the gain can be completely disabled in two ways.

(1) The color killer disables the gain by saturating Q_4 during black-and-white broadcasts.

(2) Blanking during the 3.58 MHz burst is accomplished by momentarily cutting off Q_3 during the horizontal-retrace time.

22.4 COLOR DEMODULATORS

As previously mentioned, the *I*, *Q* system of color demodulation was used in early color receivers. All recent color receivers use three other methods of demodulation. We will discuss all four methods in this section.

***I* and *Q* Demodulators** Refer to Fig. 22-9 for the phase relationship between *I*, *Q*, and 3.58 MHz signals. The figure shows that the 3.58 MHz oscillator signal must be delayed 57 degrees in order to be in phase with the *I* color sidebands. This demodulation can take place only when the phases of the *I* color sidebands and the oscillator signal are identical. For *Q* color sideband demodulation, the oscillator phase must be delayed $57 + 90 - 147$ degrees. These delays are not shown on the block diagram of Fig. 22-7. Only the 90 degree relationship between the two 3.58 MHz signals is mentioned. Many times, the locations of required delays (phase shifts) are not obvious in a schematic. Each stage of the bandpass amplifier or 3.58 MHz amplifier circuits contributes a little to this phase shift. By the time each of these signals arrives at the demodulator, we may find phase shifts are required which do not agree at all with the theory.

The important thing to keep in mind is this: proper demodulation of a color sideband requires a 3.58 MHz signal exactly in phase with that sideband.

***R-Y*, *B-Y* Demodulator** Figure 22-14 shows a solid-state *R-Y*, *B-Y* demodulator. This particular design is used in a Motorola color receiver.

Q_1 is the last stage of a two-stage, color-bandpass amplifier. This output stage utilizes an emitter follower for a low-output impedance. The low impedance is required, because Q_2 and Q_3 are driven at their emitter terminals. Q_2 is the *R-Y* demodulator, and Q_3 is the *B-Y* demodulator. The 3.58 MHz carrier is injected at the bases of Q_2 and Q_3. The mixing of the carrier with the color sidebands is thus performed across the emitter-base junctions of these two transistors. Since these junctions have nonlinear characteristics, the mixing process will produce difference frequencies, that is, the *R-Y* and *B-Y* signals.

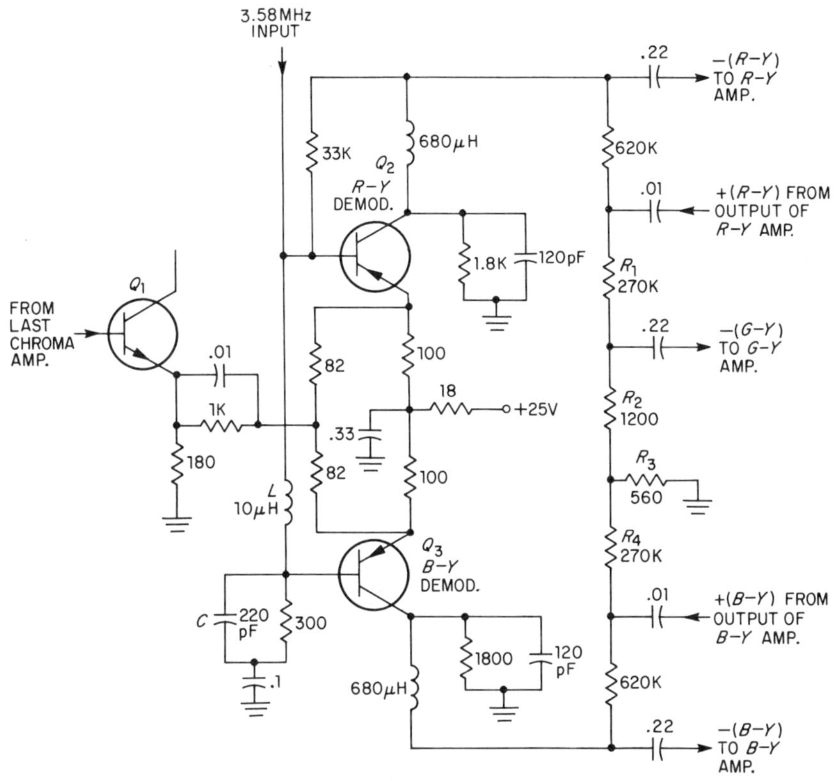

Fig. 22-14. A solid-state R-Y, B-Y color demodulator. (*Courtesy of Motorola.*)

These new signals are amplified and appear at the collectors of Q_2 and Q_3.

Referring to Fig. 22-9, we see that the R-Y sidebands can be demodulated if the 3.58 MHz carrier is delayed 90 degrees. The carrier input coming into Fig. 22-14 already has this 90 degree phase shift. Figure 22-9 shows that the B-Y color sidebands require an additional 90-degree phase shift for the 3.58 MHz carrier. This additional shift is performed by L and C in the figure.

The 120 picoFarad capacitors and 680 microhenry chokes connected to the collectors of Q_2 and Q_3 are used to attenuate the 3.58 MHz carrier. The zero to 0.5 MHz R-Y and B-Y signals are allowed to pass. These signals go directly to the R-Y and B-Y amplifiers, which in turn drive the picture-tube grids. In Section 22.2, we noted that the G-Y signal can be derived from the R-Y and B-Y signals. The exact relationship between these three signals is $-(G\text{-}Y) = 0.51(R\text{-}Y) + 0.19(B\text{-}Y)$. We need merely to obtain a fraction of the interted R-Y signal and add this to a fraction of the inverted B-Y signal. This is done in Fig. 22-14 by the R_1-R_2-R_3 voltage divider at the output of the R-Y amplifier. The R_3-R_4 voltage divider is used on the output of the B-Y amplifier. These two signals then drive the G-Y amplifier.

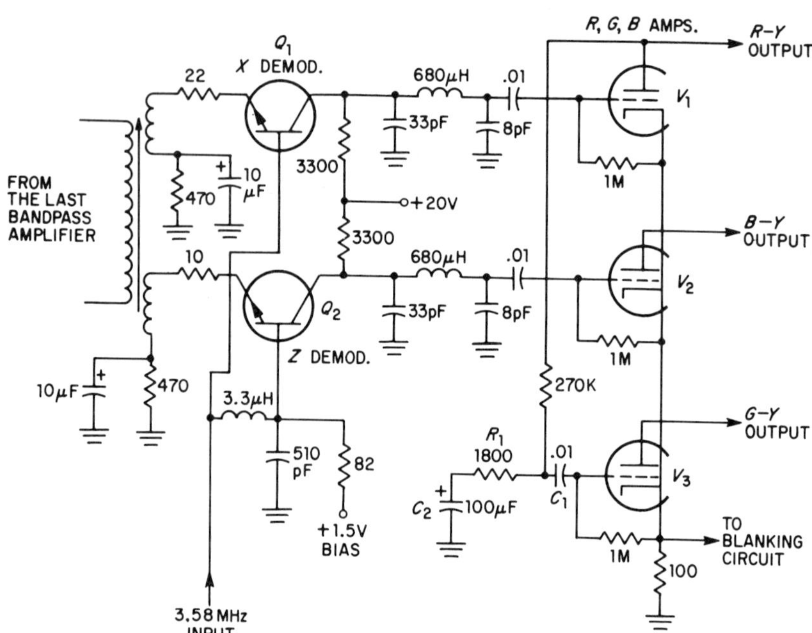

Fig. 22-15. A tube and transistor *X, Z* demodulator. (*Courtesy of Philco.*)

X and Z Demodulator A tube and solid-state diode *X, Z* demodulator is shown in Fig. 22-15. In some respects, the operation of this circuit is similar to the *R-Y, B-Y* demodulator of Fig. 22-14. The color sidebands are again applied to the emitters of the two demodulator transistors. The color subcarrier is brought to the bases of the same transistors. However, the phase difference between these two subcarriers is only 57.5 degrees instead of the 90 degrees used in the *R-Y, B-Y* demodulator. Since the phase difference between the carriers is not exactly 90 degrees, the output at the collectors of Q_1 and Q_2 is close to, but not exactly, *R-Y* and *B-Y*. These signals are applied to two grids of a three-triode, common-cathode circuit. This circuit generates the *G-Y* signal at the plate of V_3 by utilizing the signal on the common cathodes as the input to V_3 and grounding its grid through R_1, C_1, and C_2. This signal on the common cathodes also corrects the slight error in the *R-Y* and *B-Y* signals, so that we have pure *R-Y* and *B-Y* at the plates of V_1 and V_2. The 270-kilohm resistor between the plate of V_1 and the grid of V_3 also produces a slight correction in the purity of the *G-Y* signal.

The filter circuits at the collectors of Q_1 and Q_2 remove any residual 3.58 MHz carrier that may be present, in order that it will not reach the picture tube. Otherwise, a beat note between it and the 4.5 MHz sound carrier will develop an objectionable interference pattern on the screen.

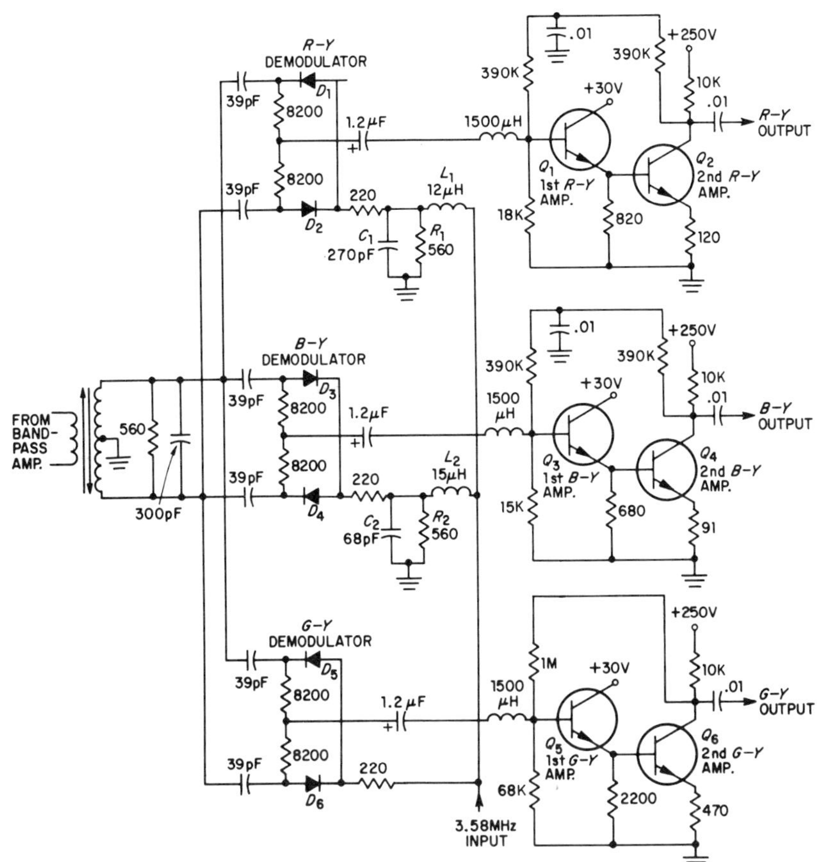

Fig. 22-16. Color demodulators and amplifiers which separately handle the *R-Y*, *B-Y*, and *G-Y* color sidebands. (*Courtesy of RCA.*)

The *R-Y*, *B-Y*, and *G-Y* Demodulator One might ask: why not simply use a separate demodulator for each of the 3 signals and avoid the confusing separation of *G-Y* from *R-Y* and *B-Y*? This is possible and is commercially done, but it is not as straightforward as one might first think. Figure 22-16 shows a system where each color signal is separately demodulated. In this circuit, the demodulators are simply diode phase detectors, almost identical to the phase detector in the horizontal AFC system. This is the same system illustrated in block diagram form in Fig. 22-11. We will refer to both illustrations to determine where each phase shift and function takes place in Fig. 22-16.

The 57° phase shift in the bandpass amplifier takes place in stages prior to the input transformer. The 180° phase shift for the color sidebands going into the *B-Y* demodulator is accomplished by reversing the *B-Y* diodes. Note that D_3 and D_4 are connected opposite to D_1 and D_2 (or D_5 and D_6).

The 57° phase shift for the carrier going to the B-Y demodulator is performed with R_2, L_2, and C_2. Likewise, the 147° phase shift for the R-Y demodulator is accomplished by R_1, L_1, and C_1. The output from each demodulator is obtained at the junction of the two 8200-ohm resistors in its input network. This output signal passes through a filter composed up of a 1.2 microFarad capacitor and a 1.5 millihenry choke and which possesses a bandpass from 20 Hertz to 500 kHz. These output signals drive R-Y, B-Y, and G-Y amplifiers which we will discuss in the next section of this chapter.

22.5 THE *R-Y, B-Y,* AND *G-Y* AMPLIFIERS

The R-Y, B-Y, and G-Y output signals from most color demodulators are not strong enough to drive a tri-gun color picture tube directly. One or two stages of additional amplification are required. Some receivers also use these amplifiers to generate the G-Y signal from the R-Y and B-Y signals. In this section we will examine three approaches to these amplifiers.

Separate Amplifiers The most straightforward approach to the R-Y, B-Y, and G-Y amplifiers is shown in Fig. 22-16. Since the three color signals are separately and completely generated ahead of these amplifiers, there is no interaction between these stages. Each amplifier is a simple two-stage wideband, transistor circuit. Feedback is provided in each stage by the simple expedient of not by-passing the emitter resistors. This improves the stabilization and the bandwidth of the stages. The bandwidth must be at least 0.5 MHz. To compensate for the differences in picture tube phosphor efficiencies, it is necessary to provide gains for each color-output stage. This variation of gain is provided mainly by the value of the emitter resistors for each color amplifier. Note in Fig. 22-16 that the B-Y amplifier will have the highest gain and the R-Y amplifier a lower gain, with the lowest gain amplifier being the G-Y stages. Note that +250 volts is used as the collector voltage for the output stages. This is necessary to develop the high color-video voltages required to drive the picture-tube grids. Special, high-voltage transistors must be used to accommodate these high dc and ac voltages.

***G-Y* Generated at Output** In Section 22.4 we noted that $-(G\text{-}Y) = 0.51(R\text{-}Y) + 0.19(B\text{-}Y)$. Many receivers use a straightforward application of this equation to generate the G-Y signal. It must be done after the R-Y and B-Y amplifiers. This provides the phase inversions required by the minus sign. Usually a simple resistor divider will provide 0.51 $(R\text{-}Y)$ and another resistor divider will produce $0.19(B\text{-}Y)$. The G-Y signal is then amplified the same as the R-Y and B-Y signals.

Figure 22-17 shows a straightforward approach to the method described above. The $0.51(R\text{-}Y)$ signal is produced by the voltage divider

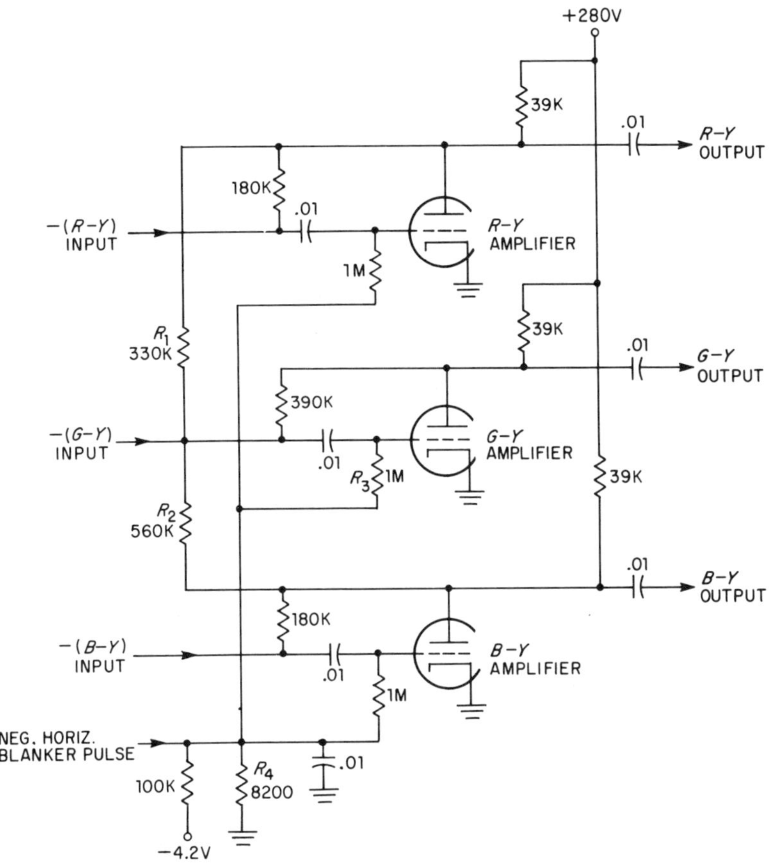

Fig. 22-17. *R-Y, B-Y,* and *G-Y* amplifiers, where the *G-Y* signal is obtained from the outputs of the *R-Y* and *B-Y* amplifiers. (*Courtesy of RCA.*)

formed by R_1 and R_3. Likewise, 0.19(*B-Y*) is developed by R_2 and R_3. R_4 is required in order that a blanking voltage may be inserted into all three amplifiers during horizontal retrace, to blank out the burst. The resistor from each plate to each input serves to introduce negative feedback. This stabilizes the stage gain and widens the circuit bandwidth.

The *G-Y* Generated by a Common Cathode The *R-Y, B-Y,* and *G-Y* amplifiers using a common cathode connection were shown in Fig. 22-15. The discussion accompanying that figure described its operation. This is a fairly common method of obtaining *G-Y,* and it can also be achieved with transistors. Note that it is no more complicated than the amplifiers shown in Fig. 22-17. However, its operation is not quite so obvious since each tube has inputs on both the grids and the cathodes.

22.6 LUMINANCE (VIDEO) AMPLIFIERS

The video amplifiers in color-TV receivers differ from those in monochrome receivers in several respects. In this section we will briefly

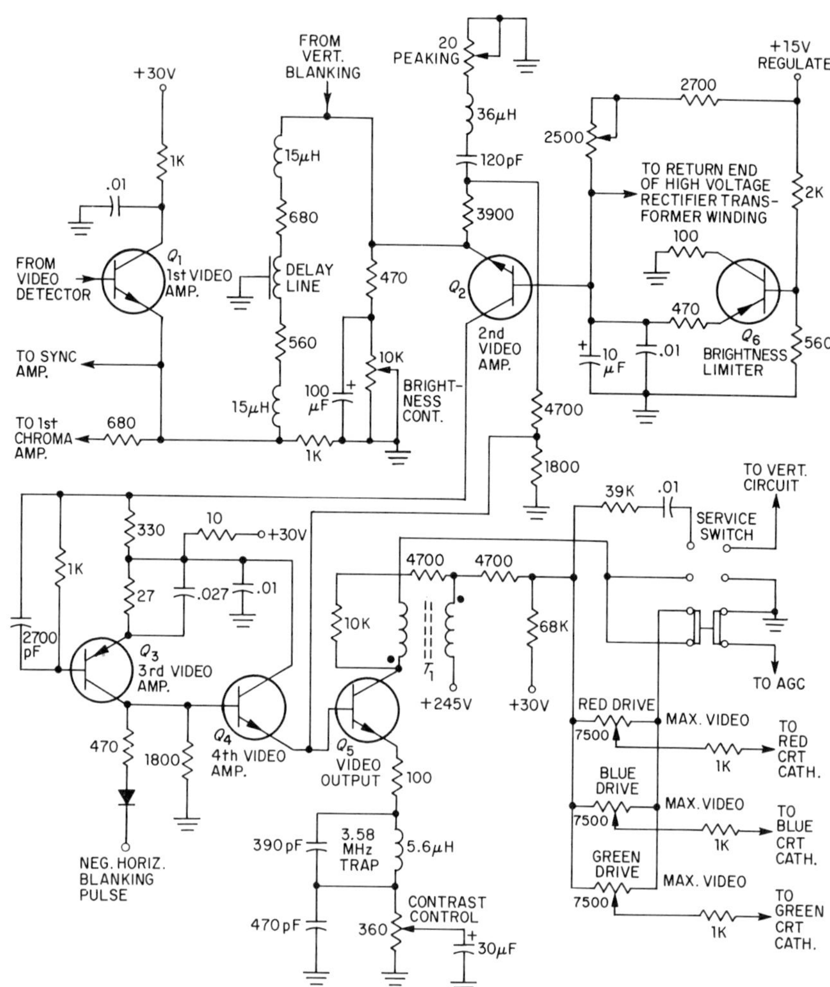

Fig. 22-18. A 5-stage high quality video amplifier for a large screen color receiver. (*Courtesy of RCA.*)

describe some of these differences. Figure 22-18 shows an example of a high quality solid-state color-TV video amplifier. This circuit was selected for discussion, because it shows quite clearly many of these differences between color- and monochrome-video amplifiers.

The first video amplifier, Q_1, performs three functions. Since it is an emitter follower, it has a low-output impedance which enables it to drive three different stages from its emitter terminal. One stage is the sync amplifier. The second stage is the first chroma amplifier connected to the first video amplifier through a 680-ohm resistor. The third output from Q_1 goes to the delay line. Note the elaborate circuitry on both sides of the delay line, designed to insure a good impedance match. If this is not done, the delay line will produce reflections which cause an effect similar to ghosts on the picture tube.

The vertical-blanking signal is injected into the video signal at the delay line output. From here, the signal goes to the emitter of a common-base amplifier, Q_2. This type of amplifier has a low-input impedance to help match the delay line impedance. Several other interesting things happen at the Q_2 emitter terminal. A series RLC circuit connects this point to ground and helps shape the frequency response curve of the video amplifier between 2 and 3 MHz. The only other frequency response adjustment in the 5-stage video amplifier is the 3.58 MHz trap in the emitter circuit of the video-output stage.

Picture tube brightness is controlled at the emitter of Q_2. This is done by adjusting the bias for all succeeding stages with a 10 kilohm potentiometer. Since all stages are dc coupled right up to the picture tube, this shift in the Q_2 bias level will adjust picture-tube brightness.

Another item worth noting at the emitter of Q_2 is the feedback from the emitter of the fourth video amplifier. Thus Q_2, Q_3, and Q_4 form a dc coupled feedback amplifier. This will stabilize the bias point of these stages as established by the brightness control. The brightness limiter stage, Q_6, also stabilizes the brightness setting by another feedback method. It samples the current going into the high-voltage rectifier and adjusts the brightness (at the base of Q_2) according to the magnitude of this high-voltage current. This will prevent the picture tube from drawing excessive beam currents on bright scenes.

The horizontal-blanking voltage is injected into the video signal at the junction of Q_3 and Q_4. A large negative pulse turns Q_4 off during the horizontal-retrace interval.

The video-output stage utilizes a common emitter amplifier. The gain of this circuit can be adjusted by varying the impedance from its emitter to ground by means of the contrast control. The lower this impedance, the higher the gain. At 3.58 MHz, the parallel-tuned trap looks like an extremely large impedance. This lowers the gain to almost zero at that frequency. The contrast control adjusts the stage gain by connecting a by-pass capacitor at various points along a 360-ohm potentiometer. Minimum gain (minimum contrast) is obtained when the by-pass capacitor is at the ground side of the potentiometer. At this point, the full value of the potentiometer is in the emitter circuit and maximum degeneration occurs.

The circuits in the output of Q_5 require explanation. The output load for Q_5 consists of peaking transformer T_1, the two 4,700-ohm resistors, and the three paralleled drive controls. The two windings of T_1 are mutually coupled in such a way as to provide high-frequency peaking. The video signals are applied to the red, blue, and green cathodes from their drive controls, which are adjusted to compensate for phosphor efficiency inequalities.

The service switch has three positions. Normal viewing utilizes the lower position. If the switch is moved from this normal position, then the wire labelled AGC kills the RF section of the receiver. Thus, in the switch's middle position only a raster, with no video, appears on the picture tube. This is useful for purity and convergence adjustment. The third position on the switch kills the vertical oscillator-buffer amplifier. This position is useful for adjusting the tilt of the yoke and the gray scale, since only a horizontal line will appear across the center of the picture tube.

22.7 COLOR-SYNC SECTION

This section of a color receiver provides a 3.58-MHz carrier to the color demodulators. The carrier must be exactly in phase with the original carrier at the studio in order to reproduce accurately all the colors. The 3.58-MHz burst on the back porch of each horizontal-sync pulse is a sample of that carrier with the proper frequency and phase. The color-sync section must produce a carrier which matches the frequency and phase of these bursts.

An abbreviated block diagram of the color-sync section of a color receiver is shown in Fig. 22-19. The input to this section is obtained from some point in the luminance system or from some other point where the chroma signal is complete, for example, the bandpass (chroma) amplifiers. What the color-sync section is primarily designed to obtain is the color burst. It is this burst which contains information concerning the proper frequency and phase of the color subcarrier. The 3.58 MHz burst takeoff point varies in different receivers, but it is always from some point where the full chroma signal is present. In some receivers, the takeoff point might be at a resonant coil or transformer in the plate circuit of one of the video amplifiers. In another receiver, it might be at one of the chroma amplifiers.

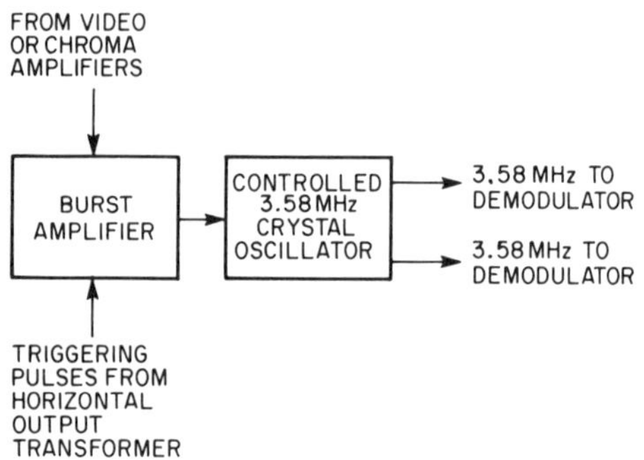

Fig. 22-19. An abbreviated block diagram of the color-sync section of a color receiver.

The voltage developed by the takeoff circuit is applied to a burst amplifier which is gated by a pulse obtained from the horizontal-output transformer. That is, the stage is cutoff except when gated by the pulse, and during this interval the color burst should be active in the system and hence pass through the burst amplifier. At the end of the gating pulse, the burst amplifier is again cutoff.

At the output of the burst amplifier, the color burst may be applied either to a crystal-ringing circuit or to an automatic phase control (APC) circuit. Some of the early receivers employing *I*, *Q* demodulation systems used a crystal-ringing circuit, but nearly all present-day receivers use an automatic phase-control circuit. Both systems are described here.

Crystal Ringing Circuit This system uses a quartz crystal which, when excited by the color burst at the start of each horizontal line, will continue to "ring" or oscillate at its natural frequency of 3.58 MHz for the duration of the line. A typical circuit is shown in Fig. 22-20. The burst from the burst amplifier activates the quartz crystal and, because of its extremely high *Q*, it continues to oscillate with very little decrease

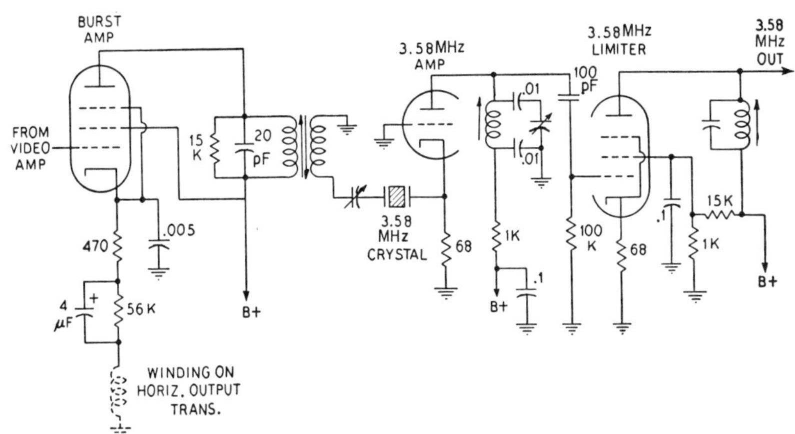

Fig. 22-20. The partial schematic circuit of a crystal ringing system.

in amplitude until the next burst arrives. The trimmer in series with the crystal can change its resonant frequency by several hundred Hertz and thus take care of crystal tolerances.

The stage following the crystal is an amplifier stage, and the stage beyond that is generally a limiter to smooth out variations in the output of the ringing circuit. The output from the limiter may be used as one of the 3.58 MHz driving voltages for the *I* or *Q* demodulators, while the same output, passed through a 90-degree phase-shift network, will provide the reference voltage for the other demodulator.

Automatic Phase-control System A second approach to the development of a 3.58-MHz subcarrier whose frequency and phase are locked in to that of the color burst is by means of an automatic phase-control system. This system, illustrated in Fig. 22-21, is common in present-day receivers. It closely resembles the horizontal AFC systems currently employed in black-and-white receivers. That is, the frequency and phase of the color burst are compared with the frequency and phase of the signal developed by a free-running, 3.58 MHz oscillator. If a difference exists between the two, a corrective dc voltage is developed and applied to a reactance modulator which is connected across the

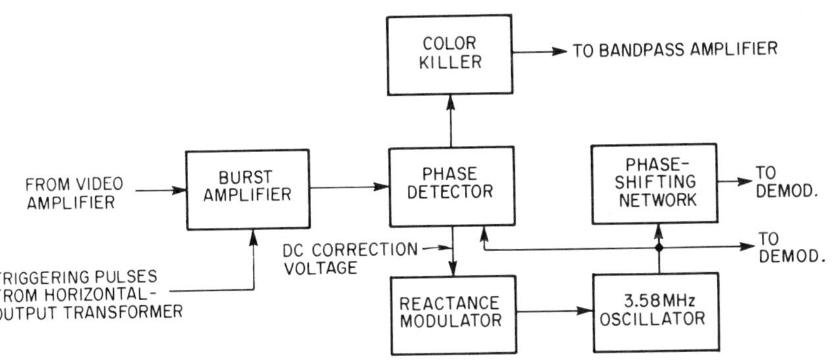

Fig. 22-21. A color-sync section using an automatic phase-control system.

resonant circuit of the oscillator. The reactance modulator, in response to the correction voltage, varies its impedance in such a way as to bring the oscillator frequency in line with the color burst. The oscillator output is fed to a 3.58-MHz tuned circuit from which signals for the phase detector and the two color demodulators are obtained. We will discuss the automatic phase-control system in more detail in following sections.

22.8 BURST AMPLIFIER (COLOR SYNC AMPLIFIER)

As illustrated in Figs. 22-19, 22-20, and 22-21, a circuit is required which extracts the 3.58 MHz burst from the chroma signal. The burst is then amplified in this stage. The circuit should not respond to the rest of the chroma signal; hence, the amplifier is turned on only during the horizontal-retrace time. One might think that the horizontal-sync pulse would also be amplified during this time. However, this does not happen because the burst amplifier is sharply tuned to 3.58 MHz. Pulses or waveforms which have frequency components other than exactly 3.58 MHz are not amplified.

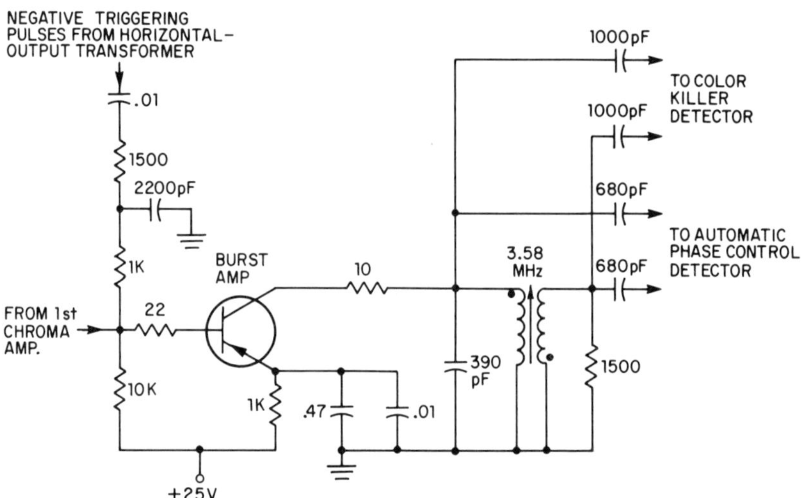

Fig. 22-22. A typical solid-state burst amplifier. (*Courtesy of Motorola.*)

A typical solid-state burst amplifier is shown in Fig. 22-22. The input to this circuit is from the output of the first chroma amplifier. Also arriving at this same point is a negative pulse during each horizontal-retrace period. During this pulse, proper bias is established so that the transistor will amplify the 3.58 MHz burst. At other times, the stage is biased slightly below cutoff. No ac signal is amplified during these times.

The rest of this burst amplifier is quite straightforward in operation. The transistor is connected with the emitter ac grounded to achieve the largest possible gain. The output transformer is tuned to 3.58 MHz.

Capacitors at the transformer primary pass the bursts to a color-killer detector and an automatic phase-control detector. The opposite phase of each signal is also provided to these circuits by the secondary of the transformer.

22.9 CRYSTAL CIRCUITS

The color demodulators require an accurate and stable 3.58 MHz carrier for proper operation. The oscillator which generates this carrier is a major contributor to the quality of the color picture. High stability is important since slight tint errors are discernible by even the most unskilled viewer. A crystal controlled oscillator is therefore mandatory. The stability of simple *LC* or *RC* oscillators is much less than that required for a color receiver, 3.58 MHz oscillator.

A solid state 3.58 MHz crystal oscillator is shown in Fig. 22-23. This differs from the crystal-ringing circuit previously discussed, in that

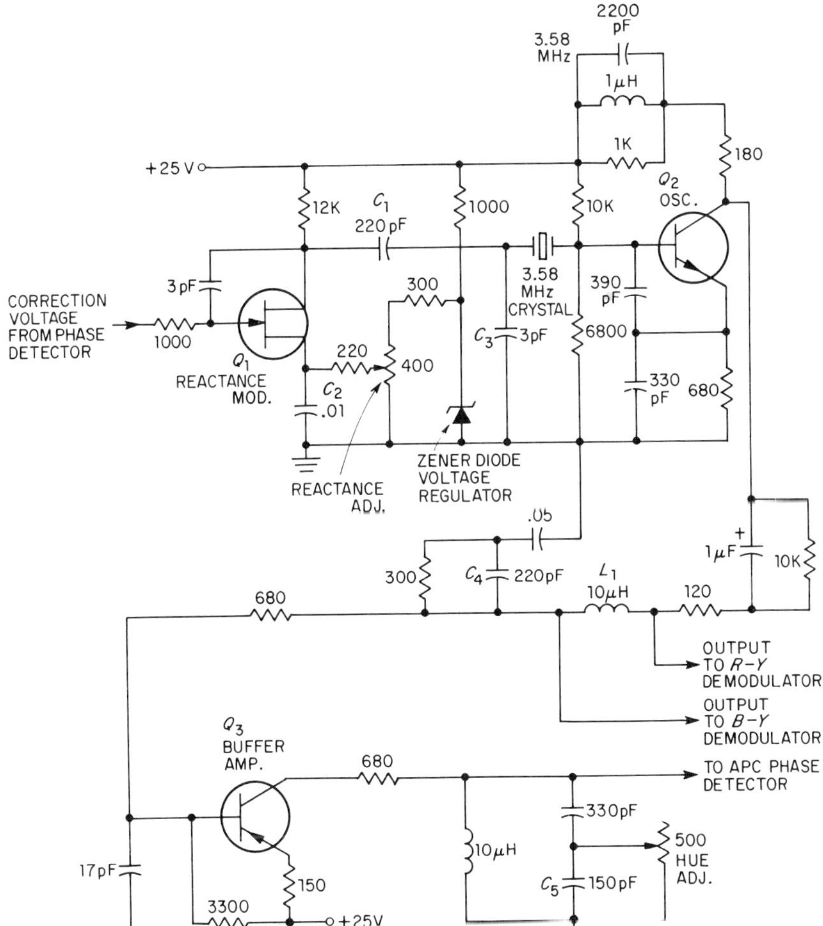

Fig. 22-23. A solid-state 3.58 MHz crystal oscillator with a reactance modulator and buffer amplifier. (*Courtesy of Motorola.*)

this oscillator is not shock excited, but receives continuous feedback and provides a constant amplitude output. The actual oscillator stage, Q_2, is a form of the well-known Colpitts oscillator. It is characterized by two series capacitors from base to ground, with the emitter tied to the junction of the two capacitors. (The pure or non-crystal Colpitts oscillator has an inductor across these capacitors, that is, from base to ground.) The variation shown in Fig. 22-23 is called the Colpitts-Pierce oscillator, where a crystal takes the place of the inductor. However, in this case the crystal must go through C_1, Q_1, and C_2 to return to ground. A parallel path is also offered by capacitor C_3.

By passing the crystal return through Q_1, a method of controlling the actual crystal oscillator resonant frequency is achieved. Typical crystals used in color-TV receivers have a sensitivity of about 160 Hertz/picoFarad. That is, if the capacitance of the C_1, C_2, C_3, and Q_1 circuit changes by 1 picoFarad, the crystal oscillator resonant frequency changes by 160 Hertz.

Let us pause for a moment to see how Q_1 changes the capacitance of the C_1, C_2, C_3 circuit. Q_1 is a field-effect transistor which behaves as a voltage-controlled resistor. Depending on the voltage at its gate terminal, its resistance between the drain and source terminals will vary from a few hundred ohms to many megohms. If the reactance adjustment potentiometer is set such that 5 volts is placed on the source electrode of Q_1, then 10 volts on the gate of Q_1 will turn it completely on. Under these conditions, the resistance between the drain and the source may be as low as 100 ohms. This small resistance can nearly be neglected in the calculations. Thus, C_1 in series with C_2 looks like C_1 only because C_2 is so large. Placing C_1 in parallel with C_3 we get $220+3=223$ picoFarads.

Suppose a zero voltage is applied to the gate of Q_1. This will turn Q_1 completely off. C_1 has no return to ground and the equivalent capacitance of the circuit becomes $0+3$ pF $=3$ pF. The field-effect transistor reactance modulator thus has a sensitivity of approximately 220 pF/10 volts $=22$ pF/volt. Combining this with the crystal sensitivity of 160 Hertz/picoFarad, the overall sensitivity is

$$\frac{160 \text{ Hertz}}{\text{picoFarad}} \times \frac{22 \text{ picoFarads}}{\text{volt}} = 3{,}250 \text{ Hertz/volt.}$$

The modulator-oscillator circuits are therefore very sensitive to small changes of voltage from the phase detector. The modulator performs very efficiently in keeping the 3.58 MHz oscillator exactly in phase with the burst frequency.

The rest of Fig. 22-23 is fairly straightforward. The output of the oscillator stage is parallel tuned to 3.58 MHz so that a harmonic and distortion free 3.58 MHz carrier is produced. The R-Y demodulator carrier is the first signal output. After a 90-degree phase shift through L_1 and C_4, the *B-Y* demodulator carrier is produced. The carrier then goes to a buffer amplifier which drives the APC phase detector. The hue control (in the output of Q_3) adjusts the phase of this output by varying the effective capacitive reactance of C_5. The oscillator phase and frequency are controlled when this output is compared with the burst frequency. The oscillator phase thus is dependent on the setting of the hue control. As mentioned previously, the oscillator phase determines the color selection properties of the demodulators.

Varactor Reactance Modulator The field-effect transistor circuitry including C_1, C_2, and C_3, of Fig. 22-23 can all be replaced by a device called a varactor diode. All semiconductor diodes possess a given junction capacitance when reverse biased. As the reverse bias gets larger, the junction capacitance becomes smaller. Varactor diodes are optimized for this phenomenon. A relatively small voltage change across these diodes causes an appreciable capacitance change, when compared to standard diodes. Also, these diodes possess a good Q (that is, $1/2\pi fC$ is large compared to the diode series resistance at 3.58 MHz).

A reactance modulator using a varactor diode is shown in Fig. 22-24. Compared to Fig. 22-23, the modulator is quite simple. This is because the variable component is now a capacitive reactance instead of a resistance. This eliminates the need for several other capacitors, as required in Fig. 22-23. The control voltage is impressed across the varactor through a 100 k resistor and consists of a positive voltage (for reverse biasing the varactor) plus any correction voltage from the phase detector. Since the varactor is in the ground return of the crystal, it helps determine the oscillator frequency, by varying the effective capacitance across the crystal.

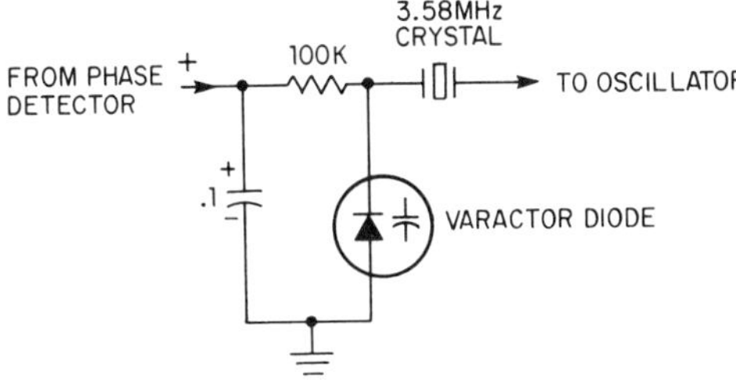

Fig. 22-24. A reactance modulator which uses a varactor diode.

22.10 PHASE DETECTOR AND COLOR KILLER

Two circuits in the automatic phase-control system remain to be discussed. These are the phase detector and color killer shown in the block diagram of Fig. 22-21. A detailed schematic of typical circuits for these functions is shown in Fig. 22-25. These circuits are discussed together here, because in most receiver schematics they are next to each other, and at first glance, they look identical. Both detectors use the 3.58 MHz burst signal and the 3.58 MHz signal from the oscillator as inputs. Both circuits are phase detectors. They compare the phase of the two 3.58 MHz input signals and generate a dc output proportional to their phase difference. The real difference in the two circuits is the destination of the dc output signal.

The output from the phase detector is fed to the reactance modulator (see Figs. 22-23 and 22-24). At that location the dc correction signal controls the frequency of the 3.58 MHz oscillator in such a way that the

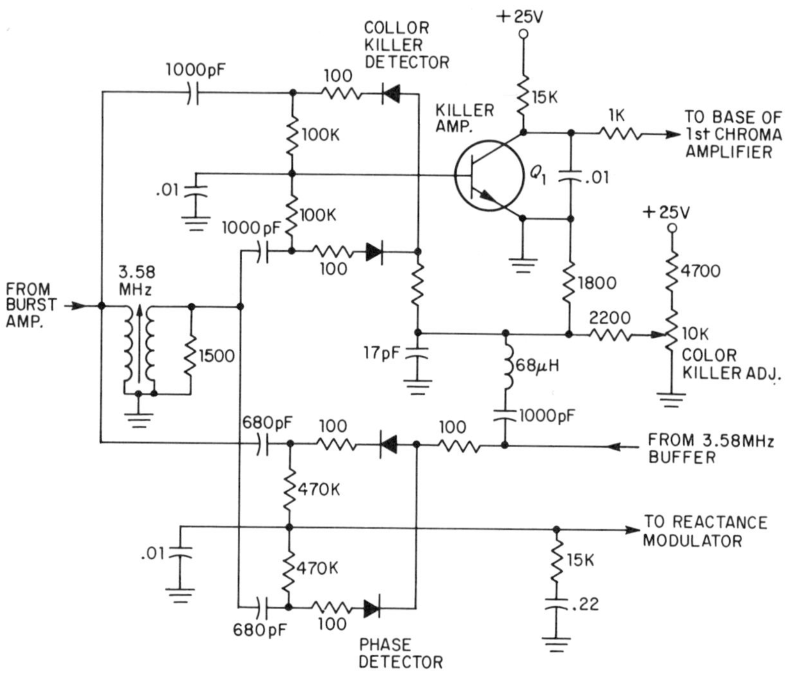

Fig. 22-25. Phase detector and color-killer circuits. (*Courtesy of Motorola.*)

burst and oscillator frequencies are identical. Thus, if one attached a VTVM to this dc voltage, a positive or negative voltage would be measured. The magnitude and polarity of this voltage would depend on the resonant frequency of the crystal oscillator compared to the burst frequency.

The output from the color-killer detector has the same polarity as the phase detector. However, its magnitude is slightly different because of different time constants. Note the function of the color-killer potentiometer. It is adjusted so that Q_1 is biased up to a point just slightly below cutoff. If no bursts are present in the video signal, then a large phase error is detected. This causes a positive bias to add to the bias mentioned above and Q_1 turns on. This sends a negative bias to the chroma amplifiers, turning them completely off.

If color bursts are being received, the color-killer detector does not turn Q_1 on. In this case the chroma amplifiers operate normally.

22.11 AUTOMATIC-TINT CONTROL

One of the most common complaints of color-TV owners is the changing picture-flesh tones on their screen. It seems that they must frequently readjust their sets to eliminate green or purple people. When a commercial comes on, or when a station changes to another camera, the whole readjustment cycle must be repeated. This is not usually the fault of the receiver; frequently the TV station is at fault. Only a very slight change in the phase of the incoming chroma signal causes a significant flesh-tone change. Changes in the yellow-orange-red portion of Fig. 22-9 are the most objectionable. Flesh tones use combinations of these colors. Variations in the green and blue side of Fig. 22-9 often go unnoticed.

One approach to stabilizing the flesh color has been developed by Magnavox. Other systems are available, but we will discuss the Magnavox approach here. This system, called the automatic-tint control (ATC) corrects color errors within ± 30 degrees of orange. Orange is 57 degrees lagging behind the burst—approximately halfway between yellow and red. Thus, most of the sector from yellow to red is under the control of the ATC. Colors outside this range are unaffected.

The ATC has a 3 position front-panel switch labelled Off-Partial-Full. On the rear of the set, there is a potentiometer called the preference control. The viewer first places the ATC switch on "Full" and adjusts the preference control for the correct flesh tone. Then the ATC switch is set on "partial" for normal viewing. If the preference control is set to produce an angle θ (between red and yellow), then whenever color signal information is received within ± 30 degrees of orange, such color signals are partially phase shifted toward θ. If the ATC switch is in the "Full" position, then the color signals within ± 30 degrees of orange would be fully phase-shifted toward θ. This might reduce the range of the red, orange, and yellow colors. The partial position is recommended for normal viewing. This will still correct the tint over the same ± 30 degree range; however, the correction voltage is only about half the value of the voltage in the full position and avoids reducing the flesh-tone color range.

Figure 22-26 shows the automatic-tint control block diagram. Before the addition of the ATC to this receiver, the chroma signal passed from the first chroma amplifier to the second, with the color control potentiometer located between the two stages. With the ATC modification, this signal path is still maintained, but a portion of the chroma signal is sampled, sorted out for its yellow, orange, and red content and then is also applied to the second chroma amplifier to correct the off-color chroma signal tint. The output of the second chroma amplifier-mixer (to the demodulators) is a phase-corrected signal, which provides the proper flesh tones.

The operation of this system will now be explained. (Refer also to the simplified schematic of Fig. 22-27. The *X*-phase (*X*-*Z* demodulators are used) 3.58 MHz subcarrier is applied to a phasing network, to which is also connected the preference control. The *X*-phase subcarrier is phase locked to the 3.58 MHz burst, and its phase will remain constant regardless of any possible phase errors in the chroma signal. By means of the preference control, this reference phase is adjusted by the viewer (as explained above), until the desired flesh tones appear on the screen. At this point and assuming the chroma signal does not phase shift with

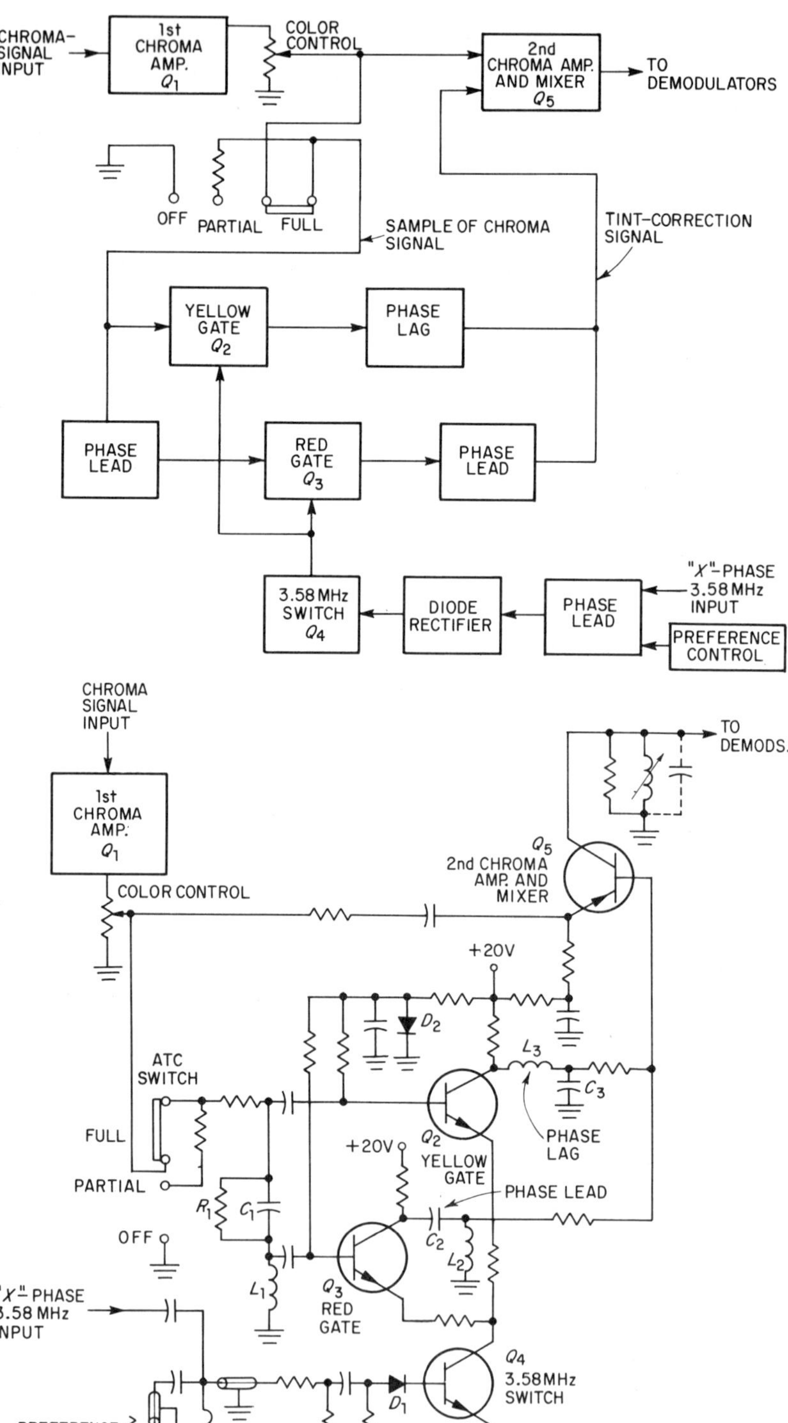

Fig. 22-26. A block diagram of one approach to automatic-tint control. (*Courtesy of Magnavox*).

Fig. 22-27. A simplified schematic of the Magnavox automatic-tint control circuit.

regard to flesh tones, the tint-correction signal (see Fig. 22-26) to the second chroma amplifier is such as to maintain the desired flesh tones. (Initially, the ATC switch is put into the "off" position and the hue and color controls adjusted for a normal color picture.)

The adjusted X-phase 3.58 MHz signal then passes through a diode rectifier, D_1, which permits only the positive half cycles to be applied to the base of the 3.58 MHz switch (Q_4), an NPN transistor. This transistor is biased so that it will conduct only on the applied positive half cycles of the 3.58 MHz signal. The red and yellow gate transistors, Q_3 and Q_2, have their emitters connected to the collector of Q_1, so that these gates can be turned on only during the period that Q_1 conducts on the positive half cycles of the applied 3.58 MHz signal to its base. Note, in Fig. 22-27, that phase shifters consisting of R_1, C_1, and L_1 are connected to the inputs (bases) of Q_2 and Q_3. The effect of these phase shifters on the input chroma signal to the bases is such that a red chroma signal produces the maximum output from the red gate (Q_3) and a low output from the yellow gate. Conversely, a yellow chroma signal will produce maximum output from the yellow gate and low output from the red gate.

If the flesh tones become too red, the output of the red gate increases. This increased amplitude red chroma signal passes through a phase-leading network, consisting of C_2 and L_2. The phase-leading network, causes this originally red chroma signal to be shifted toward yellow (see Fig. 22-9). This now yellowish signal is applied to Q_5, where it mixes with the normal straight-through chroma signal. The result of this mixing process in Q_5 is a reduction of the red color saturation and a return to the reference flesh tones.

If the flesh tones become too yellow, the output of the yellow gate increases and this increased output passes through a phase-lagging network, consisting of L_3, C_3 to the base of Q_5. The phase-lagging network causes this originally yellow chroma signal to be shifted toward magenta (between red and blue). This magenta signal is now applied to Q_5 and mixes with the straight-through chroma signal, where it causes a reduction of the yellow color saturation and again, a return to the reference flesh tones. Note that the normal flesh tones are in the vicinity of the orange region and that the corrections operate on both sides of that region which extend toward yellow on the one side and magenta on the other side.

If the flesh tones are normal, as originally set up by the hue and color controls (with the preference control set to maintain this situation), the correction signals from the red and yellow gates cancel and the chroma signal passes through Q_5 unchanged.

Diode D_2 clamps the forward bias of Q_2 and Q_3 to about 0.6 volts. This is done to make certain that the low-level chroma signals will be sampled by the yellow and red gates when the switch (Q_4) turns each transistor on with the positive peaks of the rectified 3.58 MHz wave.

22.12 TROUBLES IN COLOR CIRCUITS

Whenever a problem occurs in the color circuits, it is usually fairly simple to determine whether color or monochrome circuits are at fault. Four video signals arrive at the color picture tube—luminance, red, green, and blue. If one of these is missing or distorted, a few manipulations of the front-panel controls (on several color and black-and-white stations) will isolate which of the four signals are at fault. In this section, we will briefly summarize the effect on the picture-tube screen when each of the color circuits malfunction.

Bandpass (Chroma) Amplifier All color signals must pass through this section of the receiver before being processed by the color circuits. Often the burst amplifier signal is obtained after the color signal has passed through one chroma amplifier, while the full chroma signal continues on through a second stage (see Fig. 22-12). Under these conditions, if the first chroma stage becomes inoperative, what will we see on the screen? The picture will appear black-and-white since no chroma signal will arrive at the CRT. Also, the color killer will sense that no color signal is present, and it will turn the chroma stages off.

If the chroma amplifiers are only partially defective, then all the colors will be affected. The monochrome signal will not be affected. This can be observed by turning the color control off.

Some chroma amplifiers, such as the one illustrated in Fig. 22-13, utilize a controllable phase-shifting stage. By controlling the phase, one actually controls the hue of the signal. If this type of stage becomes defective, then the hue adjustment would be affected. However, hue adjustment troubles in most color receivers indicate trouble in the 3.58-MHz circuits.

Color Demodulators At this point, the colors start separating. If trouble is experienced in the *R-Y* demodulator, then the red signal on the CRT will be affected. Similarly, a *B-Y* demodulator problem is obvious on the screen as a blue problem. If *G-Y* is derived from *R-Y* and *B-Y*, then a green problem might also originate from a fault in either demodulator. If red, green, and blue problems all exist at the CRT, the demodulators are most likely not at fault as these stages usually affect one or two colors at a time.

The preceding statements can also be said for problems in *X-Z* demodulators. One merely identifies red color problems with a possible *X* demodulator problem and blue color problems with a potential *Z* demodulator problem. Green color problems might originate in either demodulator, together with a red, or blue color problem.

Luminance-video Amplifiers Since video amplifiers were discussed briefly in this chapter we will comment upon them here, as regards any possible problems. Their problems will only be noticed in the black-and-white portion of the picture. By turning down the color control the only signal left at the CRT is the luminance signal. If the problem is still present, then the video-amplifier section may be at fault.

Color Sync Section Troubles in the burst amplifier, 3.58 MHz oscillator, or phase detector will affect all color signals equally. These stages will not have any effect on the monochrome signal. Hue adjustment problems usually originate in these stages.

If the oscillator loses sync with the bursts, then the colors will float through the picture giving a rainbow effect. Loss of oscillator sync may be the result of a faulty burst amplifier, reactance modulator, oscillator, or phase detector. A poor signal from the RF, IF, video detector, and chroma stages may also cause loss of color sync.

Color Killer It is quite apparent when this stage experiences troubles. If it is inoperative, one will see colored snow in black-and-white scenes. If it never turns off, all color programs will appear only in black-and-white.

REVIEW QUESTIONS

1. Draw a block diagram of a color-television receiver using a tri-gun picture tube.
2. Which sections of the color block diagram would not be found in a monochrome receiver? Explain why in each instance.
3. What precautions must be observed with respect to sound takeoff in a color receiver?
4. With which portion of the incoming signal is the color-sync section specifically concerned? How does it use this information?
5. Trace the path of the monochrome portion (*Y* signal) of a color transmission, from the second detector to the picture tube.
6. What is the function of the chrominance section of a color receiver?
7. Explain why the *I*, *Q* system is not utilized in recent TV receivers.
8. Explain what the color and hue controls do in the chrominance section.
9. How can the hue control be effective in either the 3.58 MHz circuits or the chroma circuits?
10. Explain how the *G*-*Y* signal is derived in Fig. 22-14.
11. Why is a 3.58 MHz trap needed after the color demodulators?
12. How does the hue control shown in Fig. 22-23 control the phase of the 3.58 MHz carrier?
13. Explain how the reactance modulator of Fig. 22-23 controls the oscillator frequency.
14. Name at least one advantage of the varactor-reactance modulator.
15. What does the automatic-tint control (ATC) circuit of Fig. 22-26 do to the blue signals? Why?

23 Typical Monochrome and Color-TV Receivers

23.1 INTRODUCTION

For our discussion of a modern television receiver we will use a Sylvania all solid-state color receiver. We will discuss the complete receiver in block diagram form, with the monochrome and color sections treated separately. Alignment and adjustment information for TV receivers in general, will be discussed. Finally, alignment of the Sylvania color set will be given.

Figure 23-1 shows the complete block diagram, and the schematic Fig. 23-2 is a fold-out . This allows you to look at the schematic while you are studying the various sections of the receiver.

23.2 THE MONOCHROME SECTION

In this section we will follow the monochrome signal through the block diagram of Figure 23-1. Monochrome circuits are those which are needed to produce a monochrome signal, but many of these circuits are also used for amplifying the color signals.

The UHF and VHF tuners are shown on the block diagram as being on two separate chassis. The antenna imputs for the tuners are not displayed. There are two separate AGC inputs to the tuners, and the UHF AGC is amplified. The UHF signal reduced to IF is fed through the VHF tuner to the first video-IF amplifier. Four stages of video IF amplification are included.

Your attention is called to the AFC closed feedback loop consisting of the tuners, IF stages, AFC drive, AFC IF, and AFC detector. The purpose of this feedback loop is to control the local oscillator frequency in the VHF and the UHF tuners, and to prevent it from drifting. Early models of color receivers sometimes produced undesirable color changes (and sometimes a complete loss of color) due to a slight drift in the local oscillator frequency. Furthermore, without the AFC circuit, a slight misadjustment of the fine-tuning control by the customer often resulted in unsatisfactory color pictures and an unnecessary service call.

The sound takeoff point on the block diagram is between the fourth IF amplifier and the video detector. Note that a separate sound IF-detector diode is used, since the signal does not pass through the video detector. The 4.5 MHz amplification, limiting and FM detection is accomplished in an integrated circuit, and the audio signal output drives a power amplifier and speaker.

Returning to the video chain in the block diagram; following the video detector there are three stages of *Y* video amplification and a video driver. The contrast control is between the first and second video amplifiers. The delay line is shown between the second and third video amplifiers. The *Y* video driver simultaneously operates the red, green, and blue output amplifiers which, in turn, simultaneously operate the three cathodes of the color picture tube to produce a monochrome picture.

The sync takeoff point is after the first video amplifier. The noise gate system, which inverts the noise pulses and cancels them, is fed from the video detector through a noise gate and to the sync separator. This circuit prevents noise spikes from triggering the sweep oscillators.

Keyed AGC is used in the receiver (see Fig. 23-5). The sync pulses for this system are taken from the first video amplifier. A pulse input from the horizontal-output transformer keys this circuitry so that only the sync pulses are delivered to the AGC amplifier. The output of the AGC amplifier goes to the first pix IF, and also to the VHF tuner. An additional stage of amplification is used to process the AGC voltage for the UHF tuner.

The sync separator delivers sync pulses to the vertical oscillator which is followed by two stages of drive in the vertical-output stage. Notice that vertical blanking is employed to shut the picture-tube beam off during retrace. Vertical blanking is accomplished by feeding a blanking pedestal to the video amplifier (Q_{212}).

The sync separator also delivers sync pulses to the horizontal oscillator. An AFC circuit is used to improve horizontal-sync stability. The horizontal-output stage delivers a signal to the blanker which cuts off the horizontal retrace by sending a signal to the video amplifier (Q_{212}). This is especially important in color receivers, since the color burst occurs on the back porch of the horizontal-blanking pedestal. If the picture tube is not cut off during the horizontal retrace the burst signal may produce a colored vertical line on the picture-tube screen. The horizontal-output stage also provides the high voltage for the picture tube. A conventional flyback circuit is used. A boost B+ voltage is also obtained. The brightness limiter and HV protect circuit prevent picture-tube damage from excessive voltage or drive. These circuits also serve to limit x-ray radiation from the tube.

The power supply consists of two separate systems. One produces a "high voltage" *B+*, and the other produces a low voltage B+.

Having discussed the block diagram in terms of the monochrome signal, we will now look at some of the individual circuits in greater detail.

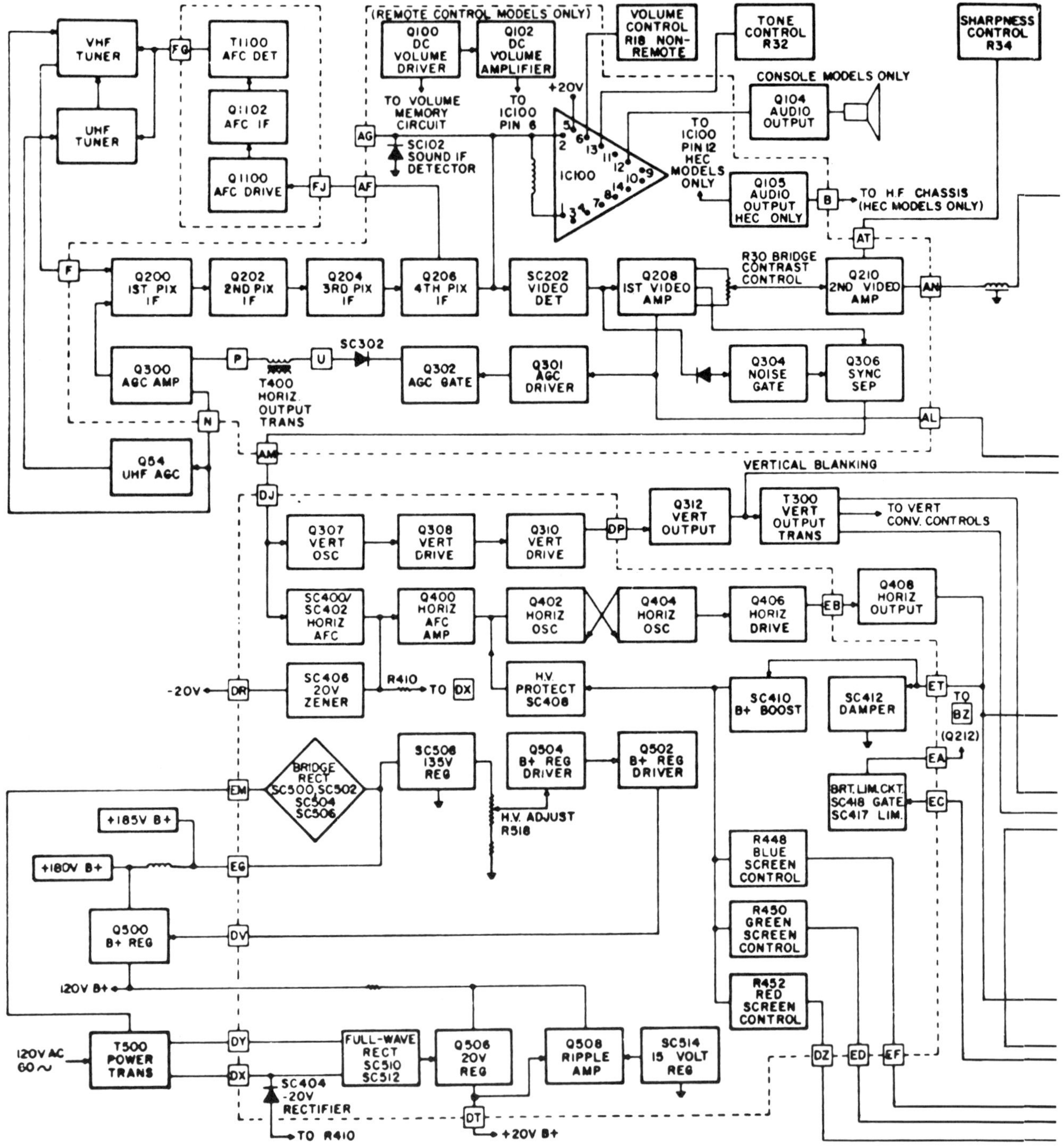

VHF TUNER
UHF TUNER
FG
T1100 AFC DET
Q1102 AFC IF
Q1100 AFC DRIVE
FJ
(REMOTE CONTROL MODELS ONLY)
Q100 DC VOLUME DRIVER
Q102 DC VOLUME AMPLIFIER
TO VOLUME MEMORY CIRCUIT
TO IC100 PIN 6
+20V
VOLUME CONTROL R18 NON-REMOTE
TONE CONTROL R32
SHARPNESS CONTROL R34
CONSOLE MODELS ONLY
Q104 AUDIO OUTPUT
AG
SC102 SOUND IF DETECTOR
AF
IC100
TO IC100 PIN 12 HEC MODELS ONLY
Q105 AUDIO OUTPUT HEC ONLY
B
TO H.F. CHASSIS (HEC MODELS ONLY)
AT
F
Q200 1ST PIX IF
Q202 2ND PIX IF
Q204 3RD PIX IF
Q206 4TH PIX IF
SC202 VIDEO DET
Q208 1ST VIDEO AMP
R30 BRIDGE CONTRAST CONTROL
Q210 2ND VIDEO AMP
AN
Q300 AGC AMP
P
U
SC302
T400 HORIZ. OUTPUT TRANS
Q302 AGC GATE
Q301 AGC DRIVER
Q304 NOISE GATE
Q306 SYNC SEP
N
AL
AM
Q54 UHF AGC
VERTICAL BLANKING
DJ
Q307 VERT OSC
Q308 VERT DRIVE
Q310 VERT DRIVE
DP
Q312 VERT OUTPUT
T300 VERT OUTPUT TRANS
TO VERT CONV. CONTROLS
SC400/ SC402 HORIZ AFC
Q400 HORIZ AFC AMP
Q402 HORIZ OSC
Q404 HORIZ OSC
Q406 HORIZ DRIVE
EB
Q408 HORIZ OUTPUT
-20V
DR
SC406 20V ZENER
R410
TO DX
H.V. PROTECT SC408
SC410 B+ BOOST
SC412 DAMPER
ET
TO BZ
(Q212)
EA
BRIDGE RECT SC500,SC502 SC504 SC506
EM
SC508 135V REG
Q504 B+ REG DRIVER
Q502 B+ REG DRIVER
H.V. ADJUST R518
BRT. LIM. CKT. SC418 GATE SC417 LIM.
EC
+185V B+
+180V B+
EG
R448 BLUE SCREEN CONTROL
Q500 B+ REG
DV
R450 GREEN SCREEN CONTROL
120V B+
R452 RED SCREEN CONTROL
120V AC 60 ~
T500 POWER TRANS
DY
DX
FULL-WAVE RECT SC510 SC512
Q506 20V REG
Q508 RIPPLE AMP
SC514 15 VOLT REG
DZ
ED
EF
SC404 -20V RECTIFIER
DT
TO R410
+20V B+

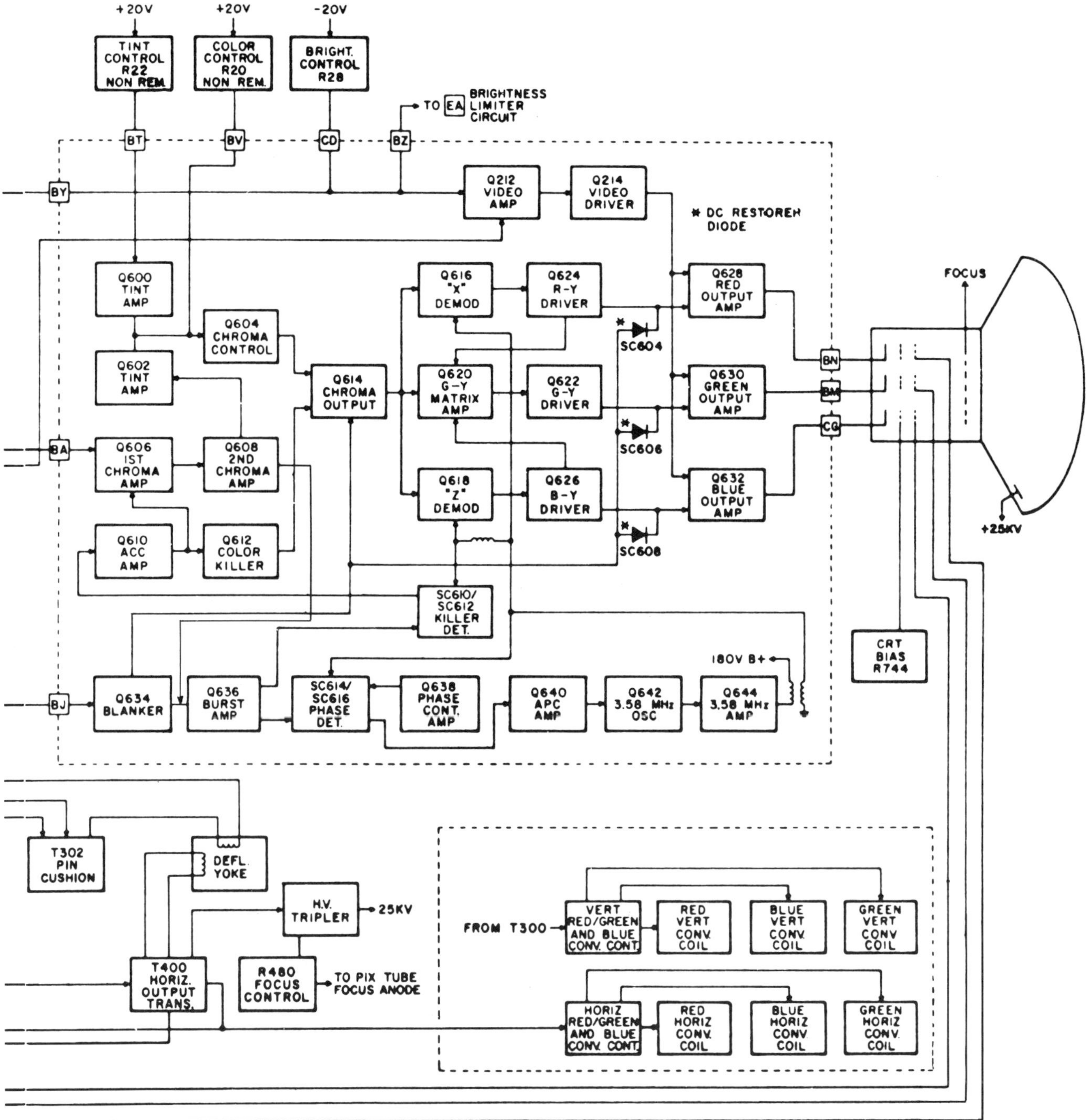

Fig. 23-1. Block diagram of the Sylvania all solid-state color receiver. (*Courtesy of Sylvania Electric Products, Inc.*)

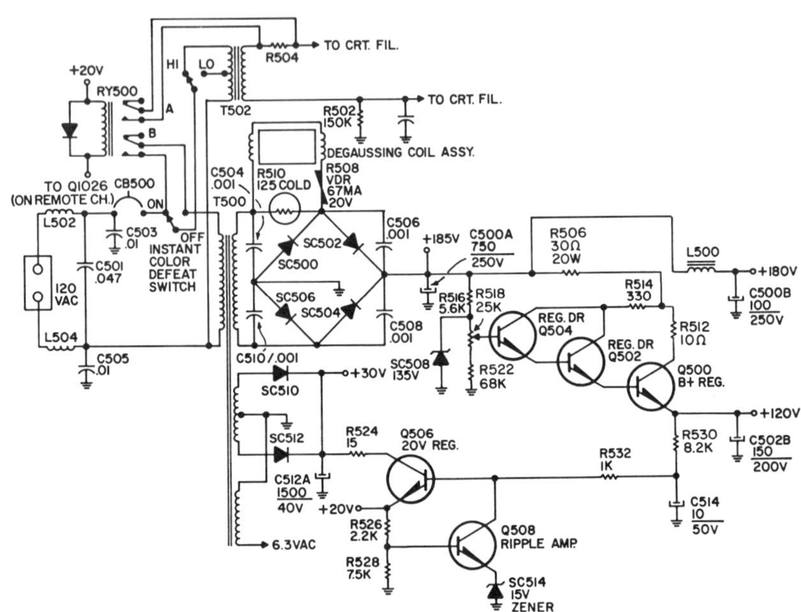

Fig. 23-3. The 120V and 180V power-supply circuits. (*Courtesy of Sylvania Electric Products, Inc.*)

The 180-volt and 120-volt Supplies (See Fig. 23-3.) The bridge rectifier output charges $C_{500}A$ to the peak ac voltage, placing 185 V across voltage divider R_{516}, R_{518}, and R_{522}. A zener diode (SC_{508}) regulates the voltage drop across R_{518} and R_{522} to 135 volts. Variable resistor R_{518} adjusts the forward bias to the regulator drivers Q_{504}, Q_{502}, and the $B+$ regulator Q_{500}, and sets the emitter voltage of the regulator Q_{500}. The emitter of Q_{500} supplies the +120 V dc buss which is the regulated high B+ for the receiver.

Should the power source voltage drop, the bridge output charges $C_{500}A$ to a lower peak voltage, and the divider network voltage is lower. However, the voltage across zener diode SC_{508} remains constant. Only the voltage drop across R_{516} decreases. This holds the voltage across R_{518} and R_{522} constant. Since the emitter voltage of Q_{500} must also remain constant, the effect of lower power source voltage is seen as a lower collector to emitter voltage on Q_{500}, but the 120 V emitter voltage remains unchanged.

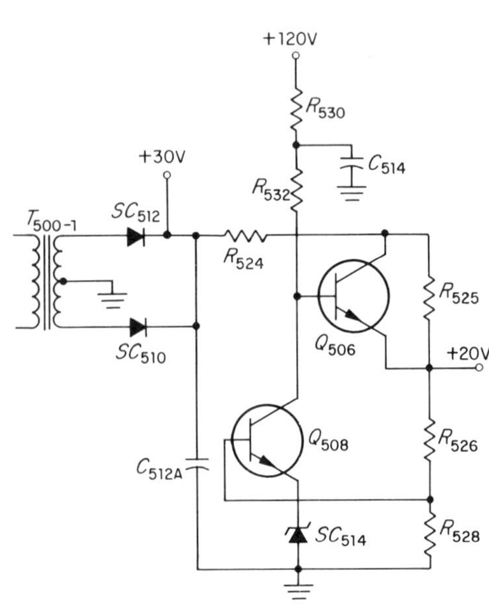

Fig. 23-4. The 20 V power supply.

The 20 V Power Supply The low voltage supply, illustrated in Fig. 23-4, utilizes a full-wave rectifier network consisting of SC_{510} and SC_{512}, and filter capacitor $C_{512}A$, which is the 30 V source for Q_{506} (20 V regulator). The collector of Q_{506} is connected to the 30 V source through R_{524}.

The bias network for the base voltage of Q_{506} (which consists of R_{530}, R_{532}, Q_{508}, and zener diode SC_{514}) is connected from the regulated 120-volt bus to ground. The forward bias to Q_{506} is developed across Q_{508} and SC_{514}. Transistor Q_{508} compares the voltage seen at the center of the divider with the +15 V dc at the junction

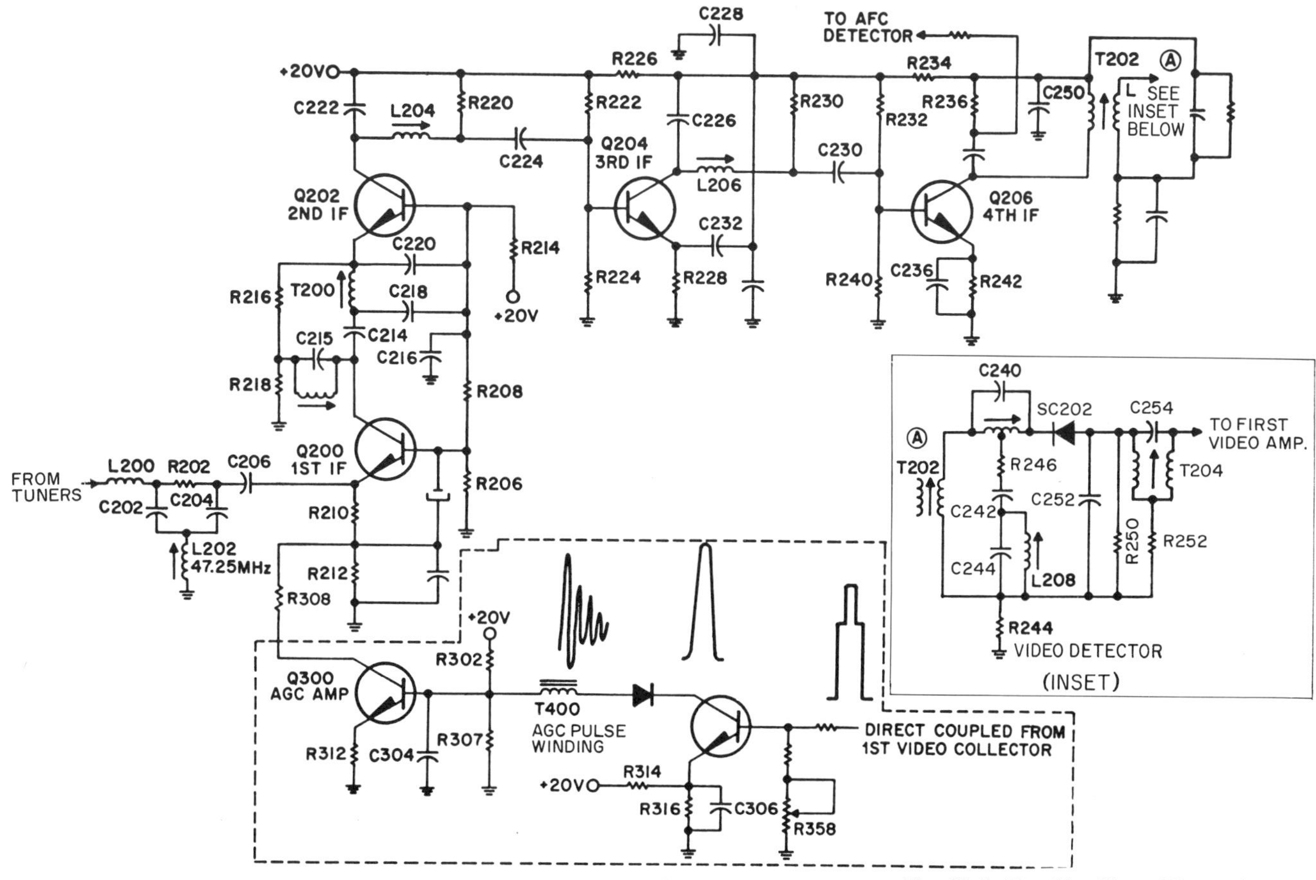

Fig. 23-5. The video-IF amplifier section.

of R_{526} and R_{528}. If the compared voltage is too high, Q_{508} conducts harder, drawing more current through R_{532}. This lowers the voltage on the base of Q_{506}, and consequently, lowers the voltage on the emitter of Q_{506}. Any tendencies (due to load or voltage input variations) toward 20 V source variations are regulated by comparing the output voltage with the zener (SC_{514}) voltage, and using the error signal to control regulator Q_{506}.

Video-IF Section and Video Detector The video-IF section is illustrated in Fig. 23-5. It employs four video-IF amplifier transistors; Q_{200}—first PIX IF; Q_{202}—second PIX IF; Q_{204}—third PIX IF; and Q_{206}—fourth PIX IF.

The RF signal from the tuner is applied to the emitter of Q_{200} through an impedance-matching coil (L_{200}), and an RC network (R_{202}, C_{202}, and C_{204}) with an adjacent sound trap coil (L_{202}). The trap coil is tuned to 47.25 MHz, which sets the upper limits of the IF bandpass. This keeps the sound and picture carrier energy of the tuned band high, and the adjacent channel sound carrier level low.

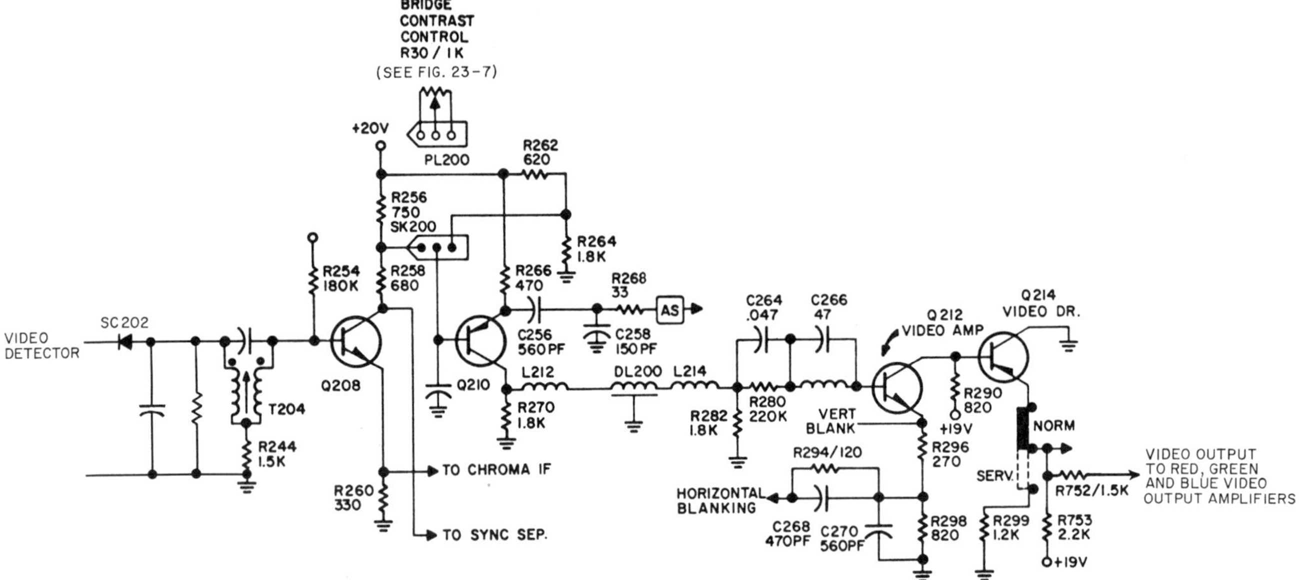

Fig. 23-6. The video amplifier. (*Courtesy of Sylvania Electric Products, Inc.*)

The output of Q_{206} (fourth PIX IF) is coupled through T_{202} (Video Output Transformer) to SC_{202} (Video Detector Diode). The 41.25 MHz sound carrier is trapped out by T_{202} and L_{208}. The signal is then fed into the cathode of SC_{202}, producing a video signal with negative sync pulses ($-Y$).

The Video Amplifier Figure 23-6 shows the video amplifier. Following SC_{202} (video detector), the signal inversion takes place at each successive amplifying stage (Q_{208}—first Video Amp, Q_{210}—second Video Amp, Q_{212}—Video Amp) resulting in a $+Y$ signal at the output of Q_{212}. Since there is no signal inversion at Q_{214} (Video Driver), the result is a $+Y$ signal at the emitter output of Q_{214}.

Horizontal and vertical pulses are fed to the emitter of Q_{212} providing blanking signals to the CRT. By inserting the blanking pulse in the low signal level stages, low pulse power is required. As additional amplification from Q_{212} and Q_{214} increases, the pulse amplitude for blanking the color CRT beam also increases.

The input drive to Q_{210} is taken from the bridge contrast-control circuit shown in Fig. 23-7. The advantage of this control system in a dc video amplifier chain is its ability to control contrast without upsetting the average dc voltage level in the video system, and hence, the brightness level at the CRT. The bridge circuit consists of R_{256}, R_{258}, Q_{208}, R_{270}, R_{262}, R_{264}, and the contrast control R_{30}. The contrast control is low impedance and utilizes the equipotential dc voltage across it to maintain the average dc level through the succeeding video-amplifier stages constant. The voltage divider R_{262} and R_{264} resistance ratio is about 3:1 dividing down the 20 V $B+$, placing about 1/4 of 20 V

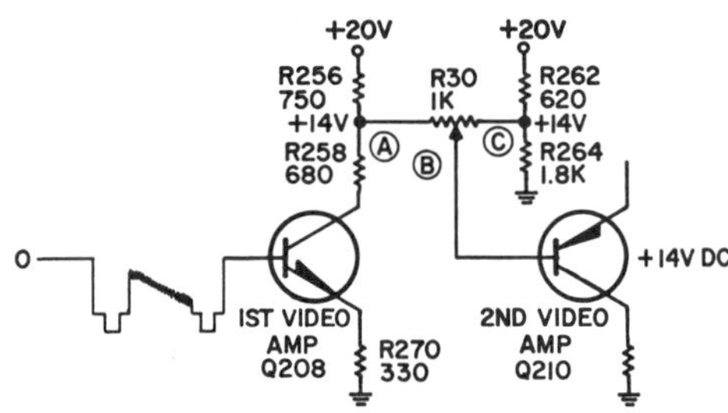

Fig. 23-7. The bridge-contrast control. (*Courtesy of Sylvania Electric Products, Inc.*)

across R_{262} and 3/4 of 20 V across R_{264}. The voltage at the junction of R_{262} and R_{264} is a constant voltage between 14 and 15 volts positive.

The video amplifier Q_{208} is biased on by the voltage divider network in its base. This forward voltage produces a 14 V drop at the junction of R_{256} and R_{258}. Now, both ends of the contrast control are equipotential and no dc current flows through R_{30}, the bridge load (Contrast Control). However, R_{30} and the resistors R_{264} and R_{262} form a video voltage divider across which the video-signal variations are present. When the contrast control slider is at point A, maximum video drive is fed to Q_{210}. On the other hand, when the slider is at point C, the video drive is minimum. However, the dc voltage on the control arm does not vary. Thus, no dc shift occurs in the video amplifiers and the brightness level remains steady.

Sound Section You can follow this discussion of the sound circuit by referring to the fold-out schematic (Fig. 23-2) in the back of the book. The output of Q_{206} (fourth PIX IF) is applied to sound detector diode SC_{102} for 4.5 MHz sound-IF detection to avoid interference in the video detector and amplifiers. The 4.5 MHz IF is filtered through a low-pass filter (L_{102} and R_{120}) to remove all 40 MHz IF frequencies, and is then impedance-coupled to the input of IC_{100}. This integrated circuit provides amplification of the 4.5 MHz IF frequencies, FM limiting, FM detection, and stage of the audio signal voltage amplification. The resultant audio signal from IC_{100} is applied to Q_{104} (Audio Output) and then to the speaker system. On console models with high fidelity, the signal from IC_{100} is applied to Q_{105} (Audio Driver) and from there to the jack plate on the hi-fi chassis.

The AGC System The AGC circuit, illustrated in Fig. 23-8, is a closed-loop regulator system controlling RF and IF signal gain. Its purpose is to maintain relatively constant video output over a wide range of input-signal levels. The system consists of a two-stage gate-amplifier circuit utilizing flyback pulses and sync-tip voltages for regulating the AGC amplifier's impedance.

A positive horizontal flyback pulse is applied (through diode SC_{302}) to the collector of the AGC gate transistor Q_{302}. Simultaneously, the base of Q_{302} receives a positive voltage from the sync portion of the video signal at the collector of the first video amplifier. During this time the AGC gate transistor will conduct current which will result in negative dc voltage at the filter capacitor C_{304}. The magnitude of this voltage is proportional to the current through the gate transistor, and hence, to the amplitude of the sync pulses at the collector of the first video amplifiers, Q_{208}. This dc voltage controls the bias (hence, gain) at the tuner and the controlled IF stages. The potentiometer R_{358} together with R_{320} and the gate emitter bias network R_{314}, R_{316} set the AGC threshold voltage.

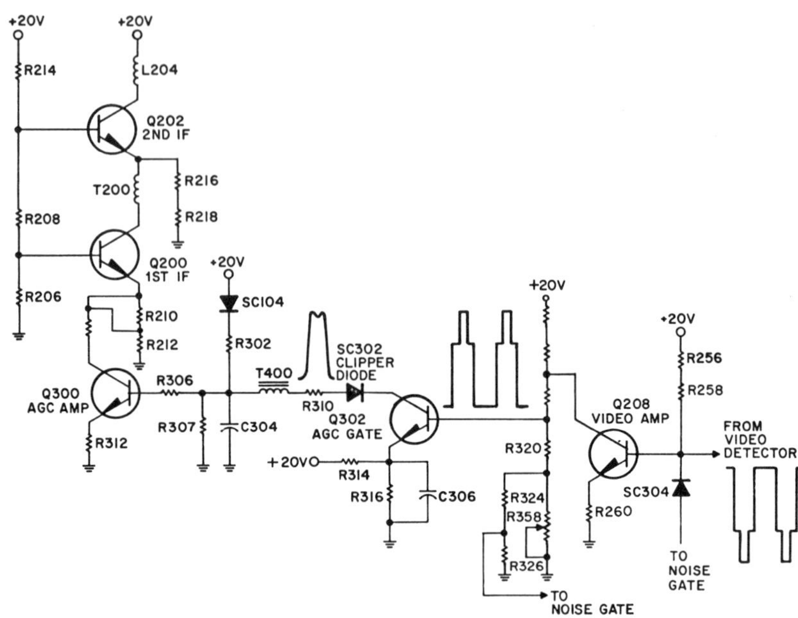

Fig. 23-8. The AGC system. (*Courtesy of Sylvania Electric Products, Inc.*)

With no AGC developed, the AGC amplifier transistor Q_{300} is biased into saturation through the resistors R_{306}, R_{302}, and R_{307}. The voltage at the junction of R_{307} and R_{302} is 8 V which gives maximum tuner gain.

With Q_{300} in saturation, the emitter current through Q_{200} and Q_{202} (series connected) will be determined by resistor R_{210}. This has a relatively low value, thus the IF stage will operate at maximum current, hence maximum gain.

When negative AGC bias is developed and added to the existing bias at the base of Q_{300}, the current through Q_{300} is reduced, thus reducing the current and the gain of Q_{200} and Q_{202}. With Q_{300} cut off by negative AGC bias, the gain of Q_{200} is low and determined by resistor R_{212} (R_{210} is low resistance.) Transistor Q_{202} is controlled similarly by R_{218}. Further change of AGC voltage will control only the gain of the tuner. See Fig. 23.5.

AFC and Tuning Control As shown in the receiver schematic, Fig. 23-2, the—AFC circuit obtains its signal from the fourth video-IF stage (Q_{206}). The tuner AFC (Automatic Frequency Control) circuit employs transistor Q_{1102} (AFC IF/Detector) and Q_{1100} as the driver. The Discriminator Circuit is composed of T_{1100} (Discriminator Transformer), diodes SC_{1100}, SC_{1102}, filter circuit R_{1108}, C_{1112}, R_{1110}, C_{1114}, and the voltage matric R_{1116} and R_{1118}. The sole purpose of the AFC circuit is to electrically control changes in oscillator frequency and IF drift by feeding correction voltage from the discriminator to the oscillator varactor located in the tuner.

The video-IF carrier 45.75 MHz from Q_{206} (fourth PIX IF stage) is applied to the input of Q_{1100}, then to Q_{1102}. Transformer T_{1100} is tuned

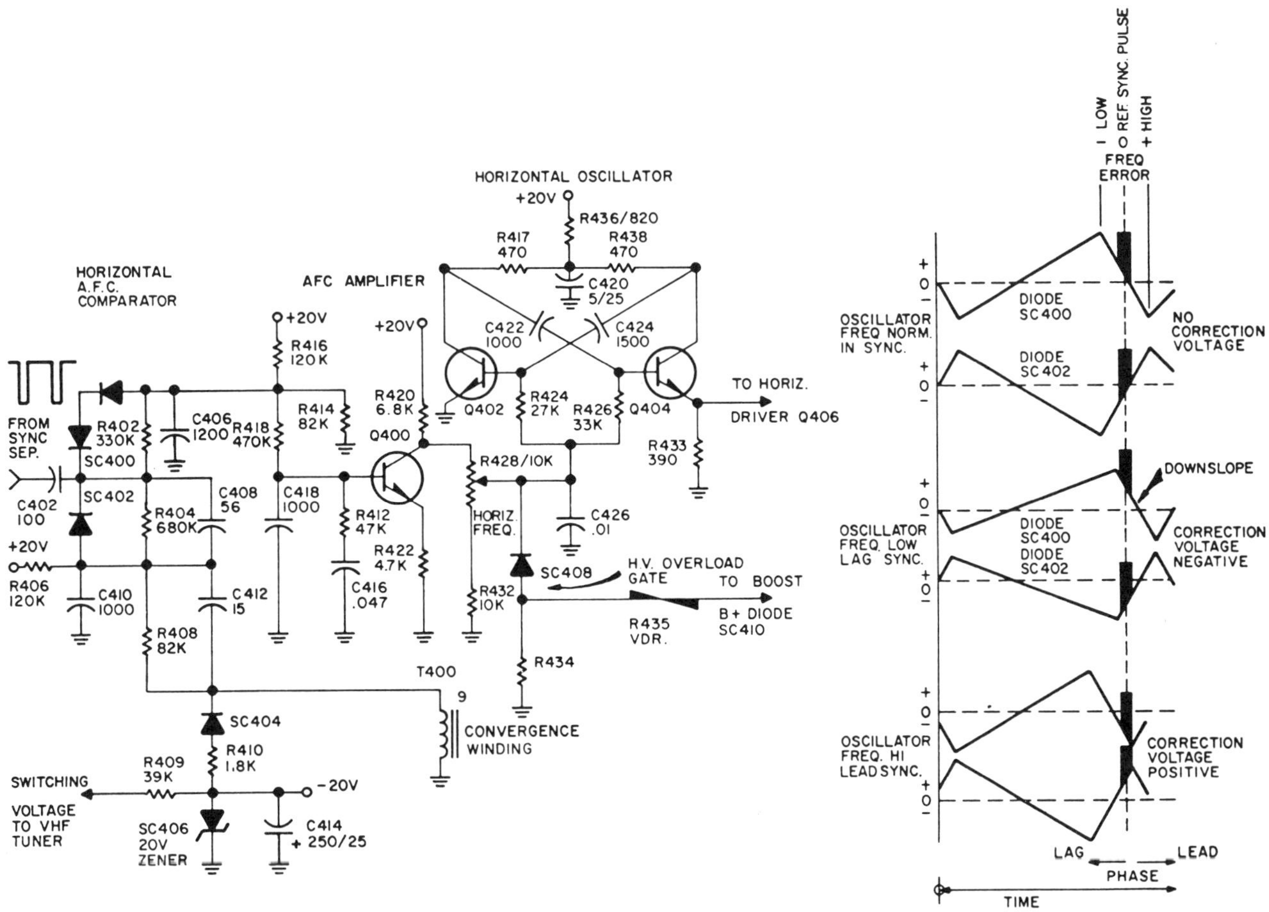

Fig. 23-9. The horizontal oscillator and AFC circuit. (*Courtesy of Sylvania Electric Products, Inc.*)

to 45.75 MHz, the video-IF carrier frequency. Should the AFC detector detect a frequency change, the matrix voltage across R_{1116}, R_{1118} changes, producing a correction voltage at their junction. This dc voltage changes the varactor-diode voltage, which in turn. brings the local oscillator frequency back to the correct value.

At the correct fine tuning, with the pix carrier at 45.75 MHz, no output voltage is developed in the discriminator. If the RF oscillator frequency is increased for any reason, the pix carrier frequency will increase by the same amount. This would develop negative voltage, which when added to the tuning voltage, would reduce the oscillator frequency, thus bringing the tuning back to the acceptable frequency. With the oscillator frequency reduced, positive correcting voltage is developed, thus compensating for the fine-tuning error.

Horizontal Oscillator and AFC (See Fig. 23-9) The horizontal multivibrator frequency is controlled by the amplified output from an unbalanced comparator circuit. consisting basically of diodes SC_{400}, SC_{402} back to back and the associated waveshaping circuits C_{412}, R_{408}, and C_{410}.

Negative polarity horizontal-sync pulses are passed through C_{402} into the unbalanced comparator from the single-ended sync separator Q_{306} (see the complete schematic). Diodes SC_{400} and SC_{402} compare the horizontal sync pulse to a sawtooth voltage applied to their anodes. The sawtooth is developed by integrating a flyback pulse in the waveshaping circuit R_{408}, C_{412}, and C_{410}.

The sawtooth wave-form is the anode voltage for each diode, and the horizontal-sync pulse drives the cathode of each diode. When an unbalance occurs in the comparator diodes, one diode conducts more than the other, producing an output voltage.

When the oscillator frequency runs normal, the sawtooth repetition rate and the horizontal-sync pulse are in sync, because the sync pulses sit on a sawtooth ac reference. Now, both diodes conduct equally, resulting in no correction voltage. However, when the oscillator frequency is lower than normal, or the oscillator phase leads the sync pulse, diode SC_{400} conducts more than SC_{402}. This unbalanced conduction results in a negative correction voltage from the comparator. Should the horizontal-oscillator frequency run high or lead the horizontal-sync pulse diode, SC_{402} conducts more than SC_{400}, causing the unbalance to produce a positive voltage from the comparator circuit.

The comparator-output voltage is fed to Q_{400}, a class *A* dc amplifier (the *AFC* amplifier), where the comparator-output voltage either aids or bucks the AFC amplifier's forward bias. When the comparator's output voltage is negative, the forward bias of Q_{400} is lowered. This raises the collector voltage of Q_{400} and also the voltage drop across R_{428} and R_{432}. This increases the voltage across R_{424} and R_{426}, and produces a more rapid discharge of C_{422} and C_{424}, thereby increasing the oscillator frequency.

A positive output voltage from the comparator circuit aids the forward bias at the base of Q_{400}. The AFC amplifier conduction increases, lowering the collector voltage and the voltage across R_{428} and R_{432}. This decreased voltage, appearing at frequency control R_{428}, slows down the discharge of C_{422} and C_{424}, resulting in a lower oscillator frequency.

Horizontal Driver Figure 23-10 illustrates the horizontal driver and output circuit. The horizontal-oscillator multivibrator in the previous stage develops a square-wave pulse across resistor R_{433}. When this pulse is applied to the horizontal-drive transistor (Q_{406}), it switches from cutoff to saturation. This effectively grounds the primary of T_{402} through Q_{406}. Current flows from B+, through the primary of T_{402}, and through Q_{406} to ground. A pulse is coupled to the horizontal-output transistor Q_{408}.

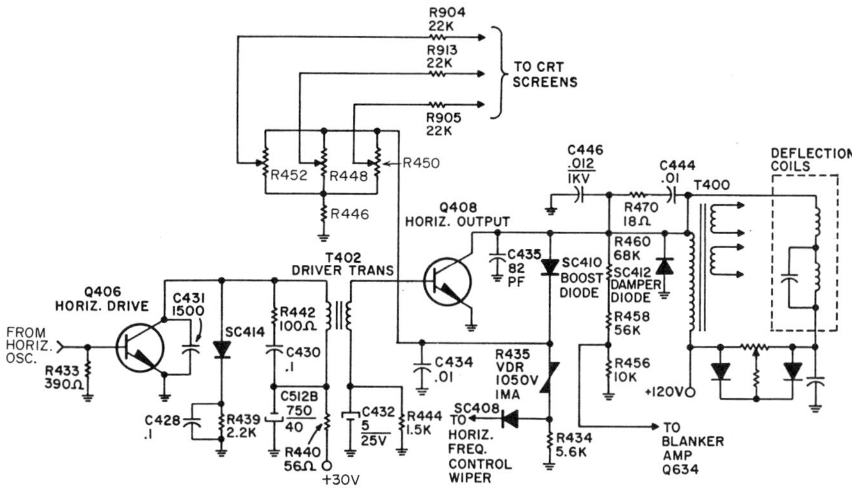

Fig. 23-10. The horizontal driver, horizontal output, damper, and boost voltage circuits. (*Courtesy of Sylvania Electric Products, Inc.*)

The driver protection diode (SC_{414}) and capacitor C_{428} form a low-impedance network for reactive voltages appearing across the primary of the T_{402} during the turn off of transistor Q_{406}. As soon as the positive voltage on the collector of Q_{406} increases over the voltage on C_{428}, SC_{414} conducts and charges C_{428}. Resistor R_{439} keeps discharging C_{428} so that SC_{414} conducts again at the following turnoff of Q_{406}, thus protecting Q_{406} from voltage surges. The series *R-C* circuit comprised of R_{442} and C_{430} forms a damping link for all higher frequency ringing in T_{402}. A filter network (R_{440} and C_{512B}) is used in the collector circuit of Q_{406}. This filter decouples the transistor collector, and also filters the 120-Hz ripple on the 30 V B+. The voltage drop in R_{440} results in a voltage of about 13 V across C_{512B}.

Horizontal Output The driver transformer T_{402} (Fig. 23-10) couples the switching signal from the driver transistor (Q_{406}) to the horizontal-output transistor base (Q_{408}). This positive-going pulse turns on Q_{408}, causing it to conduct into saturation. The collector tap on transformer T_{400} is grounded when Q_{408} is saturated, and this causes current to increase through the primary winding and deflection coils. Inductive reactance opposes the current increase through the yoke coils and the T_{400} primary winding. The magnetic field develops linearly in the yoke coils moving the CRT beam from the raster center to the extreme right.

When transistor Q_{408} gets turned off, the magnetic field in the flyback and the deflection coils forms a resonate circuit. This way the decreasing yoke current flows into C_{446} until it reaches a value of zero. At this point the electron beam has returned to the raster center. Consequently, when C_{446}, which is now charged, discharges again into the yoke, current in the yoke rises but is now negative, and therefore, brings the beam to the extreme left of the raster.

Damper Diode (See Fig. 23-10.) The voltage which started rising when Q_{408} was turned off and which reaches a maximum when the beam returned to the raster center is now back to zero again. The moment this voltage becomes negative, damper diode SC_{412} starts conducting, dissipating the stored energy. Thus the yoke current slowly drops to zero, bringing the beam back to the raster center.

B+ Boost Circuit The high energy pulse developed by T_{400} turns on boost diode SC_{410}, applying the pulse to C_{434}. This way C_{434} becomes charged to a dc voltage which is very close to the peak collector voltage of Q_{408}, and this voltage supplies the picture-tube screen controls.

HV Overload Gate Should the voltage level across the parallel-resistance network comprised of R_{435}, R_{434}, R_{448}, R_{450}, R_{452}, and R_{446} increase drastically, the VDR impedance decreases, placing a higher voltage across R_{434}. This increase in voltage turns on the HV overload gate diode SC_{408}, causing an increase in voltage at the horizontal control, (see Fig. 23-9) forcing the horizontal multivibrator into a higher running frequency. The horizontal-output transistor Q_{408} is turned on for shorter periods of time, and therefore, the peak currents and peak voltages in the output stage decrease. Also, the input pulse into the tripler, shown in Fig. 23-2, becomes lower and reduces the second anode voltage.

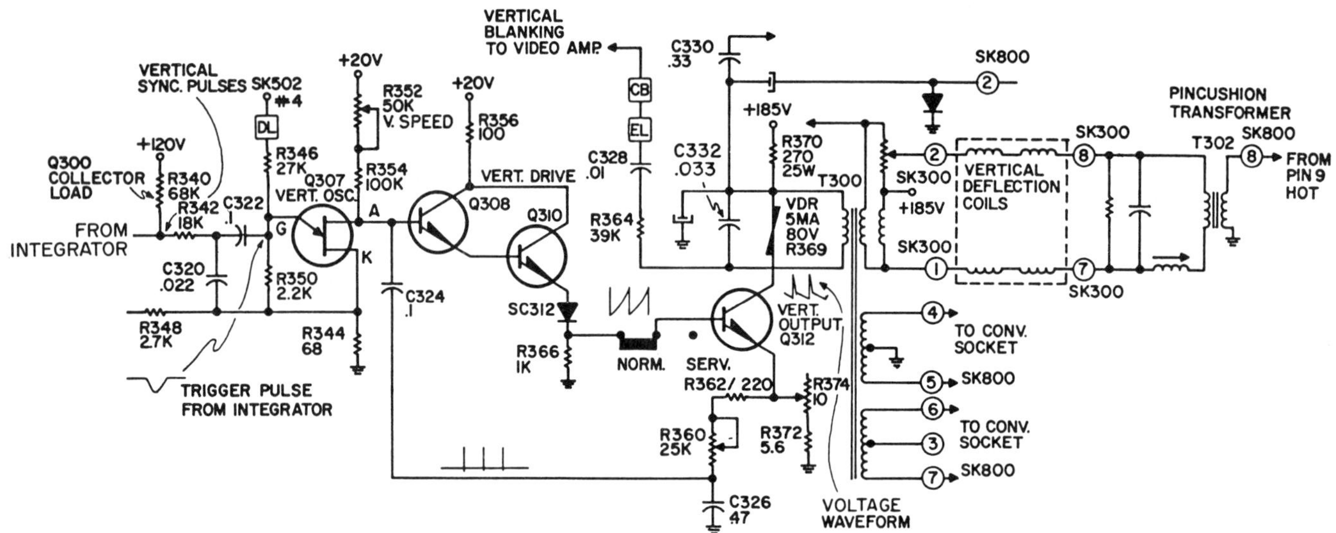

Fig. 23-11. The vertical oscillator and vertical-sweep system. (*Courtesy of Sylvania Electric Products, Inc.*)

Vertical Sweep and Sync Systems The vertical-sweep oscillator, illustrated in Fig. 23-11, is controlled by a negative-going pulse which is developed in the receiver integrator circuit. This pulse is fed to the gate of the programmable unijunction transistor Q_{307} through capacitor

C_{322}. The negative-going timing pulse controls the discharge of C_{324} and C_{236} through Q_{307}, thus forming the drive sawtooth fed to Q_{308}.

The programmable unijunction transistor can be better understood if it is viewed as a diode gate and a current generator. No current flows in the unijunction until the diode is forward biased. When the anode voltage is made 0.6 V higher than the gate voltage, current will flow in the diode and a current flows in the cathode. The programmable unijunction will remain conducting until the anode current falls below that needed to maintain conduction.

When the vertical-serrated pulse appears at the integrator circuit, the pulses are counted, discharging C_{320} and producing a negative pulse. Capacitor C_{322} couples the negative pulse to the gate of Q_{307}, lowering the gate voltage to below the anode voltage allowing C_{324} and C_{336} to discharge through Q_{307}. When the capacitors have discharged, the unijunction no longer conducts and C_{324} and C_{326} begin to charge through R_{352} and R_{354}. The relatively long time constant of this R_3C circuit produces a linear voltage ramp. When the capacitors once again allow the anode to be 0.6 V higher than the gate voltage, the unijunction will fire and the preceding process will repeat itself.

The vertical driver is connected in a Darlington configuration (a double emitter follower) which drives the vertical-output transistor Q_{312}. The feedback signal amplitude and shape is controlled by the adjustment of R_{360}. By combining the drive voltage developed by the charge and discharge of C_{324} with the shaped feedback voltage, the driving signal linearity fed to Q_{308}, and likewise, the current waveform from Q_{312} to the vertical transformer T_{300}, are improved.

The vertical-hold control adjusts the anode voltage of Q_{307}, hence the firing voltage. Adjusting more resistance into the circuit lowers the firing point and increases the oscillator frequency. Decreasing the resistance raises the firing point, slowing the oscillator.

The height control R_{374} adjusts the emitter voltage of Q_{312}. Adjusting R_{374} to less resistance in the emitter circuit increases the current gain and also increases the deflection current and magnetic field in the yoke coils, thus producing additional vertical deflection.

23.3 THE CHROMA SECTION

Before discussing the circuitry of the chroma section, we will follow the color signal through the block diagram of Fig. 23-1. The luminance portion of the chroma signal is derived from the video driver Q_{212} which delivers its signal to the red, green, and blue cathodes simultaneously. This signal is delayed by the delay line between the second and third video amplifiers.

The chroma signal is taken from the output of the first video amplifier (Q_{208}) and delivered to the first and second chroma amplifiers (Q_{606}

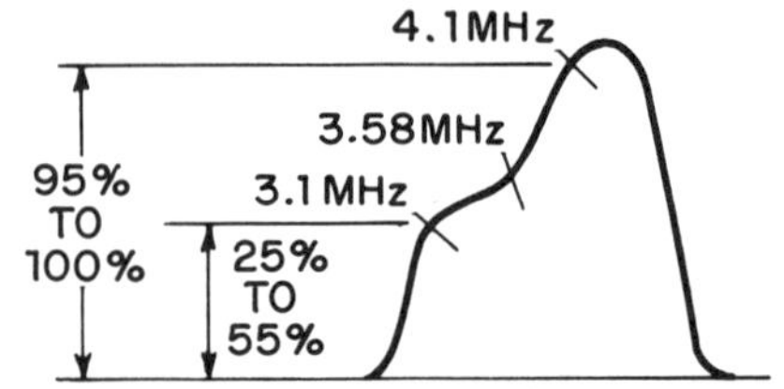

Fig. 23-12. Response curve taken at the output of the second chroma amplifier. (*Courtesy of Sylvania Electric Products, Inc.*)

and Q_{608}). These amplifiers are also known as *color IF* or *color bandpass* amplifiers. They amplify and pass only that portion of the signal that is related to color reproduction. Figure 23-12 shows a typical response curve for a chroma bandpass amplifier. The increase in the amplitude of this response curve at the upper-frequency end compensates for the drop off of the video-IF response curve at the color subcarrier point.

Returning to the block diagram of Fig. 23-1, the output of the second chroma amplifier passes through a tint control amplifier (Q_{602}) in which the phase of the signal is corrected for proper color reproduction. Transistor Q_{602} is a phase-shift amplifier, and the amount of phase shift is controlled by the magnitude of its collector voltage, which in turn, is controlled by a tint amplifier (Q_{600}). Tint amplifier Q_{600} receives its signal either from a remote control circuit, or from a variable resistor connected from B+ in sets that are not remote controlled. In either case the amount of dc voltage delivered to this tint amplifier (Q_{600}) determines the amount of phase shift of tint amplifier Q_{602}.

The output of Q_{602} is delivered to a chroma control (Q_{604}) which determines the amount of gain of the color signal. The chroma control amplifier is simply a broadband amplifier that is gain adjustable.

The chroma's output amplifier (Q_{614}) delivers the color signal simultaneously to the X demodulator, Z demodulator, and $G-Y$ matrix amplifier. The automatic-gain control (called automatic-color control or ACC when it is located in the chroma section) is obtained from a color-killer detector comprised of SC_{610} and SC_{612}. The color-control signal is amplified by Q_{610}, the automatic color-control amplifier. The output of Q_{610} determines the gain of the first chroma amplifier, and also determines whether the color killer will allow the chroma-output amplifier (Q_{614}) to conduct. During reception of monochrome signals, Q_{614} is cut off in order to prevent noise colors on the screen of the picture tube.

The 3.58 MHz oscillator is controlled by an automatic-phase control network which compares a feedback signal from the output of the 3.58 MHz amplifier (Q_{644}) with the burst from burst amplifier Q_{636}. The output signal from the oscillator is delivered to the X and Z demodulators where it is combined with the output from the chroma amplifier. The output of these demodulators drives the $R-Y$ and $B-Y$ drivers. Note that the $G-Y$ matrix amplifier obtains its signal from the $R-Y$ and $B-Y$ amplifiers, and from the chroma amplifiers.

Having traced the signal through the block diagram, we will now discuss some of the individual circuits.

Tint Control (See Fig. 23-2.) The electronic tint control uses two transistors, Q_{600}, as a dc amplifier, and Q_{602} as a phase-shift amplifier. The two amplifiers are emitter-follower and common-emitter configurations, stacked in series to B+, permitting Q_{600} to control the Q_{602} collector voltage.

Chroma tint may be changed by changing the dc voltage at the terminal *BT*. The dc voltage at terminal *BT* for a non-remote set is a variable voltage source, provided by R_{22} (Tint Control) connected to B+ and ground, and for a remote set is a memory module, providing the dc voltage. Variable resistor R_{22} is located on the receiver control board (not shown).

Color Control The phase corrected chroma signal is coupled from the Tint Amplifier to the chroma control (Q_{604}) (see Fig. 23-2). Amplifier Q_{604} gain is adjustable. This transistor is biased at 0.3 V by the forward drop of SC_{602}. This diode also compensates for gain variation in Q_{604} due to temperature variations. With no voltage at terminal *BV*, Q_{604} is biased at zero emitter current. Therefore, the transistor does not conduct and no signal appears at the collector of Q_{604}. Applying a dc voltage at terminal *BV*, forward biases this transistor, and we get chroma out at the collector. By changing the dc control voltage at *BV*, the emitter current can be varied, which varies the beta of the transistor, thus varying the color output at the collector of Q_{604}.

The dc voltage at terminal *BV*, as explained in tint operation, is, for non-remote sets, a potentiometer, acting as a variable voltage source, and for a remote set, a memory module providing the color control voltage at the terminal *BV*. The color control, like the tint control, is located on the control board.

ACC and Color Killer Figure 23-13 shows the ACC circuitry. The composite video signals from the emitter of Q_{208} (first Video Amplifier) are applied to Q_{606} (first Chroma Amplifier) through C_{616} and L_{603}.

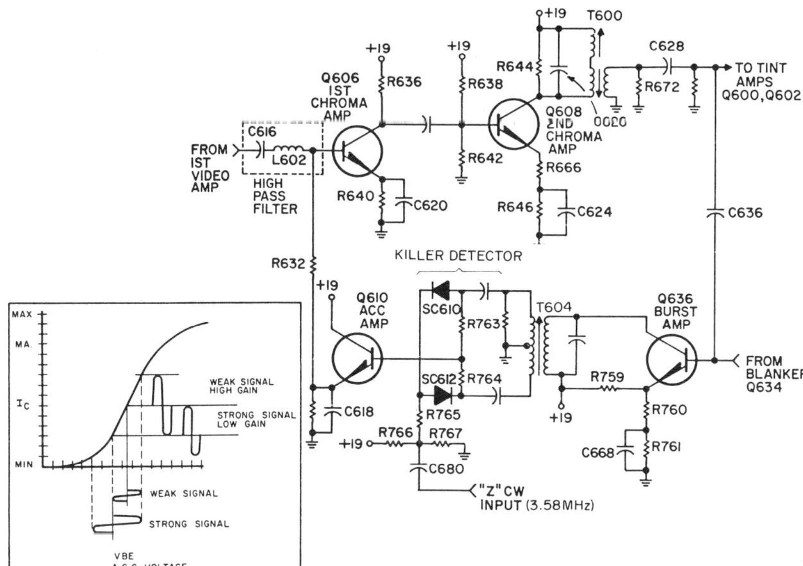

Fig. 23-13. The automatic color-control circuit. (*Courtesy of Sylvania Electric Products, Inc.*)

Q_{606} is automatic chroma gain controlled by SC_{610}, SC_{612} (Killer Detector), and Q_{610} (ACC Amplifier). The Killer Detector Diodes, T_{604}, and associated components compare the burst amplitude of the Z, CW signal (3.58 MHz phase used at the Z demodulator). When no burst is present, SC_{610} and SC_{612} both conduct equally, and no difference voltage is developed at the R_{763}, R_{764} junction. The voltage as this junction is the sum of two voltages:

(1) Voltage developed across R_{767}.

(2) Voltage developed due to the conduction of SC_{610}. In the presence of burst, if the oscillator is locked, then SC_{612} does not conduct, but SC_{610} does conduct and develops a negative voltage.

Therefore, voltage at the base of Q_{610} will decrease in the presence of burst. The higher the burst amplitude, the smaller the dc voltage at the base of Q_{610}. This voltage, through emitter follower Q_{610}, is applied to bias Q_{606} (first Chroma Amplifier). Reduction in this voltage means reduction in the emitter current of Q_{606}, thus reducing the beta, which decreases the gain of this stage. This way, the burst amplitude and the chroma amplitude are regulated.

As shown in Fig. 23-2, the Q_{610} emitter voltage is also applied to the base of Q_{612} (Color Killer). During the presence of burst, the less positive voltage at the emitter of Q_{610} (ACC Amplifier) acts as a "turn off" bias for Q_{612}. However, during the absence of burst, Q_{610} will conduct harder and therefore will forward bias Q_{612}.

***X* and *Z* Demodulators** (See Fig. 23-2.) In order to demodulate the chroma sidebands, the Z and X demodulators provide synchronous detection of these signals with the 3.58 MHz reference oscillator injection voltage, which is fed into the emitters of Q_{616} and Q_{618}. The phase of the 3.58 MHz applied to the Z demodulator is shifted approximately 90 degrees by coil L_{608} and the capacitor C_{650}. The actual shift is selected to provide the most accurate color presentation. The 3.58 MHz signal injection into the emitter of Q_{616} and Q_{618} will cause pulses at a 3.58 MHz rate in their collectors.

When the chroma signal is applied to the bases of the demodulators, the phase and amplitude of the chroma signal will determine the conduction levels and collector voltage changes in each demodulator. When the two signals, chroma and 3.58 MHz reference, are in phase across the emitter-base junction, the transistor conduction is lowered, raising the collector voltage toward B+. If the chroma signal and the 3.58 MHz reference signal are out of phase, heavy conduction will occur, and produce low collector voltage levels. If the two signals are 90 degrees out of phase, the collector voltage will be held at the level developed by the 3.58 MHz reference signal. These demodulator variations are direct coupled to the bases of Q_{624} ($R-Y$ Driver) and Q_{626} ($B-Y$ Driver).

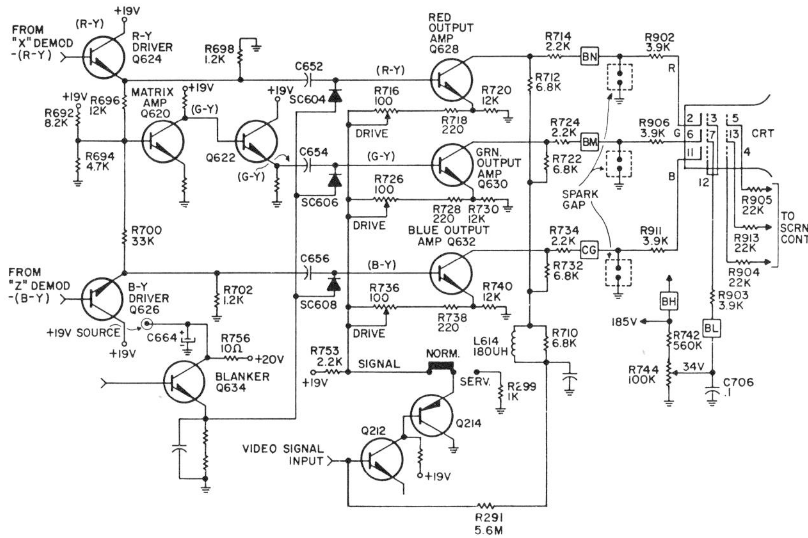

Fig. 23-14. The color drive circuitry. (*Courtesy of Sylvania Electric Products, Inc.*)

Drivers and Color Output Figure 23-14 shows the color drive and color-output circuitry. The emitter of Q_{214} (Video Driver) provides a $+Y$ signal to the emitter circuit of Q_{628}, Q_{630}, and Q_{632} (Red, Green, and Blue Output Amplifiers). By using emitter drive to Q_{628}, Q_{630}, and Q_{623}, there is no signal inversion. At this point, if the chrominance signal is absent, the $+Y$ signal is fed to the color *CRT* cathode. However, when the chrominance signal is present, color difference signals are developed by the *X* demodulator and *Z* demodulator and applied to Q_{624} $(R-Y$ Driver) and Q_{626} $(B-Y$ Driver), respectively. Q_{624} and Q_{626} are emitter-follower configurations feeding Q_{628} (Red Output Amplifier) and Q_{632} (Blue Output Amplifier), their respective signals.

The $G-Y$ signal is formed from a portion of the $-(R-Y)$ and $-(B-Y)$ signal. The signal across emitter resistor R_{698} and R_{702} are coupled through attenuating resistors R_{696} and R_{900}, and matrixed across R_{692} and R_{694}. Since the resistance of R_{696} is approximately 1/3 R_{700}, then R_{700} feeds the lesser signal to the matrix amplifier. Since R_{696} attenuates a $-(B-Y)$ signal and R_{700} attenuates a $-(R-Y)$ signal, the resultant voltage mix at the input of Q_{620} $(G-Y$ Matrix Amplifier) is a $+(G-Y)$ signal. Signal inversion occurs in Q_{620} resulting in a $-(G-Y)$ signal at the input of Q_{622} $(G-Y$ Driver). Q_{622} is an emitter-follower configuration, signal polarity is maintained, and the $-(G-Y)$ signal is fed to the base of Q_{630} (Green Output Amplifier). This color difference signal and the monochrome signal matrix across the base-emitter junction of Q_{620}.

When the $+Y$ signal drives the Q_{630} emitter, Q_{630} conduction is controlled as though a $-Y$ signal was fed to its base, therefore, this $+Y$ signal is not inverted. When the $-(G-Y)$ signal at the base of Q_{630} is amplified, the signal is inverted to a $+(G-Y)$ signal. Thus, the $-Y$ in the $+(G-Y)$ signal is cancelled by the $+Y$ monochrome signal, and the resulting signal is the individual green signal drive to the CRT. Identical matrix action takes place with the *R* and *B* output amplifiers.

Blanker Circuit and Burst As shown in Fig. 23-2, the base of Q_{634} (Blanker) receives positive pulses from a voltage-divider network across Q_{408}. Operating as an emitter follower, it provides positive pulses to Q_{636}, the burst amplifier, to diodes SC_{604}, SC_{606}, and SC_{608} to establish the dc levels in Q_{628}, Q_{630}, Q_{632}, and to Q_{614} as a blanking pulse to remove the burst signal.

Brightness Limiter Circuit The Brightness Limiter Circuit (see Fig. 23-2) is a closed loop regulator circuit controlling the CRT beam current. Its purpose is to protect the CRT by restricting its emission level, thereby increasing the CRT life. The limiter circuit controls the forward bias of Q_{212} (Video Amplifier) and hence the dc level through Q_{214} (Video Driver), Q_{628}, Q_{630}, Q_{632} (Color Output Amplifiers) to the CRT cathodes. Horizontal pulses are injected into the junction of SC_{418} and R_{468}. The peak-to-peak pulse across R_{468} (Sampling Resistor) is directly proportional to beam current. Diode SC_{418} (Brightness Limiter Rectifier) rectifies the negative position of the pulse and develops a negative voltage across the filter network C_{436}, R_{466}. SC_{417} (Brightness Limiter Gate) is reverse biased until the beam current reaches approximately 700 microamperes, which corresponds to approximately —11 V at the anode of SC_{417}, set by divider R_{462}, R_{464}, off the regulated —20 volts. The sampling resistor R_{468} is so chosen that the voltage on the cathode side of SC_{417} has increased to —11.5 volts at this time. Diode SC_{417} goes into conduction and the negative voltage at its anode starts to increase with beam current, thereby reducing the conduction of Q_{212} (Video Amplifier). From here on, the controlling action becomes continuous.

The 3.58 MHz Oscillator The 3.58 MHz oscillator shown in Fig. 23-15 is a modified Clapp configuration using the 3.58 MHz crystals series resonant operating mode. This provides a low impedance feedback path from collector to base and mechanically controls the oscillator by its vibrating period. In addition, the capacitors C_{694} and C_{696} form part of the feedback circuit to sustain oscillation. As long as the feedback loop gains in unity or better, the oscillator will sustain operation.

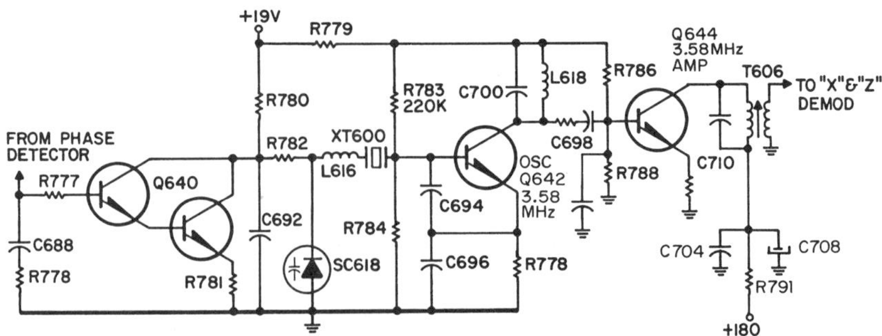

Fig. 23-15. The 3.58 MHz oscillator. (*Courtesy of Sylvania Electric Products, Inc.*)

Transistor Q_{640}, a Darlington amplifier, actively controls the varactor SCz_{18} capacity through its forward bias and dc correction voltage from the phase-comparator circuit. The oscillator load is a resonant tank circuit L_{618} and C_{700} tuned to 3.58 MHz. The output is capacity coupled through C_{698} to Q_{644}, which is a 3.58 MHz buffer stage required to develop additional 3.58 MHz drive amplitude. The collector supply for this stage is the 180 volts bus. R_{791} and C_{708} comprise a filter that decouples the 3.58 MHz energy from the bus. Capacitor C_{704} is an additional bypass capacitor placed across C_{708} (a 30/150 V electrolytic) Its purpose is to neutralize any capacitor lead inductance and assure complete 3.58 MHz by-passing. Transformer T_{606}, the 3.58 MHz output transformer provides resonant gain to the output signals and is the coupling means to the X and Z demodulators.

The 3.58 MHz Oscillator Phase-control Circuit Phase control of the 3.58 MHz reference oscillator is accomplished by comparing the oscillator signal with the burst at the A.P.C. detector. Figure 23-16 shows the

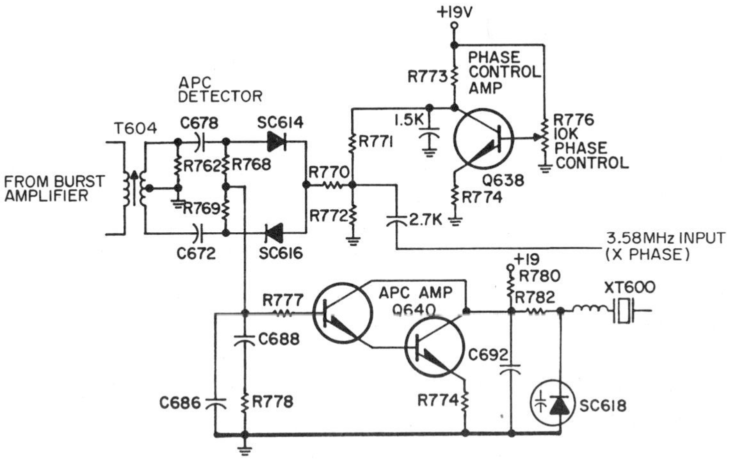

Fig. 23-16. The automatic phase control circuit. (*Courtesy of Sylvania Electric Products, Inc.*)

circuit. The voltage at the junction of R_{768} and R_{769} consists of two voltages.

(1) The dc voltage provided by phase control amplifier Q_{638}.

(2) The correction voltage provided by the A.P.C. detector, derived by comparing the phase and frequency of the 3.58 MHz oscillator to the incoming burst. With no burst, or the oscillator running in phase with burst, this voltage is zero.

The first voltage (which can be varied by R_{776}) is to bias the A.P.C. amplifier Q_{640}, so that with no burst (*i.e.*, zero correction voltage) the oscillator free-running frequency is close to the burst frequency. The second voltage (*i.e.*, the correction voltage), adds to or subtracts from the first voltage, depending on whether the oscillator is leading or lagging the burst phase, and thus decreases or increases the collector

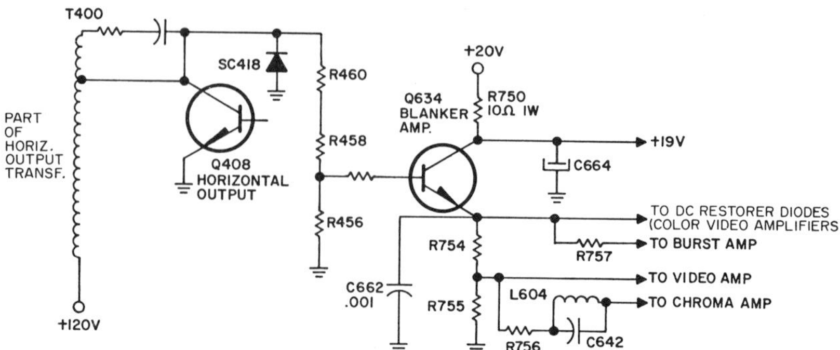

Fig. 23-17. The blanker circuit. (*Courtesy of Sylvania Electric Products, Inc.*)

voltage of Q_{640}. The change in collector voltage of Q_{640}, changes the capacity of the varactor diode SC_{618}, and this in turn changes the series resonance of the 3.58 MHz crystal, pulling the frequency and phase of the oscillator towards the burst frequency and phase.

Transistor Q_{638} stabilizes the collector voltage of Q_{640}, for temperature drift.

Blanker Amplifier The blanker-amplifier circuit, illustrated in Fig. 23-17, operates as an emitter follower providing positive pulses to the chroma amplifier, burst amplifier, color-video amplifiers, and the video amplifier.

The divider network consisting of R_{460}, R_{458}, and R_{456} steps down the flyback pulse to about 25 volts at the base of the blanker. It is clamped to +19 volts and shaped by the load consisting of R_{754}, R_{755} and C_{662}. L_{604} and C_{642} resonate at 3.58 MHz and prevent any chroma signal getting into the blanker pulse. Resistors R_{757} and R_{756} perform as dividers for burst suppression and gating.

23.4 ALIGNMENT OF THE RECEIVER—EQUIPMENT REQUIRED

For the proper alignment of monochrome-television receivers, the following basic pieces of test equipment are required: a cathode-ray oscilloscope, a wideband sweep generator, an RF-signal generator, a high-impedance voltmeter, and a marker-signal generator that is capable of indicating specific frequency points on the test pattern swept out on the oscilloscope screen. The latter may be available as a separate instrument, or it may be incorporated in the sweep generator.

Oscilloscope The cathode-ray oscilloscope is a necessary piece of test equipment among the serviceman's electrical testing apparatus. One of its greatest uses is that of observing waveforms of different voltages and frequencies in the receiver. This provides the technician with a positive means of determining rapidly exactly what is occurring at all points in the circuit under test. It eliminates guesswork and permits accurate adjustments to be made until the correct operating conditions are attained. For a television receiver, satisfactory images are observed only if the various intervening circuits are functioning properly.

Oscilloscopes used for sweep alignment of IF stages, or to check the sweep response of video amplifiers do not require a wideband response for their vertical amplifiers. In fact for such uses, the vertical amplifier response does not have to exceed 10 kHz. The reason for this is that we are not viewing such frequencies as 45 MHz, or even 3 MHz directly on the scope screen. Imagine the difficulties in designing an oscilloscope to directly display 45 MHz. It is frequently not realized by the technician, that for the above mentioned uses, the scope is not required to reproduce these high frequencies but rather the very low frequencies contained in the pattern repetition rate, which is normally 60 Hz or 120 Hz, plus the harmonics of this fundamental rate required to reproduce the outline of the response curve. Such harmonics rarely, if ever exceed about 10 KHz. This means that particularly when servicing monochrome receivers, a relatively inexpensive oscilloscope, will suffice.

The actual need for a wideband oscilloscope, such as one which will accept signals extending to 4 or 5 MHz, exists only for certain special measurements. One such requirement is when it may be desirable to directly view the 3.58 MHz burst signal on the back porch of the horizontal-blanking pulse. Another and less stringent requirement occurs when it may be desirable to examine the exact shape of the horizontal- and vertical-sync and blanking pulses. In this latter case, an oscilloscope response of somewhat less than 1 MHz will normally be adequate. The technician would be well advised to carefully consider his needs before investing an excessive amount of money in an ultra-sophisticated oscilloscope he may not really need. An oscilloscope which offers "triggered sweep" would be nice to have, but is far from essential for the TV technician.

The term "vectorscope" may sound rather mysterious and complex, but actually any scope can be a vectorscope and there is no requirement to purchase a special scope for this purpose. A vectorscope is sometimes used by the technician to view and help to align the overall color circuits. It is also of considerable aid in troubleshooting some of the color circuits. In order to use any scope as a vectorscope, it is only required to have access to one vertical CRT deflection plate and one horizontal-deflection plate. It is also necessary to be able to disable the internal sweep. The two CRT plates mentioned previously must be disconnected from any scope circuits, and there should be no input signal into the scope vertical amplifier. Now the $R-Y$ signal from the TV set is connected directly to the top CRT vertical plate, and the $B-Y$ signal is connected directly to the CRT horizontal right plate. The TV chassis is also connected to the scope ground. When the scope and TV set are energized, a vector pattern will appear on the scope, if a color-bar generator is connected to the TV set. Such a color-bar generator (described later) would be connected directly to the antenna terminals, since its output is a

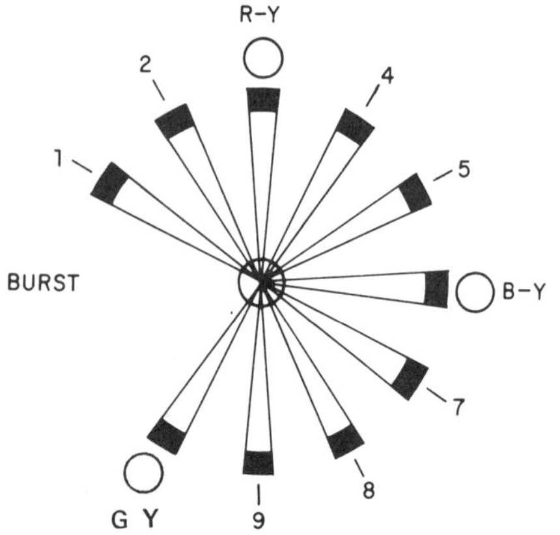

Fig. 23-18(A). An ideal vectorscope display. This display permits the phase relationship of the color signals to be studied.

modulated RF which can be selected to the frequencies of Channels 2 through 6.

Figure 23-18 (A) shows a typical vectorscope pattern with the various phases marked for identification. The proper functioning of the color demodulators, as well as certain other color circuits can be quickly checked with the aid of the vectorscope pattern.

An oscillogram of two lines of the rainbow color pattern, such as that displayed on the vectorscope in Fig. 23-18 (A), is shown in Fig. 23-18 (B). The colors as identified in part B of the figure can be correlated with part A, as both indicate 10 colors and in the same order, going clockwise in part A.

Some manufacturers of test equipment have designed oscilloscopes which conveniently perform the dual functions of a normal oscilloscope and a vectorscope, by means of simple switching operations. One such scope is illustrated in Fig. 23-19.

Sweep Generators Because of the wide bandwidths that are peculiar to television receivers, the familiar signal generator, in which only one frequency is available at any one time, is not suitable, by itself, for receiver alignment. With a single frequency entering the circuit, it is possible to determine only one point on the frequency response curve. To do this for a 6-MHz or even a 4-MHz band would require too much time for ordinary service, where time is an important factor in determining the cost of the job. To meet the special requirements imposed by television receivers, special sweep generators, similar to the unit illustrated in Fig. 23-20, have been developed.

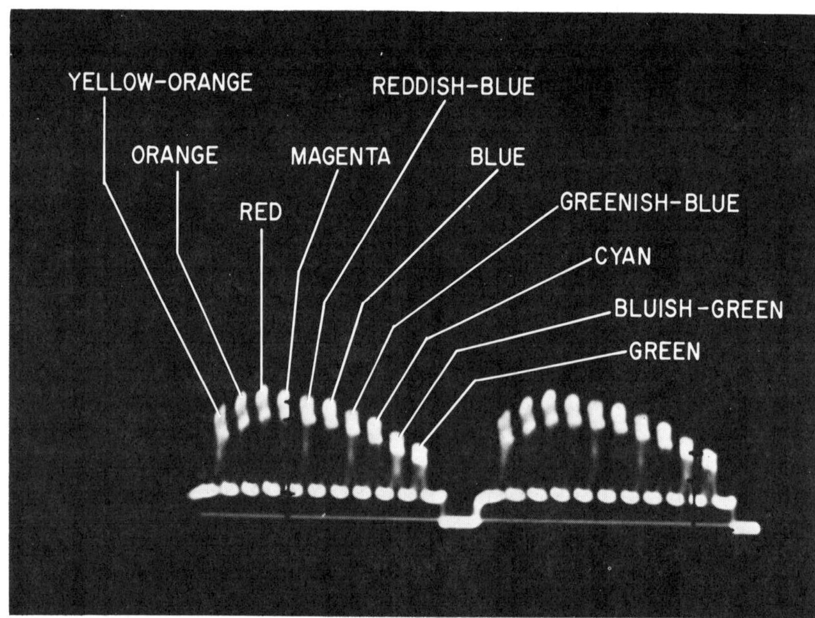

Fig. 23-18(B). Two lines of rainbow color bars as seen on a conventional oscilloscope.

These generators are designed to provide an output signal that sweeps across a range of frequencies and continuously repeats this sweeping at a rate of 60 or 120 Hertz. If this type of signal is applied to the video-IF system of a television receiver, and an oscilloscope is connected across the output of the system (at the video detector, for example), then the pattern produced on the scope screen will represent the response curve of that system.

The range of frequencies obtainable from a sweep generator varies with the instrument. Some are designed to cover only the video-IF frequencies, say from 39 to 48 MHz. The one shown in Fig. 23-20 covers this and all UHF and VHF frequencies as well. It also provides color-IF frequencies; 3.58 MHz $\pm$ 0.5 MHz. In all of these instruments, there is a sweep-width control which permits adjusting the range of frequencies swept out at any setting of the tuning dial. For example, suppose the generator is set for 43 MHz. Then the amount by which the output signal sweeps above and below 43 MHz can be varied from zero (when there is no sweep and the output frequency is 43 MHz) to, say, $\pm$5 MHz. This means the output signal varies periodically from 38 MHz (43 MHz $-$ 5 MHz) to 48 MHz (43 MHz $+$ 5 MHz). For TV circuit alignment, sweep generators are indispensable.

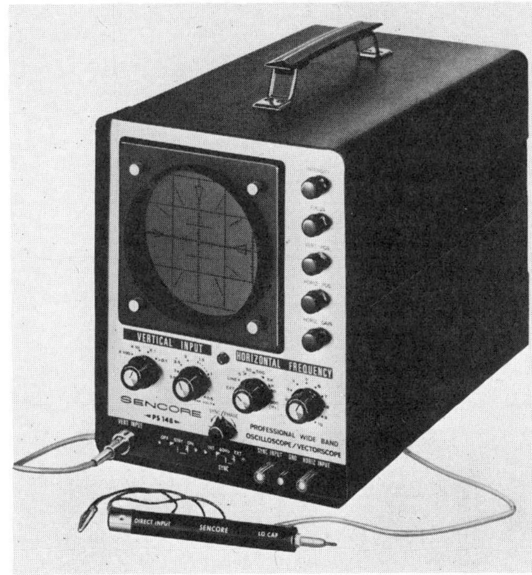

Fig. 23-19. A combination oscilloscope and vectorscope. (*Courtesy of Sencore, Inc.*)

Single-signal RF Generators Although the television receiver employs wideband tuning circuits, the conventional amplitude-modulated RF signal generator is not entirely without application. In the receiver just analyzed, some of the IF single-tuned circuits are peaked

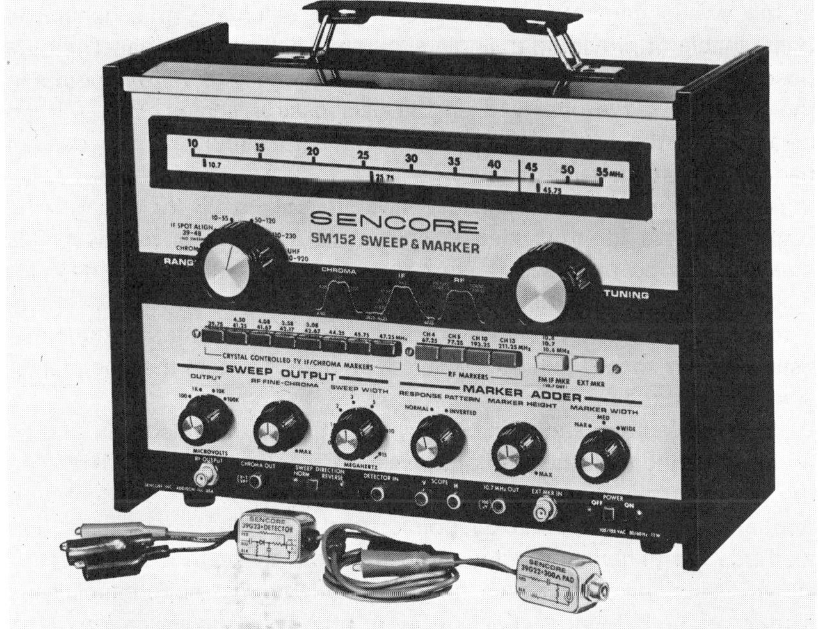

Fig. 23-20. A TV-sweep generator used for monochrome and color sets. All required marker frequencies are available. Post-injection markers are utilized. (*Courtesy of Sencore.*)

with such a signal generator. This is also true of the trap circuits. Signal generators which generate one frequency at a time may also be useful for displaying marker points, as will be seen presently, and for testing the local high-frequency oscillator of the receiver.

High-Impedance Voltmeters The vacuum-tube voltmeter has always been a very handy instrument to have around, and with television receivers it becomes even more important. The vacuum-tube voltmeter, when properly constructed, has a negligible loading effect on the circuit across which it is placed. In this respect it gives a truer indication of the conditions in the circuit under test. Furthermore, by the use of special auxiliary probes, it will measure dc voltages as high as 50,000 volts and RF voltages having frequencies of several hundred megaHertz. With the conventional 1,000-ohms-per-volt meter it is impossible to do this. Finally, the vacuum-tube voltmeter is practically immune to overloading, a feature for which even experienced technicians frequently are grateful.

A disadvantage of a vacuum-tube voltmeter is that it requires voltage from the power line for its operation. As with other vacuum-tube systems, transistors have replaced the tubes in the vacuum-tube voltmeter. The transistor which performs most nearly like the vacuum tube is the FET. It has a high-input impedance comparable to the tube.

Figure 23-21 shows a field-effect multimeter (FET meter) which may replace the vacuum-tube voltmeter in the shop. Since they are transistorized, FET meters can be battery operated which makes them useful for both bench testing and servicing in the home.

Fig. 23-21. A high-impedance portable FET multimeter. (*Courtesy of Leader Electronics Corp.*)

Marker Signals This final piece of apparatus may be either incorporated in the sweep generator, or supplied by an external signal generator. It is capable of providing a single accurately calibrated signal. The purpose of a marker signal is to indicate the frequency at various points in the response curve observed on the oscilloscope screen. This will aid in adjusting the tuning slugs in the resonant circuits to the desired bandpass characteristics.

As an example, consider the response curve of Fig. 23-22, which is the response curve for the video-IF system of a television receiver. This curve would be observed if we connected across the video-detector load resistor. What we desire to do, once we obtain the response curve, is to determine the frequencies at various points to insure that the curve rises where it should and falls where it should. It is here that the marker signal is required.

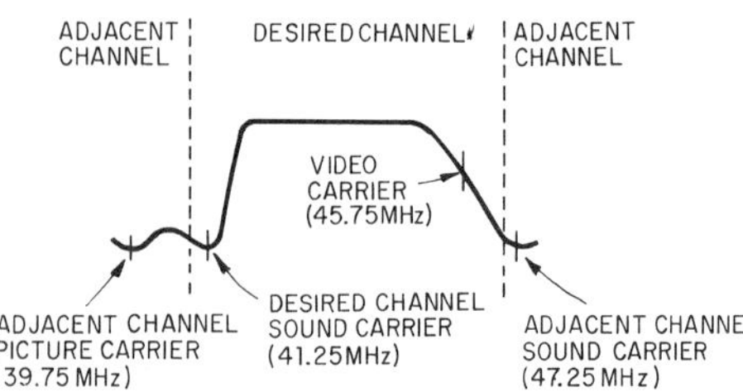

Fig. 23-22. The overall video-IF response curve of a television receiver.

If a manufacturer states in his service data that the video-carrier IF is 45.75 MHz and the other end of the response occurs at 41.75 MHz, then the curve obtained should be checked for the position of these two frequencies. To obtain marker points on the oscilloscope screen, two methods are generally employed. In the simplest method, the sweep-signal generator contains an internal oscillator that superimposes its signals on the IF being swept out (40 to 50 MHz). The indication of the marker point in the visible pattern is either a slight wiggle or else a dip in the curve at this point (see Fig. 23-23). Note that while two marker points are indicated in Fig. 23-23, only one is seen at a time. First the marker oscillator is set to 45.75 MHz and its position noted on the curve, and then it is set to 41.75 MHz and its position checked again.

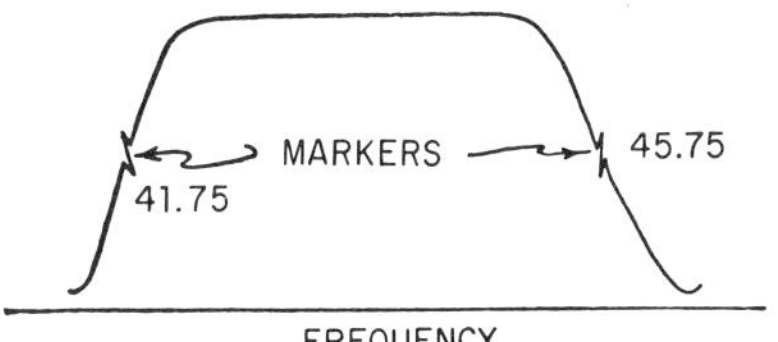

Fig. 23-23. The use of marker points for indicating definitely a frequency on a response curve.

The mentioned video-IF band limits, 41.75 MHz and 45.75 MHz, represent the entire 4 MHz that can be employed to transmit the details of the televised scene. Some portable receivers are designed to pass only 3 MHz in the IF amplifiers. In these instances, reference to the manufacturer's instructions will quickly indicate the band limits, and the marker frequencies can be changed accordingly.

If the sweep generator does not contain an internal device for supplying the marker points, these can be obtained by the following method. Take an RF signal generator and place its output leads in parallel with those of the sweep generator, using a 50 pF isolating capacitor* in the signal lead of the marker generator. Set the frequency of this second oscillator accurately to one of the frequencies that is to be checked on the response curve, say 45.75 MHz. With the equipment turned on, a wiggle (or pip) will appear on the overall response curve at 45.75 MHz. Note whether the response at this point is that indicated by the manufacturer in his service manual. Now change the marker frequency to 41.75 MHz and note where this appears on the response curve. In this particular receiver (Fig. 23-2), 45.75 MHz represents the video-IF carrier, and 41.25 MHz the sound-IF carrier. Other frequencies that should be checked are the trap frequencies (see Fig. 23-22). These include 41.25 MHz, the sound carrier of the same channel and 47.25 MHz, the lower adjacent channel sound carrier. At each of these points the response should be very low. It is advisable not to turn the amplitude of the indicating signal generator too high, but to keep it as low as possible to still obtain a marker line.

The method of marking the sweep just described has the disadvantage that the marker signal must be fed through the receiver circuitry along with the sweep signal. This same disadvantage occurs when using the sweep generator with a built-in marker system. The marker signals passing through the tuned circuits of the receiver tend to distort the response, and this often results in an unrealistic display of the overall receiver response curve on the oscilloscope. To avoid this, a system of adding markers which will not distort the response curve has been devised. This system is called the "post-injection marker system." (The sweep generator illustrated in Fig. 23-20 utilizes this type of marker system.)

* It may be found that attaching the marker generator directly to the sweep generator lead will cause the response curve to alter its shape. If this occurs, try inserting an isolating resistor of 10,000 ohms in the signal line of the marker together with the 50 pF capacitor. The value of the capacitor is not critical and values between 20 pF and 200 pF have been suggested by various manufacturers.

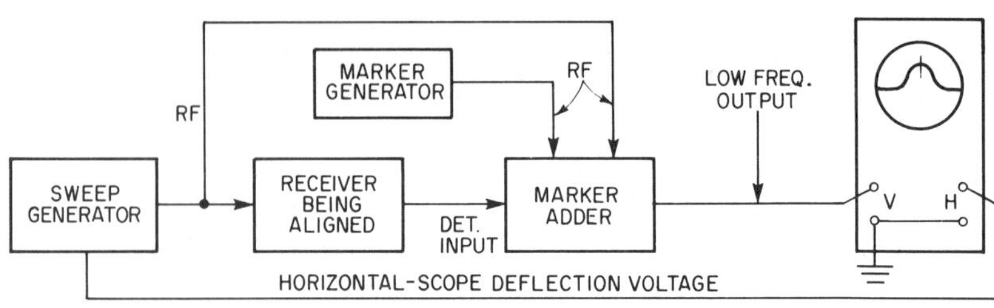

Fig. 23-24. Test setup for using the post-injection marker adder. Although shown as separate blocks here, the marker circuits may be in the sweep-generator cabinet.

The post-injection marker system is illustrated in the block diagram of Fig. 23-24. The RF output of the sweep generator is fed to the receiver input and also to the input of the marker adder. The RF marker (at the selected frequency, or frequencies) is also fed to the input of the marker adder. In the adder these two RF frequencies are heterodyned to produce a marker at an audio-beat frequency. Since the sweep generator produces the horizontal-scope sweep in synchronism with the RF sweep, markers will appear on the scope base line, even if the receiver is not connected. Note that the markers do not pass through the receiver circuits and so cannot distort the response curve no matter how great their amplitude may be. In a sweep generator, such as shown in Fig. 23-20, markers may be displayed either singly or simultaneously.

Note in Fig. 23-24 that the output of the receiver is fed to a detector input of the marker adder. Thus the response curve of any part of the RF or IF system may be examined directly and displayed on the scope. The detected response curve (in the marker adder) is added to the marker signals, and both are then displayed on the scope simultaneously when the receiver is connected and operating.

If a vertical marker is displayed on the edge of a response curve slope as shown in Fig. 23-23, it may be difficult to determine exactly where the marker appears. For this reason, some sweep generators, which incorporate the above marker system, have a provision for displaying the markers either horizontally or vertically.

23.5 COLOR-TV GENERATORS

Figure 23-25 shows a color-TV generator that is very useful for troubleshooting in color circuitry, and also for convergence adjustments on the picture tube. The features found in this generator are described in the following paragraphs. Some of these features may be obtained in individual units of test equipment, but whether they exist in individual test equipment or are combined into a single unit like the one illustrated in Fig. 23-25, their basic functions are the same.

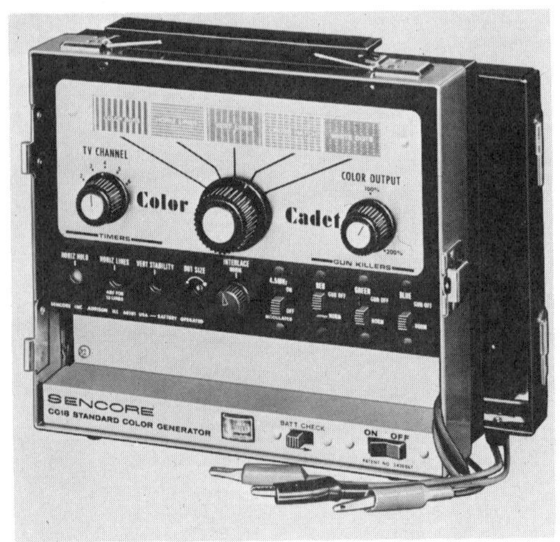

Fig. 23-25. A color generator for a color setup. (*Courtesy of Sencore.*)

Dot Generator Dot generators are used in the color picture-tube setup specifically for the purpose of making convergence adjustments. When the dot generator produces a pattern of individual dots on the screen, convergence controls can be adjusted so that the red, green, and blue

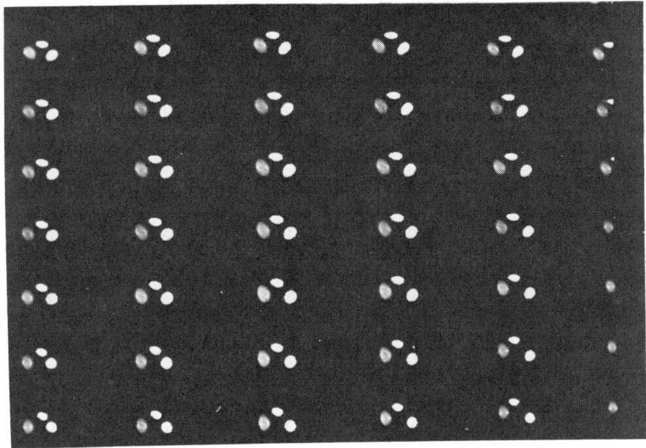

A

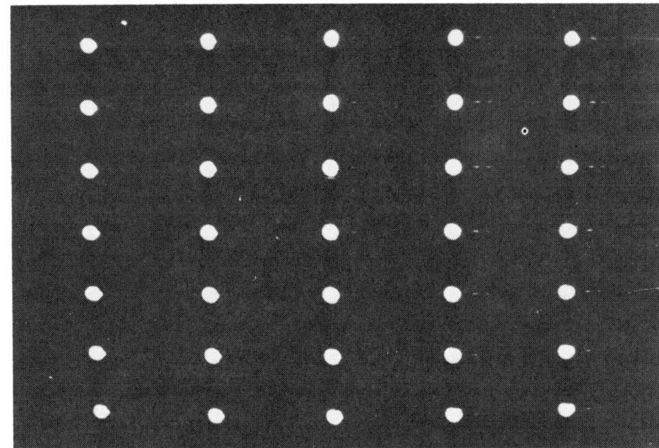

B

Fig. 23-26. (A). A section of a color picture tube with all three beams misconverged is shown. (B). The same section after convergence. Note that the three-dot groups have merged into single dots.

regions are superimposed, thus producing a white dot pattern. The white dot pattern indicates that the beams are properly converged.

Figure 23-26(A) is a photo of a section of a color-tube screen, when all three beams are misconverged and a standard dot pattern is being fed into the receiver. The same section of this color tube after convergence is illustrated in Fig. 23-26(B). Note that in all cases the three-dot groups have merged into a single dot, indicating correct convergence.

Cross-hatch Generator The display on a color-picture tube must be adjusted for vertical linearity and horizontal linearity. These adjustments are the same as for those made on a monochrome-picture tube. For color pictures there is also the need for vertical and horizontal convergence, and pin-cushion correction. The cross-hatch generator can be used for making all of these adjustments.

In order to check that there is no distortion of the raster, the cross-hatch generator feeds a signal to the receiver. This signal causes vertical and horizontal lines on the screen, and the lines form squares on the picture-tube screen. Since the squares consist of thin bright lines, the cross-hatch generator is also useful for determining if the picture tube is properly converged.

On some color generators, there is a special feature which permits a single dot or cross to be moved around the face of the screen to check dynamic convergence at the vertical and horizontal edges of the picture tube.

Color-bar Generator The display for a *keyed rainbow generator* is characterized by a black dividing line between each of the colors on display. This divides the rainbow into individual color bars on different hues, with each hue being associated with a specific phase of the chroma signal. The keyed rainbow generator can be used for setting up color receivers. It is much more useful than an unkeyed color generator, because it permits troubleshooting in different sections of the receiver by using an oscilloscope to trace the signal.

A disadvantage of the keyed rainbow generator is that the signals for the various hues are generated at the same amplitudes. This is undesirable because the three primary colors should be displayed with signals at different amplitudes to account for the fact that the eye is not as sensitive to some colors as it is to others. The fact that the signal amplitudes are the same means that the colors do not all have the same relative brightness. This makes it more difficult to interpret the pattern during setup.

Some technicians use an *NTSC generator* which produces the same range of hues, but has the additional advantage that the hues are generated with the amplitude required for making all of the colors the same brightness. However, the NTSC rainbow generator is more expensive than the keyed rainbow generator, and is not as widely used.

23.6 ALIGNMENT PROCEDURE

In a color-television receiver a complete alignment consists of aligning the *sound, video-IF, tuner, and color sections*. The alignment of monochrome receivers is essentially the same except that there is no color section. We will discuss the alignment procedure* for the Sylvania color receiver shown in Fig. 23-2. This is a fairly standard procedure, and will apply with slight modifications to most color receivers. It is always a good idea before aligning any receiver to check the manufacturer's specifications, and it is especially important to check the characteristic curves shown by the manufacturer as being ideal. It is not always possible to obtain these ideal curves during alignment, but they will give you a better idea of what you are trying to accomplish during the alignment procedure.

Video-IF Alignment and Trap Adjustments The manufacturer's recommended procedure for aligning the video-IF stages in the receiver are shown in Chart 23-1 which follows this section. To get the maximum benefit of this discussion, and also of the following alignment procedures, use the fold-out schematic of the receiver (Fig. 23-2) to locate the various components and test points.

Step 1 of the video-IF alignment and trap adjustment (Chart 23-1) describes the procedure for checking the combined frequency response of the mixer and the first two video-IF stages. You will note that the alignment setup notes state that the horizontal driver (Q_{406}) should be removed during this procedure. The reason for this is that Q_{406} drives the horizontal-output stage (Q_{408}) which operates into the flyback

* Information furnished by Sylvania Electric Products, Inc. and reproduced by permission.

transformer. The keying pulse for the automatic gain-control circuit comes from this flyback transformer. Note that the winding between pins 2 and 3 of T_{400} (Fig. 23-2) goes to terminals *U* and *P* of the video-IF stage.

When you remove the keying pulse, the keyed AGC circuit becomes inoperative, and therefore, you must apply a dc bias base voltage to make up for the lost AGC. That is the reason for applying +2.0 volts to tie point *P* on the IF board as required in Step 1.

The sweep-generator cable, shown in Fig. 1 of Chart 23-1 provides a low, 75-ohm impedance input to the UHF, IF input of the VHF tuner. This simulates the normal connections. Refer to the tuner schematics of Fig. 23-27 to locate the tuner connection points referred to in the following discussions.

The response curve shown in Fig. 3 of Chart 23-1 is the response of the mixer stage of the tuner and the first pix IF stages.

Step 2 of the alignment procedure in Chart 23-1 is similar to Step 1. An isolation capacitor (Fig. 4 of Chart 23-1) has been added to the cabling between the sweep generator and the tuner, and the sweep generator now feeds the output point of the VHF tuner. The response curve of Fig. 5 is for the first stage of IF in the receiver, but not the mixer. Note that this curve is relatively flat between 42.6 MHz and 45.75 MHz. If adjustment is necessary in this stage, then Step 1 should be repeated to make sure that the response curve has not been distorted by the adjustment.

If the response curve of Fig. 5 cannot be duplicated with reasonable similarity during the alignment procedure, then it will be necessary to align the coils and traps which are used for shaping the overall response curve. This procedure is described in Step 3. A signal generator is used for these adjustments instead of a sweep generator. The 41.25 MHz and 47.25 MHz traps are adjusted to obtain the steep sides on the skirt of the response curve of Fig. 5, and the 44.5 MHz coils are adjusted to obtain the correct top shape on the curve. After these adjustments have been made, Steps 1 and 2 should be repeated.

A VTVM is used as an indicator for proper adjustment in Step 3. It is connected to the output of the first video amplifier stage. At this point the signal has passed through the detector, and only the modulation from the generator signal is available at point *AL*. The modulation frequency is well within the frequency response of VTVM or FET meters.

The overall response curve of the IF section between the tuner output and the first video amplifier is shown in Fig. 6 of Step 4 in Chart 23-1. Note that the curve of Fig. 6 is a typical video-IF response curve for color-television receivers. The video-IF alignment procedures of Chart 23-1 are similar to those provided by other manufacturers of color receivers. Of course, the actual adjustments will vary from set manufacturer to set manufacturer, but the general alignment procedure is similar.

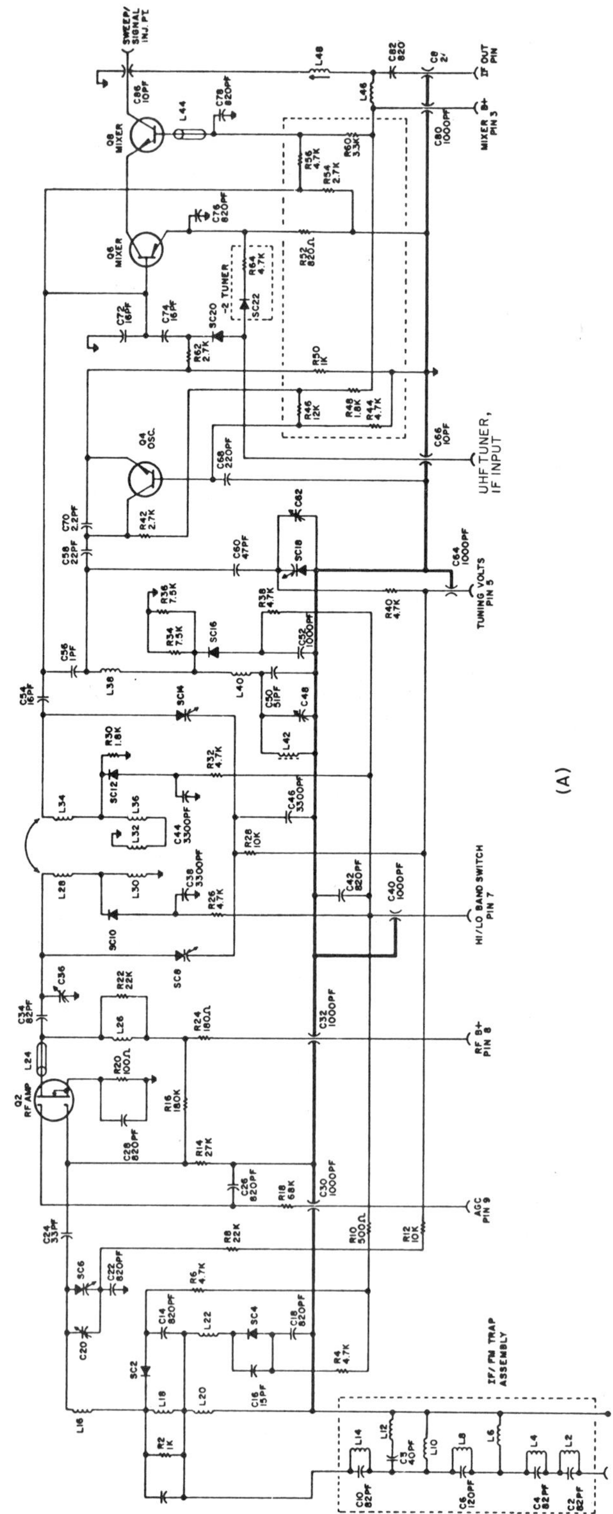

Fig. 23-27. The tuners for the Sylvania all solid-state chassis. (The schematic for the complete receiver is fold-out Fig. 23-2.) (*Courtesy of Sylvania.*)

Fig. 23-27 (A). Sylvania solid-state VHF tuner.

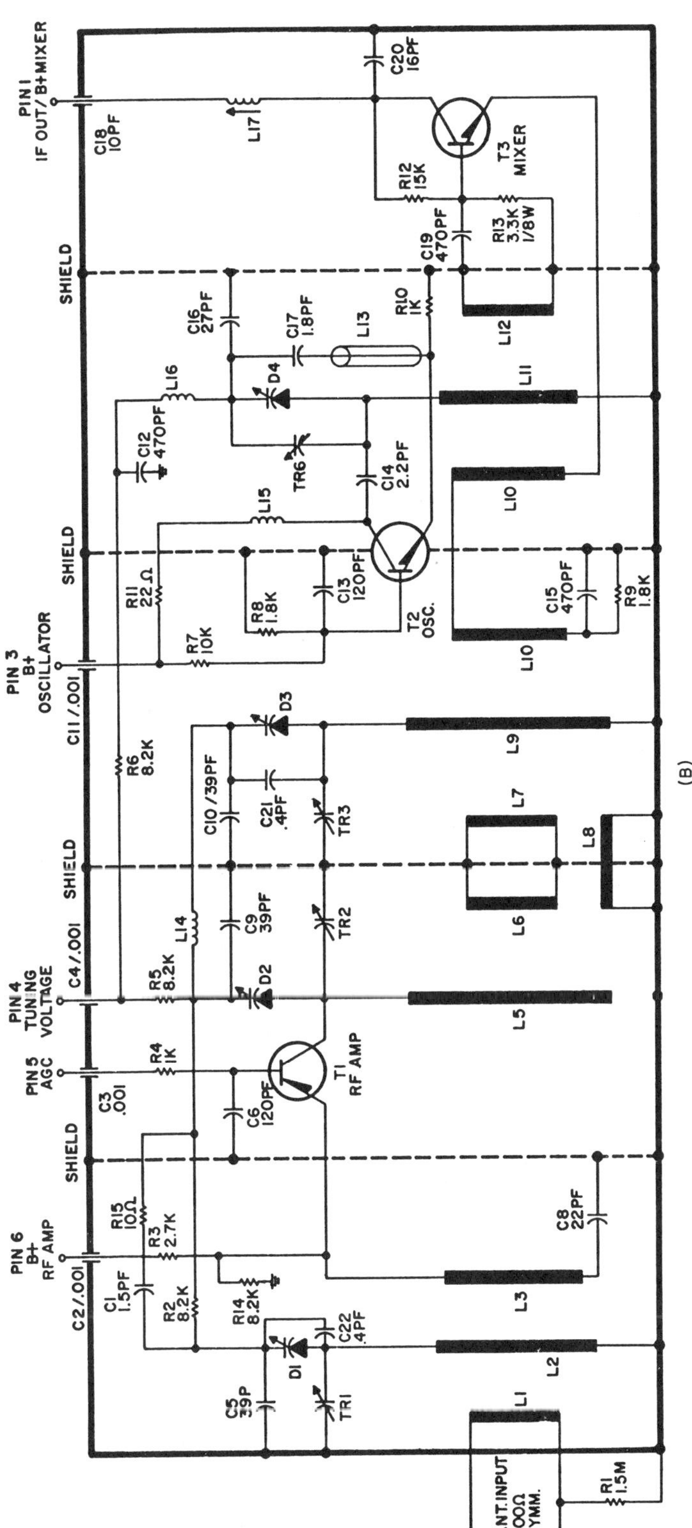

Fig. 23-27 (B). Sylvania solid-state UHF tuner.

ALIGNMENT PROCEDURE

PRELIMINARY INSTRUCTIONS

1. Line voltage should be maintained at 120VAC.
2. Keep marker generator coupling at a minimum to avoid distortion of the response curve.
3. Do not use tubular capacitors for coupling sweep into receiver. Disc ceramics are best.
4. For best results, solder the sweep generator ground to chassis, do not use clips.
5. Sweep generator "hot" lead must make good electrical contact at all points given under TEST EQUIPMENT HOOK-UP.
6. Use scope gain for maximum peak-to-peak response curve on the scope rather than sweep generator.
7. Test equipment (tube type only) should warm up for spproximately 15 minutes before alignment.
8. Bias values specified must be maintained during alignment to insure proper results.
9. To eliminate Vertical circuit interference, switch "Normal Service" switch on rear of chassis to "Service" position.
10. Markers should be kept as small as possible and should be crystal controlled.
11. Check detector response for linearity using a square wave generator at 40MHz and 50MHz.
12. To energize the E01-13,-14 HEC chassis when isolated from the AM/FM Amplifier system (PL104 - TV audio output plug and PL532 - TV power plug disconnected) apply 120VAC, 60Hz to pins 2 and 4 of PL532 - TV power plug. NOTE: There will be no audio.

VIDEO IF ALIGNMENT AND TRAP ADJUSTMENT

STEP	ALIGNMENT SET-UP NOTES	TEST EQUIPMENT HOOK-UP	ADJUST
1	Remove Horiz. Driver Q406. Apply +2.0V Bias to tie point [P] on IF board. Rotate AGC control R358 fully counterclockwise. 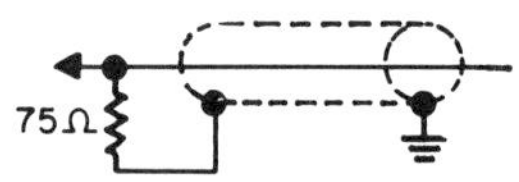 75Ω TO UHF IF INPUT ON VHF TUNER Figure 1	SWEEP GENERATOR - Through a cable shown in Figure 1 to UHF IF input on VHF tuner. Set generator to 44.5MHz with 10MHz sweep. MARKER GENERATOR - Loosely coupled to sweep generator lead. OSCILLOSCOPE - Through network shown in Figure 2 to tie point [Z] on IF board. Calibrate oscilloscope for .01 Volt peak-to-peak output. 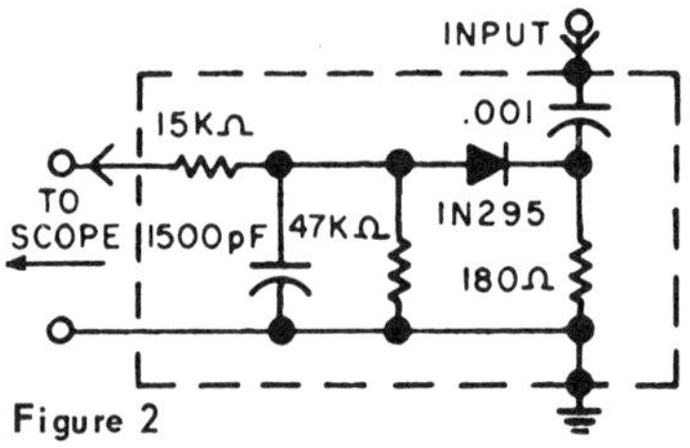 INPUT 15KΩ .001 TO SCOPE 1500pF 47KΩ IN295 180Ω Figure 2	Adjust tuner IF output coil on VHF Tuner to obtain response curve shown in Figure 3. 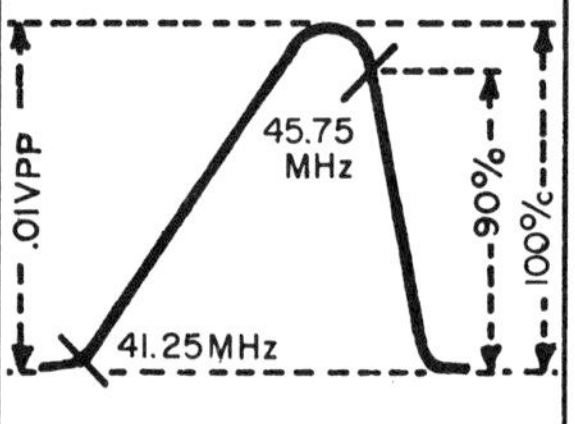 .01VPP 45.75 MHz 90% 100% 41.25MHz Figure 3
2	Same as step 1. 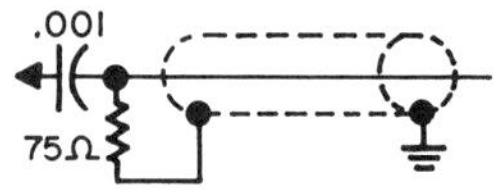 .001 75Ω TO SWEEP/SIGNAL INJECTION POINT ON VHF TUNER (RF TEST POINT) Figure 4	Same as step 1, except: SWEEP GENERATOR - Through .001 capacitor shown in Figure 4 to SWEEP/SIGNAL INJ. PT. on VHF tuner. 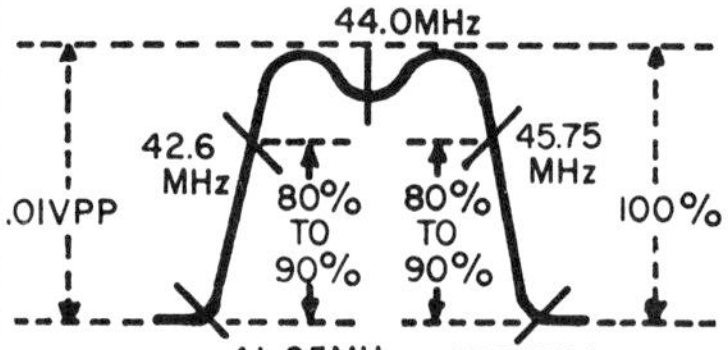 44.0MHz 42.6 MHz 80% TO 90% 80% TO 90% 45.75 MHz .01VPP 100% Figure 5 41.25MHz 4725MHz	[L202] for minimum 47.25MHz marker. [T200] top and bottom core to obtain response curve shown in Figure 5. Readjust tuner IF output coil if necessary. Repeat for optimum performance.
3	Same as step 2. Step 3 - Necessary only when alignment is extremely far out of adjustment.	SIGNAL GENERATOR - Through a .001 capacitor shown in Figure 4 to SWEEP/SIGNAL INJ. PT. on VHF tuner. Set output for maximum signal (RF frequencies are 44.5MHz, 41.25MHz and 47.25MHz.) VTVM - Through a 33K resistor to tie point [AL] on IF bd.	[L204], [L206] and bottom core of [T202] for MINIMUM POSITIVE deflection on VTVM (44.5MHz) Set [L208] for max. inductance (core approx. center of coil). Top core of [T202], [L208] (toward top of coil) for MAXIMUM POSITIVE deflection on VTVM (41.25MHz) Readjust [L202] slightly for MAXIMUM POSITIVE deflection on VTVM (47.25MHz).

Chart 23-1

ALIGNMENT PROCEDURE (CONT'D)

VIDEO IF ALIGNMENT AND TRAP ADJUSTMENT (CONT'D)

STEP	ALIGNMENT SET-UP NOTES	TEST EQUIPMENT HOOK-UP	ADJUST
4	Same as step 2. Figure 6	SWEEP GENERATOR - same as step 2. MARKER GENERATOR - same as step 2. OSCILLOSCOPE - Through a 30K resistor to point [AL] on IF board. Calibrate oscilloscope for 3V peak-to-peak.	[L204] [L206] [T202] Bottom core, in above order for maximum output. Retouch [L206] to position the 45.75MHz marker, and [L204] for 42.6MHz marker to obtain a rounded response as in Figure 6.
5	Same as step 4.	Same as step 4.	Readjust [T202] Top core, [L208] (41.25MHz), [L202] (47.25MHz), for proper marker location as shown in Figure 6. Readjust [L204], [L206] and [T202] Bottom core if necessary.

Chart 23-1 continued

(Charts 23-1 through 23-6 reproduced by permission of Sylvania.)

Alignment of the Sound Section Chart 23-2 explains the procedure for aligning the sound section of the television receiver shown in Fig. 23-2.

The video carrier and the sound-IF signals heterodyne in the video detector circuit and produce a 4.5 MHz beat signal. If this signal is allowed to pass through the video amplifier section, it will produce a herringbone pattern on the screen. Therefore, a trap is inserted in the video section which is tuned to reject 4.5 MHz signals.

Step 1 of Chart 23-2 indicates that a strong signal from a local station should be tuned, and that the fine-tuning knob should be adjusted until the 4.5 MHz pattern is visible in the picture.

The response of most VTVM's and FET meters to a 4.5 MHz signal is very poor, and therefore, the demodulation probe shown in Fig. 7 must be used to measure the dc amplitude of the 4.5 MHz wave.

The sound trap (T_{204}) is adjusted for minimum voltmeter reading, thus indicating that the 4.5 MHz signal has been reduced to its lowest value. It is also a good idea to observe the herringbone pattern on the screen during this adjustment to make sure it has been minimized. However, it is not a good practice to try to make this adjustment without the aid of a VTVM or FET meter, because small changes in signal sync cannot be observed as easily on the screen of the television receiver as they can be with the use of a meter.

After the 4.5 MHz trap has been properly adjusted, the next step is to apply an FM signal to the sound-IF section. This procedure is described in Step 2. With the receiver tuned off station, the FM signal generator feeds its signal to the integrated circuit and the tuned circuits are adjusted for maximum output signal.

STEP	ALIGNMENT SET-UP NOTES	TEST EQUIPMENT HOOK-UP	ADJUST
1	VHF tuner set to channel receiving strong signal. Rotate tuning knob until 4.5 is visible in picture.	VTVM - Through network shown in Figure 7 to tie point [AL] on IF board. 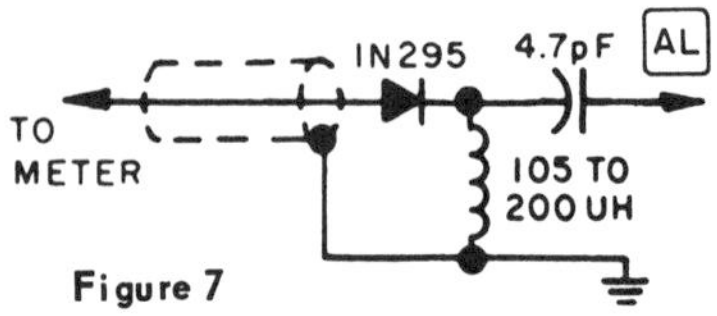 TO METER; IN295; 4.7pF; [AL]; 105 TO 200 UH Figure 7	[T204] for minimum deflection on VTVM.
2	Apply jumper lead from pin 6 of IC100 to chassis ground.	FM SIGNAL GENERATOR - Through network shown in Figure 8 to tie point [AG] on IF board. Set generator to 4.5MHz preferably crystal calibrated or controlled with at least 50 millivolts output. Set FM deviation at ±25kHz. OSCILLOSCOPE - Vertical input to pin 8 of IC100 on IF board to chassis ground. 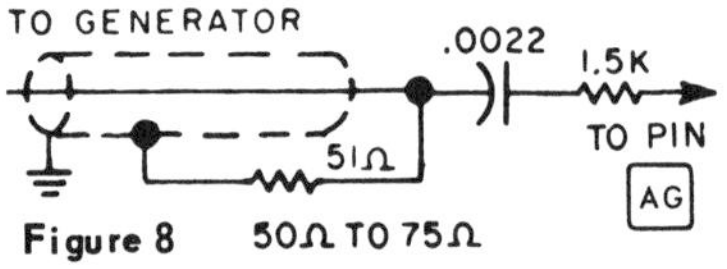 TO GENERATOR; .0022; 1.5K; TO PIN [AG]; 51Ω Figure 8 50Ω TO 75Ω	[L104] For maximum sine wave amplitude on oscilloscope. Decrease signal input until sine wave output is approx. 50% of the original amplitude. Adjust [L100] for maximum output. Further decrease the signal input and readjust [L100] for maximum output.
3	Same as step 2.	FM SIGNAL GENERATOR - same as step 2. OSCILLOSCOPE - Connect to pin 12 of IC100.	Reduce signal input and check sine wave output to insure audio driver stage is operative.
4	Remove all test equipment leads, etc. Connect antenna and check receiver sound reception on a strong local station. NOTE: Refer to Audio DC Limiter Control in ADJUSTMENTS Section.		

Chart 23-2

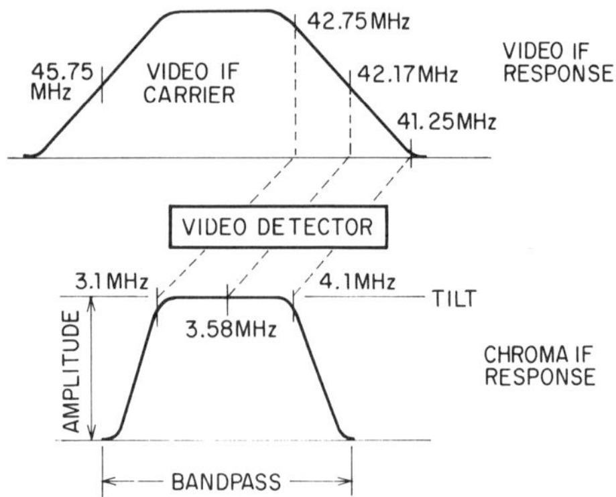

Fig. 23-28. A chroma bandpass curve and its relationship to the receiver response curve. (*Courtesy of Sylvania.*)

Alignment of Chroma-bandpass Amplifiers Before studying the chroma-alignment procedure, it will be helpful to review the characteristics of the chroma signal in the color receiver. Figure 23-28 shows a section of the receiver response curve in which the chroma signal lies. Ideally, the bandpass amplifier will allow only the chroma signals between 3.1 MHz and 4.1 MHz to pass through the chroma section. That is the purpose for aligning this section—to make sure that the chroma amplifiers will pass chroma signals correctly, above and below the color subcarrier frequency.

As shown in Step 1 of Chart 23-3, the horizontal oscillator is disabled during chroma bandpass alignment. Also, the 3.58 MHz oscillator is disabled. Test point *BJ* is grounded to disable the blanker circuit. (The blanker circuit drives the chroma-output amplifier Q_{614} to cutoff during the retrace period.) Test point *P* is grounded on the IF strip to stabilize the AGC amplifier output during this adjustment.

The sweep generstor signal is applied to the input of the second chroma amplifier through terminal *BC*. Figure 9 of Chart 23-3 indicates the required probe. Note that the center frequency of the sweep must be 3.58 MHz (the center frequency of the bandpass amplifier) with a sweep width of only 3 MHz. As described in Step 1 of Chart 23-3, the oscilloscope is connected to the output of the second chroma amplifier during the adjustment. The response curve displayed is only for the second chroma amplifier in the color-bandpass section.

In Step 2 of the chroma-alignment procedure the sweep generator feeds its signal to the input of the video-IF section of the receiver. The oscilloscope is connected to show the overall response curve of the IF and color-bandpass amplifier section with this setup. The signal that you will observe at tie point *DX* is the detected sweep-generator signal, which has passed through the video-IF, video-detector and chroma-bandpass sections. Note that the inserted 45.75 MHz signal from the signal generator serves as a video-IF carrier and is essential, if the resultant video frequencies are to be obtained from the output of the video detector. The video-IF carrier heterodynes against the various IF frequencies in the video detector, and results in the production of the video frequencies. Without the inserted IF carrier, the output of the video detector would consist only of the very low sweep-rate frequencies and the low harmonics making up the output-response curve. The desired, overall response curve is shown in Fig. 10 of Chart 23-3.

An alternate method of chroma-bandpass alignment is also shown in Chart 23-3. This method is not as good as the one described above since that one enables you to view the overall video-IF and chroma-bandpass response, whereas the alternate method only views the response of thc chroma-bandpass amplifiers alone. In the alternate method a video-frequency sweep generator is used instead of an IF sweep. One advantage of the alternate method is that it makes it possible to easily align the chroma amplifiers in cases where the alignment is badly off. This would bc difficult to do, if you were passing the sweep signal also through the IF stages of the receiver. In the alternate method, the video-sweep generator and markers are fed to the input of the first chroma amplifier. The oscilloscope remains connected to the output of the chroma-output amplifier. Alignment is performed as shown on the chart for the response curve desired and is illustrated in Fig. 14 of the chart.

Color-automatic Frequency and Phase-control Adjustments The automatic frequency and phase control circuit (AFPC) is also sometimes called the automatic phase control circuit (APC). The function of this circuit is to lock the receiver 3.58 MHz color sub-carrier oscillator to the same frequency and phase as the burst. When this is accomplished the receiver sub-carrier oscillator is exactly in step with the transmitter sub-carrier oscillator, the condition required to properly reproduce the original colors at the transmitter. Any variation between

CHROMA ALIGNMENT

STEP	ALIGNMENT SET-UP NOTES	TEST EQUIPMENT HOOK-UP	ADJUST
1	A. Remove Horiz. Osc., transistor Q406. B. Remove 3.58 Osc., transistor Q642. Ground TP [BJ] on Chroma Board, & ground TP [P] on IF Board. Apply +1.7VDC bias to TP [BD] on the chroma board. Adjust Color Killer fully CW. Adjust AGC fully CW. Adjust color control for +6.5V at TP [BV]. Adjust tint control for +6.5V at TP [BT]. 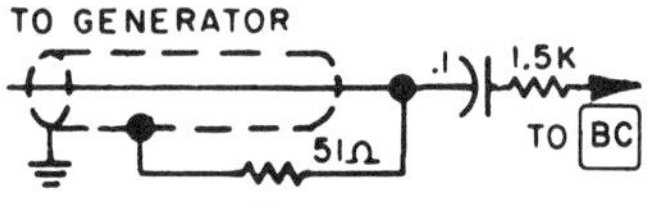 Figure 9	SWEEP GENERATOR - Through network shown in Figure 9 to test point [BC] on Chroma Board. Set generator to 3.58MHz with 3MHz sweep. MARKER GENERATOR - Insert in parallel with sweep generator to provide frequencies of 3.1MHz, 3.58MHz, 4.1MHz. OSCILLOSCOPE - Through Video detector probe shown in Figure 11 at C628 & C636 junction on Chroma Board. Calibrate scope to .1V full scale deflection. 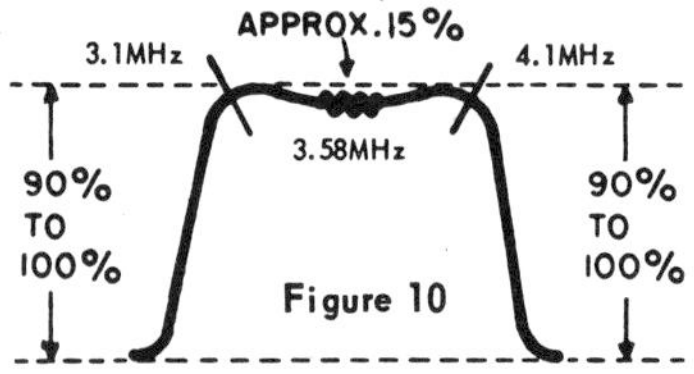 Figure 10	[T600] Top and bottom core for response curve shown in Figure 10. 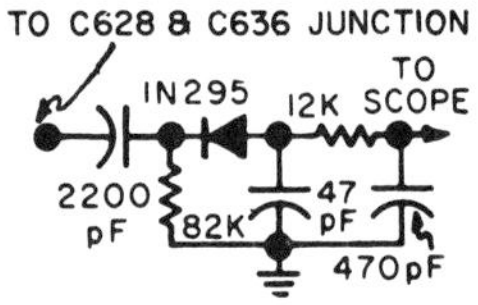 Figure 11
2	Same as step 1, except remove ground jumper and apply +2VDC bias to TP [P]. 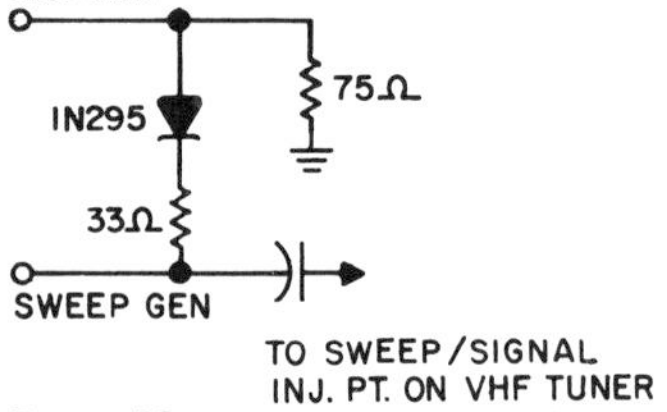 Figure 12	SIGNAL GENERATOR - Set to 45.75MHz. SWEEP GENERATOR - Set to 3.58MHz, sweep 3MHz. The outputs of both generators through network shown in Figure 12. To SWEEP/SIGNAL INJ. PT. on VHF tuner. VTVM - Point [AL] on IF board. OSCILLOSCOPE - Through network shown in Figure 11 to tie point [BX] on Chroma Board. 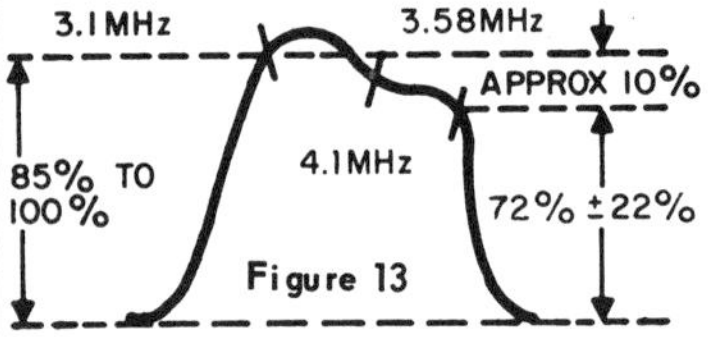 Figure 13	Set SWEEP GENERATOR to minimum for no signal output. Note DC READING ON VTVM, then increase generator output until a 1.5V reduction is indicated on VTVM. (If this 1.5V cannot be obtained, set generator to maximum and if necessary, reduce the +2V bias at tie point [P] until the voltmeter reads at least .5V DC change.) [T602], for maximum amplitude of the 3.58MHz marker, as shown in Figure 13.

ALTERNATE CHROMA ALIGNMENT FOR STEP 2

STEP	ALIGNMENT SET-UP NOTES	TEST EQUIPMENT HOOK-UP	ADJUST
2	Same as step 1.	SWEEP GENERATOR - same as step 1, except to the base of Q208. MARKER GENERATOR - same as step 1. OSCILLOSCOPE - same as step 2. 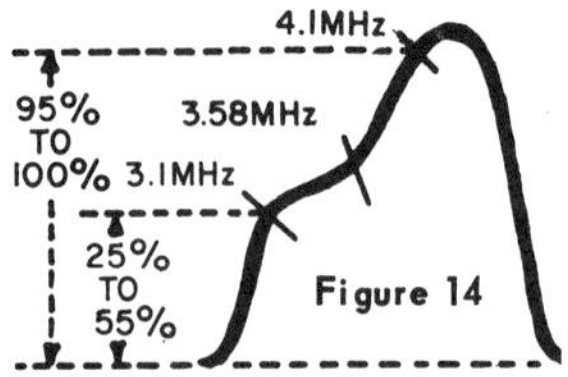 Figure 14	Adjust [T602] for maximum peak 3.58MHz marker as shown in Figure 14.

Chart 23-3

the receiver and transmitter sub-carrier oscillators will result in the reproduction, at the receiver, of incorrect colors. See Chart 23-4, which follows.

In Step 1 of Chart 23-4, one of the procedures required is to ground test point *BU*. This removes the input to the burst amplifier and permits the 3.58 MHz receiver, crystal oscillator to "free-run," without the influence of the color AFC circuit. At this time, the collector circuit of Q_{644}, the 3.58 MHz amplifier, is tuned (T_{606}) for maximum 3.58 MHz output.

In Step 2 of Chart 23-4, the ground is removed from point *BU*, thus allowing the color burst to be amplified and compared with the oscillator signal in the phase detector. The adjustment of T_{604} is made to assure that it is correctly tuned, and that the killer and phase detectors will operate properly.

In Step 3 of Chart 23-4, the color-phase control (R_{776}) adjustment is set so that the display produced by the color-bar generator is motionless on the screen. This indicates that the 3.58 MHz oscillator phase is properly adjusted. Note that during the adjustment in Step 3, tie point *BU* is again grounded to disable the automatic-frequency control. For this reason it may not be possible to get the color bars to stand still on the screen, but if they drift at all, the drift should be very slow.

When the tie point at point *BU* is ungrounded as required in Step 4, the color bars should be securely locked on the screen. The patterns observed on the oscilloscope in Step 4 (Fig. 15 of Chart 23-4) are taken at the *R-Y* output. Moving the tint control back and forth actually shifts the phase of the color signal. Remember that this is the color signal from the color-bar generator. The sixth bar on a color bar display is in the blue region, and that is why it should be cancelled at the *R-Y* output.

In Step 5 the wave-forms and amplitudes of all three color-section outputs (Fig. 16 of Chart 23-4) are checked with an oscilloscope. Of course, if the color oscillator is not in proper phase and frequency, these patterns cannot be obtained.

After the automatic-frequency and phase-control circuitry has been adjusted, the final step—that is, Step 6 of Chart 23-4—is to adjust the color-killer threshold adjustment. Ideally, the color killer should prevent any signals from passing through the color amplifiers when a monochrome signal is being received. The reason for this is that random noise signals would cause color flecks (called confetti) on the screen of the picture tube when a black-and-white picture was being displayed, provided that the noise signals could pass through the color-bandpass amplifiers. The purpose of the color killer is to automatically cutoff the color-bandpass amplifiers during monochrome reception. Step 6 simply explains how the color killer is adjusted to the threshold level.

COLOR AFPC ALIGNMENT

STEP	ALIGNMENT SET-UP NOTES	TEST EQUIPMENT HOOK-UP	ADJUST
1	Set Color Control for +6.5VDC at TP BV. Set Tint Control for +6.5VDC at TP BT. Set Color Killer fully clockwise. Set Phase Control R776 at mid-range. Ground TP BU.	COLOR BAR GENERATOR - To receiver antenna terminals. OSCILLOSCOPE - Across secondary of T606, 3.58MHz output transformer. OR AC-VTVM - Through 4.7K resistor across secondary of T606.	T606 for maximum amplitude on oscilloscope. OR T606 for maximum reading on AC VTVM.
2	Same as step 1, but remove ground from tie point BU.	COLOR BAR GENERATOR - same as step 1. VTVM - To tie point BD.	T604 for minimum DC reading on VTVM. Make certain 3.58MHz oscillator is running and locked in. (1.7V or less)
3	Color Control - Same as step 1. Tint Control - Same as step 1. Killer Control - Same as step 1. Phase Control R776 - Same as step 1. Ground tie point BU.	COLOR BAR GENERATOR - Same as step 1.	R776 for zero beat on picture tube (color bars stand still on screen or drift slowly).
4	Color Control - Same as step 1. Tint Control - Same as step 1. Killer Control - Same as step 1. Remove ground from BU.	COLOR BAR GENERATOR - Same as step 1. (use low lever color signal) OSCILLOSCOPE - To test point BN on Chroma Board. NOTE: Very little adjustment (if any) is required. If more than 1/2 turn is required in either direction, repeat step 2 AFPC Alignment and step 2 of the Chroma Alignment.	T604 so that when tint control is rotated from one extreme to the other, there is a minimum of + and - 30 deg. from nominal phase. Return tint control to mid-range. 6th bar on R-Y waveform should be cancelled. See Fig. 15.
5	Tint Control - Same as step 1. Killer Control - Same as step 1.	OSCILLOSCOPE - To test points BN, BM, CG, respectively.	Check for proper matrixing. Waveforms and amplitudes should conform to Fig. 16.
6	Tint Control - Same as step 1. Set channel selector to channel receiving snowy raster with no color transmission.	No test equipment.	R660 Killer threshold control clockwise until snow on screen appears colored; then rotate R660 counterclockwise until color in the snow disappears. Check on a color program to assure setting of R660 is not killing on color.

Chart 23-4

COLOR AFPC ALIGNMENT (CONTINUED)

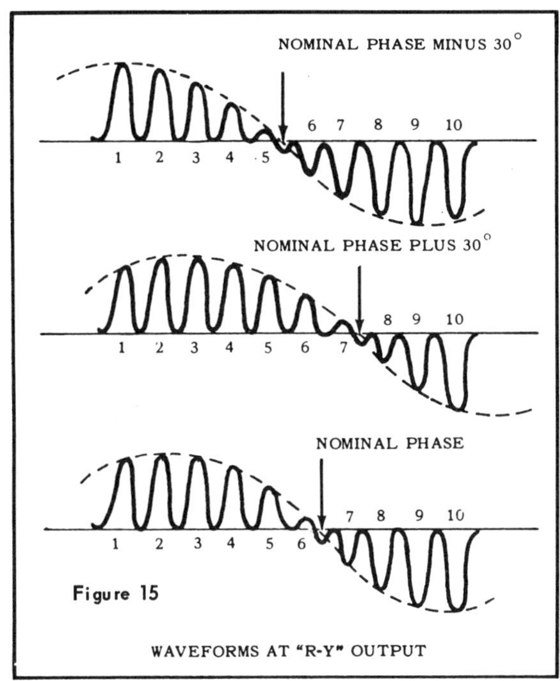

Figure 15

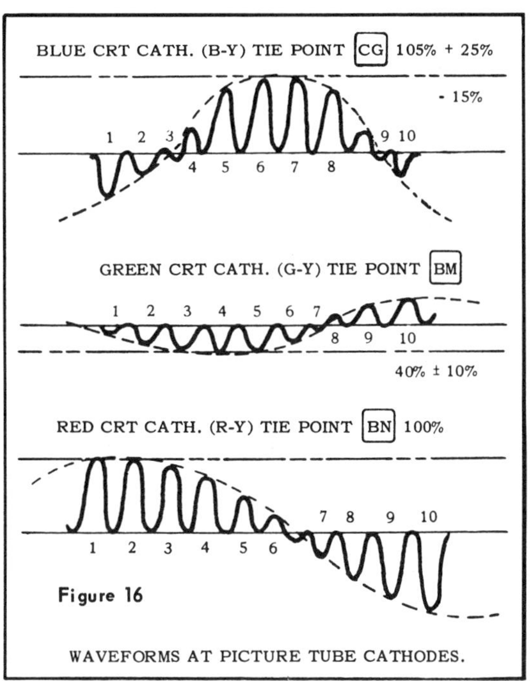

Figure 16

Chart 23-4 continued

Horizontal Deflection (See Chart 23-5.) The horizontal oscillator in the receiver of Fig. 23-2 has automatic-frequency control, and this is typical of all horizontal oscillator circuits used in television receivers. As with other horizontal AFC circuits, the automatic-frequency control circuit is adjusted by disabling the sync-pulse signal and setting the oscillator frequency so that it is approximately correct. This is the purpose of Step 1 in Chart 23-5. Note that point *DJ* is grounded, and this is the input point for the horizontal oscillator synchronizing signals. The horizontal-hold control is set to approximately the center of the range, and then the horizontal-frequency control (R_{428}) is set so that the free-running frequency of the horizontal oscillator is approximately correct. When the ground at *DJ* is removed, as required in Step 2, the synchronizing signal locks the oscillator.

Step 2 gives the procedure for adjusting the final anode voltage for the color-picture tube. This adjustment is very important, since an improper high voltage will not only result in poor picture quality and focus but also can result in the production of undesirable x-rays.

Convergence Adjustments (See Chart 23-6.) As discussed previously in Chapters 14 and 18, a color-picture tube must have its three beams properly converged, if both monochrome and color pictures are

HORIZONTAL DEFLECTION

STEP	ALIGNMENT SET-UP NOTES	TEST EQUIPMENT HOOK-UP	ADJUST
1	NORM/SERV Switch in normal position. Tune receiver to a local station and synchronize picture. Short test point [DJ] on the deflection board to ground. Center Horizontal Hold Control [R28] . Increase brightness to maximum.		[R428] Horizontal Frequency Control to bring both sides of picture vertical.
2	Same as step 1, except remove jumper from test point [DJ] on the deflection board. Reduce brightness to minimum.	VTVM - Through high voltage probe to picture tube anode.	Monitor line voltage for 120 VAC. Adjust [R518] , H.V. Adjust, for 25KV on voltmeter.
3	Readjust focus, height, and vertical linearity controls for proper focus and vertical size.		

Chart 23-5

to be correctly reproduced. (Refer to Fig. 23-26 for dot patterns which indicate mis-convergence and correct convergence conditions. Refer also to Chapters 14 and 18 for discussions of convergence.)

The general convergence procedure consists of six general steps which are given at the top of Chart 23-6.

(1) *The degaussing procedure* is given in Step 1 of the chart and needs no further comment here.

(2) *Static Convergence.* In the procedure for static convergence a dot generator is used to provide the display on the picture-tube screen. Red, green, and blue convergence magnets are adjusted until the dots are superimposed and white, indicating that the red, green, and blue color triads are being struck simultaneously by the proper electron beams, in the center area of the screen.

The magnets around the neck of the tube that are used for making convergence adjustments are shown in Figs. 1, 2, and 4 of Chart 23-6. Figure 3 of Chart 23-6 is a very important illustration because it shows in which direction the dots can be moved by each magnet. Note that the red and green magnets move the red and green dots at an angle of 45°, while the blue magnet can move the blue dots vertically and horizontally. When the set is out of convergence, three different color dots may be discernible, and as the correct adjustment of each magnet is made, the dots become superimposed to produce a pure white dot.

(3) *Purity Adjustments.* The purity adjustment is normally made with only the red gun in operation. The other two guns are turned off by adjusting the blue and green screen controls. In some receivers, switches are available to turn the individual guns on or off. Regardless of how the blue and green guns are disabled, only the red gun should now be conducting, and the screen should display a pure red color.

— INSTALLATION OF NEW COLOR TELEVISION RECEIVER (SET-UP) —

Be certain the following adjustments have been made before starting convergence procedure: H.V. Adjust, Horizontal, Focus, Height, Vertical Linearity, Width, Electrical Center and AGC.

— GENERAL CONVERGENCE PROCEDURE —

The following is an outline of the complete step by step Convergence procedure.

1. DEGAUSSING - Demagnetize shadow mask with degaussing coil.
2. STATIC CONVERGENCE - DC converge all three colors in center area of screen.
3. PURITY ADJUSTMENT - Adjust purity ring for red center screen area, adjust yoke position for red outer area.
4. COLOR TEMPERATURE SET-UP - Setting black and white brightness tracking.
5. DYNAMIC CONVERGENCE - Dynamically converge all three colors. Check horizontal, focus, height, vertical linearity, width, electrical centering and AGC.
6. REPEAT STEPS 2, 4, 5 for best results.

---

1. DEGAUSSING

A. Adjust receiver for a black and white picture. Observe picture for good black & white reproduction over all areas of the screen.
B. If color shading is evident, demagnetizing (degaussing) of the receiver is necessary.
C. This chassis has a built-in degaussing coil and usually all that is required to degauss the receiver is to turn the receiver off for approximately 20 minutes, and then on.
D. When manually degaussing the receiver, move a degaussing coil slowly around the front faceplate of the picture tube and around the sides of the receiver; then slowly withdraw the coil to a distance of at least six (6) feet from the receiver before disconnecting the coil from the AC source. NOTE: Degaussing should be done only after receiver is placed in the position where it will be viewed and remain.

2. STATIC CONVERGENCE

A. Adjust receiver for a normal black and white picture.
B. Receiver should be in the position in which it will be operated.
C. Connect a DOT/CROSS-HATCH GENERATOR with an RF output to antenna terminals of receiver. Adjust red, green, and blue magnets (shown in Figure 1) and the blue lateral magnet (shown in Figures 2 and 4) to attain convergence of dots in the center of picture tube screen. The direction of movement of the dots using these magnets is shown in Figure 3. Lateral movement of the blue dot is accomplished by rotation of the blue lateral magnet adjustment.

3. PURITY ADJUSTMENT

NOTE: BEFORE ATTEMPTING TO ADJUST RECEIVER FOR PURITY, THE SET MUST BE TURNED ON FOR AT LEAST 15 MINUTES AND BRIGHTNESS MUST BE AT NORMAL LEVEL.

A. Tune receiver to a blank channel having a white background.
B. Adjustment for purity is most accurate while observing one color (preferably red). Decrease the Blue (R448) and Green (R450) screen controls to minimum by turning fully counterclockwise.
C. Set contrast control to minimum, brightness to normal viewing level.
D. Loosen yoke adjusting nuts and slide yoke back against convergence magnet assembly. Rotate purity ring magnets around neck of picture tube and/or spread the individual ring tabs (See figure 2) for uniform red field area at center of screen.
E. Follow steps A, D, E, F as outlined under "COLOR TEMPERATURE SET UP" in section 4.
F. Readjust static magnets if necessary. See step 2.
G. Repeat step B under "PURITY ADJUSTMENT".
H. If the red field area at center of raster moves off center when static magnets were readjusted, the purity rings must be set again.
I. When the center of raster is red and static convergence is correct, move yoke forward and position for best overall red screen.
J. Retighten yoke adjustment nuts. Proceed with complete "COLOR TEMPERATURE SET UP", and observe all three beams.

4. COLOR TEMPERATURE SET-UP

A. Move service switch to SERVICE position.
B. Set brightness and contrast controls to mid range.
C. Set CRT Bias Control (R744) to minimum (counterclockwise) position.
D. Rotate channel selector to a free channel, rotate Contrast, Brightness, Color and the 3 screen controls to minimum (full counterclockwise) position.
E. Individually adjust screen controls so that line is just visible in the center of picture.
F. Return service switch to NORMAL position.
G. Rotate channel selector to strongest channel in the area.
H. Advance Contrast and Brightness for normal picture viewing.
I. If insufficient brightness is noticed at high brightness setting, repeat above procedure, then readjust CRT bias control (R744) for adequate brightness level after completion of screen adjustments.
J. Adjust green, red, and blue drive controls so that the bright parts of the picture are similar to, but slightly less blue, than a typical black and white receiver.

5. DYNAMIC CONVERGENCE

NOTE: MAKE CERTAIN STATIC CONVERGENCE IS COMPLETED BEFORE ATTEMPTING DYNAMIC CONVERGENCE. CHECK HORIZONTAL, FOCUS, HEIGHT, VERTICAL LINEARITY, WIDTH, ELECTRICAL CENTERING AND AGC.

A. (On some models) the Convergence board assembly is designed so that adjustment can be made from the front of the receiver. Loosen the screw(s) securing assembly to cabinet. Remove assembly and mount on top rail of cabinet with controls facing front of receiver. Secure with screws to prevent movement.
B. Adjust receiver for a normal black and white picture.
C. Connect a DOT/CROSS HATCH GENERATOR with an RF output to antenna terminals of receiver.
D. Switch to cross-hatch pattern with the following steps referring to Figure 5 while making adjustments.

6. Repeat Static Convergence, Dynamic Convergence, and Color Temperature Set-Up for optimum results.

Chart 23-6

STEP	ADJUST	FOR
DECREASE BLUE SCREEN CONTROL TO MINIMUM BY TURNING FULLY COUNTERCLOCKWISE.		
1	R810 - Top R/G Vert.	For convergence of Red and Green center vertical lines at top of screen.
2	R804 - Bottom R/G Vert.	For convergence of Red and Green center vertical lines at bottom of screen.
3	Repeat steps 1 and 2 for best convergence of Red-Green vertical center line from top to bottom.	
4	R806 - Top R/G Horiz.	For convergence of Red and Green horizontal lines at top center of screen.
5	R800 - Bottom R/G Horiz.	For convergence of Red and Green horizontal lines at bottom center of screen.
6	Repeat steps 4 and 5 for best convergence of Red and Green horizontal lines at top and bottom center of screen.	
7	L802 - Right R/G Vert.	For convergence of Red and Green vertical lines on right half of screen.
8	R822 - Left R/G Vert.	For convergence of Red and Green vertical lines on left half of screen.
9	L800 - Right R/G Horiz.	For convergence of Red and Green horizontal lines on right half of screen.
10	R818 - Left R/G Horiz.	For convergence of Red and Green horizontal lines on left half of screen.
FOLLOW STEPS A, D, E, F AS OUTLINED UNDER "COLOR TEMPERATURE SET-UP" IN SECTION 4.		
11	R812 - Top Blue Horiz.	For equal displacement of horizontal blue lines in center of screen from top to bottom.
12	R802 - Bottom Blue Horiz.	For equal displacement of blue horizontal lines at top and bottom of screen on vertical center line.
13	T800 - Right Blue Horiz.	For convergence of center blue line from middle to right side of screen.
14	R824 - Left Blue Horiz.	For convergence of center blue line from middle to left side of screen.
Step 15 applies only to those chassis which use the TOROIDAL DEFLECTION YOKE (on some models only).		
15	L806 - Left & Right Blue Vert.	For convergence of left and right blue vertical lines.
16	Repeat all steps for best possible convergence, remove test equipment and check on air signal.	

SLIGHT READJUSTMENT OF THE BLUE STATIC AND BLUE LATERAL MAGNET MAY BE NECESSARY. FOR OPTIMUM CONVERGENCE REPEAT THE ENTIRE PROCEDURE.

NOTE: L804 - Blue horizontal center phase is factory adjusted and MUST NOT be adjusted during the convergence alignment. Adjust only if replacement of coil becomes necessary. If coil is replaced, proceed through the convergence alignment, then connect an oscilloscope (signal lead) to test point S of convergence board, ground lead to chassis and adjust L804 until slope notch of wave shape is at 50% of amplitude. See illustration. Repeat convergence alignment for optimum performance.

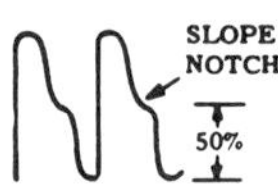

Chart 23-6—continued

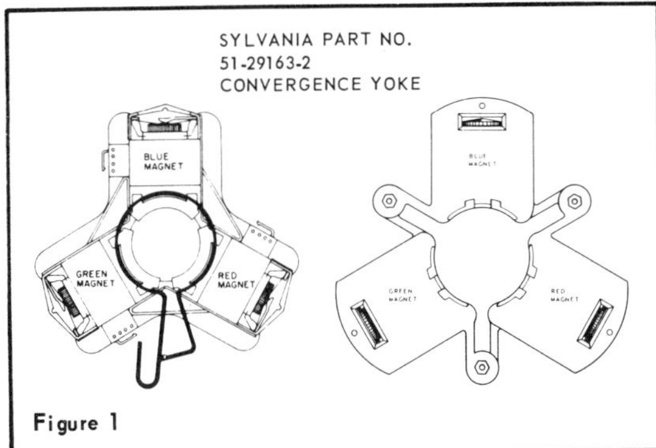

Figure 1

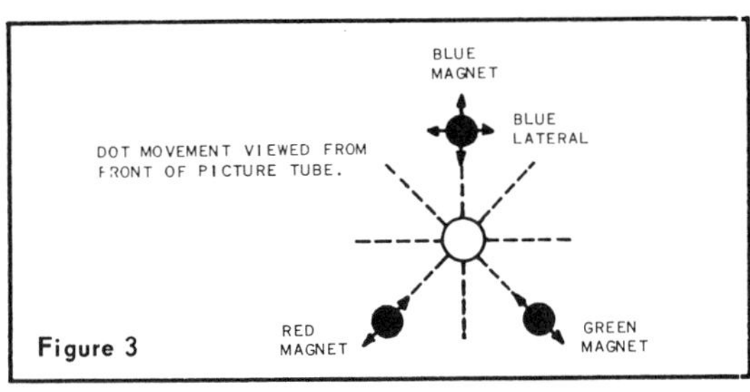

Figure 3

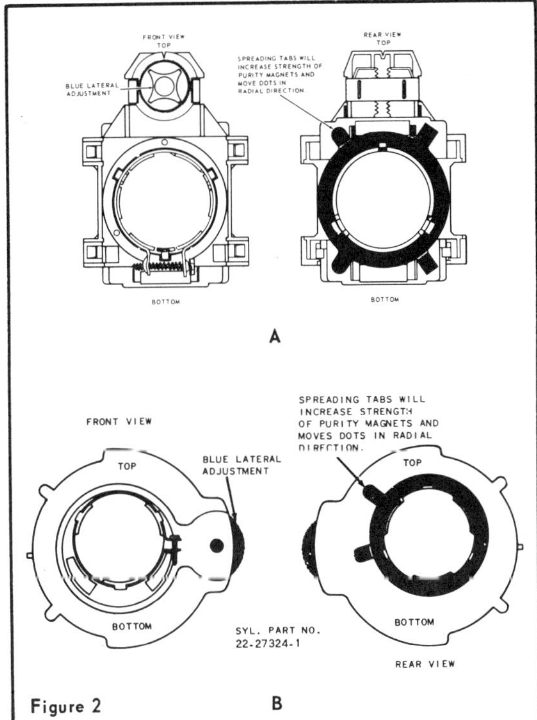

Figure 2

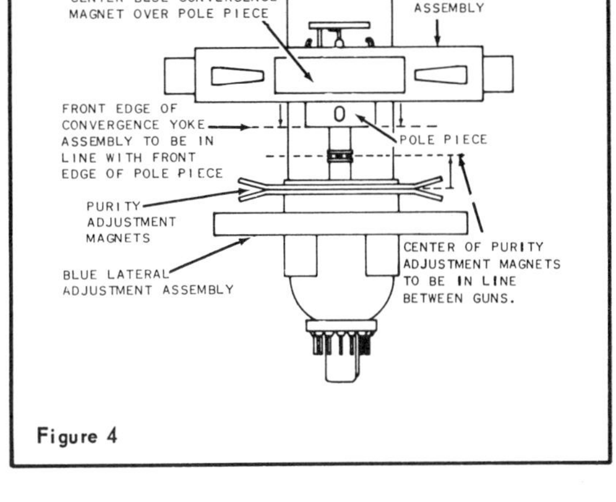

Figure 4

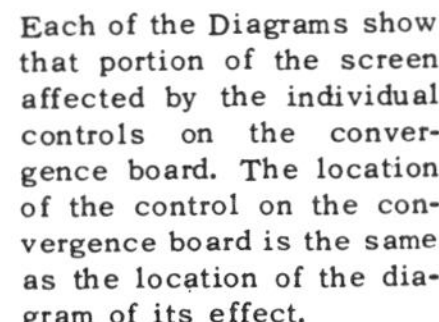
Each of the Diagrams show that portion of the screen affected by the individual controls on the convergence board. The location of the control on the convergence board is the same as the location of the diagram of its effect.

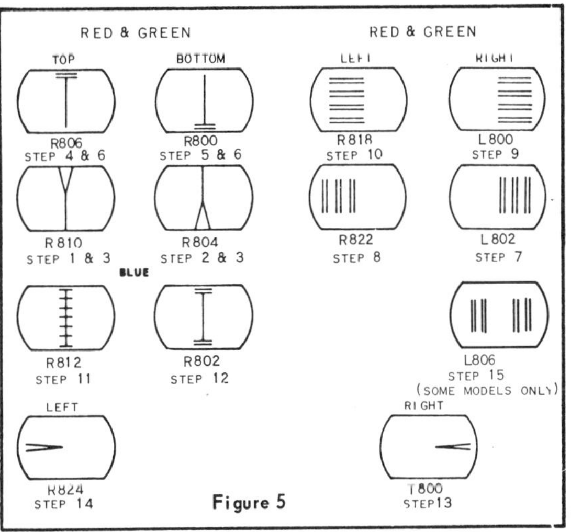

Figure 5

Chart 23-6—continued

(4) *Color Temperature and Dynamic Convergence.* The color temperature (gray scale) and dynamic convergence adjustments are self explanatory as described in Chart 23-6. Keep in mind that the color temperature or gray-scale adjustment permits black-and-white pictures to be displayed properly. Dynamic convergence adjustment assures that convergence is proper at the edges of the screen as well as at the center.

REVIEW QUESTIONS

1. In the block diagram of Fig. 23-1, what signals are at the output of the first video amplifier?
2. Using the block diagram of Fig. 23-1, trace the closed-loop circuit for the automatic-frequency control of the local oscillator in the tuner.
3. Why is a keying pulse from the flyback transformer used in the AGC circuit?
4. How is the 25 KV picture tube voltage obtained for the receiver shown in Fig. 23-2?
5. Which circuits receive a signal from the blanker? (Use the block diagram of Fig. 23-1.)
6. Which circuit is controlled by the color killer? (Fig. 23-1.)
7. Where is the delay line in the television receiver of Fig. 23-2?
8. Explain how the monochrome signal produces a picture on the picture tube. Use the block diagram for your discussion.
9. What methods of regulation are used in the low-voltage power supply in the circuit of Fig. 23-2?
10. What is an advantage of an FET meter over a VTVM?
11. What is an advantage of using an oscilloscope with a triggered sweep?
12. Explain what is meant by the expression "the post injection marker system of alignment."
13. What is an advantage of the post injection marker system over the alternative?
14. What is the first step in aligning a color, closed-loop automatic-frequency control circuit?
15. What is the procedure—with regard to AGC—when aligning the video-IF stages?
16. Draw a response curve for a color (chroma) bandpass amplifier.
17. What is a vectorscope used for?

Troubleshooting Television Receivers 24

24.1 INTRODUCTION

Television receivers are critical mechanisms that require accurately performing circuits, if the receiver is to operate correctly. The indiscriminate replacement of component parts may cause more harm than good and should be avoided. Careful adherence to manufacturer's values is especially important if correct performance is to be maintained. For example, hold controls permit some variation of the oscillator's frequency, but the limits are fairly narrow. A wide discrepancy between the values of the replacement parts and those specified by the manufacturer will usually render synchronization impossible. The same general situation holds true also, for other receiver sections.

The first step in troubleshooting a television receiver is to observe the symptoms. For a monochrome receiver the display (or lack of display) on the picture-tube screen and the sounds (or lack of sounds) from the speaker are very valuable information for quickly locating a fault. In a color receiver the display of or lack of colors on a color program and the condition of a monochrome display are important troubleshooting clues.

The visual and aural symptoms will frequently permit the technician to isolate the defective section and stage. The next step is to isolate the defective component. The tube or transistor is normally checked first. If it is then found to be in good condition, the voltage and resistance measurements are used to pinpoint the defective component.

Television troubleshooting, then, is a process of going from general information or symptoms to the location of defective sections, defective stages, and finally to the defective component(s).

24.2 GUIDEPOINTS FOR TROUBLESHOOTING-MONOCHROME RECEIVERS

A certain amount of common sense coupled with a good understanding of the television receiver system are needed for fast efficient troubleshooting. As an example, suppose a receiver exhibits a distorted picture accompanied by a distorted audio output. This situation immediately tells the technician that the defect is located in a circuit through which both signals pass, and it directs attention to the stages preceding the separation point. On the other hand, when only one signal appears distorted at the output, then obviously this distortion must have occurred in a circuit dealing solely with this signal. If only the audio is defective, then all attention is centered on the audio stages following the point of separation. If only the image is distorted, only the video stages beyond the separation point need be examined. With these simple facts in mind,

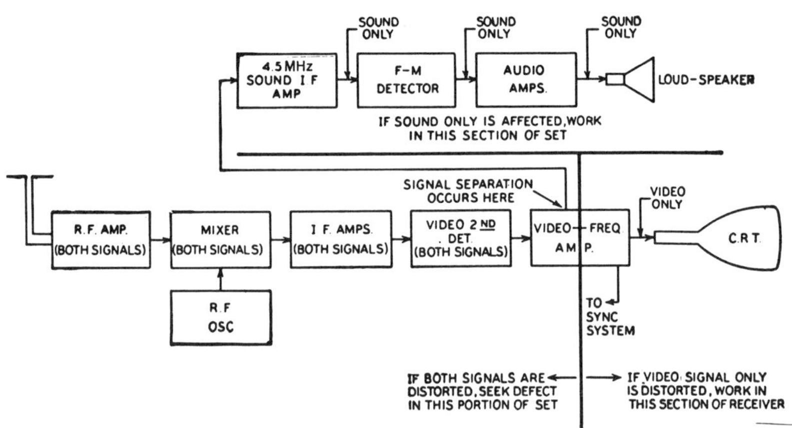

Fig. 24-1. The block diagram shows how the symptoms can be used for locating the defective receiver sections.

tracking down troubles in television receivers can be simplified considerably.

Figure 24-1 shows a block diagram of an intercarrier monochrome receiver. Note that the sound is separated from the video after the detector. (In early model split-sound receivers the sound takeoff point was at the mixer or first IF amplifier.) The block diagram separates the symptoms into three categories: distorted sound with normal video, distorted video with normal sound, and both sound and video distorted. The diagram is also applicable for no sound with normal video, no video with normal sound, and no sound or picture.

Once the defect has been traced to a specific section of the receiver, the next step is to analyze that particular section with a view toward further localization. This brings us to the block diagrams shown in Figs. 24-2 and 24-3. In the audio system, the general breakdown consists of the IF amplifiers, the FM detector, the audio amplifiers, and the loudspeaker. In the video system we have the video-IF amplifiers, the

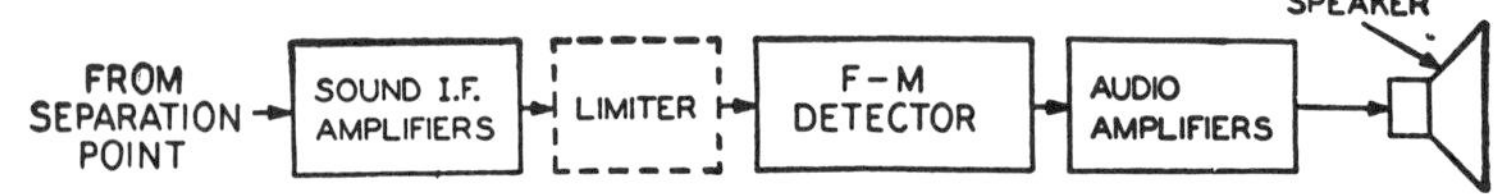

Fig. 24-2. A block diagram of the audio section of a television receiver.

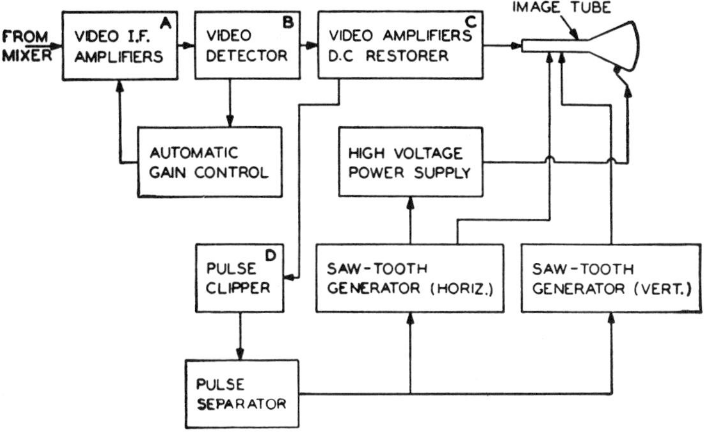

Fig. 24-3. A block diagram of the video section of a television receiver.

video detector, video amplifiers, and the vertical- and horizontal-sweep systems. The high-voltage power supply for the CRT is located at the output of the horizontal-sweep system.

Once the difficulty has been traced to a particular system in the receiver, we are in a position to conduct a further analysis of our defective receiver. In the video system, for example, breakdown of the signal path in sections *A*, *B*, or *C* will prevent the video signal from reaching the cathode-ray tube. If the sound-separation point is prior to the breakdown, sound will be unaffected. If it is after the breakdown, sound will also be lost. In either instance, the picture-tube screen will contain a raster, but no sound. The scanning raster is due to the sweep oscillators which continue to function because they are free running, relaxation oscillators. Even if they are not being triggered, they will still oscillate.

As another example, suppose the circuit opens up in section *D*. The sync pulses will be prevented from reaching the sweep oscillators; consequently the oscillators will not be controlled by the incoming signal. On the other hand, image signals are reaching the cathode-ray tube. The visual result is a scrambled picture caused by the various sections not being placed on the screen in proper sequence. The audio section will be unaffected.

These are some of the many clues that the technician may receive each time a set becomes defective. In this chapter we will undertake the analysis of many common clues encountered in defective television receivers. The recognition and interpretation of these signposts will help the technician in 90 percent of his work.

RF System A defect in the RF section of a television receiver will affect both the sound and the video outputs. If the signal is prevented from passing through completely, no sound will be heard from the loudspeaker and no image will be seen on the screen. However, what will be visible is a scanning raster (see Fig. 24-4).

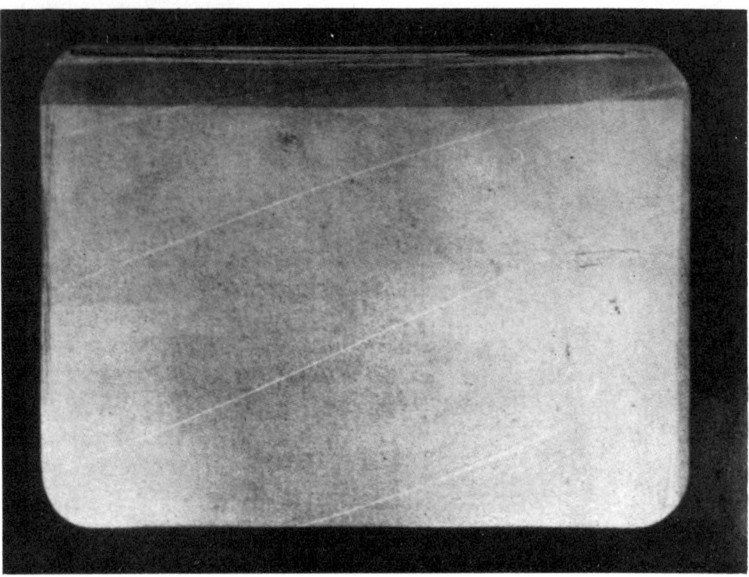

Fig. 24-4. A scanning raster.

Several signal tests are useful in indicating whether or not a signal can get through the front-end stages. A rough test, but one which will reveal whether or not a signal can get through is to turn the contrast control up and momentarily short across the antenna transmission line at the receiver-input terminals. Bursts of noise will be heard in the speaker (even with FM) and flashes of light will appear in the scanning raster. If the signal cannot get through the RF stages, these indications will not be obtained. Admittedly this is a rough test, but it is surprising how well it works.

One problem which might arise is how to distinguish between a defect in the front-end stages and the complete absence of any signal reaching the television receiver. The latter may occur if there is a break in the transmission line leading from the antenna to the receiver. A simple method for checking the front-end stages is to use an AM signal generator. Connect the instrument across the input terminals and set the generator frequency to a value about 1 MHz above the video-carrier frequency for that channel. Turn on the AM modulation in the generator. If dark bands appear across the screen of the image tube, then a signal is able to pass through the RF stages. The signal generator will not produce any indication on the screen if the receiver RF oscillator is inoperative.

If further tests indicate that the tubes or transistors are all right, but the set is completely dead on all channels, then a voltage and a resistance check will be needed to reveal the source of trouble. Start by measuring the B+ and AGC voltages to the tuner. Check the manufacturer's specifications to make sure that these voltage readings are correct. If the receiver has an automatic fine-tuning system, be sure to check the voltage at the tuner AFT terminal.

Fig. 24-30 Improper adjustment of the color set fine-tuning control can result in the color sub-carrier beat pattern shown here.

Fig. 24-31 Insufficient color amplitude, which may be the result of an improperly adjusted color amplitude control or a defective bandpass amplifier stage, may result in a weak color display like the one shown here.

Fig. 24-32 Too much color amplitude will produce a picture with oversaturated colors. An improperly adjusted color amplitude control or a defective automatic color-control circuit may cause this condition.

Fig. 24-33 The improper adjustment of the hue control (which is also called the tint control) results in this picture with an overall green shading. This may also be caused by incorrect phasing in the color demodulation circuits. A defect in an automatic tint-control circuit can also produce this general effect. The overall color might also be red or blue, or some intermediate shade.

Fig. 24-34 An example of a color tube which shows poor purity. Note the large areas of color.

Fig. 24-35 Colored confetti can result from low (or no) gain in the RF amplifier.

Fig. 24-36 The appearance of the screen of a color receiver, with a color-bar generator input, when there is a 60-Hz hum in the signal circuits.

Fig. 24-37 Ringing or close-spaced ghosts are seen here at the trailing edge of each color bar. This is a circuit-caused defect and is not due to multi-path reception.

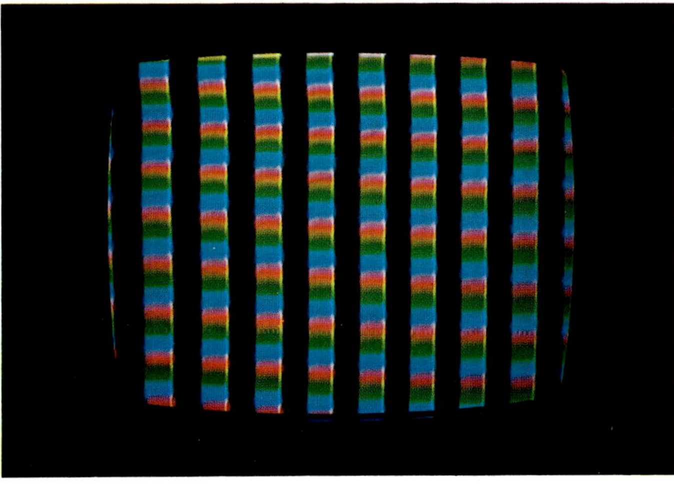

Fig. 24-38 A color-bar generator pattern when color sync is lost. (Monochrome-horizontal sync is normal.)

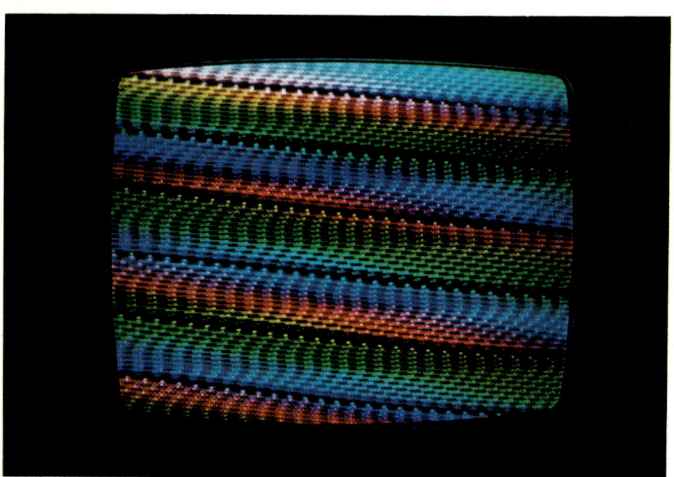

Fig. 24-39 The effect on a color-bar pattern, when both monochrome (horizontal) sync and color sync are absent.

Fig. 24-40 Appearance of the color-bar signal when there is no *G-Y* signal present. Note the complete absence of green coloring.

The Video Circuits The next group of circuits through which the signal passes are the video circuits. These include all of the video IF stages, the video detector, and the video-frequency amplifiers. Let us consider each in turn.

Failure in the video IF system will cause the image to become distorted or to disappear entirely from the screen, but it may or may not affect the sound. In addition to tubes or transistors, any of the other components—resistors, capacitors, and coils—can be the cause of failure. To locate the defective unit, several methods of approach are possible. First, there is voltage checking. With a voltmeter measure the voltages of the various IF stages at the elements of the tubes or transistors. Then compare these with the manufacturer's values. This method is effective and frequently successful.

Incidentally, if it is discovered that all the B+ voltages in a system are low, trouble in the B+ system is indicated, and the trouble is not necessarily in the stages where the voltage measurements are being made. A filter capacitor in the power supply may be leaky, or the rectifier may be defective, or a shorted by-pass capacitor may be at fault. In an effort to determine the one branch containing the defective component, your job now may be the tedious one of disconnecting the various branches leading off the B+ supply.

Voltage tests are most useful in detecting open resistors, shorted capacitors, and the incorrect AGC voltages. This latter voltage is most important, because an improper AGC voltage can completely disrupt the set operation. Incorrect AGC voltages may be caused by defective tubes, transistors, diodes, or other parts.

A defect that will not be found by voltage checks (such as an open capacitor or a shorted coil) may be brought to light by injecting a signal into the circuit. Take an AM generator and connect it to the input of the last video IF amplifier. (This stage is just before the video second detector.) Set the frequency to the mid IF value and turn on the modulation. If this stage is functioning (and all the others that follow it), then black-and-white horizontal bars will appear across the face of the picture. The generator probe can be moved back, stage by stage, until the point of failure is located. This method is simple, requires only an AM signal generator, and is readily carried out.

If you have a sensitive oscilloscope and a demodulator probe, you may be able to trace the video signal as it travels through the IF system. Connect the probe to the vertical-input terminals of the oscilloscope. With the oscilloscope in its most sensitive condition, touch the probe tip to the input of the first video IF amplifier and see if you obtain the video-signal wave-form on the scope screen. A normal indication is shown in Fig. 24-5. (The same waveform with positive-sync pulses is also acceptable. The phase of the pattern depends upon the probe circuit and the number of vertical amplifiers in the oscilloscope.) If you obtain

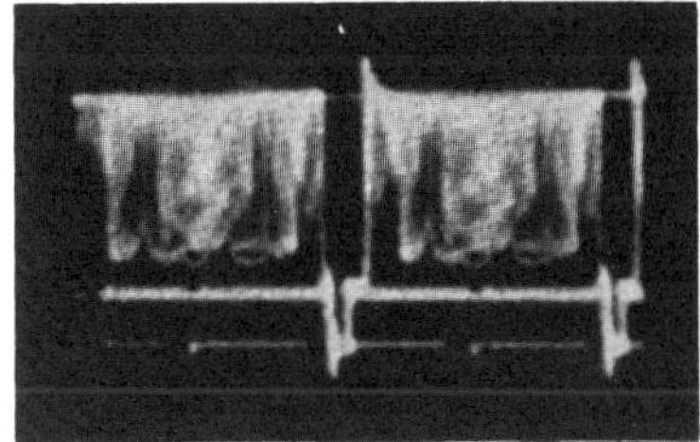

Fig. 24-5. The appearance of a normal video signal as seen on an oscilloscope screen.

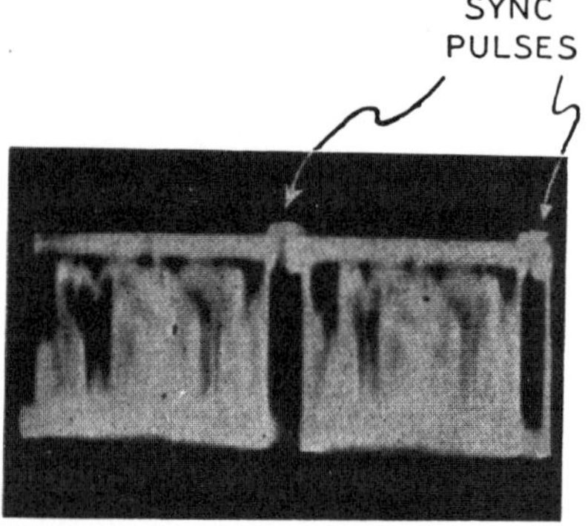

Fig. 24-6. A video signal in which the sync pulses have been partly compressed. Note that the video signal extends up to the level of the pulses.

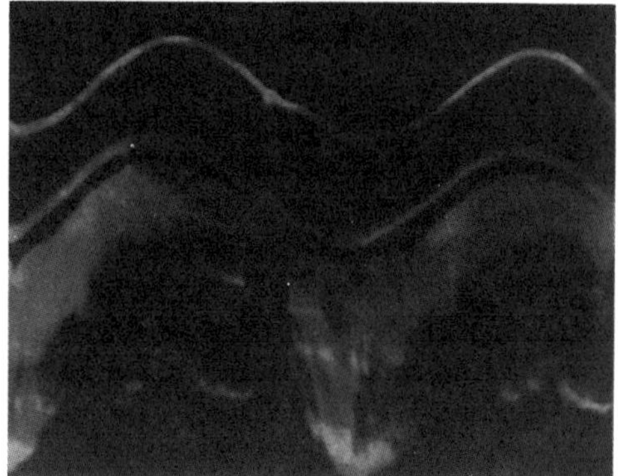

Fig. 24-7. 60 Hertz hum in the video signal.

a normal indication, move on to the input of the next IF amplifier, and so on. Continue until you lose the signal, or it becomes distorted.

When you reach the video-second detector, the detector probe is no longer required. Detector action will be furnished now by the video detector. Signal-tracing can be very effective, provided you have a sensitive scope and can distinguish between a normal-appearing video signal and a distorted one. Overloading in any of the video IF amplifiers as well as ac hum in the signal is clearly brought to light by this method (see Figs. 24-6 and 24-7). Signal tracing with an oscilloscope is used extensively in the video amplifiers (those beyond the video detector) and in the vertical- and horizontal-sweep systems.

In the video detector, a completely defective diode will prevent any signal from passing through it. Defects such as a weak picture lacking in contrast, an unstable vertical or horizontal lock-in, or intermittent operation are symptoms of a defective diode detector.

Following the video second detector are the video amplifiers. Unless a video amplifier becomes completely inoperative, in which case no image at all is obtained on the screen, indications of other defects will be evident only by their effect on the image. In a video-frequency amplifier stage, the following defects may be found:

(1) A defective low-frequency compensating network.
(2) A defective high-frequency compensating network.
(3) Improper voltages at the tube or transistor electrodes.
(4) An inoperative tube or transistor or a defective component.

When the low-frequency compensation network is defective, the background shading of the image becomes darker and the larger objects in the image "smear" (see Fig. 24-8). Check by-pass capacitors, load,

Fig. 24-8. A visual indication of poor low-frequency response in a video system. (*Courtesy of RCA.*)

dropping, and bias resistors. Capacitors are highly vulnerable components, and may be shorted or open. A fast method of checking for open capacitors is to shunt a suspected unit with another capacitor of equal value that is known to be good. Also useful are in-circuit capacitor and transistor checkers which reveal a defective unit without removing it from the circuit.

Smearing can also occur when the bias voltage is incorrect. Hence, measure the bias voltage at each video-amplifier stage. Leaky interstage coupling capacitors may cause incorrect-bias voltage. Low collector or plate and screen voltages produce smearing because under these conditions the transistor or tube is readily overloaded. Look, too, for a load resistor that has increased considerably in value. A tube or transistor with a low gain is another cause of smearing.

A more difficult defect to detect is the loss of fine detail caused by poor high-frequency response of the video stages. High-frequency compensation is provided by the series and shunt-peaking coils in the video amplifiers.

The peaking coils are frequently shunted by a resistor to prevent them from sharply increasing the amplifier gain at their resonant frequency. Should this resistor increase substantially in value or perhaps open, its shunting effect would be removed and transient oscillations would develop in the coil and in its distributed capacitance whenever the signal frequency fell within this range. The oscillations would appear on the screen as ghost lines (or multiple lines) after any small object or sharply defined line or edge in the picture. Their presence is readily noticeable in test patterns. This effect, also known as "ringing" can be distinguished from ghost signals by the fact that the various lines are evenly spaced. Also, each successive line becomes progressively fainter. Ringing may also be caused by replacing a defective peaking coil with one whose value is not correct for that circuit.

If the signal is not passing through a video amplifier, as is indicated by no image on the screen (only a scanning raster), then a quick check to locate the inoperative stage is in order. Apply a 400 Hertz audio signal, obtained from an audio-signal generator, across the load resistor of the video-second detector. Alternate black-and-white bars will appear across the screen if the video amplifiers are working. If the screen remains blank, move the audio generator toward the cathode-ray tube, one stage at a time, until the defective amplifier is found. Voltage, resistance, transistor and tube checks will then quickly reveal the defective component.

Another approach to video-amplifier troubleshooting when no signal is reaching the picture tube is to tune in a station, then follow the video signal from the second detector to the point where the break occurs. The oscilloscope is an ideal instrument for this purpose. Place the vertical-input leads of the scope across the output-load resistor of

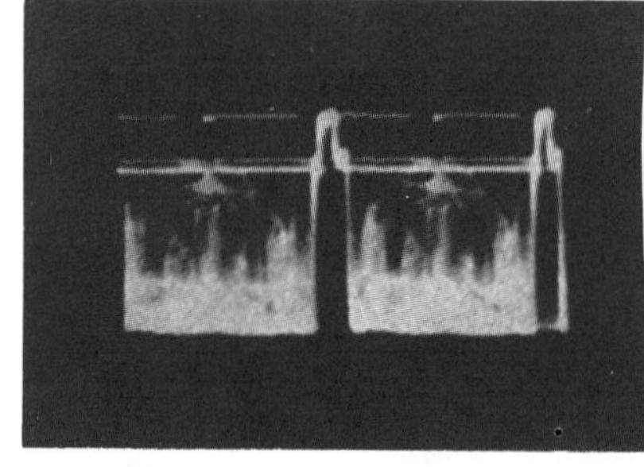
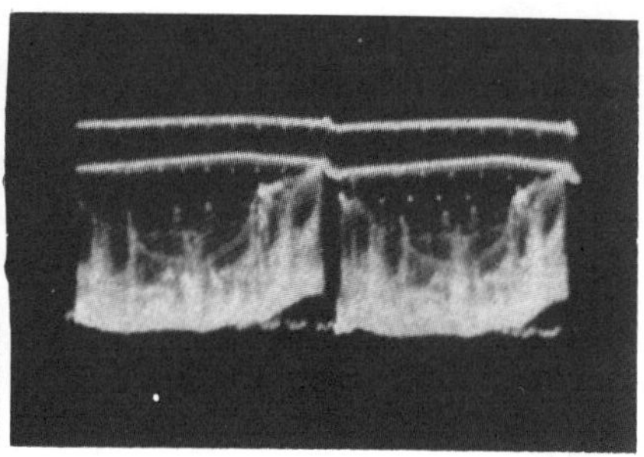

Fig. 24-9. Typical line (left) and field (right) patterns of video signals.

the second detector. Set the scope-sweeping frequency either to 30 Hertz (to observe two fields) or to 7875 Hertz (to observe two lines). Typical field and line patterns of video signals are shown in Fig. 24-9. Whether the polarity of the observed signals is as shown in the figure or reversed depends upon how the detector is connected. In any event, it is the presence of the signal that is important and not its polarity.

Once the signal is observed at the output of the second detector, it can be traced to each succeeding video amplifier until it disappears or becomes distorted. Where this occurs represents the location of the defect, and voltage and resistance measurements should then bring it to light.

Additional Video-amplifier Defects Dimness in a picture, or one that lacks good contrast, can be caused by a weak video-amplifier or improper dc operating voltages. A dim picture may also stem from a defective picture tube or one having an excessively negative grid-bias voltage.

On the other hand, excessive contrast—when the picture is quite dark and the grays are missing—may be caused by a defective tube or transistor, bias which is incorrect, a leaky coupling capacitor, a shorted cathode by-pass capacitor, or an improperly functioning dc restorer, if one is employed. Too strong a signal from the video-second detector, due perhaps to a defective AGC system, will also produce the same symptom.

A negative picture (Fig. 24-10) represents a greater aggravation of the conditions that lead to excessive contrast. In this case, the amplifier is overdriven, resulting in the reversal of picture values, as is indicated in the illustration. Picture tubes often cause this trouble; they may be either quite gassy or have an internal short.

There is usually a 4.5 MHz trap between the video detector and the picture tube. The purpose of this trap is to eliminate the 4.5 MHz beat frequency obtained by combining the video- and sound-IF frequencies in the nonlinear video detector stage. If the trap is out of adjustment, a 4.5 MHz beat pattern will appear in the picture. The beat pattern will also appear if any of the components in the trap becomes defective.

Fig. 24-10. A negative picture in which all the tonal values are reversed.

24.3 SERVICING PICTURE-TUBE CIRCUITS

The proper presentation of an image on the screen of a monochrome-picture tube depends upon the following conditions: First, a signal being present at the grid (or cathode) of the tube; second, the correct dc voltages being applied to the various electrodes; third, the neck-mounted coils and magnets being properly positioned; and finally, sufficient deflection power being available. Failure of any one of these conditions will either distort the picture or result in its absence. Let us consider each condition separately to see what its effect will be on the picture.

No Video Signal In the absence of a signal, only a raster will be seen on the screen. The raster indicates that the high voltage plus all the other voltages are operating normally. It will indicate also that the horizontal- and vertical-deflection systems are delivering the necessary deflection currents to the yoke. And from the sharpness of the raster lines and the absence of any shadows over the screen, you will know that the focus action and ion trap are functioning properly.

Loss of High Voltage The surest indication of high-voltage failure is the appearance of a perfectly blank screen. If a blank screen is accompanied by normal-sound output, then we know that the low-voltage power supply is operating and we can concentrate on the high-voltage system. However, a blank screen and no sound are more likely caused by a defective low-voltage supply which should be tackled first.

High-voltage power-supply failure will cause a completely dark screen but there may be other reasons for the same condition. If the filament of the picture tube is lit, check the high-voltage system. If a kilovoltmeter is available, measure the output of the high-voltage power supply. If the voltage is low or missing, replace the high-voltage rectifier. As a final step, make resistance (continuity) checks in the high-voltage system. Be sure to discharge the high-voltage capacitors first!

Not only the high-voltage transformer and the cathode-ray tube may be at fault here, but in addition, the horizontal deflection system may be defective. To determine where the trouble exists in these areas, first measure the voltage at the output of the high-voltage supply with a kilovoltmeter. If the voltage is zero, determine whether the horizontal deflection system is operating by checking the waveform at the input of the horizontal-output amplifier.

If the waveform and its peak-to-peak value are normal, check the waveform at the cathode or emitter. A normal condition here will indicate that the tube or transistor is functioning as it should. Concentrate now on the horizontal amplifier-output circuit, particularly the output transformer, the damper and its associated components, and also the high-voltage rectifier circuit.

The absence of any deflection waveform indicates that the trouble exists in the horizontal-deflection system. This circuit is best checked with an oscilloscope. Note the waveforms at various points, and compare them with those given by the manufacturer. Note that a distorted waveform may still cause a high voltage to be produced, although this voltage will be low. A distorted deflection waveform will visibly affect the horizontal linearity of any image appearing on the screen.

It is well to distinguish between a dark screen, which is due to no high voltage, and a screen containing a scanning raster but no image. The latter difficulty, when accompanied by a normally functioning audio system, indicates a defective video system. In this case, the horizontal deflection system and the high-voltage power supply are both operating satisfactorily, as revealed by the appearance of the scanning raster.

Defective Picture Tube The picture tube is undoubtedly the most expensive single item in the television receiver. Because of this, the owner of the set is understandably very much concerned with its condition. Fortunately, these tubes are ruggedly built and most of them can be expected to last 24 months or longer. However, defects in them do arise and it behooves the television technician to be familiar with them.

The brightness of the raster (and the picture) depends upon the number of electrons striking the fluorescent screen. This, in turn, depends upon the bias between the cathode and control grid and upon the number of electrons emitted from the cathode. If a tube is improperly constructed, or after it has had long use, it is possible that less than the normal number of electrons will be emitted from the cathode. The result will be a dim picture, even for advanced settings of the brightness control.

A tube with low emission should generally be replaced. However, it has been found that raising the filament voltage above its normal value will frequently cause enough additional electrons to be emitted to restore the brightness of the picture. It can be expected, of course, that in time the emission will again decrease below a usable value, but in the meantime useful service is being obtained from a normally unusable tube. Picture-tube brighteners—which are actually step-up transformers—are commercially available.

An incorrectly-positioned ion trap can also be responsible for reduced brightness. Owners who tamper with their sets may also bring about a reduction in brightness by moving the ion trap. Or, in moving a set about the house, the ion trap magnet may have been jarred out of position. Or, an inexperienced technician might have had difficulty arranging the deflection yoke and centering device to obtain a shadowless picture, and might have "solved" the problem by shifting the ion-trap magnet from its optimum position.

When a picture tube becomes gassy, several things can happen. It may develop negative pictures (see Fig. 24-10). Negative pictures may be due to a bad picture tube or to an excessively strong signal. If the signal is the cause, then negative pictures should be obtained only on the channel where this signal is being received. On all other channels, normal pictures should be seen. If the tube is at fault, of course, all stations will be affected.

A picture tube which is somewhat gassy may sometimes cause picture-blooming, a condition which causes the picture to expand in all directions as the brightness control is turned up. Blooming is accompanied by loss of focus and sometimes by complete loss of picture and raster. The blooming or spreading out of the picture is due to a decrease in the high voltage applied to the tube. If the tube is gassy, it may be drawing an excessive amount of current, and this drain on the high-voltage supply can cause a considerable reduction in high voltage.

Another source of picture-tube trouble is a cathode-to-heater short. In sets where the cathode is externally connected to the filament, there will be no noticeable effect. However, in many television-receiver circuits, the brightness control is placed in the cathode circuit of the picture tube (see Fig. 24-11). One side of the heater is usually grounded. Hence, when a cathode-to-heater short develops, the cathode is placed at ground potential, and no variation of the brightness control will be able to alter this.

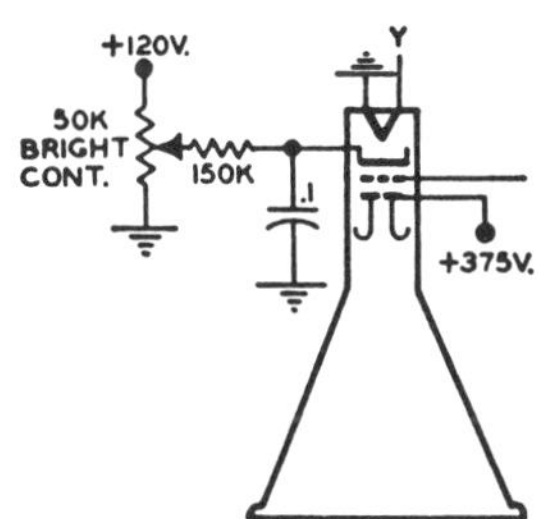

Fig. 24-11. A brightness control in the cathode circuit of a picture tube.

Picture Distortions due to Defective Picture-tube Components
Associated with the picture tube are the deflection yoke, a centering

magnet, and perhaps an ion trap and a focus coil. Improper placement of any of these components will have an adverse effect on the picture. By the same token, any defect within these components will also tend to distort the picture or even prevent it from appearing on the screen.

When a picture is not in proper focus the cause may be: a defect in the electron gun (in electrostatically focused tubes), the application of an incorrect voltage to the focus electrode (again, in electrostatically focused tubes), an improperly adjusted focus control, a defective focus coil (in electromagnetically focused tubes), or a resistive change in the focus circuit. When a focus coil is employed, it must have a certain

Fig. 24-12. Improper placement of ion trap.

Fig. 24-13. The shadow around the outer edge of the pattern is caused by a deflection yoke which is not as close to the cone of the tube as it will go. (*Courtesy Sylvania News.*)

amount of current flowing through it. Too much or too little current will produce defocusing. If the set uses a permanent-magnet focus unit, poor focus indicates improper placement of the magnet or possibly a weakened magnet. In Fig. 24-12, the ion trap is out of position. In Fig. 24-13, the deflection yoke is not as close to the cone or bulb of the picture tube as it should be. Figure 24-14 reveals distorted images

A

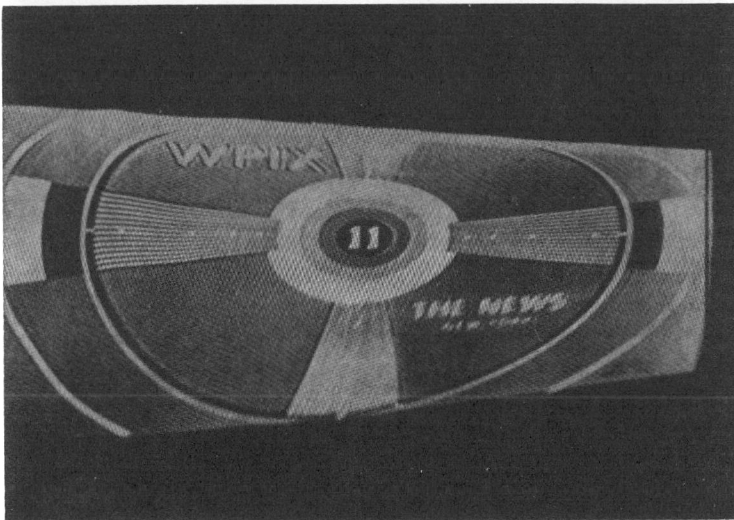

B

Fig. 24-14. Distorted images which stem from the deflection yoke. (A) Distortion as produced by a short across part of the horizontal deflection yoke windings. (B) Short in vertical section of yoke windings.

which stem from the same source—the deflection yoke. A shorted turn in the deflection coil may be the cause. Thermistors (resistors that change resistance value when their temperature is changed) are used in some yokes to compensate for changes in yoke resistance when the yoke temperature changes. If the thermistor becomes defective the symptoms shown in Figs. 24-14(A) and 24-14(B) may be the result.

Besides the two sets of windings in the deflection-yoke housing, there will be found damping resistors (across the vertical windings) and a small capacitor (across one of the horizontal windings). Should any of these components become defective, picture distortion will be produced. A typical appearance of the image in this case is shown in Fig. 24-15, in which several ripples at the left side of the screen cause the pattern to appear wrinkled.

Fig. 24-15. Ripples caused by an open or wrong value of capacitor across half of the horizontal winding of the yoke. (*Courtesy of Sylvania News.*)

24.4 SERVICING LOW-VOLTAGE POWER SUPPLIES

Since the low-voltage power supply of a television receiver is, in many respects, similar to the power supply in a radio receiver, the same types of troubles will be encountered in both. The following discussion will discuss what effects can be expected from defective components.

Tubes A rectifier tube is designed to pass a considerable amount of current. In many television receivers these tubes are pushed to the limit of their capacity. If, after being in use for some time, the emission level of the rectifier tube decreases somewhat, it will have an immediate effect on the value of voltage developed. Lowered voltages can lead to dim pictures, unstable picture hold-in, decreased volume, and a decrease in set sensitivity, to name several of the more obvious results. It is frequently difficult from a number of symptoms to point directly to the power supply and say that here lies the cause of the trouble. However, when the trouble cannot be corrected by rectifier-tube replacement, then a check should be made of the B+ voltage to determine if this possesses its normal value.

Semiconductor Rectifiers Semiconductor rectifiers are widely used in television receivers, because they will perform the same function as vacuum-tube rectifiers but with greater efficiency. They are small in size, rugged, comparatively cool in operation, and have a long, useful life. However, from time to time, these units may become defective and will require replacement.

These semiconductor rectifiers are sometimes tested by measuring their forward and reverse resistances. This is done by first removing one side of the rectifier from the circuit. The procedure is to place the ohmmeter leads across the diode and note the resistance, then reverse the ohmmeter leads and again measure the resistance. The readings will normally show greatly different values in each direction. However, what these resistance values are will depend on the ohmmeter voltage and scale. Because of this, ohmmeters should be used *only* to determine whether current can flow through the rectifier. They should never be used to gauge how well the unit will perform.

Whenever B+ voltage is below normal and no other obvious defects exist, many technicians simply substitute other rectifiers for the ones in the receiver. In performing the changeover, the following precautions should be observed:

(1) The soldering iron or solder should not be brought into contact with the semiconductor rectifier body. Also, the iron should not be applied to the soldering terminals any longer than necessary.

(2) Mount the rectifier in the same manner as its predecessor and see that it receives adequate ventilation.

Electrolytic Capacitors Electrolytic capacitors in the low-voltage power supply provide filtering. These capacitors will have to be replaced if they become defective.

When making a visual inspection of a receiver during any servicing job, the technician should check the filter capacitors for signs of corrosion, dripping, or scaling around the base of the capacitor. The capacitor may still work perfectly, but these signs may indicate that the capacitor is about to fail and should be replaced. The failure of a filter capacitor usually produces excessive hum (in the speaker), low-volume output, and heavy black bars on the face of the screen.

Effects of Shorted or Leaky Filter Capacitors. A shorted filter capacitor can cause the rectifier to burn out. It can cause the fuse to blow if the circuit is fused.

With a leaky filter capacitor, the voltage output of the power supply may be low, there is hum in the speaker, dark horizontal bars appear on the pictures, and the transformer and filter chokes may overheat.

Effects of Open-Filter Capacitors. An open filter capacitor produces most of those symptoms caused by leaky capacitors except the heating of parts. It may lower the B+ voltage. This is especially true if the capacitor in question is the input capacitor of the filter network.

Open-Filter Chokes. One visual effect of an improperly filtered power supply is shown in Fig. 24-16. The distortion in Fig. 24-16 is due to a 60-Hz ripple in the power supply voltage, reaching the horizontal-deflection system. This may occur if one half of the full-wave rectifier becomes inoperative. The curvature at the edge of the image represents one cycle of a 60-Hz sine wave laid on its side. If the ripple voltage is 120 Hz (open filter capacitor) then the number of "bends" is doubled. The visual effect of ripple in the vertical-deflection system is an alternate spreading and crowding of the image in the vertical direction.

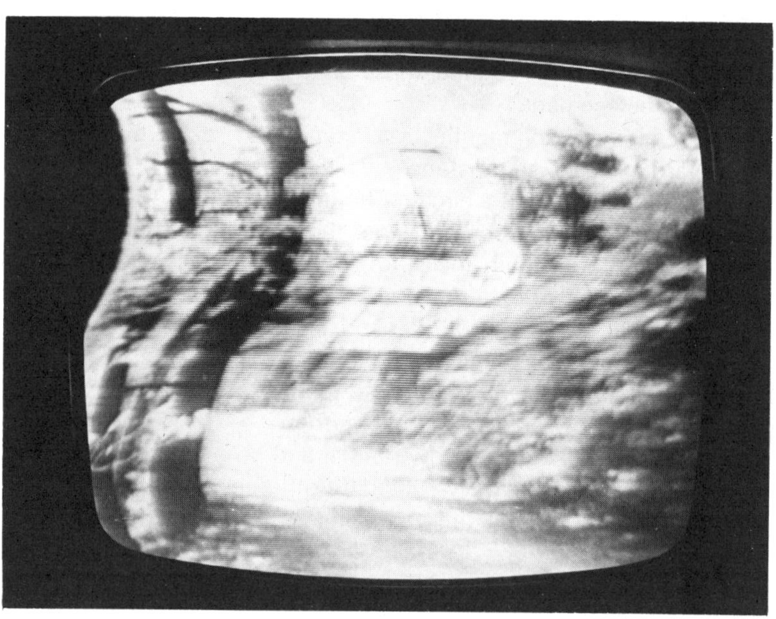

Fig. 24-16. Visual effect of poorly filtered power supply. The distortion is due to 60-Hz ripple in the voltage reaching the horizontal-deflection system.

Sound *versus* Filter Ripple Sound signals reaching the cathode-ray tube produce an effect which is similar in certain respects to ac ripple. Each defect causes black bars to appear across the screen, but those caused by the sound voltages are more numerous and their intensity changes in step with the amplitude of the applied audio (see Fig. 24-17). The black bars produced by hum in the power supply seldom exceed two, are much wider than the audio bars, do not vary in intensity with the sound, and are visible on all channels.

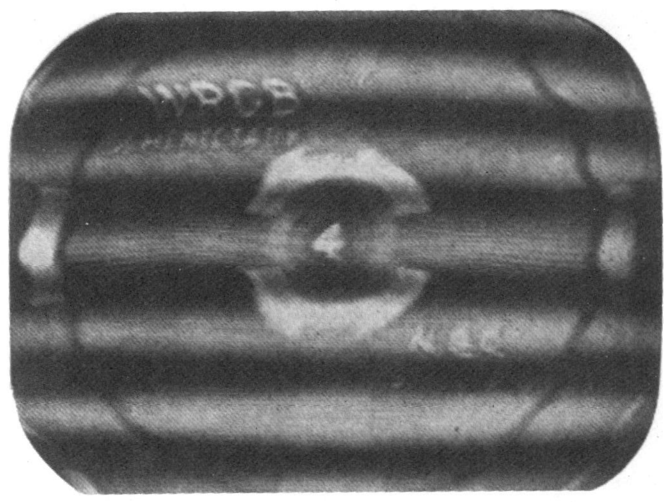

Fig. 24-17. The visual effect of sound voltages in the video system.

Sound voltages reaching the cathode-ray tube may arise from two sources: improper alignment of the trap circuits in the video-IF system or an incorrect setting of the fine-tuning control. As a first step toward removing the sound bars, adjust the fine-tuning control. In a receiver functioning normally, this adjustment should be enough to remove the sound bars from the image without distorting the sound output. If the sound bars cannot be removed or the sound becomes fuzzy when they are removed, then the alignment of the circuits should be checked.

Transformerless Receivers Television receivers which use semiconductor rectifiers do not always use a power transformer, which means that the technician must be careful not to ground the chassis of such receivers unless he is certain that the side of the power line to which the chassis is connected is also at ground potential. Unless this precaution is observed, it is quite possible that a fuse will be blown and any equipment connected to the receiver may be damaged. For this reason, isolation transformers should always be used when servicing receivers without power transformers.

24.5 SERVICING HIGH-VOLTAGE POWER SUPPLIES

A good indication of high-voltage failure is the appearance of a dark screen. If the dark screen is accompanied by normal sound, then the low-voltage power supply is probably operating normally, and we can concentrate on the high-voltage system. However, a blank screen and no audio is generally due to a defective low-voltage supply.

As a first step in checking the high-voltage supply, measure the voltage at the end of the second-anode connector (after the connector has been removed from the picture tube). If there is no high voltage, a likely source of trouble is the high-voltage rectifier. Substitute another rectifier and see if the voltage returns. Also check the damper.

In the flyback type of supply, a check should be made with an oscilloscope to determine whether the proper deflection voltage is reaching the input of the horizontal-output amplifier. If this wave does not have the proper shape and peak-to-peak amplitude, work back toward the sweep oscillator to find out why the proper wave is not being developed.

If the drive voltage is normal, a check should be made of the horizontal-output tube or transistor. The important thing is to determine whether the trouble is arising from the horizontal-output circuit or from insufficient drive on the output amplifier. This will then direct you to the proper circuit.

Within the high-voltage system, the usual causes of failure are the rectifier (tube or semiconductor diode), the damper, the filter capacitor, and the series filter resistor. If the capacitor shorts, there will be no high voltage. A common trouble developed by the series filter resistor is an increase in value. This does not remove the high voltage, but it does decrease it and produces an effect known as blooming. As the brightness control is turned up, the picture begins to lose brightness and starts expanding. It is possible for the picture to disappear altogether at some high setting of the brightness control.

A rectifier with a low output can also be responsible for blooming.

24.6 SYNC-SEPARATOR STAGES

The sync-separator stages are located between the vertical- and horizontal-sweep systems and the video system.

A defect in the sync-separator stages of a television receiver will almost always cause both vertical and horizontal sections to fall out of sync.

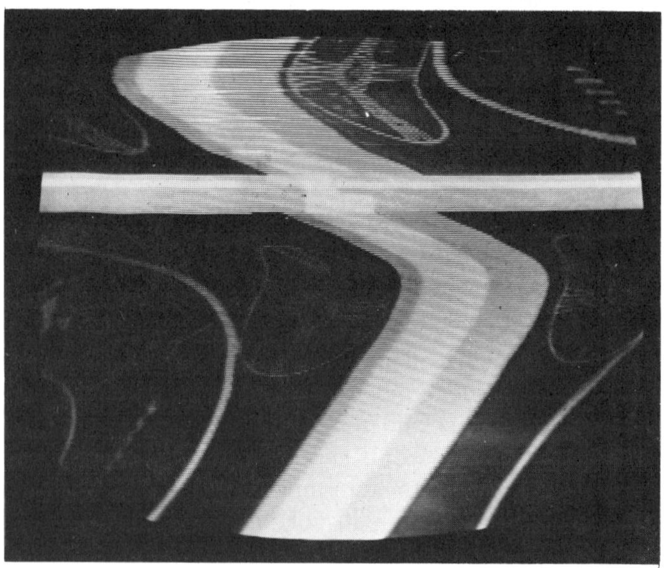

Fig. 24-18. Loss of vertical and horizontal lock-in. (*Courtesy of H. W. Sams and Co.*)

The picture on the screen will be found to roll both vertically and horizontally (see Fig. 24-18). While it may be possible to lock the picture in momentarily by adjusting the vertical- and horizontal-hold controls, the picture will soon fall out of sync again.

The best way to check through a sync circuit in search of a defect is by tuning in a signal and then using an oscilloscope to check pulse waveforms at the input and output of each of the sync stages. The point at which to start is the input to the sync-separator section. The waveform here will generally be the composite video signal (see Fig. 24-19(A)). At the output of the sync separator the vertical and horizontal waveforms will appear as shown in Fig. 24-19(B). (The oscilloscope sweeping rate used to observe the vertical-sync pulses is 30 Hertz; the horizontal sync pulses, 7,875 Hertz.) Note how much the video signal has been suppressed, or even eliminated, while the sync pulses are being amplified.

Horizontal Vertical

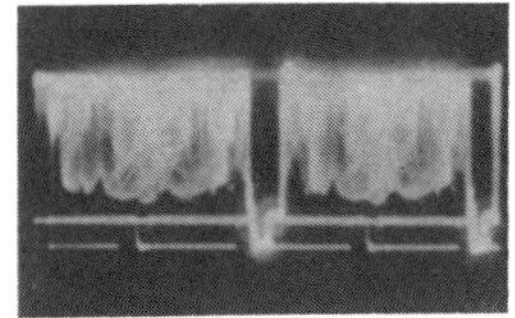

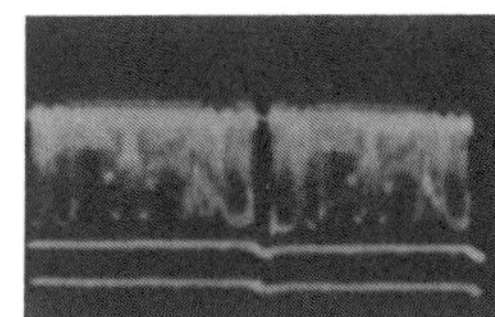

A- At Input To Sync-Separator System.

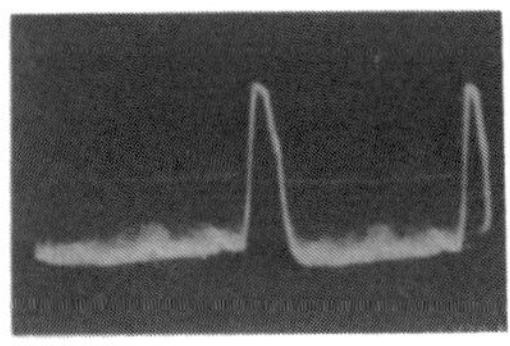

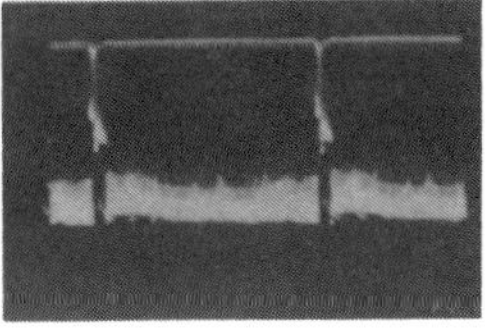

B- At Output Of First Sync Separator.

Fig. 24-19. An illustration of how the sync-separator stages separate the sync pulses from the video signal. (A) At input to sync separator system. (B) Output of sync-separator .

When observing sync pulses on an oscilloscope screen, the technician will find that the horizontal pulses stand out clearer and more distinctly than the vertical pulses. One reason for this is that the horizontal pulse is simpler in structure than the serrated vertical pulse. Also, the horizontal-sync pulse occupies a greater proportion of a line than the vertical sync pulse does of a field. Hence there is more of the pulse to be observed when the scope-scanning rate is set to the proper value. These facts are borne out by the oscillograms of vertical- and horizontal-sync pulses shown in Fig. 24-19.

It is important to observe carefully the composite video signal which is applied to the input of the sync-separator section. If, for example, it is

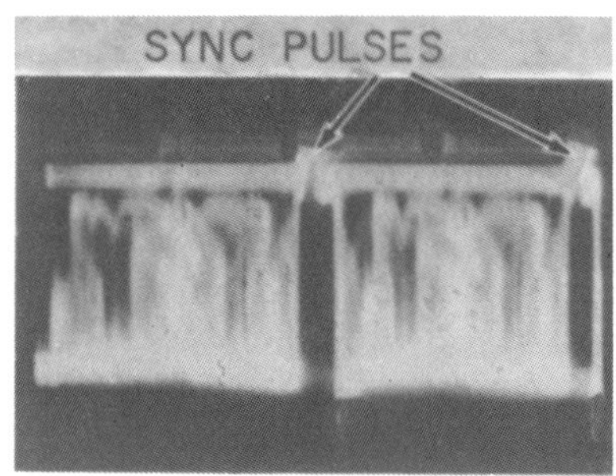

Fig. 24-20. A video signal in which the sync pulses have been partly compressed. Note that the video signal extends up to the level of the pulses.

found that the sync pulses have been compressed (Fig. 24-20), and it is difficult to keep the picture in sync, then it can be surmised that it is not the sync-separator stages that are at fault, but the preceding video system. It is quite possible that the defect exists in the AGC network, wherein the controlled RF and IF amplifiers are being permitted to operate at higher than normal gain because of incorrect bias. This can readily lead to overloading with subsequent sync-pulse compression. When this signal reaches the sync separators, there is a considerable amount of video signal operating at nearly the same level as the sync pulses, and consequently it becomes impossible to effect a clear-cut separation. In the vertical system, this will show up as unstable or critical hold in. The picture will have a tendency to roll. In the horizontal system, the automatic frequency-control network may hold the picture more firmly in place. However, the hold-in range will undoubtedly be smaller than normal. Also, a bend may appear at the top of the picture. Vertical lines or objects in the picture will be found to curve to the right or to the left. In extreme cases, the top of the picture will "flag-wave," that is, will move rapidly from side to side.

If it is found that sync pulses are being compressed or clipped entirely before they reach the sync-separator stages, then the path of the video signal should be traced back to the video-second detector with the oscilloscope. Sync-limiting or compression may occur at almost any point from the RF amplifier to the sync takeoff point. A defective tube or transistor, a defective AGC system, a signal which is too strong, incorrect operating voltages or component values can all be responsible. Examine the signal at the video-second detector. If it looks normal, then the preceding RF and IF stages are operating normally. If the sync pulses are absent or have been compressed, the trouble exists prior to the video-second detector. Check the voltages at the video-IF tubes or transistors first. Pay particular attention to improper AGC voltages since these are a frequent source of trouble. Defective tubes or transistors may be the source of trouble. The troubleshooting procedure should now follow along the lines previously outlined for video-IF amplifiers.

When poor sync action has been traced to the sync-separator stages, the oscilloscope is employed first to localize the defective stage. Thereafter voltage and resistance checks will be required to isolate the component.

24.7 VERTICAL-DEFLECTION SYSTEM

Difficulties in the vertical-deflection system are perhaps the easiest to analyze because the signal voltages developed here deal only with the vertical sections of the image. There are no high-voltage power supplies associated with this system, such as we find in the horizontal-deflection system. When only the vertical-deflection system is affected, analysis of the source of the defect is generally simple and straightforward.

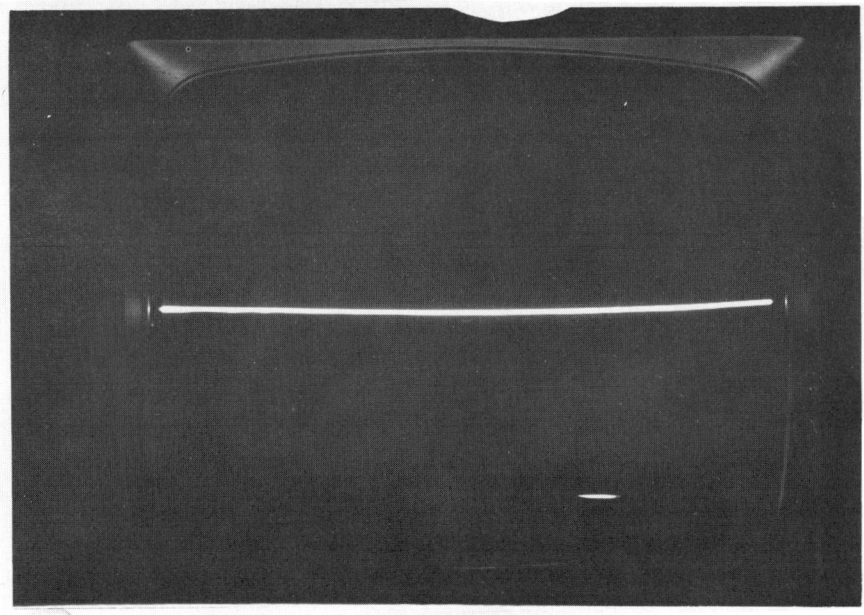

Fig. 24-21. Complete loss of vertical deflection.

(1) The most positive indication of complete failure of the vertical-deflection system is the appearance of a narrow, horizontal line on the cathode-ray-tube screen (see Fig. 24-21). The first components to check are the vertical oscillator and amplifier tubes or transistors. If these are working, check the vertical-sweep waveforms with an oscilloscope, starting at the vertical-oscillator stage and progressing forward to the vertical-output amplifier. Once the defective stage is located, voltage and resistance checks should reveal the defective component. Keep in mind, when making the foregoing tests, that an open in the windings of the vertical-deflection yoke coils or vertical-output transformer is also a possibility.

(2) The picture may fail to lock-in vertically (see Fig. 24-22). Ordinarily, if the vertical-hold control is rotated, a point will be found where the frequency of the oscillator is brought close enough to the incoming vertical-synchronizing pulses to permit lock-in. The picture then becomes stable. However, if the oscillator is not functioning properly, or the vertical-synchronizing pulses are not reaching the oscillator, then lock-in will not occur. When the vertical-hold control proves ineffective, make the following tests:

a. Check the waveform at the input to the vertical-sweep oscillator. Determine whether the pulses reaching the vertical sweep oscillator are sufficiently strong to maintain control. It is quite possible that the path from the sync-pulse separator to the vertical oscillator does not transmit the full vertical pulse. Defective coupling capacitors, open resistors, or components which have changed value appreciably may be the causes.

Fig. 24-22. The effect of an incorrect setting of a vertical-hold control. The vertical sync and its equalizing and blanking pulses create the pattern between the portions of the picture.

Resistors can be checked by measurement and capacitors by substitution. Check, too, the amplitude of the complete video signal at the point where the sync-pulse separation occurs. The image on the screen will also help determine whether sufficient signal strength is present.

b. If the foregoing test produces normal results, check the waveform at the output of the vertical-sweep oscillator. Note whether variation of the vertical-hold control has any effect on the frequency of the observed waveform. A lack of such frequency variation may indicate an open resistor in the circuit containing the hold control. Check the vertical oscillator resistors and capacitors to find the defective component.

c. Sometimes poor vertical lock-in comes from a completely unsuspected source. In one set brought into the shop, the picture could not be held vertically. All tubes and suspected parts were checked and found to be all right. However, when the technician started repositioning some of the wires near the vertical-sweep oscillator, it was found that a lead from the vertical system had moved near the filament leads of another tube. The 60-Hertz field radiated by the filament wires induced sufficient voltage in the vertical sweep-oscillator circuit to trigger this circuit prematurely, resulting in lock-in instability. When the offending filament leads were dressed away from the vertical-sweep oscillator, the set returned to normal operation.

(3) The picture may "bounce" when the set is jarred. In this case check all the tubes and connections in the vertical-synchronizing system for microphonics. Merely by tapping each tube, while it is in the

set may be inconclusive. The best test is to replace each tube, in turn, with one known to be good.

Incidentally, too strong a signal or an incorrect setting of the contrast control may produce a jumpy picture. When this condition is present, the tubes or transistors are driven to saturation, thus compressing or otherwise distorting the synchronizing-pulse waveforms. The result is poor sync control of the sweep oscillators. Since the horizontal-sweep system employs some form of automatic control, noise disturbances or a weak signal may affect the vertical-sweep system to a greater degree than the horizontal-sweep system.

(4) The height of the picture may be insufficient. In a receiver functioning normally, adjustment of the height control will produce the proper picture height. Inability to obtain this result may be caused by one or more of the following conditions:

a. Weak vertical-output-amplifier tube or transistor.

b. Incorrect voltages on vertical-oscillator and vertical-amplifier tubes or transistors.

c. Low-line voltage.

d. Improper placement of the deflection yoke.

Items c and d will affect the width of the image, too. If the width is normal, these items can be disregarded.

(5) The picture may be compressed at the top. This is an indication of poor linearity of the vertical-deflection voltage. The vertical-trapezoidal deflection wave is developed in the output circuit of the vertical-sweep oscillator, amplified by the vertical-output amplifier, and then applied to the vertical-deflection coils. If this wave does not have the proper form, if parts of it curve or bend more than they should, then the electron beam in the cathode-ray tube will not travel down at an even rate. The visual result will be a vertical bunching of lines in some sections of the image and the spreading apart of them in others. This is known as "poor linearity." When the image displays this type of distortion, the trouble is in the vertical sweep and amplifier circuits. This assumes, of course, that the vertical linearity control has been properly adjusted.

A precise vertical linearity check can be made with an audio oscillator. Place the receiver in operation but have it tuned so that no signal is being received. This will leave the screen with a blank raster. Now connect the audio oscillator across the load resistor of the video-second detector and set the frequency to 660 Hertz. As the output of the generator is turned up, a series of alternate black-and-white stripes will appear (see Fig. 24-23). If the bars are evenly spaced, we know that the scanning of the screen is linear. On the other hand, a nonlinear scanning rate will cause the bars to bunch together at some points.

The figure 660 was suggested because it is an integral multiple of the 60-Hertz vertical scanning rate. Almost any multiple of 60 may be used, although it is desirable to have at least ten or more bars on the screen.

Fig. 24-23. Black-and-white horizontal bars for checking vertical linearity.

The audio signal was injected at the video-second detector, because this same signal will also lock-in the vertical-sweep system and produce a steady pattern. The sync signals are almost always taken from some point beyond the second detector, and in this way part of the injected signal will reach the vertical-sweep oscillator and lock it in.

(6) Inability to center the image vertically.

a. *Electromagnetic-Focus Picture Tubes.* Test for an open resistor in the centering network if such a circuit is employed. If the position of the focus coil or magnet is the sole method of centering in the receiver, check this. Where a centering potentiometer is employed, measure the voltage across this control. If it is low, measure the voltage of the low-voltage power supply. A low voltage at either point will produce a restricted centering range. Check the position of the ion trap if there is one. Finally, if all these tests do not reveal the trouble, check the cathode-ray tube itself. The electron gun may become tilted because of a sudden jarring of the tube, although this is quite rare.

b. *Electrostatic-focus picture tubes.* In these tubes, the component that governs the picture position is a magnetic centering device, the arms of which must be properly oriented with respect to each other. Check the manufacturer's service manual for the proper procedure to follow. Also examine the position of the ion trap if there is one.

(7) Vertical foldover in a picture might be considered as an aggravated case of nonlinearity. In form, the picture will appear somewhat as shown in Fig. 24-24. The bright horizontal stripe across the bottom of the image represents the point where the scanning beam stopped moving downward. This condition generally arises from some defect between the output circuit of the vertical oscillator and the output stage. Thus, a leaky coupling capacitor between the oscillator and the output amplifier

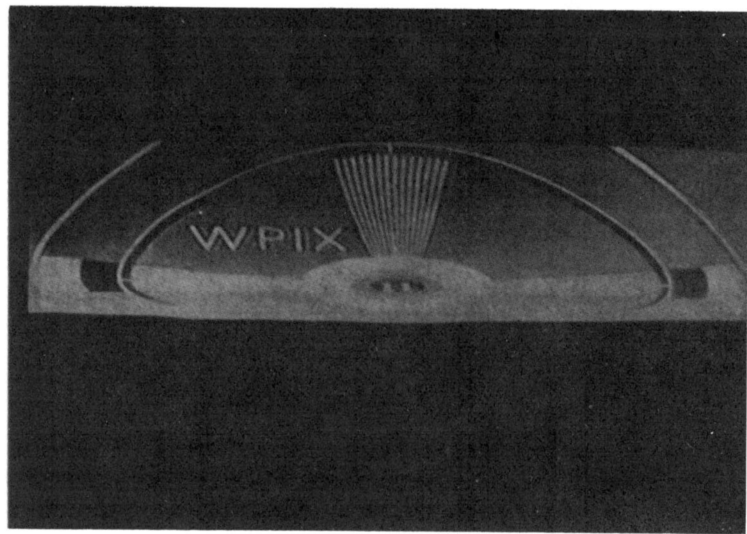

Fig. 24-24. An illustration of vertical foldover. (*Courtesy of Sylvania Electric Products, Inc.*)

or a substantial decrease in the value of the bias resistance may often be responsible. It might also be wise to check the filter capacitor in the cathode leg of the output amplifier. In tube receivers, a gassy output tube is also a distinct possibility and a new tube should be tried.

Sometimes a condition will be obtained in which there is a bright horizontal line across some intermediate point in the picture. This is usually due to heater-to-cathode leakage in the vertical-output tube.

(8) Shorted turns in the vertical-output transformer will result in a loss of picture height. It may sometimes give the same indication that insufficient driving voltages give to the vertical-output amplifier. This type of trouble is usually difficult to detect because it reveals itself neither to normal voltage or to resistance checks. If the shape and the peak-to-peak amplitude of the deflection wave are correct at the input to the vertical-output amplifier and if the amplifier appears to be operating normally, then the output transformer is a logical suspect.

Measuring the peak-to-peak value of a wave can be a very efficient method of servicing, especially in the sweep sections of a television receiver. Make it a habit to measure this value at the input of the output amplifier. Then compare the value with that recommended by the manufacturer.

Fig. 24-25. Vertical keystoning. (*Courtesy of Sylvania Electric Products, Inc.*)

Shorted turns in the vertical-deflection coils will produce the trapezoidal or keystone-shaped pattern shown in Fig. 24-25. Which side is wider depends upon which section of the coils contains the short. This condition is called a *keystone raster*.

24.8 HORIZONTAL-DEFLECTION SYSTEM

Distinctive waveforms are produced at each point in the vertical- and horizontal-deflection systems, and the reception of a signal does not appreciably alter the shape of these waves.

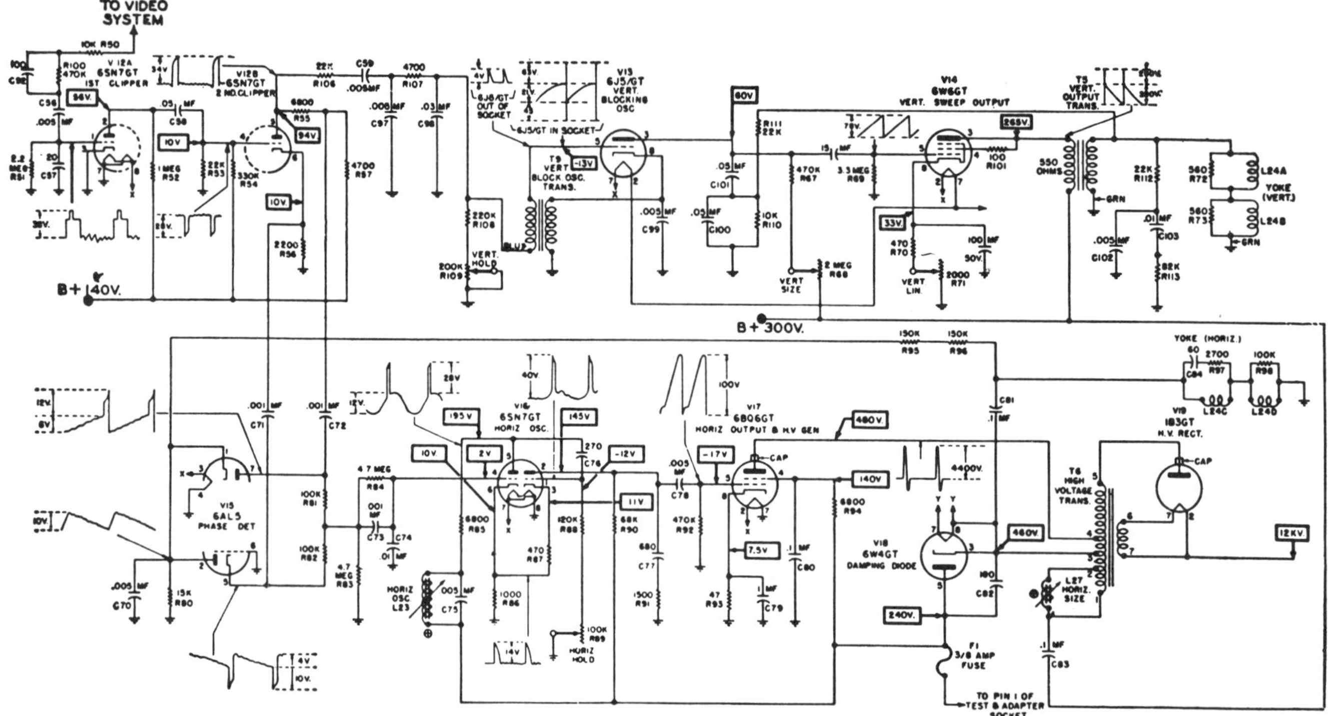

Fig. 24-26. The sync separator and the vertical and horizontal-sweep systems of a television receiver, together with the normal waveforms found in these circuits. (*Courtesy of Motorola.*)

The best troubleshooting method is to compare the wave shapes of the voltages in the defective receiver with the corresponding waveforms given by the manufacturer in his service manual. For example, refer to the waveforms shown in Fig. 24-26 (see also Chapter 23 for waveforms of an all solid-state receiver). These are the proper waveforms present in the vertical and horizontal circuit of this receiver when everything is operating normally. The peak-to-peak voltage values are also important in waveform checking, and these should be carefully noted. When the wave amplitudes are found to be appreciably smaller than recommended by the manufacturer, tubes and the B+ power supply voltages should be checked. On the other hand, distorted or improperly shaped waves usually indicate defective capacitors and/or resistors in the circuits. Transistor sweep circuits have waveforms similar to the ones shown in Figure 24-26 (see Chapter 23 for these).

When trouble is traced to the deflection systems, check the tubes or transistors first. If these test all right, then the next job is the waveform check. Take an oscilloscope and connect the grounded vertical-input terminal to the receiver chassis. Connect a test prod to the other vertical-input terminal. Then, starting at the output of the synchronizing oscillator, check the input and output waveforms of each amplifier, working toward the deflection coils of the cathode-ray tube. At the point where the waves disappear or are not in their proper form, voltage and component checks should be made to determine the reason for the change or disappearance of the waves. It is desirable to use a ruled plastic mask for the oscilloscope screen and note approximately the peak-to-peak voltage values of each of the waveforms checked. Variations of from 10 to 15 percent from the recommended values can be accepted since the adjustment of various controls can readily affect the wave amplitude by this amount.

Incorrect Horizontal Frequency Incorrect horizontal frequency is indicated when the picture assumes the distorted appearance of Fig. 24-27. In the figure, the image appears to slip in a horizontal direction.

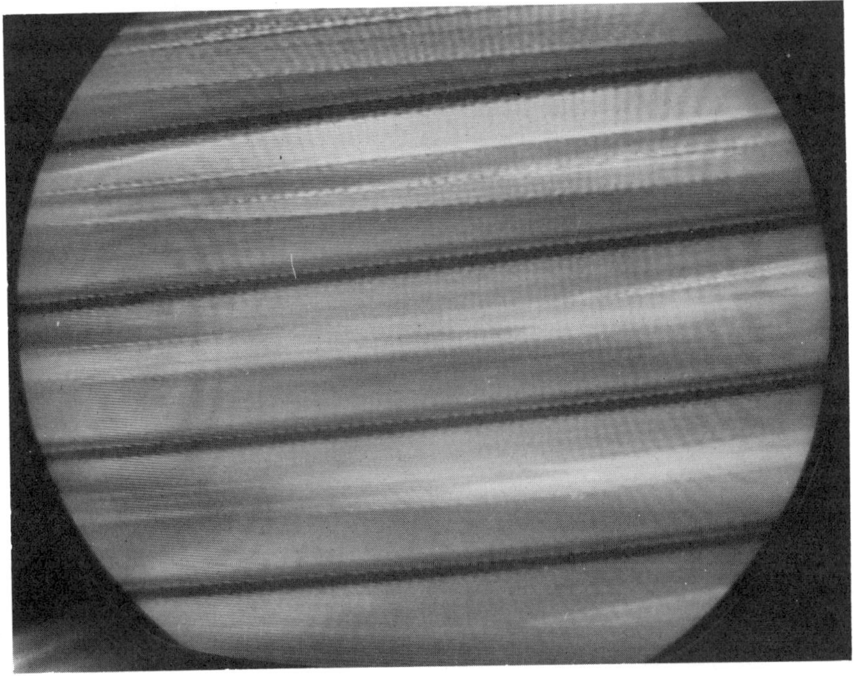

Fig. 24-27. Incorrect horizontal frequency (locked-in).

If the slippage is great enough, it results in several images overlapping, and no detail can be clearly distinguished. Slow rotation of the horizontal hold control may bring the oscillator back ot the correct operating frequency at which one stationary image is again visible. Failure of the hold control to correct the oscillator frequency generally indicates a faulty component in this oscillator or in its automatic-frequency control network.

Other Horizontal Defects A fairly common type of picture defect, and one which gives many technicians a considerable amount of trouble, is horizontal nonlinearity. The picture may be compressed or elongated at the left- or right-hand sides (see Fig. 24-28). There may be a series of ripples across a part of the screen (see Fig. 24-15), or the left side of the picture may possess one or more closely spaced dark lines (see Fig. 24-29).

Fig. 24-28. An image possessing horizontal nonlinearity. (*Courtesy of RCA.*)

Fig. 24-29. The visual effect of Barkhausen oscillation.

To track down the part of the horizontal circuit most likely to contain the seat of the trouble, the technician must recall something that was learned previously: the first part of the horizontal beam travel is controlled by the damper (and its circuit), while the remaining part of beam travel across the screen is controlled by the horizontal-output amplifier.

From this it can be assumed that if the picture is impaired on the left-hand side, the defect is most likely to exist in the damper circuit. If it is the right-hand side of the picture which is distorted, the most likely place to look for the defect is in that portion of the horizontal-sweep system extending from the horizontal oscillator up to and including the horizontal-output stage.

Thus, from the foregoing reasoning, horizontal foldover caused by a defective damper or vertical light and dark stripes all quite naturally fall at the left-hand side of the picture because they arise from damper circuit defects. Also at the left side, may be found the dark stripes of Barkhausen oscillations. These are spurious oscillations set up within the horizontal-output amplifier tube after the tube has been cut off. (This problem does not occur with television receivers having transistorized-sweep circuits.) Barkhausen oscillations may be cured by replacing the horizontal-output tube or by placing a magnet near the top of the tube. Foldover or compression at the right-hand side of the picture may be produced when there is excessive drive to the horizontal-output stage.

The most frequent defects in the damper stage include the damper diode, any capacitors in the damper circuit, and the linearity coil.

In the horizontal output stage, check the following:

(1) The input signal to the output stage (shape and amplitude).
(2) Operating voltages.
(3) Leaky coupling capacitor at the stage input.
(4) A bad horizontal output tube or transistor.
(5) Incorrect bias.

In most instances the general location of the defect, as revealed by the section of the picture which is affected, is fairly well-defined. However, because a B+ boost voltage developed by the damper stage is fed back to the horizontal-output section (and sometimes to other sections in the horizontal system as well), a certain amount of interaction between the various sections is bound to occur. For example, changes in the B+ boost filter-network components may have an effect on both sides of the picture, although the left-hand side will be the section principally affected. Or, insufficient driving voltage applied to the output stage will have its greatest effect on the right-hand side of the picture. But, since the B+ boost voltage developed depends upon the drive voltage, the left-hand side of the picture will suffer, too. Thus, while the circuits directly associated with each part of the picture will have their greatest effect on that section, the close relationship between all circuits will produce disturbances in other sections of the picture as well.

There is still another feature of horizontal-sweep system operation that is worth noting and this is the peak-to-peak amplitude of the driving voltage fed to the grid of the output-amplifier tube. The width of the picture is governed by this voltage as well as the high-voltage for the picture tube and the B+ boost in the damper circuit. It is an important check point in the horizontal-sweep system and measuring its value early in the servicing analysis is strongly recommended.

24.9 SERVICING AFC CIRCUITS

The purpose of the AFC circuit in the horizontal-sweep system is to keep the horizontal oscillator locked-in with the incoming sync pulses. When this system is operating properly, random noises and momentary disturbances will have no visible effect on the horizontal-sync stability;

the picture will remain securely in place. However, when the system is not operating normally, it may slip out of synchronization occasionally, or the setting of the horizontal-hold control will be critical, or the picture may refuse to be locked-in at all.

Failure of the AFC system to maintain the picture in synchronization may arise from a defect in the control circuit or from a defect in some prior circuit. If the latter is true, then the sync pulses reaching the AFC control stage are distorted in some manner or they may even be missing altogether.

If it is determined that the proper pulses are reaching the AFC circuit, then any reason for sync instability must be caused by faulty operation of the control circuit itself. To localize the source of the trouble, several methods of approach are open to the technician. As a start, he can measure the voltages within the control circuit. This lends itself quite readily to the location of fairly obvious defects caused by resistances, which are either open or have changed radically in value, and leaky capacitors. Tubes are not mentioned because it is assumed that these will receive first attention whenever work is done on *any* tube circuit.

Another approach to the service problem is by waveform checking within the AFC circuit. To check the waveform effectively, the technician should check the correct waveforms as given in the manufacturer's literature. Slight variations in the circuit will frequently cause some modifications in the wave shapes developed therein.

Another item to check in your examination of an AFC system is whether the dc control voltage is being fed to the AFC controlled stage. Connect a VTVM between the dc control path and chassis. Then, with a signal coming in, slowly rotate the horizontal-hold control from end to end. In a properly operating system, the needle on the VTVM should swing back and forth in step with the hold-control variation. This indicates that the control network is developing corrective voltages to offset the changes in the horizontal-sweep frequencies produced by rotating the hold control. Failure to observe these voltage swings on the VTVM indicates that either no dc control voltage is being developed or that the amount developed is not teaching the controlled stage.

Changes in component values are frequently the cause of instability in AFC systems. Be especially mindful of this fact when making voltage and resistance checks.

24.10 GUIDEPOINTS FOR TROUBLESHOOTING COLOR RECEIVERS

Difficulties in obtaining the proper color display on the screen of a color-television receiver may be caused by one or more of the following: *improper adjustment of applicable receiver controls, improper color system alignment*, and *defective color circuits.*

Improperly-adjusted Controls In early color television receivers one of the most frequent causes of poor color display was in the improper adjustment of the receiver's customer-operated controls. The manufacturers have reduced this problem considerably by the use of closed-loop feedback systems which automatically adjust the gain, phases, and frequency of various critical sections in the receiver, and thus assure the proper color display.

The fine-tuning control of the receiver must be properly adjusted in order for the tuner and IF sections to be able to pass the color signals. A misadjusted fine-tuning control can result in the complete loss of color display. Also, improper adjustment of the fine tuning can result in a color subcarrier beat being displayed on the screen. This latter effect is shown in Fig. 24-30.

The brightness and contrast control settings are important for getting the right ratio of light to dark areas. If the contrast control setting is too low, the picture will look washed out. Most viewers prefer to adjust the brightness and contrast controls with a monochrome picture (with the color-amplitude control turned off) so they can get the proper ratio between the light and dark areas.

If the color-amplitude control of the receiver is improperly adjusted, the colors will be either too weak as shown in Fig. 24-31, or too saturated (strong) as shown in Fig. 24-32. The proper setting of the color amplitude control can usually be best made after the hue (or tint) control is adjusted with a fairly low color-amplitude control setting.

An important preliminary test can be made with the color amplitude control to determine if there is trouble in the color-bandpass amplifier section. Rotate the color knob from its OFF position (producing a monochrome picture) to its full ON position (producing highly saturated colors). There should be no appreciable change in the picture tint when the color control is adjusted throughout its range. A change in tint is easiest to recognize by watching the flesh tones. A change in tint may cause the flesh tones to shift from their normal color toward yellow or toward red. A shift in tint with color control adjustment may indicate that the color-bandpass section should be aligned, or that a component in that section is defective.

The hue control varies the phase relationship in the color signal and therefore determines the correct coloring of the picture. Figure 24-33 shows an example of the picture with the improper hue control setting. The setting has turned the picture toward the green vector and thus gives a green shade to the overall display. Remember that the hue control is properly adjusted when the flesh tones are correct.

The adjustments of the horizontal and vertical hold controls of a color receiver are the same as for those in a monochrome receiver. Since misadjustment of the color-receiver controls can result in an unsatisfactory picture, the first step in trouble-shooting a color-television receiver should be to make the proper control adjustments, including the monochrome controls.

Improper Color Setups The term *setup* in color television refers to the process of making adjustments for purity, convergence, and gray scale. (The procedure for making the color setup was described in Chapter 23.)

If a receiver has poor purity; that is, if a purity adjustment is needed, the symptom will be a large colored area on the screen which does not change from picture to picture. This symptom is also present when degaussing of the picture tube and mounting hardware is needed. It may also indicate that the automatic degaussing circuitry is not working (see Fig. 24-34).

Color fringes around the objects displayed on the screen generally indicate that the picture tube is out of convergence. Improper dynamic convergence is indicated by fringes near the edges of the screen.

If it is not possible to obtain the proper black-and-white shades in a monochrome picture, the problem is likely to be improper gray scale (color temperature) adjustment. Remember that the color section is inoperative during a monochrome picture, and improper gray scale is not a problem that can be traced to the color section.

24.11 DEFECTIVE MONOCHROME CIRCUITS

If a receiver will not properly reproduce a color picture, the problem is not necessarily in the color section. In referring to the block diagram of Fig. 23-1, you will see that the color (and monochrome) signals must pass through the tuner, IF stages, video detector, and first video amplifier stages before they actually reach the color section. Proper operation of some of the stages in the color section is also dependent upon a blanking signal coming from the blanker. This blanker receives its signal from the horizontal-sweep section of the receiver. The color portion of the color picture is superimposed upon the monochrome picture which passes through the delay line in the video amplifiers. Thus, when analyzing the trouble in a color-TV receiver, it is important to understand that the difficulty may lie in a monochrome circuit rather than in the color section. The more common problems in monochrome circuits which may affect the color portion of the picture are discussed by the following.

Colored Snow Pattern (Confetti) Trouble in the tuner can produce snow in a color picture. If the color killer adjustment is normal, the snow will pass through the color section and the display will be a colored confetti on the screen. Thus, color confetti (shown in Fig. 24-35) is not necessarily a problem in the color section, but rather, it is due to a loss of amplification in the tuner RF section. If the RF gain is normal in the tuner and confetti occurs on the screen during reception of a monochrome picture, the trouble can be presumed to be in the color-killer section.

Normal Sound and Color Picture· no Monochrome Picture While the color elements are presumed to be normal in this condition, you must note that there can be no fine detail present in the picture, since this is ordinarily supplied by the *Y* signal. From the above symptoms, it should be obvious that the trouble exists in circuits which follow the sound and chroma takeoff points. Since there are different schemes for handling the *Y* signal in various receivers, the receiver schematic must be checked to see what circuits, following the sound and chroma takeoff points, exist in your particular receiver. In many receivers, after the takeoff points, the *Y* signal passes through additional video amplifier stages, including a delay line, and is applied to the three cathodes of the color tube. Thus, these stages and also the delay line would be suspect.

Hum Bars in Monochrome and Color Pictures This effect on a color-bar picture is illustrated in Fig. 24-36. The picture shows the effect of 60-Hz voltages which have been introduced into the sync and video circuits. Note the single cycle of bend in the vertical bars caused by hum in the horizontal sync. Also note the single horizontal color bar caused by hum in the video circuits. If the hum frequency was 120 Hz, the frequency of these effects would be doubled.

In a vacuum-tube, or hybrid receiver, 60-Hz hum effects may be caused by heater-to-cathode leakage in a tube. This fault cannot cause 120-Hz effects, since the filaments are heated from a 60-Hz source. In transistor receivers, this cause for a hum problem, of course, is absent. However, in any type of receiver, hum problems can be caused by power supply defects. A failure of one of the rectifiers in a full-wave rectifier can produce 60-Hz hum, while a defective power-supply filter capacitor, may produce 120-Hz hum effects.

Raster only; no Sound and no Monochrome or Color Picture Present The presence of a raster indicates the operation of the power supply and of the horizontal- and vertical-deflection systems. The lack of snow generally indicates that the problem follows the mixer stage of the tuner and is therefore most likely in the IF or AGC sections. Since there is no sound, as well no picture, it follows that the problem exists prior to the sound takeoff point (or at it). In color sets, sound is tapped off at the last common IF stage, but always prior to the video detector. Thus, the usual source of this trouble would be found in either the IF or AGC sections. However, a completely defective mixer stage in the tuner might also be a cause for this problem and should not be overlooked.

Best Color and Monochrome Pictures do not Track The meaning of this trouble symptom is that the tuner fine-tuning control must be adjusted to different settings to obtain the best color and monochrome pictures. The most likely cause of this trouble is a defective or misaligned

IF system, including the mixer stage of the tuner. A good way to begin the troubleshooting for this problem is to check the frequency response of the tuner and IF section, as described in Chapter 23. Any appreciable variation from normal alignment must be corrected before proceeding further. It is important to remember that IF response may be severely affected by a regenerating IF stage. A good test for the presence of regeneration is to sweep the tuner and IF stages and observe the overall response curve. While doing this, change the AGC voltage (with an external bias supply). If the response curve changes appreciably when varying the AGC bias, regeneration is present. Some causes of a regenerating IF stage are: defective neutralizing circuit, open by-pass capacitor, and incorrect tuning of resonant circuits.

Ringing in Monochrome and Color Picture Ringing, or close-spaced ghosts, are illustrated in the color-bar picture of Fig. 24-37. Note the ringing effect at the trailing edge of each color bar. This effect is easily distinguished from ghosts caused by multipath reception. This type of ringing is most easily localized by means of a sweep generator. The response of the IF and video amplifiers is checked separately with regard to their response curves. If the ringing is caused by a defect in the receiver circuitry, it usually causes serious distortion of the response curve. Some defects which may cause ringing are: regeneration of an IF stage, excessive high-frequency compensation in a video-amplifier stage, and a broken lead in the antenna transmission line or tuner input.

Color Picture Displaced Horizontally from Monochrome Picture This symptom can occur only if there is a phase difference between the color and monochrome pictures. One obvious cause of this trouble is the delay line. As discussed in Chapter 23 and elsewhere in this book, the function of the delay line (in the video (*Y*) amplifier section) is to delay the *Y* signal about 1 microsecond so that it will coincide with the chroma signals, in phasing. A defective delay line will cause the monochrome picture to either lead or lag the color picture by a small amount, depending upon the type of defect.

If the alignment is improper in either the monochrome or chroma sections, signal phase changes may occur which can cause delays or advances of the color picture with respect to the monochrome picture. If this problem is suspected, check the alignment of the monochrome and color sections, as described in Chapter 23. Remember that the alignment of any tuned circuit stages will be seriously affected by the presence of regeneration. This problem may also be caused by a defective transmission line, an incorrect impedance match between the transmission line and the receiver, or a broken lead in the transmission line.

24.12 DEFECTIVE COLOR CIRCUITS

In connection with the following discussions, refer both to the block diagram of Fig. 23-1 and to the schematic diagram of Fig. 23-2.

No Color Picture: Monochrome Picture and Sound Normal This symptom indicates that the problem exists only in the chroma circuits, since a normal monochrome picture and normal sound is present. The following are the circuits to check:

(1) *Color Killer.* The threshold control may be set incorrectly, or the circuit may be defective. The color killer (Q_{612}) will hold the Chroma Output (Q_{614}) cut off in the absence of a color picture (color burst signal), in normal operation. When a color picture is present, it should permit Q_{614} to operate normally. Note that the operation of the color killer depends upon the correct functioning of Q_{610}, the ACC amplifier, which in turn is operated by the output of the killer detector (Q_{612}). The killer detector operation requires correct signal inputs from the burst amplifier (Q_{636}) and the 3.58 MHz amplifier (Q_{644}). Thus, if there is a suspected problem of killer malfunction, all of the mentioned circuits may have to be checked for proper operation.

(2) *3.58 MHz, Subcarrier Oscillator.* If the subcarrier oscillator signal does not reach the X and Z demodulators, color video cannot be produced from the transmitted chroma signals. In this case, a color broadcast will be reproduced only in monochrome. In the block diagram of Fig. 23-1, the subcarrier oscillator proper is shown as Q_{642}. However, note that this is followed by Q_{644}, the 3.58 MHz amplifier. Defects in either of these two stages will prevent the subcarrier from being applied to the two demodulators. The trouble may be easily localized by checking with a wide-band scope. Voltage measurements may also be useful, as well as resistance measurements in the suspected circuits.

(3) *Chroma-amplifier Section.* In Fig. 23-1, note that the chroma signal is picked off from the first video amplifier (Q_{208}), as a composite video signal, and is applied to the first chroma amplifier (Q_{606}). Because of the restricted bandpass characteristics (about 3 MHz to 4 MHz) of this amplifier and the ones following, only the chroma-sideband signals are passed through Q_{606}. From this point, these signals are passed through the second chroma amplifier, (Q_{608}), the tint amplifiers (Q_{602}, Q_{600}), the chroma control (Q_{604}) and the chroma output amplifier (Q_{614}). From this last chroma amplifier, the chroma signals are applied to the X and Z demodulators and to the $G-Y$ matrix amplifier. Since the chroma signals pass through all of the mentioned chroma and tint amplifiers before being applied to the demodulators, it is obvious that a major defect in any of these amplifiers would prevent the passage of the chroma signal and would result in the reproduction of only a monochrome picture from a color transmission.

In localizing problems in the mentioned circuits, the best approach would be through the use of a color-bar generator and oscilloscope. With these instruments, it would be a relatively simple matter to check the progress of the signal from stage to stage. When the defective stage has been located, then checks of the transistors (or tubes, where applicable), as well as voltage and resistance checks will make it possible to locate the defective parts.

Note: In the preceding discussions we have been assuming a complete loss of color under all signal conditions. However, it is possible that color will appear only in the presence of a very strong received signal. This indicates only a partial malfunction of the circuits previously described in numbers 1, 2 and 3. The best approach to localizing this type of problem is also by means of a color-bar generator and oscilloscope. However, in this case, it is particularly important to monitor the peak-to-peak values of the waveforms at each stage, since this will enable you to most easily isolate the defective stage. If you refer to the waveforms associated with the schematic of Fig. 23-2, you will get a very good idea of what to expect. However, remember that these waveforms are only for the particular set shown and may be somewhat different for other receivers. Thus, it is most important to obtain the correct manufacturer's information for each receiver you will be checking. The general methods for alignment and for the connection of the test equipment to be used, however, will be very similar to that given in Chapter 23, and this may be used as a guide regardless of the receiver involved.

Abnormal Color Intensity When we speak of color intensity, we are not referring to the hue (or tint) of the color, but rather of its strength (technically, its saturation). In describing this problem as "abnormal color intensity," we mean a situation where the color intensity is too great or too weak and that the condition cannot be corrected by adjustment of the color-intensity control. While it is possible for other conditions in the receiver to cause this condition, we are here restricting this discussion to the possible malfunctioning of the ACC (automatic-color control) circuit. The ACC circuit controls the gain of the first chroma amplifier (Q_{606}) in a manner very similar to AGC operation. The functioning of the ACC circuit is described in Chapter 23 and should be reviewed if necessary. The gain of the first chroma amplifier varies (normally) in proportion to the amount of ACC voltage applied to its base, which, in turn, is proportional to the amplitude of the incoming burst signal. Malfunctioning of the ACC system is best determined by checking the dc voltages throughout the system. These dc voltages should be carefully compared against the ones provided by the manufacturer. Note that the ACC system includes the killer detector, the ACC amplifier, the first chroma amplifier, the second chroma amplifier, and the burst amplifier. This is a closed-loop system and ACC problems could exist in any of the aforementioned stages.

The manufacturer will sometimes provide ACC system voltages for no-color input as well as color-input conditions from a color-bar generator. It is necessary to check the various stage voltages under both conditions, in order to localize the problem. As in troubleshooting AGC problems, it may be necessary, at times, to place a manually controlled dc voltage at the base of the first chroma amplifier and remove the voltage from the ACC amplifier. This will make it possible to isolate the problem between faults in the chroma amplifiers and faults in the remainder of the ACC loop.

Other possible symptoms of problems in the ACC loop are drifting values of color intensity and intermittent changes in color intensity. The isolation procedures here are the same as outlined previously, except that long-term monitoring is usually required.

Improper Color Sync It is possible for the color sync alone to be improper, or for both the color and monochrome sync to malfunction. Both of these conditions are discussed here:

(1) *No Color Sync: Monochrome Sync Normal.* This condition is shown in Fig. 24-38. (The input signal is from a color-bar generator.) The first step in isolating this problem is to be certain that the defect exists solely in the chroma circuits and does not originate in monochrome circuits. This is readily determined by an oscilloscope check at the input to the chroma section (Q_{606}). With a color-bar generator providing the color signal, check the waveform and amplitude of this input signal against the manufacturer's information. Any serious deviation from specifications, indicates that the cause of the trouble precedes the chroma section.

Remember that proper color sync depends upon the exact frequency of the 3.58 MHz subcarrier oscillator, as compared to the transmitted color burst. Any circuit defects which affect the frequency of the subcarrier, or which seriously affects its amplitude will affect the color sync. Referring to the block diagram of Fig. 23-1, we note that the color sync may be affected by the blanker (Q_{634}), the burst amplifier (Q_{636}), the phase detector, the phase control amplifier (Q_{638}), the APC amplifier (Q_{640}), the 3.58 MHz oscillator (Q_{642}), and the 3.58 MHz amplifier (Q_{644}). An oscilloscope is invaluable in localizing problems in the color-sync system. Defects in the operation of the blanker or burst amplifier are readily detected by means of scope waveforms. Operation of the 3.58 MHz oscillator and amplifier can also be quickly checked with a wide-band scope. The remaining color-sync circuits can best be checked by dc voltage and resistance measurements. It is important to note that the color-sync system is a closed-loop, automatic-phase control system, so that a defect which seems to be in one particular circuit may actually be in an associated circuit. However, careful waveform observations, coupled with dc voltage and resistance measurements, which are correlated with the manufacturer's specifications will make it possible to isolate the trouble. Since transistors (or tubes, where used) are a prime suspect, be sure to check these first with an in-circuit transistor tester, before spending possibly needless time on other circuit checks.

(2) *No Color Sync: no Monochrome Sync.* The appearance of this problem (color-bar generator input) is shown in Fig. 24-39. It is unlikely that both the color-sync and monochrome-sync circuits would fail simultaneously. If both kinds of sync are out together, this is usually caused by a defect in the monochrome-sync circuits. The color burst is situated on the back porch of the horizontal-blanking pulse and is gated in the color-sync circuits by a horizontal-keying pulse. Therefore, unless the monochrome-horizontal sync is in the exact required frequency and phase, the color sync will also suffer. Troubleshooting for defective horizontal-sync action has been covered previously in this chapter and need not be repeated here. In the unlikely event that fixing the defect in the horizontal-sync action, does not also correct the color-sync operation, refer to part 1 of this discussion.

Incorrect Hues This problem may occur in the form that all of the hues may be incorrect, or that only one or more hues may be incorrect. This problem is, of course, best checked with the aid of a color-bar generator.

(1) *All Hues Incorrect.* The correct reproduction of all color hues is dependent upon the phase of the 3.58 MHz subcarrier oscillator signal applied to the color demodulators. This phase must be synchronized with that of the transmitted color-burst signal. If the phase of the subcarrier signal is shifted, all of the colors on a color-bar display will be changed from normal and this will be readily evident on the color-tube display. Such a condition may be caused by poor alignment of the bandpass (chroma) amplifiers, as described in a prior section, and this cause will not be further pursued here, except to note that apparent poor alignment may not be caused by incorrect tuning only, but may also be the result of defective circuit components. In any case, a sweep-response curve will reveal the fact that a response defect exists, and it then remains to isolate the problem.

An important clue in analyzing this problem is to check and see if all the colors are presented in the correct sequence, but not in the correct position on the display. (We are of course assuming that such an obvious fault as a misadjusted hue control is not the cause of our problem.) If the sequence of colors is correct, but their position incorrect, this indicates that the subcarrier or chroma phase is wrong. In order to determine if the entire color sequence is correct, it may be necessary to adjust the hue control, since some of the color bars may be off screen. In the circuit of Fig. 23-1, incorrect phasing may be caused by malfunctions of the tint amplifier (Q_{600}, Q_{602}), or in the tint control (R_{22}) circuit proper. This condition may also be caused by malfunction in the subcarrier oscillator proper (Q_{642}) or its associated circuits. These associated circuits are the automatic phase control system which includes the 3.58 MHz amplifier (Q_{644}). The automatic-phase control system includes the burst amplifier (Q_{636}), the phase detector, the phase control amplifier (Q_{638}), the APC amplifier (Q_{640}), the 3.58 MHz oscillator (Q_{642}) and the 3.58 MHz amplifier (Q_{644}). Correct alignment of all tuned circuits involved here is a must and this should be checked first if there are no immediately obvious defects in the APC system. The next step would be to carefully check all transistors for defects, which are not necessarily of a major nature. If in doubt, replace the transistor with a known good one. The next step (if the defect has not yet been found) is to make a careful check of all dc voltage and resistance values and to compare the readings against those furnished by the manufacturer. Unfortunately, an oscilloscope is not of much value generally, when looking for defects which only cause signal-phase shifts.

In Fig. 23-1, the phase control amplifier (Q_{638}) depends upon the setting of a potentiometer for its proper operation (see Fig. 23-2). This setting, as well as the various dc voltages associated with the stage must be carefully checked.

One or More Hues Incorrect When only a single hue (red, green, or blue) is incorrect (missing), the trouble will naturally be found in a circuit which handles only that particular color. In Fig. 23-1, this problem may be caused by a defective color-tube gun, or by a defective output amplifier (Q_{628}, Q_{630}, Q_{632}), or a defective driver (Q_{624}, Q_{622}, Q_{626}). By the use of a color-bar generator and scope, the defective channel can be readily identified and the defective stage localized. Although not used in this receiver, some color sets employ three separate color demodulators. In this event, a failure of any one of the demodulators will also produce a condition where only one of the basic colors is absent from the picture.

In the block diagram of Fig. 23-1, there are two color demodulators (X and Z). The third color ($G-Y$) is derived from a proportion of the outputs of the two demodulators. In the event that one of the two color demodulators becomes inoperative (for example, the X demodulator), we would not only lose that particular color (red), but also the red component of chroma-video signal which is applied to the $G-Y$ matrix amplifier (Q_{620}). Thus, the only color which would be correct would be the $B-Y$, or blue. This type of trouble is easily isolated by means of a color-bar generator and scope. In this case it would be found that there was no output signal from the X demodulator or the $R-Y$ driver.

If the $G-Y$ matrix amplifier were inoperative, the red and blue (and mixture) colors would be normally present, but there would be no green contribution to the color picture. This situation is shown in the color-bar photo of Fig. 24-40. Note the complete absence of the color green.

REVIEW QUESTIONS

1. Explain how a monochrome receiver can be separated into sections for the purpose of troubleshooting.
2. How does a color receiver differ from a monochrome receiver as far as troubleshooting is concerned?
3. If a color receiver has no sound, no monochrome picture, or color picture, but it has a bright raster, what section(s) may be at fault?
4. What may cause color confetti to be displayed on a color-receiver screen?
5. What causes problems in tracking between monochrome and color pictures in a color receiver?
6. What are some of the causes of a loss in high voltage in a monochrome receiver?
7. What is a "negative picture" and what may cause it?
8. What causes a keystone raster?
9. Explain how the hue control in a color receiver is adjusted.
10. If the IF section is badly misaligned in a color receiver, list at least three symptoms which may be present.
11. Explain two reasons for the presence of a 920 kHz signal in the picture of a color set.
12. What are two causes of a loss of high-frequency response in a TV picture?
13. What are two reasons for a complete loss of sync in a TV set?
14. If there is a shorted turn in a deflection yoke coil, what will be the effect on the picture?
15. There are bright, vertical lines at the right side of a TV raster. What may cause this?
16. Describe some causes of horizontal-sync instability.
17. Explain in general how some of the monochrome circuits may affect color reproduction.
18. If the 3.58 MHz oscillator is not at the correct frequency, how will this affect the color picture?
19. Discuss some of the possible results of a defective chroma amplifier.
20. A color TV set, showing a color-bar pattern, has the color bars in the correct sequence, but in the wrong location. What might cause this problem?
21. If one hue of a color-bar pattern is missing, what circuit(s) might be at fault?

Index

NOTES

NOTES

NOTES

NOTES

NOTES

NOTES

NOTES